Humanizing Digital Reality

Klaas De Rycke · Christoph Gengnagel
Olivier Baverel · Jane Burry
Caitlin Mueller · Minh Man Nguyen
Philippe Rahm · Mette Ramsgaard Thomsen (Editors)

Humanizing Digital Reality

Design Modelling Symposium Paris 2017

énsa-v
école nationale supérieure
d'architecture de versailles

Springer

Editors

Klaas De Rycke
Ecole Nationale Supérieure d'Architecture de Versailles
Versailles
France

Christoph Gengnagel
University of the Arts
Berlin
Germany

Olivier Baverel
Ecole des Ponts ParisTech
Champs sur Marne
France

Jane Burry
Swinburne University of Technology
Melbourne
Australia

Caitlin Mueller
School of Architecture + Planning
Massachusetts Institute of Technology
Cambridge, MA
USA

Minh Man Nguyen
ENSA Paris-Malaquais
Paris
France

Philippe Rahm
Philippe Rahm architects
Paris
France

Mette Ramsgaard Thomsen
The Royal Danish Academy of Fine Arts, Schools of Architecture, Design and Conservation
Copenhagen
Denmark

ISBN 978-981-13-4915-7 ISBN 978-981-10-6611-5 (eBook)
https://doi.org/10.1007/978-981-10-6611-5

Softcover re-print of the Hardcover 1st edition 2018

Printed on acid-free paper

This Springer imprint is published by Springer Nature
The registered company is Springer Nature Singapore Pte Ltd.
The registered company address is: 152 Beach Road, #21-01/04 Gateway East, Singapore 189721, Singapore

Foreword

The National School of Architecture at Versailles (ENSA-V) and the Design Modelling Symposium have come together because they share the same vocation, the same commitment that of offering an innovative terrain to explore the potential of the digital revolution through education and research in architecture and urbanism.

The DMS will support the transformation of the school and aims to accentuate the path that so clearly defines ENSA-V: ensure fruitful links between education, research, and a deliberate return via digitalization to a control and an intensified exploration of the material. The creation of a fablab focusing on the new paradigm "digitalization as a return to the material" will be launched. It will be run by teams concerned with innovation, made up of students, teachers, researchers, and practitioners.

Together with the DMS, ENSA-V will inaugurate these new spaces—hospitable, flexible, and open—destined to provide new national and international educational perspectives, able to flourish at all levels and weaving close ties between patrimony and innovation, between heritage and a vision of the future.

ENSA-V will also inaugurate, with the DMS, a new educational vision which incites teachers, researchers, and students to focus on three unifying themes—"plan/non-plan," "acceleration/deceleration," and "living in the world". Extending over three years, they will encourage a community in movement, interdisciplinary and cosmopolitan, to meet to devise scholarly contributions, to envisage innovative experiments, and to construct audacious hypotheses. From today, ENSA-V's objective is to provide this community with the resources necessary to develop, to flourish, and to fully disseminate its work.

In anticipation of the first manifestations of this three-year strategy, this work will illustrate the powerful opportunity for areas of exploration that crosscutting architectural and digital disciplines provide when together they devise a hope for human activities.

Jean-Christophe Quinton

Workshop at ENSA Versailles

The original version of the book was revised:
For detailed information please see Erratum.
The erratum to the book is available at
https://doi.org/10.1007/978-981-10-6611-5_56

Acknowledgements

For their advises, we thank Andreas Kofler and Abigail Bachelor.
For their kind understanding, we thank Bettina Pinks, Lennart, and Arvo De Rycke.
For their precious help, we thank Thomas Charil and Solène Assouan.
For their helpful support, we thank Baptiste Fizelier and Quentin Rihoux.
For their kind support, we thank Alexandre Labasse and Carlos Pérez.
We would like also to thank our sponsors Bollinger+Grohman, WoMa, Sofistik, Ecole des Ponts ParisTech, Konbini, Design By Data, and especially Autodesk.

We hereby thank the ENSA-V's BDE for their assistance in the logistics of the event.
Thanks also to the ENSA-V Kfet for taking care of meals, coffees, and refreshments.
We also thank Architectonique, the ENSA-V Junior Enterprise, for having relayed to the local professionals.
For their kind participation, we thank La Maréchalerie Centre d'Art Contemporain, Grolsch and Makery.

Finally, we thank ENSA-V for hosting and supporting the Design Modelling Symposium Paris 2017.

Design Modelling Symposium Paris, 16/09–20/09/2017

International Scientific Committee

Oliver Tessmann, TU Darmstadt
Tobias Wallisser, LAVA, Berlin
Kathy Velikov, University of Michigan, Ann Arbour

Advisory Committee

Ollivier Baverel, ENSA Grenoble/ENPC
Jane Burry, Dean of Design at Swinburne University, Melbourne
Klaas De Rycke, ENSA Versailles
Christoph Gengnagel, UDK Berlin
Caitlin Mueller, MIT Cambridge
Minh Man Nguyen, ENSA Paris Malaquais/Digital Knowledge
Philippe Rahm, ENSA Versailles
Mette Ramsgard Thomsen, CITA Copenhagen

Organizing Committee

Klaas De Rycke, ENSA Versailles
Minh Man Nguyen, ENSA Paris Malaquais/Digital Knowledge
Dagmar Rumpenhorst-Zonitsas, Daglicious Coordination, Berlin

Main Sponsors

AUTODESK.

énsa-v
école nationale supérieure
d'architecture de versailles

Co-sponsors

BOLLINGER + GROHMANN
Ingénierie

SOFiSTiK

DESIGN BY DATA

École des Ponts
ParisTech

WoMa
FABRIQUE DE QUARTIER

Introduction

Digital and physical human environments interfere more and more. Mobile devices, drones, cars, robots, smart production methods, intelligent cities and—following Antoine Picons observations in his text further in this book—even digitized odors are constituting new layers on top of our daily lives and are becoming much more common layers in an ever more digitized world. They seem to alter our regular physical interactions and surrounding physical or digitized world. Even without looking as far back as the middle ages, but just thirty years ago at the dawn of the World-Wide Web we might question how far the reality is from becoming digital and how much the digital influences reality?

The changes that we are considering are inherently human, being man-made and expressing an unbridled belief in constant progress. This positivist belief focuses primarily on action and overlooks a multitude of aspects; It disregards the real role and place of human (inter)action, of human and digital (inter)action, how far the digital can and will change the real/reality, how much we can control our reality with(in) the digital, and how much of it is actually still controlled by humans?

Much has been written and debated about the influence CAD, and data has had on buildings, on city planning, on fabrication and its processes. We could invert the question asking instead to what extent are humans still in control of our digital/real surroundings?

In the line of the positivist movement some 150 years ago, anthropocentrism seems to be a logical continuation of progress where humans control the built and unbuilt environment and are central in bringing and implementing solutions. In contemporary reality, ever better computers emulate and control ever better natural phenomena evolving to a hoped for all-encompassing matrix for future cities and future relationships.

Technological advances seem a culmination of this positivist idea. Nowadays, preachers and technological zealots and specifically proponents of AI even predict the removal of humans from the equation. Building on that thought, we can ask ourselves how far our reach as humans really goes? Do the complex algorithms that we use for city planning nowadays live up to expectations and do they offer sufficient quality? Are they an extension of ourselves? Are they self-controlled? How much data do we/they (the computing power) have and can we control? Are current inventions reversing the humanly controlled and invented algorithms into a space where humans are controlled by the algorithms?

Are processing power, robots, and algorithms of the digital environment and construction in particular, not only there to rediscover what we already know or do they really advance the fields of construction and architecture?

The Design Modelling Symposium is an interdisciplinary platform to explore recent developments, their meanings and place in the environment of architecture, engineering, and art. This year's conference will try to offer some answers and exchange on the following questions;

1. What is big data? What does it teach us and in what fields? Is it preconceived or only processed data? How is data flowing, toward or away from something? Intelligent cities? How is city-planning changed by the data knowledge, what are the current and future algorithms running the current and future towns? What smarter elements are they suggesting and how "smart" should cities become? What are its flaws?
2. What are the benefits and the possibilities in design with robotics? Are they just a tool or a goal in themselves? Should we look for ways by which the very technical approach of robotic design could define future design solutions? Is there a new archi-botic paradigm?
3. How much can we construct nature? What is nature/material; a perfectly controlled algorithm which can be used/abused as one wishes? What role does geometry play? Is the data crunching, digital design, and digital fabrication merely a help in statistical empirical and descriptive science? Is it merely a numerical transcript of natural phenomena? Does it just support—but in continuous faster loops—exploring and exploiting natural possibilities for constructing or can it go beyond that point and help uncover new ways and perhaps new laws of nature? Can it be treated as a behavioral and statistical science proactively informing building processes?
4. What is data sharing, workflow, collaborative? How digital tools create an environment that can help teams designing projects? Is it disruptive constructive or linear? How do we inform the process and then control the output until the physical result?
5. How emulation can influence the project? What is the relationship between hypothesis, analysis, physical testing, post-rationalization, and when can we fully predict the reactions in the real world? How can post-construction measurements inform real-time projects?
   As computational fluid dynamic analysis and the gathering of data through drones and other real-time data is gaining ground, how is the digitalization of the weather, of the total physical environment, of real-time data influencing design?

This initial approach to the position of man in the digital era leaves us deliberately with only questions. We felt that over the course of 2017 and with the approaching symposium in September, these open-ended questions would be slowly answered by—in chronological order—the members of the board, by the keynote speakers, by the scientific contributions, and finally by the event of the symposium itself with workshops, speeches, and conversations.

The first step was taken by the board. The board works both as a think tank, as a guardian of the researched content and of quality and as a moderator.

The first step was to organize the questions into different topics. The topics then should be given boundaries—or a framework—within which possible answers (hypothesis/theories) can be treated and organized. The hypotheses should come from the participants, namely the keynote speakers, the contributors of scientific papers, and the participants at the symposium.

The framework of the topics is described in more detail by the responsible topic leaders later in the book.

We found that for the five topics, we could further reorganize them under three overarching global themes;

A. **Design and Modeling of Matter**

- material practice (Mette Ramsgaard Thomson, Christoph Gengnagel)
- structural innovation (Olivier Baverel, Caitlin Mueller)

B. **Design and Modeling of Data**

- data farming (Tomas Diez, Klaas De Rycke)
- data shaping cities (Jane Burry)

C. **Design and Modeling of Physics**

- thermodynamic practice (Philippe Rahm)

Each topic is debated over half a day during the symposium. Specific chosen keynote speakers help to conceive the topics.

From 148 very interesting and high-level scientific submissions, 45 were finally chosen to deliver possible answers to the initial questions.

The symposium is the final step. It started with a series of questions, evolved to defining topics, delivering a framework and initial definitions, being tested in scientific papers and finally being debated at the symposium.

About this book, the book tries to follow an empirical research approach. There is an original question based on some observation (hypothesis-intro), followed by induction or the formulation of hypothesis (topics), further to deduction (experiments and rephrasing by the keynotes) and then to the testing (scientific papers). This should normally be finalized with an evaluation or a general conclusion. Since empirical research is observation based and has no simple final value but rather a field of possibilities or rather probabilities, the book leaves it open to the conference participants to formulate any conclusion. This way the book perfectly falls in line with this year's overall theme; what shall each human become in the vast area of data treatment in the fields of architecture and engineering?

All of the available data on these topics will be gathered on a digital and physical platform which we hope will enhance further discussion and—who knows—progress!

Klaas De Rycke
Minh Man Nguyen

Contents

Part VI Scientific Contributions

Contributors

Yasmine Abbas Paris College of Art/Neo-nomad, Paris, France

Sigrid Adriaenssens Princeton University, Princeton, USA

Fernando Porté Agel Wind Engineering and Renewable Energy Laboratory (WIRE), Ecole Polytechnique Fédérale de Lausanne (EPFL), Lausanne, Switzerland

Tristan Al-Haddad Georgia Tech, Atlanta, USA; Formations Studio, Atlanta, USA

Pantea Alambeigi Royal Melbourne Institute of Technology, Melbourne, VIC, Australia

Pierre André Ecole des Ponts ParisTech, Champs-sur-Marne, France

Audrey Aquaronne Ecole des Ponts ParisTech, Champs-sur-Marne, France

Dorit Aviv School of Architecture, Princeton University, Princeton, NJ, USA

Phil Ayres School of Architecture, Centre for Information Technology and Architecture, The Royal Danish Academy of Fine Arts, Copenhagen, Denmark

Carlo Bailey New York, NY, USA

Rainer Barthel Technical University of Munich, Munich, Germany

Olivier Baverel Laboratoire GSA-Géométrie Structure Architecture (Ecole Nationale Supérieure D'architecture Paris-Malaquais), Laboratoire Navier (UMR 8205), CNRS, Ecole Des Ponts ParisTech, IFSTTAR, Université Paris-Est, Marne la vallée, France

Joseph Benedetti F-O-R-T Ingénierie, Lille, France; Ecole Nationale Supérieure d'Architecture et de Paysage de Lille, Villeneuve d'Ascq, France

David Benjamin The Living, an Autodesk Studio, New York, NY, USA

Christopher Beorkrem University of North Carolina at Charlotte, Charlotte, NC, USA

Louis Bergis B+G Ingénierie Bollinger + Grohmann S.a.R.L, Paris, France

Shajay Bhooshan Block Research Group, Institute of Technology in Architecture, ETH Zurich, Zurich, Switzerland

Philippe Block Block Research Group, Department of Architecture, Institute for Technology in Architecture, ETH Zürich, Zurich, Switzerland

Philippe Bompas Elxir Sasu, Paris, France

Maite Bravo Institute of Advanced Architecture of Catalunya IAAC, Barcelona, Spain

Maurizio Brocato ENSA Paris-Malaquais, Paris Cedex 06, France

Jane Burry Swinburne University of Technology, Melbourne, VIC, Australia

Ignacio López Busón AA MA Landscape Urbanism, MAPS, AAVS Shanghai and Turenscape Academy, Shanghai, China

Edouard Cabay Institute for Advanced Architecture of Catalonia, Barcelona, Spain

Miguel Vidal Calvet Foster + Partners, London, England; Computational Design, Master in Architectural Management and Design (MAMD), IE University, Segovia, Spain

Corneel Cannaerts KU Leuven Faculty of Architecture, Ghent, Belgium

Jean-François Caron Laboratoire Navier (UMR 8205), CNRS, Ecole Des Ponts ParisTech, IFSTTAR, Université Paris-Est, Marne la vallée, France

Paul Casson Multipass, Newton in Furness, Cumbria, UK

Damiano Cerrone Spin Unit, Tallinn, Estonia

Stephanie Chaltiel Institute of Advanced Architecture of Catalunya IAAC, University Polytechnic of Catalunya UPC, Barcelona, Spain

Ashkan Cheheltan School of Architecture, Centre for Information Technology and Architecture, The Royal Danish Academy of Fine Arts, Copenhagen, Denmark; Berlin University of the Arts, Berlin, Germany

Canhui Chen Royal Melbourne Institute of Technology, Melbourne, VIC, Australia

Angelos Chronis Institute for Advanced Architecture of Catalonia, Barcelona, Spain

Kenn Clausen 3XN Architects and GXN Innovation, Copenhagen, Denmark

Jeroen Coenders TU Delft, The Hague, The Netherlands; White Lioness technologies, Amsterdam, The Netherlands

Zack Xuereb Conti Singapore University of Technology and Design, Singapore, Singapore

Ann Cosgrove New York, NY, USA

Jean-François Coulais ENSA-Versailles, Paris, France

Christof Crolla The Chinese University of Hong Kong, Sha Tin, Hong Kong

Pierre Cutellic AREA/CPMK, Paris, France; ETH Zurich, Zurich, Switzerland

Ashley Damiano University of North Carolina at Charlotte, Charlotte, NC, USA

Renaud Danhaive Massachusetts Institute of Technology, Cambridge, MA, USA

Daniel Davis New York, NY, USA

Cyril Douthe Ecole des Ponts ParisTech, Marne la Vallée, Champs-sur-Marne, France; IFSTTAR, Champs-sur-Marne, France

R. Duballet Laboratoire Navier, UMR 8205, Ecole des Ponts, IFSTTAR, CNRS, UPE, Champs-sur-Marne, France; XtreeE, Immeuble Le Cargo, Paris, France

Billie Faircloth Kieran Timberlake, Philadelphia, USA

Al Fisher University of Bath, Bath, UK; University of Bath, London, UK

Michael U. Hensel Oslo School of Architecture and Design, Oslo, Norway

Anja Jonkhans University of Applied Arts, Vienna, Austria

Sam Conrad Joyce Singapore University of Technology and Design, Singapore, Singapore

Xavier De Kestelier Hassell Studio/Smartgeometry, London, UK

Axel Kilian Princeton University, Princeto, USA

Tomas Diez Ladera IAAC, Barcelona, Spain

Linxue Li Tongji University, Shanghai, China; ATELIER L+, Shanghai, China

Steve De Micoli Institute for Computational Design and Construction, University of Stuttgart, Stuttgart, Germany

Philippe Morel ENSA Paris-Malaquais, Paris, France; EZCT Architecture & Design Research, Paris, France

Jean-Rémy Nguyen B+G Ingénierie Bollinger + Grohmann S.a.R.L, Paris, France

Yusuke Obuchi The University of Tokyo, Tokyo, Japan

Federico De Paoli Foster + Partners, Riverside, London, UK

Klaas De Rycke B+G Ingénierie Bollinger + Grohmann S.a.R.L, Paris, France; ENSA-Versailles, Paris, France

J. Dirrenberger Laboratoire PIMM, Ensam, CNRS, Cnam, Paris, France; XtreeE, Immeuble Le Cargo, Paris, France

Alexandre Dubor Institute for Advanced Architecture of Catalonia, Barcelona, Spain

Jefferson Ellinger University of North Carolina at Charlotte, Charlotte, NC, USA

Philipp Eversmann Eversmann Studio, München, Germany

Jan Friedrich University of the Arts, Berlin, Germany

Christoph Gengnagel University of the Arts, Berlin, Germany

Evan Greenberg Bartlett School of Architecture, UCL, London, UK

Manfred Grohmann School of Architecture, University of Kassel, Kassel, Germany; Bollinger + Grohmann Ingenieure, Frankfurt am Main, Germany

Sean Hanna UCL IEDE, Central House, London, UK

Philippe Hannequart Laboratoire Navier (UMR 8205), CNRS, Ecole Des Ponts ParisTech, IFSTTAR, Université Paris-Est, Marne la vallée, France; Arcora, Groupe Ingérop, Rueil-Malmaison, France

Mary Katherine Heinrich School of Architecture, Centre for Information Technology and Architecture, The Royal Danish Academy of Fine Arts, Copenhagen, Denmark

Denis Hitrec University of Ljubljana, Ljubljana, Slovenia

Roland Hudson Universidad Piloto Colombia, Bogota, Colombia

Ryan Hughes Aarhus School of Architecture, Aarhus, Denmark

Marco Ingrassia Institute for Advanced Architecture of Catalonia, Barcelona, Spain

Michael-Paul "Jack" James University of North Carolina at Charlotte, Charlotte, NC, USA

Ewa Jankowska-Kus B+G Ingénierie Bollinger + Grohmann S.a.R.L, Paris, France

Jeroen Janssen AKT II, White Collar Factory, London, UK

Sawako Kaijima Singapore University of Technology and Design, Singapore, Singapore

Farshid Kardan Wind Engineering and Renewable Energy Laboratory (WIRE), Ecole Polytechnique Fédérale de Lausanne (EPFL), Lausanne, Switzerland; Laboratoire GSA-Géométrie Structure Architecture (Ecole Nationale Supérieure D'architecture Paris-Malaquais), University Paris-Est, Paris, France

Alireza Karduni UNC Charlotte, Charlotte, NC, USA

Ole Klingemann Structure and Design|Department of Design, Faculty of Architecture, University of Innsbruck, Innsbruck, Austria

Jan Knippers Institute of Building Structures and Structural Design (ITKE), University of Stuttgart, Stuttgart, Germany

Benjamin S. Koren ONE TO ONE, New York, NY, USA

Johannes Kuhnen Design-to-Production, Erlenbach/Zurich, Switzerland

Riccardo La Magna Konstruktives Entwerfen Und Tragwerksplanung, Berlin University of the Arts, Berlin, Germany; ITKE—University of Stuttgart, Stuttgart, Germany

David Andres Leon Centre for Information Technology and Architecture, School of Architecture, The Royal Danish Academy of Fine Arts, Copenhagen, Denmark

Julian Lienhard Structure GmbH, HCU Hamburg, Hamburg, Germany

Taavi Lõoke Estonian Academy of Arts, Tallinn, Estonia

Richard Maddock Foster + Partners, Riverside, London, UK

Areti Markopoulou Institute for Advanced Architecture of Catalonia, Barcelona, Spain

Anna Mavrogianni Bartlett School of Architecture, UCL, London, UK

Achim Menges Institute for Computational Design and Construction, University of Stuttgart, Stuttgart, Germany

Caitlin Mueller Massachusetts Institute of Technology, Cambridge, MA, USA

Kristjan Männigo Spin Unit, Tallinn, Estonia

Danil Nagy The Living, an Autodesk Studio, New York, NY, USA

Paul Nicholas School of Architecture, Centre for IT and Architecture, Royal Danish Academy of Fine Art, Copenhagen, Denmark

Jens Pedersen Aarhus School of Architecture, Aarhus, Denmark

Michael Peigney Laboratoire Navier (UMR 8205), CNRS, Ecole Des Ponts ParisTech, IFSTTAR, Université Paris-Est, Marne la vallée, France

Sven Pfeiffer Institute for Architecture, TU-Berlin, Berlin, Germany

Nicole Phelan New York, NY, USA

Paul Poinet Centre for Information Technology and Architecture (CITA), KADK, Copenhagen, Denmark

Mary Polites AA MArch Emergent Technologies, MAPS, D&I College, Biomimetic Design Lab (BiDL), Tongji University, Shanghai, China

Mariana Popescu Department of Architecture, Institute for Technology in Architecture, ETH Zürich, Zurich, Switzerland

Kåre Stokholm Poulsgaard Institute for Science, Innovation and Society, University of Oxford, Oxford, UK

Renee Puusepp Estonian Academy of Arts, Tallinn, Estonia

Jonathan Rabagliati Foster + Partners, Riverside, London, UK

Mette Ramsgaard Thomsen School of Architecture, Centre for IT and Architecture, Royal Danish Academy of Fine Art, Copenhagen, Denmark

Aurel Richard Institute for Advanced Architecture of Catalonia, Barcelona, Spain

Katja Rinderspacher Institute for Computational Design and Construction, University of Stuttgart, Stuttgart, Germany

Matthias Rippmann Department of Architecture, Institute for Technology in Architecture, ETH Zürich, Zurich, Switzerland

Dagmar Rumpenhorst-Zonitsas Daglicious Coordination, Berlin, Germany

Moritz Rumpf School of Architecture, University of Kassel, Kassel, Germany

Jonas Runberger Chalmers Department of Architecture and Civil Engineering, Gothenburg, Sweden

Virginia San Fratello San Jose State University, San Jose, USA; Emerging Objects, Oakland, USA

Eric Sauda UNC Charlotte, Charlotte, NC, USA

Markus Schein School of Art and Design, University of Kassel, Kassel, Germany

Fabian Scheurer Design-to-Production, Zürich, Switzerland

Eike Schling Technical University of Munich, München, Germany

Maxie Schneider University of Arts Berlin, Berlin, Germany; Springer Heidelberg, Heidelberg, Germany

Bob Sheil The Bartlett School of Architecture, UCL, London, UK

Paul Shepherd University of Bath, Bath, UK

Walter Simone PREVIEW/UFO, Grenoble, France; ENSA-Versailles, Versailles, France

Vasily Sitnikov KTH Royal Institute of Technology, Stockholm, Sweden

Maria Smigielska Zurich, Switzerland

Hanno Stehling Design-to-Production GmbH, Erlenbach, Zurich, Switzerland

James Stoddart The Living, an Autodesk Studio, New York, NY, USA

Seiichi Suzuki Institute of Building Structures and Structural Design (ITKE), University of Stuttgart, Stuttgart, Germany

Tom Svilans Centre for Information Technology and Architecture (CITA), KADK, Copenhagen, Denmark

David Tajchman Architectures David Tajchman, Paris, France, Tel Aviv, Israel; Technion IIT, Haifa, Israel

Martin Tamke Centre for Information Technology and Architecture (CITA), KADK, Copenhagen, Denmark

Eric Teitelbaum School of Engineering and Applied Science, Civil and Environmental Engineering, Princeton University, Princeton, NJ, USA

Oliver Tessmann TU Darmstadt, Darmstadt, Germany

Edoardo Tibuzzi AKT II, White Collar Factory, London, UK

Sylvain Usai Design-to-Production GmbH, Erlenbach, Zurich, Switzerland

Minh Man Nguyen ENSA Paris Malaquais/Digital Knowledge, Paris, France

Tom Van Mele Block Research Group, Department of Architecture, Institute for Technology in Architecture, ETH Zürich, Zurich, Switzerland

Kathy Velikov University of Michigan, Ann Arbour, USA

Rodrigo Velsaco Universidad Piloto Colombia, Bogota, Colombia

Petras Vestartas School of Architecture, Centre for Information Technology and Architecture, The Royal Danish Academy of Fine Arts, Copenhagen, Denmark

Emmanuel Viglino Arcora, Groupe Ingérop, Rueil-Malmaison, France

Lorenzo Villaggi The Living, an Autodesk Studio, New York, NY, USA

Tobias Wallisser LAVA, Berlin, Germany

Ginette Wessel Roger Williams University, Bristol, RI, USA

Han Yu MLA, Landscape Architect, Kuala Lumpur, Malaysia; SOM, Shanghai, China

Dale Zhao The Living, an Autodesk Studio, New York, NY, USA

Mateusz Zwierzycki School of Architecture, Centre for Information Technology and Architecture, The Royal Danish Academy of Fine Arts, Copenhagen, Denmark

Part I

Material Practice

Tamke M., Baranovskaya Y., Holden Deleuran A., Monteiro F., Fangueiro R, Stranghöhner N., Uhlemann J., Schmeck M., Gengnagel C., Ramsgaard Thomsen M.: Bespoke materials for bespoke textile architecture, IASS annual symposium 2016 “spatial structures in the twenty-first century”, at Tokyo 2016

Current design practice in architecture and engineering is undergoing radical changes. The ability to integrate advanced simulation in the early design phase and live sensor data from the environment of site, production, or material, fundamentally changes the act of design from one of pure projection to one of calibration of behavior. The speculative and creative process of design now engages tools that change the way we understand performance across the scales of environment, structure, element, and material giving us the ability to conceive new hybridized structural morphologies and rethink and invent their underlying material practices.

Christoph Gengnagel
Mette Ramsgard Thomsen

Computational Material Cultures

Achim Menges(✉)

Institute for Computational Design and Construction, University of Stuttgart, Stuttgart, Germany
achim.menges@icd.uni-stuttgart.de

Fig. 1 Pavilion's envelope consists of wooden lamellas formed by an intricate network of bent and tensioned segments. This self-equilibrating system physically computes the shape of the pavilion during assembly on site. *Institute for Computational Design (Achim Menges) and Institute of Building Structures and Structural Design (Jan Knippers), ICD/ITKE Research Pavilion 2010, University of Stuttgart, 2010*

Architecture provides the material context within which most of our everyday life unfolds. As a material practice, it effectuates social, cultural, and ecological relevance through the articulation of the built environment. This articulation is intrinsically tied to the processes of intellectual and physical production in which architecture originates: the processes of design and materialization. Today, the reciprocal effects of these two processes on each other can be seen through a different lens, and computation constitutes a critical factor for this contemporary reassessment of the relation between the generation and the materialization of form and space.

On the one hand, computation enables architects to engage facets of the material world that previously lay far outside the designer's intuition and insight. On the other, it is increasingly understood that—in its broader definition—computation is not limited to processes that operate only in the digital domain. Instead, it has been recognized that

K. De Rycke et al., *Humanizing Digital Reality*,
https://doi.org/10.1007/978-981-10-6611-5_1

material processes also obtain a computational capacity—the ability to physically compute form. When seen together, these two aspects suggest that we are now in a position to rethink the material in architecture through the computational. As the material ambience emanating from architecture represents a critical constituent of material culture, this essay seeks to inject this notion—usually reserved for historic thought—with a projective capacity by introducing a design approach that integrates materiality and materialization as active drivers.

Design Computation: Material Integration

Over the last two decades, digital processes have had an unprecedented impact on architecture. The computer has pervaded all aspects of the discipline, from the inception of design at an early stage, to the management of building information, all the way through to fabrication and execution on site. However, the underlying conception and logic of established processes more often than not remained largely unchallenged during this adaption of new technologies, rendering them a mere computerized extension of the well known. In areas that are primarily concerned with an increase in productivity, efficiency, and accuracy, this may be expected, if not particularly satisfying intellectually. But in design, with its intrinsic striving for innovation that is in sync with technological and cultural developments, it is surprising to see how often digital processes have been absorbed in the discipline without questioning traditional modes of conceiving form, structure, and space.

A striking example of this languishment is the primacy of geometry in design that has dominated architectural thinking since the Renaissance and has not changed much in the transition from the manually drawn to the digitally drafted, parametrically generated, or computationally derived. From a methodological point of view, this deeply entrenched prioritization of the generation of geometric shape over the processes of material formation imbues most digital design approaches with a deeply conventional touch, even when well camouflaged in exotic form and exuberant articulation.

If, in contrast, we begin to view the computational realm not as separate from the physical domain, but instead as inherently related, we can overcome one of the greatest yet most popular misconceptions about the computer in architectural circles, namely that it is just another tool (Kwinter 2012). We need to embrace the computer as a significant technological development, one that offers the possibility of a novel material culture rather than just another architectural style. In the same way as "the early moderns used the telescope and the microscope, to engage aspects of nature whose logic and pattern had previously remained ungraspable because they were lodged at too great a remove from the modalities of human sense," as Sanford Kwinter aptly puts it (Kwinter 2011), today architects can employ computation to delve deeper into the complex characteristics of the material world and activate them as an agency for design. In other words, through computation material no longer needs to be conceived as a passive receptor of predefined form as in established approaches, but instead can be rethought and explored as an active participant in design.

Of course, material-specific design is by no means an extraordinary thought or even a new idea. In fact, most architects would probably claim that their design decisions are directly linked to intended materiality. But usually the relation between form, space, structure, and material is locked in the aforementioned hierarchy and follows an established set of preconceived rules. These "do's and don'ts" are assumed to correspond with a constructional "can and can't," and in the case of the modernist's truth to materials, a "should and must not" that carries a strong moral overtone that still resonates in many architectural schools. Most often, these implicit design conventions are expressed typologically, whereby material characteristics are thought to directly relate to a set of constructional, spatial, or structural typologies.

Probably the most famous example of such an assumption, still frequently glorified by architectural practitioners and academics alike, is Louis Kahn equating the will of brick with the structural typology of the arch. Today, one would hope, we can steer clear of such a one-sided conversation based on prejudged interpretation. When listening to brick, computation would have been a good hearing aid to gain a more differentiated, multifaceted, and open-ended understanding of material characteristics and their latent design potential. Instead of relating material property to constructional form based on direct and linear rules, and in contrast to the unquestioned application of preconceived construction-handbook knowledge, computation enables us to employ material behaviors and materialization processes as a truly explorative agency in design, in which novel material, structural, and spatial effects may originate.

Design Agency: Materiality and Materialisation

Questioning the conventional hierarchy of form generation and materialization, as well as the established typological approach to material-oriented design, has been a central area of study at the Institute for Computational Design (ICD) at the University of Stuttgart (Menges 2012). While this research aims at contributing insights to disciplinary concerns of architecture, the integrative nature of the approach requires interdisciplinary collaboration with various partners from engineering and natural sciences, most frequently with the university's Institute of Building Structures and Structural Design (ITKE). Over the last few years, these investigations have been conducted and tested through a series of full-scale pavilions and demonstrator buildings. Several examples of the related research and design works are presented below.

One initial area of research focused on employing the relatively well-known yet architecturally largely unexplored material behavior of elastic (de)formation. This simple form of material computation—the self-forming output of an elastic curve based on the input of the application of a given force at one support point—allows for spatially articulating initially planar elements while at the same time increasing the capacity of such a bending-active structure (Lienhard et al. 2011). As even this simple material behavior cannot be conventionally drawn or modeled, it has only rarely been employed in architecture. However, today's technical possibility to compute form in unison with material characteristics enables tapping into the fascinating design potential that elastic material behavior may offer both spatially and structurally.

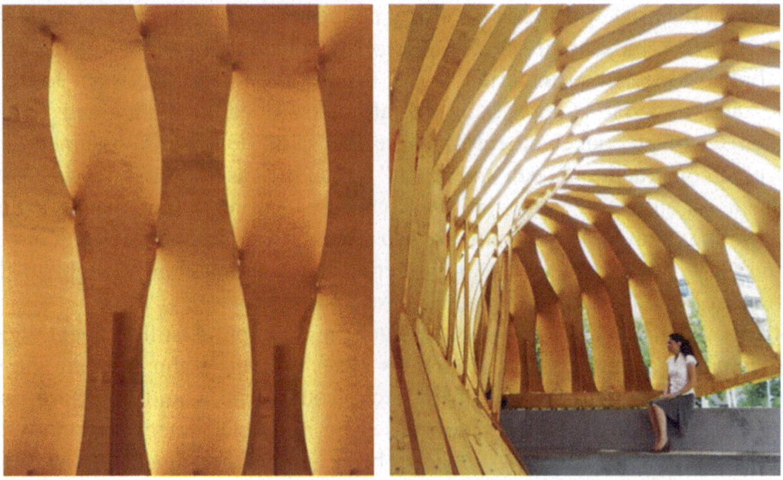

Fig. 2 *Left* The undulations of the thin wood lamellas, which are at the same time the pavilion's skin and structure, lead to differentiated direct and indirect illumination that accentuates the spatial depth. *Right* The residual stresses embedded in the building elements during assembly, which are a decisive factor in the structural capacity of the system, are experienced through the sinewy delicacy of the extremely thin wood lamellas

The ICD/ITKE Research Pavilion 2010 pursued such an investigation through an intricate wood lamella system (Fleischmann et al. 2012). During assembly, these initially planar elements are connected at differential length intervals so that they form tensioned and elastically bent regions along one wooden strip that are locked into position by the adjacent strips. The distribution of the joint points between the strips oscillates along the torus shape, and this morphological irregularity results in both significant global stability and a distinctive articulation of the pavilion's envelope, which is at the same time skin and structure. Here, form and material are inherently and inseparably related, and this not only applies to the design process, but also the construction procedure where even on site the material physically computes the shape of the pavilion.

The resulting intricate network of bent and tensioned segments that form this self-equilibrating structure is perceived through the sinewy delicacy of the extremely thin wood lamellas. The residual stresses, which are embedded in the strip elements during assembly and are a decisive factor in the structural capacity of the system, form part of the visual and spatial experience through the varying undulations of the envelope, which at the same mediates a gentle transition from direct to indirect illumination that accentuates the depth of the toroidal space. The pavilion offers a glimpse of the design potential dormant in even the simplest material elements, and how this can be teased out as formerly unexplored architectural possibilities when focusing the computational process on material behavior rather than on geometric shape. In the case of this pavilion, material is no longer just a passive receptor of predefined form, but rather becomes an active generator of design.

The profound impact of integrating the characteristics of material behavior and materialization processes in computational design thinking and techniques also allows for enriching material systems that have hitherto been considered "amorphic" with novel morphological and tectonic possibilities. Amorphic refers to materials that are seen as "shapeless" and thus require an external shaping device such as a mold or formwork. Concrete is a familiar example, but also fibre-composite materials such as glass- or carbon-fibre reinforced plastics (GRP/CRP) are commonly understood in this way by architects, designers, and engineers alike. Based on a higher-level integration of computational design, simulation and fabrication, the ICD and ITKE have investigated an alternative approach to conceptualizing and constructing fibre-composite systems in architecture that no longer relies on elaborate molds or mandrels (Menges and Knippers 2015). The goal of the study is twofold: on a technological research level, it aims to reduce the considerable effort, waste, and investment involved in the fabrication of molds, which currently render the applications of GRP and CRP systems only suitable for the serial production of identical building elements or for application in projects with extraordinary budgetary means. On a design research level, the investigation at the same time seeks to question the common conception of brous composites as amorphic by minimizing the need for external molds and thus teasing out the "morphic" character of the material itself, enabling the study of architectural articulation that unfolds from the self-expression of the fibres.

The research has led to a series of further research pavilions that are all based on constructional principles culled from the vast pool of biological composite systems. This remains a strongly tangible quality of the resulting architectural morphologies, as does the integrative approach to the computational generation and robotic materialization of the fibrous forms. For example, in the ICD/ITKE Research Pavilion 2012, the computational approach allowed for choreographing the interaction of sequentially applied fibres through a robotic fabrication process (Reichert et al. 2014). Here, only a simple, linear scaffold is required for the lament winding, as the initial fibre layers become an embedded mold. During the production process, the application of fibres in conjunction with the prestress induced by the robot continuously (de)forms the system so that the nal shape emerges only at the very end.

Fig. 3 *Left* The form of the pavilion's composite shell gradually emerges through the interaction of fibres applied on a minimal scaffold during a robotic filament-winding process. Computation allows for understanding this complex material behavior in design and enables the strategic differentiation of fibre layout, organization, and density of fibres, resulting in novel fibrous tectonics. *Right* The pavilion's skin structure is an extremely thin composite shell that does not require any additional support elements. In the translucent glass fibre surfaces, the black carbon rovings provide a distinctive visual reference to the intricate interplay between the fabrication- and force-driven fibre arrangements

Computation not only allows understanding and deploying this complex fibrous behavior in design, it also enables the strategic differentiation of fibre layout, organization, and density of fibres. In the resulting translucent composite surfaces of the 2012 pavilion, the black carbon rovings provide a distinctive visual reference to this intricate interplay between the fabrication- and force-driven fibre arrangements. While the constructional logic is revealed in this way, it avoids a simple and singular reading. Very different to the typical, glossy gel coat finishes stemming from molding processes that dominate our experience of these materials, here the carbon rovings form a deep skin with a rich, layered texture. This surface texture, as well as the overall morphology and resulting novel fibrous tectonics emerge from the computationally modulated, material formation process.

More recently, the research has been expanded toward cyber-physical production systems, in which the fabrication machine is no longer dependent on receiving a comprehensive and finite set of manufacturing instructions, but instead has the sensorial ability to gather information from its fabrication environment and changes its production behavior in real time (Menges 2015). Here, machine and material computation become fully synthesized in an open-ended process. In the ICD/ITKE Research Pavilion 2014–15, this approach allowed for gradually hardening an initially soft—and thus continuously deforming—inflated envelope by applying fibres on the inside. Eventually, a structurally stable state was reached so that the internal air pressure could be released and the pneumatic envelope changed into the pavilion's skin. In daylight, there remains only a subtle trace of the fibrous structure on the reflective envelope from the outside, which transforms into an expressive texture when illuminated from within at night. On the inside, the initial softness of the roving bundles remains tangible in the fibres' texture, which strongly contrasts with their actually hardened state. This evokes at the same time a strong sense of transparency and even airiness, as well as a stringy leanness and tangible tautness of the extremely lightweight structure.

Fig. 4 *Left* The interior of the pavilion reveals the intricate carbon-fibre structure that articulates the spatial surface and at the same time provides the structural support for the transparent—and initially inflated—ETFE envelope. *Center* Computational design, simulation, and fabrication enable a synthesis of structure and skin that is perceived differently on the interior and exterior of the pavilion. During the day, only a subtle trace of the fibrous structure is visible on the reflective envelope, whereas the stark contrast between the transparent skin and the black carbon is strongly perceived on the inside. *Right* At night, the constructional logic of the cyber-physical design and fabrication approach remains tangible in the distinctive architectural articulation of the pavilion

Emerging Material Culture

The projects introduced above begin to suggest how material performance and architectural performativity can be synthesized in ways that go far beyond a trite truth to materials and related fixed and singular structural and spatial typologies. The computational convergence of the processes of form generation and materialization enables new modes of architectural speculation and experimentation that will contribute to the definition of a truly contemporary, computational material culture, which also constitutes an important facet and ambition of Parametricism 2.0. A humble indication of the potential richness of such an integrative design approach may be given by the pavilion examples illustrated here, which all stem from one coherent body of design research yet display a considerable variety in formal, spatial, and structural articulation.

References

Fleischmann, M., Knippers, J., Lienhard, J., Menges, A., Schleicher, S.: Material behaviour: embedding physical properties in computational design processes. In: Menges, A. (ed.) 2 Material Computation, March/April (no 2), pp. 44–51 (2012)

Kwinter, S.: The computational fallacy. In: Menges, A., Ahlquist, S. (eds.) Computational Design Thinking, pp. 211–215. Wiley, London (2011)

Kwinter, S.: Cooking, Yo-ing, Thinking. Tarp: Not Nature, p. 108. Pratt Institute, New York (2012)

Lienhard, J., Schleicher, S., Knippers, J.: Bending-active structures: research pavilion ICD/ITKE. In: Nethercot, D., Pellegrino, S., et al., (eds.): Proceedings of the International Symposium of the IABSE-IASS Symposium, Taller Longer Lighter. IABSE-IASS Publications, London (2011)

Menges, A.: Material resourcefulness: activating material information. In: Menges, A. (ed.) 2 Material Computation, March/April (no 2), pp. 34–43 (2012)

Menges, A.: The new cyber-physical making in architecture: computational construction. In: Menges, A. (ed.), 2 Material Synthesis, September/October (no 5), pp. 28–33 (2015)

Menges, A., Knippers, J.: Fibrous tectonics. In: Menges, A. (ed.) 2 Material Synthesis. September/October (no 5), pp. 40–47 (2015)

Reichert, S., Schwinn, T., La Magna, R., Waimer, F., Knippers, J., Menges, A.: Fibrous structures: an integrative approach to design computation, simulation and fabrication for lightweight, glass and carbon fibre composite structures in architecture based on biomimetic design principles. CAD J. **52**, 27–39 (2014)

Part II

Structural Innovation

Photograph credits: *top*: © Block Research Group, ETH Zürich, *bottom*: © Iwan Baan

Structural Innovations often occur at the intersection of several aspects of construction: technology, form, structure, materials, and forces. For example, Felix Candela associated a specific form (The Hypar which is a doubly ruled surface) with a specific technology (formwork constructed with straight boards) to design and construct innovative thin shell structures with an economy of means. On the other hand, Frei Otto associated a specific type of form (anticlastic doubly curved surfaces) with a specific method to find the equilibrium of forces (The Force Density Method) to generate innovative tensile structures. In contemporary practice, numerical parametric tools provide designers a deeper understanding of the interplay between all of these critical design issues and allow for rapid exploration and testing of a multidimensional design space. Modern design methods such as computational graphical static, grammar-based design, and construction aware design also provide insight to help the designer to explore complex design scenarios. These continuously evolving methods and computational tools are promising, and they provide a rich paradigm to facilitate structural innovation in the future.

Olivier Baverel
Caitlin Mueller

Make Complex Structures Affordable

Jean-François Caron[1(✉)] and Olivier Baverel[1,2]

[1] Navier Laboratory (ENPC/IFSTTAR/CNRS), Ecole des Ponts ParisTech, 6/8 avenue Blaise Pascal, 77455 Marne la Vallée, France
caron@enpc.fr
[2] Laboratoire GSA, École nationale supérieure d'architecture de Grenoble, 60 Avenue de Constantine, 38036 Grenoble, France

Navier laboratory, a joint research unit between the Ecole des Ponts ParisTech, IFSTTAR, and CNRS, gathers about 50 permanent scientists, 120 Ph.D.'s and general skills in the mechanics and physics of materials and structures, in geotechnics, and their applications to, in particular, civil engineering and petroleum geophysics. For last 15 years, Navier has chosen to reinforce the links between mechanics of materials, structural engineering, and applied mathematics at a very high level to explore new paths and propose building innovations. Three main keys are explained and illustrated with examples. The first key is a deep knowledge about materials and especially about innovative ones which often look still after their tailor-made purpose. A second one is a rigorous mathematical management of shapes and geometry to rationalize complex situations in a fully integrative way, including cladding and connections for example. And finally, if one can't stay blind to the new digital and technological prospects which penetrate all the economic sectors, as robots and 3D printing, what are the relevant digital ideas for building?

Several recent scientific results and prototype are presented in the following as proofs that combining all these aspects may help for the design of complex structures making them more affordable.

Think Mechanics of Material

Ph.D. of Saskia Julich, Cyril Douthe, Lina Bouhaya, Frederic Tayeb, Lionel Du Peloux, Natalia Kotelnikova, Philippe Hannequart, Sina Nabaei, Nicolas Bouleau

To design a structure, the specific behavior of the material that will be used plays a key role. It is always important to remember that the first bridge in iron looked very similar to the one in masonry. It is only after few decades of development that the steel found its own structures such as Pratt truss… New materials in the building industry such as composite materials are often copying unadapted steel solutions (truss, bolted connections, I or T beams…), or even more trendy, some 3D printing solutions for concrete without rheological consideration. Several examples of innovative use of material to design structures are investigated in our laboratory. A first example concerning smart material alloys is the topic of a DMSP presentation [P. Hannequart] and is relied to sun-shading device. It is impossible to propose relevant systems without modeling and testing the highly nonlinear

K. De Rycke et al., *Humanizing Digital Reality*,
https://doi.org/10.1007/978-981-10-6611-5_2

behavior of SMA, since its behavior is coupled with the structure response. A second example concerns composite materials and an innovative way to use them. Composite materials must be used according to their specific characteristics (lightness, flexibility, etc., and even if classically considered as weakness), and by looking for new adequate designs. For instance, elastic gridshells have been made of wood because it is the only traditional building material that can be elastically bent with large deformations without breaking. But glass fiber reinforced plastics (GFRP) have higher elastic limit strain (1.5% at best for GFRP and 0.5% for wood), and their Young's modulus also is higher (25–30 GPa against 10 GPa for wood). One can expect the buckling load of a gridshell in GFRP to be 2.5–3 times higher than one made of wood.

Four full-scale prototypes of composite gridshells have been built by Navier laboratory. The two first ones were built on the campus of Ecole des Ponts ParisTech and were tested under several loading conditions in order to investigate the behavior of gridshell structures and to compare with the numerical models. Detailed results of these tests can be found in Douthe et al. (2006, 2010) and Tayeb et al. (2013). The behavior of the prototype is very close to numerically performed simulations. After that, and several small prototypes, two major gridshells built to house people have been recently made. The first one for the Solidays festival (June 2011, Baverel 2012, Fig. 1 above) and the last one built to temporarily replace the Creteil Cathedral (February 2013, Du Peloux et al. 2013, Fig. 1 below, right and left). Both of them were built in collaboration with the consultant engineering office T/E/S/S and the building contractor VIRY. The dimensions of these structures are quite similar: around 7 m high, 25 m long and 15 m wide. They are constituted of about 2 km of pultruded unidirectional tubes (polyester resin and glass fibers) with a Young's modulus of 25 GPa and a limit stress of 400 MPa. The available length and diameter of the tubes are, respectively, 13.4 m and 41.7 mm; the wall thickness of the tubes is 3.5 mm.

Fig. 1 Composites gridshell, Solidays 2011 (*middle*), Créteil 2013 *below*, (*left* and *right*)

Only a scientific approach of this kind of innovative material (see also Kotelnikova-Weiler et al. 2013) had permit these realizations. And it is possible to go further. For example, using composite profiles with anisotropic sections (different from square or axisymmetric ones) generates bending/torsion coupling and consequently richer freeform. We propose such development in Lefevre et al. (2017). And why not a fully elastic prestressed footbridge like we proposed in Caron et al. (2009)?

Constructibility Is Mathematics

Ph.D. of Romain Mesnil, Nicolas Leduc, Xavier tellier, Robin Oval, Yannick Masson

The last decades have seen the emergence of new tools for architects; these tools mainly based on Nurbs allowed to produce nonstandard architectural shapes, and unfortunately these shape were often costly economically and environmentally in terms of the amount of materials needed for fabrication and due to the complexity of the assembly. Designers find often themselves helpless with the geometrical complexity of these objects. Furthermore, the available tools dissociate shape and structural behavior, which adds another complication.

To tackle the problem, some companies propose a post-rationalization of the shape proposed by the designer in order to solve some construction aspects such as planar facet or node without torsion… This post-rationalization could be time consuming and often needs an expert to run it.

The research carried out at Navier takes the mathematical point of view based on invariance under geometrical transformations and studies several strategies for fabrication-aware shape modeling. In other words, we provide tools that give shape that are natively fabrication-aware with no need of heavy post-rationalization technic. Three technological main points have been identified and correspond to three independent contributions.

a. The repetition of nodes is studied via transformations by parallelism. They are used to generalize surfaces of revolution. A special parametrization of molding surfaces is found with this method. The resulting structure has a high node congruence (Fig. 2, left). Freeform generated as an isogonal molding surface has only few different types of node (Mesnil et al. 2015).

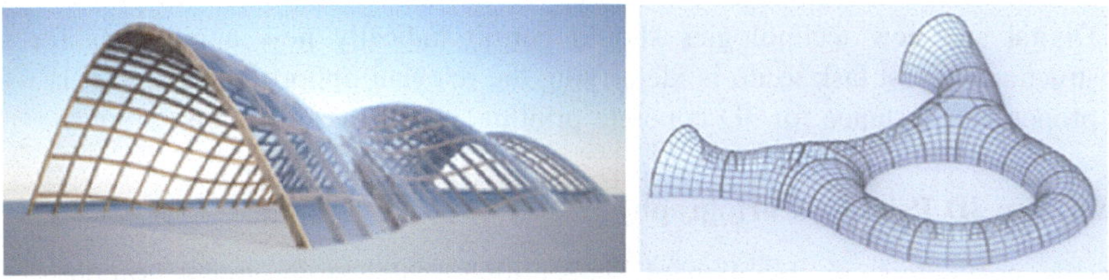

Fig. 2 *Left* Freeform generated as an isogonal molding surface. *Right* A geometry with a complex topology modeled with cyclidic nets

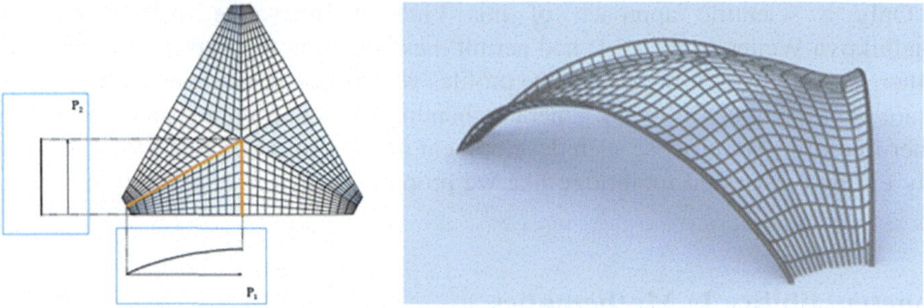

Fig. 3 *Left* Decomposition of a complex mesh into simple patches. *Right* The corresponding lifted mesh, a Marionette Meshes with a singularity

b. Cyclidic nets are then used to model shapes parametrized by their lines of curvature. This guarantees meshing by planar panels and torsion-free beam layout. The implementation of several improvements were done, like doubly curved creases, a hole-filling strategy that allows the extension of cyclidic nets to complex topologies, and the generation of a generalization of canal surfaces from two rail curves and one profile curves (Fig. 2, right) (Mesnil et al.).
c. An innovative method inspired by descriptive geometry is proposed to generate doubly curved shapes covered with planar facets. The method, called marionette technique, reduces the problem to a linear problem, which can be solved in real time (Fig. 3).

A comparative study shows that this technique can be used to parametrize shape optimization of shell structures without loss of performance compared to usual modeling technique. The handling of fabrication constraints in shape optimization opens new possibilities for its practical application, like gridshells or plated shell structures. The relevance of those solutions has been demonstrated through multiple case studies (Mesnil et al. 2015, 2017a, b).

How May Digital and Robot Help?

Ph.D. of Romain Duballet, Vianney Loing, Tristan Gobin, Nicolas Ducoulombier, Pierre Margerit

Digital and new technologies should permit radically new approaches for the construction. A first task to do is identifying the relevant opportunities, and it is what we propose for instance for 3D concrete printing, or for mason robots.

Innovative 3D Printing Cartography

Specific parameters are highlighted—concerning scale, environment, support, and assembly strategies—and a classification method is introduced. We denote a given set of building systems by enumerating the parameters it is concerned with:

Fig. 4 *Left* An example of an a^4 assembly type, (*center*) an example of a s^1 situation, and *right* a a$x_0^1 x_e^1 e^2 a^2 s^4$all application (from Democrite French project, 2015, Xtree, ENSAM…)

$$x_0^n x_e^m e^i a^j s^k \ldots$$

where p^n stands for the nth version of parameter p. The two parameters x_0 and x_e are scale parameters, respectively, about the printed object and the extruded paste, while the parameters e, a, and s, respectively, concern printing environment, assembly, and support. For instance: x_O^0 concerns printed object of size less than a meter, typically a connection, while x_O^1 a constructive element of size around 1–4 m, typically a beam, column, or slab. x_e^2 is for a thickness of the printed layer between 5 and 30 cm, e^0 means On-site (direct printing), while e^2 means Prefab factory (indirect printing), a^4 an assembly of external element(s) during printing (Fig. 4a) and s^1 a printed support, left in place (Fig. 4b). A wall prototype from Ensam and Xtree, $x_0^1 x_e^1 e^2 a^2 s^4$ on Fig. 4c.

A map of the different approaches and their associated robotic complexity will be proposed in the presentation and detailed in a paper under review. We have compared existing works (13 teams) thanks to this classification and decided a first development relevant with our devices (6 axis robots), a $x_0^1 x_e^1 e^2 a^4 s^3$ system of collaborative printing.

"$x_0^1 x_e^1 e^2 a^4 s^3$", an Innovative Insulated Printed Wall

Constrained masonry, the assembly technique of breeze blocks and mortar restrained in a reinforced concrete frame, is a very popular building system, especially for individual and collective housing, for it is at once cheap, fast, and easily implemented. From a purely mechanical point of view, it is however quite inefficient. In the case of a one- or two-storey house, the need in mechanical resistance for the wall itself, considering the presence of the reinforced concrete frame, is indeed far lower than the breeze blocks/mortar system can provide. The main role of this staking is in fact to allow solid continuity between the concrete frame, for bracing purpose and to act as separating wall. Up to a limit, it could be said that the mortar between the blocks is the only needed element to provide resistance. Such considerations lead us to the idea of assembling insulating blocks instead of breeze blocks, leaving the mechanical role to the mortar in between, and getting thermal performances in addition. The new system for the wall is now a generalization of the previous one: a continuous spatial structure

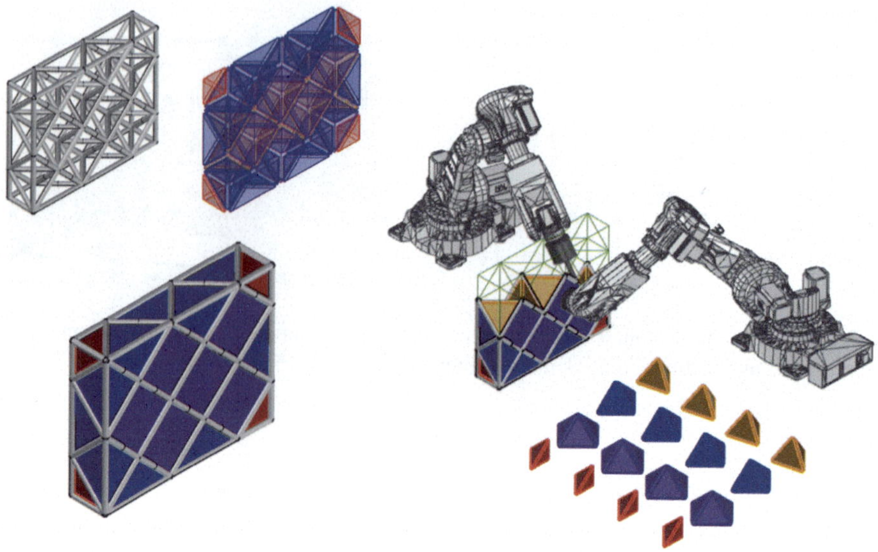

Fig. 5 Concept of fabrication of the insulated printed wall

in mortar, and thermic insulation in the negative space. The key aspect of the technique is to assemble specifically shaped insulating blocks by printing a mortar joint at the edges location. The mortar is extruded through a nozzle controlled by a robotic arm, as described in Gosselin et al. (2016). The mortar acts as joint for the insulating blocks, while they act as printing support for the mortar. It can be decomposed (Fig. 5) into the following steps, insulating blocks fabrication (from polystyrene panels thanks to a hot-wire robotic cutting process), insulating blocks assembly (by a pick and place robot) which form a layer of printing support, and mortar extrusion. The mortar is extruded by a printing system (Mesnil et al. 2015). This system brings some shape constraints for the blocks, and they must form a space tessellation, the edges of which will form a mortar space truss.

Smart Mason Robots

The aim of this work is to show the potential of robotics arms and computer vision to autonomously assemble masonry walls where all the ashlars can have different shapes. We collaborate here with Imagine, another research laboratory of Ecole des Ponts ParisTech, expert in computer vision, 3D reconstruction and scene understanding, machine learning, and optimization.

For the technique to be easily and robustly usable on a construction site, we only rely on simple 2D cameras rather than 3D scanners. Pose (location and orientation) estimation of the ashlars according to the robot is performed through the training of a convolutional neural network trained on a dataset of synthetic images (He et al. 2015). Synthetic datasets are created and a networks trained to estimate the position and orientation of one specific ashlar among others ashlars on the same image. A wall with

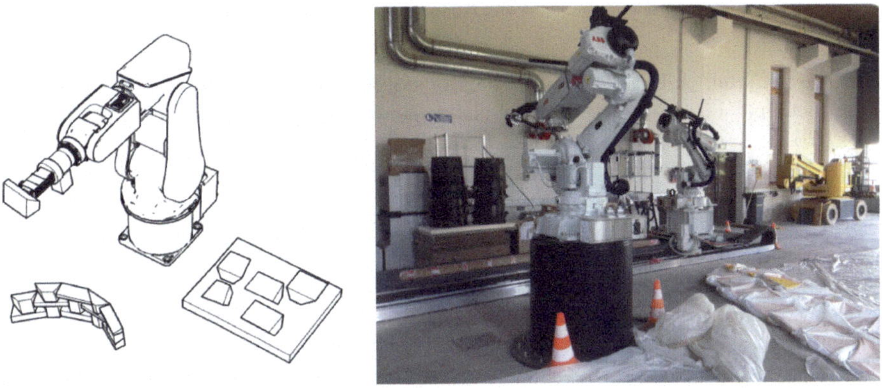

Fig. 6 *Left* Abeille-like curved wall and mason robot. *Right* Navier robotic devices

ashlars whose shapes are all slightly different could be therefore easily built, for example, an Abeille-like wall (Fig. 6) with different ashlars with interesting mechanical properties (bending behavior, seismic or dissipative applications) that we study from a theoretical and numerical point of view (finite element, discrete media models). First experimentations are in progress in our new collaborative platform Buildin'lab (Fig. 6 right). This approach therefore opens a new way to think effective masonry structures.

Conclusion

This short paper gives a brief description of some researches carried out for construction in the Navier laboratory during the last 15 years. The paper focuses on proposals that make complex structures more affordable. Several examples are shown such as composite gridshells, complex shapes natively fabrication-aware, 3D printing of concrete or smart mason robot for complex masonry. The main specificity is to go from theoretical aspects to scale:1 prototypes, through material testing, numerical simulations, worksite aspects and also by mastering the numerical workflow to go from the concept to the construction. The strength of the team is to be composed of experts in mechanics of materials, structures, discrete differential geometry, and now robotics. The development of a new collaborative platform building lab in the Ecole des Ponts Paristech including several robotics solutions and digital learning will be a chance to explore new challenges with our academic and industrial partners.

References

Baverel, O., Caron, J.-F., Tayeb, F., du Peloux, L.: Gridshells in composite materials: construction of a 300 m^2 forum for the Solidays' festival in Paris. Struct. Eng. Int. **22**(3), 408–414 (2012)

Caron, J.-F., Julich, S., Baverel, O.: Selfstressed bowstring footbridge in FRP. Compos. Struct. **89**(3), 489–496 (2009)

Douthe, C., Baverel, O., Caron, J.F.: Form-finding of a grid shell in composite materials. J. Int. Assoc. Shell Spat. Struct. **47**(150), 53–62 (2006)

Douthe, C., Baverel, O., Caron, J.F.: Gridshell structures in glass fibre reinforced polymers. Construc. Build. Mater. **24**(9), 1580–1589 (2010)

Du Peloux, L., Tayeb, F., Baverel, O., Caron, J.F.: Faith can also move composite gridshells. In: Proceedings of the International Association for Shell and Spatial Structures (IASS) Symposium (2013). Gridshell Struct. Const. Build. Mat. **49**, 926–938 (2013)

Gosselin, C., Duballet, R., al.: Large-scale 3D printing of ultra-high performance concrete—a new processing route for architects and builders. Mater. Des. **100**, 02–109 (2016)

He, K., et al.: Deep residual learning for image recognition. Arxiv.Org **7**(3), 171–180 (2015)

Kotelnikova-Weiler, N., Douthe, C., Hernandez, E.L., Baverel, O., Gengnagel, C., Caron, J.-F.: Materials for actively-bent structures. Int. J. Space Struct. **28**(3–4), 229–240 (2013)

Lefevre, B., Tayeb, F., du Peloux, L., Caron, J.-F.: A 4-degree-of-freedom Kirchhoff beam model for the modeling of bending–torsion couplings in active-bending structures. Int. J. Space Struct. (on press) (2017)

Mesnil, R., Douthe, C., Baverel, O., Léger, B., Caron, J.-F.: Isogonal moulding surfaces: a family of shapes for high node congruence in free-form structures. Autom. Constr. **59**, 38–47 (2015)

Mesnil, R., Douthe, C., Baverel, O., Léger, B.: Generalised cyclidic nets for the modeling of complex shapes in architecture. Int. J. Archit. Comput.

Mesnil, R., Douthe, C., Baverel, O., Léger, B.: Marionette mesh: from descriptive geometry to fabrication aware. Adv. Archit. Geom. (2016)

Mesnil, R., Douthe, C., Baverel, O., Léger, B.: Buckling of quadrangular and Kagome gridshells: a comparative assessment. Eng. Struct. **132**(3), 337–348 (2017a)

Mesnil, R., Douthe, C., Baverel, O., Léger, B., Caron, J.-F.: Structural morphology and performance of plated shell structures with planar quadrilateral facets. J. IASS (2017b)

Tayeb, F., Caron, J.F., Baverel, O., Du Peloux, L.: Stability and robustness of a 300 m^2 composite. **49**, pp. 926–938 (December 2013)

Part III

Data Farming

Fig. 13.—*Improved Corn and Seed Drill.*

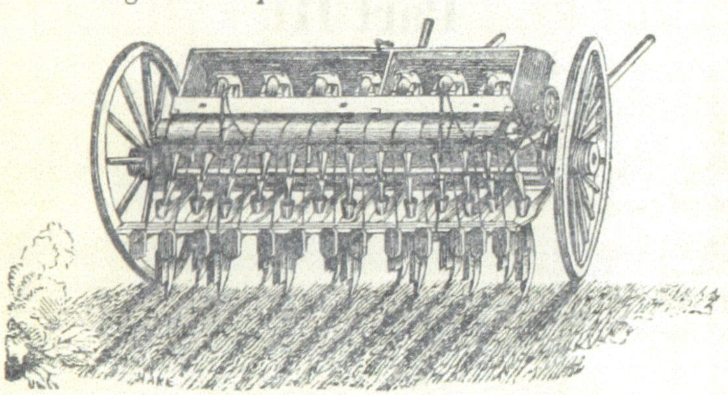

From "farming implements, their various uses and recent modern improvements compared with the old machines." By F. D. P, page 51

Agriculture is arguably the most transformative technology humankind created before entering the digital age. It allowed the control of natural environments, the creation of complex systems and networks to provide food (energy) supplies for humans, the creation of the sedentary human, and finally of the modern age of cities. The industrialization process of the last 150 years incorporated the mechanization of agriculture, the industrialization of farming, collecting, harvesting, and conservation techniques and processes all powered by fossil fuels (energy). Automation seems to be the next paradigm for agriculture systems, and many other systems that rely on technology today. It contains a set of principles that could be used as a metaphor for our approach to the latest form of resources: data, which is the result of the digital revolution in computation, communication, and fabrication. Bits or data is becoming a precious resource that fuels the emergent global economy, based on information management and knowledge production and distribution. This explains partially the growing and already extensive interest in digital design, or designing with digital tools. It has been part of and is still growing in the discussion in the architectural field for more than 20 years and has seen rapid changes in the last decade with the rise of smartphones, bigger data storage, and faster computing power. We propose in this chapter a discussion on the design of digital systems and processes: farming, harvesting, collection or distribution of bits, storage and conservation, use and interpretation in multiple scales, and levels of engagement. Data is often described as the "new oil",[1] obviously from the extractive and centralized paradigm of the twentieth century. Our approach is generative and distributive, understanding that it will be in the twenty-first century when the digital age reaches its full potential.As in the food chain, data can be farmed in different ways from small systems at the domestic scale through low-cost and DIY sensors, to big data scrapped by server farms in which algorithms extract patterns from human interactions through digital platforms. Just as food farmers need manual or automated tools, seeds, or tractors, data farmers need to have access to create and choose their own tools for data-driven design at different scales: from sensors to algorithms, to servers, to interpretation and visualization tools. Data is a resource that is produced by many agents in a distributed manner; there is not just one source and one result. It is collected, stored, and processed in decentralized networks that are part of the complex infrastructure behind the bits that organize information. Data is a tool to generate knowledge, to understand patterns, and to program behaviors that are transforming the way we design the inhabitable world.Designers, urbanists, architects, and artists integrate datasets into design and creative processes. Design outputs are determined by the use of data, turned back into another layer of information and by processing algorithms to another set of automatically created (re)interpretations. Digital, parametric, and generative design processes start from the integration of data and computational tools, forming a workflow in which decisions are taken between machine and human:

[1] Read also the Economists printed edition of 6th May 2017 on "The world's most valuable resource".

Illustration by David Parkins, in “The world’s most valuable resource is no longer oil, but data”, The Economist, 6 May 2017

Turning atoms into bits: capturing data from the physical world to the digital world

Interpreting datasets to turn them into information

Informing design decisions by programmed processes

Using bits to create atoms: creating a design output to intervene in the physical world

Transforming the way we read the digital world through design

Transforming the way we take decisions and finally to let decisions be made by (automated) computed interpretation

We, as designers facing these changes, should be asking ourselves: How much we are influenced by the quality of the data and the interpretation processes that are used to inform our decisions? How can we design our own design processes to maintain human input? How can we create our own data inputs, data interpretation algorithms, data storage, and distribution infrastructure?

In this session, we want to question the tools, the processes, and the political issues behind data farming as a new (re)generative design process. One permits design systems that enable the distribution of value as a core principle, and one creates a positive impact and is embedded in the environment and society.

Klaas De Rycke
Tomas Diez

Data Morphogenesis

Kasper Jordaens(✉)

Solution Designer at imec, De Krook, Miriam Makebaplein 1, 9000 Gent, Belgium
kasper.jordaens@imec.be

How can we shape data so that it helps us as designers. How can seemingly random noise evoke emotions and look nice? Apparently designers can guide data to take pleasing shapes. I will elaborate in this paper on how to look at data, strategies to process it, and why live interaction can be key to better understand what the data means. We will also look at how designers can make use of data processing to support a creative process.

Introduction

Data Data Data

In the last two years, we as humanity have produced more data than we ever did in our entire history. It is just one way of telling an audience that we do actually produce a lot of data. What are we supposed to do with all that data? Let me contextualize a bit.

First of all the process of creating data was made almost automatic, a picture here, a filter there, a multimegapixel camera in every pocket, and unlimited resources to store them…

This makes it really easy to create data. If it is automatic, is it that important? Of more importance is the fact that we (can) share this data that we can keep it (if not using snapchat) and that we can create and access it anywhere anytime.

On top of that, we create metadata, so the data about the data is added to the pile. Actually, this process is similar to how we used to create data, and store in our carbon-based brain but there is a difference.

Filter Data

Because writing the above paragraph feels like, take everything into account because FOMO and because YOLO, you need to know everything and you need to know it instantly… It's all about context. Context is a filter; it takes care of us. It consists of noise, filtering out surrounding conversations for example. And our brains are well trained to take only the essentials and leave peripheral conversations unparsed. The Internet trains our brains otherwise. Submit to the algorithm. The Internet will parse the context for you. Do not try to alter the stream… This is the completely opposite strategy of our own brain.

K. De Rycke et al., *Humanizing Digital Reality*,
https://doi.org/10.1007/978-981-10-6611-5_3

I pose that you have to filter your own stream… take care of your own data. This is what I am saying. We have been producing loads of data in all our history. What's changed is…

1. we started recording it
2. we added machines to the conversation
3. we started filtering the data on other man's behalf
4. we started denying peers access to the data we are creating ourselves
5. we started losing control of our own context…

The context we operate in is data. How can we deal with this context without losing our mind or without losing control?

Strategies of Taking Control

Capture Data

Lots of data is surrounding us. We can capture our surroundings ourselves to be more informed, or, as a first step toward being more informed, to have more data. Data can be transformed in intel, but it starts with data. It is step one. Recent technological progress made this simple. We can sense data automatically and we can process it automatically and we can connect the data to any other target.

Merge Data

Making sense of data is hard for machines. We as humans are trained to do so. Think of "the conversation in a bar" example. This is an extremely hard task for a computer. Machines still need a hand when handling multiple data streams finding out which is the relevant one Facebook's stream is the ultimate data merge (for now) for your social network, merging all data from your friends and their friends. To give you the most relevant and for investors most financially rewarding information, they process a lot of data. This is quite an achievement, but not enough as the efforts to merge this data with the context are far from sufficient to provide a good experience. Strategies for merging data ourselves are experimental at the moment. The algorithms required are not open, and writing these strategies yourselves involves a lot of effort and endurance.

From Chaos to Information

So that we took control of the data, what can we do with it? How can we deal with it and learn from it?

Interpret Information

Data by itself is meaningless (even after we filtered it). Eventually when we look at data before you can even think about it, it is just signals firing some neurons in your brain. You have to give data a meaning yourself. Ordering impulses and structure them is a very basic skill of the human brain. But to do this requires an enormously complex organic chemical-electronic device, the brain. In 1943, a model was devised that was the basis of all the neural network computation models to follow[1] McCulloch & Pitts were on the right track; their insights paved the way for nowadays' digital neural networks. The way we need to train these software neural networks is also similar to how we believe humans learn. Without elaborating on this and challenging this process, want to show a little experiment I did to illustrate what these networks are capable of.

I created an image consisting of 1 million random pixels using a standard graphics processing library (ImageMagick) by running: convert-size 1000 × 1000 xc: +noise Random noise.png

The result looks like this (Fig. 1).

Fig. 1 Random noise

I fed this data into a deepdream cluster at https://deepdreamgenerator.com. This is based on a neural network that was built to recognize objects, but people started abusing it to generate images, by reversing the process, essentially asking "What does this look like" instead of "What is this".

---

[1] LOGICAL CALCULUS FOR NERVOUS ACTIVITY https://classes.soe.ucsc.edu/cmps130/Spring09/Papers/mculloch-pitts.pdf.

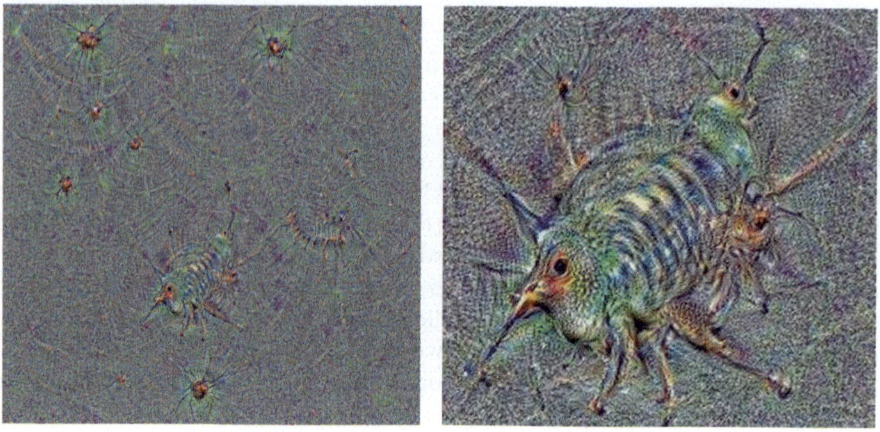

Fig. 2 Deep dream rendition + detail

Now if you feed it random pixels like I just did, it comes back with (Fig. 2).

Remarkably enough deepdream interprets random data into things our own brain can somehow place into the bigger picture. The process itself is pretty well explained online[2] but the point here is that even random data can (seem to) mean something, if processed correctly. Meaning is arbitrarily chosen, in this case because of the training data it was fed must have looked like birds or insects or something.

Trying to understand patterns in seemingly random data is difficult, but big business. Stock market trading seems to be all about it, and all wrong as well from time to time. But analyst has been trying to make patterns of the numbers and to influence the numbers the other way around by interpreting the context. This is true for financial newspapers who came at least a day late with their insights in 1929 as with algotrading where algorithms compete to get insights milliseconds before the competition does. Everyday more and more data is added to a stack, and big data algorithms try to take a shot at predicting this seemingly random data. Humans nor computers can tell the difference between random (e.g., random walk or Brownian noise) data and the market data and this is true even though the market data (Dow Jones e.g.) does not even follow the standard deviation.[3] So how do computers or humans make sense of this data as it does not even follow the most basic law of statistics. To start with it is always an interpretation. This is true for any data. 35 °C can be hot, but it is rather cold for a sauna. So what we do is add a ruleset. Rules can be "are we in a sauna" or "are we trying to boil an egg". These rules help to get meaning out of the data. Computers are very strict in abiding rules; they will (at least for now) stick to what they were programmed to do. This makes it impossible for a computer to react to a data event it was not programmed to react to. But the moment a data event outside the computer's

[2] https://web.archive.org/web/20150703064823/http://googleresearch.blogspot.co.uk/2015/06/inceptionism-going-deeper-into-neural.html.

[3] The Misbehaviour of Markets—2004 MandelBrot & Hudson pp. 185–187.

"normal" comes in it's stuck. You get a flash crash[4] or some other difficult to manage event. In the 2010 flash crash, trillions of dollars were lost in minutes because computers were not programmed to respond to an event like this. Whatever the cause was (numerous official institutes, press and academics tried to untangle the data of that day, but there is no real consensus) it is clear that computers cannot respond to events they are not trained to respond to. This is how Kasparov tricked a battery of chess computers into losing to him in 1985 (12 years before deep blue beat Kasparov). The computers back then were not powerful enough to brute force all combinations and Kasparov exploited this at some point. This is human against machine. Today, we humans lose from chess computers (and Go, Jeopardy, checkers, etc.). Unless we get assistance from a machine. Even if that assistant machine is much less powerful than the opponent computer. Just for a reference on how complex the chess problem is:

> A player looking eight moves ahead is already presented with as many possible games as there are stars in the galaxy.[5]

But apparently if we use the right tools, humans can amplify their cognitive states to grasp a part of this.

Interacting with Information

So computers help us to turn chaos into information, but unless the universe the computer has to interface with is completely internal (fully known and bound) interaction is required. The word interaction implies some form of real-time info flowing back and forth, influencing the state in both directions. A feedback loop that keeps going.

In the last couple of years, I have had some chances on studying this behavior either by going to conferences on data both creative and commercial (e.g., visualized or resonate[6]) and academic, for example, at ICLC where the systems for transforming data in a live interactive environment are being discussed and researched.[7] (http://iclc.livecodenetwork.org) Also through experiments as a computer-based performance artist in sound and visuals where I aim to maximize the interaction between the performers and the data both with classical rock-band setup or 100% computer-based flows. A restriction I put into place is that I do not want to just program music, I want to play the music and visuals that I program. This way, when the universe reacts, you can respond to this. So being able to bend the rules, I just set for the computer or even completely reprogram it, based on the data the computer and the context give back.

[4] https://en.wikipedia.org/wiki/2010_Flash_Crash.

[5] From the book *Chess Metaphors*, Diego Rasskin-Gutman.

[6] Meeting people like Karsten Schmidt (@toxi) who wrote https://medium.com/@thi.ng/evolutionary-failures-part-1-54522c69be37 or Amy Sterling (@amyleesterling) who made me see the neuroscience part of data or seeing people like Jer Thorpe (@blprnt) talk heavily influenced my thinking concepts on data and computing.

[7] Leading the way here in being creative with a live interface to the algorithms for creating music is Alex McLean (@yaxu) who created tidal and makes music with it https://slab.org/publications/.

Fig. 3 Pattern game

An interesting tool for starting live reprogramming is pure data.[8] It allows to interrupt dataflows and change a running program on the fly, and it gives you visual feedback on while programming. If you are not afraid of text-based environments you should try SonicPI[9] which is a common entry point for music coding. There are many different livecoding environments out there what they have in common is that they allow reacting on a running program, without interrupting it.

Set Some Rules

This is a picture I kept from a game my son was playing (see Fig. 3). He asked me if I could see a pattern. I pointed out that I see diagonals from alternating colors. He however constructed it using another pattern. Make sure no two colors would touch in a straight line. So we had the same result from two different rules. Another interesting thing is that he constructed this using a simple rule, and patterns started emerging. This is known as cellular automatons, and they have been part of computer science for a long time. Although they can return pretty amazing patterns in their own universe make, if you then make them interactive they are a great source of inspiration for pattern-based designing. For a one-dimensional system (using only black and white, and the immediate neighboring cells as input) 256 rules are possible. See here[10] for a plot of all the 256 rules for one-dimensional CAs. Some of them are great examples of how simple rules render complex results and can be used as a seed for further explorations. As an example take rule 30 (Fig. 4):

[8] http://puredata.info/.

[9] SonicPI was created by Sam Aaron (@samaaron) who is evolving the computer as an instrument, bringing the power of computing to people as easy and expressive as possible.

[10] https://plato.stanford.edu/entries/cellular-automata/supplement.html.

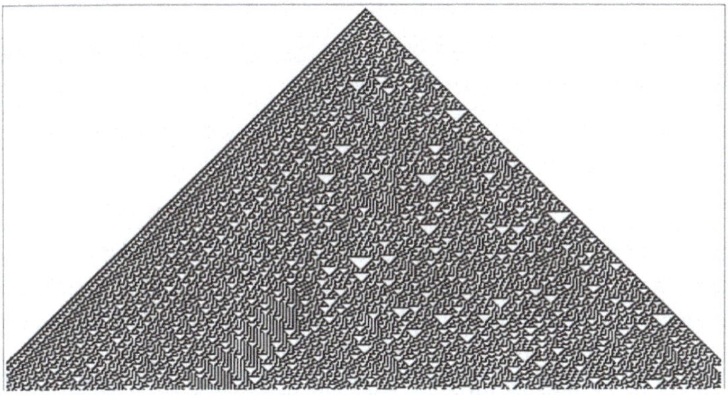

Fig. 4 Rule 30 from "a new kind of science"

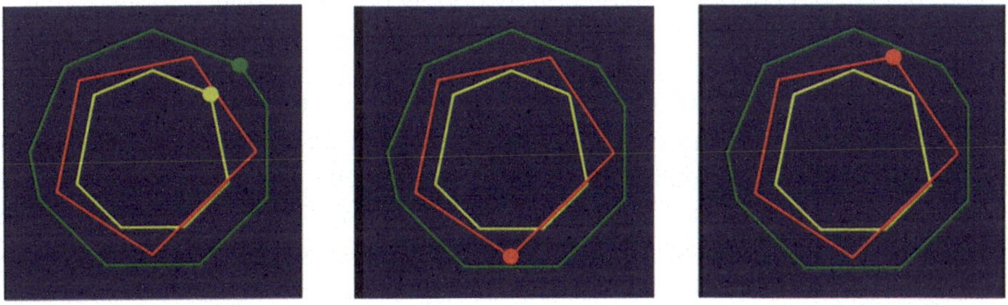

Fig. 5 Visual representation of euclidean rhythms as found on YouTube (https://www.youtube.com/watch?v=vMMcNlqzikw)

Cell-based plotting of "rule 30" renders seemingly random patterns interleaved with structured patterns as demonstrated in the image above. But in this image there is no interaction possible, the context is what it is. Variations on the rule render different and surprising results.

Sometimes we have been using these rules without knowing. Different Euclidian Rhythms can be generated using a very simple formula. The results of this formula have been part of traditional music for a very long time. Nobody knew the rules, but different cultures played music following this simple rule.[11] They just felt the pattern (Fig. 5).

Bend the Rules

Music, and certainly traditional music which always had to be played live required participants to be interactive with the rules. They had to feel and interact with their music and their context.

[11] The Euclidean Algorithm Generates Traditional Musical Rhythms Godfried Toussaint∗ School of Computer Science, McGill University.

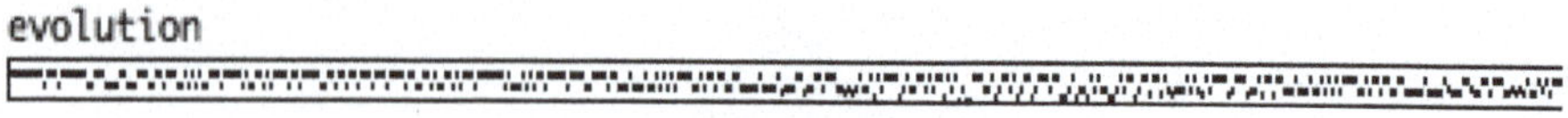

Fig. 6 This strip is part of a live rulebend pattern from a cellular automata I wrote in pure data. It was played live to respond to my input while sticking to rules I put in place before, resulting in unpredictable yet "conforming" structures

Rules can be bend and by changing and moving constraints you get different results. By sticking to the rules, you just get ordinary programs running resulting in regular morphogenesis like the rule 30 plot above. By making the rules interactive, and not strictly stick to them, you can keep interesting aspects resulting from using a rule-based pattern generator combined with the live aspects. For example, you can modulate rules using live input from connected data sources (Fig. 6).

Conclusion

Now even with the simplest rules that have been bend in the most fantastic ways, it boils down to what you can do with it. The data is still an intermediate result. In the end, you want to shape something. To do this, you steer the morphogenesis. You need to guide the data and place it in a new context where it can flourish. As example of this I like to conclude with a project that is a data visualization of the same rhythm from a record, just two measures of a whole record. But it is combined with the context so it renders different results each time it is plotted. The context is the paper you plot on, the pens you use, and an arbitrary number extracted from the entropy of the computer. Three hundred covers have been plotted this way, and not one is the same, although families can be seen (Fig. 7).

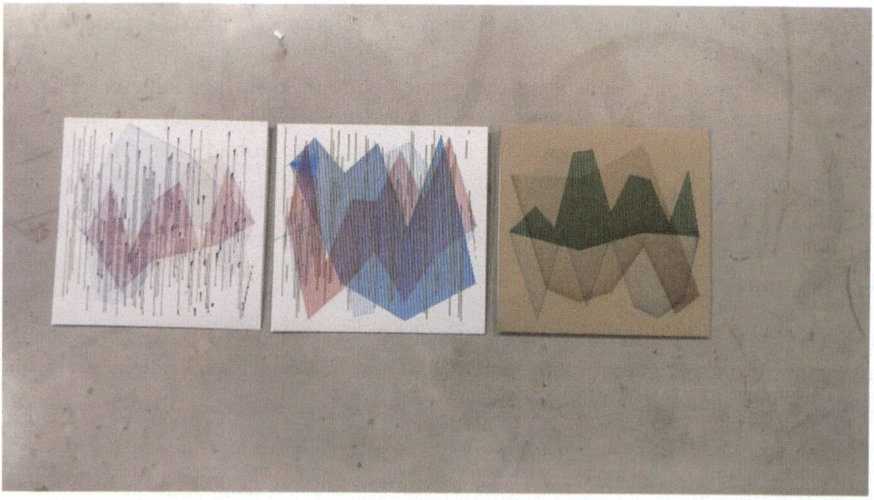

Fig. 7 Triangle Yur - Sondervan | Cover art algorithm kaosbeat

Generating form, whether sound, image or shape or even function is always a matter of getting data, processing it, adding the right amount of context and see if it fits in that context. It can remain flat and just add a layer of fun to massive layers of data. It can also be a deep dive into unknown territories and results. The data remains something to be processed and without guidance it is lost. With some guidance, it can surpass expectations and add additional layers to our direct thinking and environment. Without losing ourselves in doom scenarios, the computing power of today just adds a further possibility of working with a massive gathering of data. The place of a guide—the human—is unmistakably central both in steering toward results and in enhancing already existing results. This paper is meant to inspire, give some direction, a manifesto for data with a soul.

Mutually Assured Construction

Usman Haque(✉)

Umbrellium, London, UK
usman@umbrellium.co.uk

I am going to talk about designing participatory systems. I will talk about some of my work over the last 15 years, some of the things I noticed along the way, and some of the things I would like to work more on future. In particular, I will talk about a design strategy I'm calling *mutually assured construction*.

Fig. 1 Clockwise from *top-left*: Pachube, burble, listening & assemblance

But first a reminder: In the 1970s and 80s, the idea of "mutually assured destruction" was pretty central to cold war conflict management. It was a doctrine that essentially said that, since *I* will fire my nuclear weapons if *you* launch a nuclear attack on me, and since *you* will fire yours if *I* attack you (and since either outcome results in total annihilation), therefore, neither of us has any incentive to attack the other (or, for that matter, any incentive to disarm of course!). You could spend thousands of game theory hours examining this dynamic, but the essential point was that the condition helped bind together our futures and assured that we didn't destroy each other.

I have been looking at how you take that dynamic a step further, albeit at a much smaller scale. I am interested not just to agree that we *won't destroy each other*, but

K. De Rycke et al., *Humanizing Digital Reality*,
https://doi.org/10.1007/978-981-10-6611-5_4

more to use the consequences of apparent paradoxes or contradictions *to be positively constructive together*. The frictions to cooperation exist at every scale you might look at, even when the benefits of cooperation seem so self-evident; my interest, as a designer and more specifically as a designer of participatory systems [PDF], is in figuring out how to deal with such frictions effectively, to structure participation in order to account for them, and even thrive on them. Mutually assured construction is essentially a set of design strategies for building, acting, and deciding a future together, without requiring consensus on that future.

If you are interested in the structures of participation, the question of design, and more specifically who designs, is a tricky one: because the extent to which a system is participatory is partly also the extent to which it is not centralized around one single designer. The dilemma is how you design for participation, when being a designer means to a certain extent making decisions on behalf of others.

The way that I have dealt with this dilemma (after much angst!) is by realizing that no matter what design act we make in this world, there is always someone, or some group that makes decisions about that act and that get affected by the decisions—you cannot get away from the fact that you will make designs/decisions/distinctions that impinge upon other people. What is important, however, is to ensure that the decisions you make, and the designs you make, *open up* the set of possibilities rather than constricting the set of possibilities—and even better that the decisions/distinctions themselves are open to rescripting, repurposing, redeciding, and reappropriating by others. Here, I often refer to Heinz von Foerster, and his Ethical Imperative [PDF]: "Act always so as to increase the number of choices".

This, I think, is a fundamental concept in participatory design: to accept that there is a designer, perhaps a meta-designer, making decisions, but to question constantly how many individual decisions involved in the deployment or manifestation can be made by others instead, either now or in the future. *And not to be precious about these.* Matthew Fuller and I wrote about this at length in the Urban Versioning System. Such an approach often results in complex initiatives, that are hard to describe and difficult to bound—initiatives that are necessarily described differently by different people. This means there is no definitive, authoritative description—the description is owned by many. I realize this flies directly in the face of philosophies that say design is about clarifying or solving problems (which assumes the world is knowable and solvable) or that design is about simplifying (which assumes that simplicity is desirable and achievable). I will leave critique of those to another time (though I like what Jack Schulze had to say about it).

But why is a participatory approach in design so important right now? For me it is pretty pragmatic. We are faced with a number of potential crises that are in many ways inter-related.

Our democratic institutions look increasingly creaky because voting outcome is essentially being affected more by the number of people that don't vote than those who do vote.

Our environmental infrastructures have to respond to the conflicting impacts of climate change, mass migration, and the fickle boundaries of geography.

Fig. 2 Scenes from "democratic" decision-making (UK, USA, Colombia)

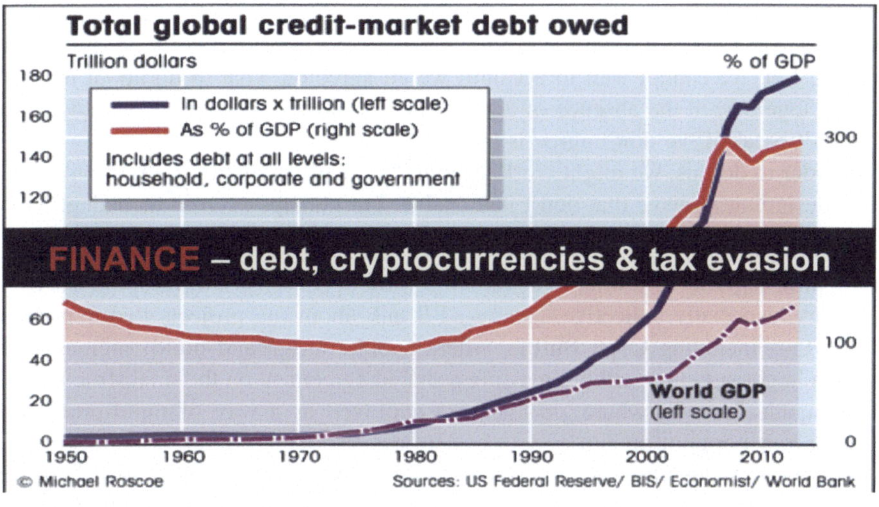

Fig. 3 Total global credit-market debt owed—US Federal Reserve/BIS/Economist/World Bank

Fig. 4 Covert surveillance, NSA hack, Volkswagen Emissions Scandal, Tesla beta-testing dangerous features on the road, telco-hacking by teenagers, fake news—these are all part of the "smart city"

And in the midst of all this, technological solutionism sees propositions, largely by Silicon Valley corporations, that through their "smartness" throw open arms toward mass-surveillance, mass-hacking, mass-deception, mass-insecurity and mass-delusion.

Any of these, but especially all of them together, mean that we are faced with needing to redesign, radically, our everyday lives in the near future. How are we going to do that? Who is going to design that future? Do we outsource the fundamental decisions to Silicon Valley corporations, technologists, and algorithms who see humans as problems to solve? Or do we take ownership of our combined and collective futures? No single voice or even small selection of voices is going to resolve all these complexities. Yet, we cannot wait to act until we all agree on what to do about them. So working together, in the absence of agreement, is essential; designing systems so that they don't break if we don't agree is crucial.

In order to do this, we might work through "mutually assured construction", which means designing systems that don't just enable, but perhaps even in their deployment *require*:

working together: learning to collaborate without consensus, developing a sense of agency, especially collective agency—I'll talk about my projects that experiment with this, including Open Burble, where people design and build fragments of a much larger structure without needing a shared agreement on the final structure; and Flightpath Toronto, where zip-lines are deployed as a way of rapid-prototyping urban transportation in situ.

deciding together: figuring out how to build a shared responsibility for a collective future—I'll talk about Natural Fuse, in which a network of connected plants enables a community to balance energy consumption and the collective carbon footprint;

Fig. 5 Projects clockwise from *top-left*: Burble, Flightpath Toronto, Natural Fuse, Winds-of-Fukushima (by Seigo Ishino), VoiceOver, Cinder

and Cinder, in which students at a new school interact with an augmented reality cat in order to make decisions about resource allocation (cat food) based on the building's solar panel productivity.

acting together: embedding accountability and a sense of collective accomplishment that we can actually achieve something by working and deciding together—understanding that empathy is about listening more than just mere sharing. I'll talk about VoiceOver, which saw a radically public communication infrastructure developed and deployed with a small community in northeast England; and what I saw in the Pachube community following the radiation disaster at Fukushima; as well as a more recent project, the Urban Innovation Toolkit).

This is not about crowd-sourcing to find the "best ideas" for the future. I am arguing that the only way we will *have* a future is by working together on complex projects that embrace the messiness of our conflicting desires and imaginations that reinforce the notion that we can collaborate even when we don't agree on everything, and that enable us through variations on Ulysses pacts to design and create a collective future. The outcome would not just be that we have a future, but that our future is one in which we have necessarily learned how to coexist, co-create, and co-evolve.

As a designer, I don't have a clear idea of that future, or how to construct it. Instead I am working on ways that we can together build a shared memory of a possible future, so that we can decide together whether and how we move toward it.

References

https://en.wikipedia.org/wiki/Mutual_assured_destruction
http://www1.udel.edu/johnmack/frec406/game_theory.html
https://www.haque.co.uk/papers/notesonthedesignofparticipatorysystems_eng.pdf
http://pespmc1.vub.ac.be/books/Foerster-constructingreality.pdf
http://uvs.propositions.org.uk/
http://www.core77.com/posts/13905/design-is-not-about-solving-problems-13905
https://www.amazon.co.uk/Save-Everything-Click-Here-Technological/dp/1610393708
https://roarmag.org/magazine/mass-surveillance-smart-totalitarianism/
https://www.theverge.com/2017/5/12/15630354/nhs-hospitals-ransomware-hack-wannacry-bitcoin
https://en.wikipedia.org/wiki/Volkswagen_emissions_scandal
https://www.theguardian.com/technology/2016/jul/06/tesla-autopilot-fatal-crash-public-beta-testing
http://www.salon.com/2016/12/03/fake-news-a-fake-president-and-a-fake-country-welcome-to-america-land-of-no-context/
http://umbrellium.co.uk/initiatives/citizen-engagement-spectacles/#open-burble
http://umbrellium.co.uk/initiatives/flightpath/
http://umbrellium.co.uk/initiatives/natural-fuse/
http://umbrellium.co.uk/initiatives/cinder/
http://umbrellium.co.uk/initiatives/voiceover/
http://umbrellium.co.uk/initiatives/pachube/
http://umbrellium.co.uk/initiatives/urban-innovation-toolkit/
https://en.wikipedia.org/wiki/Ulysses_pact

Seven Short Reflections on Cities, Data, Economy and Politics

These Are a Series of Short Essays on the Future of Technology, Economy and Society

Tomas Diez(✉)

IAAC, Barcelona, Spain

Essay Number 1: The City of Cities

It is not a secret that cities are the biggest creation of mankind, where most problems concentrate, where most national budgets are being spent, where more than 50% of the global population already live, and we will find the opportunities to create new models for our economy and society of the future. For many years, authors, architects, economists, movements, urbanists, and even artists have proposed their model of cities; now, we have a family of different approaches: the Garden City of Ebenezer Howard, the Bioregional City of Peter Berg, the Polycentric City, the Green City, the Ecological City of Richard Register and Paul Downton, the Open City, the Smart City now being developed by Google Sidewalk Labs in New York, the Sharing City of Neal Gorenflo. Models that try to reflect and map the understanding of what the city is under different principles connected with technology, philosophy, strategies, approaches to governance and more. The city is actually a city of many cities that contain as many visions as people inside them, everyone is a city and master of his/her own destiny (https://en.wikipedia.org/wiki/Otto_Neurath), or we live in one single city (http://news.yale.edu/2015/08/25/yale-architecture-exhibition-takes-global-problem-envisioning-city-7-billion). The city is a multiscalar system of systems (http://www.eamesoffice.com/the-work/powers-of-ten/), connected to different networks at subatomic or universal scales. The complex task of understanding a city requires a permanent shift of scale, resilience, and adaptation, without fixing ideas and models, but taking this understanding to concrete actions that are driven by a vision that might not have a form, name, or shape, but that assumes the role of the city as the place to offer the best environment for man and women to thrive, without compromising resources, or rely on others exploitation.

Essay Number 2: Monocultivation

Hundreds of years ago human agriculture made possible the excess of production, which leads to accumulation of goods, the concentration of population in towns (which would become cities), and the end of the hunter-gatherer. Few hundred years later, economy is based on the flow of real and fictional money that simplifies the value of

K. De Rycke et al., *Humanizing Digital Reality*,
https://doi.org/10.1007/978-981-10-6611-5_5

assets, skills, people, resources, and almost every single element of our reality. Money has become a mean and an end itself, the ultimate resource of our time, real, and fictional. If agriculture transformed dramatically the way humans inhabited this planet, money monoculture is threatening life itself. Our economy assumes that we have a limitless planet in order to look after one objective, no matter what: we need to cultivate money, more money. Money monoculture is possible thanks to the control over the access to information (the Internet is being sequestered in case you did not know), and concentration of the means for production: energy, agriculture, and objects/tools, which allow humans to survive, and better interact, with their habitat. The management (sequester) of physical assets and natural resources is articulated by other abstractions in the form of legal systems, economic laws, and models that we have invented recently, backed by nations and corporations. If means of production are democratized and made accessible to everyone and our digital information is protected to be owned by us, we would be challenging the foundations of the current economy, politics, and social structures as never before.

Essay Number 3: The New Capitalism Is Data Driven… Digital and Totalitarian

We could simplify traditional capitalism by explaining that it was created thanks concentration to the access to means of production in few hands lead society to structure itself into classes; it might start with agriculture thousands of years ago, but industrial revolution made it more obvious during the last couple of centuries. Factory owners were the economic forces driving late nineteenth and early twentieth centuries. Today capitalism has mutated and evolved, factories are marginal, and those forces are now the ones controlling the access to information and scientific knowledge. Holding control of information and scientific knowledge allows any organization capable of concentrating it to control our economy, as it has been happening during the last 20–30 years, when digital technologies turned the whole world market into a fiction (although there are proven studies that it is just a programmed series of algorithm with identified patters, more like a video game for white male adults), which is using democracy as a management body that keeps banal dialogues through media (TV, news, social media, etc.), while the real battles are fought away from our sight, just as Edward Snowden showed us in the award winning Citizen Four documentary produced by Laura Portras. Snowden not only raised awareness, but also took action together with private Manning, by releasing thousands of documents that prove a denounce as reality, with no serious consequences than for themselves and the ones that collaborated to release the truth about the invasion of privacy by a new alliance between the military, government and corporations, a trinity that is a threat for the future of civilization in this planet. We could have a hopeless and apocalyptic view of the world when we check Donald Trump's tweet feed, or see Venezuela's president dancing in TV while students are being killed by official security forces, or observe how Le Pen turns into a political force in France, and Facebook or Google are profiting billions (or trillions) out of managing the emptiness and nothingness where we express ourselves. Yes, our current democratic, economic, and social scenarios are not the best.

Essay Number 4: The Paradoxical Now

The polarization of politics responds to the nature of our current transition period, which might last years, decades, or an entire century. This transition will produce winners and losers, as it happened a 100 years ago, and more than 500 years ago. Although it looks like we are repeating history and be condemned to it, it is only up to us not to do it, and build on top of it by taking the best from it. William Gibson used to say that the future is here, but it is not evenly distributed: The challenges of our times are not about developing the next big futuristic technology, instead, we have to find out how we will give technology a different purpose beyond sustaining a model that only seeks monocultivation. Check Silicon Valley VC fever to make useless technology extracting money from voluntarily uninformed population. As in every transition, we live surrounded by paradoxes and contradictions, in which *the old* and *the new* overlap with each other. Our values and ethics are challenged everyday, ideology dissolves fast, or tries to survive, no matter what. We claim to be saving the planet, while we are mining it until exhaustion using coltan from Congo, aluminum from Australia, meat from Brazil, sneakers from Vietnam, mobile phones from China; while moving materials and products thousands of miles until they get to our hands, used, and disposed; while burning million of years petrified dinosaurs to have a warm bedrooms and living rooms, or to move our cars and planes. We live in a beautiful world with many good things too, we invented it, and we can reinvent it and make it even better, anytime we want.

Essay Number 5: Is this Going to Change?

There are new indicators showing a change of this tendency toward local production and manufacturing, powered by distributed networks and accessible/affordable new technologies. China's city factories are looking for alternative countries to establish their production plants caused by social and economic pressure by Chinese workers. The proposed projects for the construction of the Nicaragua canal will support a new pattern of the supply chain in the global market: Goods are traveling slower than before, production is moving back to countries, in a slow pace. While the Panama canal can allow ships with certain size to go through it esclusas, the truth is that due to the reduction of the demand of transport, larger cargo ships are needed in order to reduce the cost per container. The Nicaragua canal will allow large cargo ships (five times bigger than the ones going through the Panama canal) to navigate from the Pacific Ocean to the Atlantic Ocean, with a considerable effect in the ecology of Central America *. We hope the process of relocalising production is more radical and makes shipping containers obsolete.

- *Note or reference:* http://worldif.economist.com/. *In the section What if launched by The Economist claims that the shipping container traffic is slowing, as production of consumer-goods is getting closer to centers of consumption (cities, townships, regions); this is one of the main reasons to have a Nicaragua canal: the need of bigger containers ("five times the container-carrying capacity of those that currently traverse the Panama Canal") to go from the Pacific to the Atlantic, which allows to reduce the cost per container shipped by suppliers.*

When did we turn our cities into parasites, killed our local industry, and decided that products should be "made in China"? Answer is simple: when cheap products based on cheap production and maximum profit became our religion. The problem is that we did not only loose production capacity in terms of infrastructure, we have lost the knowledge of how things are made, we do not care since we just have to use and dispose, and this is the base of contemporary culture. But things flow, as Zygmunt Bauman (https://en.wikipedia.org/wiki/Zygmunt_Bauman) states, we have a constant flow of resources, products, money, trash, polymers, rivers, oceans, air streams, clouds… all inside the same planet make the trash we dispose come back to our dining table, and the misery caused in countries that make our products possible is embedded in the things we buy with money earned in jobs we do not like at all. Flows create constant change, and that is the base of our contemporary life, that is, what creates tension between the established and accommodated class and the redistribution of resources and power. Things are going to change, they are changing, the question is how we want them to change.

Essay Number 6: The Big Questions

Purpose, meaning, and ownership are keywords to keep in mind when talking about the future. The conversation is not really about VR, AI, AR, ML, robotics, quantum computers, automation, big data, or synthetic biology. Instead, we need to ask ourselves: What and who are these technologies serving for? who decides what to do with them? and how much I really know about them? These questions motivate individuals, communities, and organizations to collaboratively propose and build new ways to own and use technology, to put it to the service of humans and the planet, not only to survive, but to transcend in harmony with our living systems. At least that is the aspiration, we do want to invent the future, not only to predict it (as Lincoln would say), but to make it more accessible, and respond to the biggest challenges of our times, which are mainly social and environmental.

Essay Number 7: The Future Is Already Here

We already have the capacity to produce energy using solar technologies plugged to home batteries, or grow food at domestic and local scale using synthetic biology to cultivate our own meat, or we could produce anything we need with endless recycling materials using digital fabrication technologies in neighborhoods. Our current economic, political, legal, and social structures are struggling to keep the control on everything. The current “operating system” running the world is not ready to support the democratization of production, and the mass distribution of everything. We are in a transition period, that is, opening a unique opportunity to make technology more accessible to everyone in order to increase the resilience of communities and individuals, and break the dependence on intermediaries. If we boot a new “productive operating system”, it would mean that millions of these intermediaries controlling the distribution of wealth will lose their advantages: We might see bankers applying to get the universal basic income, as dystopian as it might sound. The extended global economic crisis is making evident that money is losing its value thanks to the emergence of fluid infrastructures for fluid economies, based on trustable and transparent assets. Blockchain technologies are disrupting the creation of “money” by allowing anyone to be their own bank, or build new digital institutional infrastructure. In few years, we will be able to have the computer power and data storage of a Google or Facebook server farm in the size of a home appliance thanks to quantum computers, being able to store the Internet (or parts of it) locally. In this context, we need an optimistic view of the rather challenging transition we are living today and make it operative to build the future we want.

Part IV

Data Shaping Cities

Approximately 58,000 college students are homeless; ALEXIS BENVENISTE, AOL.COM, Jul 24, 2015 12:00 p.m. https://www.aol.com/article/2015/07/24/approximately-58-000-college-students-are-homeless/21213755/

The NSA and Alphabet having given us reason to fear the facelessness of data gathering and its application. Their whistle-blowers have shown us that we do not always have real access to our own data or agency in its use. But in trade, there is almost no category in cities that remains completely untouched by access and digital interaction with "big data." Supermarkets track the behavior of customers hawkishly to relate sales data to product placement and visual merchandizing. For these organizations, shelf location has long been a commodity to trade alongside the groceries. Which of us would now travel in a car without reference not only to GPS routing but to real-time feedback on the traffic conditions on alternative routes and expected travel times? Most evidence points to a correlation between data openness and prosperity. But are the design of the public fabric and infrastructure, and are the social and environmental imperatives in cities as well served by the abundance and availability of data as commerce? What does the data fail to tell and which, and whose, data is not flowing into the system? Homelessness is a condition often cited as invisible. While there is some understanding of the numbers of people living on the streets or seeking shelter in the temporary refuges of NGOs, the couch surfers and young families surviving in temporary or cramped, and unsuitable conditions are estimated to be a much larger number but unseen. Who else remains unrepresented? Is there equity of access to representation and benefit from data?

Another question is whether in design and design modeling for cities, our legacy systems are yet making use of newly abundant open data or the opportunities to share data between stakeholders in meaningful ways. In 2015, a joint report by the RIBA and Arup "Designing with Data: Shaping Our Future Cities" made three recommendations: improve coordination between government departments; digitize the planning process; and get governments and urban planning experts to work together. To flow, data needs the human and digital system conduits to interconnect. It seems that at least as recently as two years ago in the UK the digital pipelines still did not connect government departments or the human capital contributing to the shaping of city fabric. The uptake of information modeling in building has been slow and riven with technical and organizational challenges, how much more complex is shared information and data modeling across stakeholder interests at the scale of the city? Where are the Precinct information models that combine infrastructure, services, landscape, transport design, feedback on microclimate, lighting, drainage?

How is data shaping our cities, and how should or could it be contributing to more environmentally sensitive, human-centric urban spaces? With the automation promised by the combination of big data and machine learning, what are the humanizing trends in virtual reality modeling of cities?

This section opens with Antoine Picon's beautiful exposition of the shift from the cleansing influence of modernism to a celebration of sensory richness and diversity in the contemporary sentient city. Fábio Duarte and Carlo Ratti explore applications for the exponentially expanding photographic data in Google Street View when machines analyze the qualities of different precincts from the photos.

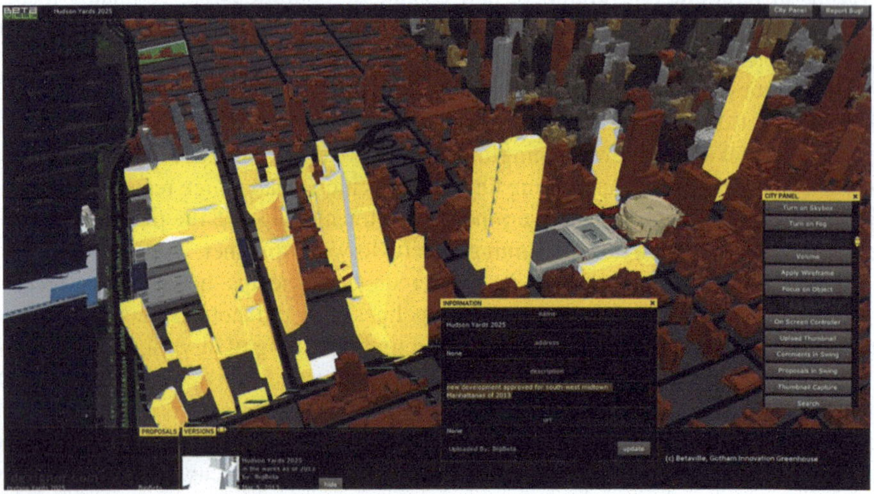

Arup and RIBA, designing with data: shaping our future cities; published in Dexigner November 10, 2013 https://www.dexigner.com/news/27057

The research papers cover topics that range from the integration of microclimatic data into urban design, to stimulating pedestrian behavior and gaming to increase public participation in design; from models for social integration, to using machine learning in the design of workspaces. Data and model relations shared between collaborating consultants in the design process for a ramp and a facade contrast with new work to simulate pedestrian movement and the use of twitter data to follow the movement of individuals in the city. Design modeling researchers are taking on the significant challenge of data shaping cities.

Jane Burry

What Big Data Tell Us About Trees and the Sky in the Cities

Fábio Duarte(✉) and Carlo Ratti

Massachusetts Institute of Technology, Cambridge, USA
{fduarte, ratti}@mit.edu

Since Google Street View (GSV) was launched in 2007, its cars have been collecting millions of photographs in hundreds of cities around the world. In New York City alone, there are about 100,000 sampling points, with six photographs captured in each of them, totaling 600,000 images. In London, this number reaches 1 million images. The GSV fleet now also includes bicycles, trolleys (for indoor spaces), snowmobiles, and "trekkers" (for areas inaccessible by other modes). Using the images to fly over the Grand Canyon, visit historic landmarks in Egypt, discover national parks in Uganda, or circulate through the streets of Moscow, although great experiences, explore only the most immediate and visual aspects of the images. Such an overwhelming abundance of images becomes much more interesting when we consider them as a rich source of urban information.

Researchers in the fields of computer sciences and artificial intelligence have been applying computer vision and machine learning techniques to interpret GSV images. Very few of them move beyond the technical aspects of deciphering these images to explore novel ways to understand the urban environment. The few examples include the detection and counting of pedestrians (Yin et al. 2015) or the inferring of landmarks in cities (Lander et al. 2017). Still, most of this research is either based on small subsets of GSV data or presents a combination of techniques in which the participation of humans is required:

Fig. 1 Computer vision process

At the Senseable City Lab, we have been using computer vision and machine learning techniques to analyze full datasets of GSV images in order to understand urban features in ways that would take too long or be financially prohibitive for most cities using human-based or other technological methods. We started by looking at the trees and to the sky. Exposure to greenery and natural light is essential to human well-being,

K. De Rycke et al., *Humanizing Digital Reality*,
https://doi.org/10.1007/978-981-10-6611-5_6

outdoor comfort, and climate mitigation. Therefore, quantifying green areas and light exposure in different parts of the city will inform better urban design as well as environmental and public health policies. By using GSV data with computer vision techniques, we demonstrate the value of bringing big data to the human level, to the tangible aspects of urban life.

Usually, street trees are quantified and characterized using field surveys or other technologies such as high spatial resolution remote sensing. These techniques depend on intensive manual labor, specialized knowledge, and ad hoc data acquisition. Although satellite imagery analysis gives accurate quantification and characterization of green areas in cities, the technology has two critical caveats for urban dwellers: firstly, it looks at the city from above, not from a person's perspective. Satellite imagery does not show greenery at the street level, which is the most active space in the city and where people see and feel the urban environment. Secondly, larger green areas are highlighted in detriment to the relatively sparse street greenery. However, visits to parks and urban forests do not happen frequently and the benefits of these areas are felt at a large scale, whereas street trees are part of citizens' daily experience and have immediate positive effects on people's lives. We are not dismissing such techniques, but finding ways to take advantage of the huge amount of standardized visual data freely available of hundreds of cities to propose a human-centric and comparable assessment of street greenery.

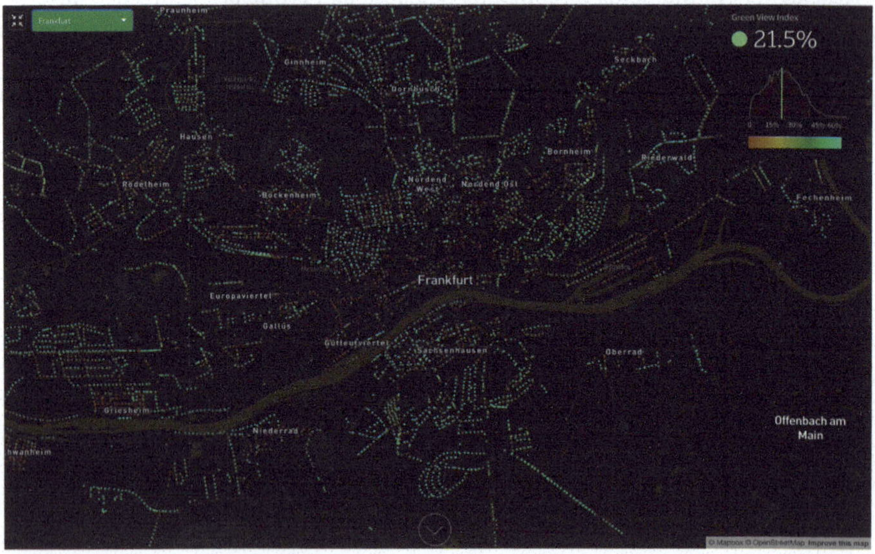

Fig. 2 Treepedia in Frankfurt

Using large GSV datasets composed of hundreds of thousands of images per city, Li et al. (2015) and Seiferling et al. (2017) calculated the percentage of green vegetation in streets, using computer vision techniques to detect green pixels in each image and subtract geometric shapes. With a few computational steps, what is left from this subtraction is greenery. Since the GSV data acquisition procedure is standard, these methods allow us to calculate street greenery in dozens of cities around the world and to compare them—using what we called the green view index.[1]

By avoiding the pitfalls of creating "algorithmic sorting of places" (Shapiro 2017), which automates the attribution of social values onto visual aspects of an image, the analysis of large visual datasets with the same computer vision techniques across different cities and countries has the power to become a civic tool, by which citizens can compare street greenery in different cities and neighborhoods and demand adequate measures from public authorities.

A recent work (Li et al. 2017) has applied similar techniques to measure the sky view factor in cities. The sky view factor is usually understood as "the ratio between radiation received by a planar surface and that from the entire hemispheric radiating environment" (Svensson 2004: 203), varying from 0 to 1. In cities, it can used to quantify the degree of sky visibility within urban canyons, by which one can infer the exposure to natural light in each site, for instance. A common technique to measure the sky view factor is to capture fisheye images with special cameras. Again, as this technique is time consuming—and therefore financially prohibitive for most cities—even when it is done, it usually covers only part of the city. We have been using computer vision algorithms to analyze GSV panorama images in order to optimize the process, cover the entire city, and make such analysis more accessible.

Besides using sky view factor as an indicator of local environmental conditions, at the Senseable City Lab we are exploring using it in order to optimize urban infrastructure. One example is optimizing energy-saving programs in public areas. Cities have been converting their traditional street lights into LED technology, which consumes less energy and save cities millions of dollars per year—the 26 million street lights in the USA consume more than $2 billion in energy, and the greenhouse gas emissions they generate is comparable to 2.6 million cars. However, in most cities, even in those converting streetlights to LED, unless lampposts are equipped with photosensors, all streetlights turn on automatically at the same time, in some cases varying daily according to the astronomical sunset. Applying computer vision techniques to analyze dozens of thousands of GSV images, we can determine the sky view factor at each data point and match them with the nearby streetlights. By accounting for buildings and trees blocking the adequate amount of lighting required in each point of the city, it would be as if we had hyperlocal sunsets close to each streetlight and could determine the optimal time to turn on the lights, which would save energy and money to cities at an aggregate level. Using this highly granular information, we could optimize existing infrastructures without adding another layer of devices, but rather by using data which is already available.[2]

The underlying research question is how not to take data at face value but instead by the intrinsic information they hold about how cities work and how citizens live in the urban environment. A GSV image is more than simply a combined photograph if you analyze it with the appropriate tools. In both cases discussed here—street greenery

---

[1] Treepedia project is available at http://senseable.mit.edu/treepedia.

[2] We are grateful to Ricardo Álvarez and Xiaojiang Li for some of the ideas discussed here; and to Lenna Johnsen for revising the paper.

and sky view factor—it is possible to imagine that soon such large amount of visual data will be collected more frequently and in many more cities. Furthermore, with more sensors deployed in urban infrastructure, embedded in personal mobile devices, and soon in driverless cars, we can foresee all this data available in real-time maps, which will help to design actuations at the local level as well as enable the creation of worldwide urban dashboards that would show multiple cities in a comparative way. Making sense of the sizeable quantities of data that is already generated in and about our cities will be key to creating innovative approaches to urban design, planning, and management.

References

Lander, C., Wiehr, F., Herbig, N., Krüger, A., Löchtefeld, M.: Inferring landmarks for pedestrian navigation from mobile eye-tracking data and Google Street View. In: Proceedings of the 2017 CHI Conference Extended Abstracts on Human Factors in Computing Systems—CHI EA'17 (2017)

Li, X., Zhang, C., Li, W., Ricard, R., Meng, Q., Zhang, W.: Assessing street-level urban greenery using Google Street View and a modified green view index. Urban For. Urban Greening **14** (3), 675–685 (2015)

Li, X., Ratti, C., Seiferling, I.: Mapping urban landscapes along streets using google street view. In: Patterson, M. (ed.) Advances in Cartography and GIScience, Lecture Notes in Geoinformation and Cartography. DOI: 10.1007/978-3-319-57336-6_24 (2017)

Seiferling, I., Naikc, N., Ratti, C., Proulx, R.: Green streets—quantifying and mapping urban trees with street-level imagery and computer vision. Landscape Urban Plann. **165**, 93–101 (2017)

Shapiro, A.: Street-level: Google Street View's abstraction by datafication. New Media Soc. 146144481668729 (2017)

Svensson, M.K.: Sky view factor analysis—implications for urban air temperature differences. Meteorol. Appl. **11**(3), 201–211 (2004)

Yin, L., Cheng, Q., Wang, Z., Shao, Z.: 'Big data' for pedestrian volume: exploring the use of Google Street View images for pedestrian counts. Appl. Geogr. **63**, 337–345 (2015)

Urban Sensing: Toward a New Form of Collective Consciousness?

Antoine Picon(✉)

G. Ware Travelstead Professor of the History of Architecture and Technology, Harvard University Graduate School of Design, 48 Quincy Street, Cambridge, MA 02138, USA
apicon@gsd.harvard.edu

Traditional Planning and the Purification of the Sensory Experience

Cities have always been placed where the senses are constantly solicited. If as historian of art Michael Baxandall writes, "living in a culture, growing and learning to survive in it, involves us in a special perceptive training,"[1] cities figure among the primary educators of civilizations. But this education presents negative counterparts. On streets and squares, sight, sound, smell, touch, and taste have a lot to process, too much sometimes. The amount of information with which the senses are confronted can prove overwhelming. Above all, the urban sensory experience is not without drawbacks. From visual chaos to foul smells, and from loud noises to corrupted food, the city can at times be trying, even unbearable. The poet Juvenal already drew a severe picture of Ancient Rome. French writer Louis-Sébastien Mercier has remained famous for his apocalyptic evocation of late-eighteenth-century Paris with its stench and noise. Numerous journalists and novelists have criticized the dire conditions of life in the popular districts of Victorian London or early twentieth-century New York. From such a perspective, the need to purify the urban experience from too much or disastrous sensory stimulation appears justified. Indeed, urban planning and design have had a long history of efforts at sensory purification.

One could begin such a history with the Renaissance ambition to reform completely the urban visual experience by eliminating the complex and tortuous urban sequences of the medieval city. Visual reform was to remain a constant dimension of urban planning and urban design, from Renaissance and baroque attempts to make the city more regular to the modernist project of entirely rational and geometrically rigorous new urban compositions. From the nineteenth century onwards, these endeavors were, however, balanced by a growing sense of the value of the urban historic heritage. Through novels like Victor Hugo's celebrated *Hunchback of Notre-Dame* published in 1831, Romanticism contributed to the reevaluation of the picturesque value of ancient districts. The monumental legacy of past centuries led also to evolutions such as the reintroduction of color, after the Greeks and the Gothic, in contemporary buildings.

[1] Baxandall (1985).

K. De Rycke et al., *Humanizing Digital Reality*,
https://doi.org/10.1007/978-981-10-6611-5_7

Fig. 1 Gustave Doré, "Dudley Street, Seven Dials", from Gustave Doré and Blanchard Jerrold, *London: A Pilgrimage*, London, 1872

The visual experience of the modern city thus appeared as a contested field full of tensions and even contradictions.

In modern metropolises, attempts were also made to purify other sensory experiences. Some of them succeeded beyond expectations. Smell, for instance, was gradually disciplined through measures ranging from the banishment of traditional stinking activities such as tanning to the construction of sanitation infrastructure. Although new sectors such as the chemical industry could also produce unpleasant odors, none could compare with the stench of some districts in preindustrial cities. The evolution was all the more radical in that it was rooted in a profound change in the sensibility of individuals toward odors, which has been admirably documented in the French case by historian Alain Corbin in his 1982 book *The Foul and the Fragrant: Odor and the French Social Imagination*[2].

In other domains, the outcome was less univocal. Food was, for instance, framed by more and more demanding regulations and standards that contributed to the disappearance of certain traditional culinary practices. In addition, food distribution was rationalized through institutions such as central markets that possessed a clear infrastructural dimension. But these evolutions were counterbalanced by the effects of the

[2] Corbin (1986).

Fig. 2 Le Corbusier, contemporary city for 3 millions inhabitants, 1922, Fondation Le Corbusier

transportation revolution, which brought new products to the table of the urban consumer.

Regarding hearing, the results were especially ambiguous. On the one hand, modernity was obsessed with the reduction of urban noise and with a longing for the purity of the sound experience, which led to new techniques and practices of musical performance and recording, as evoked by Emily Thompson in her pioneering study of the subject.[3] On the other hand, large cities remained places of diverse and intense auditory stimulation, and this far beyond any attempt made to regulate levels of noise. Like the visual complexity of historic heritage, the complex soundscape of cities had its defenders who were quick to oppose it to the deadly silence of modern districts and facilities. The 1967 movie *Playtime* by French director Jacques Tati offers a brilliant variation on this theme.

Despite these mixed and often contested results, the trend was overall to purify the urban sensory experience. Nowhere was such a trend as conspicuous as in the sprawling suburbs of the USA. Suburban life promoted such a purification, which was seen as a prerequisite for the standardization of lifestyles. From the perspective of its advocates, the gentle and rarefied suburban sensory experience contrasted favorably with the chaotic sensory landscape of city centers. The contrast had often to do with

[3] Thompson (2002).

social and racial prejudices, the noise and smell of other populations being assimilated to mere nuisances.

Rich, Diverse, and Common Sensory Experience as a Project

The past decades have seen a dramatic inversion of this trend. Books, journals, and exhibitions now insist on the importance of a rich and diverse sensory experience in the overall quality of urban life. During 2005–2006, for instance, the Canadian Center for Architecture presented an exhibition on this theme.[4] The subject went on to become the main theme of the French contribution to the 2010 Shanghai Expo curated by architect Jacques Ferrier.[5] Following the rising interest in everything pertaining to food, from organic agriculture to reality television shows centered on cooking, the journal *Log* devoted one of its 2015 issues to food as a founding dimension of urban life.[6] The subject had already been broached by sociologist François Ascher, who had envisaged the eater as the emblematic figure of contemporary "hypermodernity."[7]

Architecture is part of this movement. The renewed importance of the senses can be traced in its recent production through a phenomenon like the so-called "return" of ornament, which modernism had condemned in the name of intellectual rigor.[8] From Herzog & de Meuron's to Farshid Moussavi's buildings, contemporary ornament is often associated with a heightened sensory dimension as well as with the desire to challenge the divide between the senses, to blur for instance the distinction between sight and touch. Indeed, realizations like Herzog & de Meuron's De Young Museum in San Francisco display a tactile quality without equivalent in modernist architecture.

It is easy to interpret this evolution in the context of an increased competition between cities all over the world and the emphasis put by economists, sociologists, and other specialists of the urban on an "economy of knowledge" based on the skills of an elite of creative individuals who want to be stimulated rather than dampened by their urban environment. The "creative class," to use Richard Florida's expression, appears of strategic importance in city development, and the fulfillment of its expectations seems the best way to attract its members, from researchers to entrepreneurs.[9] If "thriving cities connect smart people" according to urban economist Edward Glaeser, they need arguments to convince these people to come and stay.[10] Thus, the richness and diversity of the urban sensory experience is not only about enjoyment; it represents an asset that needs to be cultivated along with infrastructural realizations that make life easier.

---

[4] Zardini (2005).

[5] Leloup et al. (2010).

[6] *Log* 34, Spring/Summer 2015.

[7] Ascher (2005).

[8] Moussavi and Kubo (2006), Picon (2013).

[9] Florida (2002).

[10] Glaeser (2008). See also Glaeser (2011).

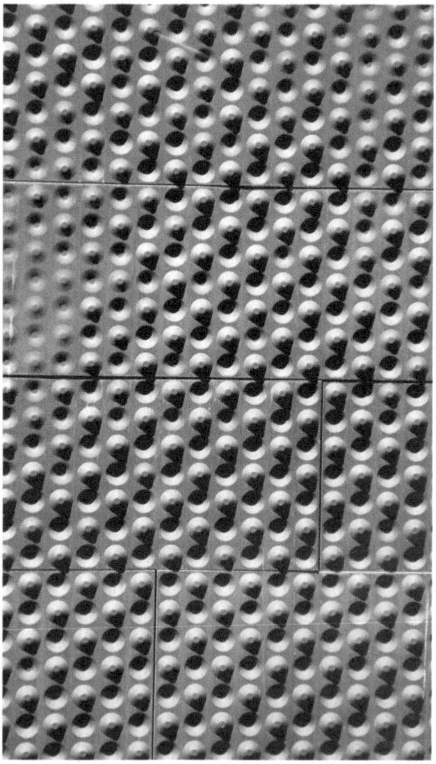

Fig. 3 Herzog & de Meuron, de Young Museum, San Francisco, 2005

Now, catering to the needs of a highly skilled elite in a context of generalized urban competition is not enough to explain the rise of the sensory in contemporary urban practices and debates. It is striking to observe how the urban sensory experience has become a concern for all sorts of constituencies, from the broad audiences attracted by reality television shows based on culinary contests to the more rarefied patronage of high-end restaurants. Music, fashion, and design also concern extremely diverse segments of the urban population. Sensory experience no longer appears as inevitable as in former times, when purifying it was on the agenda of politicians and planners, nor does it seem to be a privilege reserved for the happy few. Now, it represents something akin to a common, a right for every inhabitant of the city.

This right may be interpreted differently from one group to another. Advocates of the environment would like the city to offer as many nature-based experiences as possible. From their perspective, the air is never pure enough; there are never enough trees and meadows interspersing the urban fabric. Birds all of kinds should be able to find a home in cities. Their desire is often in contradiction with the wishes of the proponents of an artificially luxurious urban environment, for whom there should be no upper limit to the number of fashion retail stores and gourmet restaurants. The senses have become a new contested field where our common urban future is taking shape among unavoidable tensions and contradictions.

Besides its vast appeal, the role played by digital technologies constitutes another novelty in the evolution of urban sensing. The full realization of what this implies is just beginning to dawn upon us. When the modern computer appeared in the late 1940s to early 1950s, it was, as suggested by its name, understood as a mere machine for computing. Then, thanks to Norbert Wiener and other proponents of cybernetics and artificial intelligence, the new machine began to be interpreted as a powerful aid for reasoning, an aid that would perhaps one day surpass its human creator in logical and deductive power. With the digital revolution—that is, the massive spread of information and communication technologies and their interference with every aspect of life, both at the individual and the collective level—it has become clear that our senses are now reshaped.

We can no longer see without being decisively influenced by digital cameras: They are like substitute eyes, whether standalone or integrated into our telephones. With these instruments, we have got used to images that seem to be perfectly focused regardless of the distance. We listen to MP3s, and we wear perfumes that have been synthesized with the help of sophisticated software that allows us to combine fragrances in the same way that designers juggle with forms and colors using their graphics programs. The list of ways that our sensory relationship to the physical world has changed is endless. Zooming in and out now seems almost as natural as blinking. The scale of phenomena often becomes problematic in that we can shift almost constantly from the infinitely small to the infinitely large. The American architects Charles and Ray Eames, who were trailblazers in thinking about the relationship between computers and architecture, had already glimpsed this evolution in their 1977 documentary, *Powers of Ten*, which offered a vertiginous voyage from galaxies to atoms; Google Earth has since made this banal. The boundary between the abstraction of numbers and pure sensation is becoming blurred; music in digital formats testifies to this. We are entering a new world of sensory givens marked by the effects of synesthesia. Forms appear to be becoming tactile, and sounds gustatory. Neuroscience is exploring the biological and cognitive foundations of this new world, within which the blind can be made to "see" in many different ways, such as by engaging the sensitivity of their skin or tongue. When it tries to blur the distinction between vision and touch, architectural ornament takes after this more general synesthetic trend.

The digital revolution also brings to sensing the possibility of sharing the sensory in ways that were still unthinkable a couple of decades ago. Whereas traditional sensing was a unique experience difficult to communicate—hence the crucial importance of the arts in such a communication—such is no longer the case in the digital age. Indeed, code can be transmitted with precision, and image or sound can travel from terminal to terminal and be appreciated in the same way by multiple users. This ability to share reinforces the notion that urban sensing constitutes a common, something that does not belong to one social category or another but must be considered as a shared resource.

Toward the Sentient City

Digitally augmented in many cases, and easy to communicate online, the urban sensory experience seems to blur the distinction between the individual and the collective, just like the billions of highly personal moments and emotions that people share with others on social networks. From this perspective, the development of the practice of taking a photograph of the meal one is about to consume and posting it immediately on Facebook could very well be indicative of a much broader evolution. We no longer want to experience the city and its pleasures alone. We want this experience to contribute to a larger awareness of what is going on in cities, our city or the cities of our "friends."

Both individual and collective, urban sensing could be constitutive of a new way to understand the evolution of the urban. From the sensory city, it seems that we are gradually passing to the sentient city—that is to say, to a new form of collective intelligence very different from the traditional crowd insofar as it does not negate individual behavior but rather integrates it, with its specificities, within a larger frame in which endless differences and infinite nuances can coexist. The success met by the various research projects and experiments led by MIT's Senseable City Lab is probably linked to the way it has foreseen and accompanied this evolution.[11] Going further, one might be tempted to assimilate such evolution to a process of rising self-awareness. The sentient city might very well represent a first step toward a truly smart city, a city that would not only prove more efficient and livable but be also synonymous with new ways to perceive, deliberate, and ultimately decide about one's collective future[12].

There is perhaps no better allegory of the link between sensing and self-awareness than the story portrayed by the philosopher Étienne Bonnot de Condillac, the leader of the Sensualist school in Enlightenment France, in his *Traité des sensations* (Treatise on the Sensations) of 1754. Here, he describes a statue endowed with the potential to become intelligent but deprived of the ability to feel. By successively investing it with a sense of smell, hearing, taste, sight, and finally touch, the philosopher evokes the statue's gradual awakening to the world of sensations, soon followed by ideas. Each sense brings its own set of discoveries, but the real tipping point is when the statue acquires the sense of touch and thus discovers the existence of obstacles beyond itself that put up a resistance to it, and it simultaneously realizes that it possesses a body

[11] See the Lab's site: http://senseable.mit.edu, consulted on 25 February 2017.

[12] We have developed this approach in Picon (2015).

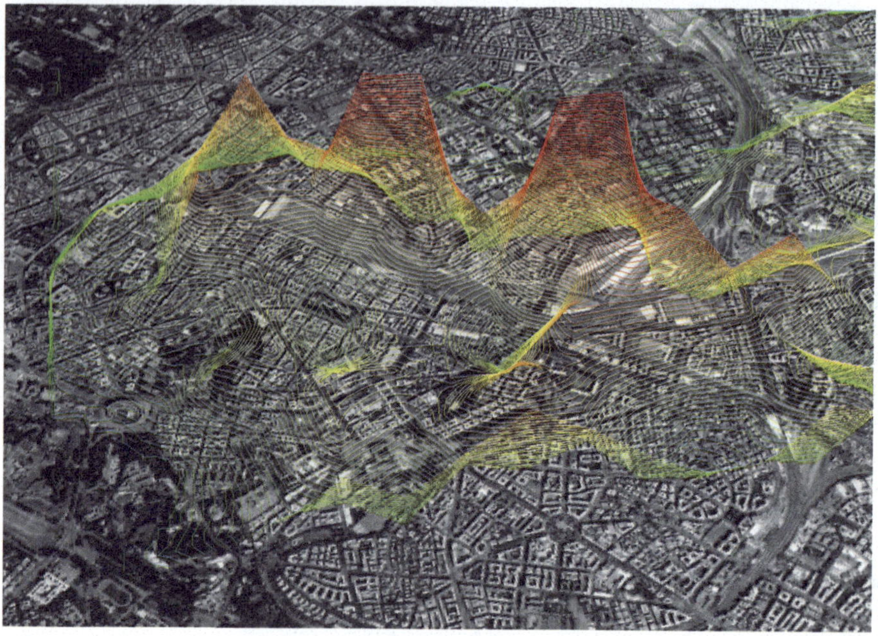

Fig. 4 Senseable City Laboratory, MIT, Real Time Rome 2006

which interacts with them. Sensations generate ideas and ideas progressively lead to consciousness. In Condillac's apologue, the discovery of the inertia and impenetrability of matter finally give the statue access to true consciousness: an awareness of itself as an entity, that is, at once linked to the world and distinct from it.

If I were to attempt a transposition of this story to our contemporary urban problems, the real challenge of today's urban sensing lies not so much in its intensity and profusion. Contrary to traditional planners, we not only accept these sensory qualities: We embrace them as a true source of riches. The challenge might have more to do with determining what could be the source of the resistance that Condillac identifies as the true starting point of the process leading to self-awareness. In other words, what kind of resistance, or rather opacity and inertia, can we identify and relate to in an increasingly mobile world, a world in which sensations can be manipulated more and more easily, using a mix of physical and digital tools, and leaving us often perplexed? The difficulty we experience so often today when we try to identify "real" images, maps for instance, as opposed to digitally edited ones, is typical of this growing perplexity.[13] Can we trust our senses without being assured that something can resist our power of manipulating our own sensory experience? Another way to put it would be to refer to the traditional notion of materiality and the need to redefine it now that atoms of matter are increasingly hybridized with bits of information. Without opacity and inertia, without materiality envisaged as a resistance to our capacity to scheme, urban sensing could

[13] Kurgan (2013).

Fig. 5 Musée Grévin, Palais des Mirages

end up transforming itself into a labyrinth of misguided perceptions, a fascinating shared sensorium analogous to the halls of mirrors that one finds in places like the Musée Grévin in Paris. Groups that enter them see their reflection indefinitely replicated—but this is an illusion. Urban sensing is an opportunity to reach a new stage of collective consciousness; let us not transform it into a trap leading to collective delusion.

References

Ascher, F.: Le mangeur hypermoderne. Odile Jacob, Paris (2005)

Baxandall, M.: Patterns of Intention: On the Historical Explanation of Pictures, p. 107. Yale University Press, New Haven and London (1985)

Corbin, A.: The foul and the fragrant: odor and the French social imagination (originally published in Paris, 1982) (trans. Miriam Kochan). Harvard University Press, Cambridge (1986)

Florida, R.: The Rise of the Creative Class: And How it's Transforming Work, Leisure, Community, and Everyday Life. Basic Books, New York (2002)

Glaeser, E.: Thriving cities 'connect smart people'. Harvard Gazette, 13 March 2008, http://news.harvard.edu/gazette/story/2008/03/thriving-cities-connect-smart-people/, consulted on 21 February 2017 (2008)

Glaeser, E.: Triumph of the City: How Our Greatest Invention Makes Us Richer, Smarter, Greener, Healthier, and Happier. Penguin Press, New York (2011)

Kurgan, L.: Close Up at a Distance: Mapping, Technology, and Politics. Zone Books, New York (2013)

Leloup, M., et al.: Pavillon France, Shanghai Expo 2010, Jacques Ferrier Architectures, Cofres Sas. Archibooks, Paris (2010)

Log 34, Spring/Summer 2015

Moussavi, F., Kubo, M. (eds.): The Function of Ornament. Actar, Barcelona (2006)

Picon, A.: Ornament: The Politics of Architecture and Subjectivity. Wiley, Chichester (2013)
Picon, A.: Smart Cities: A Spatialised Intelligence. Wiley, Chichester (2015)
Thompson, E.: The Soundscape of Modernity: Architectural Acoustics and the Culture of Listening in America, 1900–1933. MIT Press, Cambridge (2002)
Zardini, M. (ed.): Sense of the City: An Alternative Approach to Urbanism, exh. cat. Canadian Centre for Architecture, Montreal; Lars Müller, Baden (2005)

Part V

Thermodynamic Practice

How Representation's Tools Have Changed the Architectural Form From Alberti' Perspective to Computational Fluid Dynamics

In the 1980s, a historic revolution took place in architectural offices, which transformed radically the work of architects and then its objects. For hundreds of years, the modes of conception and representation of architecture, i.e., the real tools with which the architect works—pencil, T-square, set square, rule, compass, and paper, had remained virtually unchanged. It is a strong specificity of the architect, as of the musician, not to work directly with the material which ultimately materializes his work, unlike the painter or the sculptor, but upstream, with other tools, without any in relation to the construction, by drawing it, describing it, according to graphic and written codes. A building being ultimately in three dimensions, the architect, to simplify the task, to communicate his intentions and measures to the workers who will later be in charge of building, decomposed the volumes in two dimensions representable on paper with a pencil, according to horizontal and vertical planes called plan, section, and elevation. The dimensions and description of the materials can be measured or read on these. The forms of architecture which the general public calls "square", that is to say, the most common architectural form, made of straight lines and junction of walls at right angles, were evidently the predominant shapes of the architect because of the use of the T-square, facilitating the drawing of straight lines and the square, the 90° angle, and the simplicity of its use. The tools to draw curves or to obtain other shapes were more complicated to
make—plastic helped during the twentieth century to create templates, French curves or flexible ruler—and more complicated to use: One had to be very focused to follow the curve with the pencil, with a 50% chance to miss. These complex forms then implied greater efforts on the part of the workers to carry them out. On the construction site, the lead line for placing the vertical lines and a tight wire to locate the horizontal lines remained the most practical and cost-effective means of building, resulting in the profusion of a "square", ordinary and simple architecture.

At the beginning of the nineteenth century, the philosopher Georg Wilhelm Friedrich Hegel criticized architecture, perhaps for that reason, that of producing forms that ultimately depended only on the obviousness and constraints of its wooden tools of representation, and the surrender of the construction to the natural forces, that of gravity, which causes a wall to be straight, since it would fall if it was leaning, the shape of the sloping roof, since the rainwater would stagnate and infiltrate inside if it was flat, the bottom walls of the building being thicker than the top walls since they have to bear more weight, the window which should not be too large because otherwise the weight of the wall above the void of the window will cause the lintel to collapse, etc. Hegel thus described architecture as the poorest, most imperfect art, because human imagination is most strongly constrained by the difficulty of representing volumes or of measuring a curve, by the force of gravity which makes all the walls and columns to be vertical, by the wind that makes the wall exist as a shelter, by the sun and snow that induce the roof, by the length of the tree trunks that determines the length of the rooms, because the length of the wooden beam will ultimately determine the position of the walls and therefore the size of the room, without finally being able to do much in the face of these physical constraints of nature, if not to bring some order.

It is nevertheless through the invention of new techniques of representation and construction that architectural forms have changed during the course of history, such as those who took place during the 1980s with the introduction of computer science. One notable precedent is that of the invention of perspective in the early fifteenth century, which revolutionized the mode of representation of architecture until then confined to plan, section, and elevation. It is important to note that this revolution does not originally take place in the final field of architecture, its object and its end, which would be built construction, but in the field of its means, its tools, those of the design and representation on paper, upstream of construction by workers. And because the architect actually uses the means of the painter and the draftsman to work, that is to say drawing, that for this reason artists were often at the same time painter and architect, a revolution in painting caused a major shift in architecture. History tells us that Filippo Brunelleschi demonstrated in 1425 on the baptistery of St. John in Florence, Italy, a new mode of representation in drawing on a glass panel, perfectly superimposed with reality. He used a hole to look with one eye through a drawing of the baptistery that is reflected on a mirror placed in front of the real baptistery, allowing to perfectly superimpose reality and representation. He thus validated a mode of geometrical representation in drawing that of perspective with a center point, which is today called a vanishing point on which all the lines parallel to the direction of the gaze converge. In fact, what Brunelleschi is putting into practice is a popular saying, "Well-targeted for a one-eyed man", who says that when one shoots far from the target, relying on the fact that by closing one eye, we see the world no longer in three dimensions, but in two dimensions, as on a plane, as a table, which finally makes that, as the universal dictionary of 1690 writes: "one sees better, farther and straighter with one eye than when the two are used together." This will result in the picturesque image of the painter, his palette of colors in the left hand, closing an eye, and holding out the brush before him to measure the angles and dimensions of the objects placed in front of him. The invention of perspective is thus first of all the idea of closing one eye to deprive oneself of the stereoscopic binocular vision in three dimensions in order to flatten all that are seen, having all depths on the same plane, the same two-dimensional surface, and thus being able to measure all the lengths, all the angles, and all the shapes in a single plane. This mode of reception of reality in two dimensions, by closing one eye, was immediately transformed into a pictorial production mode with the first paintings in perspective of the Italian painter Masaccio for Chapel Brancacci in Florence finished in 1427. Before that, painting represented little or only in a succession of planes, without perspective, in the manner of a superposition of architectural elevations placed one behind the other. And architecture proceeded with the same difficulty to represent the volume.

By introducing perspective into painting, as a new means of representation, the consequence was a major change in the art of painting, which has persisted for many centuries. But this also led to an upheaval in architecture and town planning that used the same tools of representation as those of painters. The art of architecture is transformed at the same time that the art of painting is changed by the invention of perspective. Thus, the painter and architect Leon Battista Alberti, 10 years before writing his treatise of architecture De re aedificatoria in 1449, establishes in his book of theory of painting De picture (1435) the geometric principles of drawing in perspective, prescribing the graphic principles to represent the world in perspective. A new type of architecture document appeared that of perspective henceforth associated with the general presentation of an architectural project, alongside plans, sections and elevations, still valid today. But this new mode of drawing was not merely a mode of representation. It also became a mode of conception, embodied in the three urban paintings of the Città ideale of the end of the fifteenth century, where we see the profound modification of the way of conceiving the plan of a city and to draw architecture that was generated by the invention of perspective. It thus marks the origin of urbanism as a discipline, for one can plan the city according to depths, points of view, and vanishing points. The invention of perspective has first upset the mode of representation of the architect and then the real world. In designing the plan of a city or a building according to perspective, the architect begins to organize the arrangement of the volumes in depth, to arrange them according to the central point, generating new compositions, perspectives, symmetries, axes, ordered according to a point of view that begins to modify urbanism from the Renaissance, first with the village of Pienza by Rosselino or the Piazza Ducale of Vigevano built in 1492, then all European cities during the following centuries. The invention of perspective as a means of representing reality thus transformed the fabrication of reality afterward, with a never-ending success in the design of new town plans, squares and streets, and buildings until today. One of the first compositions based on the perspective that of the Capitol of Rome by Michelangelo in 1537 with a composition where the center point is an ancient equestrian statue will become the reference of the French squares of the seventeenth and eighteenth centuries, for example, the Place des Victoire or the Place Vendôme in Paris. The three urban paintings of the Città ideale at the end of the fifteenth century offer the three models of urban compositions based on perspective: one where the center point is occupied and blocked by an object (which will be used, e.g., in the compositions of the urban area of Rome in the Piazza del Popolo by placing an obelisk at the end of the horizon), one where the center point meets a never-ending horizon (which will be taken over for the Park of Versailles or the Salt Institute of Louis Kahn), and one which combines the first two models, leaving through the object that occupies the center a hole allowing the gaze to continue further (following the example of the triumphal arches on the Champs Elysees).

As a result of the invention of perspective, the architecture is also transformed, with the church of Santa Maria dei Miracoli or the Scuola Grande di San Marco by Pietro Lombardo of 1489, whose openings or stairs play with effects of perspective, unimaginable before the development of the technique by Alberti. And posterity shall be continued onto this day.

What is thus observed is that one of the most important changes in urban and architectural forms and composition that of the Renaissance was not played out in reality, at the final level of its constructed objects, but in the representation, in a transformation of the tools and means of the architect, on paper, with a ruler and a pencil. In the world of ideas, it has also not played out transcendent or philosophical concepts that would have been upstream, as it is sometimes understood, but thanks to the empirical discovery, on paper, of geometric and mathematical rules, making it possible to reproduce reality. The discovery of a new way of representing the world in two dimensions, the one-eyed world, if you like, put on paper, eventually coincided with urban models. The great urban perspectives, the magnificence of the break-throughs and squares of the Renaissance, the Baroque and the Neoclassicism, ultimately have less to do with a symbolic program that would have been imagined upstream than with a geometric technique of representation on paper, implying the centrality of perspective, the effects of symmetry, and the repetition of elements in the manner of colonnades, which made it possible to take a good measure of the effect of depth. The resulting order and magnitude is the consequence of perspective and not its cause. And it is to this same type of revolution in the means that we are witnessing from the 1980s.

The arrival of computers with Computer-Aided Design (CAD) software in architectural offices during the 1980s was first used to save working time by freeing the architect from the tedious load of technical drawings by hand, first in pencil, then in ink, and in its corrections involving the use of ink on the layer. Tracing paper itself, used at the beginning of the twentieth century by architects instead of paper, had already made things easier. By scratching bad lines with a razor blade, it was possible to correct a part of the drawing without having to redraw everything, as it was the case with paper. In the first instance, computer science had only this practical role of CAD (computer-aided drawing), allowing to draw faster, to be able to correct and modify the plans without difficulty and without limit, the sections and the elevations, graphic documents still relevant in the presentation of an architectural project. It was certainly because of the perspective that we began to wonder about the use of CAD. Indeed, it is necessary to build the building completely in 3D to be then able to walk inside or outside and to choose a point of view that will be found advantageous for the project. But it is also by walking in 3D in the building modeled on the computer that we begin to perceive things that we had not planned, qualities or defects that we can attenuate or accentuate by modifying the plan and section. But going a step further, some architects wondered whether the computer simply did not offer the possibility of conceiving the architecture directly in perspective, in three dimensions, without having to go first by the plan and the section, i.e., reversing the usual process of designing the project that begins with the plan, followed by the section and finished by the perspective. This is what the American architect Greg Lynn and the other teachers of the Paperless studio created in 1993 at the University of Columbia in New York at the initiative of its director Bernard Tschumi. In this new revolution of means of representation of architecture, new forms have emerged as a result of immediate work in 3D. In a lecture, the architect Alejandro Zaera-Polo recalls that working directly in three dimensions, in the space without the gravity of the computer screen, made him realize that in 3D, the plane and the section did not exist, because it allowed surface continuity between horizontality and verticality, and even the latter had no reason to exist that the planes could be biased, bent, freely deformed in she space of the screen that a floor could become a wall, without any link to the flatness in two dimensions of the sheet of paper. The plane and section suddenly appeared as an archaic means of representation of the volumes, a codified simplification of the two-dimensional architecture separating the horizontal planes of the vertical sections without any reason on different layers or sheets of paper. No opportunities to work and properly represent space and volume were easily accessible. Alejandro Zaera-Polo thus naturally explained the appearance of these continuous surfaces which deform from the horizontal to the vertical through bias, because it was no longer limited to a representation split in plan on one side and in section of the other. Greg Lynn and the Paperless Studio questioned the T-square, set square, and the "square" shapes that these instruments induced, the resulting straight lines and planes. Computer software for 3D modeling, with functions such as NURMS (non-uniform rational mesh smooth), made it possible to work freely on curved surfaces, distort plans and create what Greg Lynn called "Fold". Endless deformations of the surfaces, as well as "Blobs", deformed volumes with soft and organic shapes, with no net breakage, no straight angle or line. A few years ago, architect Frank Gehry was the first to use 3D modeling software from aeronautics, but without inventing new shapes, to build the complex and curved shapes of a fish sculpture for Barcelona. Later, in an opposite approach to the Paperless studio in New York, he began to scan his models of crumpled paper with completely random shapes, so that he could then transcribe the otherwise inconceivable forms according to coordinates and geometric lines constructions that could be constructed later, following the example of the Guggenheim of Bilbao.

After the form, computing was applied to the calculations of the structures supporting the buildings, calculating and simulating geometrically the descent of loads in a wall or a volume, allowing to remove material in places where no or few loads are transmitted, as in the example of the Tod's Omotesando building in Tokyo by Toyo Ito with engineer Mutsuro Sasaki.

The latest computer revolution is the use of multiphysics software, also known as computational fluid dynamics (CFD), initially designed for mechanics, electricity to study flow, friction, velocities, fluid turbulence, allowing to design and draw urban plans and buildings according to the behavior of the air, its velocity, its temperature, according to principles of convection, conduction, pressure. The mass plan of our project for the Jade Eco Park is based on three urban simulations of winds flowing into a new neighborhood of the city of Taichung in Taiwan built on the site of an old airport. By simulating the passage of fresh wind from the north into the new neighborhood, we were able to identify areas where it will naturally be cooler in the park, corresponding to the areas where the wind is strongest and therefore the most refreshing, acting as a natural fan. This first simulation carried out by the German Transsolar office allowed us to draw a first mass plan that was superimposed on two other simulations: the wet wind from the southwest and the air pollution fluxes from the neighboring roads. These three models, heat, humidity, and air pollution, have thus directly generated the mass plan, its forms and then its uses, in a linguistic reversal of urban and architectural composition where "The form and The function follow the climate".

To illustrate the application of CFD software in architectural design, we present two urban-scale projects for the Jade Eco Park in Taichung and at the architectural level for the convective apartment project in Hamburg.

Urban Design

The ambition of the Jade Eco Park in Taichung, Taiwan, is to give back the outdoors to the inhabitants and visitors by proposing to create exterior spaces where the excesses of the subtropical warm and humid climate of Taichung are lessened. The exterior climate of the park is thus modulated so to propose spaces less hot (more cold, in the shade), less humid (by lowering humid air, sheltered from the rain and flood), and less polluted (by adding filtered air from gases and particle matters pollution, less noisy, less mosquitoes presence). The design composition principle of the «Taichung Jade Eco Park» is based on climatic variations that we have mapped by computational fluid dynamics simulation (CFD): Some areas of the park are naturally warmer, more humid, and more polluted, while some of them are naturally colder (because they are in the route of cold winds coming from the north), dryer (because protected from the southwest wind providing humidity of the see in the air) and cleaner (faraway from the roads). In theses last naturally cooler, less humid, and less polluted microclimates, we increase the coolness, the dryness, and the clean for creating more comfortable spaces for the visitors. Beginning with the existing climatic conditions as a point of departure, we have defined three gradation climatic maps following the results of three computational fluid dynamics simulations. Each map specifically corresponds to a particular atmospheric parameter and its variation of intensity thought out the park. The first one corresponds to variation on the heat on the site, the second one describes the variations in humidity in the air, and the third one the intensity of the atmospheric pollution. Each map shows how the intensity or strength of the respective atmospheric parameter is modulated through the park. By doing so, the maps keep areas within the park from reaching excessive natural conditions while making the experience of changes in climate much more comfortable in the areas where we will reinforce the coolness, the dryness, the clean. The three maps intersect and overlap randomly in order to create a diversity of microclimates and a multitude of different sensual experiences in different areas of the park that we could freely occupy depending the hour of the days or the month in the year. At a certain place, for example, the air will be less humid and less polluted but it will still be warm, while elsewhere in the park, the air will be cooler and dryer, but will remain polluted. The three climatic maps vary within a gradation, which ranges from a maximum degree of uncomfortable atmospheric levels that usually exist in the city (maximum rate of pollution, maximum rate of humidity, maximum rate of heat) to areas that are more comfortable where the heat, the humidity, and the pollution are lessened. Climatic Lands.

According to the computational fluid dynamics simulation (CFD), we have identified the coldest, driest, cleanest areas of the park where these existing climatic qualities are increased by densifying the number of climatic devices and trees with specific climatic properties to cool, to dry, to clean the air. These eleven most comfortable areas are designated as the Climatic Lands and are denominated Coolia, Dryia, and Clearia according to their specific climatic specification. The Coolia are the most colder areas in the Park. The Dryia are the dryest ares, and the Clearia are the areas with the less polluted air. The eleven Climatic Lands contain all the activities and programs. The three Climatic Paths link all the same types of Climatic Lands together.

Coolia

The Cool Lands (Coolia) are the most cooler areas of the park. They are defined by a high density of cooling trees and cooling devices. There are four Cooling Lands named as Northern Coolia, Western Coolia, Middle Coolia, and Southern Coolia. Inside these four Coolia (Cooling Lands), visitors can find all the equipments for Leisure.

Dryia

The Dry Lands (Dryia) are the most dryer areas of the park. They are defined by a high density of drying trees and drying devices. There are three Dry Lands named as Northern Dryia, Eastern Dryia, and Middle Dryia. Inside these three Dryia (Drying Lands), visitors can find all the equipments for Sports.

Clearia

The Clean Lands (Clearia) are the areas of the park with the less polluted air. They are defined by a high density of depolluting trees and depolluting devices. There are four Clean Lands named as Northern Clearia, Eastern Clearia, Middle Clearia, and Southern Clearia. Inside these four Clearia (Clean Lands), visitors can find all the equipments for Family Activities.

Jade Eco Park, Taichung, Taiwan

Philippe Rahm Architectes, Mosbach Paysagistes, Ricky Liu & Partners

Client: IBA Hamburg, Germany/Partners: Transsolar, Böllinger & Grohmann, fabric.ch

Masterplan Composition

Heat map on site

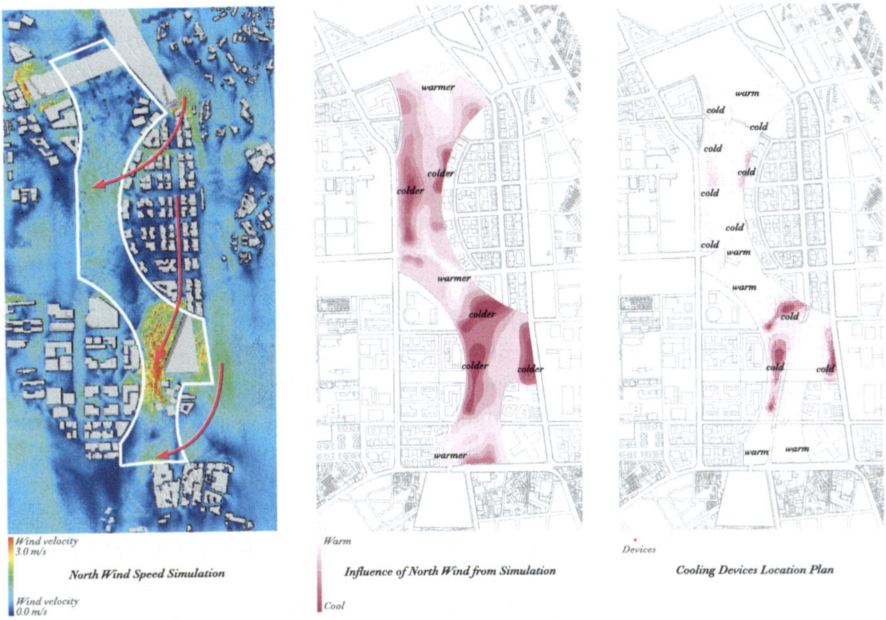

North Wind Speed Simulation

Influence of North Wind from Simulation

Cooling Devices Location Plan

Architecture

The design of the IBA Hamburg condominium building is based on the natural law of Archimedes that makes warm air rise and cold air drop. Very often in an apartment, a real difference of temperature can be measured between the floor and the ceiling, a difference that could sometimes even be 10 °C. Depending on our physical activities and the thickness of our clothes, the temperature doesn't have to be the same in every room of the apartment. If we are protected by a blanket in bed, the temperature of the bedroom could be reduced to 16 °C. In the kitchen, if we are dressed and physically active, we could have a temperature of 18 °C. The living room is often 20 °C because we are dressed without moving, motionless on the sofa. The bathroom is the warmest space of the apartment because here we are naked. Keeping these precise temperatures in these specific areas could economize a lot of energy by reducing the temperature to our exact needs. Related to these physical and behavioral thermal figures, we propose to shape the apartment into different depths and heights: The space where we sleep will be lower, while the bathroom will be higher. The apartment would become a thermal landscape with different temperatures, where the inhabitant could wander around like in a natural landscape, looking for specific thermal qualities related to the season or the moment of the day. By deforming the horizontal slabs of the floors, different heights of the spaces are created with different temperatures. The deformation of the slabs also gives the building its outward appearance.

In thermodynamics, energy transfer by heat can occur between objects by radiation, conduction, and convection. Convection is usually the dominant form of heat transfer in gases. This term characterizes the combined effects of conduction and fluid flow. In

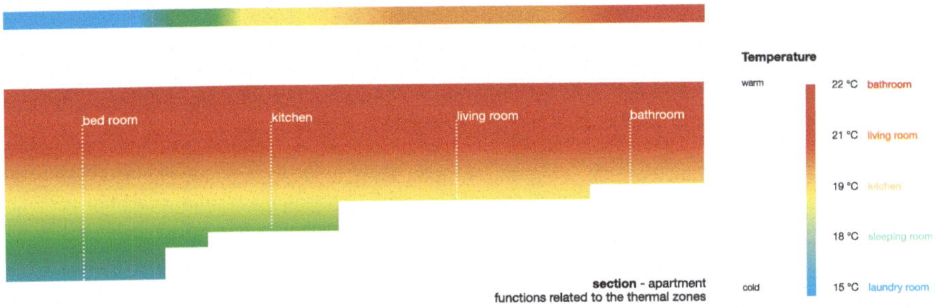

convection, enthalpy transfer occurs by the movement of hot or cold portions of the fluid together with heat transfer by conduction. Commonly an increase in temperature produces a reduction in density. Hence, when air is heated, hot air rises, displacing the colder denser air, which falls. In this free convection, gravity and buoyancy forces drive the fluid movement.

Convective apartments, Hamburg, Germany
Philippe Rahm Architectes
Client: IBA Hamburg, Germany/Partners: Fabric | Ch, Arup/Epfl/Werner Sobek/Weinmann Energies

Philippe Rahm

On the Nature of Thermodynamic Models in Architecture and Climate

Nadir Abdessemed(✉)

Consultant for Climate, Energy and Sustainability in the Built Environment at Transsolar, Curiestrasse 2, 70193 Stuttgart, Germany
abdessemed@transsolar.com

Most of the disciplines engaged in architecture and design describe our built environment with models. Be it science or engineering, art or architecture, philosophy or economics, models help us understand and inform a future world shaped by designers. They are an essential element of our thinking and allow us to connect the physical world with a cognitive process. The purpose of creating a model may lie on a spectrum from finding "truth" to storytelling. It may range from answering practical question to aesthetics, from optimizing resources to more effectively use time and money needed to finding answers to a question. Models may be of physical nature or virtual, they may require a computer, and sometimes only be a conversation involving an analogy.

However, models have various aspects in common, one could say, it is the following common ground that could define the nature of a model. They typically consist of (a) a context (b) a problem, and (c) a method. I would like two follow up with two examples from my work as a climate engineer at Transsolar: Our context could be a fully transparent gallery space of a museum as illustrated in Fig. 1, the problem could then be the associated greenhouse effect and questions could be raised about thermal comfort within this space. A method to address these questions could be a thermodynamic computer simulation to predict the natural flow flushing the space with cold outside air.

The context could also be our longing for experiencing the (im)materiality and spatial environment of a landscape of clouds, a cloudscape. The associated problem could be questions related not only to materiality, but also to space, the visual perception, and obviously to the difficult accessibility of clouds. How does it feel to enter a cloud? How much latent energy is absorbed in a cloud? The method could be an experimental installation at an architectural exhibition, providing access to a cloud on a human scale (Fig. 2).

The latter, an experimental approach to modeling, comprises installations, mock-ups or prototypes, experiments, physical tests, and observations, while the former example represents models of a virtual nature, like a computer simulation, intelligent algorithms to optimize isolated aspects of the design or just a thought experiment written with pen and paper.

I am intrigued by the power of models, but also fascinated by the extent to which models are sometimes used with little virtue. The famous Hammer–Nail problem, also known as the Law-Of-The-Instrument, refers to a boy, equipped with only a hammer, who will see a nail in every problem. Ironically a model in itself—this imagery

K. De Rycke et al., *Humanizing Digital Reality*,
https://doi.org/10.1007/978-981-10-6611-5_8

Fig. 1 Context: Academy Museum of Motion Pictures Los Angeles. Architect Renzo Piano building workshop, Genova. Climate engineering of the glass sphere surrounding a public and open gallery space by Transsolar. © RPBW

Fig. 2 "Model of a cloud". Cloudscapes by TRANSSOLAR and Tetsuo Kondo Architects, Biennale di Venezia, 2010. Photo © Tetsuo Kondo Architects

represents how reliance one tool may limit our perception of a problem boundary due to the lack of appropriate alternatives.

So, What Is a Good Model?

Being involved in many different challenges during various stages of design, I often ask myself: "What is a good model?", and I would like to take the opportunity of the Design Modelling Symposium to share my thoughts while attempting to answer that question from the perspective of a designer facing practical, everyday problems. I will do this using two examples from our recent work on Renzo Piano's Academy Museum of Motion Pictures and the Cloudscapes installation by Transsolar and Tetsuo Kondo that will highlight the qualities which I believe make a good model.

1. **Simplicity**

A model could be understood as a reduction of complex (in most of our cases physical) systems to a more simplified set of parameters that may be of particular interest. In that sense, the model should be only as complex as necessary, and as simple as possible.

Simplicity is one of the qualities of Cloudscapes that could be understood as a *model of a cloud*. There are fairly complex physical mechanisms taking place in clouds. They are typically based on transient, i.e., time-dependent processes and develop into spatially three-dimensional formations with sometimes dramatic or spectacular sceneries. Meteorologists categorize clouds into complex sets of classifications with Latin naming conventions such as Cirrus Fibratus Radiatus or Altocumulus Stratiformis Perlucidus—just to get a sense of the various properties that may describe a cloud specie.

However, what all clouds have in common despite their complexity is the presence of warm humid air and cold dry air, interacting with each other and separating space through condensed droplets. Only this reduced set of aspects of an otherwise complex, real cloud, has been physically created in Cloudscapes and made accessible to visitors.

While the cloud is floating through a three-dimensional space, the stratification defining the space below, in between and above, is a one-dimensional phenomenon (as temperature only varies in height, not in plan). The model of the cloud is furthermore independent of time (one may call it quasi-steady-state), while experiencing the cloud inherently involves the concept of time as part of our human perception. To physically observe and appreciate a real cloud through a model, scales are distorted (certainly flying through a real cloud would involve an altitude change of several hundred meters). As such, we could say that the amplified set of interesting phenomena has to do with stratification of cold and warm air, the rise of humidity (contrary to common belief humid air is lighter than dry air), the limited ability of air to absorb water, and varying transparency of condensation droplets in the air.

A certain simplicity also lies in the computer simulation describing the aerodynamics in the Academy's gallery space shown in Fig. 4. While we are interested in temperatures within that semi-open space, initially, only winds have been addressed. In particular, the momentum transferred between a given amount of air particles in a simplified geometry sheds light onto the already complex mechanisms that would take place in a swirling flow. Effects having to do with solar radiation, buoyancy (warm air rising), gusts of winds and turbulences, humidity, the presence of people and internal structures, to name a few, have been isolated from the problem, in order to come to valuable conclusions related to the most important questions within the design process.[1] This leads me to the next quality of a good model:

2. **Asking the Right Questions**

A good model is employed in the context of the **right questions**. It is less a matter of giving the right answers, as we often don't a priori know what lies ahead. Posing the right question is the basis for choosing the right model raising our understanding of an unknown situation to the next level.

Some of the right questions during the design of Cloudscapes were the following: In what different ways can one experience a cloud? What perspective do we take on? The answers to these questions lead to the ramp designed by Tetsuo Kondo Architects, bringing people from below the cloud, through the cloud, above the cloud, experiencing not only visual separation, but also layers of heat and different humidity. While I am exploring ways of visualizing three-dimensional object such as a cloud (not only of condensation droplets but also of information) with my dear friend and colleague Silvia Benedito from the GSD, it turns out that a cloud is typically experienced from either below, or above, or from inside as we all know from a landing plane. The only experience that could be made of all three layers at one time is outside the cloudscape, separated by a transparent film as created in an altered version of the Cloudscapes installation at the Museum of Contemporary Art in Tokyo in Fig. 3.

But why are we interested in a gallery space enclosed in a transparent glass dome in the middle of Los Angeles? Certainly, we are talking about the sheltering characteristics balanced with what one could refer to as an adverse greenhouse effect. So the ultimate question is: How can we create a protected space of comfort while ensuring connection to the outside world. The question originally would not evolve around conditioning systems nor materials, and neither included form nor aesthetics. These, one could argue, follow the fundamental wish to create a comfortable, sheltered space within an otherwise uncomfortable environment. Various comfort models can be used to simulate how a person would feel in an environment with unconventional temperatures, movement of air (as shown in Fig. 5), solar radiation, humidity, radiation of warm surfaces, even clothing or physical activity of a cliché LA audience arriving from the beach with different expectations for thermal comfort in a somewhat outdoors, yet sheltered space.

[1] We could go further: The well-understood, even simpler, two-dimensional archetypal "flow over a cavity" certainly has similar air flow features, although I refrain from speculating here on the physical mechanisms in the gallery context.

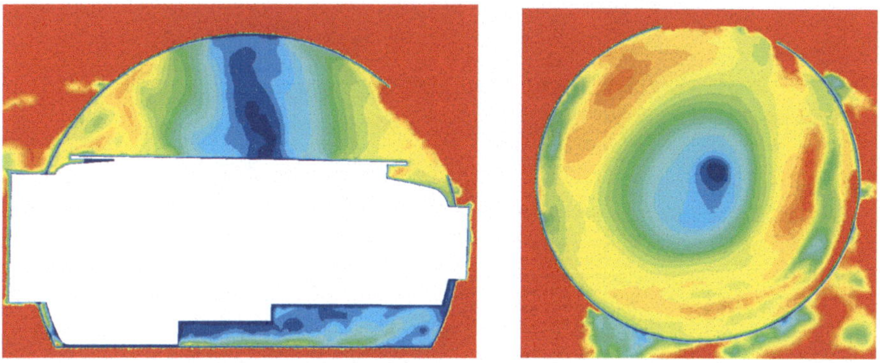

Fig. 3 Cloudscapes by TRANSSOLAR and Tetsuo Kondo Architects 2012, Museum of Contemporary Art Tokyo MOT, perspective from outside allows to visually connect to all three layers of the cloud: below, in between and above. Photo © Tetsuo Kondo Architects

3. **Discovery**

A good model allows us to discover some form of truth if the model represents reality (to an acceptable degree of accuracy). Clearly, there is a physical correlation between humidity rising to a height above people's heads, but only until it encounters a layer of warm dry air. The atmosphere under the ceiling is so warm that its density is even lower than that of the cloud, forming a natural barrier for the water in the form of vapor that now balances between two dryer and clearer spheres. The correlations we observe are universal to humid air and form the basis of the fundamental physical mechanisms related to density of vapor/air mixtures. Similar dynamics have been exploited in the design of an art school in the humid climate of Houston, Texas, where natural ventilation relies on the fact that humid air rises naturally. While these correlations are explored by climate engineers and "makers" of the cloud, there is a high discovery potential from an architectural point of view: interaction with natural daylight, definition of space, and the reaction of visitors, for instance.

In our museum case, truth or discovery is directly related to finding answers to the question of people's comfort. Figure 5 shows the amount of colder outside air displacing the warm air of our "greenhouse" reducing overall temperatures. Quantifying the ventilation effects, potentially including the exposure to solar radiation not only informs the design, e.g., of height, orientation, and size of openings to the outside, but also increases our confidence in the design process. While visitors in our physical model are of course real people that could be asked about their experience, the ones in a virtual model are not, at least initially—only one aspect in the classic discussions about the value of experiments versus that of computer simulations. The faculties of experimentalists and simulation experts typically agree on collaboration and the next example explores an interesting effect that could be observed in both, virtual and physical models.

Discovery often raises new questions that naturally where not evident to the designer in the beginning. These may be associated to surprising findings in the design process, leading me to the last quality of a good model.

Fig. 4 Simple: model using a simplified set of unknowns to address the question interaction between winds and building form. Surprise: unexpected vortex of air driven by wind tangential to the opening. *Red* indicates higher and *blue* lower air velocities. Vertical (*left*) and horizontal (*right*) section

4. Surprise

Employing models sometimes generates surprising moments. Surprises are often perceived as undesirable realities of research and design. They are typically associated to risk, or throwbacks in the process, diverting us from supposed progress. Every surprise calls for an explanation, challenging our past way of thinking about a situation and dragging us out of our intellectual comfort zone. I believe there is a danger in ignoring surprising findings, for often surprises can be inspiring, hence a good model should allow and even more enforce and unleash surprising moments.

In our gallery context as shown in Fig. 5, the "swirl of air", reminiscent to the pattern of a tornado is certainly not an effect we would expect from winds tangential to the dome opening. What forces are driving this "tornado"? Do we observe a similar relative flow on a carrousel? Can we enhance this vortex and generate a breeze giving

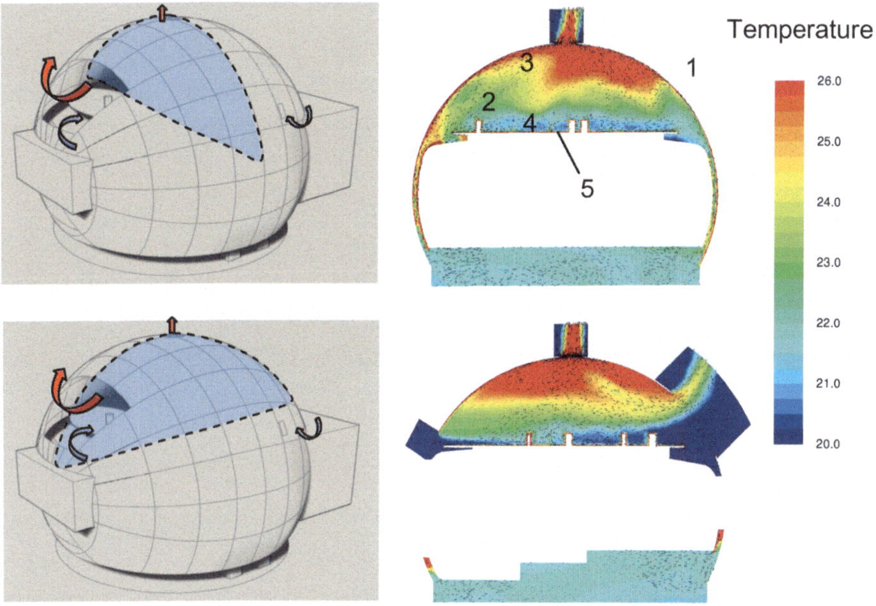

Fig. 5 Section through the transparent dome surrounding the semi-external gallery space. CFD model describing the three-dimensional flow (temperature stratification and velocity field) due to solar radiation and buoyancy effects

thermal comfort to visitors in summer? And how can we protect from it in winter? Surprises raise new questions and inspire design.

A vortex can also be experienced as a surprising moment in Cloudscapes. A group of children was asked to engage with the cloud to test its stability. As they started running in a circle, a rotating flow developed, creating a negative pressure in the "eye of the vortex", locally dragging the cloud down to the ground like a tornado; a phenomenon that would stabilize itself right after the circular dance would stop.

Clearly, cause and effect between air movement and people are reversed in our two examples, and the underlying forces are very different. However, it gives rise to the idea that a vortex-like air movement is an important phenomenon on different levels and scales of architecture.

While simplicity, relevance, discovery, and inspirational surprise are certainly qualities that could be seen in isolation, it is the combination of these dimensions that shape an interesting narrative around our creative processes in the context of models. Like asking about the qualities of a good model, we could ask, what is the material a good story is made of. From Homer to Hollywood, there are many qualities shared between a tale and a model. Arguably, the ultimate virtue of a good model and a good story is the idea of progress—on a physical and a meta-physical level.

The Tool(s) Versus The Toolkit

Chris Mackey[1,2(✉)] and Mostapha Sadeghipour Roudsari[1,3]

[1] Ladybug Tools, Boston, USA
{chris,mostapha}@ladybug.tools
[2] Payette Associates, Boston, USA
[3] University of Pennsylvania, Philadelphia, USA

As the architectural practice continues into the digital age, most designers agree that new computational technologies should be harnessed for the betterment of their buildings and workflows. However, there are currently several competing philosophies for how such technologies should be best integrated into the design process. Until recently, much of today's practice has found itself leaning toward one of two camps: one that allows for a distributed yet disconnected approach and another that strives toward a centralized methodology. While each philosophy has its strengths, they both suffer from significant limitations.

Perhaps one of the clearest examples of the distributed and disconnected approach is the current state of environmental performance software, which has seen the introduction of countless new specialized tools over the last few decades. The goals of these distinct applications range across the board from daylight/glare modeling, to HVAC sizing, to full-building energy simulation, to thermal comfort forecasting, to embodied carbon estimation, to structural member sizing, to envelope insulation evaluation, to condensation risk mitigation, to acoustic modeling, to stormwater/rain collection management (Moe 2013). The list goes on and, in order to engage the full range of topics necessary for good environmental design, practitioners frequently find themselves recreating models in each of these different interfaces. As a result, designers are typically unable to account for all of these criteria in the scope of their projects and this additive approach of piling more tools onto the process becomes over-complicated and inefficient.

Many have recognized this trend in recent years and have attempted to respond to the issue. Perhaps the clearest example of this is the rise of building information modeling (BIM), which anticipates a streamlined design process by having all design team members put their data into a single model built with one software package (Negendahl 2015). While such BIM models can be useful for organizing and documenting final designs, their sheer size can make them inflexible and difficult to iterate upon. The more that is added to a model, the harder it becomes to change or test out new ideas with it. If one were to support all of the previously listed environmental analyzes with a single central model, each model element would have a huge number of properties and geometry types associated with it. This is because the data that is needed for one type of study, such as energy modeling, is often very different than that needed for another study, such as structural member sizing. So adding/changing any given element of such a central model would either be time consuming and unfriendly

K. De Rycke et al., *Humanizing Digital Reality*,
https://doi.org/10.1007/978-981-10-6611-5_9

to iteration or would result in messy, poor data for certain types of studies as practitioners add building elements to satisfy only one objective at a time.

Accordingly, neither the disjointed set of tools nor the "one tool to rule them all" is a suitable solution to our dilemma of software integration. As with many dualistic situations that contrast two extremes, the best route is often a middle one that harmoniously balances the two. So what would such a harmonious balance look like for our problem of software integration? One might notice that both the disconnected and the centralized approaches suffer from the same philosophical fallacy—they focus on the tools themselves as the solution rather than the workflows or interconnection between tools. Instead of a single all-powerful tool or a disjointed set of tools to address our contemporary dilemma, a cohesive toolkit that seeks enhanced workflows between software might be far more effective. Unlike the centralized method, a toolkit would have the flexibility to engage different objectives whenever they become relevant. In other words, one does not have to specify all properties/formats of a given building element at once but can wait until one is ready to use the pertinent tool in the kit. Yet, unlike the disconnected approach, the reference to tools as part of a "kit" means that they are expected to work together in a continuous process, allowing the work done with one tool to be cleaned, formatted, and passed onto another with minimal loss of relevant data (Fig. 1).

This notion of a "toolkit" is arguably what has made visual programming languages (VPLs), such as Grasshopper (McNeel and Associates 2017) and Dynamo (Autodesk 2017), so successful in addressing software integration issues in contemporary design practice (Negendahl 2015). The fact that these VPLs break down the functions of software into discrete "components" or "nodes," makes them the literal embodiment of toolkits. Each component within a VPL has its own inputs and outputs, essentially acting as an individual tool within a larger script or workflow. As a result, users can customize their workflows based on the arrangement of components in their scripts, enabling them to engage different issues as they become relevant and experiment with new creative workflows as unique situations arise on projects (Tedeschi 2014). Furthermore, the ability to output data at different points along a VPL workflow allows plugins intended for different purposes to easily pass relevant information between one another. While VPLs are one of the more obvious examples of toolkits in contemporary

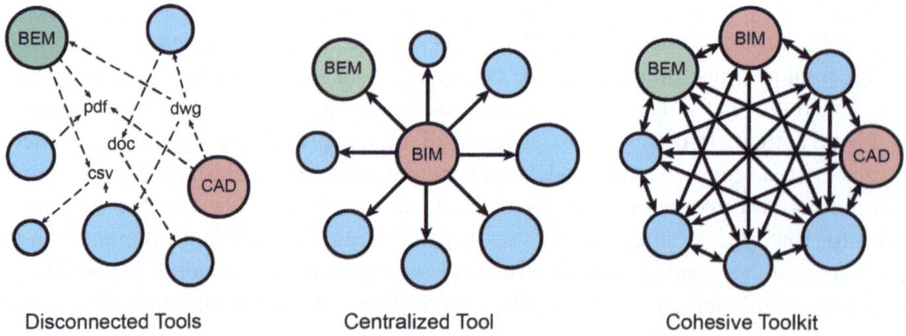

Fig. 1 Diagrams of software integration methods

practice, the general sentiment that all software should work together can also explain why some computer applications have integrated more successfully than others with contemporary practice. Accordingly, with the goal of a toolkit in mind, the rest of this article will define key guiding principles and features that make software a part of a toolkit and therefore a particularly useful element of contemporary design processes. These "principles of the toolkit" are derived from the author's' experience developing the "Ladybug Tools" plugins (Sadeghipour Roudsari and Mackey 2017) for the aforementioned VPL interfaces. While the authors attribute much of the success of this project to these principles, it should be noted that not all must be fulfilled for a given software to act as part of the toolkit and there can be multiple pathways to addressing each principle. Ultimately, it is hoped that this list will assist both practitioners who are seeking to identify software that can be used in their toolkits as well as software developers looking to make their projects behave with this "toolkit" functionality.

"Do One Thing and Do It Well"

Perhaps the most important feature of any software seeking to be a part of a kit is that it performs one task (or a few related tasks) exceptionally well. Many of us know from experience that our most valuable and continually-used tools are often simple in presentation. For example, many attribute the early success of Google to its founding engineers' focus on making a fast and well-indexed search engine rather than adding extraneous news, weather, and advertising images (Williamson 2005). The same can be said of many plugins in software ecosystems like those surrounding Grasshopper and Dynamo. When a large number of plugins exist within the same community, they are often forced to focus on a particular task in order to define a niche for themselves within their ecosystem. The more developers that there are in a given software environment, the stronger the need to differentiate oneself and the more intense the speciation. Even Ladybug Tools, which many people see as an umbrella for several different types of studies, has a clear boundary that defines what it does well. Specifically, this is "analysis related to climate/weather data" and, while there are many other tools related to good environmental design (like optimization algorithms, building structural solvers, and tools for creating generic charts), Ladybug Tools does not include these. Instead, if you need this functionality, we recommend that you use other tools in your "kit" that are better suited toward these tasks, like the Octopus optimization plugin (Vierlinger 2017), Kangaroo form-finding plugin (Piker 2017), or just export your data to Excel to make some generic charts. This focus on one particular task is essentially a software developer's recognition of the tool's place within a larger toolkit. As a result, a tool that is intended to be a part of a kit follows the first tenant of the Unix philosophy (Weber 2015), having fewer extraneous features and instead focusing efforts on its primary stated purpose.

Build Interoperability with Other Tools

While it is important for software in a toolkit to focus on performing one task well, the suggestion that one "use another tool in your kit" is of little help if one cannot export one's work to such other tools. For this reason, it is critical that any software seeking to participate in a toolkit possess BOTH import and export capabilities to a variety of other formats. This is particularly relevant given that many software companies prefer to focus on importing data from a wide range of formats and neglect the development of export functionality. This is understandably the result of traditional competitive economics as companies feel that it is better to keep users within their own interfaces rather than letting them export and roam to competitor software. Yet, the adherence to this thinking often ends up hurting such software projects more than it helps since the time that could have been spent building export capabilities is instead devoted to adding features that mimic competitor functionality. Because such mimicked functionality is usually never as good as another piece of software that is dedicated to the task, there is a missed opportunity to add the most value to their work. This competitive mindset also makes it important to highlight that good interoperability not only includes the export to generic file types, like PDFs for drawings or gbXML for energy modeling, but also allows direct export to more specific formats when possible, like Illustrator for drawings and OpenStudio (NREL 2017) for energy modeling. Such direct exporting affords the smallest loss of relevant data in translation and allows the software to perform more successfully within a broader toolkit. Understanding that this interoperability is critical for a functioning toolkit, it is clear why plugins developed for VPLs excel as members of toolkits. Such plugins take this concept to the extreme by allowing any relevant data to be connected/exported from one plugin to another. Of course, a plugin that outputs standardized data types is likely to be more successful at achieving this interoperability and this brings us to the next guideline in the list.

Use Standardized Open Formats for Data Transfer

While interoperability with existing major software is critical for any toolkit, a good member of a kit also anticipates its compatibility with possible future extensions of itself and insertions of custom data at different times during its use. For this reason, the way in which data is passed between the different utilities of a tool can be just as important as the tool's overall ability to export to other platforms. This is particularly relevant as many programs use compiled proprietary file formats that are only readable by computers and cannot be easily translated to human-readable data, such as text, numbers, and geometry. While these compiled formats have some important uses in compressing data, their use in proprietary schema can severely limit a tool's ability to be extended and to "talk" to other tools. As such, the use of open text-readable standards like ASCII and UTF-8 greatly increases a software's usefulness within a toolkit. Furthermore, the use of standard file types for storing data, like CSV for tabular data or JSON/XML for object-oriented data, also helps enable cross-compatibility. Finally, VPL components that pass standardized text-readable streams of data between

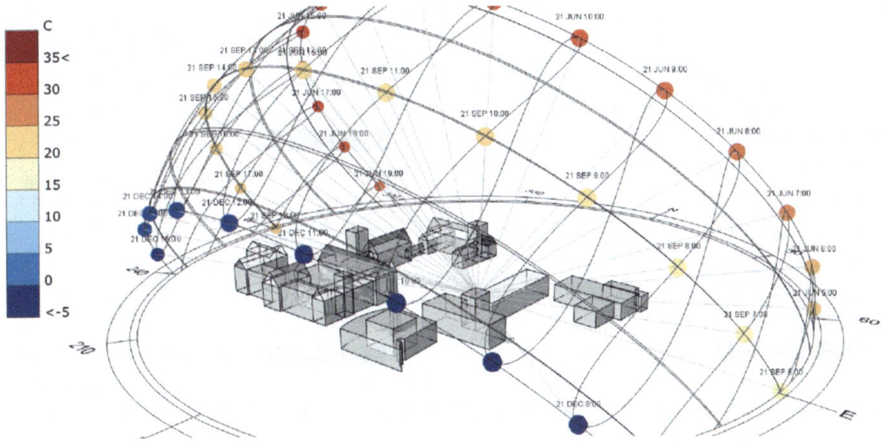

Fig. 2 An example of annual hourly data usage in a Ladybug visualization: the suns of a sunpath are colored with hourly temperatures, denoting which parts of the sky should be blocked/shaded for thermal comfort

their components will more easily facilitate integration with other plugins and data sources. Perhaps the best example of this within Ladybug Tools is the structure used for annual hourly data, which can originate from several sources including downloaded climate data and annual building energy simulation results. Nearly all Ladybug visualization components can accept this hourly data as an input (Fig. 2) and both the standardization of this data across the plugin and the human-readable format of this data are critical to the success of Ladybug Tools. Specifically, this annual hourly data is derived from the standard format of .epw climate data and consists of a text header followed by numerical values for every hour of a year. This simple format enables both easy math operations to be performed on the numerical data while also providing text instructions for the specific components that make use of it. The fact that this data is human-readable also means that, if a user has hourly data coming from any other source outside of the plugin (like another piece of software or recorded empirical data), this can be directly input into Ladybug to visualize and analyze it. Other examples of standardized, human-readable formats in Ladybug Tools include text formats borrowed from its underlying simulations engines, like the Radiance standard for daylight materials and the EnergyPlus standard for full-building energy materials (Fig. 3).

Modularize the Tool

The success achieved through the use of standard, text-readable formats initially depends on a tool being modularized into discrete elements that can pass this standardized data back and forth. The more modularized that a tool is, the more locations that exist for people to input/export custom data, build extensions on top of the tool, and connect it to other software. From this principle, we can understand that VPL

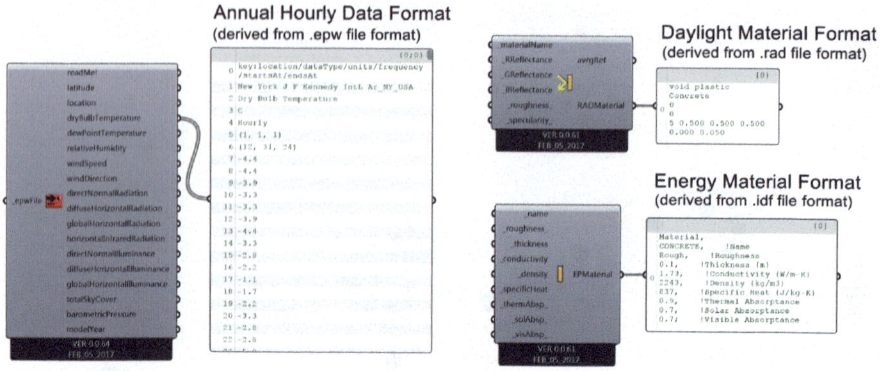

Fig. 3 Standardized, human-readable data formats in Ladybug Tools

plugins will be more successful at integrating into toolkits if they break down their functions into more and more components or nodes. This is something that Ladybug Tools takes to heart since it is very rare to run an entire study with a single component. For a daylight simulation alone, one has separate components for geometry, materials, sky types, "recipes" (or simulation instructions), and result-processing (Fig. 4). This modularization ultimately allows for a much higher degree of customization and potential integration with other tools than would be possible if these processes were wrapped in a single component. It is also important to highlight that software does not necessarily have to exist in the form of VPL components in order for it to be modularized. The vast majority of software in the world achieves a modularization by breaking down all its capabilities into a well-documented Application Programming Interface (API) or Software Development Kit (SDK). The most "toolkit-like" of these APIs make use of a principle known as object-oriented programming, which essentially divides the functions of software into several different "objects," each with properties that can be set and operations that can be performed on it (Kindler and Kriv 2011). These "objects" can refer to anything and, as an example, the Rhinoceros CAD

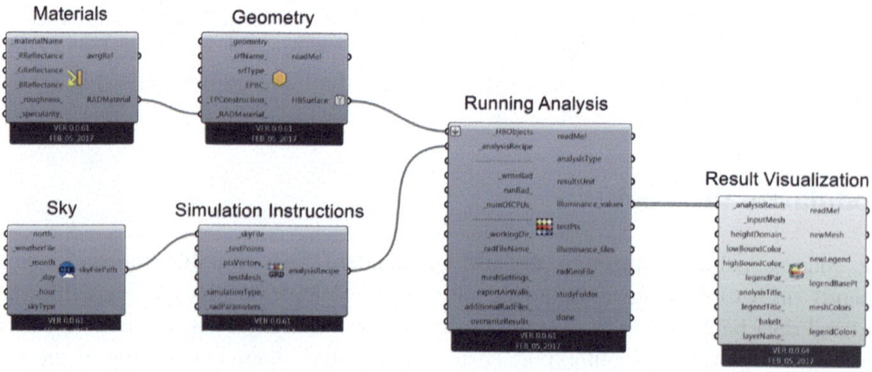

Fig. 4 Modularization of daylight simulation in Ladybug Tools

software (McNeel 2017) includes several object types that one might readily recognize (like points, curves and surfaces) as well as objects that are less obvious (like an object for the viewport or an object for document settings). The more modularized, object-oriented, and well-documented such APIs are, the easier it is to build extensions off the software and connect it to other tools. Accordingly, VPL plugins that maximize their number of components and APIs that break down software into many objects tend to be more successful at operating within toolkit schemas. Given this principle, it is important to recognize that, the more components or objects that a tool is broken into, the steeper the learning curve is to mastering the software. For this reason, there is need for one final principle of the toolkit.

Make It Easy to Start but Impossible to Master

Much like the instruments of any craftsperson, software toolkits are most successful when there is an art to mastering them. However, if such kits are too difficult to use from the start, new entrants can feel discouraged and will find it hard to engage. For this reason, the most successful toolkits follow a philosophy that was perhaps best summarized by the founder of Atari when describing their most popular video games—they are "simple to learn but impossible to master" (Bogost 2009). Following this mantra, the intense modularization, customizability, and exposing of options within a toolkit must be balanced with plenty of defaults for these options. Within Ladybug Tools, this manifests itself in the form of components that have large numbers of inputs but only a small number of them that are actually required to run the component. For example, the Ladybug sunpath has over 15 inputs, which allow for a high degree of customization, yet only a single input (the location) is necessary to produce the familiar solar graphic (Fig. 5).

This large number of defaulted inputs along with a visual standard to communicate which inputs are required or defaulted (using dashes _before or after_ input names) helps new users of Ladybug Tools navigate the capabilities of the software. In addition to default values, having a low number of required components for a given operation will further make a toolkit "easier to start." Within Ladybug Tools, this is best illustrated by the fact that only three components are necessary to run a full-building energy simulation. Yet, users can engage with over 100 other components in order to add meaningful aspects to this energy simulation (like window geometries, wall/window constructions, and energy-saving strategies like natural ventilation, shade, and heating/cooling efficiency upgrades). Structuring software like this enables new entrants to rapidly arrive at a tangible results and, with an understanding that these quickly-achieved results are far from perfect, users will be inspired to delve further into the toolkit. If done well, new entrants will find themselves quickly advancing through the kit and educating themselves as they go. Together, this forms a community of masters, new entrants, and many in between, who can help each other reach deeper into the kit through online discussion forums (Fig. 6). Eventually, masterful users may find a route all of the way to the core functions of the software and it is for this reason that many of the software packages that are most devoted to the notion of a toolkit are also

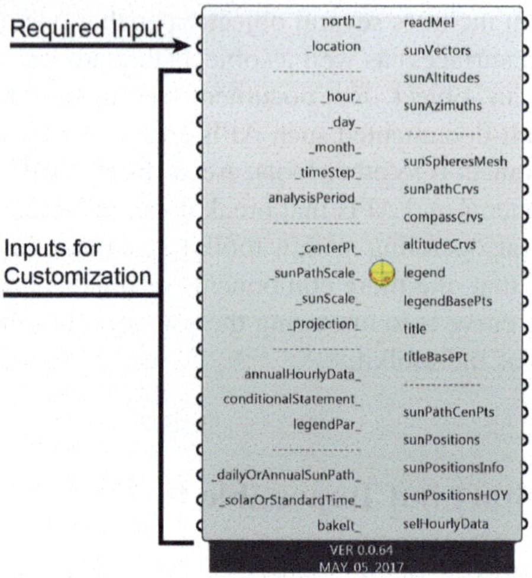

Fig. 5 One required input at over fifteen customizable inputs on the Ladybug Tools sunpath

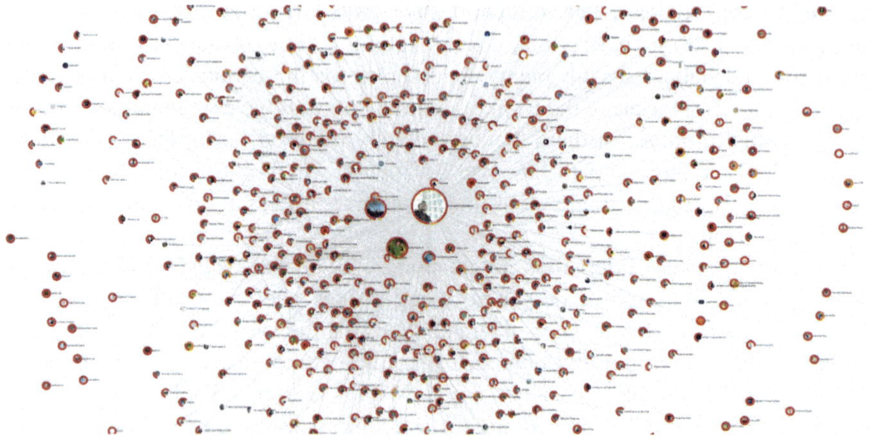

Fig. 6 Snapshot of activity on the Ladybug Tools community forum (2017)

open source. Allowing everyone to view a tool's source code is an invitation for people to master it and, while not all software needs to be open source to participate in a toolkit, one can typically use this to identify software that intends to be participate as such.

By following the principles listed above, software can better meet the needs of today's architectural practice by enabling BOTH the flexibility to engage different iterative studies AND the integration that is needed to make coherent, coordinated designs. When we look at many of our buildings today, they increasingly resemble complicated collections of separate systems—structure, envelope, air conditioning, fire protection, electrical, interior furnishings, etc. In this separation of systems, we increasingly overlook opportunities for synergies between them, such as when part of the envelope can act as a structural system. This separation can also lead to misinformed decisions, such as removing exterior shade to save construction costs only to pay for it in a larger air conditioning system that can remove the higher solar gain. These missed opportunities are a direct result of our design thinking that is shaped by poorly integrated tools. If we are to have elegant, coherent building designs in this contemporary era, we must first address these underlying issues in our design workflows. By moving toward software toolkits instead of disconnected or centralized tools, our vision of elegant designs can more quickly become the reality around us.

References

Autodesk: Dynamo. http://www.dynamobim.org (2017)

Bogost, I.: Persuasive Games: Familiarity, Habituation, and Catchiness. Gamasutra (2009)

Kindler, E., Krivy, I.: Object-oriented simulation of systems with sophisticated control. Int. J. General Syst. 313–343 (2011)

McNeel, R., & Associates: Grasshopper. Algorithmic modeling for Rhino. http://www.grasshopper3d.com/ (2017)

Moe, K.: Convergence: An Architectural Agenda for Energy. Routledge (2013)

National Renewable Energy Laboratory (NREL): OpenStudio. https://www.openstudio.net/ (2017)

Negendahl, K.: Building performance simulation in early design stage: an introduction to integrated dynamic models. Autom. Constr. **54**, 39–53 (2015)

Piker, D.: Kangaroo. http://kangaroo3d.com/ (2017)

Sadeghipour Roudsari, M., Mackey, C.: Ladybug tools. http://www.ladybug.tools/ (2017)

Tedeschi, A.: AAD Algorithms-Aided Design. Parametric Strategies Using Grasshopper. Edizioni Le Penseur (2014)

Vierlinger, R.: Octopus. http://www.food4rhino.com/app/octopus/ (2017)

Weber, S.: The Success of Open Source. Harvard University Press (2005)

Williamson, A.: An evening with Google's Marissa Mayer (2005)

Capturing the Dynamics of Air in Design

Jane Burry(✉)

Spatial Information Architecture Laboratory (SIAL), Master of Design Innovation and Technology (MDIT), RMIT University, GPO Box 2476, Melbourne, VIC 3001, Australia

By contrast with science and unraveling the mysteries of the natural world, architecture seems in many ways inherently focused on end points and the production of a solid and enduring presence in the world rather than mechanisms and dynamism per se. Yet architecture and design have been continually invoked in relation to order-in-the-universe, sometimes for their intentionality in producing an ideal outcome, sometimes as the schema of a system. The simile seems as useful to the divine universe—Thomas Aquinas' simile between God and the role of the architect and Calvin's "Great Architect"—as to the scientific universe.[1] Stephen Hawking and Leonard Mlodinow no doubt adopt the title *The Grand Design* with a note of irony, and in reference to the struggle between Darwinism and tenets of Christianity.[2] However, we might take this adoption of the design metaphor, this time for the *mechanisms* of quantum mechanics in a continuously unfolding universe, to reinforce that the inseparability of mechanism and teleology is as fundamental in design as it is in nature.

By its very definition, design is the expression of an intention, or an end point and the process of getting there. But as post-empirical science leads us back to the natural world of Heraclitus[3]—a world of restless and relentless change—how can human design and artifice embrace this dynamism rather than idealizing a static foil to a tumultuous world? The six or seven decades of digital computation has only amplified the ambition to extend the process of representing a design outcome into representing *mechanisms* to shape such ends. Computation has proffered models of process where influential inputs from the design context are linked to outcomes in a manner analogous to the evolution and genotyped development of species and phenotypic development of their organisms. Designers seek to sculpt architecture like land form in response to forces of climate and geophysics and engage with "riddles of form…problems of morphology" like trees and clouds[4].

We harbor a dream of closing the loop as the natural world seems to in its ever-altering ebb and flow, to be able to create almost naturalistic dynamic feedback systems in which the impact of one state of a design is tested in some abstraction of the

---

[1] Thomas Aquinas, Summa Theologiæ (1265–1274 unfinished).

[2] Stephen Hawking and Leonard Mlodinow, The Grand Design, New York: Bantam Books, 2010.

[3] Heraclitus of Ephesus, 535–475 BC, a pre-Socratic Greek philosopher who wrote that you could never step into the same river twice.

[4] D'Arcy Wentworth Thompson. *On Growth and Form, the Complete Revised Edition*. New York: Dover Publications Inc., 1992 (1917), 10.

K. De Rycke et al., *Humanizing Digital Reality*,
https://doi.org/10.1007/978-981-10-6611-5_10

Fig. 1 The dynamics of sensual atmospheres: The use of multiple scanning lasers enabled visualizing air through several planes, in parallel or perpendicular to each other. *Credit* Malte Wagenfeld

Fig. 2 The dynamics of sensual atmospheres: visitors observing and interacting with the fine fog produced by the installation atmospheric sensitivity. Aesthetics of Air, RMIT Gallery, April–May 2011. *Credit* Malte Wagenfeld

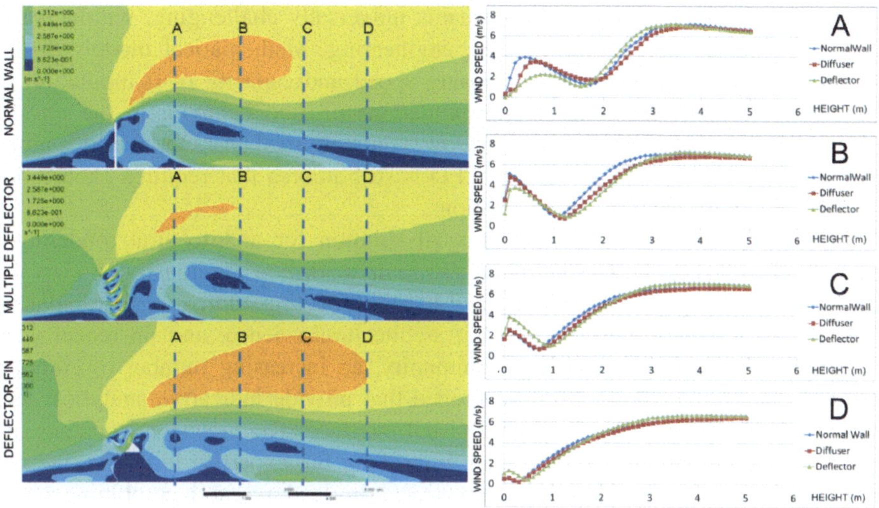

Fig. 3 Wind deflection by low urban wind screens. *Credit* Rafael Moya Castro

world and its "performance" fed back to the designer or to the system to orchestrate the next.[5]

There are some acute problems in architecture that demand novel working paradigms and tools effectively combining design, mathematical and technical expertise. Three of these are the engagement with fluid dynamics, thermodynamics, and sound in the architectural design process. For some years we have been investigating how mathematical and design knowledge can be integrated effectively in design at an early, synthetic and fundamental level to explore the extreme, challenging case of the dynamic interaction of air and architecture. The aim has been not to research fluid dynamics or acoustics per se but to find out how the architect can bring design and mathematical knowledge together within the design process to lead to design innovation that addresses the problems and opportunities of airflow and the transit of heat and sound in air in architectural design.

Most of the major design challenges in the built environment, whether interior or exterior, involve air, its movement, its temperature, its quality in terms constituent gases and airborne pollutants, its potential for energy generation, wind loading and drag.

Passive and low energy modes of transport and renewable energy generation engage the wind, but the built environment less so. Yet paradoxically, the very statics of build urban fabric, its static nature, makes its interaction with ever-changing turbulent air so much more complex to interpret, model, and exploit in any consistent way than mechanisms and conveyances that move.

[5] These first three paragraphs are adapted from first publication by Melbourne Books in the introduction to the book Jane Burry ed. 2013 Designing the Dynamic: High Performance Sailing and Real-time Feedback in Design.

Fluid dynamics is a subject that remains universally challenging, within mathematics, physics, biology, earth sciences, engineering. Mathematical models of fluid flow are developing rapidly with increasing speed and capacity of computation but remain extreme generalizations or approximations of fluid behavior in the physical world, for instance, treating it as laminar flow, while physical air in the world is almost universally turbulent.[6] For this reason, CFD is still an area that requires expert judgment and sensitivity analysis for evaluation.

CFD modeling is not a one-size fits all solution for architectural modeling of airflow. Back in 2011, The *Designing the Dynamic* workshop[7] demonstrated to us very clearly the need for physical analogue and hybrid modeling addressing particular and unique design research questions. Within architectural science, and, in recent years, increasingly in the design modeling community, an increasing number of computational tools and plugins have been developed that provide indicative analytical feedback at the early design stage through a visual interface. The precision, accuracy, and meaning in the outputs of such tools are improving but expertise in setting the conditions and interpreting the outputs remain critical. In-depth research combining design, analogue and digital simulation, and feedback in early design can lead to novel and apposite design paradigms and models.

The increased accessibility of electronic microsensing, and more designer-friendly software in the building physics domain has made real-time feedback in design on wind, air movement, temperature, acoustics in a designer's studio a reality. There has been an opportunity to discover a real place for rapid mixed, or complementary reality feedback in design. Such feedback systems facilitate the iterative design of highly tuned systems, whether semi-open enclosures to modulate auditory experience in open work areas, modest urban installations to alter local microclimate, or sculptural 3D tiling system to introduce thermal lag in hot arid climates. The work has included full-scale enduring prototype spaces that also provide a site for the longitudinal study of their performance in the context of use.

What are the killer applications for this new facility with rapid mixed, complementary, and augmented reality for performance-driven design? One situation which has been under intense academic scrutiny since at least the 1970s is the user experience of the open-plan office. Initially the way that human density, light levels, openness affected work satisfaction and social interaction attracted attention.[8] As personal computers became standard office fare, heat loads from CRT screens presented a new issue, in many places addressed through ubiquitous air conditioning year-round and "controllable" sealed perimeter office spaces, which also provided some white masking sound, reducing the disturbance and privacy issues of conversation and phone calls in open work areas. Increasingly mobile working facilitated by subsequent waves of technology development in personal and mobile computing gave much more variable

[6] Malte Wagenfeld, 2013. The dynamics of sensual atmospheres. In: *Designing the Dynamic: High-Performance Sailing and Real-time Feedback in Design*, Melbourne books, Melbourne, Australia.

[7] Jane Burry, ed. 2013 Designing the Dynamic: High Performance Sailing and Real-time Feedback in Design, Melbourne Books.

[8] Oldham, Greg R., and Daniel J. Brass. "Employee reactions to an open-plan office: A naturally occurring quasi-experiment." *Administrative Science Quarterly* (1979): 267–284.

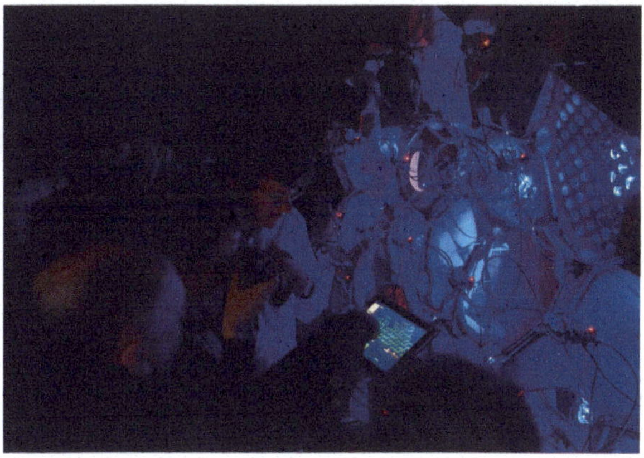

Fig. 4 Sensory detective installation at SG 2016 Gotenborg: visualizing thermal interactions in real time with your phone or tablet

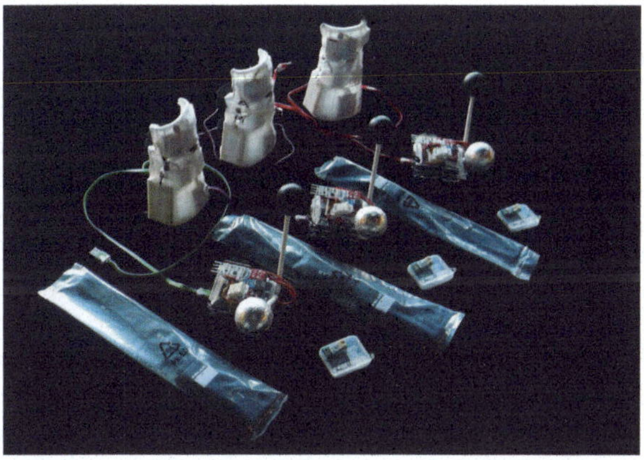

Fig. 5 Daniel Prohasky's system to explore the thermal environment within buildings: wearable node-based deployable wireless sensing, wearable biometric monitoring, wearable thermal ambience measurement. *Credit* Daniel Prohasky

densities of workers in space and the concept of the hot desk and more recently, activity-based working.

The nature of shared work spaces has morphed continually in response to many factors during the twentieth and twenty-first century. First an increasing preponderance of office-based work with increasing automation in many primary and secondary industries increased office worker density, with the transition from cellular office layouts to more open plan working from mid-century onwards, a progressive expansion of this model to an increasing number of sectors. Steep computer adoption from the 1980s onwards resulted in substantial heat loads in office spaces that accelerated the adoption of ubiquitous air conditioning in all sorts of climates and sealed spaces in response. The

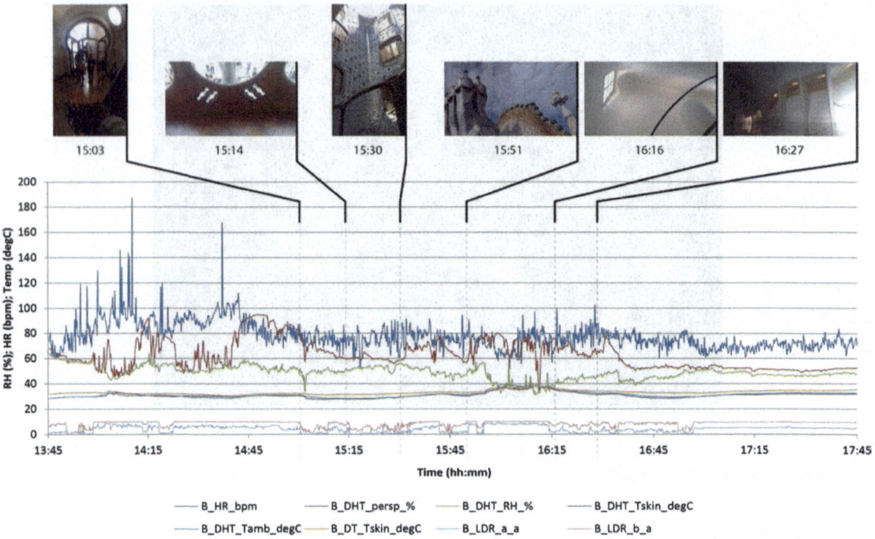

Fig. 6 Observing and monitoring indoor microclimates with physical sensing and wireless technology: Casa Battlo—Barcelona

Internet and mobile technologies have increased the mobility of workers in diverse roles.

The shift from considering open plan as a means to higher density of static desks and workers to differentiating spaces on an activity basis invites the investigation of human auditory experience and the variable sensation of air movement and temperature differentiation within a single open space. Sitting for long periods at a desk is coined "the new smoking"[9] for its manifest long-term health costs from orthopedic maladies to obesity and increased risk of heart disease and other issues exacerbated by physical inactivity. Similarly, evidence is building for the hazards of lack of natural light, unmonitored indoor air quality, stress of distraction through inappropriate acoustic zoning for focused work, while the benefits of open offices compared to the defensible cellular model for increased interaction, communication, and team building are lauded.[10] The health benefits of movement and change during office-based work provide the case for investigating new approaches to spatial, atmospheric, and auditory design in open work areas.

Regardless of which phenomenal interaction with architecture is investigated, this is research with the human at the center. While the auditory research draws on methods and metrics from acoustic science and engineering, it is the actual human experience in all its multisensory complexity that is at stake: speech privacy in open work

---

[9] Dr James Levin, Mayo clinic is credited with coining "sitting is the new smoking."

[10] Thomas J. Allen and Peter G. Gerstberger, 1973, A Field Experiment to Improve Communication in a Product Engineering Department: The Non-Territorial Office Human Factors: *The Journal of the Human Factors and Ergonomics Society* Volume: 15 issue: 5: 487–498.

environments and the opportunities that the age of advanced manufacturing offers for a more refined architecture to sculpt or mediate sound experience. Adaptive thermal comfort—the idea that a ubiquitous uniform thermal condition is not the panacea for indoor living leads to the need for ways to deepen our understanding of what people experience, how air movement and thermal variation affect individuals in various undertakings and environments, and how greater variation and individual control might return to interiors that are nominally set to a uniform temperature and air speed; nominally, because air will respond to in chaotic and turbulent ways to every difference, it encounters even within air-conditioned buildings.

Thirdly a renewed interest in a more conducive, stimulating, and variable work setting excites the need for better ways to capture how people are using space and how they respond to spatial changes in that space. Data captured through tracking and semi-automated complex systems analysis and real-time visualization throws up rich feedback to both space users and their managers. It offers the unexpected connections from inference from selective filtering of big data that simple observation cannot.

A second critical interface between architecture and air is the building façade. The combination of climate and microclimate provides an endlessly rich variety of contextual challenges to maximize passive architectural performance for cooling, heating, breathing, and tempering the atmospheric conditions inside the building. A third is the behavior of wind in urban space and the dimensionally small but formally fine-tuned

Fig. 7 Mini-wind tunnel with multiple sensing rig installed at the private microclimates cluster at SG 2014 Chinese University of Hong Kong. *Credit* Daniel Prohasky

Fig. 8 Whole sensory detective cluster installation at SG 2016, Chalmers, Gotenborg: cellular wall with Daniel Prohasky's novel multisensing wireless units in each cell mapping changing microclimate across the surface

architectural interventions that can transform the human experience by sculpting the wind.

In each of these applications, we have found that critical to a deep designly understanding of the phenomenal architectural interaction is a panoply of tools and approaches spanning digital and analogue simulations, real time or otherwise intuitively engaging analytical feedback and the opportunity to make design changes and rapidly witness their influence on the complex atmospheric and atmosphere-borne phenomena.

Part VI

Scientific Contributions

Stone Morphologies: Erosion-Based Digital Fabrication Through Event-Driven Control

Steve De Micoli, Katja Rinderspacher(✉), and Achim Menges

Institute for Computational Design and Construction, University of Stuttgart, Keplerstrasse 11, 70174 Stuttgart, Germany
{steve.de-micoli,katja.rinderspacher}@icd.uni-stuttgart.de

Abstract. Recent research on the introduction of cyber-physical systems in manufacturing have demonstrated the potentials of a shift from a linear design-to-fabrication approach to an interconnected process utilizing real-time sensory feedback and interactivity to link design and computation with the physical process of making (Menges 2015). Apart from the immediate benefits of self-monitoring and networked manufacturing systems, such technological advancements present the opportunity for novel design practices and material exploration. The presented research investigates the potentials of using event-driven control as a fabrication process based on the morphodynamic forces of erosion to produce an architectural prototype. The project aims to investigate how erosion-based processes, together with an integrated control system, can expand a material's morphospace (Menges 2013).

Keywords: Erosion-based fabrication · Procedural fabrication · Event-driven programming · Integrated numerical control

Introduction

The work presented in this paper culminates a body of work which was developed within the scope of the authors' individual doctoral research and investigated through two student workshops and the fabrication of a 1:1 prototype. The project is contextualized within two topics of research in the field of digital fabrication: indeterminate fabrication processes and numerical control systems.

Indeterminate Fabrication Processes

Inspired by non-standard formation processes in nature, this research investigates the potential of erosion-based processes in the context of a novel computational design and fabrication process. While projects such as "Augmented Materiality" by Johns (2014) or "Erosion Machine" by Paine (2005) have dealt with indeterminacy in fabrication, the presented project seeks to instrumentalize erosion as a form generating design and fabrication methodology and to exploit its novel application possibilities.

Central to the development of the system is an extensive body of material cataloguing and prototyping to acquire a thorough understanding of the erosion process and

K. De Rycke et al., *Humanizing Digital Reality*,
https://doi.org/10.1007/978-981-10-6611-5_11

the resultant surface resolution to build a material knowledge similar to the innate material understanding a craftsperson develops over decades of material interaction and form generation with a specific medium (DeLanda 2015).

The presented system emerges through the material's heterogeneous composition and results in a highly articulated surface resolution whose complexity is much finer and detailed than the size of its fabrication tool and tooltip. In contrast to standard fabrication processes, the erosion-based methodology is highly indeterminate which necessitates an integrated workflow and real-time feedback to react to the emergent material behavior during fabrication.

Numerical Control and Human Machine Interaction (HMI)

Recent modes of making, introduced through real-time sensing and actuation allow for fabrication processes to be integrated into early design stages, transforming it into a generative tool for design. This disruptive paradigm has the potential to shift the making of buildings from a computerized production to computational making (Menges 2015).

Within such human-machine systems, as system complexity increases, the human operator's task moves from physical labor to cognitive behavior (Gorecky et al. 2014). Similarly, within the context of this project, a sound understanding of the process of erosion and its causal behavior becomes inherent to the development of cognitive strategies and the utilization of the system as a design tool.

Project Context and Technological Diffusion

The project took place in Malta, which contrary to other disconnected microstates, has a rich industrial heritage of manufacturing and craft. Notwithstanding, small island states tend to be inherently disadvantaged in today's economic climate due to high transportation costs, intellectual dearth and diseconomies of scale in manufacturing and production (Briguglio 1995).

The project itself becomes an exploration into the capacity to overcome such limitations through disruptive technologies and the introduction of novel paradigms in integrated computational design and digital fabrication combined with the utilization of local resources.

Furthermore, this project seeks to understand whether the integration of fabrication processes through direct numerical machine control within the CAD environment would allow a user, with no particular skill in fabrication, to use and misuse the technology, extending the range of formative and performative potentials embedded within the material itself and also the users' own cognitive design capacity.

Erosion-Based Formation Processes

The forces of natural erosion subject the surface of the earth to an ever-shifting formation process. These morphodynamic processes (Embleton-Hamann et al. 2013) condition the surface of the earth based on the mobilization, transportation and

Fig. 1. Erosion-based limestone fabrication process. ©ICD, K. Rinderspacher

Fig. 2. Detail of erosion morphology as the calcium-based fragments impede the water. ©ICD, K. Rinderspacher

deposition of material by a flowing medium, such as water, wind or ice and result in the remarkable land-formations visible all over the world (Ahnert 2015). The presented research investigates the potential for using flowing water as a form-generating fabrication methodology (Figs. 1 and 2).

Material Characteristics

A high-pressure waterjet operating on 25–200 bar is used to erode a specific type of Maltese Globigerina Limestone. The range of variation in erosion energy of the water stream is notably important, since it is directly linked to the resulting material morphology. The stone, locally known as "Soll", is a transitional layer between the stronger "Franka" Globigerina Limestone and the softer strata of green sand and blue clay (Pedley et al. 1976). "Soll" Globigerina Limestone is composed of various hard, calcium-based marine fossil fragments and a high percentage of softer Globigerina cells, planktonic foraminifera, which are compacted together through the process of sedimentation (Pedley et al. 1976).

The pressurized water gradually removes the softer globigerina cells during fabrication, while being repelled and diverted by the harder calcium-based fossil fragments. As a result, an intricate topology of calcium-capped peaks and deep troughs emerges as a direct mapping of the erosion, as well as the stochastic fossil layering embedded within the individual stone.

The integration of such a process into standard design and fabrication protocols confronts a designer with considerable conceptual and technical challenges. The ephemeral nature of the erosion process fluctuates through energy flow and material transfer and is as a result in constant modification. The organizational and functional relationship between process, material and resulting form is interconnected, time-dependent and results in an extensive range of scales.

The capacity for a material to take on new form is conditioned by a dialectic between its internal composition and its ability to absorb and in some cases, retain external forces (DeLanda 2004). A vast spectrum of tooling typologies has led to highly predictive and precise processes in the industry. In such standard fabrication systems (Kolarevic 2003), both additive and subtractive, the final resolution of the machined material or surface is directly related to the size of the tooltip.

In contrast, the presented fabrication methodology exploits material heterogeneity and process indeterminacy and utilizes them as generative drivers (Menges 2012) by engaging directly with the material's inherent makeup.

Erosion Characteristics

Central to the project is the formalization of the multifaceted and complex materialization this erosion-based fabrication process unfolds. A comprehensive set of experiments and prototypes are utilized to determine the specific properties, topological characteristics and material behavior and finally to explore the morphological differentiation that result from changes in erosion parameters. The most primary and influencing factors are the nozzle type and variation in water pressure. While the nozzle type influences the erosion and its outcome through a specific distribution angle, a variation in water pressure controls the erosion-intensity and is directly linked to the emergence and proportion of peaks. Since the magnitude of erosion significantly depends on its time duration, its control is determined by the number of passes and the machine's feed-rate (Figs. 3, 4 and 5).

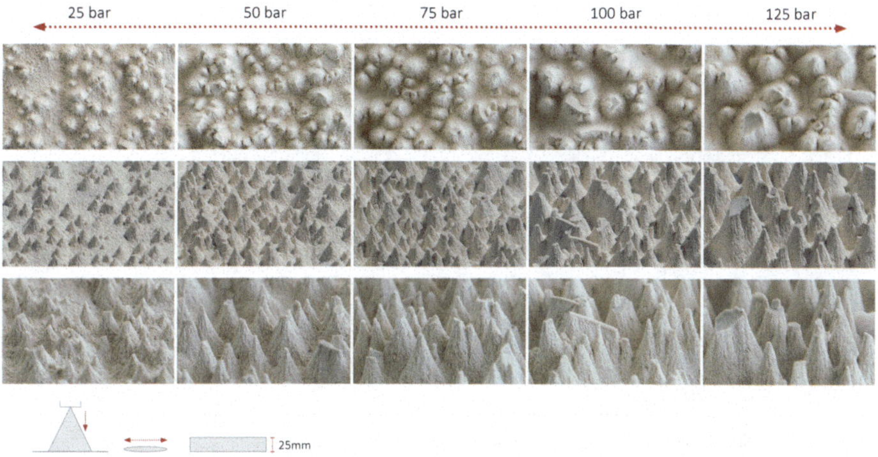

Fig. 3. Table showing the relation between erosion-intensity and peak formation. ©ICD, K. Rinderspacher

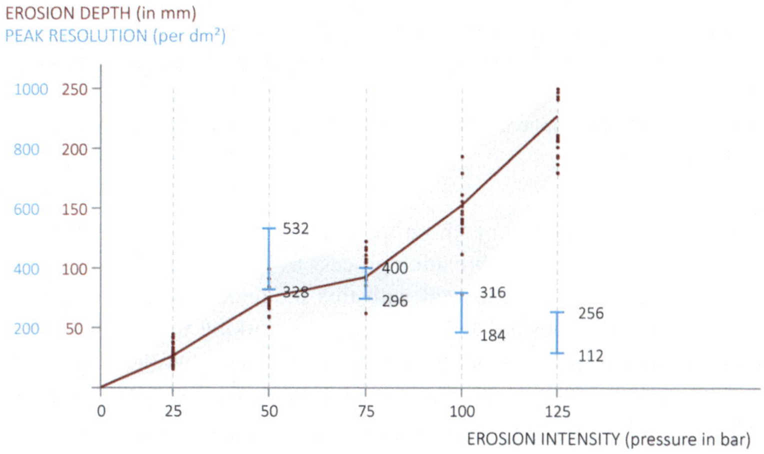

Fig. 4. Table showing the relation between erosion intensity, erosion depth and peak resolution. ©ICD, K. Rinderspacher

Fig. 5. Limestone sample displaying results from a gradation of water pressure. ©ICD, K. Rinderspacher

The extensive cataloguing of these parameters was critical for a successful implementation of the proposed fabrication process as it delivered insight into the range of material effects that can be achieved with this methodology, but also provided an empirical foundation and consolidated knowledge in working with such a non-standard fabrication process. This process and material catalogue became the design palette and mediator helping the designer to engage with such an indeterminate method of fabrication.

Machine Augmentation, Event-Driven Control and Fabrication Interactivity

Non-linear Numerical Control and Material Heterogeneity

The development of Numerical Control (NC) as a formal language provides manufacturers the ability to translate descriptive geometry into programmable mechanic movements. Characterized by the dichotomy for simplified Human-Machine Interfaces (HMI) and complex programming of extensive fabrication strategies, industrial CNC machines have become synonymous with "black-box" and highly inflexible systems.

While being apt for industry, allowing for increased sub-specialization, the inability to engage with the unidirectional and fragmented machining process creates a significant hindrance for the generative making of architecture as the fabrication process remains isolated and detached from the design process-forfeiting the opportunity for the former to actively engage with the form-making process.

Machine Augmentation

A disused CNC plasma cutter is sourced locally and reconditioned. The water-jet lance replaces the plasma end-effector while other necessary precautions are taken to ensure all electronics and motors are isolated from the oncoming water. Typical to industry standard CNCs, a proprietary software interfaced between the operator and machine control.

To overcome the fragmented workflow common to CAD/CAM processes, an open source micro-controller running a G-code parser is used to bypass the main machine-controller and parse stepper-motion signals directly to the motor drivers. This provides the user the ability to carry out all machine control, including the execution of programs within the CAD environment, allowing design decisions to be executed in reaction to, and throughout the fabrication process. The function of homing the machine or programming a toolpath becomes as intuitive and instant as drawing a line on screen.

Machine Control to Cognitive Strategy

Within human-machine systems, unskilled users initially place great intellectual effort on learning control skills to acquire the ability to create cognitive and goal-oriented strategies (Anzai 1984). One crucial factor in the development of the system architecture was the students' involvement during the workshops. This demanded a system of control which did not rely on advanced knowledge in CAD or CAM. The resulting integrated control bypasses the need for a deep understanding in G-code programming language and allows an unskilled user to begin to create generative erosion strategies which engage with the material's morphology rather than being overcome by the tedious sequence of setting up an executable machine program.

In addition to machine control, the presented fabrication method requires a secondary level of control skills specific to the erosion process and the relations of causal behavior on the material (Fig. 6).

An event-driven system is particularly suitable as it relies on an open system of cause-and-effect, whilst retaining flexibility for instant user-based decision making. Furthermore, this hybrid system provides the right framework to embed cognitive rule-based strategies of both machine control and empirically-derived erosion material-outcomes.

The machine control consists of a series of handler modules divided into two main categories; operational modules and feature modules. Embedded within these handlers are small preset methods consisting of both toolpath control as well as erosion sequences. The operational modules handle regular machine control as well as generic uniform erosion routines such as a horizontal walker which consists of a series of

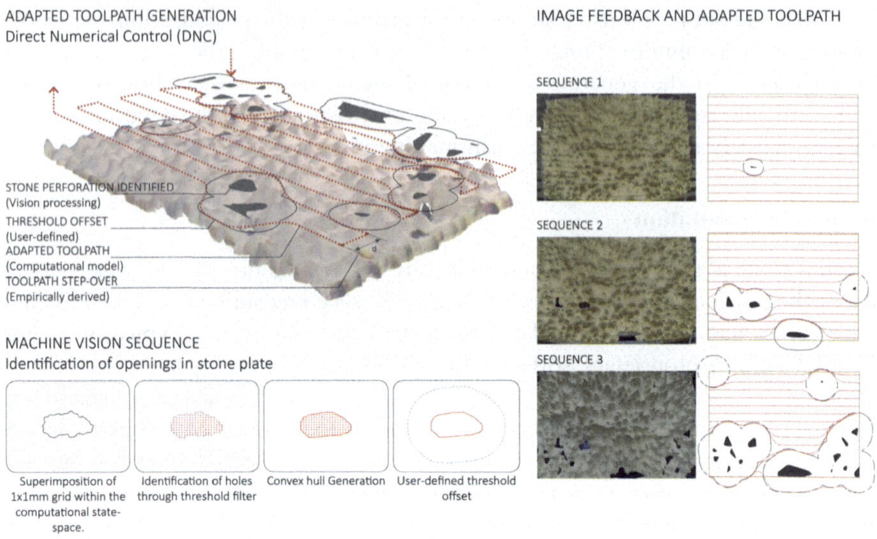

Fig. 6. A machine-vision sequence identifies openings in the stone and adjusts machine toolpaths accordingly. ©ICD, S. De Micoli

parallel passes over the entire face of the stone. The distance of the lance to the stone, toolpath step-over and machine feed rate are all embedded within the method to relieve the user from unnecessary system complexity. The other handler methods are written to allow the user to react to emerging features within the stone and enhance material removal in selected areas.

Prototype Fabrication

The above-mentioned research findings are carried out and brought into a larger architectural context with a physical 1:1 prototype. The wall, being the quintessential archetype of a free-standing masonry structure is used to test the applicability of the fabrication process on a larger scale and to display the material system's ability to function as a load-bearing architectural structure. It is composed of 36 individually eroded 300 × 200 mm limestone plates and supported by a 30 × 50 mm C-section aluminum frame. Contrary to traditional masonry work, the stones' perpends are purposely made to align vertically - similar to cladding systems. Horizontal braces are introduced to tie the wall laterally in-between each course. These T-sections also provide some additional support for the stone edges to be mounted on and evenly distribute the stone's load across the entire horizontal length (Fig. 7).

A computational model is developed to allow the user to design the wall's differentiated topology in its entirety through the identification of varied pressure zones. A computer simulation visualizes the anticipated material behavior, discretizes the wall into individual units, each containing its unique water-pressure zoning, and calculates the initial erosion toolpaths.

Fig. 7. The prototype wall is a 1.80 m × 1.20 m structure comprised of 36 individual 25 mm thick limestone plates (*left*). Detail of wall (*right*). ©ICD, K. Rinderspacher

While the limitation of a 3-axis machine constrained the body of work to a planar surface, similar to natural erosion, the erosive energy initially affects the surface however begins to penetrate into the depths of the material engaging with its inherent performative capacities for load-bearing and the redirection of force-flow as material is iteratively removed. A separate series of laboratory compression tests investigated the eroded stones' load bearing-capacity which set a limit to the amount of erosion it could withstand. In the case of the prototype, all individual stones are initially eroded until the critical threshold where small holes emerged is reached. From this stage forward, a detailed monitoring of the material behavior proved to be crucial. The main control loop iteratively calls a user-selected erosion strategy. After each sequence, a handler is called to capture data from the stone, extract key information and feed it back into the computational model. As the erosion process is essentially subtractive, toolpaths are recalculated to adapt to the stone's status and account for areas which might be detrimentally fragile and compromise structural integrity (Figs. 8, 9 and 10).

Two feedback systems were tested for the prototype wall; a 3D scanning system and an image-based system. While the 3D data provided insight into differentiated stone thickness, hole distribution and peak heightfield mapping, the hardware limitation in point-cloud resolution resulted in a relatively coarse 3D mesh. Due to the above and the wet environment of the fabrication process an alternative solution utilized a web-cam in a waterproof housing to stream an image directly into the CAD program. Machine vision analysis is carried out and used to directly map out stone perforation. The toolpaths recalculate their trajectories allowing the fabrication system to adapt to the real-time sequential erosion of the stone whilst not undermining its capacity for structural performance.

Fig. 8. Detail of limestone wall illustrating topological diversification and material perforation. ©ICD, K. Rinderspacher

Fig. 9. Close-up of wall displaying the surface articulation resulting from a gradation of water pressure. ©ICD, K. Rinderspacher

The advantage of the event-driven process is that bottom-up material exploration together with machine control can be embedded within the handler modules allowing the user to influence high-level design decisions such as controlling the level of material porosity, without having to concern oneself with structural integrity or complex machinic procedure.

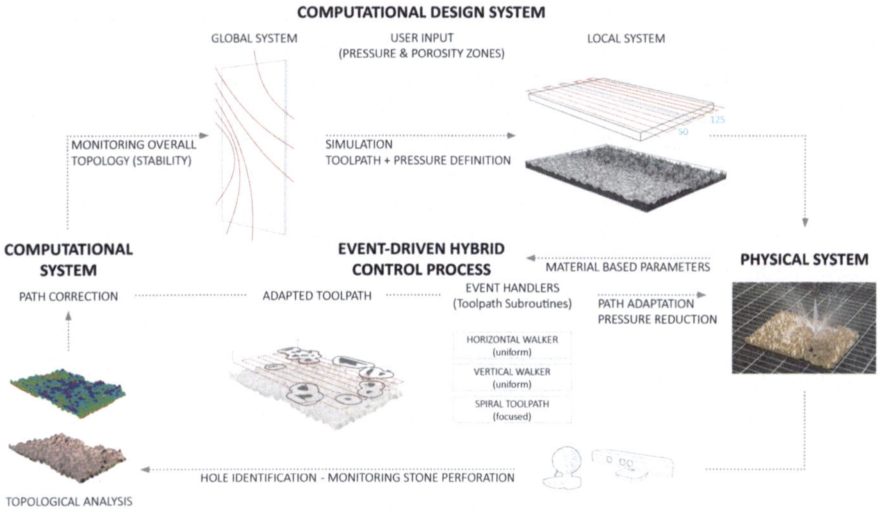

Fig. 10. Computational hybrid-control system integrating the physical fabrication and the digital model through real-time visual sensory and adaptation of machinic toolpaths. ©ICD, S. De Micoli and K. Rinderspacher

This paper presents a novel erosion-based fabrication process utilizing event-driven control and the integration of computational design and fabrication utilizing local resources. In-of-itself it also provides a framework and methodology in which integrated machine control and design thinking can spur novel material morphologies specific to local materials and hardware. While the fabrication of the prototype showed the scalability of the material system and fabrication setup, further explorations into potential multi-performative aspects of the process-based high-resolution surface texture, such as acoustic deflection and moisture transpiration brought about by the limestone´s natural ability to retain and release moisture, are currently investigated. In addition, the development of a water recycling system would be critical for the production of larger scale prototypes. While the 3-axis CNC machine limited the fabrication to a planar limestone module, a 5-axis machine will allow the three-dimensionality of the fabrication method to be explored with larger limestone blocks.

Conclusion

The presented prototype exploits the possibilities erosion-based fabrication processes present as form-generating methods in an architectural context. The imprecision of the erosion-based process and the heterogeneous material logic of the Globigerina Limestone results in a highly indeterminate fabrication process. By integrating these characteristics as generative drivers for design, the presented fabrication methodology unfolds highly articulated and complex surface structures and a novel topological diversification which is, in contrast to standard fabrication systems, much finer than its fabrication tool. The implementation of such non-standard processes shapes a novel material identity by extending the possibilities of high-resolution surface variation and unfolding a diverse material morphology which is unique to its formation process.

The synthesis of physical experiments and real-time feedback in machine control offers a great potential for further exploration within the design field as it transforms Human-Machine-Interaction into a digitally enhanced Human-Material-Interaction. The augmentation of standard tools within a confined environment provokes the necessity for further investigation on the potentials for the revisiting of various material and fabrication processes within local industry and to explore disruptive interventions to achieve novel material practices.

Acknowledgements. The authors would like to thank Irina Miodragovic Vella, Faculty for the Built Environment, University of Malta, M. Eng. Anja Mader, Institute for Building Structures and Structural Design, University of Stuttgart, Lara and Mathias Droll.

References

Ahnert, F.: Einführung in die Geomorphologie, 5th edn. UTB GmbH, Stuttgart (2015)

Anzai, Y.: Cognitive control of real-time event-driven systems. Cogn. Sci. **8**(3), 221–254 (1984)

Briguglio, L.: Small island developing states and their economic vulnerabilities. World Dev. **23** (9), 1615–1632 (1995)

DeLanda, M.: Material Complexity. In: Leach, N., Turnbull, D., Williams, C. (eds.) Digital Tectonics, pp. 14–21. Wiley, Chichester (2004)

DeLanda, M.: The new materiality. In: Menges, A. (ed.) Material Synthesis—Fusing the Physical and the Computational, Architectural Design, vol. 85(5), pp. 16–21 (2015)

Embleton-Hamann, C., Wilhelmy, H.: Geomorphologie in Stichworten III. Exogene Morphodynamik. Schweizbart Science Publishers, Stuttgart (2013)

Gorecky, D., Schmitt, M., Loskyll, M., Zuhlke, D.: Human–machine-interaction in the industry 4.0 era. In: 12th IEEE International Conference on Industrial Informatics, INDIN (2014)

Johns, R.L.: Augmented materiality: modelling with material indeterminacy. In: Gramazio, F., Kohler, M., Langenberg, S. (eds.) Fabricate, pp. 216–223. gta Verlag, Zürich (2014)

Kolarevic, B.: Digital production. In: Architecture in the Digital Age—Design and Manufacturing. Reprint, pp. 30–54. Taylor & Francis, New York and London (2003)

Menges, A.: Morphospaces of robotic fabrication: from theoretical morphology to design computation and digital fabrication in architecture. In: Brell-Cokcan, S., Braumann, J. (eds.) Robotic Fabrication in Architecture, Art and Design, pp. 28–61. Springer, New York (2013)

Menges, A.: Material computation: higher integration in morphogenetic design. In: Menges, A. (ed.) Material Computation—Higher Integration in Morphogenetic Design, Architectural Design, vol. 82(2), pp. 14–21 (2012)

Menges, A: The new cyber-physical making in architecture: computational construction. In: Menges, A. (ed.) Material Synthesis—Fusing the Physical and the Computational, Architectural Design, vol. 85(5), pp. 28–33 (2015)

Paine, R.: Erosion Machine. http://roxypaine.com/machines (2005). Accessed 2017/06/10

Pedley, H.M., House, M.R., Waugh, B.: The geology of Malta and Gozo. In: Proceedings of the Geologists Association, vol. 87(3), pp. 325–341 (1976)

Designing Grid Structures Using Asymptotic Curve Networks

Eike Schling[1(✉)], Denis Hitrec[2], and Rainer Barthel[1]

[1] Technical University of Munich, Arcisstr. 21, 80333 München, Germany
eike.schling@tum.de
[2] University of Ljubljana, Zoisova Cesta 12, 1000 Ljubljana, Slovenia

Abstract. Doubly-curved grid structures pose great challenges in respect to planning and construction. Their realization often requires the fabrication of many unique and geometrically-complex building parts. One strategy to simplify the fabrication process is the elastic deformation of components to construct curved structures from straight elements. In this paper, we present a method to design doubly-curved grid structures with exclusively orthogonal joints from flat and straight strips of timber or steel. The strips are oriented upright on the underlying surface, hence normal loads can be transferred via bending around their strong axis. This is made possible by using asymptotic curve networks on minimal surfaces. We present the geometric and structural fundamentals and describe the digital design method including specific challenges of network and strip geometry. We illustrate possible design implementations and present a case study using a periodic minimal surface. Subsequently, we construct two prototypes: one timber and one in steel, documenting bespoke solutions for fabrication, detailing and assembly. This includes an elastic erection process, by which a flat grid is transformed into the spatial geometry. We conclude by discussing potential and challenges of this methods, as well as highlighting ongoing research in façade development and structural simulation.

Keywords: Asymptotic curves · Minimal surfaces · Elastic deformation · Gridshells

Introduction

There have been a number of publications and design strategies aiming to simplify the fabrication and construction process of doubly-curved grid structures. Therein, we can distinguish between discrete and smooth segmentations (Pottmann et al. 2015). One strategy to achieve affordable smoothly curved structures relies on the elastic deformation of its building components in order to achieve a desired curvilinear geometry from straight or flat elements (Lienhard 2014). Consequently, there is a strong interest in the modelling and segmentation of components that can be unrolled into a flat state, such as developable surfaces (Tang et al. 2016a). Recent publications have given a valuable overview on three specific curve types—geodesic curves, principal curvature lines, and asymptotic curves (Fig. 3)—that show great potential to be modelled as

K. De Rycke et al., *Humanizing Digital Reality*,
https://doi.org/10.1007/978-981-10-6611-5_12

Fig. 1. Grid structure based on asymptotic curves: The model is built from straight strips of beech veneer. All joints are orthogonal. *Image* Denis Hitrec

developable strips (Tang et al. 2016a). Both geodesic curves and principle curvature lines have been successfully used for this purpose in architectural projects (Schiftner et al. 2012). However, there have been no applications of asymptotic curves for load-bearing structures. This is astounding, as asymptotic curves are the only type which are able to combine the benefits of straight unrolling and orthogonal nodes (Fig. 1).

In this paper we present a method to design strained grid structures along asymptotic curves to benefit from a high degree of simplification in fabrication and construction. They can be constructed from straight strips orientated normal to the underlying surface. This allows for an elastic assembly via their weak axis, and a local transfer of normal loads via their strong axis. Furthermore, the strips form a doubly-curved network, enabling a global load transfer as a shell structure (Schling and Barthel 2017).

In Section “Fundamentals”, we describe the geometric theory of curvature and related curve networks, and present the strained gridshells of Frei Otto, as our primary reference for construction. In Section “Design Method” we introduce our computational design method of modelling minimal surfaces, asymptotic curves and networks, and the related strip-geometry. In Section “Design Implementation”, we illustrate our method in a typological overview of shapes and implement it in a design study of a pavilion. In Section “Construction”, we discuss the fabrication, construction details and assembly by means of two prototypes, in timber and steel, and give insights into the load-bearing behavior. The results are summarized in Section “Results”. We conclude in Section “Conclusion”, by highlighting challenges of this method, and suggesting future investigations on structural simulation and prototype development.

Fundamentals

Geometry

Curvature. The *curvature of a curve* is the inverse of the curvature radius. It is measured at the osculating circle of any point along the curve (Fig. 2 left).

To determine the *curvature of a surface*, we intersect it with orthogonal planes through the normal vector. The two section-curves with the highest and lowest curvature are perpendicular to each other and indicate the principle curvature directions (Fig. 2 middle). The two principle curvatures k_1 and k_2 are used to calculate the Gaussian curvature ($G = k_1 \times k_2$) and mean curvature ($H = (k_1 + k_2)/2$). If k_1 and k_2 have opposite orientations, i.e. their osculating circles lie on opposite sides of the surface, then the Gaussian curvature has a negative value. This surface-region is called anticlastic or hyperbolic. If k_1 and k_2 have opposite orientations and the same absolute value, then the mean curvature is zero. Surfaces with a constant zero mean curvature are called minimal surfaces. They are a special type of anticlastic surfaces that can be found in nature in the shape of soap films.

To measure the *curvature of a curve on a surface*, we can combine the information of direction (native to the curve) and orientation (native to the surface) to generate a coordinate system called the Darboux frame (Fig. 2 right). This frame consists of the normal vector **z**, the tangent vector **x** and their cross-product, the tangent-normal vector **y**. When moving the Darboux frame along the surface-curve, the velocity of rotation around all three axes can be measured. These three curvature types are called the geodesic curvature k_g (around **z**), the geodesic torsion t_g (around **x**), and the normal curvature k_n (around **y**) (Tang et al. 2016b).

Curve networks. Certain paths on a surface may avoid one of these three curvatures (Fig. 3). These specific curves hold great potential to simplify the fabrication and construction of curved grid structures (Schling and Barthel 2017). *Geodesic curves* have a vanishing geodesic curvature. They can be constructed from straight, planar strips tangential to the surface. *Principle curvature lines* have a vanishing geodesic torsion—there is no twisting of the respective structural element. They can be

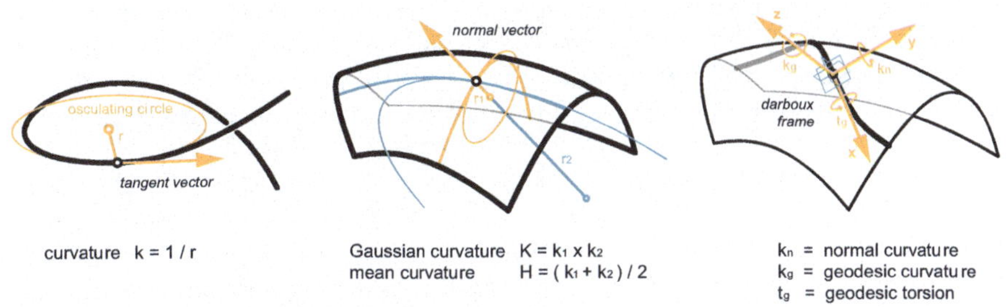

Fig. 2. Definitions of curvature. *Left* Curvature of a curve is measured through the osculating circle. *Middle* The curvature of a surface is calculated through the two principle curvatures k_1 and k_2 as Gaussian or mean curvature. *Right* There are three curvatures of a curve on a surface. Geodesic curvature k_g, geodesic torsion t_g and normal curvature k_n

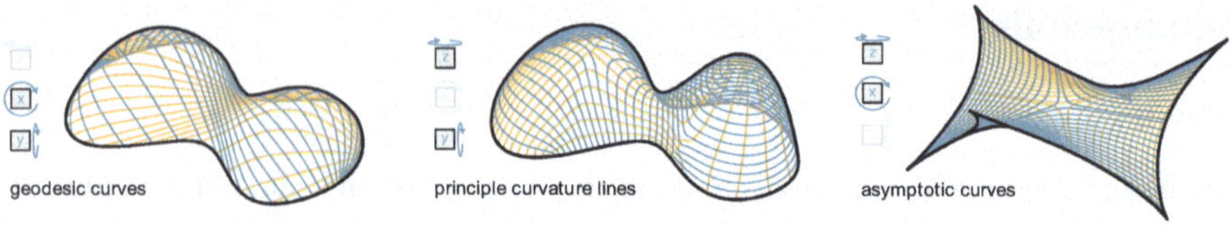

Fig. 3. Surface-curves have three curvatures: geodesic curvature (z), geodesic torsion (x), and normal curvature (y). For each of them, if avoided, a related curve type exists: geodesic curves, principle curvature lines and asymptotic curves

fabricated from curved, planar strips, and bent only around their weak axis. Their two families intersect at 90°. *Asymptotic curves* have a vanishing normal curvature, and thus only exist on anticlastic surface-regions. Asymptotic curves combine both geometric benefits: They can be formed from straight, planar strips perpendicular to the surface. On minimal surfaces, their two families intersect at 90° and bisect principle curvature lines.

The design of related curve-networks is not trivial. *Geodesic curve networks* tend to vary in density and intersection angles. The designer may choose their start and end point but not their path. *Both principle curvature networks and asymptotic curve networks* consist of two families of curves that follow a direction field. The designer can only pick a starting point, but cannot alter their path. If the surface is locally planar (or spherical), the quadrilateral network forms a singularity with higher or lower valence.

Construction

Our construction method (Section "Construction") is based on the strained timber gridshells of Frei Otto (Fig. 4). This paradigm utilizes elastic deformation to create a doubly-curved lattice structure from straight wooden laths.

Otto's curve network is subject to all three types of curvature (as described in Section "Geometry"). Thus, the laths need to be bent and twisted in all directions. The

Fig. 4. Multihalle Mannheim by Frei Otto, 1975. This strained timber gridshell is formed from elastically-bent timber laths. *Image* Rainer Barthel

doubly-symmetrical profiles must be appropriately flexible in bending and torsion (Schling and Barthel 2017). Otto's lattice grid is assembled flat and subsequently erected into a doubly-curved inverted funicular shape. Once the final geometry is reached, the quadrilateral network is fixed along the edges and braced with diagonal cables (Adriaenssens and Glisic 2013).

Design Method

Minimal surface. A minimal surface is the surface of minimal area between any given boundaries. In nature such shapes result from an equilibrium of homogeneous tension, e.g. in a soap film. The accurate form is found digitally in an iterative process by either minimizing the area of a mesh, or finding the shape of equilibrium for an isotropic pre-stress field.

Various tools are capable of performing such optimization on meshes, with varying degrees of precision and speed (Surface Evolver, Kangaroo-SoapFilm, Millipede, etc.). They are commonly based on a method by Pinkall and Polthier (2013).

The Rhino-plugin TeDa (Chair of Structural Analysis, TUM) provides a tool to model minimal surfaces as NURBS, based on isotropic pre-stress fields (Philipp et al. 2016).

Certain minimal surfaces can be modelled via their mathematical definition. This is especially helpful as a reference when testing the accuracy of other tools.

Asymptotic curves. Geometrically, the local direction of an asymptotic curve can be found by intersecting the surface with its own tangent plane.

We developed a custom VBScript for Grasshopper/Rhino to trace asymptotic curves on NURBS-surface using differential geometry. For this, the values and directions of the principal curvature (k_1, k_2) are retrieved at any point on the surface. With this information, we can calculate the normal curvature k_n for any other direction with deviation-angle α (Fig. 6 left) (Pottmann et al. 2007, p. 490).

$$k_n(\alpha) = k_1(\cos\alpha)^2 + k_2(\sin\alpha)^2. \tag{1}$$

To find the asymptotic directions, the normal curvature must be zero, $k_n(\alpha) = 0$. Solving for α results in:

$$\alpha = 2\pi - 2\tan^{-1}\sqrt{\frac{2\sqrt{k_2(k_2 - k_1)} + k_1 - 2k_2}{k_1}}. \tag{2}$$

By iteratively walking along this asymptotic direction and calculating a new α at every step, we can draw an asymptotic curve on any anticlastic surface. Our algorithm uses the Runge–Kutta method to average out inaccuracies due to step size. On minimal surfaces, the deviation angle α is always 45° (due to the bisecting property of asymptotic curves and principle curvature lines).

In the case of meshes, we use EvoluteTools to find the asymptotic curves. Both EvoluteTools and our VBScript were checked for accuracy by comparing their results with a parametrically defined asymptotic network on an Enneper minimal surface. Depending on the quality of the mesh or NURBS surface and the step size, a sufficient accuracy for design and planning can be achieved.

Network design. When designing an asymptotic curve network, we take advantage of the bisecting property between asymptotic curves and principle curvature lines (Pottmann et al. 2007, p. 648). By alternately drawing each curve and using their intersections as new starting point, we can create an "isothermal" web with nearly quadratic cells (Fig. 5) (Sechelmann et al. 2012). Combining these two networks allows us to benefit from both their geometric properties simultaneously for substructure and façade, allowing for cladding solutions with tangential developable strips or planar quads (Pottmann et al. 2007, p. 680).

Strip geometry and fabrication. A developable, i.e., singly-curved, surface-strip is defined by straight rulings. If this strip is orthogonal to the reference surface and follows a given surface-curve, the rulings are enveloped by all planes that contain the normal vector **z** and tangent vector **x** of the Darboux frame (Fig. 6 left). The vector of these rulings **r** is calculated via the equation

$$\mathbf{r} = k_g \mathbf{z} + t_g \mathbf{x}, \tag{3}$$

where k_g is the geodesic curvature and t_g is the geodesic torsion. For asymptotic curves, k_g is measured simply via its osculating circles. To calculate t_g we use the two principle curvatures (Tang et al. 2016b):

$$t_g = \frac{1}{2}(k_2 - k_1)\sin 2\alpha, \tag{4}$$

where α is constant 45° on a minimal surface. The rulings of a developable strip along asymptotic curves are not necessarily perpendicular to the surface. Thus two orthogonal strips commonly have a curved intersection (Fig. 6 middle). In the worst case of a vanishing geodesic curvature, the rulings are parallel to the tangent vector, which makes modelling and construction of a developable surface-strip impossible.

In our method, the strip geometry is defined with the normal vectors **z**, which allows for straight intersections and well-defined strip surfaces (Fig. 6 right). As a consequence, some twisting of the structural strips needs to be considered. This

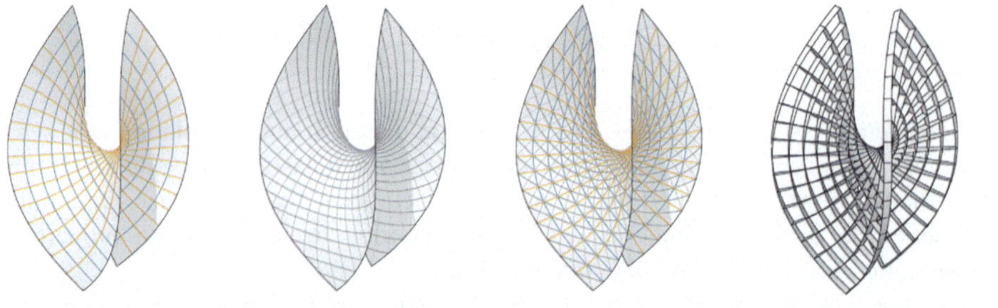

Fig. 5. Enneper surface with **a** asymptotic curves **b** principle curvature lines **c** web of both networks **d** strip model of the asymptotic network

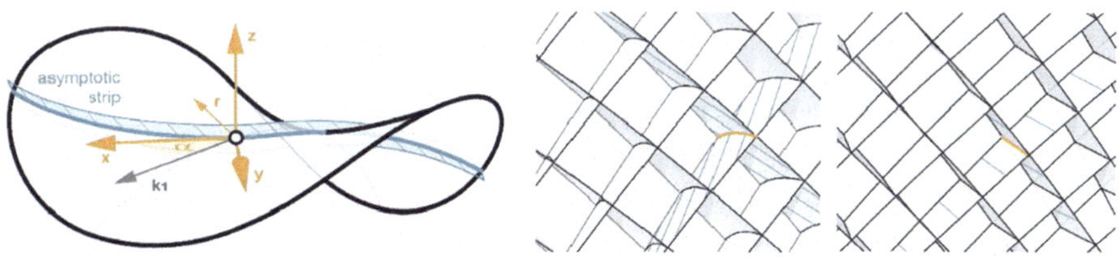

Fig. 6. *Left* A developable strip along an asymptotic curve and orthogonal to the surface, is define by rulings r, which are generally not parallel to the normal vector z. *Middle* This results in curved intersections between the strips. *Right* In our method, the strips are defined by the normal vectors z, to ensure straight intersections. Consequently, the strips are not truly developable, but twisted

deviation from a truly developable strip is essential to realizing a simplified construction.

The node to node distance, measured along the asymptotic curves, is the only variable information needed to draw the flat and straight strips. They are cut flat and then bent and twisted into an asymptotic support structure.

Design Implementation

Design strategies. Even though minimal surfaces can only be designed through their boundaries, there is a wide range of possible shapes and applications. Examples shown in Fig. 7 display how varying boundary conditions influence the surface and asymptotic network. Boundary-curves may consist of straight lines (a), planar curves (d), or spatial curves (b). Straight or planar curves are likely to attract singularities (a, d). A well-integrated edge can be achieved by modelling a larger surface and "cookie-cutting" the desired boundary. A minimal surface can be defined by one (a, b), two (c), or multiple (d) closed boundary-curves. Symmetry properties can be used to create repetitive (c) and periodic (e) minimal surfaces. The Gaussian curvature of the design surface directly influences the geodesic torsion of asymptotic curves, the density of the network and the position of singularities. A well-balanced Gaussian curvature will produce a more homogenous network.

Case study. We applied this method in a design studio, Experimental Structures, to develop the concept for a research pavilion. The design shape is based on the periodic minimal surface, Schwarz D, which can be described within a simple boundary along six edges of a cube. This basic cell is repeatedly copied and rotated to form an infinitely repetitive surface. Finally, this surface is clipped with an inclined block to cut out the desired shape. The entire network is a repetition of the curve-segments within one-sixth of the initial cubic cell (Fig. 8).

Despite the high level of repetition, the pavilion displays a complex and sculptural shape. A model ($2 \times 2 \times 1$ m) was built from strips of beech veneer at a scale of 1:5 to verify the fabrication and assembly process (Fig. 9).

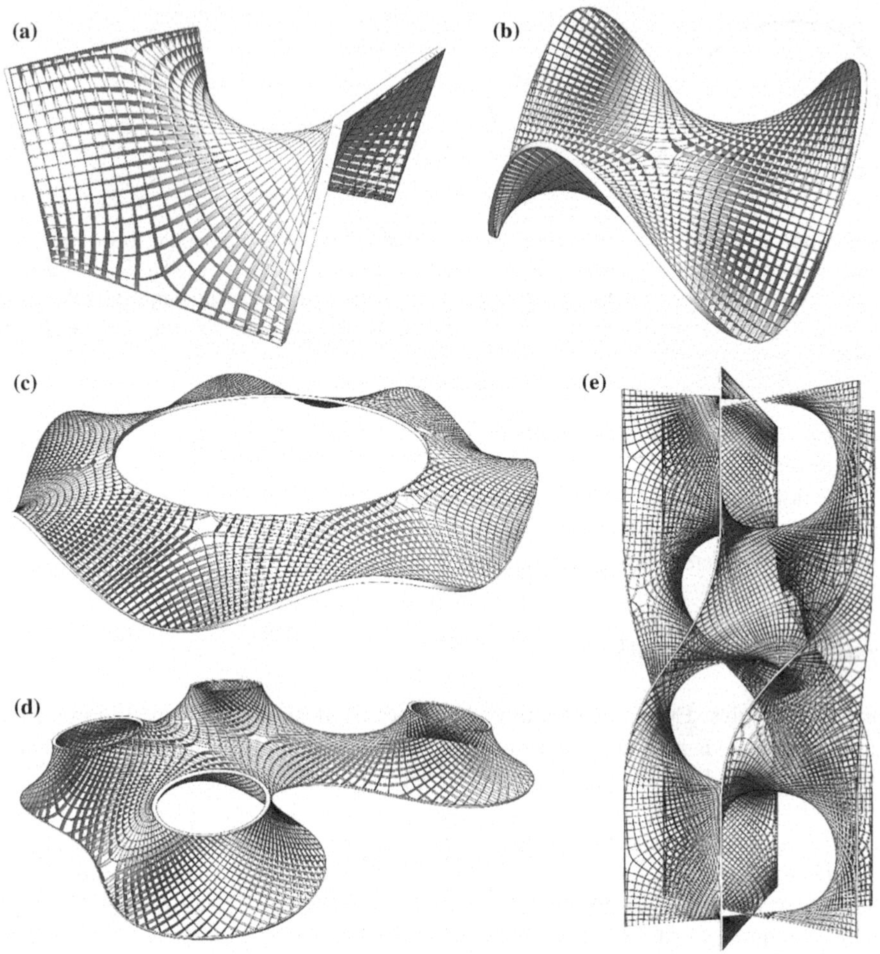

Fig. 7. Overview of asymptotic strip networks on minimal surfaces. **a** One polygonal boundary, creating a saddle shaped network with singularities appearing along the edges. **b** One spatially curved boundary, creating a surface with three high and low points and a network with central singularity. **c** Two boundary curves creating a rotational repetitive network with regular singularities. **d** Multiple boundaries creating a freely design minimal surface with four high points. **e** Variation of a singly-periodic Sherk's two minimal surface, with six interlinking boundaries

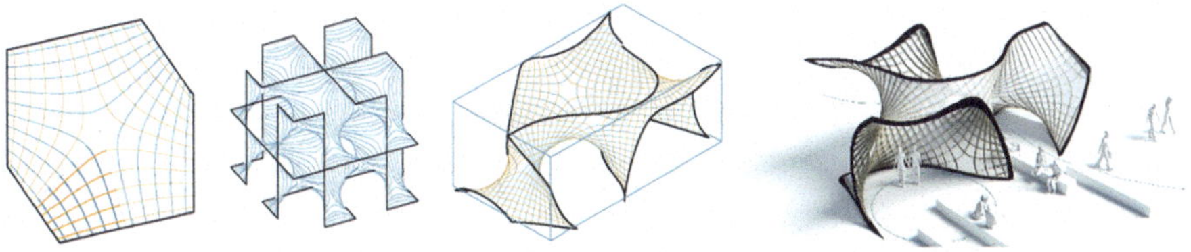

Fig. 8. Design implementation using a Schwarz D periodic minimal surface

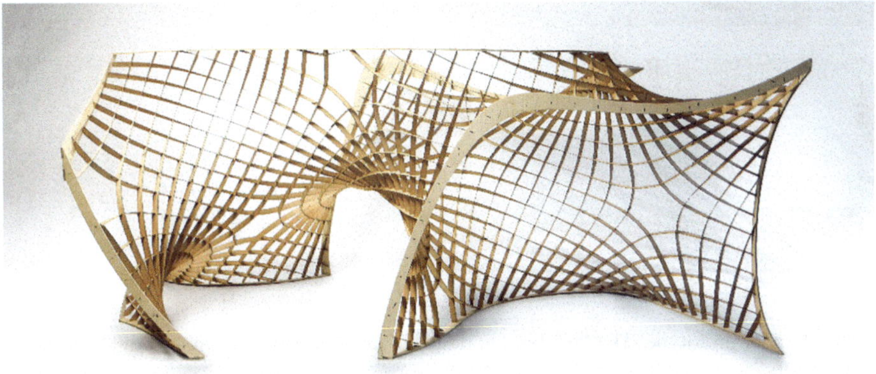

Fig. 9. Model of the design implementation. The structure is built from straight timber strips of beech veneer with exclusively orthogonal nodes. *Image and model* Denis Hitrec

Construction

Curvature and bending. Our construction process follows the reference of Frei Otto's strained gridshells (see Section "Construction"). The strips are fabricated flat and subsequently bent into their spatial geometry. As asymptotic curves admit no normal curvature, no bending in the strong axis of the strips is necessary during assembly. Due to the geodesic torsion, there is a certain amount of twisting of the lamellas. Additionally, the geodesic curvature results in bending around the z-axis.

When choosing the profiles, the section modulus and thickness need to be adjusted to the maximum twist and minimal bending radii to keep deformation elastic. At the same time, the profiles need to provide enough stiffness to resist buckling under compression loads. These opposing factors can be solved by introducing a second parallel layer of lamellas. Each layer is sufficiently slender to easily be bent and twisted into its target geometry. Once the final geometry is fixed, the two layers are connected with a shear block in regular intervals to increase the overall stiffness similar to a Vierendeel truss.

This technique was applied in the construction of two prototypes, in timber and steel, each with an approx. 4×4 m span (Figs. 10 and 11).

Timber prototype—spatial construction. For the timber prototype, the two asymptotic directions were constructed on separate levels out of 4 mm thick poplar

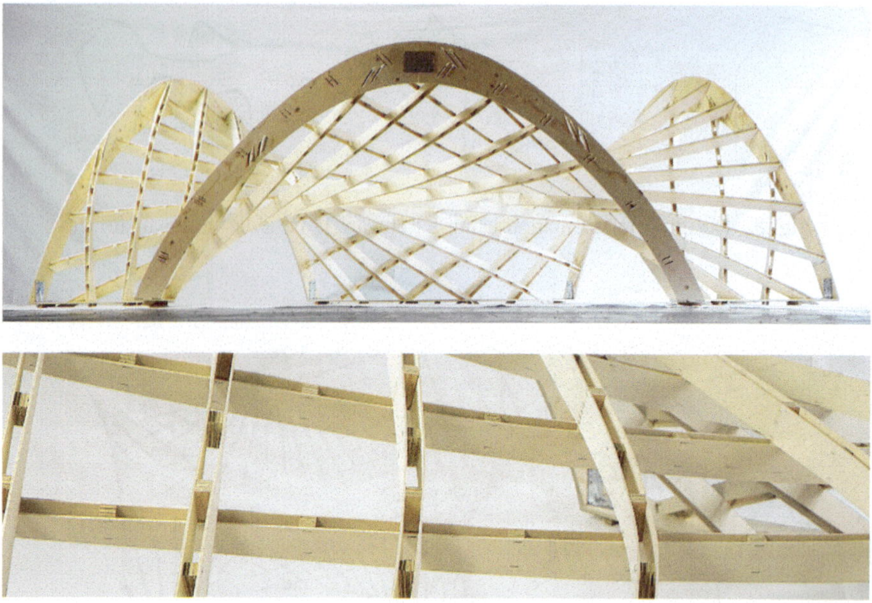

Fig. 10. Timber prototype. The lamellas are doubled and coupled to allow for low bending radii and high stiffness. *Image* Eike Schling

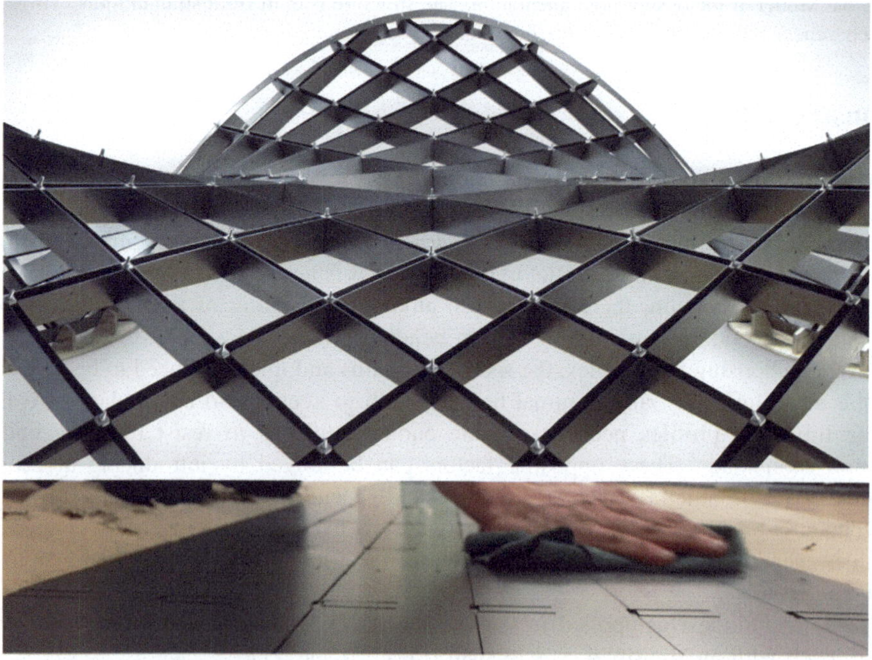

Fig. 11. Steel prototype. The straight lamellas are doubled and coupled to allow for low bending radii and high stiffness. *Image* Eike Schling

plywood strips. This allowed for the use of continuous, uninterrupted profiles. The upper and lower level were connected with a square stud, enforcing the orthogonal intersection angle. This rigid connection could only be fitted if all elements were curved in their final geometry. Consequently, this prototype had to be erected spatially using framework and edge beams as temporary supports. The height of the edge profiles was determined by their intersection angle with the lamellas, creating a dominant arched frame.

Steel prototype—elastic erection. The steel prototype was built from straight, 1.5 mm steel strips. Both strip families interlock flush on one level. Therefore, the lamellas have a double slot at every intersection (Fig. 11). Due to a slot tolerance, the joints were able to rotate by up to 60°. This made it possible to assemble the grid flat on a hexagonal scaffolding. The structure was then "eased down" and "pushed up" simultaneously and thus transformed into its spatial geometry (Fig. 12) (Quinn and Gengnagel 2014). During the deformation process, a pair of orthogonal, star-shaped washers were tightened with a bolt at every node, enforcing the 90° intersection angle.

Fig. 12. Assembly process of the steel prototype, showing the elastic transformation from flat to curved geometry. *Image* Denis Hitrec

Fig. 13. Silhouette of the strip model from beech veneer

Once the final geometry was reached, the edges were fitted as tangential strips on top and bottom. This edge locks the shape in its final geometry, generates stiffness and provides attachments for the future diagonal bracing and façade (Figs. 13, 14 and 15).

Structural behavior. The structural behavior of asymptotic grids is greatly dependent on the overall shape and support of the structure. Our initial investigations have observed a hybrid load-bearing behaviour of both a grillage and a gridshell (Figs. 16 and 17).

The strip-profiles are orientated normal to the underlying surface, allowing for a local transfer of normal loads through bending via their strong axis. This is especially helpful to account for the local planarity of asymptotic networks (due to their vanishing normal curvature) and to stabilize open edges.

On the other hand, the strips form a doubly-curved network, enabling a global load transfer as a gridshell (Schling and Barthel 2017). For this, the quadrilateral grid needs to be appropriately braced via diagonal cables. The edge configuration adds additional stiffness by creating triangular meshes.

The elastic erection process, results in residual stresses inside the curved grid elements. Additional compression forces, originating from a membrane load-bearing behavior, increase the bending moment around the weak axis of these curved elements. The strategy of doubling and coupling lamellas (see Section "Curvature and bending") is therefore essential to control local bucking.

Finally, it needs to be said, that the principle stress trajectories of a shell constitute the optimal orientation for compression and tension elements in a respective grid structure. In our method, however, we choose to follow a geometrically optimized orientation along the asymptotic directions.

Fig. 14. Close-up of the strip model from beech veneer

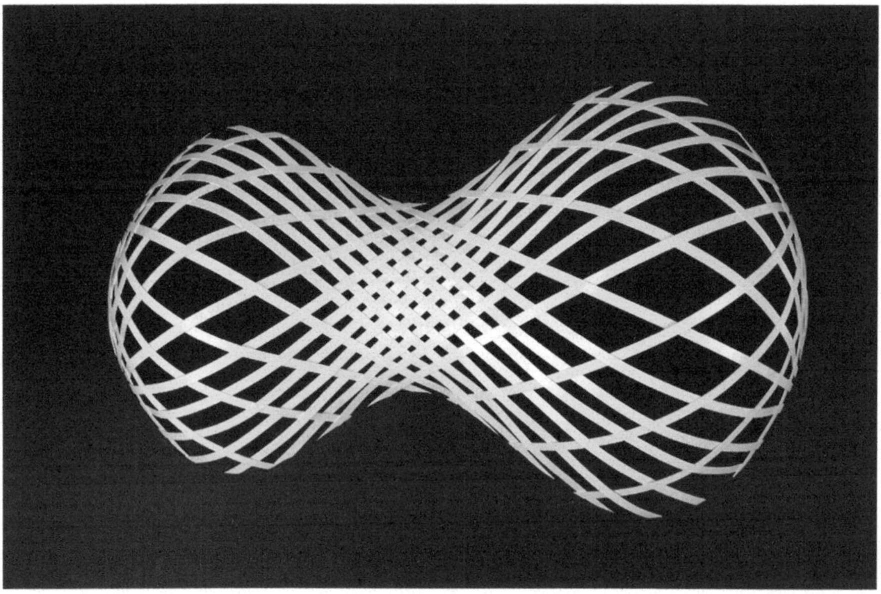

Fig. 15. Strip model of a geodesic curve network

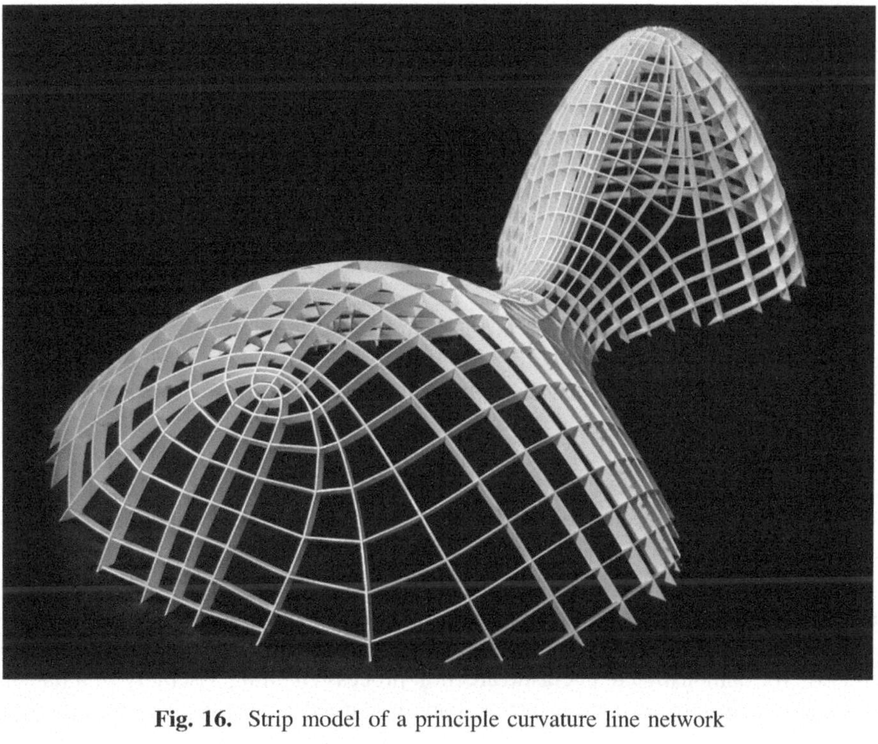

Fig. 16. Strip model of a principle curvature line network

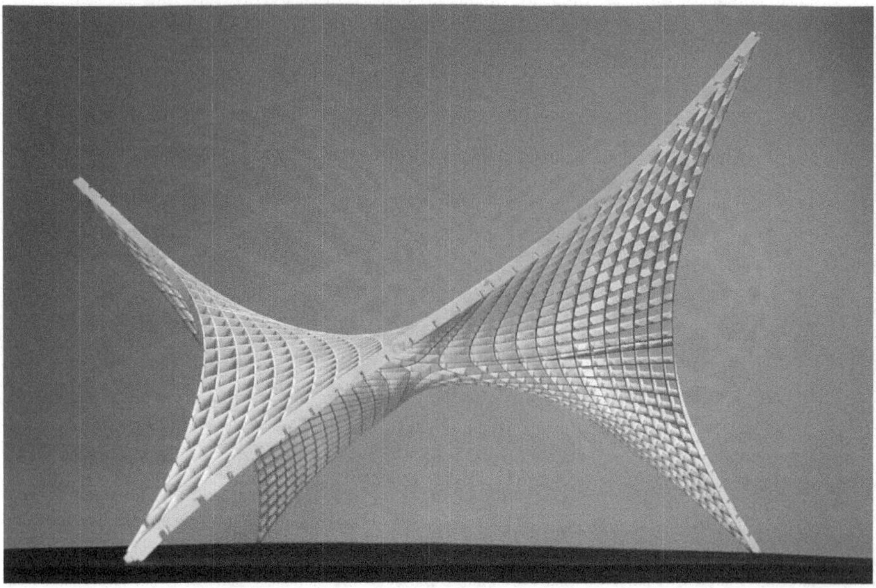

Fig. 17. Strip model of an asymptotic curve network

Results

We compared the geometric properties of three specific curve networks: geodesic curves, principle curvature lines and asymptotic curves.

We identified that only asymptotic curves are able to combine the benefits of straight unrolling and orthogonal nodes. They can be formed from straight strips perpendicular to the underlying anticlastic surface. This way, they resist loads normal to the surface by bending in their strong axis. On minimal surfaces, asymptotic curves intersect at 90°, which allows the use of identical nodes throughout the structure. The bisecting property with principle curvature networks offers further geometric advantages for the façade.

We developed a custom VBScript that can trace asymptotic curves on anticlastic surfaces with sufficient accuracy for design and construction. A wide spectrum of design solutions was visualized in a typological overview of strip networks and implemented in a case study for a pavilion.

We model the strip-network along the normal vectors, to ensure a well-defined geometry and simplified construction with identical and straight intersections, thus deviating from a true developable geometry. This measure results in some twisting of the profiles.

Both twisting (geodesic torsion) and bending (geodesic curvature) have to be considered when choosing profiles for construction. We have presented a strategy of doubling and coupling the bent structural elements to achieve sufficient stiffness of the final grid. We demonstrated an elastic erection process enabling assembly in a flat state and a subsequent transformation into a spatially-curved geometry. The findings were verified in the realization of two prototypes: One in timber and one in steel, each with a span of 4 × 4 m. The structural behavior of asymptotic grids was discussed along the basis of both a grillage and a gridshell.

Conclusion

An analytical approach to both geometry and material properties is required to achieve a symbiosis of form, structure and fabrication. Even though the design freedom is limited to the choice of boundary curves, there are a wide range of design solutions. Additionally, asymptotic networks offer an individual aesthetic quality. Our design method may be applied in all scales from furniture design to stadium roofs. Structurally, asymptotic gridshells show great potential, as they combine the benefits of upright sections with a doubly-curved grid. Hence, loads can be transferred locally via bending, and globally as a shell structure.

We are continuing to investigate the structural behavior of asymptotic structures on the basis of a grillage and a shell, comparing grid orientations, shapes and supports. Simultaneously, we are developing a workflow to compute the residual stress of the initial bending and torsion through the local geodesic curvature and geodesic torsion, without simulating the construction process. Another ongoing development is the implementation of constructive details: This includes cable bracing and façade systems using planar quads, developable façade strips and membranes.

Acknowledgements. This paper is part of the research project, Repetitive Grid Structures, funded by the Leonhard-Lorenz-Stiftung and the Research Lab of the Department of Architecture, TUM. Both the geometric background as well as the digital workflow were developed in close collaboration with the Department of Applied Geometry and the Center for Geometry and Computational Design, Prof. Helmut Pottmann, TU Wien. The plugin TeDa was provided and supported by Anna Bauer and Bendikt Phillipp from the Chair of Structural Analysis, TUM. Further support in MESH modelling was granted by Alexander Schiftner and the Evolute GmbH in Perchtoldsdorf, Austria. We would like to thank Matthias Müller, locksmith at the Technisches Zentrum, TUM, as well as Thomas Brandl and Harry Siebendritt of the Brandl Metallbau GmbH&Co. KG in Eitensheim for their extensive support in steel fabrication.

References

Adriaenssens, S., Glisic, B.: Multihalle Mannheim. Princeton University, Department of Civil and Environmental Engineering. http://shells.princeton.edu/Mann1.html (2013). Accessed 9 May 2017

Lienhard, J.: Bending-Active Structures. Form-Finding Strategies Using Elastic Deformation in Static and Kenetic Systems and the Structural Potentials Therein. ITKE, Stuttgart (2014)

Philipp, B., Breitenberger, M., Dàuria, I., Wüchner, R., Bletzinger, K.-U.: Integrated design and analysis of structural membranes using isogeometric B-rep analysis. In: Computer Methods in Applied Mechanics and Engineering, vol. 303, pp. 312–340 (2016)

Pinkall, U., Polthier, K.: Computing discrete minimal surfaces and their conjugates. Exp. Math. **2** (1), 15–36 (2013)

Pottmann, H., Asperl, A., Hofer, M., Kilian, A.: Architectural Geometry. Bentley Institute Press, Exton (2007)

Pottmann, H., Eigensatz, M., Vaxman, A., Wallner, J.: Architectural geometry. In: Computers and Graphics. http://www.geometrie.tuwien.ac.at/geom/fg4/sn/2015/ag/ag.pdf (2015). Accessed 9 May 2017

Quinn, G., Gengnagel, C.: A review of elastic grid shells, their erection methods and the potential use of pneumatic formwork. In: De Temmerman, N., Brebbia, C.A. (eds.) MARAS 2014. Ostend, WIT Press, Southampton, pp. 129–143 (2014)

Schiftner, A., Leduc, N., Bompas, P., Baldassini, N., Eigensatz, M.: Architectural geometry from research to practice: the eiffel tower pavilions. In: Hesselgren, L., et al. (eds.) Advances in Architectural Geometry, pp. 213–228. Springer, New York (2012)

Schling, E., Barthel, R.: Experimental studies on the construction of doubly curved structures. In: DETAIL structure 01/17, Munich, Institut für Internationale Architektur-Dokumentation, pp. 52–56 (2017)

Sechelmann, S., Rörig, T., Bobenko, A.: Quasiisothermic mesh layout. In: Hesselgren, L., et al. (eds.) Advances in Architectural Geometry, pp. 243–258. Springer, New York (2012)

Tang, C., Bo, P., Wallner, J., Pottmann, H.: Interactive design of developable surfaces. In: ACM Transactions on Graphics, vol. 35, No. 2, Article 12, New York (2016a). http://www.geometrie.tugraz.at/wallner/abw.pdf. Accessed 9 May 2017

Tang, C., Kilian, M., Pottmann, H., Bo, P., Wallner, J.: Analysis and design of curved support structures. In: Adriaenssens, S., et al. (eds.) Advances in Architectural Geometry, pp. 8–22. VDF, Zürich (2016)

The Grand Hall of the Elbphilharmonie Hamburg: Development of Parametric and Digital Fabrication Tools in Architectural and Acoustical Design

Benjamin S. Koren(✉)

ONE TO ONE, New York, NY 10011, USA
koren@onetoone.net

Abstract. The paper will discuss the larger historical context of acoustic sound diffusion in concert halls and summarize the development and manufacturing of sound diffusive acoustic panels in the Grand Hall of the Elbphilharmonie Hamburg. The Grand Hall seats 2100 people and is clad in approximately 6000 m^2 of CNC-milled gypsum fibreboard panels. Each panel has a seamless, non-repetitive geometric surface pattern applied, which aims to optimally diffuse the sound throughout the concert hall and prevent long path echoes.

Keywords: Acoustics · Sound-diffusion · Parametric design · Automation · Optimization · Panelization

The Elbphilharmonie is a cultural and residential complex designed by the Swiss architectural firm Herzog and de Meuron. The building is located in Hamburg, a port-town in Northern Germany and the country's second largest city. A full 14 years after the inception of the project, it opened on January 11, 2017. Apart from a myriad of ancillary programs, such as a hotel, apartments, a public viewing platform, restaurants, cafes and a garage, the building houses three concert venues. The largest venue, the 2100 seat Grand Hall (see Fig. 1), is the subject of this paper.

Acoustical consulting for the Grand Hall was provided by Nagata Acoustics "from the design phase through construction and project completion" (Toyota 2017). The President of the company's US division, Dr. Yasuhisa Toyota, was responsible for the acoustics of the Grand Hall of the Elbphilharmonie. It was the close collaborative effort between Herzog and de Meuron's architectural and Nagata Acoustics' acoustic design which ultimately created the ideal basis for the application of innovative computational design and fabrication methods throughout the entire building process. The formal challenge posed by the hall's characteristic room shape, its vineyard style, intertwining and continuous double-curved surfaces (see Fig. 1), as well as the functional requirements of varying acoustic specifications of the interior's materials and panels, sound-diffusion and absorption, led to the development of custom parametric design tools and digital fabrication methods. These tools created the non-standard acoustic panels, substructure and seamlessly prepared them for production on CNC milling machines and assembly.

K. De Rycke et al., *Humanizing Digital Reality*,
https://doi.org/10.1007/978-981-10-6611-5_13

Fig. 1. The finished Grand Hall of the Elbphilharmonie. Photo by Michael Zapf

Sound Diffusion

Yasuhisa Toyota reflected on the state of concert hall acoustics in an interview that "the sound of space is one of the most complex engineering problems faced by the architect. People have been building structures for performances for millennia, but it's only in the last century that we've started to understand why some physical spaces sound better than others" (Bell 2014). Indeed, the scientific approach to concert hall acoustics is only about a century old, beginning with Wallace Sabine's work on Boston Symphony Hall which was completed in 1900. It wasn't until the 1970s that Manfred R. Schroeder pioneered the work on acoustic sound diffusers and there continues to be "much debate about the role of surface diffusers in concert halls" (Cox and D'Antonio 2003).

In broad terms, sound diffusion is the even scattering of sound energy in a room (see Fig. 2). Non-diffusive, reflective surfaces in concert halls can lead to a number of unwanted acoustic properties, which can be rectified, in part, by adding diffusers. A perfectly diffusive space is one where acoustic properties, such as reverberation, are the same, regardless of the location of the listener.

Discussing sound diffusion, Cox and D'Antonio (2003) identify the inherent link between the acoustics of a concert hall and the architectural design. They note that "in older, pre-twentieth century halls, such as the Grosser Musikvereinssaal in Vienna, ornamentation appeared in a hall because it was the architectural style of the day. Such walls were therefore naturally diffusing: large flat surfaces were very rare". They go on to lament the detrimental effects of reduced, modernist designs on concert hall acoustics: "in the twentieth century, however, architectural trends changed and large expanses of flat areas appeared in many concert halls. The expanse of flat surfaces can

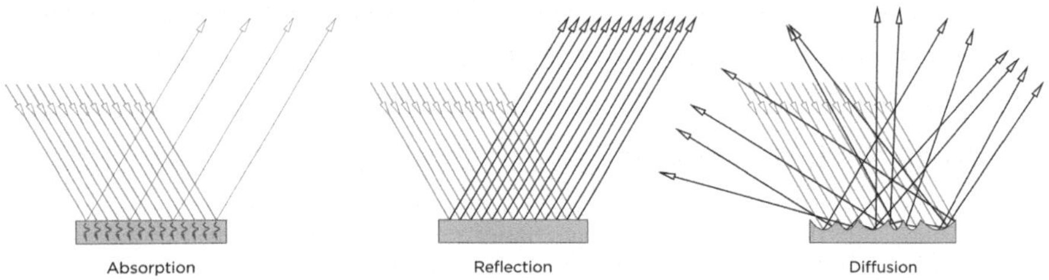

Fig. 2. Sound absorption, reflection and diffusion. Image by the author

lead to distortion in the sound heard as a result of comb filtering, echoes and other mechanisms". They conclude that "the key to good diffuser design is to find forms that complement the architectural trends of the day. The diffuser must not only meet the acoustic specification, it must fit in with the visual scheme required by the architect". The Elbphilharmonie serves as a great, contemporary example where this trend has been successfully reversed; Toyota (2017) explains that the Grand Hall was indeed conceived in "a single collaborative process of the architectural and acoustic design".

Toyota (2017) approached the design of the Grand Hall at two scales: at the macro-scale of the shape of the room and of the micro-scale of the diffusive surfaces. Oguchi (2017) of Nagata acoustics explains that "to study the basic shapes in the 'Grosser Saal' (German for "Grand Hall"), we began by using computer simulation based on the geometrical acoustics to study how the reflections would distribute in the 'Grosser Saal'". The acousticians went on to build a "1/10 scale model of the hall interior and conducted acoustical experiments in the scale mode to study details about how sound would behave in the space". The acousticians' "first step in scale model testing was to check for detrimental echoes and, if an echo was found, to solve how to eliminate it". Conducting the tests did in fact reveal "long path echoes at the stage and nearby seating. In the case of a long path echo, the options to eliminate are: changing the angles of some reflective surfaces; adding sound absorbing measures; or adding diffusing elements. The architects favoured diffusion to eliminate the long path echo as one of diffusing surface patterns which were usually expected to create soft reflections" (Oguchi 2017).

The acousticians therefore decided that the diffusing pattern should serve two functions, that of causing soft reflections and to eliminate echoes. To test the performance of the diffusing pattern, the acousticians added the pattern onto the walls in the 1/10 scale model that were causing the echoes. Once applied successfully, the acousticians went on to use scaled testing to "determine the depth of indentations necessary to achieve our desired results" (Oguchi 2017). These tests would determine the parameters that would eventually drive the parametric generation of the sound diffusing pattern described below. "In the locations where we expected the surface producing soft reflections, the depth of indentations for diffusion measures 10–30 mm. Where we aimed to eliminate echoes, the depth of indentations for diffusion measures 50–90 mm. Also, because we needed these panels to have sufficient weight to effectively reflect sound even at low frequencies, the panels were fabricated so that they have a post-shaping average density of 125 kg/m^2" (Oguchi 2017). Areas which were deemed acoustically unimportant, the density was reduced to 35 kg/m^2, including surfaces which were perforated with large holes.

Parametric Surface Generation

The variables for the sound diffusing pattern determined experimentally by the acousticians were on the level of the width and depth of the pattern, leaving considerable freedom as to its visual shape. The shape of the pattern itself was developed in close collaboration between the architects and acousticians with the aim of achieving harmony within the overall design of the project, making reference to the characteristic peak-and-valley shape of the building's roof.

A bespoke software plug-in of about 18,000 lines of code was written in VB.net to compute about one million, parametrically defined, uniquely shaped non-uniform rational basis spline (NURBS) surfaces and output them in Rhino 3D. A paper by the author of this paper presented at the Design Modeling Symposium in Berlin in 2009 outlined the tool development in detail (Cox and D'Antonio 2005). The pattern itself was initially based on the distortion of a two-dimensional, orthogonal grid of Voronoi seeds. The program allowed for random seed displacements, deletion and insertion in order to control the degree of randomness and the scale, i.e. the average width of each element of the pattern. In a subsequent step, each closed 2-D polygon of the Voronoi pattern was used as input in the 3-D formation of a parametrically defined NURBS surfaces.

The above mentioned specifications of the acousticians regarding the depth of the pattern on various surface within the hall would become the input that would drive a total of six parameters (see Fig. 5), which would control the exact geometry of each NURBS surface by defining the placement of each surfaces' control points. This allowed for a precise definition of the depth of each element and also its overall shape, ranging from harder to softer edges. All the parameters were driven using grayscale bitmap images which mapped XY coordinates from bitmap space, with brightness corresponding to the various parameter values, to each of the concert hall's wall and ceiling surfaces' UVW coordinates. In a last step, every control point of every NURBS cell was mapped topologically onto the flat, single- and double curved wall and ceiling surfaces in 3-D, with special attention given to the continuation of the pattern across seams (see Figs. 3 and 4), a quality deemed highly important to the concept and overall visual quality of the hall by the architects (Koren and Mueller 2017).

Digital Fabrication

The parametrically defined pattern was generated to be continuous, not only across panel edges (see Fig. 3) but also to cover the wall and ceiling surfaces continuously (see Fig. 4). Subsequently, the hall's interior had to be subdivided into about 10,000 unique panels. As each panel is unique, with regard to its overall shape as well as the surface pattern applied to it, an additional software program of about 20,000 lines of code was developed to automate the 3-D planning and digital production of the panels from CNC-milled gypsum fibreboard (see Figs. 6 and 7), as well as to optimize the

Fig. 3. CNC milled diffusive pattern. Photo by the author

Fig. 4. Surface transitions across seams. Image by the author

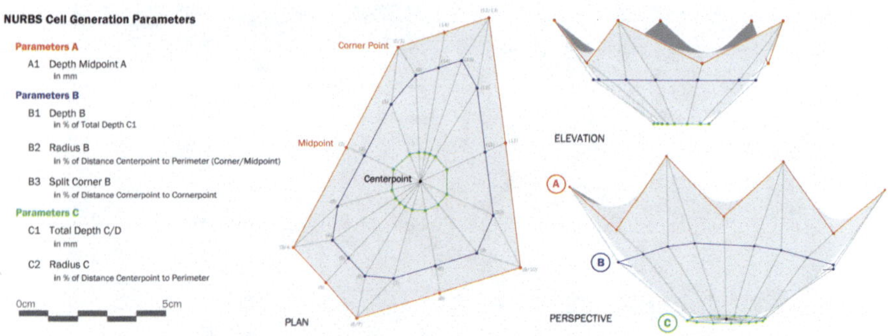

Fig. 5. Parametrically defined NURBS surface. Image by the author

Fig. 6. CNC milling of sound diffusing pattern. Photo by Peuckert

acoustic surface's substructure (see Figs. 8 and 9). Most of the panels had to meet the aforementioned weight per unit area of 125 kg/m^2, with acoustically inactive panels being reduced down to 35 kg/m^2.

Gypsum fibreboard was chosen for a combination of reasons: once milled for its appealing, natural looking architectural qualities, but also because the material's high volumetric density of 1500 kg/m^3 made it an ideal choice for meeting the high weight requested by the acousticians. The material was also hard enough to retain the shape of the diffusive pattern once milled, while still being able to be milled with a diamond cutter.

Fig. 7. Manufactured unique panels in the off-site workshop. Photo by Peuckert

Fig. 8. Ceiling panels being assembled onto the substructure on site. Photo by the author

Fig. 9. Backside view of the ceiling's substructure. Photo by the author

Since the specific gypsum product chosen was only produced in up to 40 mm thickness, most of the panels had to be build up in several layers, up to 200 mm in total, glued and mechanically fixed together with screws, in order to achieve the desired weight and accommodate the curvature of the panels. The architects defined a precise and intricate network of gap lines, which, not unlike the sound diffusing pattern itself, was meant to be continuous across the hall's larger surfaces (see Fig. 3).

Because of the varying degrees of complexity in edge conditions, it was inevitable to employ a 5-axis milling machine in the manufacturing of the panels. The curvature of the front surface was achieved by keeping the backside of each panel planar, while the front would be milled to shape. For each panel, the edges had to be digitally generated, the fixings had to be placed and a groove along the entire perimeter, for the placement of a sealing band, had to be positioned exactly 5 mm below the lowest point of the sound diffusing pattern. In addition, mechanical fixings to secure the glued layers were placed and, most importantly, the generated diffusing pattern was assigned to each panel.

Each raw panel was prepared to size. The panels were CNC milled (see Fig. 6) first from the back, which included the 5-axis formatting of the edges and the placement of the holes for fixing the substructure and for mechanically securing the glued layers. Then each panel was flipped, repositioned on the machine and milled from the front, which included a stage for 3-axis milling of the sound diffusing pattern using a ball-end cutter, milling in parallel tracks spaced at a fairly large centre distances. This resulted in a rough, final surface texture that would also exhibit the characteristic peak-and-valley motive down at the scale of the trails left by the milling head.

Once the panels were milled, they were lacquered from both sides using a clear lacquer. Lacquering was done not for visual reasons but to harden and protect the grainy surface of the milled material. Part of the substructure, profiles standardized in length at 100 mm intervals, was prefixed to the back of each panel using a combination of five standard screw lengths and standard washers of ten different thicknesses, which allowed varying the depth of penetration from 25 to 90 mm at 1.5 mm increments by combining different screw and washer types. The panels were packed for shipping. Once they arrived on site, each panel was manually installed (see Fig. 8). Simple details of the substructure (see Fig. 9) allowed for panel adjustments with three degrees of freedom, allowing them to be fitted with a 5 mm gap between panels, and at the required precision.

Evaluation

According to acoustic tests carried out by Nagata Acoustics according to ISO 3382-1, an international standard for measuring acoustic parameters in performance spaces, the reverberation time of the occupied hall measures 2.3 s (Oguchi 2017) and perfectly meets the requirement for classical music. Toyota (2017), however, rightly observes that "critical assessments and judgments about the excellence or failings of a concert hall's acoustics do not come from people reading data sheets with reverberation time and other numerical measurements of physical properties". When the concert hall opened on January 11, 2017 (see Fig. 10), assessments of the Grand Hall's acoustics

Fig. 10. The Grand Hall's Opening on January 11, 2017. Photo by Michael Zapf

are startlingly similar, by critics and performing musicians alike. The hall is ultimately praised for its extreme clarity, balance and transparency of sound.

Martin Piechotta, timpani player of the Mahler Chamber Orchestra (2017) observes that "the sound is, on the whole, very *transparent*, *balanced*, and *beautiful*", while his colleague Matthew Sadler, notes that "the sound of the orchestra coalesces onstage and in the auditorium but it is a startlingly *transparent blend* in which *every individual voice is heard*". Kennincott (2017) of the Washington Post agrees, noting that "there is no golden aura, but there is *fantastic clarity* and *spatial presence*". The critic for the Economists (2017) also agrees, noting that the sound is "*balanced* and *warm* with *absolute clarity of detail*". Pointing out aural similarities with Frank Gehry's Disney Concert Hall, Swed (2017), critic for the Los Angeles Times, praised that from his "seat in the front, the effect was *pure magic*. The singers seemed to radiate from some mysterious place inside the walls".

The Grand Hall's clarity, it should be noted, is corroborated by the tests carried out by Nagata. The objective measurement of clarity for music, C_{80}, determines how clear music sounds in a hall, i.e. if a listener can hear every separate note of a fast passage or if they blend together. C_{80} measures the ratio of early sound energy arriving before 80 ms versus reverberant sound energy arriving later, from 80 to 3000 ms, and is measured in decibels (Beranek 2004). The higher the ratio, the clearer the hall sounds. One of the reasons the clarity of the Grand Hall may be so noticeable to critics and musicians alike is that, according to Beranek (2004), halls with clarities between $-5\text{ dB} \leq C80 \leq -1\text{ dB}$ are judged best, and the Grand Hall, with a clarity factor of +0.3 dB (Oguchi 2017), lies slightly above that range. Meyer (2009), on the other hand, extends that range slightly, stating that the clarity factor should lie between $-2\text{ dB} \leq C80 \leq +4\text{ dB}$, a range the Grand Hall falls into.

Neither the assessment of critics nor musicians, nor the tests carried out by the acousticians point to any severe, detrimental acoustic phenomena, such as long path echoes or hot spots. The consensus regarding the hall's clarity by critics and musicians alike and the lack of echoes are a testament to the acousticians' successful strategic application of sound diffusive surfaces, to create soft reflections and to avoid echoes, as well as its precise and rigorous translation and execution made possible by parametric design and digital fabrication techniques.

Acknowledgements. Figures 1 and 10: Courtesy of Hamburg Marketing. Figures 6 and 7: Courtesy of Peuckert.

References

Bell, J.: The hear and now. Uncube Magazine No. 21 Acoustics (2014). Retrieved from http://www.uncubemagazine.com/magazine-21-12784433.html

Beranek, L.: Concert Halls and Opera Houses. Springer, New York (2004)

Cox, T.J., D'Antonio, P.: Engineering art: the science of concert hall acoustics. Interdisc Sci Rev **28**(2), 119–129 (2003)

Cox, T.J., D'Antonio, P.: Thirty years since 'diffuse sound reflection by maximum-length sequences': where are we now? Forum Acust **2005**, 2129–2134 (2005)

Kennincott, P.: A new concert hall in Hamburg transforms the city. The Washington Post (2017, May 15). Retrieved from https://www.washingtonpost.com/graphics/augmented-reality/what-perfect-sound-looks-like

Koren, B.S., Müller, T.: Digital fabrication of non-standard sound-diffusing panels in the large hall of the Elbphilharmonie. In: Fabricate Conference, pp. 122–129. UCL University Press, London (2017)

Mahler Chamber Orchestra: The Acoustics of the Elbphilharmonie (2017, March 8). Retrieved from http://mahlerchamber.com/news/the-acoustics-of-the-elbphilharmonie

Meyer, J.: Acoustics and the Performance of Music. Springer, New York (2009)

Oguchi, K., Dr.: Highlights of Room Acoustics and Sound Isolation Design (2017, February 25). Retrieved from http://www.nagata.co.jp/e_news/news1702-e.html

Swed, M.: What does this critic hear at the new Elbphilharmonie concert hall? The Sound of the Future. Los Angeles Times (2017, March 3). Retrieved from http://www.latimes.com/entertainment/arts/la-ca-cm-elbphilharmonie-hamburg-notebook-20170303-story.html

The Economist: A New Concert Hall is Worth the Wait, and the Cost (2017, January 21). Retrieved from http://www.economist.com/news/books/21714573-nine-years-late-11-times-its-proposed-cost-it-also-stunning-achievement-will

Toyota, Y., Dr.: Elbphilharmonie Opens in Hamburg (2017, February 25). Retrieved from: http://www.nagata.co.jp/e_news/news1702-e.html

Bloomberg Ramp: Collaborative Workflows, Sharing Data and Design Logics

Jonathan Rabagliati[1(✉)], Jeroen Janssen[2], Edoardo Tibuzzi[2], Federico De Paoli[1], Paul Casson[3], and Richard Maddock[1]

[1] Foster + Partners, Riverside, 22 Hester Rd, London SW11 4AN, UK
jrabagli@fosterandpartners.com
[2] AKT II, White Collar Factory, 1 Old Street Yard, London EC1Y 8AF, UK
[3] Multipass, Unit 5, Miller Close, Newton in Furness, Cumbria LA13 0NE, UK

Abstract. This paper examines how building collaborative workflows, sharing data, and the use of digital tools served the design team of the ramp within the new European Headquarters for Bloomberg to achieve remarkable design resolution not just at the digital design stage but through fabrication and installation. This paper examines practice-led research approaches to the sharing of digital design data and logics between designers, structural engineers, fabricators and installation teams on the Bloomberg ramp. The project could be amongst the first in the industry to use a centralised, comprehensive parametric model, shared between an architectural practice, an engineering consultant and specialised cladding contractor. It is also one of the first projects to use Virtual Reality to help refine the geometry. The use of laser scanning to analyse the as-built steel position of the ramp structure and incorporation of point cloud analysis into the workflow for setting out cladding for installation is the third example of employing digital technologies from allied fields into architecture.

Keywords: Collaborative workflows · Design logics · Sharing data

Introduction and Context

The new European headquarters for Bloomberg designed by Foster + Partners, begun in 2010 and due for completion in late 2017, provides 100,000 m^2 of new working space at the heart of the City of London. At its centre the stepped ramp, clad in bronze panels, rises six floors up the central atrium. It is designed as a place of meeting and of connection within the office, allowing people to walk abreast ascending or descending in unbroken conversation. Each flight of the ramp extends 30 m between landings. Structurally, each flight is supported only at landings and one mid-landing hanger. Unrolled in its entirety the ramp would measure almost 200 m. This paper examines how building collaborative workflows, sharing data, and the use of digital tools served the design team of the ramp to achieve remarkable design resolution not just at the digital design stage but through fabrication and installation.

This paper focuses on the nature of the collaborative workflow between architects Foster + Partners and structural engineers AKT II. It also examines how a different approach was used in sharing the parametric logics with cladding contractor Gartner. It

K. De Rycke et al., *Humanizing Digital Reality*,
https://doi.org/10.1007/978-981-10-6611-5_14

Fig. 1. Photo of Bloomberg ramp showing first flight steelwork being craned in as construction progresses floor by floor. October 2014 *Photo* Nigel Young/Foster + Partners

details the use of mathematically defined curves and VR in the design process, while examining the use of laser scanning, point cloud data and metrology in implementing the setting out and installation.

Sharing One Parametric Model

Foster + Partners sought to establish an integrated digital design relationship with structural engineers, extending the approach cultivated by Foster + Partners for Crossrail Place (Rabagliati and Huber 2014) (Fig. 1). At the design stage on the Bloomberg ramp, Foster + Partners and AKT II shared a single parametric grasshopper model. Sharing the same parametric model affords opportunities for both architects and structural engineers, but also demands rigour, discipline and accommodation. When working within the same definition, steps in logic need to be negotiated as do conventions for naming, syntax, style, scripting language and layout. Shared parameters need to be agreed as do the sequence of relationships of what drives what. Sharing relies on human relationships of trust and a high skill level of collaborators.

The benefits of sharing a digital parametric model mean that once the geometry is coordinated, updates to the geometry continue to be automatically coordinated between parties. As parameters and even geometry logics are altered, the structural model simultaneously updates. This avoids barriers to updates due to time, resources or money—reasons cited by structural engineers as barriers to updating geometry on previous projects.

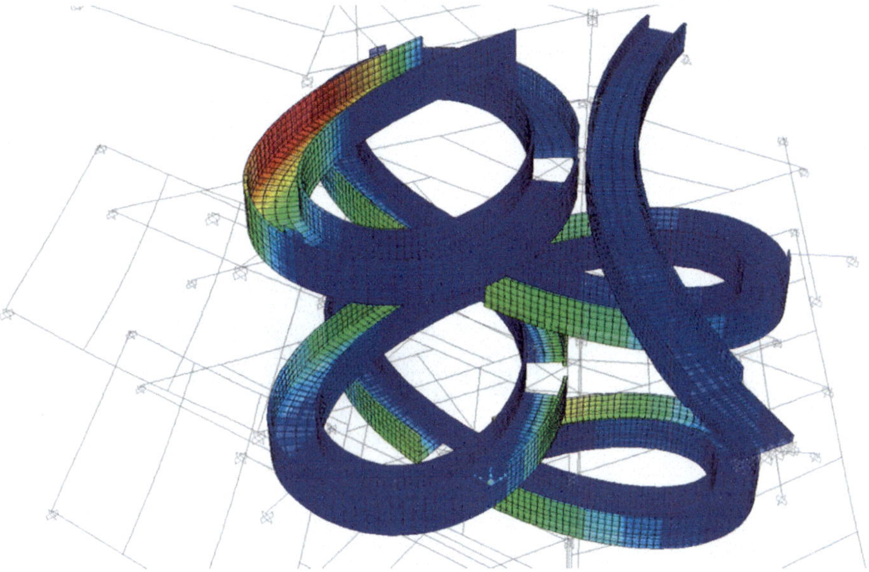

Fig. 2. Finite element analysis structural mode shape (normalised) for dominant natural frequency *Image* AKT II

Sharing a geometric definition with AKT II, permitted an accurate, fully detailed geometry with all the stiffness and material properties embedded to be used for Finite Element (FE) analysis. The Bloomberg ramp has a steel box structure whose geometry strongly influences the structural behaviour, so it was essential to have an accurate model whose dynamic behaviour could be understood in much greater detail (Fig. 2).

Synchronising the model between the design team and structural engineers frees time which would otherwise be diverted into merely maintaining 3D coordination. This allowed for more practice-based research on detailed analysis of vibration, a critical factor in the structural design. The parametric setup of geometry and interoperability with the FE analysis software allowed for an extensive assessment of dynamic response factors. Typically for a staircase, a single load case would be defined for every step and response factor results would be read for the excited step. On this project, a set of time-history load cases was created for each of the 96 steps, while simultaneously reading the results for all the steps over the full height of the ramp. This created a matrix of data that could be processed and analysed. Systematic interrogation of the results revealed some peak response levels at unexpected locations along the stepped ramp, allowing the structural engineers to precisely target specific areas to absorb critical vibration frequencies.

When architects and structural engineer share the same logic for generating geometry, the logic can be interrogated from both sides and issues can be addressed immediately. One shared model facilitates data flow between different disciplines, enhances design flexibility, increases productivity and improves level of coordination (Figs. 3 and 4).

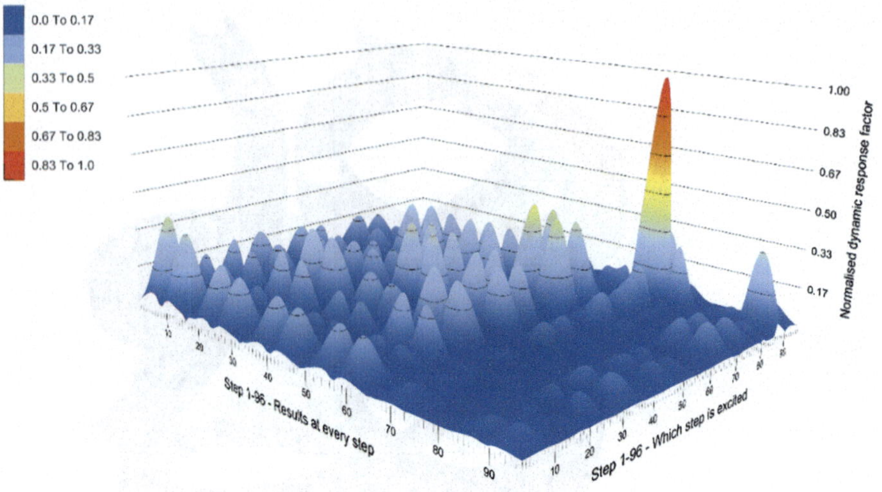

Fig. 3. 3D graph showing response factor results of dynamic vibration analysis for each step of the ramp *Image* AKT II

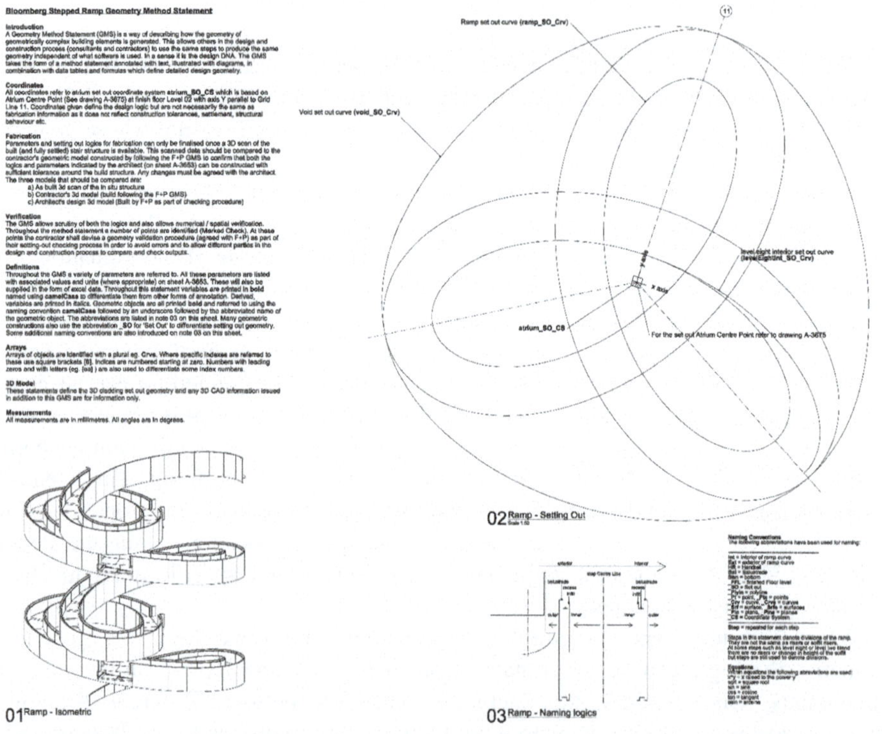

Fig. 4. Introduction page from Bloomberg ramp geometry method statement *Drawing* Foster + Partners

Geometry Method Statement

The approach to sharing the parametric model with the cladding contractor was different. Rather than sharing a digital parametric model the logics and parameters were encapsulated in a geometry method statement. This comprised of explanatory text, annotated diagrams and data tables providing step by step geometric instructions on how to construct the ramp (Peters and Whitehead 2008).

This was possible because the ramp had been specifically designed with a set of considered geometric logics. Mathematical equations describe the setting out curves in plan and the setting out curve in profile. The setting out plan curve is a three petal hypotrochoid. In plan, all the elements of the atrium are members of the same mathematical family of curves which set out the ramp, the atrium edge and the skylight.

$$\begin{aligned} x(\theta) &= (R-r)\cos\theta + d\cos\left(\frac{R-r}{r}\theta\right) \\ y(\theta) &= (R-r)\sin\theta - d\sin\left(\frac{R-r}{r}\theta\right) \end{aligned} \tag{1}$$

The curve in profile for the ramp is derived from the equation describing the cumulative distribution function of the Kumaraswamy distribution.[1] This profile curve is asymmetric to allow the hangers at the mid-flight point to be positioned slightly above the halfway height between the two landing balustrades.

$$F(x; a, b) = \left[1 - (1 - x^a)^b\right] \tag{2}$$

To understand the rationale behind this profile curve it is necessary to consider the design approach for the ramp. On typical stairs, the balustrade follows the slope of the risers and is horizontal at the landings. On the Bloomberg ramp the landings are the equivalent in length of two steps however on the soffit below the steps continue stepping across the underside of the landing, so there is a continuity from one flight through to the next. The continued rise of the balustrade through the landing means it is higher on the upper side. To reduce the disparity between the height of the balustrades at each side of the landings a slight ‘s’ curve is introduced to the balustrade in profile which allows the gradient across the landing to be less steep than the gradient around the mid flights (Fig. 5).

Utilising VR in Design of Ramp

Virtual Reality (VR) headsets were used to determine the specific parameters for this ‘s’ shaped profile curve. The aim of the architects was to achieve the maximum ‘s’ shaped curve in profile while making the ‘s’ curve virtually impossible to detect by eye. Because the plan and profile curves play off each other in three-dimensional space, the

[1] https://en.wikipedia.org/wiki/Kumaraswamy_distribution.

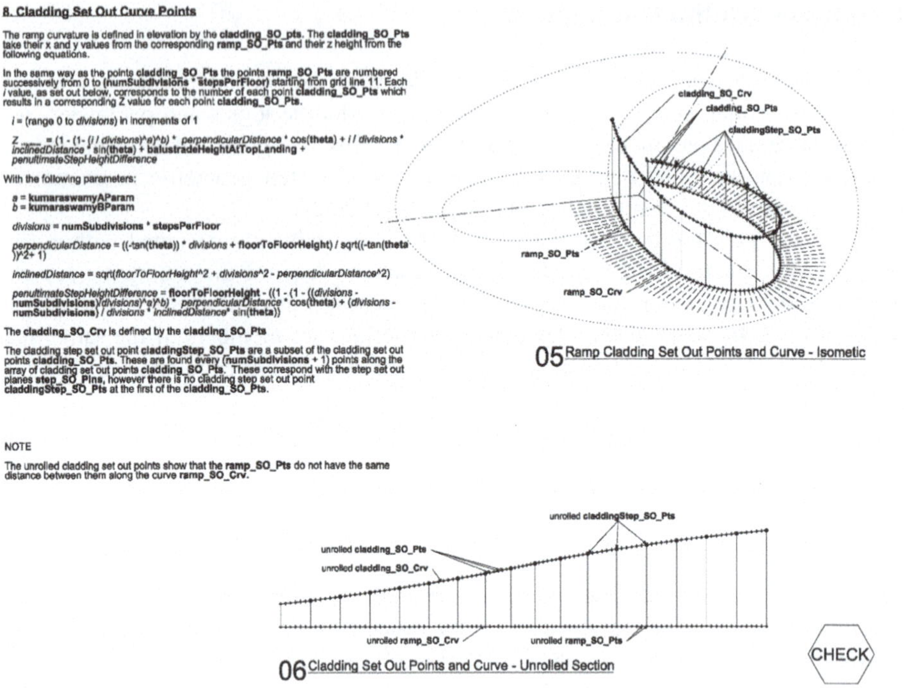

Fig. 5. Bloomberg ramp geometry method statement showing step 8 of 91 *Drawing* Foster + Partners

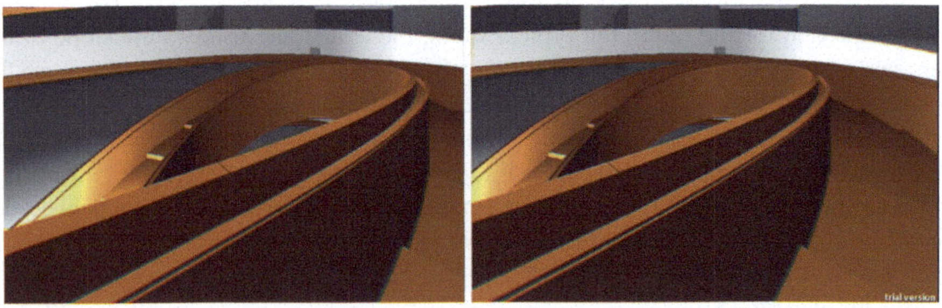

Fig. 6. Rendered motion parallax images of Bloomberg ramp used in Oculus VR headset *Image* Foster + Partners

effect of very gradual changes in curvature in profile is disguised by the curvature in plan. A model was set up so designers could don VR headsets to virtually walk up the ramp and examine views from various vantage points. The exercise looked at the 's' shaped curve from different positions on the ramp to gauge whether the 's' was noticeable at all. Different profile curves were scrutinised and the maximum 's' shaped curve that was not perceivable was chosen. This is an example of optics and human perception informing the mathematics (Fig. 6).

Using a geometry method statement allows architects to embed design logics into the cladding contractor's model. This meant as certain fabrication dimensions became established, parameters could be updated whilst still maintaining the logics of the model. The curvature is very critical to the design of the Bloomberg ramp and defining curves mathematically allowed precise curvature to be communicated independent of software. By building their model afresh the cladding contractor is required to understand the geometry and the challenges posed by the design. They take full ownership of the model. Constructing the model afresh also allows the cladding contractor to gear it towards their internal production workflows and pipelines. From the design side, the discipline of documenting the design in a geometry method statement provides a vital review process. Once the cladding geometry was rebuilt by the cladding contractor, the maximum discrepancy between the architect's and cladding contractor's 3D model was 0.004 mm.

Design to Fabrication

The geometry method statement provided the cladding contractor, Gartner, with a geometric description of all the architectural surfaces. Gartner constructed a 3D solid model that included all the bracketry and fixings enabling coordination of other services such as the sprinkler systems, which were integrated into the ramp. Gartner then sent their 3D models to the fabricator Kikukawa based in Japan who fabricated the bronze cladding panels to the correct curvature. Kikukawa was selected as they were found to be the only company capable of producing the panels to the required quality of finish and fidelity. Kikukawa could join sheets together with invisible welds for large panels that exceeded the maximum width of sheets available. The finished fabricated panels were then shipped from Japan to the UK for installation (Fig. 7).

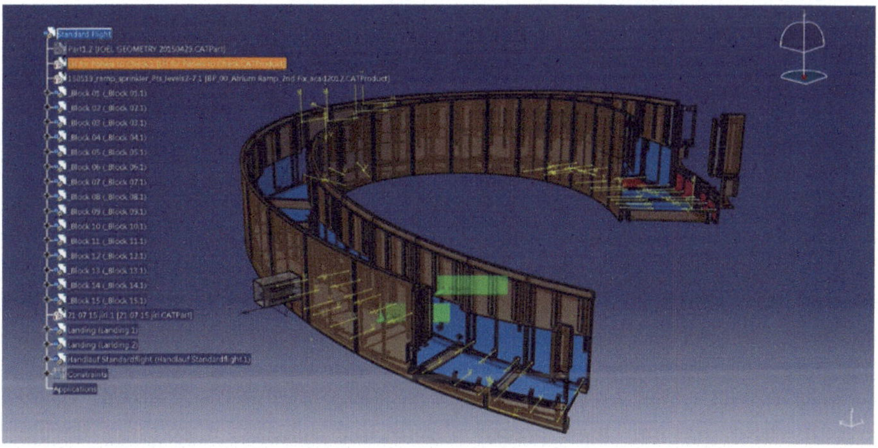

Fig. 7. Cladding Contractor, Gartner's CATIA, solid 3D model of ramp, flight 7–8 *Image* Foster + Partners

Fig. 8. Ramp Steelwork fabricated in UK by Littlehampton Welding *Photo* Nigel Young/Foster + Partners

The ramp steelwork was built entirely in the UK. The design allowed for a ±15 mm tolerance for the steel. The intention was to fit the cladding panels around the steel tolerance zone, so that any deformation of the steel would be contained within the confines of the cladding. Even before the first cladding panels were built in Japan, ramp steel was craned onto site floor by floor and stored in situ, as the main steel structure of the building was constructed. It was initially positioned on temporary supports until the concrete for each floor has been poured and the main structure has settled. The ramp steel was then welded to the main structure and the supports were removed (Figs. 8, 9 and 10).

Laser Scanning As-Built Steelwork

To verify the position of the ramp steel, Multipass, who specialise in the application of large volume laser scanning techniques to niche applications such as those found in the nuclear industry, were brought into scan the steelwork before and after removal of temporary supports. Accuracy of the laser scanning process was critical as the ramp cladding was designed around a steel tolerance of ±15 mm. The point cloud laser scan of the ramp was done from 82 locations. Scans were registered to a network of 150 laser tracker nests which were hot glued to the ramp steel and surrounding structure. The nests held 100 mm high precision reflective spheres, a solution developed specifically for the project. At least six targets are visible in each scan. In each individual Laser scan the centre points of the spheres were derived in registration software.

Fig. 9. Ramp Steelwork propped in situ on steel supports whilst surrounding floor are built 2014 *Photo* Nigel Young/Foster + Partners

Fig. 10. Point cloud resulting from laser scan of ramp steel work prior to de-propping May 2015 *Image* Multipass

The centre points were aligned to the equivalent points in the control network. The resulting size of the 82 combined scans was around three billion points.

Registered scans were cropped to remove extraneous data from the surrounding building. The remaining point cloud data representing the steel ramp was merged and duplicate data was removed to produce a single high-resolution point cloud model. It was then decimated to create a uniform point density of around 3 mm spacing between

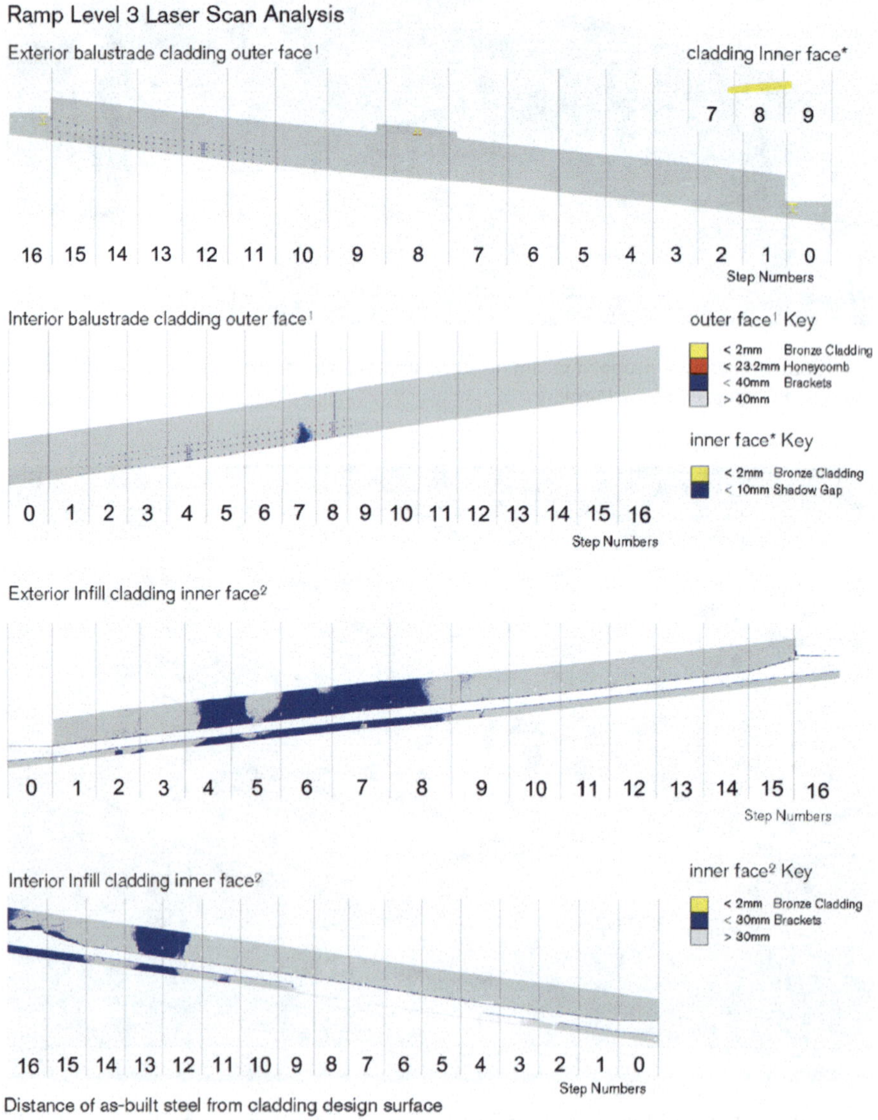

Fig. 11. Analysis of point cloud resulting from laser scan of ramp steel in situ after de-propping. showing zones where possible clashes might occur. January 2016 *Image* Foster + Partners

adjacent points. After cropping, registration, merging and optimising point density the point cloud model contained around 35 million points per flight. Around 250 million for all 6 flights, plus the bridge.

Precision of the point cloud model depends on several factors. Although the quoted precision for the Tracker instrument is better than 50 μm depending on atmospheric conditions, and the quoted precision for the Surphaser Laser Scanner is better than 400 microns at 10 m range (2017), the typical cumulative error for the point cloud data was better than 2 mm after registration. The as-built scans revealed that there were places where the surface of the steel was out by as much as 47 mm in X/Y and several areas in Z were out between 25 and 30 mm (Fig. 11).

From Point Cloud to Watertight Solid Model

Once the steel was found to be out of tolerance, and with production and shipping of bronze panels already underway, it was decided that the setting out of the panels would need to be adjusted to absorb the deviation in the underlying steelwork. An exercise was then initiated to determine a new setting out position that deviated minimally from the design geometry. This demanded a high precision solid CAD model of the as built ramp to enable clash detection against Gartner's solid model.

Turning the huge point cloud data into a watertight solid CAD model was achieved with a workflow utilising several CAD packages including Pointools, Polyworks, Geomagic and Rhino and for smaller components such as bolt fittings, with some custom in-house software. This process is usually carried out on much smaller size objects to reverse engineer parts for replacement parts from machines, turbine housings or obsolete mechanical components. In this case, the Bloomberg ramp demanded modelling a huge item with millimetre precision. The model had to include small items on the ramp such as protruding bolt heads 25 mm in diameter and 40 mm high which had to represented in the solid CAD model for inclusion in the clash detection process. This as-built solid model was also issued to the flooring contractor, who was also using Autodesk Inventor software, to allow them to manufacture and fit wooden decking to the ramp (Fig. 12).

Reconfiguring the Set Out of the Cladding Panels

The cladding fabricator's solid model was stripped down to a minimum number of critical components which were likely to come into contact or clash with the steel. The model was imported into Autodesk Inventor where mechanical constraints were applied to allow adjustment of the setting out. The setting-out curve derived from the Geometry Method Statement was adapted to produce a fit according to a series of rules established by the architects. This prioritised curvature continuity, minimising variation in shadow gaps between adjacent blocks of panels. It locked off geometry within blocks so that inner and outer balustrades of each step moved as one unit, avoiding misalignment locally.

Fig. 12. CAD model showing clash detection of steel studs with cladding brackets *Image* Multipass

The cumulative effects of small shifts between blocks allowed flights to iteratively absorb the distortion of the steel box inwards or outwards. Each iteration was double checked using colourised 3D displacement maps to indicate the location and degree of any clashes. Once proposals were agreed by the architect and client, the task of producing setting out information for installation began.

Setting Out Panels

The fixing points of the panels were set out using a two-step process. Coordinates were generated from the Inventor model producing a spreadsheet of XYZ setting out points for mounting the top brackets of the main panels. Using a control network of survey nests from the initial laser scan, a laser tracker was used to direct a hand-held target to the exact planometric position of the centres of the top bracket studs which were then welded to the steel structure. The top bracket assemblies were then mounted onto the studs. A coordinate measuring machine (CMM) was used to adjust the bracket assembly until the trunnion pin saddles/panel hanging pins were within 100 microns of nominal. The precision of the control network, laser tracker and CMM machines allowed the cladding panels to be set out with sub millimetre accuracy in 3D space (Faro Laser Tracker).

The resulting installation of bronze panels achieved the ±2 mm variation agreed in the specifications. In the context of 30 m landing to landing distance the variation from the mathematical curves at the worst case of 32 mm is imperceptible. What proved critical from a visual point of view was the alignment of the bronze panels along the inner curve of ramp inner balustrade. While most stairs are stop-start with ninety degree turns, landings and repetitive doubling back on themselves, the Bloomberg ramp

Fig. 13. First cladding panels positioned on first flight of ramp April 2016 *Image* Nigel Young/Foster + Partners

successively accelerates and decelerates maintaining curvature continuity. It emancipates its passengers from the confines of segmented flights and landings creating instead an experience of curvature which transforms the experience and flow of the entire building (Fig. 13).

Conclusion

The Bloomberg ramp employed different approaches to sharing parametric design definitions. A progressive approach to practice-led research created an opportunity for sharing of digital definition between architect and structural engineers at design stage. Use of mathematical formulae to define the geometry and design of a clear set of geometric rules captured by a Geometry Method Statement communicated the same model to cladding consultants. The process enabled parameters to adapt and meet the steel and cladding fabricator's constraints according to design logics established by the architects. Analysis of point cloud data form laser scans of as-built steel to inform setting out of cladding panels demonstrated how digital design workflows give new possibilities for fidelity in construction of complex designs. Yet the paper indicates that the human processes of building collaborative relationships, interrogating data, responding to issues as they come up and informing decision making are as vital as ever in successful delivery of construction projects.

References

Faro Laser Tracker and CMM Technical Data Sheets. http://www.faro.com

Peters, B., Whitehead, H.: Geometry, form, and complexity. In: Littlefield, D. (ed.) RIBA Publishing, London (2008)

Rabagliati, J., Huber, C., Linke, D.: Balancing complexity and simplicity. In: Gramazio, F. et al. (eds.) Fabricate: Negotiating Design & Making, pp. 45–51 (2014)

Surphaser Laser Scanner Precision Data. http://www.surphaser.com/. Accessed 01 Feb 2017

Nature-Based Hybrid Computational Geometry System for Optimizing Component Structure

Danil Nagy(✉), Dale Zhao, and David Benjamin

The Living, an Autodesk Studio, New York, NY, USA
danil.nagy@autodesk.com

Abstract. This paper describes a novel computational geometry system developed for application in the design of full-scale industrial components. This system combines a bottom-up growth strategy based on slime mould behaviour in nature with a top-down genetic algorithm strategy for optimization. The growth strategy uses an agent-based algorithm to create individual instances of designs based on a small number of input parameters. These parameters can then be controlled by a genetic algorithm to optimize the final design according to goals such as minimizing weight and minimizing structural weakness. Together, these two strategies create a hybrid approach which ensures high performance while allowing the designer to explore a wider range of novel designs than would be possible using traditional design methods.

Keywords: Design and modelling of matter · Multi-objective optimization · Generative design · Computational geometry · Additive manufacturing

Introduction

The Design Problem

The hybrid computational geometry system described in this paper was developed in partnership with a team of researchers at a large aircraft manufacturer and applied to the redesign of a partition inside a commercial aircraft (Fig. 1). The partition is the wall that divides the seating area from the galley, and the goal for the project was to reduce its weight by 50%. This weight reduction is critical to the aerospace industry to reduce fuel consumption, cost of flying, and carbon emissions.

While the partition wall may seem like a relatively simple component, it actually presents two complex structural challenges. First, the partition must support a fold-down cabin attendant seat (CAS). Unlike the partition, the CAS is not attached to the airplane's fuselage or the floor, thus the full weight of two flight attendants and the seat itself must be transferred through the partition into the aircraft's structure. Since the CAS is hanging from the partition, this creates an asymmetrical load. And to pass certification, the partition must withstand a crash test in which the weight of the CAS and its attendants is accelerated to 16 times the force of gravity (16G)—an extremely challenging structural task.

K. De Rycke et al., *Humanizing Digital Reality*,
https://doi.org/10.1007/978-981-10-6611-5_15

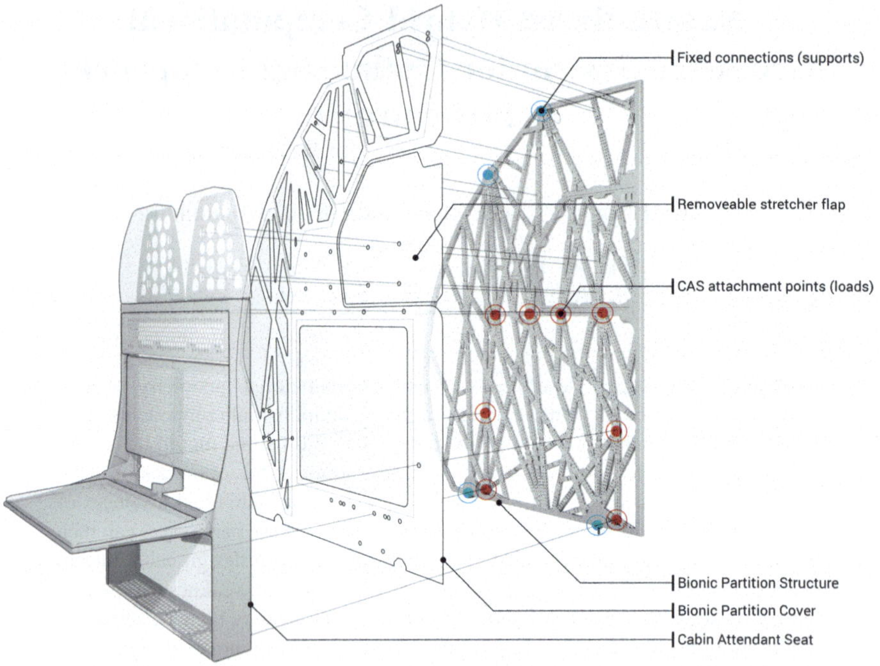

Fig. 1. Description of aircraft cabin partition design problem

Second, due to new safety regulations, the partition must include a panel called the 'stretcher flap' which can be removed to allow a stretcher carrying a sick or injured passenger to be carried around the corner from the seating area to the galley and exit. This results in a big hole in the partition which makes it difficult to route forces from the CAS directly into the aircraft's fuselage.

Due to these structural challenges, the state-of-the-art partition design is very heavy and expensive. The goal of our collaboration was to develop a computational design workflow based on natural intelligence that could leverage the potential of metal additive manufacturing to create the next generation of lightweight, strong, and affordable aircraft components.

Designing for Complexity

New technologies of metal additive manufacturing have made possible the fabrication of fully usable industrial components with complex geometries that could not be manufactured using traditional methods. Although additive manufacturing processes come with their own set of limitations in terms of what can be produced and the human labor needed to produce it, in general printing the same volume of material costs the same regardless of its formal complexity. This contrasts with traditional methods such as machining or casting, where formal complexity is often a significant aspect of the part's final cost.

These new capabilities in manufacturing have opened new design possibilities that have only barely been explored. One issue is that high-resolution formal complexity is very difficult to comprehend using traditional tools and design methods. While 3D printers can easily describe surface features to the tenth of a millimeter, there are no existing design tools which would allow the human designer to reason or design at this level of detail. To take advantage of the opportunities of these new manufacturing methods, we need new computational design tools that can assist us in the exploration of this huge space of potential designs and find the best performing solutions to our design problems.

Evolving Design

A common tool in the exploration of complex, highly multidimensional design spaces is the genetic algorithm (GA). The GA is a particularly popular example of a meta-heuristic search algorithm (Yang and Luniver Press 2010) which can explore a 'black box' parametric model to find the highest performing designs based on one or more objectives. Unlike other optimization methods based on gradient descent, GAs are completely top-down and do not require any knowledge of the model in order to mine it for the best results. This is well-suited for optimizing design models which are typically defined by a large set of geometric operations, all of which may be difficult or impossible to analytically describe or differentiate.

To utilize a GA to solve a design problem, the designer must specify a generative geometry system that describes a 'design space' of possible solutions to the problem, as well as one or more measurable goals for the design. In an engineering context, performance goals tend to be fairly straight forward, and usually involve maximizing structural performance (by minimizing stress, displacement, etc.) while minimizing the amount of material used. Thus, our main challenge in this project was to develop a unique generative geometry system which could create a wide set of possible designs in a way that could be optimized by a GA. This paper describes the system we developed, demonstrates how it was used to redesign the aircraft partition, and speculates about potential applications of such systems for future design problems.

Literature Review

The use of genetic algorithms for structural optimization is well explored in the literature, and Marler and Arora (2004) provide a good overview of the subject. This problem is usually formulated as the optimization of the topology, shape, and size (TSS) of a structural layout given a set of loading conditions and an arrangement of supports. While structural optimization is a broad field, a more specific subtopic that relates to this paper focuses on the way in which structures are represented and how they are parametrized so that they can be explored productively by a GA—in other words, how the design system's input parameters, or genotype, relates to the physical form, or phenotype, of each design iteration. Aulig and Olhofer (2016) provide a good overview of the variety of representation strategies that have been developed and break down the strategies into three basic categories: (1) grid, (2) geometric, and (3) indirect.

The first two representational strategies use a direct parameterization of geometry. In the grid strategy, the entire space of the potential structure is discretized into an orthogonal grid whose values represent either the presence of material (binary encoding) or its density (real-valued encoding) at each location. In the geometric strategy, the structure is represented directly as a collection of geometric shapes. For example, a truss structure may be represented by a collection of straight-line beams which encode the size and shape of each element.

The indirect strategy, on the other hand, uses an intermediate system to generate the structure. Two common techniques for such indirect parameterization are rule-based algorithms such as Lindenmeyer systems (Hornby and Pollack 2001) and behavioural algorithms such as cellular automata (Mitchell et al. 1996). With this approach, parameters are used to control either the rules or behaviours of the intermediate algorithm, which is then executed to create the actual structure.

Although they are more complex, indirect strategies offer several advantages over direct representations, including easier scalability to address larger design problems (Kicinger et al. 2004). In the context of design, they also offer the benefit of using a relatively small number of input parameters to define a highly complex design space with a wide variety of design solutions that may not be intuitive to the designer. Combined with an optimization process based on the GA, such complex design spaces are more likely to produce design solutions that are not only high performing but also novel and unexpected.

Because of the indirect nature of this kind of parameterization, there is a huge potential for designers to invent new types of indirect representations which are customized to address specific design problems. This paper contributes to this endeavour by describing a novel indirect representation based on a behavioural algorithm inspired by the growth of slime mould in nature.

Methodology

The following section describes the details of our generative geometry system, including the parameterized behavioural algorithm which creates each design iteration, the set of measures which determine the performance of each design, and the specification of the optimization process based on the genetic algorithm which was used to achieve the final design.

Design Model

The geometry of each partition design is created by a behavioural 'bottom-up' algorithm which is inspired by the growth of slime mould in nature. As slime mould grows, it first spreads out a dense network of connections. Then, based on where it finds food, it starts to prune the network to keep only those connections that most efficiently connect the food sources. This behaviour forms complex adaptive networks that are efficient and redundant. They are efficient because they use a small amount of "lines" to connect a set of "points" (sources of food). They are redundant because when one of the lines is damaged, the network can often "route around" the problem and keep the

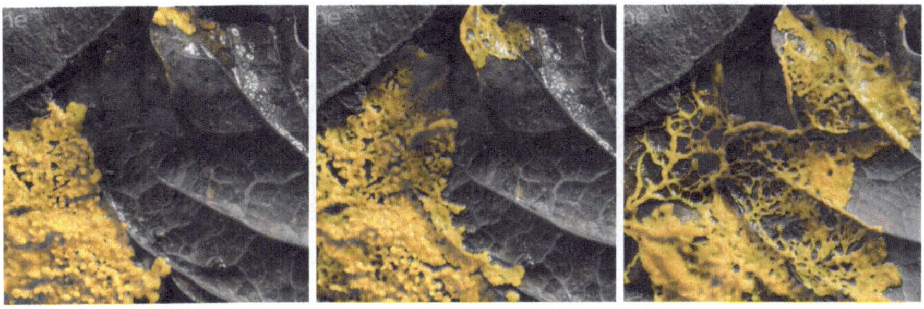

Fig. 2. Growth of slime mould over time showing initial dense network and later pruning to the most important connections

points connected (Tero et al. 2010). Overall, the living slime mould organism produces complex, highly-efficient food distribution networks based only on local behaviour (Fig. 2).

Although the geometry of slime mould growth is based on connecting points (food sources) in 2D to transport nutrients, we hypothesized that a similar logic may be beneficial for connecting points (structural attachment points) in 2D to resist structural loads. To enact a slime mould geometry system in our design model, we first specified a set of 'seed points' which consisted of the 4 support points (where the partition attaches to the fuselage), the 16 load points (where the CAS attaches to the partition), and 28 additional points sampled evenly along the partition boundary. Then, a ground structure (Bendsøe et al. 1994) of all valid structural connections is defined as the set of all straight-line segments between two seed points which do not cross the boundary of the panel. This ground structure is encoded as a graph whose vertices are the seed points and whose edges are the structural elements.

Next, a real-valued weight parameter in the domain [0, 1] is assigned to each vertex of the graph. The structural members of each design iteration are then sampled from the edges of the graph based on the following algorithm:

For each structural member (s):

1. Locate the vertex with the highest weight (w)
2. Select the edge which connects this vertex with its highest weight (w) neighbor
3. Decay the weight of both vertices by multiplying it by a decay parameter (d)

The final structural design is defined by the boundary of the panel plus the set of all selected edges (Fig. 3). Because all load points need to be connected into the structure for the design to be valid, a final step checks each load point to see if it was connected during the main growth step. If not, an additional structural member is created from the point to the closest point on the structure.

Like the slime mould, our model starts with a dense network of possible connections. The weights assigned to each seed point represent a varying quantity of food at each point, and structural pathways are selected for the design based on those that connect the highest food quantities. Just as the slime mould eats the food causing its

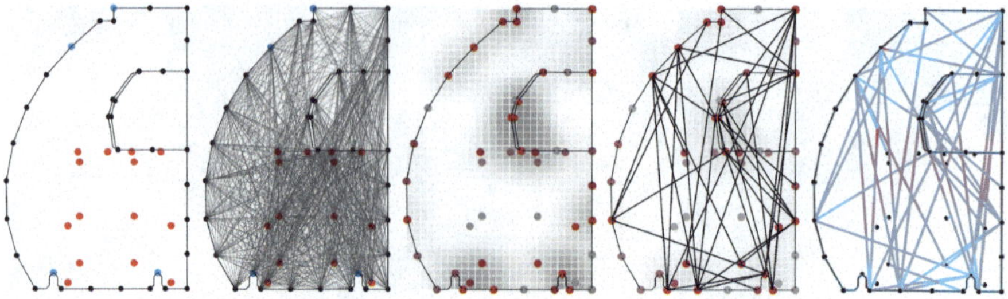

Fig. 3. Diagram of computational geometry system based on growth of slime mould

network to evolve over time, the decay factor slowly reduces the weight of each utilized seed point allowing connections to grow in other parts of the structure.

The parameters of this model are the weights (w) of the 48 seed points, plus the number of structural members (s) and the decay parameter (d). Since all the parameters are continuous, the GA is able to "learn" how to work with the growth behaviour and tune it to create better performing designs over time.

Model Evaluation

This behavioural generative geometry model can create a large variety of structural designs for the partition based on a relatively small set of input parameters. However, in order to use a genetic algorithm to evolve high-performing designs, the model must also contain a set of measures which tell the algorithm which designs are better performing. Our model uses static finite element analysis (FEA) to simulate the performance of each design under the given loading conditions. This analysis gives us a set of metrics which we can use to establish the objectives and constraints of our optimization problem:

1. Total partition weight. This should be minimized (objective).
2. Maximum displacement, which is how much the panel moves under loading. This should be less than 2 mm based on the given performance requirements (constraint)
3. Maximum utilization, which is the percentage of the maximum stress allowance of the material experienced by the structural members. This should be less than 50% based on a standard safety factor (constraint).

In addition to these structural goals and constraints, we specified an additional design objective to maximize the distribution of material (minimize the number of large holes) within the perimeter of the partition. This is to discourage designs which solve the structural loading problem while leaving large holes in the structure which may cause other problems when passengers or objects bump into the partition. This set of two objectives and two constraints completes the specification of the model (Fig. 4) and allows the genetic algorithm to automatically search the range of possible designs to find a set of valid and optimized designs.

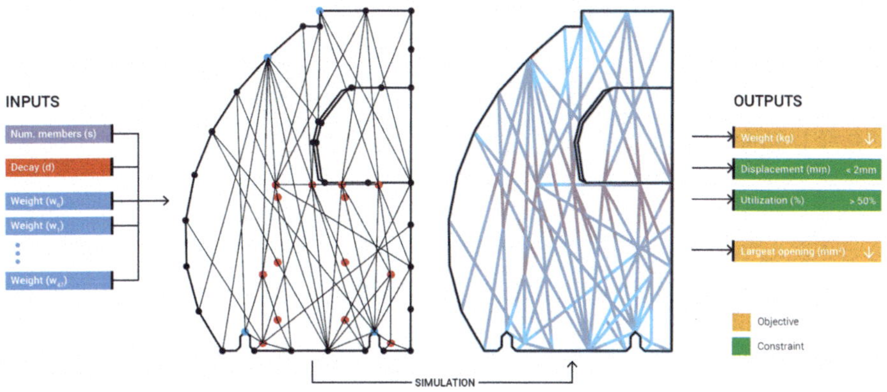

Fig. 4. Diagram of geometry system with 50 inputs, 2 constraints derived from FEA simulation, and two objectives derived from the geometry of the model

Model Optimization

Using this model, we performed an optimization using a variant of the NSGA-II genetic algorithm (Deb et al. 2002) with the following settings:

- Number of designs per generation: 200
- Number of generations: 100
- Mutation rate: 0.05
- Cross-over rate: 0.9

Once the optimization is complete, we can visualize the results by plotting them relative to the objectives of the optimization. Figure 6 shows each partition design explored by the optimization plotted as a point on a scatter plot where the x-axis represents the weight of the partition, the y-axis represents the infill factor, and the colour represents the generation in which it was created. The squares are the invalid designs based on the two constraints.

The goal of the optimization is to find designs which meet the two structural constraints with a minimum weight and a minimum number of large holes. Because these two objectives are in competition with each other, there is no single best solution (Fig. 5). Thus, the goal of the genetic algorithm is to discover the set of 'Pareto optimal' solutions that each solve the trade-off between these competing objectives in a different way. Figure 6 shows the ability of the genetic algorithm to develop subsequent generations of designs which push the 'boundary' of optimal designs closer and closer to the conceptual point of optimal performance in the lower left hand corner of the plot. The set of Pareto optimal designs are shown with a thick black outline in Fig. 6, and a subset of them is visualized above the figure.

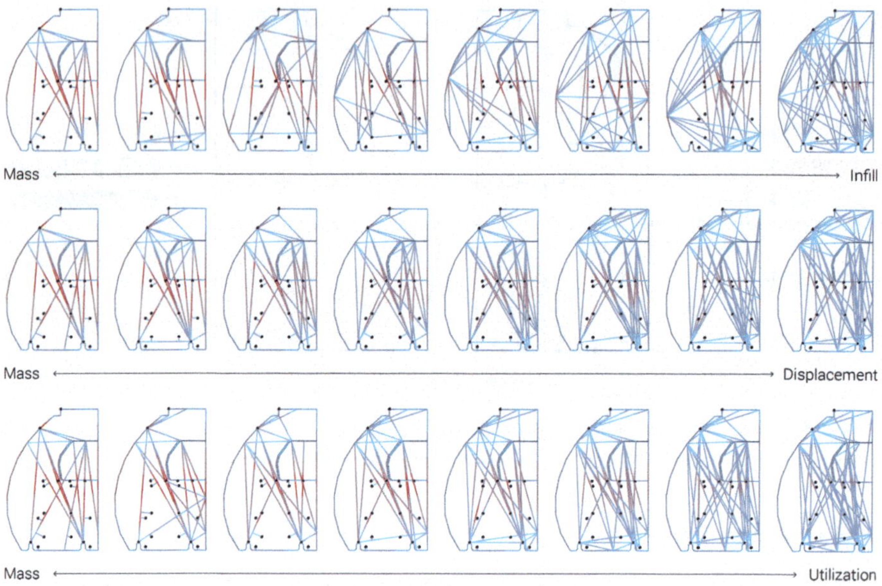

Fig. 5. Sample of designs showing trade-offs between sets of two objectives

After Optimization

Once the optimization process was complete, we selected the final design as the one which minimally met our 50% weight reduction goal while creating the least large holes in the structure (second from right in Fig. 6). The design was then further developed by breaking each of the structural members optimized through the evolutionary process into a set of smaller lattice structures. Each of these lattice beams was further optimized by changing its diameter based on the local stress distribution in the structure. This secondary optimization allowed us to fine tune the component structure so that we exactly met the performance requirement while reducing the weight as much as possible. Finally, in order to manufacture the component we needed to break it down into a set of smaller components which could fit into the bed of a metal selective laser sintering (SLS) machine (Fig. 7).

Once the parts were manufactured we assembled them into the final partition shown in Fig. 7. This part met our design goals: having the same structural performance as the state-of-the-art component with only 50% of the weight. The partition created through

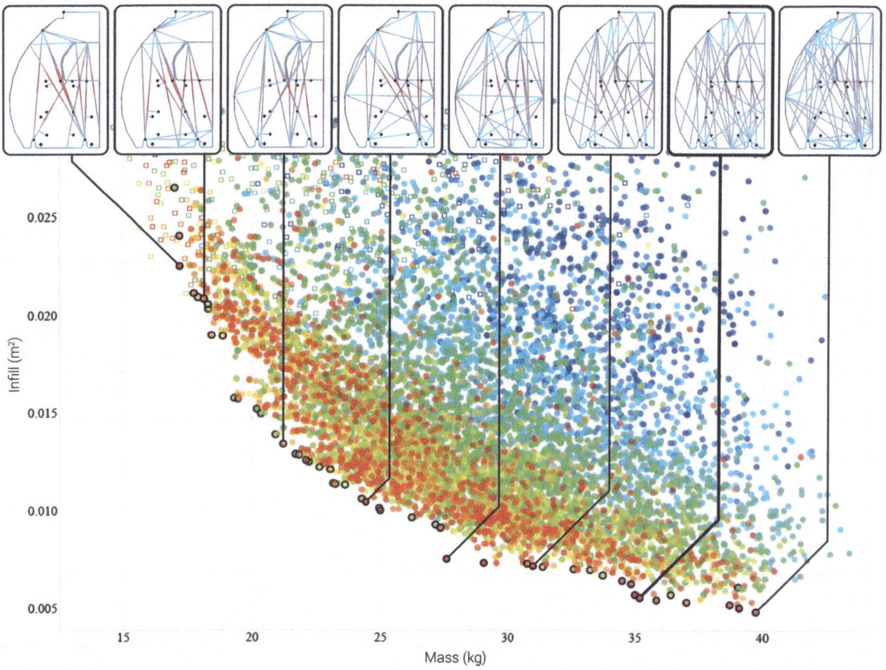

Fig. 6. All designs explored during the optimization process plotted according to the two objectives. *Colour* represents the generation in which the design was evaluated, with *blue* for earlier and *red* for later designs. Designs with a *black outline* are part of the Pareto-dominant set of optimal solutions

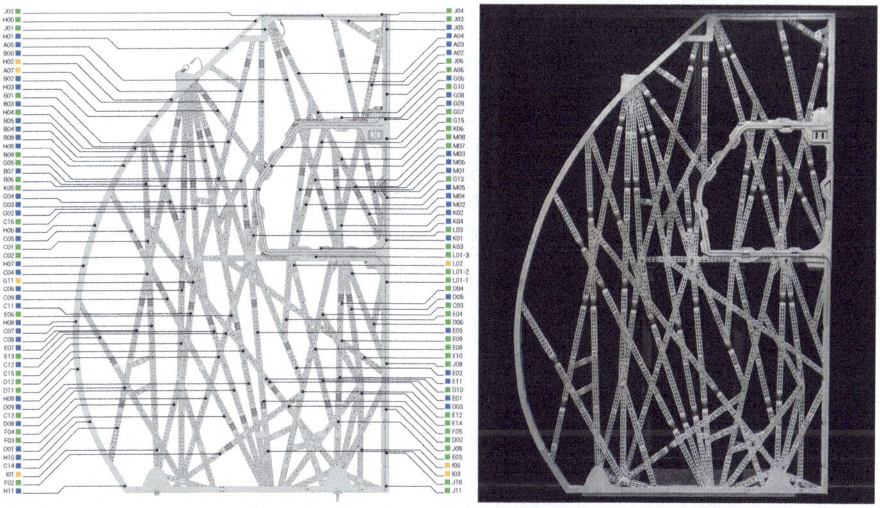

Fig. 7. Images of final design including diagram of component breakdown (*left*) and photograph of printed partition prototype (*right*)

this process is the largest aircraft component ever produced entirely through metal additive manufacturing, and is currently undergoing testing and certification which will allow it to be integrated into future airplanes for commercial flights.

Conclusions

This paper describes a novel computational design method which combines a generative geometry model based on the 'bottom-up' agent-based growth processes found in natural systems with a 'top-down' genetic algorithm for optimization. The paper also describes the application of this method toward the design of a unique industrial component which can take advantage of the formal freedoms allowed by recent advances in metal additive manufacturing. Finally, our method suggests future research into the development of other nature-based generative design systems which can leverage the power of evolutionary computing to derive unique, high-performing solutions to complex design challenges.

References

Aulig, N., Olhofer, M.: Evolutionary computation for topology optimization of mechanical structures: an overview of representations. In: 2016 IEEE Congress on Evolutionary Computation (CEC), IEEE, pp. 1948–1955 (2016)

Bendsøe, M.P., Ben-Tal, A., Zowe, J.: Optimization methods for truss geometry and topology design. Struct. Multidiscip. Optim. **7**(3), 141–159 (1994)

Deb, K., Pratap, A., Agarwal, S., Meyarivan, T.A.M.T.: A fast and elitist multiobjective genetic algorithm: NSGA-II. IEEE Trans. Evol. Comput. **6**(2), 182–197 (2002)

Hornby, G.S., Pollack, J.B.: Evolving L-systems to generate virtual creatures. Comput. Graph. **25** (6), 1041–1048 (2001)

Kicinger, R., Arciszewski, T., De Jong, K.: Morphogenesis and structural design: cellular automata representations of steel structures in tall buildings. In: Congress on Evolutionary Computation, 2004. CEC2004, vol. 1, pp. 411–418. IEEE (2004)

Marler, R.T., Arora, J.S.: Survey of multi-objective optimization methods for engineering. Struct. Multidiscip. Optim. **26**(6), 369–395 (2004)

Mitchell, M., Crutchfield, J.P., Das, R.: Evolving cellular automata with genetic algorithms: a review of recent work. In: Proceedings of the First International Conference on Evolutionary Computation and Its Applications (EvCA'96) (1996)

Tero, A. et al.: Rules for biologically inspired adaptive network design. Science (2010)

Yang, X.-S.: Nature-Inspired Metaheuristic Algorithms. Luniver Press (2010)

Enabling Inference in Performance-Driven Design Exploration

Zack Xuereb Conti([✉]) and Sawako Kaijima

Singapore University of Technology and Design, 8 Somapah Rd, 487372 Singapore, Singapore
xuereb_zack@mymail.sutd.edu.sg

Abstract. In this paper we present an approach to enable inference when coupling computational design systems and engineering simulation, in order to narrow the ambiguity of a design space to a space that is meaningful for a designer's goals. Inference is a statistical technique to draw judgement about data. The emergence of computational design systems in architecture has enabled the utilization of engineering simulation to evaluate and drive exploration of the design space. However, we argue that designers find it challenging to infer an thorough understanding of the design space when considering many variables because coupled systems are limited to one directional operations (input → output). Consequently, the qualitative control over the quality of design comes into question. In response, we present a probabilistic representation of the design-analysis system whereby, input and output variables are represented as probability distributions to enable bi-directional inference between input and output. Subsequently, the capability to infer cause from effect provides a deeper understanding about the relationships between design variables and physical behaviour. Furthermore, we discuss Bayesian networks as a statistical technique to handle inference over complex design spaces.

Keywords: Engineering simulation · Design control · Bayesian probability

Introduction

In this paper we suggest a shift in the way we approach simulation in computational design. We present a statistical approach to improve control, when design-analysis systems become convoluted.

The increased availability of engineering simulation to architectural design has facilitated a performance-conscious exploration from the very early stages of design (Shea et al. 2005). In computational design, this typically involves setting up a computational system composed of design variables and simulation analysis tools. Designers then search through a design space for solutions by manipulating the design input values and observe the quantitative output of the simulation, in a trial and error fashion or using some form of stochastic algorithm that outputs singular outcomes.

It is good to note that in this paper we direct our efforts towards the challenge involved with understanding the characteristics of a design space, rather than with the search for 'optimal' solutions, because we argue that the early stages of design are

K. De Rycke et al., *Humanizing Digital Reality*,
https://doi.org/10.1007/978-981-10-6611-5_16

mostly concerned with a process of knowledge-gain, as a means to improve the definition of the design problem.

In the architectural community we tend to utilise engineering simulation tools as a black box. This approach may be satisfactory at times however, we find that when systems become complex and the number of variables considered are many, using simulation as a black box compromises the designer's control over the system. It becomes cognitively challenging to infer and keep track of the cause and effect relationships between design variables and physical behaviour. Consequently, utilising simulation blindly, yields a shallow understanding of the complex design space.

To overcome this challenge, we argue that in order to gain a deeper understanding of the relationships between the inputs and outputs of a system, we require a different way of thinking about 'performative design'. We propose to move beyond setting up design-analysis systems limited to performance evaluation and optimisation only (see Fig. 1, left) towards systems that enable inference. In statistics, inference refers to the practice of drawing conclusions about the behaviour underlying some collected data. In this paper we discuss the potential of inference as a means to reduce the vagueness of a design space to a region that is meaningful to a designer's goals (see Fig. 1, right). In other words, we suggest a shift from using simulation as a tool limited to inputs → output operations, towards setting up a systems that enables bi-directional computations (inputs←→output) such that we can gain a deeper understanding of the cause and effect relationships between design variables and physical behaviour.

Approximating relationships between inputs and outputs immediately suggests the use of statistical methods such as polynomial regression, whereby relationships are captured into a mathematical model such that they can be exploited. However, statistical inference demands that inputs and outputs are modelled as distributions. By definition, inference is the practice of forming judgments about underlying distributions of a population of values, implying that typical regression techniques are not suitable because they are limited to a deterministic representation of the form $(y = f(x_1, \ldots, x_n))$, whose inputs $(x_1, \ldots, x_n)$ and outputs (y) only take on single values.

In response, we take on a probabilistic modelling approach, where inputs and outputs are represented as probability distributions of sampled values and simulated

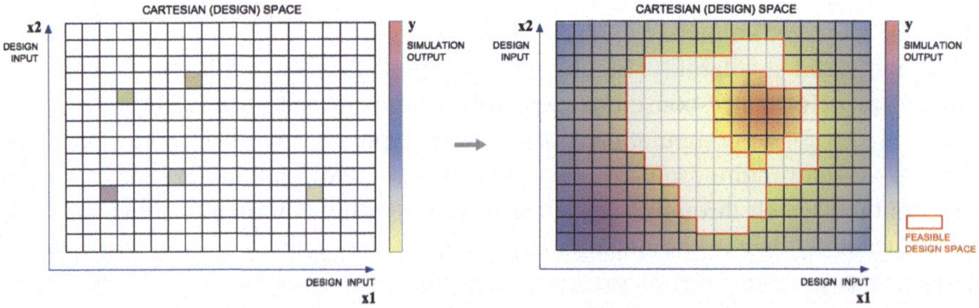

Fig. 1. A hypothetical illustration of using simulation blindly to evaluate a (discretised) design space 'manually' (*left*), and identifying feasible design spaces that are meaningful to a designer's design/performance goals (*right*)

response values, respectively. A probabilistic model does not reduce relationships between variables to mathematical relationships, instead, are represented by the likelihood of the input to cause effect on the output, in a joint probability distribution (JPD). This way, the probabilistic representation enables bi-directional inference between cause $\longleftarrow\!\!\longrightarrow$ effect, because a JPD does not distinguish between the 'right hand side' (inputs) and 'left hand side' (output), unlike classical functional representations ($y = f(x_1, \ldots, x_n)$). To summarise, our approach embeds a deterministic problem into a probabilistic framework to enable the statistical concept of inference.

In the following section we discuss and illustrate how a probabilistic approach can enable us to infer meaningful boundaries in the design space. Furthermore, we present Bayesian Networks, which is a powerful statistical technique, as a solution to perform inference over high-dimensional JPDs, i.e. complex design systems.

A Probabilistic Representation to Enable Inference

Background on Probability

In statistics, there exist two schools of taught on probability; 'frequentist probability' and 'Bayesian probability'. The frequentist interpretation views probabilities as frequencies of events (data, in our case), for example, modelling a probability distribution as a histogram. Frequentist probability is also referred to as classical statistics, which is what is taught in most school curricula. On the other hand, a Bayesian approach models variables as probability distributions, derived from some belief system or knowledge of uncertainty in the domain. For this reason, most communities are more familiar with 'data-driven' classical statistics because it is not trivial to derive a distribution intuitively. Having said that, a Bayesian representation is more suitable for complex problem solving and decision making because the representation of variables as a joint probability distribution enables the use of omnidirectional inference. In our approach, we adopt both views on probability, such that we represent the input and output model as a Bayesian probabilistic model, whose input distributions are derived through a frequentist approach, from data generated by sampling the design space.

Bayesian Probability

In Bayesian statistics, relationships between multiple probability distributions are observed as a joint probability distribution (JPD). A JPD can be defined as a representation of the probability of every possible combination of values of each variable (Binder et al. 1997). In order to illustrate this, we present a case study of a simple cantilevering structure with varying depth, described by six geometric variables (see Table 1). For example's sake we will illustrate the JPD between these variables and the maximum deflection of the structure.

A parametric frame of the structure (see Fig. 2) was modelled in Grasshopper (Rutten 2012) and then analysed using 'Millipede' (Michalatos and Kaijima 2014), which is a Grasshopper plugin for structural analysis. We discretised the frames BC, BA, and BE to emulate the varying depth of the beams (see Fig. 2). We set up a

Table 1. List of selected input variables and their corresponding ranges

	Variable	Notation	Max (m)	Min (m)
Input	X coordinate of C	C_X	−1.5	2
Input	Y coordinate of C	C_Y	−3	2
Input	Amplitude	B_pos	−1.5	2
Input	Beam depth at C	C_depth	0.3	1
Input	Beam depth at B	B_depth	1	2.5
Input	X coordinate of E	E_X	0	5
Output	Maximum deflection	max_def	–	–

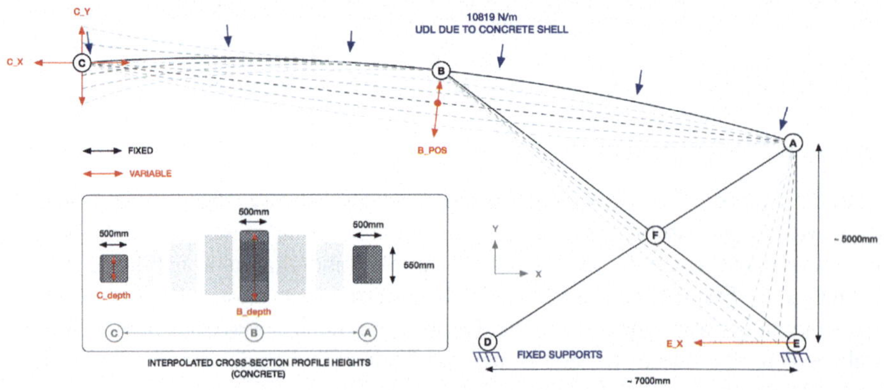

Fig. 2. Illustration of the parameterised frame geometry and the applied boundary and loading conditions

design-analysis system and ran a computer simulation experiment to generate data (values of maximum deflection), which involved running a number of simulations at various input value configurations. We made use of Sobol' sampling (Sobol 1990), to sample the design space as extensively as possible and pseudo randomly as to avoid any systematic correlations between the (independent) input variables. We then ran a parametric simulation using the generated sequences of input values as Grasshopper slider values, while recording the maximum deflection for each simulation run. For this study we selected a sample size of 4000, which is a benchmark size to secure a prediction error (RMSE) below 10%, according to a verification test we computed for six-input variable problem.

Figure 3 illustrates an example of the JPD between the depth of the beam at B and the maximum deflection from the generated data. On a side note, the shape of the distribution indicates that increasing beam depth reduces maximum deflection (as expected), while the spread of contours indicates an effect due to the other design variables on physical behaviour. Figure 3 illustrates the anatomy of the JPD where slicing at beam depth values of for example, 1.25, 1.75 and 2.25 m, reveals probabilistic relationships between the beam depth and the maximum deflection, in the form of a conditional probability functions. Conditional probability can be expressed as the

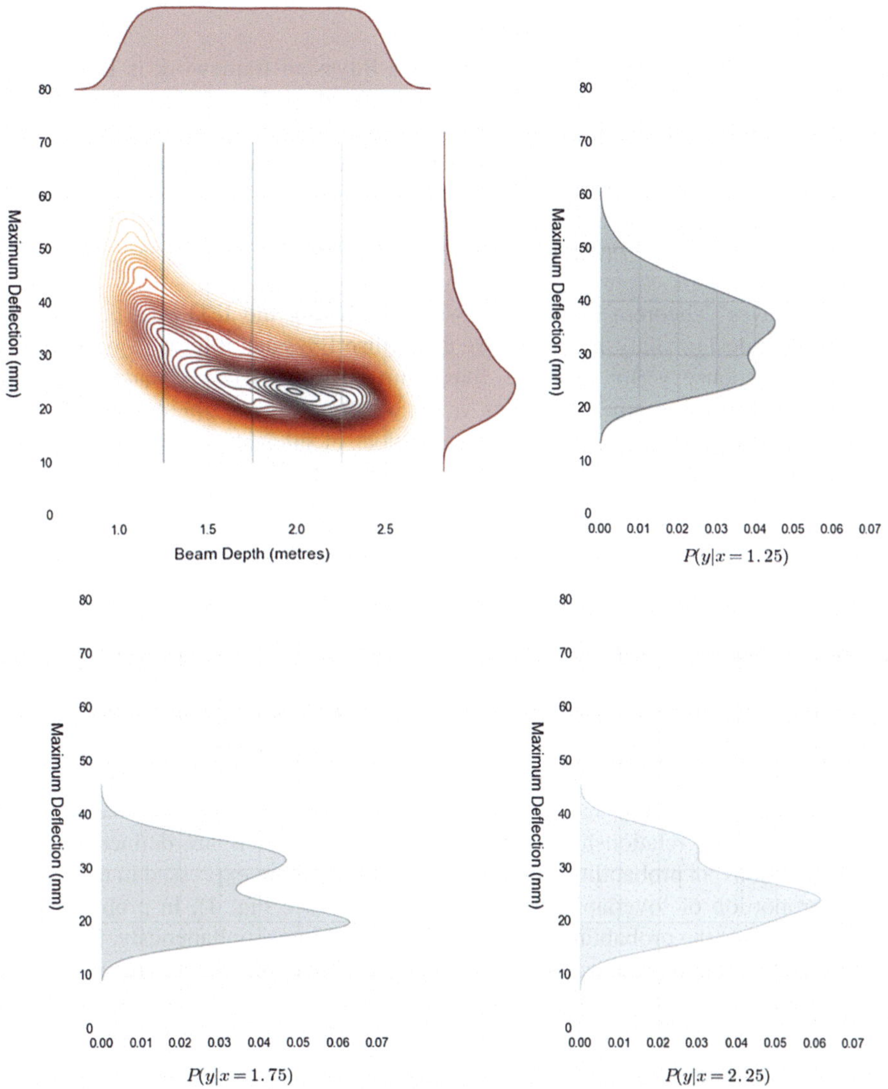

Fig. 3. Illustration of the anatomy composing the joint probability distribution between the beam depth at B (B_depth) and the maximum deflection (max_def)

probability of a variable's behaviour occurring due to another variable. In other words, rather than fitting a deterministic function to a set of points (as is practiced in classic regression), here we model a JPD.

Bayesian Inference

The key advantage of a JPD representation in a Bayesian framework is Bayes' Theorem (1763), which is what Bayesian inference is based on. Bayes' Theorem is used to estimate conditional probabilities and is defined statistically as the probability of an event (A) happening, given that it has some relationship to other event/s (B), and is given by Eq. 1.

$$P(A|\mathrm{B}) = P(\mathrm{B}|A) * (P(A)/P(\mathrm{B})) \tag{1}$$

Therefore, Bayes' Theorem implies that when modelling the JPD of a design space pertaining to a design-analysis system, we can utilise Bayesian inference to predict the engineering response y (for example, maximum deflection) (Eq. 2) and in reverse, query the values of a group X of inputs $x_1, \ldots, x_n$ (for example, the design geometric variables), when given a specific performance goal of interest (for example, keeping within an acceptable range of maximum deflection) (Eq. 3).

$$P(y|\mathrm{X}) = P(\mathrm{X}|y) * P(y)/P(\mathrm{X}) = P(y \cap \mathrm{X})/P(\mathrm{X}) \tag{2}$$

$$P(\mathrm{X}|y) = P(y|\mathrm{X}) * P(\mathrm{X})/P(y) = P(\mathrm{X} \cap y)/P(y) \tag{3}$$

As can be observed from Eqs. 2 and 3, $P(\mathrm{X}|y)$ and $P(y|\mathrm{X})$ are an identical computation, demonstrating Bayesian approach does not distinguish between inputs and outputs, thus, enabling inference of effect → cause (Conrady and Jouffe 2015).

To illustrate further the statistical concept of Bayesian inference, we can think of the Cartesian design space (bound by X and y), as a mathematical set S composed of all possible combinations of X and y values, which together make up a probability of 1, hence S = 1. While relationships in typical regression models are defined by deterministic functions, in probability, conditional relationships are expressed in terms of the relative proportion of 'overlap' between probabilities (see Fig. 4). In probability, the overlap represents the probabilities of two events occurring simultaneously, however in our case, the overlap represents the effect of one variable on the other. Therefore, using Bayes' Theorem to predict the simulation output y (Eq. 2), corresponds to the

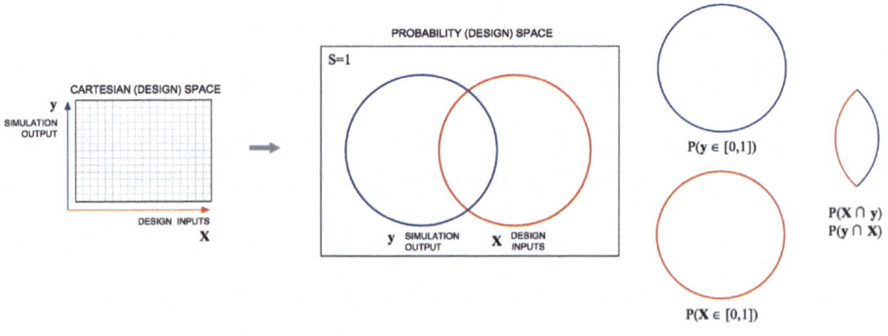

Fig. 4. Illustrative breakdown of the design space represented as a space of probabilities

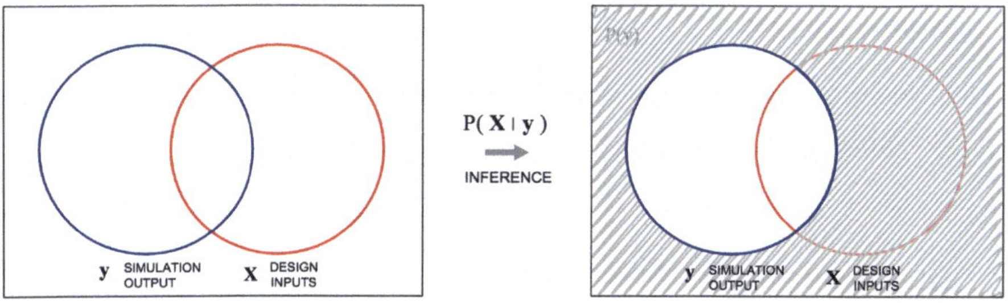

Fig. 5. Reducing the space of all probable input configurations to the ones that concern the specific output (y) being queried

proportion of X containing overlap with $y[P(y \cap \mathrm{X})/\mathrm{P(X)}]$. Similarly using Bayes' theorem to infer the values of input variables X (Eq. 3), corresponds to the proportion of y containing overlap with $\mathrm{X}[P(\mathrm{X} \cap y)/\mathrm{P(y)}]$.

When performing inference, we are therefore reducing the space of all possible outcomes to the probability space concerned only with the outcome of interest. Therefore, in terms of the design space, the operation of $P(\mathrm{X}|y)$ (which can also be denoted as $y \rightarrow \mathrm{X}$) implies that inference can reveal how the design variables (X) are effecting the physical behaviour (y), suggesting that we can immediately identify which ranges of input values are most likely to reach the performance goal of interest. In other words, we can reduce a vague design space of all possible solutions, to a more meaningful design space because it would concern only solutions that produce a feasible performance within the range of interest (see Fig. 5).

So, with reference to our cantilever example (see Fig. 2), we can identify the most likely geometric configurations that yield a low maximum deflection by:

$$\begin{aligned} & P(C\_X, C\_Y, B\_pos, C\_depth, B\_depth, E\_X | max\_def = min) \\ & \quad = P(max\_def = min | C\_X, C\_Y, B\_pos, C\_depth, B\_depth, E\_X) \\ & \quad * P(C\_X, C\_Y, B\_pos, C\_depth, B\_depth, E\_X)/P(max\_def) \end{aligned} \tag{4}$$

Performing inference in complex domains such as the example presented above, becomes challenging because the conditional relationships increase exponentially with the number of variables. This is where Bayesian Networks (BN) (Alberola et al. 2000; Pearl 1988), come into play because they can represent high dimensional JPDs very well and very compactly, thus enabling inference over multiple distributions. BNs are a type of probabilistic graphical model (PGM) that are capable of representing causal relationships probabilistically and enable the use of Bayesian inference by simply manipulating probability distributions in the network to update the other probability distributions with respect to their conditional relationships.

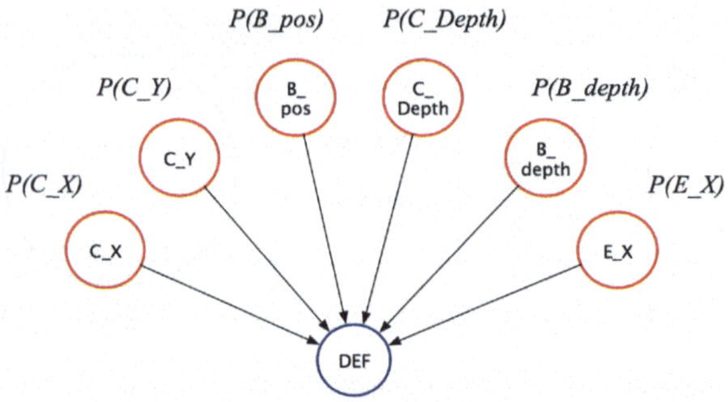

Fig. 6. Bayesian network of simulation inputs and output

Bayesian Networks to Handle Inference in Complex Domains

Background on Bayesian Networks

A probabilistic graphical model (PGM) is a type of statistical model that represents a JPD in the form of a graph structure where variables are represented as nodes, whose input is described by a probability distribution and edges between the nodes represent probabilistic relationships. Bayesian networks (BN) use directed edges to describe a causal relationship between one variable and another. Figure 6 illustrates a BN representation of the design-analysis system described in the cantilever example (see Fig. 2), where the geometric variables and maximum deflection are represented as nodes, and the directed edge structure reflects the causal direction of computation between simulation inputs and output, respectively. Bayesian networks take on only discrete distributions, therefore, all continuous input and output values were discretized respectively, into five equally spaced bins/states.

The relationships between variables, together form a network of paths, through which probabilistic information can flow. In fact, one can view a BN as a visual representation of a JPD of a group of variables (Koller and Friedman 2009). BNs convert the JPD, which essentially is the global distribution of the whole model, to local conditional distributions at each node, according to the structure of relationships (Pearl and Russell 1998). The local probability distributions of the input nodes are marginal because they do not receive a directed edge ($P(C\_X)$, $P(C\_Y)$, $P(B\_pos)$, $P(C\_depth)$, $P(B\_depth)$, $P(E\_X)$), while the local probability distribution of the output node is conditional because its' behaviour is an effect caused by the input values ($P(max\_def|C\_X, C\_Y, B\_pos, C\_depth, B\_depth, E\_X)$). In the latter case, the dependencies are quantified by conditional probability table (CPT), which essentially is a table, containing probabilities assigned to each combination of input and output values. If knowledge about the problem domain is available, the CPTs can be specified manually, otherwise learned automatically from data using supervised learning algorithms such as expectation-maximization (EM) algorithm, borrowed from machine learning. On the other hand, the topology between nodes can be encoded manually or learned automatically from data. In our case, we generated the CPTs automatically from the generated simulation data and specified the topology manually. For detailed description regarding automatic learning of conditional probabilities see Spiegelhalter (1998), while for structure topology, see Steck and Tresp (1999).

Performing Inference to Identify a Meaningful Design Space

Once the Bayesian network is set up we can explore the JPD between the geometrical variables (X) and maximum deflection (y) of the cantilever structure, using inference to predict the effect y, and in reverse identify the cause X.

The input and output distributions shown below (see Figs. 7, 8 and 9) were generated within BayesiaLab (Bayesia 2012), which is a powerful statistical software to model Bayesian networks and perform inference, besides various other statistical computations. The values on the left of each histogram indicate the probability for the respective range on the right of the histogram, to occur. Figure 7 illustrates the marginal distributions, i.e. before any inference was performed. In other words, this state corresponds to our state of knowledge about the design space during the initial stages when setting up a system.

We can specify hard probabilities to the input distributions to predict maximum deflection (for example, see Fig. 8). As a form of verification, running the simulation analysis with the following input values C_X = −0.46 m, C_Y = −1.27 m, B_pos = 1.19 m, C_depth = 0.95 m, B_depth = 1.80 m, and E_X = 0.79 m, yields max_def = 50.45 mm. On the other hand, selecting the corresponding histogram bins of the Bayesian network produces a max_def distribution with mean 51.73 mm, which is comparatively similar, as indicated in Fig. 8.

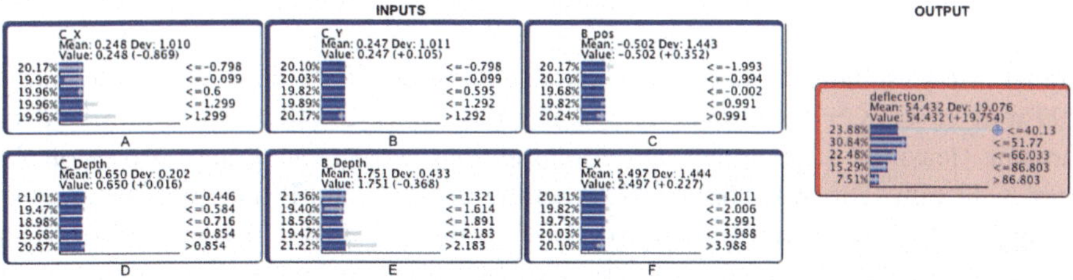

Fig. 7. Input and output distributions in their marginal state

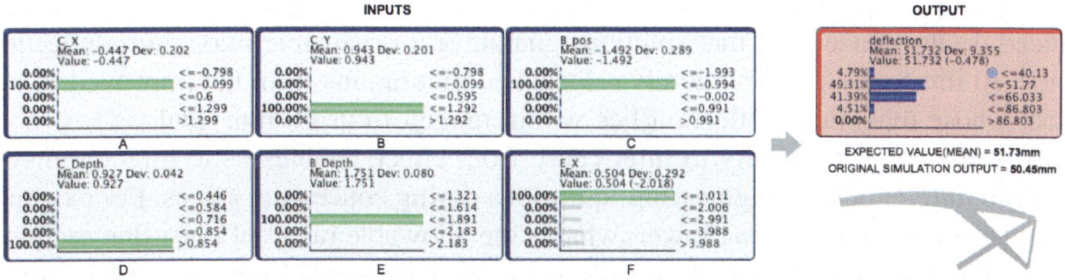

Fig. 8. Forward inference to predict simulation output

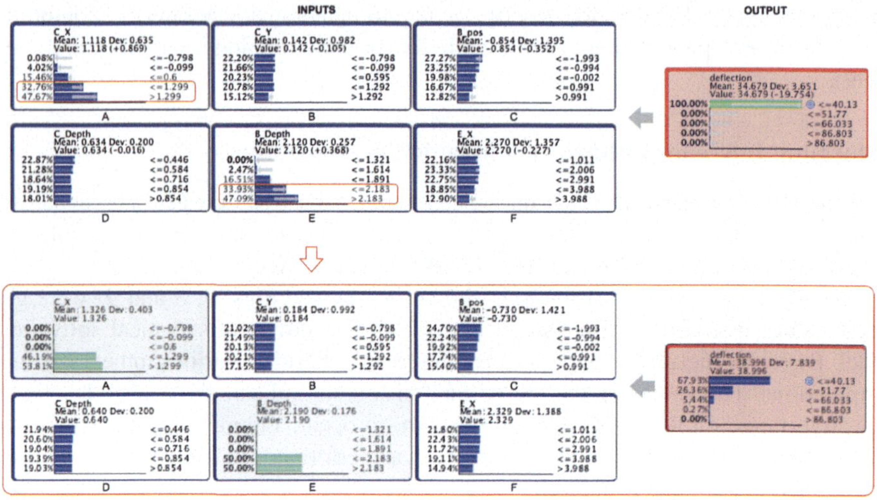

Fig. 9. Reverse inference to understand design space (*top*). Using insight to reduce design space to space relevant to our goal (*bottom*)

When it comes to performing reverse inference in Bayesian Networks, we simply need to manipulate the output distribution. For example, let us say we are interested in the range of geometric configurations that are most likely to produce a maximum deflection of the cantilever within a feasible range <40 mm. We can perform inference by simply hard-assigning 100% probability to the histogram bin containing deflection values ≤ 40.13 mm and observe the updated changes in the input distributions. The updated input distributions as shown in Fig. 9 illustrate a collective understanding how the input values can minimise deflection. More specifically, the updated input histograms in Fig. 9 indicate that if we look at the regions in the design space bound by 1.299 mm $\leq$ C_X > 1.299 mm, and 2.183 mm $\leq$ B_Depth > 2.183 mm, we can guarantee 'feasible' structural frame configurations whose maximum deflection falls below $\sim$40 mm. This was confirmed by setting the C_X and B_Depth distributions to the suggested ranges (Fig. 9, bottom). On the other hand, the almost uniform distribution shape of the remaining geometric variables indicates that minimising deflection is insignificantly sensitive to their effect, indicating a freedom of design within regions at those values.

To illustrate further, we adopted the suggested ranges as improved slider ranges in the Grasshopper parametric model. As a result, we could 'safely' search within the reduced design space such that solutions guarantee a reasonable maximum deflection. Figure 10 illustrates a few randomly selected configurations from the improved design space, whose maximum deflection lies within range of our 40 mm goal.

To summarise, the ability to infer cause from effect, enables us to directly answer the less intuitive 'design-engineering' questions during conceptual stages. For example, in the design of a tall twisting tower, what is the allowable range of twist that secures a dynamic behaviour of the structure within comfortable range? In the design of a single tower cable-stayed bridge, which are allowable aspect ratios between tower height and

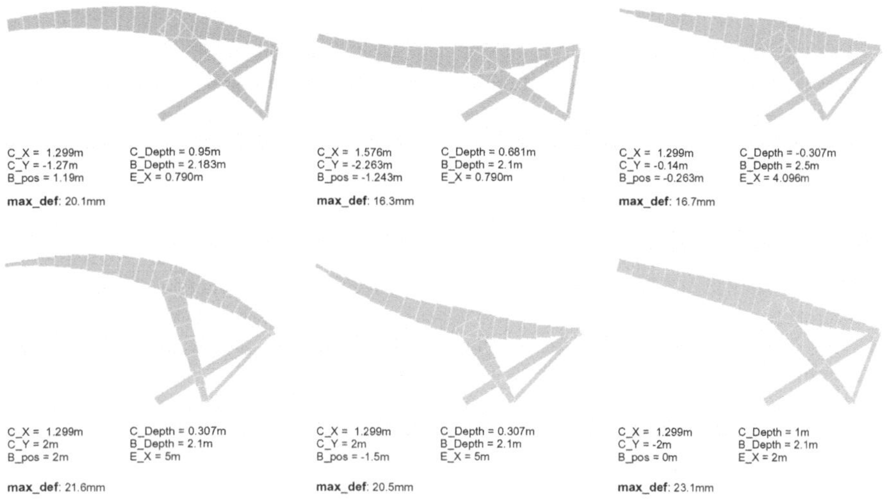

Fig. 10. Random geometric configurations from the 'reduced' design space

bridge span that produce a desirable stiffness in the bridge? Answers to these direct questions would reduce the vague design space into those regions that contain design configurations that are meaningful to desired performance goal/s.

Conclusion

The conceptual stages of real architectural design projects comprise of complex domains because they are typically characterised by multiple design variables of interest. In this paper we tackled the issue of control when exploring problem spaces defined by convoluted design-analysis systems. We emphasised that it is essential to understand the system in order to be in control over the quality of design.

We presented a probabilistic representation to enable inference in the design space such that we can identify the relationships between the inputs and the outputs of a system, thus reduce the ambiguity of the design space to a narrower space that is meaningful to a designer's goals.

The proposed framework is not limited to a particular type of engineering simulation and/or parametric approach. Furthermore, we look at the use of statistical inference in our approach to this research problem, as a translational mechanism between numerical engineering simulation and architectural design. In other words, the ability to express physical behaviour in terms of design variables can help to improve communication between the engineering and architectural domains.

Our next step will focus on communicating the inferred knowledge more graphically in terms of the 3D geometry. More specifically we envision a gradation of configurations, whereby opacity relates to probability.

References

Alberola, C., Tardón, L., Ruiz-Alzola, J.: Graphical models for problem solving. Comput. Sci. Eng. **2**(4), 46–57 (2000)

Bayes, T.: A letter from the late Reverend Mr. Thomas Bayes, FRS to John Canton, MA and FRS. Philos. Trans. (1683–1775) **53**, 269–271 (1763)

Bayesia, S.: BayesiaLab: The Technology of Bayesian Networks at Your Service (2012)

Binder, J., et al.: Adaptive probabilistic networks with hidden variables. Mach. Learn. **29**(2–3), 213–244 (1997)

Conrady, S., Jouffe, L.: Bayesian Networks and BayesiaLab: A Practical Introduction for Researchers (2015)

Koller, D., Friedman, N.: Probabilistic graphical models: principles and techniques. MIT Press, Cambridge (2009)

Michalatos, P., Kaijima, S.: Millipide (Grasshopper Plugin for Structural Analysis) (2014)

Pearl, J.: Probabilistic Reasoning in Intelligent Systems: Networks of Plausible Inference. Morgan Kaufmann, Los Altos (1988)

Pearl, J., Russell, S.: Bayesian Networks. Computer Science Department, University of California (1998)

Rutten, D.: *Grasshopper: Generative Modeling for Rhino.* Computer software, Retrieved 29 April 2012, p. 2012

Shea, K., Aish, R., Gourtovaia, M.: Towards integrated performance-driven generative design tools. Autom. Constr. **14**(2), 253–264 (2005)

Sobol', I.M.: On sensitivity estimation for nonlinear mathematical models. Matematicheskoe Modelirovanie **2**(1), 112–118 (1990)

Spiegelhalter, D.J.: Bayesian graphical modelling: a case-study in monitoring health outcomes. J. R. Stat. Soc. Ser. C (Appl Stat) **47**(1), 115–133 (1998)

Steck, H., Tresp, V.: Bayesian belief networks for data mining. In: Proceedings of the 2. Workshop on Data Mining und Data Warehousing als Grundlage moderner entscheidungsunterstützender Systeme. Citeseer (1999)

Aspects of Sound as Design Driver: Parametric Design of an Acoustic Ceiling

Moritz Rumpf[1(✉)], Markus Schein[2], Johannes Kuhnen[3], and Manfred Grohmann[1,4]

[1] School of Architecture, University of Kassel, Henschelstr. 2, 34127 Kassel, Germany
{rumpf, grohmann}@asl.uni-kassel.de
[2] School of Art and Design, University of Kassel, Menzelstrasse 13-15, 34121 Kassel, Germany
markus-schein@uni-kassel.de
[3] Design-to-Production, Seestrasse 78, 8703 Erlenbach/Zurich, Switzerland
kuhnen@designtoproduction.com
[4] Bollinger + Grohmann Ingenieure, Westhafenplatz 1, 60327 Frankfurt am Main, Germany

Abstract. The academic project presented within this paper dealt with the development and fabrication of a customized acoustic structure combining various analogue and digital design methods. The aim was to improve the acoustic situation in an office space and at the same time to represent the clients working philosophy, which is rooted in parametric design itself. The methodical framework originates from the idea of design as the resolution of "wicked" problems in the sense of Rittel and Webber (1973). It is a bottom-up approach, conceptualized to allow for a broad exploration of the design subject. Different aspects of sound have informed the process—from the conceptual stage until the detailed geometrical formulation of the structure. We used artistic experiments, designerly interpretations of sound emitting patterns, sound engineering expertise, physical sound measurement, digital simulation of sound performance properties and parametric geometry modelling. Our paper focusses on two aspects. First, the methodology developed and used to create, integrate and transform knowledge gained from different acoustic analysis and interpretation techniques into the design process. Examples will highlight benefits and difficulties encountered. Among the latter, especially aspects of acoustic simulation will be discussed. Second, we demonstrate some key aspects of the final design: the acoustic concept and the strategies used for modelling and fabrication.

Keywords: Digital design methods · Acoustic ceiling · Sound · Digital simulation · Design driver · Digital process chain · Design to production

Introduction

The initial design brief asked to develop a novel typology of an acoustic ceiling, that expresses the underlying functional and methodical aspects. Particularly the use of computational design methods had to be clearly perceivable (see Fig. 1). Consequently,

K. De Rycke et al., *Humanizing Digital Reality*,
https://doi.org/10.1007/978-981-10-6611-5_17

the use of associative modelling techniques, which relate simulation data to geometry formulation, seemed to be a natural choice.

Simulating performance aspects is common practice in architectural design. The scope of performance aspects comprises structure, light, sound and other aspects of energy and matter. The moment of utilizing simulations varies from very early design stages—where simulation is used in order to explore the design space (Wortmann 2016)—to very late phases—where it is used to prove an aspired performance of an already elaborated design.

In particular, structural simulation is well established (Tessmann 2008) for both; qualitative synthesis and quantitative analysis of solution spaces. In contrast, acoustic simulation is mostly used in the latter case, for predicting and respectively fine-tuning the acoustic performance of spaces, where it is of paramount importance (van der Harten 2011).

In his analysis of the design process of the Melbourne Recital Hall, Holmes (2013) outlines some reasons contributing to this situation:

- the lack of reliable methods for simulating absorption and diffusion,
- modelling techniques that "are reactive rather than proactive making contribution to the design process through aural experience lag behind a fluid and rapidly changing design environment",

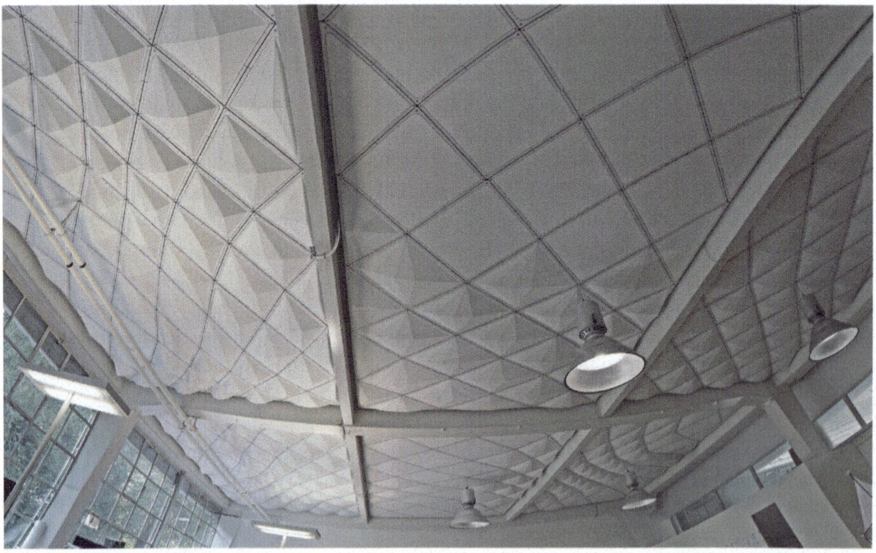

Fig. 1. Realized adapted acoustic ceiling composed out of self-similar elements—a typical property of digital process chains. Size and steepness of the modules express the proximity of sound emitting sources

- performance limitations of modelling software and simulation techniques that do not correspond well with actual hearing experience.

In short, the use of acoustic simulation as a design driver is limited due to questions of reliability, performance and suitability.

However, two acoustic parameters were identified to be valuable for different computational acoustic simulations. First, reverberation time, which is mainly dependent on material properties (absorption coefficient) and has the strongest impact on the subjective impression of acoustic quality. Secondly sound scattering is an important parameter determining a good acoustic quality, whereby a diffuse sound is regarded to be more comfortable. The scattering coefficient geometry dependent, with high frequencies being scattered through smaller geometries and low frequencies through larger geometries as described by Peters and Olesen (2010). Both parameters are represented in the concept of the realized design (see Fig. 2).

Other aspects, like subjective perception of sound phenomena, the individual notions of good acoustic quality and the aforementioned common difficulties in acoustic simulation, motivated our decision to favour a more open, patchwork-like methodical setting.

Fig. 2. Part of the realized acoustic ceiling: combinations of flat, only absorbing elements and three-dimensional elements, scattering and absorbing sound at the same time

Sound as Design Driver—Experimentation, Interpretation, Simulation

Aiming to improve the acoustic performance by a solution, aesthetically expressing its functional properties, the design process used various approaches to generate and embed knowledge from the field of sound: Artistic experiments helped to understand the perception and interpretation of sound as a spatial phenomenon expanding over time. Custom-made Grasshopper tools allowed for visualizing sound emittance to deepen this understanding. Two external consultants brought sound and acoustic engineering expertise into the project. Designerly interpretations of sound emitting patterns were implemented in different concepts. Using acoustic simulations for geometry formation as well as their benefit for evaluating variations was examined.

Digital Acoustic Simulation in Early Design Stages

Generally speaking, simulations with loose dependencies on material properties and strong dependencies on geometry are more suitable to be embedded into a design process for geometry formulation—in particular at early design stages.

One famous example from the realm of structural capacity is the inverted catenary curve, as used by Gaudi for the design of the Sagrada Familia. In this case, the geometry intrinsically represents its mechanical behaviour. "However, the interrelationship between them is inseparable; if a structures geometry is changed, its mechanical behaviour will also change." (Sasaki 2004).

This is different for acoustic simulation. Only if sound scattering is a design criterion, geometry gains a significant role (Burry et al. 2011). The Melbourne Recital Hall provides a descriptive example. The cladding of the walls and ceiling vary on three different scales according to the distribution of high, mid and low frequencies (Nichol and Orlowski 2009). Sound absorption, which is most crucial in order to reduce the reverberation time is mainly determined by material properties and not macro geometry. The absorption coefficient specifies the proportion of sound waves absorbed in relation to the proportion reflected.

Potential fuzziness of and dynamic elements (people, bookshelves, varying furniture configurations, unknown materials, etc.) in the space simulated impede the formalization of simulation parameters (see Fig. 3).

Furthermore, digital acoustic simulations are computationally heavy. They combine multiple simulation strategies (ray tracing plus image source tracing), running through several simulations which are interpolated for consistent results. This rhythm does not conform with the quick responses required in early, conceptual design stages.

In summary, we found three main obstacles for implementing an associative geometry model directly influenced by sound simulation data.

- First, the lack of a direct relation between reverberation time and geometrical shape.
- Secondly potential difficulties and flaws in formulating simulation parameters, due to a certain fuzziness of some elements of the space.
- And most obvious, there is a significant performance problem.

Our methodical patchwork to avoid these drawbacks is outlined in the following paragraph.

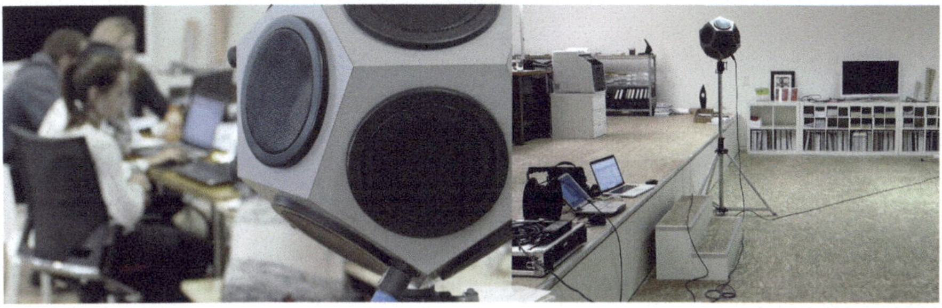

Fig. 3. Fuzzy, dynamic elements in the office space (*left*) and the setting of our measurement equipment

Patches of Experimentation and Designerly Interpretation of Sound Phenomena

Apart from linguistic, especially technically formalized information, serious design activity necessarily demands aesthetic methods of gaining knowledge. In this sense, experiments of the sound artist Wolfram Spyra helped us to access the dimension of sound as a spatial phenomenon expanding over time (see Fig. 4).

An important knowledge source for potential design strategies were two consultants, who brought engineering expertise into the project—from different, sometimes contradicting points of views, resulting in useful discussions. The pragmatic, acoustic engineer offered applicable rules of thumb for what portion of the ceiling should be used for absorption, whereas the sound engineer directly manoeuvred to the foundations of our aural perception—clearly indicating that the use of ceiling surfaces is the less favourable option.

Custom-made grasshopper tools for simple three-dimensional visualisations of sound emittance helped to deepen this understanding and to avoid being biased towards the common two-dimensional representation of frequencies. They are based on the ray characteristics of sound rather than their wave-like properties. The graphic output, depicted in Fig. 5, is driven by the geometric principle whereby the angle of reflection

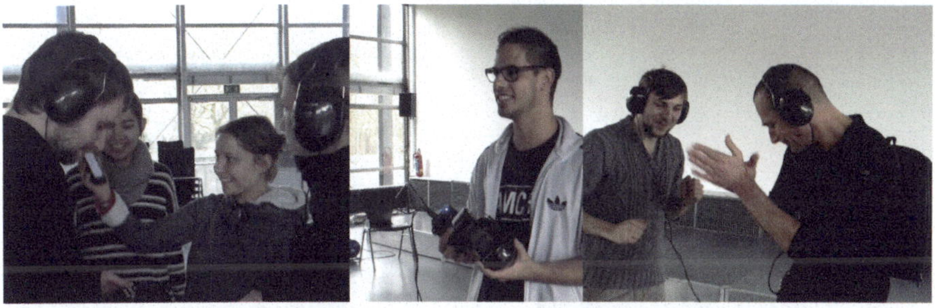

Fig. 4. Sound perception workshop about hearing sound events intermediated, via headphones, by a potentially manipulated record. Alienation effects comprise sound compression, channel changes, random variations of tone, pitch and rhythm, amongst others

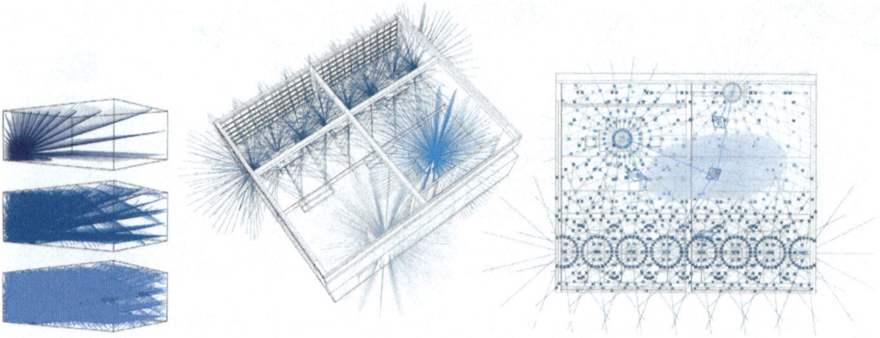

Fig. 5. Development from spatial parametric tools for the visualisation of sound emittance (*left*) to diagrams and design concepts (*centre*/*right*)

equals the angle of incidence. Arranging the digital emitters to their equivalent actual sources of noise reveals the density of primary, secondary and further. reflections onto walls and the ceiling.

Finally, these and some other approaches to understand sound phenomena resulted in various designerly interpretations of sound emitting patterns and were implemented in different designs, which exemplified and interlaced their functional concepts by formal, aesthetic means, as described in Sect. "Design Concept".

Digital Acoustic Simulation for Evaluating Variations and Validating the Design Intervention

The chosen application was Pachyderm, an open source tool for acoustic analysis and simulation (van der Harten 2013). The setup for the digital acoustic simulation (see Fig. 6 left) is analogue to the actual conducted standardized acoustic measurement, whereby six receiver positions are simulated with two speaker positions. The twelve simulated and measured results are interpolated respectively.

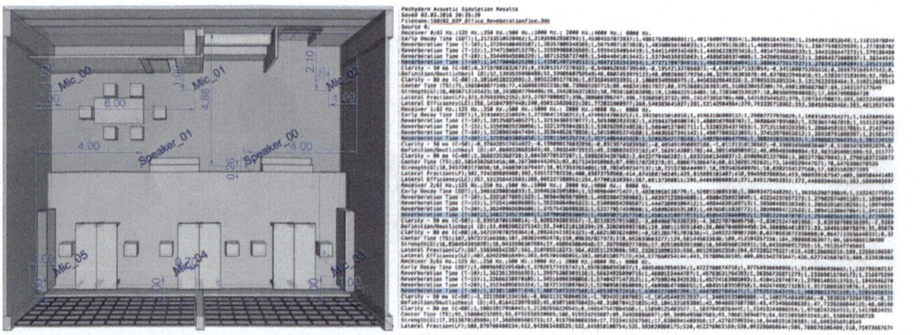

Fig. 6. Scheme of the setup for both, the actual acoustic measurement and the digital simulation, simulation results

For comparison and potentially predictions of the intervention's performance (a) the initial situation (b) a flat non-adapted version and (c) an adapted spatially unfolding version where simulated.

Since no reliable measurement of an altered spatial situation was available, an unobjectionable benchmarking of the digital simulation was not possible. However, the simulation of the original situation came rather close to the outcome of the actual measurement, showing only a slightly higher reverberation time.

The required absorption coefficients of the polyethylene acoustic felt we were using for our project was employed in the simulation, as provided by the manufacturer. They were given for the absorbing material installed with a gap of 30 mm against its background.

The simulated negligible improvement for low frequencies coincides with the small absorption coefficients for low frequencies (1% for 62.5 Hz/4% for 125 Hz/15% for 250 Hz). However, the simulated performance of the adapted acoustic ceiling deviates significantly from the actual measured results. Especially for lower frequencies the realized ceiling outperforms the simulated one. This is most probably due to the varying distance between the concrete and the acoustic felt, not covered in the technical specifications. This shows that meaningful predictions can only be made when absorption coefficients are known for the specific installation situation.

Although the simulation may lack precision in order to make reliable quantitative predictions, it can still be valuable for comparing variations qualitatively, without the necessity to build them. This is because inaccurate absorption coefficients apply for all variations containing the polyethylene felt respectively.

Comparing the simulation of two ceilings with the same area of absorption material (Table 1) reveals that the adapted ceiling outperforms the flat ceiling—particularly for lower frequencies and with a more evenly distributed improvement.

Table 1. Simulated reverberation time (T30) for the given office space

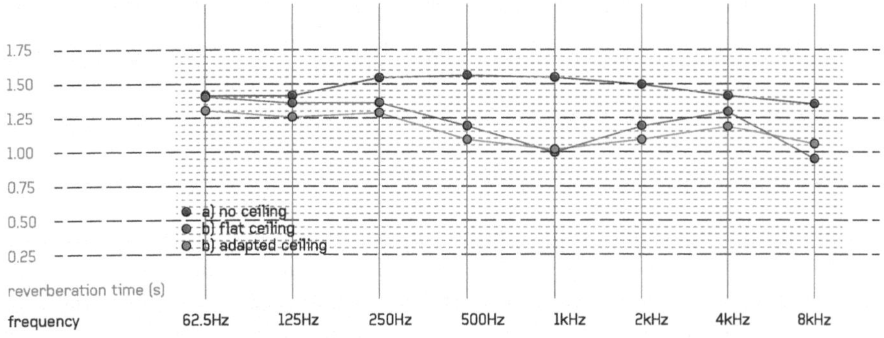

Parametric Acoustic Ceiling—Final Design and Realization

Project Context

The design location, a box-shaped office room in an old factory building has dimensions of 12/10/4 m (l/w/h). The walls, ceiling and floor are made of concrete, except for one longitudinal side which consists mainly of single layer glass in a metal frame. Structural boards (OSB) cover the floor. The interior contains furniture with nearly no absorbing material and some bookshelves.

The audibility was clearly perceivable as being poor, impeding and hindering communication. This was confirmed by a measurement of the reverberation time of 1.31 s in average. General recommendations for a good audibility and comfort in equivalent spaces range from 0.5 to 0.8 s.

Apart from improving the acoustic situation, the project brief asked to do this with a spatial structure representing the design philosophy of the client who is specialized in parametric planning and digital fabrication of non-standard architectural projects. The ambition was to build a visually appealing bespoke structure using simple means to skilfully answer a complex task.

Design Concept

The office room contains clearly located and basically immovable sources of noise: Working and meeting tables, a small kitchen area, a church in the direct neighbourhood. Even though desirable, the use of walls, floors and especially corners turned out to be impractical.

The geometrical relations are dependent on the proximity of the sound absorbing elements to the noise emitters (see Fig. 7 left). The closer they are, the smaller the units become and simultaneously they fold more and more into the space (see Fig. 7 right), generating a larger surface for sound absorption and constituting angled polygonal surfaces for sound scattering. In this way, the three-dimensional, differentiated ceiling combines absorption and diffusion and makes the acoustic situation visually perceivable.

Structural bases of the design are slender frames fabricated from white birch plywood. Lanisor™, a sound absorbing polyethylene felt, is the infill of these frames. In line with the existent subdivision of the ceiling, the overall structure is divided into six fields. The noise informed gradient is globally applied. The modules are rhombus-shaped within the fields and triangular at the edges.

Modelling and Fabrication Strategies

The acoustic ceiling should not only fit for the given situation, but it should be applicable to different rooms and environments. Consequently, an associative model for adjusting the design and adapting it to specific spatial and acoustic situations as well as for generating all 3D-representations and fabrication data was developed. The structure shows all four possible combinations of serialized/individualized and flat/folded parts.

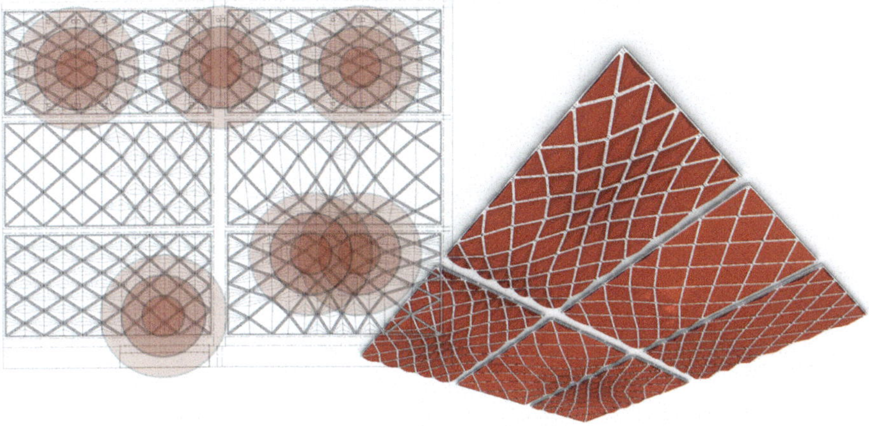

Fig. 7. Location of noise sources and distribution of modules, modules folding into the space

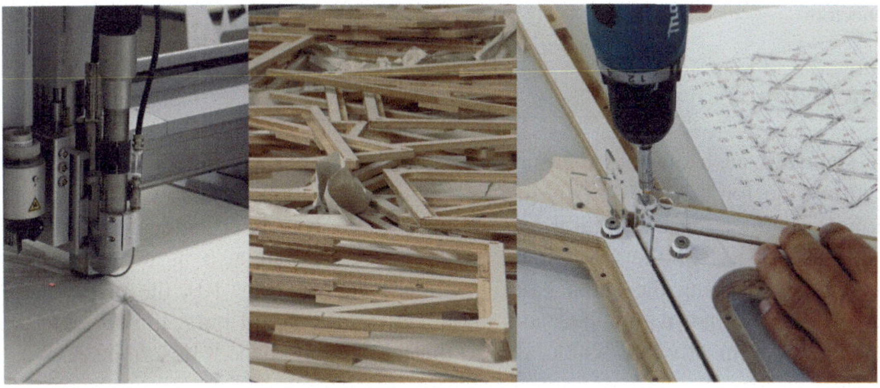

Fig. 8. Knife cutting of Lanisor©, assembly of cnc-milled frames

For the fabrication of physical models, prototypes and final parts, the project used 3-axis-milling, laser and knife cutting in 2D- and 2½ D-applications (see Fig. 8). All parts of the structure are made from materials available as standardized sheets and were designed for being processed from a single side. The infilling white polyethylene felt has a thickness of 10 mm. The white coated and natural birch-plywood for joints is 15 mm thick.

The result has been 348 modules composed of 1272 wooden frame parts, the same number of connectors and 609 pieces of the sound absorbing material Lanisor™. 354 wooden joints interconnect the frames and are the interface to the conventional building structure (see Fig. 9).

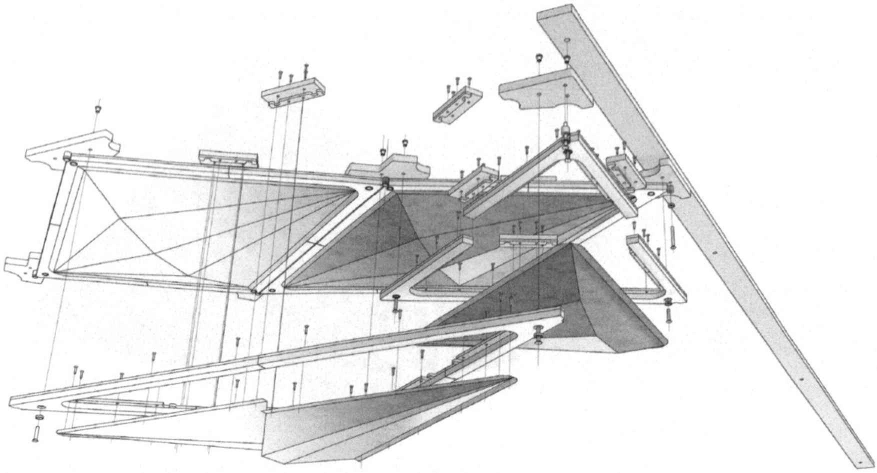

Fig. 9. Corner configuration of the final design. Spatially unfolding acoustic polyethylene felt is framed by birch plywood with white resin cover and joined by different birch plywood connectors

Conclusion

The standardized acoustical measurement revealed a drastic reduction of the reverberation time for the relevant frequency ranges (125–4000 Hz), now predominantly below 0.7 s after the installation as shown in Table 2. This coincides with the subjective impression of the content users. In this sense, but also as an eye-catching structure representing the client's design philosophy (see Fig. 10), the project was a success. Despite the overall size and amount of non-standard parts the structure was also still economically reasonable, due to the fabrication concept described above.

Table 2 Reverberation time (T30) measured (*a*) before and (*b*) after installing the ceiling

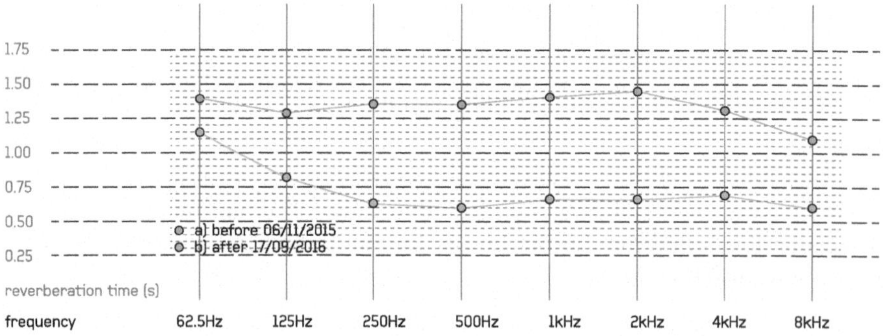

Fig. 10. Realized adapted acoustic ceiling

As a case study for the implementation of acoustic simulation into a design process chain, it provided valuable insights. Concerning reverberation time, reliable quantitative results for early design stages could not be obtained with a feasible effort. Also, the duration of the simulation would have been too long to implement it into iterative procedures. The lack of clear dependencies between simulated acoustic performance and macro-geometry was an obstacle as well.

However, these limitations led to new designerly interpretations of sound phenomena and a novel method of how the formulation of geometry can be influenced by acoustic parameters. The examination of locatable sources of noise and ray-like sound distribution turned out to be particularly successful. Digital acoustic simulations proved to be beneficial for comparing variations in a later design stage but lack precision for accurate performance predictions when installing even established sound absorbing material in a novel way.

Acknowledgements. The project was a collaboration between Design-to-Production (Zurich) with the Lab for Digital 3D Techniques at the School of Art Kassel and the Department for Structural Design at the School of Architecture at the University of Kassel.

Supervisors: Dr. Markus Schein (School of Art) Prof. Manfred Grohmann and Moritz Rumpf (School of Architecture).

Participating students: Luis Miguel Garcia Ballesteros, Joost Fähser, Niclas Garotti, Hannah Hartmann, Robert Kost, Jule Leinpinsel, Andreas Merkel, Mallika Pholchop, Robert Redel, Oliver Schaub, Katharina Schmelting, Michael Schreiner, XiaoYue Su, Oliver Waldsachs and Cynthia Ward.

References

Burry, J., Davis, D., Peters, B., Ayres, P., Klein, J., de Leon, A., Burry, M.: Modelling hyperboloid sound scattering—the challenge of simulating, fabricating and measuring. In: Gengnagel, C., Kilian, A., Palz, N., Scheurer, F. (eds.) Computational design modelling, pp. 89–96. Berlin, Heidelberg (2011)

Holmes, P.: Sonosentio—research into a new sonic design process for composing responsive spatial geometries. Melbourne (2013)

Nichol, A., Orlowski, R.: Melbourne recital centre (2009). http://www.abc.net.au/catalyst/stories/2488923.htm

Peters, B., Olesen, O.: Integrating sound scattering measurements in the design of complex architectural surfaces: informing a parametric design strategy with acoustic measurements from rapid prototype scale models. In: Schmitt, G., Hovestadt, L., Luc van Gool, L., (eds.) Future Cities—Proceedings of the 28th eCAADe Conference, Zurich, pp. 481–491 (2010)

Rittel, H., Webber, M.: Dilemmas in a general theory of planning. In: Working Papers from the Urban and Regional Development, University of California, Berkley, p. 160 (1973)

Sasaki, M.: Shape design of free curved surface shells. In: a + u—Architecture and Urbanism, Nr. 404, p. 36 (2004)

Tessmann, O.: Collaborative Design Procedures for Architects and Engineers. Norderstedt (2008)

van der Harten, A., Gulsrud, T., Kirkegaard, L., Kiel, A.: A multi-faceted study of sound diffusing elements in an auditorium. J. Acoust. Soc. Am. Seattle, Washington (2011)

van der Harten, A.: Pachyderm acoustical simulation—towards open-source sound analysis. In: De Kestelier, X., Peters, B. (eds.) Computation Works—The Building of Algorithmic Thought, London, pp. 138–139 (2013)

Wortmann, T.: Surveying design spaces with performance maps—a multivariate visualization method for parametric design and architectural design optimization. In: Herneoja, A. et al. (eds.) Complexity & Simplicity—Proceedings of the 34th eCAADe Conference, Oulu, pp. 239–248 (2016)

La Seine Musicale

A Case Study on Design for Manufacture and Assembly

Sylvain Usai(✉) and Hanno Stehling

Design-to-Production GmbH, 8703 Erlenbach, Zurich, Switzerland
{usai, stehling}@designtoproduction.ch

Abstract. *La Seine Musicale* is the latest project by architects Shigeru Ban and Jean de Gastines. Opened to the public in April 2017, this cluster of rehearsal and representation spaces sits on the Île Seguin a few kilometers to the west from the center of Paris, in place of the former Renault industrial plants. On behalf of the timber contractor Hess Timber along with the structural engineer SJB Kempter Fitze, Design-to-Production was responsible for the *digital planning* of the entire *timber structure* covering the main auditorium at the very tip of the island. Shigeru Ban's signature *hexagram pattern* covering the egg-shaped auditorium consists of 15 horizontal ring beams interconnected by 84 Diagonals (42 in clockwise and 42 in counterclockwise direction). The entire structure fits into a bounding box of 70 m length, 45 m width and 27 m height. The beams have an average cross section of 320 × 350 mm and are subdivided into 1300 individual segments. All beams sit on the same layer and interpenetrate at crossings. The façade on top of the beam structure is composed of triangular and (almost) planar hexagonal glass panels. The interface between the curved timber structure and the straight glass edges is achieved by another 3300 CNC-fabricated timber parts. The focus of this paper is on the principle of *design for manufacture and assembly*, the solutions specifically developed for this project and the way workshop pre-assembly and on-site constraints have been integrated into the parametric modelling process of a complex timber structure such as the Seine Musicale auditorium.

Keywords: Timber grid shell · DFMA · Design for manufacture and assembly · Digital fabrication

Input

For the Seine Musicale project, the defining geometric input consists of a reference NURBS surface for the outer beam level and a set of 99 reference curves for the beam axes, provided by the architect and the project engineer in a 3D-CAD model. Since all subsequent planning steps are depending on this *reference geometry*, it is critical to ensure its quality right at this early stage, in order to avoid later problems (Scheurer et al. 2013): Are the surface and axes continuous? Are the curves lying within the surface and correctly intersecting? When working with an axis network on a surface, it is also worthwhile to check for the location of the surface's seam since it can lead to discontinuities in the pulled curves (see Fig. 1).

K. De Rycke et al., *Humanizing Digital Reality*,
https://doi.org/10.1007/978-981-10-6611-5_18

Fig. 1. The timber structure under construction

Once checked, the objects are imported, sorted and named, following an unambiguous and comprehensive naming system that fits the needs of all involved parties. In this project, ring beams are numbered starting from 100 at the bottom, counterclockwise diagonals from 200 and clockwise diagonals from 300. Crossing joints are named after their intersecting beams (105×212 sits between the fifth ring and the 12th counterclockwise diagonal), longitudinal joints are counted along the beam (105#03 is the third longitudinal joint on the fifth ring). Ideally, the name should provide information about the position of the final piece while remaining short enough to be handled.

Preliminary Modeling

Due to different structural requirements between the lower and the upper part of the building, the beams' cross section height gradually decreases from 400 mm at the foot to 300 mm at the ridge while the width remains constant at 320 mm. In the model, this is reflected by four additional reference surfaces with variable offsets from the base master surface: top, upper, middle, lower and bottom. The intermediate surfaces between top and bottom are later used in the detailing process to define the lap joint levels.

The beam edges are created using perpendicular frames on the axis and intersecting these with the reference surfaces. The same perpendicular frame division is used to loft the beam volume. This construction ensures that all the beam faces are ruled surfaces, which can be *easily milled with a cylindrical tool*, and that *all edges intersect* properly at crossings, which is a critical condition for precise detailing and execution.

Preliminary joints are defined at the axis intersections and sorted into types (crossing Diagonal/Ring, Diagonal/Diagonal, Diagonal/Foot, Diagonal/Ridge…). During the modelling process, joints are to be seen as *abstract containers*. At this point, the joints only contain simple construction geometry (local beam tangents and surface axes) or analysis (angle, attributes) but will later be filled with fastener objects (screws, bolts, steel plates…) and fabrication operations (drillings, cutouts, etc.) represented as simple geometry. Joints are named after the beams they are connecting, so the link can be retrieved anytime in both directions. Additionally, since a joint connects multiple beam segments, every fastener object and fabrication operation in the joint keeps the name of the segment it affects as an attribute. The modelling and detailing process is about *creating and maintaining relations as much as geometry*. When exporting documentation or fabrication data for a beam segment, the segment collects the relevant objects and operations of all the joints it is linked to and builds the requested export data set. This abstract modelling process is lighter and allows for more flexibility and process stability than working directly with solid cut elements.

From these preliminary connections, a structural polyline model of the beam axes is created, which is used by the engineer to perform early structural calculation and prepare the detailing work. At this point, the geometry can be analyzed to *identify the critical connections*, which would be solved in priority. Typically, the boundary joints are the most difficult ones to deal with because they combine heavy load bearing and extremely diverse geometrical configuration.

Beam Segmentation

Up to this stage, the beams are continuous entities spanning all around or from bottom to top of the structure. Due to fabrication and logistic constraints they obviously have to be segmented into shorter pieces. In this project, the segmentation strategy is different for the horizontal ring beams and for the diagonal beams.

The ring beam segmentation is defined by the *transportation constraints*: anything longer than 18 meters requires a costly special convoy (the pieces were shipped from around Frankfurt to Paris by truck). The goal is then to find a harmonious division within this 18-m range while avoiding vertical alignment between consecutive levels. In the structurally most critical regions of the structure, segment lengths of up to 24 m are employed, accepting the logistical effort in order to keep cross sections and detail complexity in an acceptable range.

The diagonal segmentation strategy is directly defined by the assembly concept: the building is designed to be *self-supporting* from the beginning of the erection process. Levels are added horizontally one after the other with few punctual supports saving the space and cost of a scaffolding (although, ironically, one was then mounted by the facade team on top of the timber to set up the glass panels).

Thus, diagonals are always spanning between two consecutive rings. The crossing between ring and diagonal also acts as a longitudinal joint between diagonal segments: the ring has one lap joint on the inside for the diagonal below and one on the outside for the diagonal above, while the middle layer is running through. The diagonals overlap each other and fully enclose the ring. They are pre-assembled into cross elements before being shipped to the construction site (see Fig. 2).

Interestingly, this intricate connection is subdivided into three different joints for the engineer's model (upper longitudinal diagonal connection, middle ring/diagonal connection and lower diagonal longitudinal connection) while only one joint with three different detail types is used on the production model.

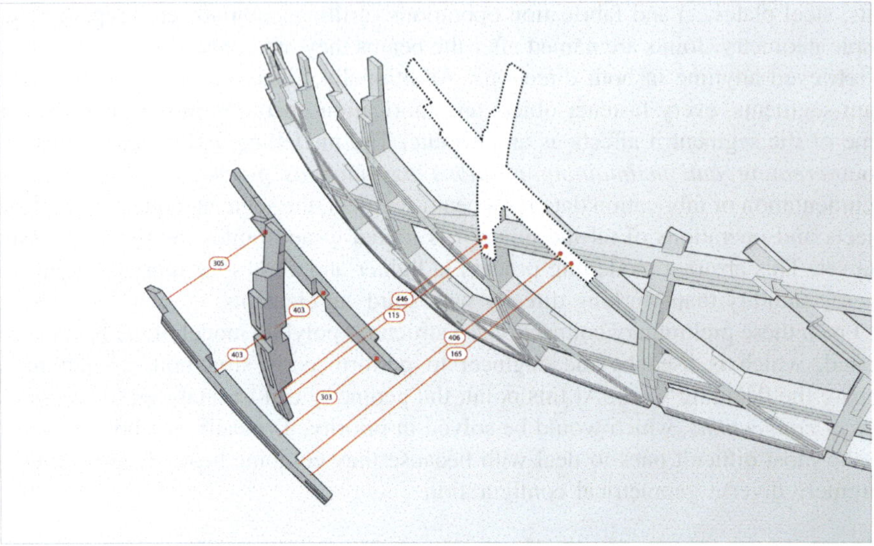

Fig. 2. Diagonal pre-assembly and combined cross- and longitudinal joint at the ring/diagonal crossings

Blanks

Once the beam segmentation is set, *blanks* are created. The term "blank" refers to the raw glue-laminated timber piece, which is milled down to the final shape of the segment. In free form projects, single curved or doubly curved blanks approximate the final piece's curvature. They are produced by bending and gluing raw lamellas on a set of pre-shaped supports.

On doubly curved blanks, there is no planar side, making it difficult to achieve *precise placement* of the blank on the CNC mill.

In this project, *positioning points* are created according to the positions of the gluing supports and the later positions of connection details in order to be *easily located* in the real world and to *avoid collisions* between the tool and the clamping equipment during milling. These points are stored in the model and passed along with the blank fabrication data to the fabricator to be precisely marked on the blank while still on the gluing supports. Later on the CNC mill, the same points can be referenced, ensuring precise positioning of the raw material on the CNC bed.

As the diagonal/diagonal crossing is a lap joint as well (see Fig. 2), one of the diagonals (the counterclockwise one) must be split in two halves along the beam direction. This differentiation is introduced during blank creation: during initial modelling of the piece and gluing of the blank, both diagonals are considered as integral elements, the only difference being the insertion of a 'dry' lamella during the gluing in the middle of the split counterclockwise diagonal (see Fig. 3). On the 'dry' lamella, glue is only applied at three distinct positions, namely both ends and the central lap

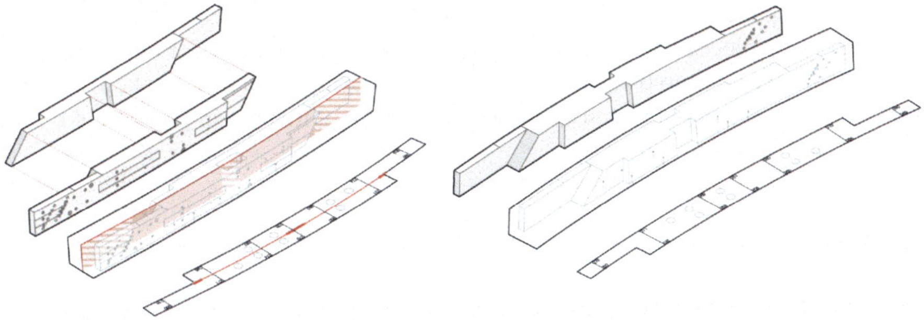

Fig. 3. Split diagonal (*left*) with blue dry lamella and its full counterpart (*right*)

joint area. When the piece is placed on the CNC mill, it splits in two as soon as the glued areas are cut away with the connection details at the end of the milling process.

This way, *only one set of fabrication data* (gluing and milling) must be produced per diagonal, sparing one third of the total effort on diagonals. It also ensures that the wood fibers on the side faces of the upper and lower part are visually continuous since they come from the same blank.

For the modelling of the blank geometry, the only implication is that the blank middle axis must strictly follow the piece's middle axis and that the lamella count in height must be an even number.

Detailing

Despite both the design surface and the beam grid being axis-symmetric, every crossing point between two beams is different, because clockwise and counter-clockwise diagonals need different detailing. These 2798 unique joints are boiled down into distinct types, in this case 8 main types and 120 sub types with minor differences. 2D detail drawings are provided by the engineer and translated into parametric systems. A spreadsheet based interface is set up with the engineer to assign the correct type to each joint and make sure that structural and production model stay synchronous. The details are then instantiated on each joint, resulting in 3D geometry featuring both fastener objects and fabrication operations.

It is very important to resist the impulse to start with the simplest situations (because they cover 90% of the project) and deal with the special cases afterward, as this leads to a huge amount of piling up detail types that need to be individually implemented and checked. Instead, the most complex situations should be solved first, leading to a parametric definition that can also cover the simpler cases. Ideally, the parametric implementation of the detail is flexible enough to allow for some variation, allowing to react to possible changes during the planning and execution process: due to the parametric nature of the detailing work, it is not such a problem to evolve an existing system if the evolution lies within the predefined solution space. In other words, it is easy to change a screw length or its distance to the border but it is harder to transform it into a steel plate…

All cutter operations are placed in the model at the reference level (i.e. without margin between pieces at crossings). The tolerances necessary for fabrication and assembly are instead stored as attributes and can be individually tweaked following feedback from workshop and construction site. The margin value in projects like this is usually around half a millimeter.

Assembly

Along with the detailed joints, every segment stores its assembly number and theoretical assembly direction. This direction verifies that there is at least one way to bring the piece into place without any collision with the already assembled surrounding.

Since a regular lap joint can only be engaged from "above" in respect to its own orientation and one ring segment is connecting to up to 11 diagonals, every single joint's assembly direction cannot be satisfied. As soon as the joint normal deviates from the segment's assembly direction, the lap joint needs to be skewed to allow assembly. This skewing can be hidden in the joint if material is taken away from the diagonal or shown as visible gaps if cut from the ring. In both cases, material is removed and the beam is weakened. So, the objective is to find the assembly direction that leads to the least possible skewing. Due to the almost spherical nature of the surface, a rotational assembly gives the best results here (see Fig. 4), because the assembly direction follows the curvature more closely along the crossings than with any lateral movement.

On the construction site, the ring segment's start is brought next to the previous piece's end and rotated into place around the longitudinal joint from the outside. As the

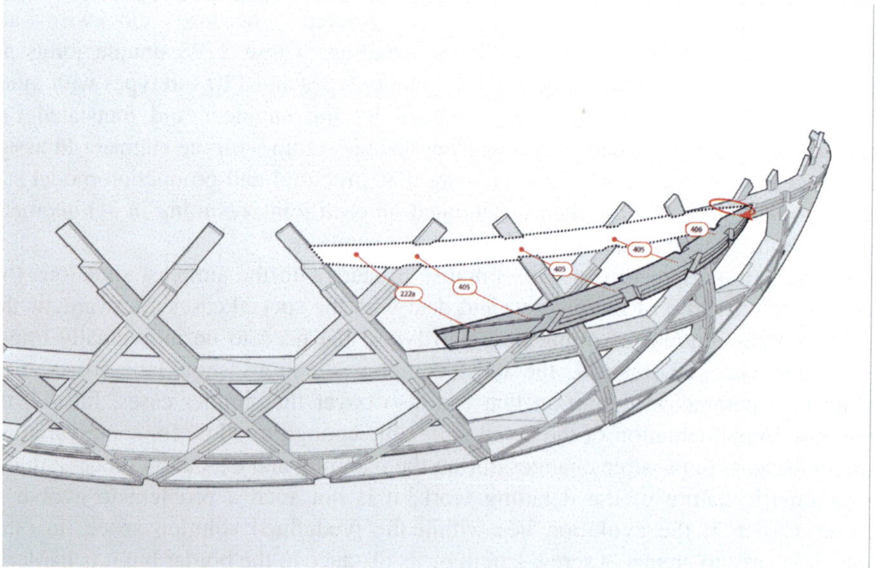

Fig. 4. Rotational assembly of a ring segment spanning over multiple lap joints

piece rotates, the lap joints of the ring successively make contact with the diagonals allowing the assembly team to engage them one after the other. Only the last segment of each ring cannot be rotated as described, as it has to connect to already assembled ring segments on both ends. It is shaped as a 'keystone' with scarf joints on both sides and is translated into place using the average direction of all its joints. To minimize possible assembly problems, the assembly order is set up so that the last segment has least curvature, as in the flat areas on the sides of the egg.

Fabrication and Pre-assembly

Before the export of fabrication data, a detailed volume of the segment is created for approval and to check the assembly direction for collisions. Once this test is passed, the fabrication operations represented in the model as lines, polylines, curves etc. are translated into BTL format as described by Stehling et al. (2014): the geometry is converted into corresponding machining operations (drillings, slots, pockets, contours…). These BTL files also contain information about the blank geometry and the previously mentioned positioning points for placement on the CNC bed.

This information is not machine specific and still needs to be processed by the timber contractor to be transformed into actual machine code. This distinction allows to optimize machining strategies during the ongoing project without having to alter the process of fabrication data generation and ensures a clear separation of responsibilities in the project.

Along with fabrication data, a workshop drawing is exported for each segment. These are automatically generated following the same parametric logic as the detailing and feature control measurements on the final geometry in plan, elevation and axonometry as well as lists and position drawings of all fasteners to be pre-assembled.

A flat plan of every level is also generated for the on-site assembly team. These are abstract representations showing the names, assembly order and detail types of the 2 rings and 84 diagonals in one level of the structure transposed into a 2D template drawing (see Fig. 5).

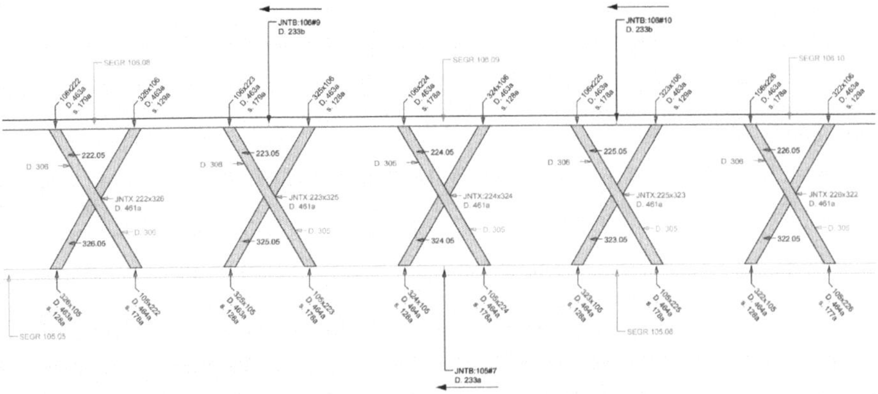

Fig. 5. Extract of a generated montage plan

Future Challenges

The digital planning and fabrication process of *La Seine Musicale* has been largely smooth. However, of course there are process steps with potential for improvement. Looking back on the project, three main challenges can be identified:

First, the definition of parametric connection details on a conceptual level. This is done by issuing a 2D drawing of one exemplary geometric configuration. While this works fine in most cases, it becomes problematic whenever exceptions such as collisions between fasteners of adjacent joints occur. The resolution of such exceptions can be an extensively manual process including a feedback loop with the structural engineer. This could be improved by *further abstraction* of the detail definition in order to react to exceptions *within the initial solution space*. For example, instead of precise positions for fasteners, a weighted range of possibilities could be issued (e.g. "place screws along this line, total force n kN"). The implementation would still try to use a predefined "preferred" position but could autonomously move or replace fasteners to a certain extent if collisions occur.

Second, the verification and approval process of parametric model and fabrication data. While the implementation of parametric details can be verified by code review and spot tests, the verification of the resulting detail geometry is still largely done by *looking at the model*. Manual checking will (and should) never be completely replaced, but an automated quality control, *implemented independently* from the logic that creates the geometry in the first place and possibly involving a feedback loop into the structural model, could help finding issues earlier and easier.

And third, the process management for individual building components between all involved companies. While most are mapping the components to their ERP (*enterprise resource planning*) system or running dedicated databases, the common interface boils down to an Excel sheet. A common platform reflecting development and approval status of all components could greatly improve productivity and process safety.

Conclusion

The concept of *design for manufacture and assembly* has already proven to dramatically increase productivity since the 1980s, in the production industry (Andreasen et al. 1988). Unfortunately, it is not yet fully acknowledged in the building industry, partly because the concept of pre-fabrication, which would be a prerequisite, has also been widely neglected.

Design-To-Production has been successful in applying these principles to a number of freeform-timber projects, like the Centre Pompidou Metz and the Nine Bridges Golf Club (also designed by Shigeru Ban), the French Pavilion for the EXPO 2015 (designed by X-Tu Architectes) and now La Seine Musicale, shown in above case study. The complex geometry of these designs cuts short any discussion on whether or not to exploit the possibilities of digital planning and pre-fabrication, including the necessary changes to the collaborative processes between designer, engineers, planners, and fabricators. Those "lighthouse projects" would simply be technically and economically impossible otherwise.

Having now gained substantial experience from these "non-standard" projects, the authors are convinced that the principle of digital pre-fabrication could and should be applied more often in the AEC industry also for "standard" projects. These methods enable to shift complexity from the construction site to the planning, pre-fabrication and pre-assembly phase and allow a much more efficient and reliable process on site, resulting in a better overall productivity. Fragmentation and inertia of the industry have been prohibitive to such a re-organization of the planning- and building processes, but this is about to change for three reasons: Internally, the accelerating use of 3D Building Information Modeling (BIM) is accompanied with a re-allocation of focus and budget towards the planning stage. Externally, the demographic development in the western countries will be requiring a shift from on-site to off-site work due to the shrinking work-force (Farmer 2016). And lastly, due to environmental concerns and lighthouse projects like La Seine Musicale, building with timber has gained a tremendous momentum over the recent years, also for large-scale housing and high-rise projects. This increasing demand is now meeting with a well-prepared timber pre-fabrication industry. To unfold the true potential of this combination, *design for manufacture and assembly* will finally become a standard term in the building industry.

References

Andreasen, M., Kahler, S., Lund, T., Swift, K .: Design for Assembly. Springer, New York (1988)

Farmer, M.: The farmer review of the UK construction labour model. In: Construction Leadership Council (CLC), UK (2016)

Scheurer, F., Stehling, H., Tschümperlin, F., Antemann, M.: Design for assembly—digital prefabrication of complex timber structures. In: Beyond the Limits of Man, Proceedings of the IASS 2013 Symposium, Wroclaw (2013)

Stehling, H., Scheurer, F., Roulier, J.: Bridging the gap from CAD to CAM. In: FABRICATE—Proceedings of the International Conference, Zürich, gta Verlag (2014)

The Design Implications of Form-Finding with Dynamic Topologies

Seiichi Suzuki(✉) and Jan Knippers

Institute of Building Structures and Structural Design (ITKE), University of Stuttgart, Keplerstrasse 11, 70174, Stuttgart, Germany
s.suzuki@itke.uni-stuttgart.de

Abstract. During early design stages of lightweight structures, enhanced user-model interactions are desired for improving shape explorability of form-finding processes. For this purpose, the common approach has been to variegate form-found geometries through the dynamic calibration of metric parameters controlling proximity and material relationship without altering the topologic model. Although this condition simplifies the problem, it also constraints the exploration of more desirable solutions. This limitation could be tackled by introducing topologic differentiation during form-finding. Therefore, form-finding is set to be topology-driven when such characteristic is activated. Unfortunately, the implementation of dynamic topologies is a complicated task requiring the complete restructuration of the entire workflow. In this paper, we present a framework for the development of a topology-driven approach for form-finding based on four basic building blocks categorised as data-structure design, discretisation, interactivity and decision-making. The study presented is placed within the context of particle-based methods and bending-active tensile hybrid structures given the large design space for shape exploration that these structures require.

Keywords: Form-finding · Dynamic topologies · Particle-based methods

Introduction

The computational form-finding of bending-active tensile hybrid (BATH) structures is challenging architectural design practices. Since the exact geometry of these lightweight structures can't be easily approximated through standard geometric modelling techniques, more interactive numerical processes of form-finding are required. The common denominator has been the exclusive use of metric parameters controlling proximity and material relationship for variegating solutions. Because of this, such design approaches are suggested to be categorised as geometric-driven. From a topological viewpoint, this means that the input geometry is set to be the same as the output geometry. While the latest simplifies the problem and facilitates convergence and stability of numerical solvers, it also highly constraints design exploration.

A natural solution to this problem is to specify a structure of navigable metric and non-metric design spaces for driving the entire form-finding process. Therefore, form-finding is set to be topology-driven when both design spaces are activated. The

K. De Rycke et al., *Humanizing Digital Reality*,
https://doi.org/10.1007/978-981-10-6611-5_19

impact of topology-driven approaches is that output geometries can be easily differentiated from input geometries by dynamically altering the connectivity of the system during form-finding. In other words, a radical increase of flexibility and design space freedom. Unfortunately, the implementation of dynamic topologies is a complicated task requiring a complete restructuration of the form-finding workflow.

The research presented in this paper is situated within a larger body of research in the field of interactive form-finding. At this stage, our study is only focused on BATH structures that are shaped from the combination of textile membranes with linear elastic rods. Therefore, the presented study intends to establish a general framework for the development of a topology-driven approach for BATH structures. All methods presented in this paper are implemented within a force-based schema developed with the Java programming language.

Background

Physically-Based Modelling Techniques for Computational Form-Finding

During the conceptual design stage, computational form-finding needs to support enhanced user-model interactions through lighter and more flexible simulations. Particle-spring systems, or in a more generalized form particle-based methods (PBM), are widely used for this purpose. Initially introduced by Reeves (1983) in the field of computational graphics, these particle models are easier to implement and allow fastest computations since numerical integration is carried locally for each particle (Debunne et al. 2001). Kilian and Ochsendorf (2005) proposed the use of PBM for form-finding funicular arcs and vaults. Since then, several efforts have been conducted for extending PBM logics within digital design workflows (Senatore and Piker 2015; Piker 2013; Harding and Shepherd 2011; Kuijvenhoven and Hoogenboom 2012; Attar et al. 2009). Compared with the dynamic relaxation (DR) method, PBM support different dynamics and allow more sophisticated implicit integrations. For example, PBM can be implemented as a conditionally stable force-based scheme like DR where internal and external forces are accumulated in accelerations for updating velocities and positions. Yet, it can also be implemented as an unconditional stable position-based scheme where positions are directly updated by omitting acceleration and velocities. In such cases, new positions are computed by iteratively solving a set of weighted geometric constraints (Müller et al. 2007; Bender et al. 2014). Current efforts are focused on extending position-based schemes by introducing constraints derived from energy potentials for improving the mechanical accuracy of results (Bouaziz et al. 2012, 2014).

Related Work

Major studies for addressing real-time topologic changes within digital models have been conducted in the field of surgical simulators (Misra et al. 2008; Wu et al. 2015). Due to major advances, these types of simulations have rapidly expanded to more immersive virtual spaces where users have the capability to freely interact with digital models while producing real-time topologic and geometric responses (Dunkin et al.

2007). From a design perspective, however, simulating real-time topologic alterations during form-finding has not been fully explored. Few efforts are reported on this subject mainly because there is still some debate regarding the loss of computational performance when increasing the flexibility of simulations. Ahlquist et al. (2015) presented a study on tensile form-active models where topological spring-meshes are altered through active modelling, or via evolutionary algorithms. This study uses an open source PBM developed by Greenwold (2017) that implements a force-based scheme. Here, the entire database is stored through index assignation which requires a delicately update when one handles topologic changes. Another study proposed the development of a computational pipeline based on Kangaroo2 (K2) (2017) to support topological changes during form-finding (Quinn et al. 2016; Deleuran et al. 2016). In this case, topologic modelling is supported through an auxiliary graph data structure which is then used to continuously update the entire data structure of the K2 solver. Like Greenwold's library, K2 was originally designed to address common deformable body problems based on static or quasi-static topologic models.

Topology-Driven Form-Finding of BATH Structures

The main problem for addressing form-finding with dynamic topologies is the necessity to develop a methodological approach for the activation of metric and non-metric design spaces. Current form-finding workflows are linearly organized within multiple and independent stages—modeling, discretization, initialization and simulation. Recurrent operations for adding, deleting or modifying topologic models are then treated separately and require breaking the simulation. In contrast, the topology-driven approach for form-finding BATH structures presented here may be categorised as a circular and recursive process with highly interconnected stages categorised as data-structure design, discretisation, interactivity and decision-making. The proposed approach integrates two re-design stages—data-structure design and discretization—for facilitating the management and generation of recursive form-finding data. Moreover, the two remaining stages may be categorised as new within form-finding workflows since they directly derived from the use of dynamic topologies. These two last stages can be activated on the fly when the size of the model increases, and are focused on managing the rate of interactivity/accuracy of simulations and the taking of modelling decisions in complex design spaces. In the following, the four main stages are presented.

Data-Structure Design

The design of an appropriate network model is one of the key problems to address when form-finding with dynamic topologies. In current approaches, a mesh-topology is required to generate the network of particles and forces—force-topology. Even if both networks are created, it is only the latter that is passed to the numerical solver. Therefore, changes in the configuration of forces inevitably require returning to the referenced mesh-topology and re-initialize the entire force-topology network.

For dynamizing topologic models during form-finding, an evolving network model with multiple dimensions is proposed. Despite that the numerical solver needs to be slightly changed, this model enables multiple and mutable connections between a pair of nodes and several types of relationships within its dimensions. Each node of the network represents a particle, and each network space delineates a dimension. In this way, the model is composed of two dimensions being the mesh-topology network and the force-topology network (Fig. 1a). The force-topology network can be iteratively redefined by altering the configuration of the mesh-topology network, and vice versa. It is important to note that both dimensions can be a direct graphic medium to interact with the designer. Moreover, nodes and edges are labelled with unique key identifiers to improve data dynamization operations (Fig. 1b). At the force-topology network dimension, nodes and edges are extended with additional information for describing its states. Nodes are then colour-coded to display its states. While the type of force determines the state of an edge, the state of a node is defined by local network measurements and by analysing the states of its adjacent edges. The finite set of states constitutes the vocabulary of the network from where axioms are formulated. User-defined rules are then triggered by such axioms to conduct transformations on the connectivity while guaranteeing the numerical stability. An implementation of the model is presented in Suzuki and Knippers (2017).

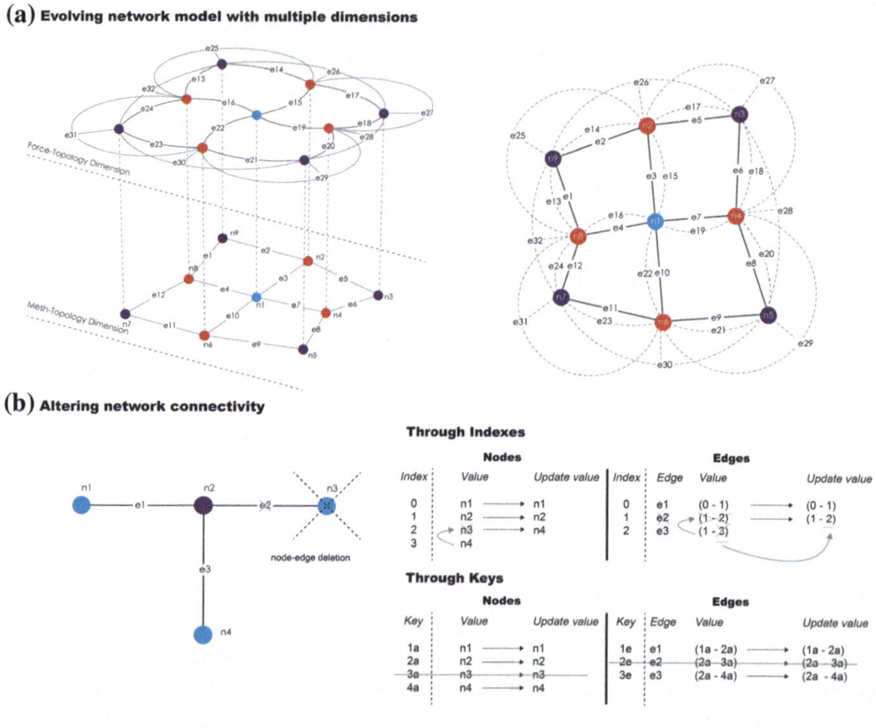

Fig. 1. Evolving network model with multiple dimensions. **a** Multigraph representation. **b** Data dynamization

Discretisation

Once data organization and management is determined, the generation of such form-finding data needs to be tackled. As previously stated, mesh-generation is a completely integrated stage within our recursive form-finding workflow. The ease of implementation and versatility of the proposed network model is based on the continuous creation of regular quadrangular meshes using a half-edge structure. Quad-meshes are preferred because its two local directions can be directly associated with weaving directions of textile membranes, and quad-faces can be easily converted into regular triangles. The problem is, however, how to generate the same topology-mesh within different and highly constrained boundary conditions which are dynamically re-defined during form-finding. In doing so, we have identified three types of recurrent boundary conditions within our form-finding workflow of BATH structures, and where specific meshing algorithms are required (Fig. 2). The main consideration for meshing is that the selected algorithms need to guaranty mesh conformity within multiple meshes and valid manifolds.

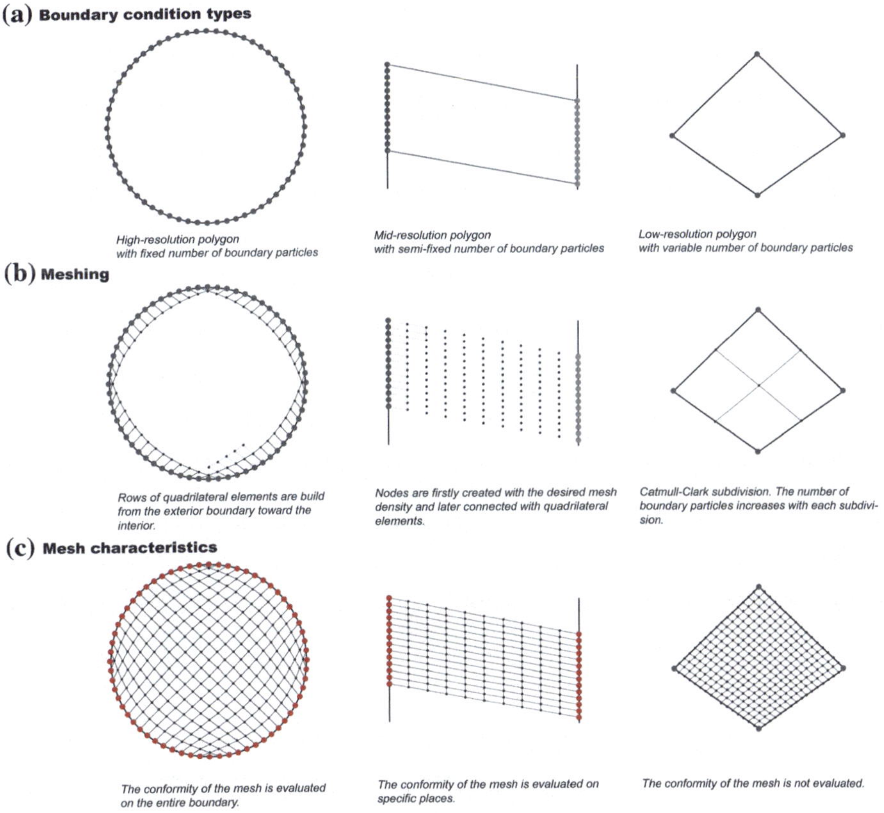

Fig. 2. Recurrent boundary conditions and meshing algorithms

The first type of boundary conditions requires the implementation of an advancing front method for iteratively "paving" rows of quad elements (Blacker and Stephenson 1991; Zhou et al. 2016). The input of the algorithm is a high-resolution polygon defined through a sorted collection of user-selected particles. The paving algorithm is selected because the number of boundary particles can't change during meshing so that mesh conformity is guaranteed on the entire boundary. The second type of conditions uses a direct method for firstly creating nodes and later connecting them through quadrilateral elements. In this case, the input is a double and equally-sized set of user-selected particles defining opposite sides of a mid-resolution polygon boundary. While the number of boundary particles at these sides can't change, internal particles are computed to satisfy the desired mesh density. The conformity of the mesh is evaluated at specific locations. Finally, for the third type of boundary conditions, a Catmull and Clark's subdivision process is implemented (1978). In this case, the number of boundary particles can change during meshing since mesh conformity is not evaluated. Therefore, a low-resolution polygonal representation can be used as input. Figure 3 shows the dynamic use of meshing algorithms within the topology-driven form-finding process of two BATH structures.

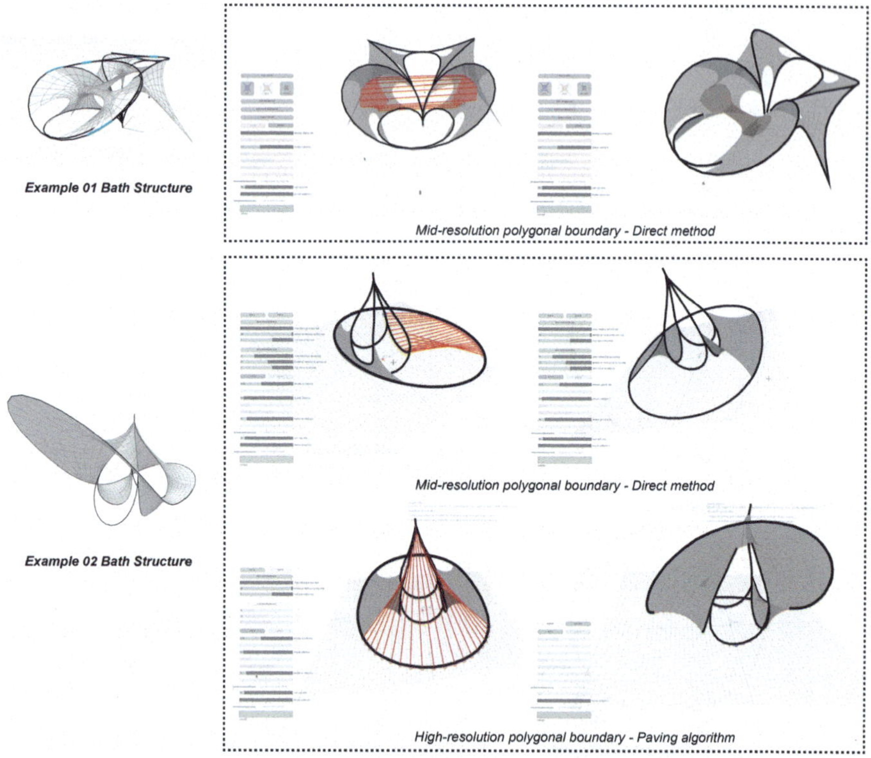

Fig. 3. Examples of dynamic mesh generation within different boundary conditions

Interactivity

The third stage is designed to address expected problems regarding the progressive diminution of interactivity when massively increasing the size of the network model. Even if the number of particles plays a key role for this, it is the type of force-element that drastically affects the number of computations per iteration. Consequently, a multi-scale strategy for modelling membranes was proposed for assisting the control of interactivity and accuracy during form-finding. This strategy is based on the construction of an element's hierarchy enabling an interactive shift of elements by "implicitly" sorting data on a quad-mesh topology.

Binary force elements, like springs and cables, are among the less expensive elements to compute. Depending on the orthogonality of the mesh topology, such formulations can simplify the simulation of membranes structures by modelling a pseudo-spring- or cable-net system. Since both models derive into rough approximations of equilibrium shapes, surface elements are required for improving the accuracy of results. Yet, surface elements like the constant strain triangle (CST) are more expensive to compute and tend to produce numerical instabilities when geometric and topologic modelling is activated (Fig. 4). What is more, these elements require a regular triangular mesh topology where each triangular face needs to have an edge parallel to the warp direction of the fibres (Barnes 1999).

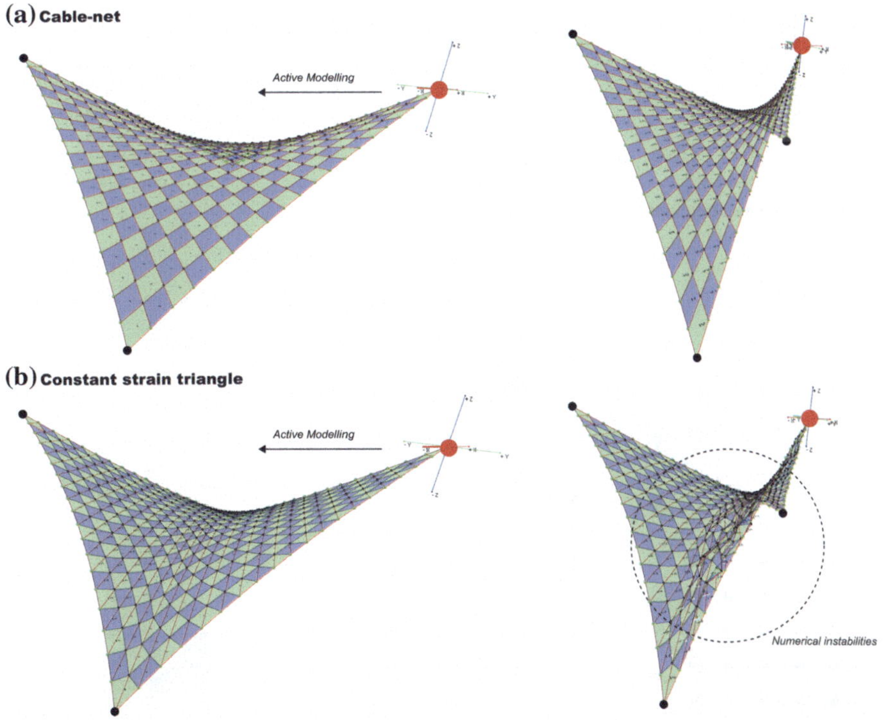

Fig. 4. Numerical instabilities of force elements when active modeling

Under these conditions, initial studies for shape and topology exploration can be conducted through simple spring/cable-net systems. As soon as a satisfactory shape is attained, the designer can dynamically shift to a surface element like CST. For the moment, prestress values need to be manually converted from one element to another. This interactive shift is only possible because the entire process is based on regular quad-meshes that can be easily converted into triangular meshes. However, the problem is that each meshing algorithm sorts data in diverse ways leading to unsuitable triangulations for CST and problems for finding weft and warp directions (Fig. 5a). To solve this, an automatic process for edge classification based on constant adjacency queries has been designed (Fig. 5b).

The process starts by randomly picking an initial face of the quad-mesh. Since the edges of the face are always cyclically sorted, we can alternate the assignation of 1 for edges aligned to the first direction and 2 for those aligned with the second direction. Because in this data structure an edge is defined by two half-edges, faces paired to each

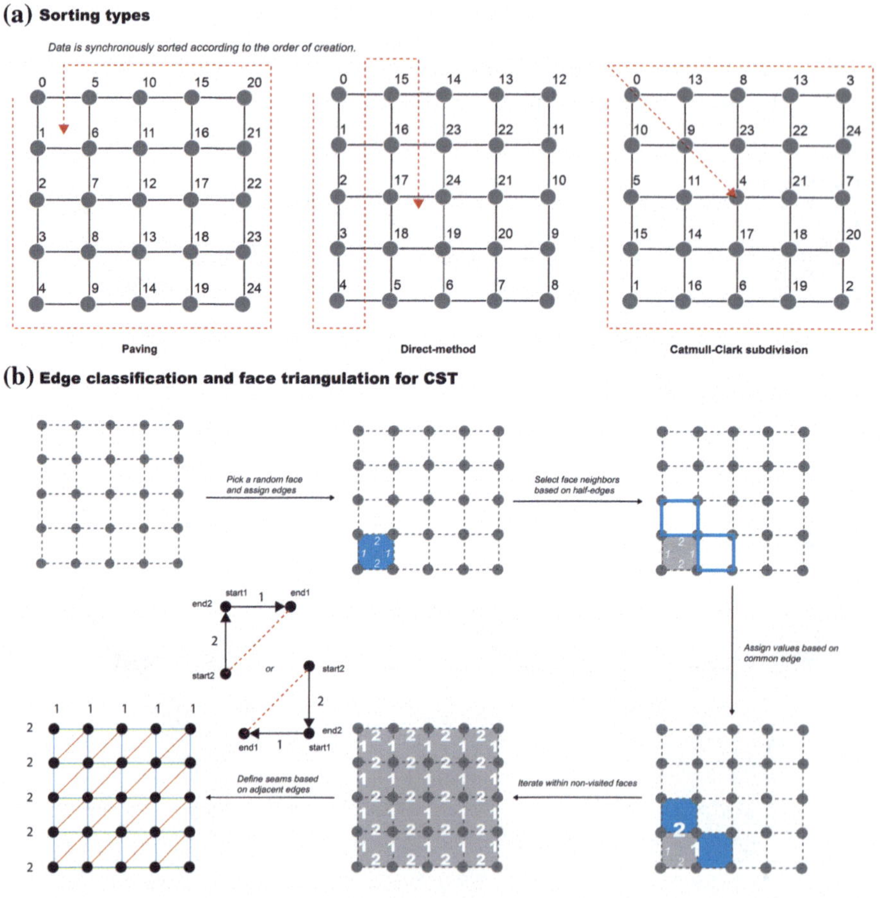

Fig. 5. Topologic adjustments for multiscale modelling. **a** Data sorting of different meshing algorithms. **b** Triangulation process for CST shifting

half-edge are searched. For each face, if all its edges are not categorized then the assignation process starts from the pre-assigned edge. This process is repeated until all edges are assigned. For triangulation, each face creates a diagonal link between the end-nodes of two face-edges assigned with a different direction. This is only possible because of the classification of edges and the cyclic order of face-edges. All diagonals are then aligned with the same direction, creating the required seams for CST. In doing so, a volatile triangular topology is created that is suitable for CST shifting. Since this process only affects the force-topology network without explicitly changing the quad-mesh topology, a reverse shifting from CST to cable/spring-net is still possible.

Decision-Making

Finally, the fourth stage is intended to address the increased amount of work required for modifying the connectivity of the network model. This problem is even worse with each transformation since decisions regarding the addition/deletion of elements need to be carried within more complex spaces. For tackling this problem, we explored the use of multi-agent based systems and machinic learning techniques.

For the first case, we proposed the design of two types of particle—parent and child—which are extended with specific behaviour. A "parent" particle flocks (Reynolds 1999) within a predefined digital space while dropping a finite set of child particles and connections within them (Fig. 6). Static targets, used to control the openness of the structure, are placed around the digital space to activate seeking and avoidance behaviours of parental particles. All the children of a single parent describe the entire network model of an elastic rod (Fig. 7a). As soon as three child particles are dropped, the elastic rod starts its deformation while the parental particle is still flocking and connected to the last child. Boundary conditions are then established by setting the initial and the last child particles as pinned supports. At the same time, child particles distinguish possible partners from different families based on distance constraints and its view range. Once the potential partners are found, tensile connections are created as shown in the example of Fig. 7b.

In the second case, a decision tree algorithm was implemented for making choices regarding the creation of rod connections between large sets of disconnected rods.

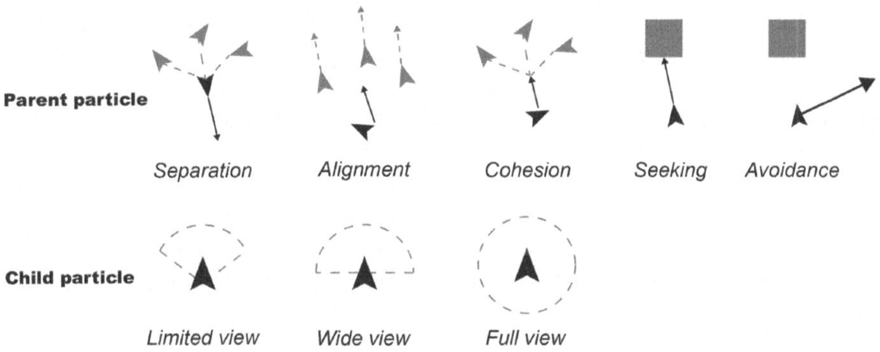

Fig. 6. Individual behaviours of programmed particles

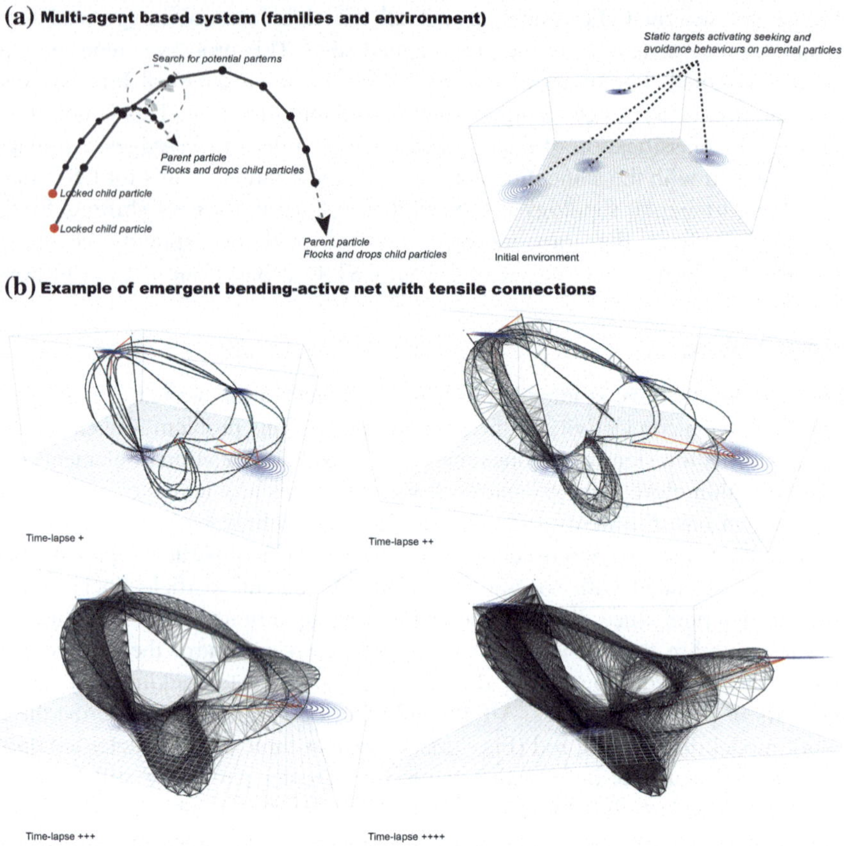

Fig. 7. Multi-agent based system. **a** Families of agents and environment. **b** Example of a BATH structure where a multi-agent based system drives the form-finding process

Through Boolean classification, a decision tree returns a positive, or negative, output by evaluating an input with a predefined value at each tree node (Russell and Norvig 2010). All predefined values are set by a user-defined sample pattern (Fig. 8a.1). Considering that each rod constitutes a directed subgraph from where a local order of particles indexes is established, the sample pattern consists of a sorted collection of those local indexes within different rods (Fig. 8a.2). A tree structure is then created from each sample pattern (Fig. 8a.3). At each tree-node, the difference of consecutive particle indexes is computed. If the difference of an input pattern is equal to the difference of the sample pattern, a positive response is triggered. Under these conditions, the goal of the algorithm is to decide whether a computer generated random pattern satisfies the path driving towards a positive decision for creating the connection. Figure 8b shows the application of the decision tree algorithm for form-finding a bending-active network.

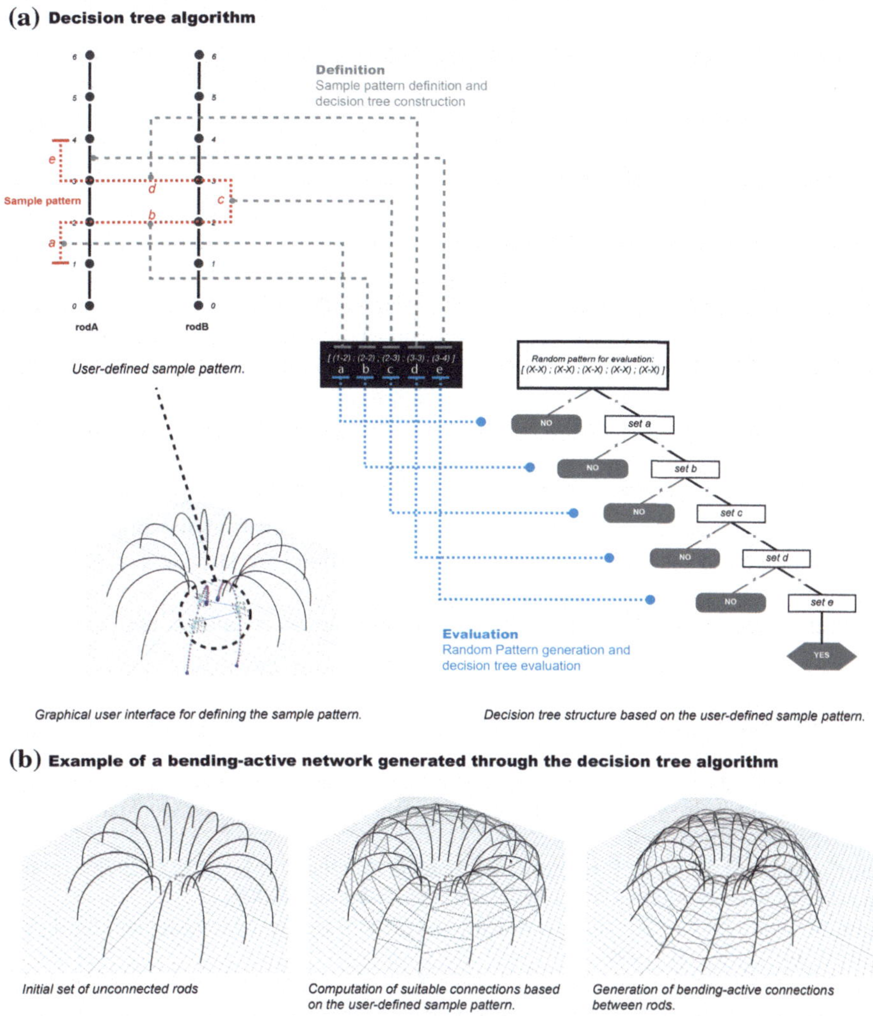

Fig. 8. Decision tree algorithm. **a** Definition of the sample pattern and construction of the decision tree algorithm. **b** Example of bending-active structure generated through the decision tree algorithm

Conclusion and Future Work

In this paper, a general framework for the development of a topology-driven form-finding approach has been presented based on four building blocks. The purpose of this approach is to enhance user-model interactions during form-finding leading towards more flexible design spaces and versatile workflows. The framework is formulated from the analysis of related problems derived from the activation of topologic modelling during form-finding. For facing these problems, we used a collection of methods that are suitable to be integrated into the design workflow. Such methods range from the design of an evolving network model with multiple dimension, the specification of mesh-types and meshing techniques, the development of a multi-scale modelling strategy, and the introduction of agency and machinic learning techniques for taking decisions in complex design spaces.

Although studies were conducted only on BATH structures, we believe that topology-driven form-finding approaches can embrace the design of more lightweight structures by adjusting and/or identifying additional building blocks. In this context, we are currently working on the implementation of a complete set of force elements and derived strategies, in addition to studies for approaching topology-driven form-finding within more immersive design spaces. The presented study has also addressed the coupling of "intelligent systems" and physically-based techniques without affecting numerical stability. Even if geometric outcomes are still too speculative, this open new research questions regarding the potential integration of autonomous systems for assisting form-finding processes in complex design spaces. The wide range of machinic learning techniques may offer promising mechanisms for controlling and optimising the addition of new elements within the network model. Yet, further efforts are still needed to understand how these mechanisms can also be used to support deletion and/or modification operations. To finish, many questions are still open regarding the theoretical and critical discussion of the nature and scope of topology-driven form-finding approaches within architectural design practices and its applicability.

References

Ahlquist, S., Erb, D., Menges, A.: Evolutionary structural and spatial adaptation of topologically differentiated tensile systems in architectural design. Artif. Intell. Eng. Des. Anal. Manuf. **29** (4), 393–415 (2015)

Attar, R., Aish, R., Stam, J., Brinsmead, D., Tessier, A., Glueck, M., Khan, A.: Physics-based generative design. In: CAAD Futures Foundation, Montreal (2009)

Barnes, M.: Form finding and analysis of tension structures by dynamic relaxation. Int. J. Space Struct. **14**(2), 89–104 (1999)

Bender, J., Müller, M., Otaduy, M.A., Teschner, M., Macklin, M.: A survey on position-based simulation methods in computer graphics. Comput. Graph. Forum **33**(6), 228–251 (2014)

Blacker, T.D., Stephenson, M.B.: Paving: A new approach to automated quadrilateral mesh generation. Int J Numer Meth Eng. **32**(4), 811–847 (1991)

Bouaziz, S., Deuss, M., Schwartzburg, Y., Weise, T., Pauly, M.: Shape-up: shaping discrete geometry with projections. Comput. Graph. Forum **31**(5), 1657–1667 (2012)

Bouaziz, S., Martin, S., Liu, T., Kavan, L., Pauly, M.: Projective dynamics: fusing constraint projections for fast simulation. ACM Trans. Graph. **33**(4), 154 (2014)

Catmull, E., Clark, J.: Recursively generated B-spline surfaces on arbitrary topological meshes. Comput. Aided Des. **10**(6), 350–355 (1978)

Debunne, G., Desbrun, M., Cani, M.P., Barr, A.H.: Dynamic real-time deformations using space and time adaptive sampling. In: Pocock, L. (ed.) SIGGRAPH 2001, pp. 31–36. ACM Press, New York (2001)

Deleuran, A.H., Pauly, M., Tamke, M., Trinning, I.F., Ramsgaard, Thomsen M.: Exploratory topology modelling of form-active hybrid structures. Procedia Eng. **155**, 71–80 (2016)

Dunkin, B., Adrales, G.L., Apelgren, K., Mellinger, J.D.: Surgical simulation: a current review. Surg. Endosc. **21**(3), 357–366 (2007)
Greenwold GitHub Homepage (2017) https://github.com/juniperoserra. Last Accessed 6 Aug 17
Harding, J., Shepherd, P.: Structural form finding using zero-length springs with dynamic mass. In: Proceedings of the International Association for Shell and Spatial Structures Symposium, London (2011)
Kangaroo HomePage (2017) http://www.grasshopper3d.com/group/kangaroo. Last Accessed 06 Aug 2017
Kilian, A., Ochsendorf, J.: Particle-spring systems for structural form-finding. J. Int. Assoc. Shell Spat. Struct. **46**, 77–84 (2005)
Kuijvenhoven, M., Hoogenboom, P.: Particle-spring method for form finding grid shell structures consisting of flexible members. J. Int. Assoc. Shell Spat. Struct. **53**(1), (2012)
Misra, S., Ramesh, K.T., Okamura, A.M.: Modeling of tool-tissue interactions for computer-based surgical simulation: a literature review. Presence Teleoperators Virtual Environ. **17**(5), 463–491 (2008)
Müller, M., Heidelberger, B., Hennix, M., Reatcliff, J.: Position based dynamics. J. Vis. Commun. Image Represent **18**(2), 109–118 (2007)
Piker, D.: Kangaroo: form finding with computational physics. Archit. Des. **83**(2), 136–137 (2013)
Quinn, G., Deleuran, A.H., Piker, D. Gengnagel, C.: Calibrated and interactive modelling of form-active hybrid structures. In: Kawaguchi, K., Ohsaki, M., Takeuchi, T. (eds.) IASS Annual Symposium, Tokyio (2016)
Reeves, W.T.: Particle systems: a technique for modeling a class of fuzzy objects. ACM Trans. Graph. **2**(2), 91–108 (1983)
Reynolds, C.: Steering behaviours for autonomous characters. In: Proceedings of Game Developers Conference, San Francisco (1999)
Russell, S.J., Norvig, P.: Artificial Intelligence: A Modern Approach. Pearson, Boston (2010)
Senatore, G., Piker, D.: Interactive real-time physics. J. Comput. Aided Des. **61**, 32–41 (2015)
Suzuki, S., Knippers, J.: Topology-driven form-finding: implementation of an evolving network model for extending design spaces in dynamic relaxation. In: Raonic, A., Herr, C., Wash, G., Westermann, C., Zhang, C. (eds.) CAADRIA2017: Protocols, Flows and Glitches, Suzhou (2017)
Wu, J., Westermann, R., Dick, C.: A survey of physically based simulation of cuts in deformable bodies. Comput. Graph. Forum **34**(6), 161–187 (2015)
Zhou, X., Sutulo, S., Guedes, Soares C.: A paving algorithm for dynamic generation of quadrilateral meshes for online numerical simulations of ship manoeuvring in shallow water. Ocean Eng. **122**, 10–21 (2016)

City Gaming and Participation

Enhancing User Participation in Design

Areti Markopoulou(✉), Marco Ingrassia, Angelos Chronis, and Aurel Richard

Institute for Advanced Architecture of Catalonia, Pujades 102, 08005 Barcelona, Spain
areti@iaac.net

Abstract. Technologies are transforming architecture into a more reactive and evolutionary organism, able to interact in real time with multiple agents such as the environment, time or user needs. Architecture moves towards the performative (Kolarevic and Malkawi in Performative architecture: beyond instrumentality, Routledge, London, 2004) or a performative instrument (Beesley and Khan in Responsive architecture/performing instruments, The Architectural League of New York, New York, 2009). The emergence of these responsive environments boosts new relations among users, architects and space. If architecture of built (or unbuilt) space can be programmed to perform, the key question to deal with, is who the actuator of such performance is. This paper engages with the idea that responsive technologies, such as Virtual and Augmented Reality and User/Gaming Interfaces, can be used by architects and urban designers as a tool for enhanced participatory design as well as a tool for evaluating existing planning or future design decisions. Two case studies are being presented, which use responsive technologies for creating participatory urban design processes. The case studies have been developed in Mumbai (use of Virtual and Augmented Reality) and Barcelona (use of Gaming Interfaces) as experimental pilot projects for acquiring qualitative and quantitative data on the process of technologically mediated design participation.

Keywords: Participatory design · Virtual reality · Augmented reality · Game design · Urban design

Research Framework

The idea of technologically mediated user empowerment in design has its roots back in the 1960s and 1970s when for the first time the capacity of professional designers to respond to the social complexity of the users was questioned. It was at that moment that the designer's community started to understand and even quantify the benefits and advantages of user's participation in the design of their spaces. From visionary architects, such as Yona Friedman and Cedric Price to pioneers of computation such as Nicholas Negroponte and the MIT Machine group, the idea of design for the people was giving way to the idea of design with the people. In this idea, high end final aesthetics are becoming obsolete, spaces are conceived as "a kind of scaffold enclosing

K. De Rycke et al., *Humanizing Digital Reality*,
https://doi.org/10.1007/978-981-10-6611-5_20

a socially interactive machine" (Hobart and Colleges 2005), the model of unique design and unique decision is questioned and the user is being placed as the protagonist that operates the various performances of the responsive "built structures integrated with computing power" (Negroponte 1975).

In Henry Sanoff's extensive documentation of participatory design methods (Sanoff 1990) we observe the importance of physical artefacts (models, games, figures) for facilitating user's participation. Such physical artefacts, although crucial for traditional design participation processes, limit the latter, since only a small community can interact with them and can collect/socialize around them. On the other hand, information and communication technologies, interfaces or user apps inhabit the "Cloud" and might be accessed by any person connected to the internet through any personal computing device. The democratization of information is bringing personal empowerment as the central theme associated with information technologies (Carr 2008). "You" were chosen in 2006 as Time magazine's Person of the Year controlling the Information Age and a series of adjectives have been used to define the empowerment of the user such as "prosumer" (Toffler 1980) or the latest "maker" (Anderson 2012).

Background

Within the scope of user empowerment and democratization of information described above, this paper aims to assess, through the two presented case studies, the potential of the technologies of information and communication in the design of architectural and urban environments. The use of gaming interfaces as a tool to envision possible scenarios for the urban environment has been investigated before. One important precedent can be found in the use of the video game Minecraft as a participatory design tool for young people, developed by UN habitat (Westerberg and Von Heland 2015) in South America, Africa and Asia. The video game allows the player to design 3d architectural and landscape elements through the addition of small cubes, working like pixels in photographs. UN-habitat intervened in low income areas, organizing workshops with young citizens asking them to interact with the digital reproduction of their neighborhood in Minecraft. The players, helped by a facilitator during 2–4 days could design and visualize possible scenarios for their urban environment per their needs and desires. The different proposals were then proposed and discussed with the local stakeholders. UN habitat's research with Minecraft has clearly demonstrated the strong potential of video games in engaging the youth with urban planning and design. In UN habitat's use of Minecraft, the focus lies on youth participation while in our case studies, the experiments were targeted to users of all ages. Also, our interventions are mostly done in situ and aim to engage the public's participation within their active urban environment.

Another example of game usage for participatory design is Block'hood developed by Sanchez (2015). Block'hood is envisioned as a crowd-sourced simulation of ecological urbanism. The game is similarly based on a rectangular voxel grid that allows the player to place blocks, each of which represent a different unit as well as mobile agent units that can circulate throughout the grid. Block'hood is again demonstrating the potential of user participation in urban design. Although Block'hood is mostly abstract and not situated within an existing urban condition it proves the significance of gaming and user participation in education and contemporary urbanism. The use of both games and virtual and augmented reality technologies in architecture is not new. Woodbury et al. (2001) have for example used games in early design education to explore the metaphorical relation of play and design. More recent examples (Otten 2014) examine the use of games in customizing mass housing systems. The use of games and virtual platforms for user empowerment and participation is though not thoroughly explored and it is certainly a fruitful research field in architecture.

Scope and Limitations

The main aim of this ongoing research project is to evaluate the potential of using gaming and virtual/augmented reality technologies to empower the user's participation in the design process. The presented case studies are all based on real urban design projects on which our interventions have been developed based on pragmatic constraints and our qualitative and quantitative analyses are informed by the actual users of the sites. Our objectives are thus focused on creating a real channel of communication with the actual users of the projects and through our digital intervention to better understand their needs and communicate their design intentions to the designers and stakeholders. It is not in the scope of this study to either create a specific methodology for participatory design or to create a quantitative or qualitative theory of the role of the user and the issues of expertise and domain knowledge in a participatory design process. Our aim is on the other hand, to develop such participatory design processes, based on pragmatic resources and real-life design problems and to try to understand and make sense of the user input.

In both case studies presented here, we have closely worked with authorities and stakeholders to create this channel of communication and we have found that although both studies are in a preliminary stage there is significant potential for further integration of user participation in these urban projects. In both case studies, extensive urban analysis and planning studies were undertaken which informed the development of the participation platforms. The design affordances offered to the users are in themselves a design input and their development has been a significant part of this study, something which is however not thoroughly investigated under the scope of this paper. The paper primarily reports on the affordances of the technological means in enabling a participation channel rather than the specific design decisions.

Case Study A: BDD Chawls, Mumbai, India

The first case study was situated in Worli BDD Chawl, a 27-ha residential district located in Mumbai, India. Through MHADA, the Housing and Area Development Authority, responsible public authority for the development of the BDD Chawls, and a private real estate office, responsible for the execution of the plan, data was selected from the existing participatory processes developed prior to our involvement in the project. This data showed numerous sessions organized with the residents of the site but very little input and decisions that the neighbors could give and take. Residents were called to choose one out of 3 design solutions that only concerned interior space distribution of the apartments such as the arrangement of the kitchen and bedroom. No other input related with the overall program planning, building form, landscape design, or other infrastructural aspects, such as energy generation, or urban agriculture, was provided to them.

The data collected in this participatory process fed the guidelines of an architectural design competition for the renewal of the area which included demolishment of the existing buildings and construction of high-rise residential towers based on the previous typologies. Our intervention involved the organization of a 3-month workshop with the students of the Master in City and Technology at the Institute for Advanced Architecture of Catalonia. Together with the students we designed an open system through a Virtual and Augmented Reality Interface that would allow the residents to visualize different design scenarios and give feedback on the design scenarios they preferred.

Methodology and Data Collection

In the first trip to Mumbai, an extensive survey of the residents allowed us to collect data on the way the users inhabit space and the citizen's desires and needs. Data ranged from timetable of working hours, to leisure and social activities, to mobility and inhabitation patterns. Based on the data collected a series of open design solutions were developed that incorporated features that were identified from the survey's data, such as community kitchens and leisure space, reconfigurable social spaces that enabled different activities like cinemas and lounges and different space configuration depending on the working and living patterns of the habitats. The modules were designed using parametric software (Grasshopper 3D) which allowed continuous design iterations and were later visualized in Virtual and Augmented Reality applications, developed using the Unity 3D development framework. The different design proposals developed were integrated within a user interface, that allowed the users to make specific design decisions for their private and public space, which would then be assessed by the stakeholders.

A second trip to Mumbai was organized to meet with the residents and allow them to test the interactive open models. We have provided them with a low-cost VR device (Google Cardboard), that allows each smartphone to be transformed into a VR headset and a tablet running an AR interface that could be overlaid over a physical scaled model of the district (Figs. 1 and 2). The VR devices allowed the users to experience in an immersive way the different design proposals and give us feedback on their preferences. The data collected from their preferences and design decisions informed the initial design proposals and provided valuable insight on the user's perspective and needs t. The residents that joined the participatory event were of different age, gender and profession.

The data collected in the first survey showed the flexibility and multiplicity of use of the spaces in the neighborhood, as well as the collective and intense inhabitation of several public spaces for ceremonies, celebrations and technical uses. At the same time, the data highlighted the shortages of water and energy in the neighborhood through the day. In contrast with this observation though, when asked if they would be interested in

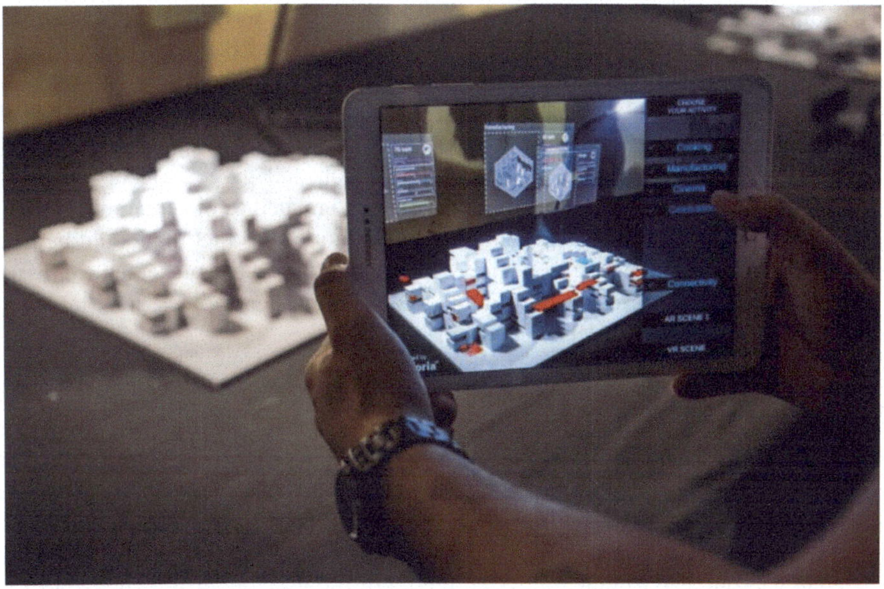

Fig. 1. Augmented reality visualizations of design proposals

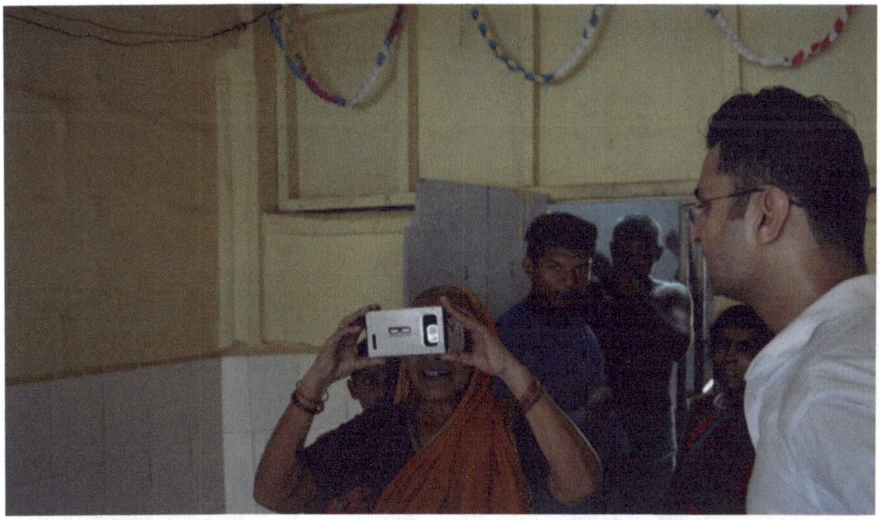

Fig. 2. Participation event in Mumbai—residents using VR devices

being provided with energy and food production systems, the citizens answered mostly negatively, showing that there is a discrepancy between the expressed needs and the design decisions of the user, which could potentially relate to lack of adequate information on the proposed solutions.

The different design solutions proposed alternative uses for the public space (such as shared kitchens, collective theatres, shared workshops) and the private spaces (such as small sustainable energy generation systems, private orchards, water recycling devices). The VR and AR interface allowed the users to inhabit and visualize the different configurations of the physical space, but was also giving them meaningful information that could allow them to take a decision and give us feedback on the preferred proposals. An interesting example is the data related to the economic value of the energy produced by the energy plant, or the amount of product that could be made in the shared digital fabrication workshop. Both data points were presented to the user in real time, allowing her to get quantitative feedback on her choices, thus inherently educating her on the design decisions taken.

Conclusions of Case Study A

Despite the small sample of users and the qualitative nature of the surveys, this first in situ and real-case experiment demonstrated both a significant potential of the participation process but also some important limitation, such as the lack of communication between the designer and the user, which relates to their expertise and knowledge gap but also to inherent social inertia of the users against a change of their urban environment. We have found that in both cases the use of virtual and augmented reality was significantly helpful in engaging the people with the design process, both due to a curiosity for the technology itself, but also due to the immersiveness of the virtual and augmented reality platforms (Fig. 3).

Fig. 3. Participation event in Mumbai—residents using VR devices

Case Study B: Poble Nou Superblock Pilot, Barcelona, Spain

The second case study was situated in the SuperBlock project of Barcelona, known in Catalan as the Superilla which is currently under development. The first part of the experiment was conducted in April 2017. The SuperBlock consists of an urban regeneration tool that is part of a plan for Barcelona, developed by the Agencia de Ecología Urbana de Barcelona, within the Urban Mobility Plan of Barcelona 2013–2018. The plan aims at shutting down two thirds of Barcelona's roads to traffic, an urgency brought on by the high levels of traffic pollution in the city. Specifically, the SuperBlock consists of a three by three "superblock" grid of Cerda's urban blocks in which the internal traffic will be reduced to residents and local business related traffic, at a lower speed, leaving the great majority of the city's traffic to circulate around the perimeter of these SuperBlocks. This intervention allows the development of new pedestrian areas and consequently new spaces for citizens. The first Superblock developed in Barcelona is known as "superillapilot" and it is located in the 22@ Innovation District, the former manufacturing area of the city and today the object of an extensive application of urban regeneration projects.

Our activity has been based on designing and testing on site a gaming interface to be used as an alternative tool for participatory design processes, allowing the citizens to create and visualize different proposals for the public space of their neighborhood, and enabling a significant data collection about their needs and desires. The interface could be freely downloaded from the internet for any device thus enabling the participation of all members of the community. A first test of the developed game and interface was organized as a collective event inside the SuperBlock at its launching event in March 2017. Residents and users of the Superblock area were all asked to play and interact with the interface, helped by a facilitator generating a first dataset about the neighborhood.

Methodology, Data Collection and Data Process

The first part of this case study was to digitally 3d model the physical space of the Superblock. That model was then imported in the interface (Fig. 4). The accuracy of the 3D model was particularly important as it allowed the users to identify their neighborhood and key areas of the site, such as their houses, local shops etc.

An initial questionnaire survey was conducted with 130 people of the neighborhood, collecting personal information such as age, gender and place of residence, as well as data about the citizens' needs and desires on the topics of mobility, social interaction, ecology and production of energy, food and goods. This data was then used to develop a gaming strategy, by designing a series of modular elements that the user could place in the virtual urban environment. Each element belongs to one of the 4 categories listed above and represents a functional program for the public space, including greenery, energy devices, alternative vehicles, leisure devices and more (Fig. 5). To demonstrate the impact of these elements on the overall neighborhood which is considered as a connected complex social system, 5 metrics were introduced to show the variation in the accessibility, economy, productivity, ecology and social

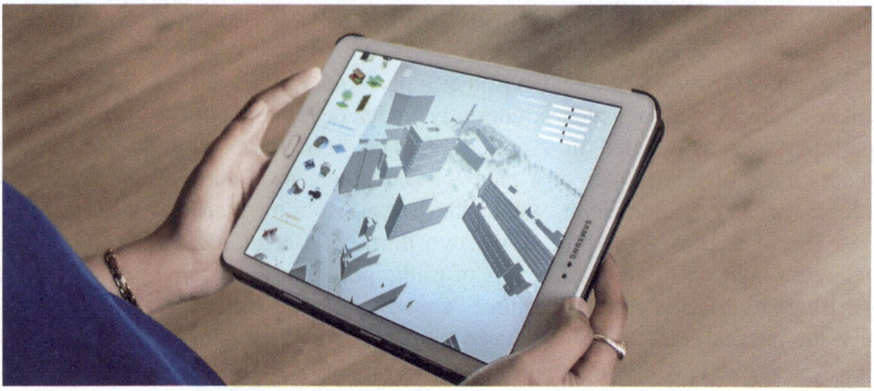

Fig. 4. The beta version of the gaming interface

Fig. 5. Example of interface elements, including bike stations, energy generators, greenery etc.

interaction. Each element placed generates a score and has an impact on the overall score of the neighborhood.

Different strategies of gamification were also used to engage the user and allow the collection of a consistent dataset. Using textual notifications, for example, the interface encourages the players to place the elements by keeping the balance between the different values. Warnings about imbalance along with recommendation pop-ups and similar features can essentially help the user understand the game, but also inherently educate her on the design affordances of the game, and consequently the design affordances of the actual project. This aspect of the game was quite interesting as it seemed to override the initially expressed needs of the user. Even if the target group was similar, the data collected at the game test on site was in cases significantly different to the data collected in the first survey. An interesting point for example is that in the first survey the majority of the neighbors (60%) expressed the desire to have more car parking spots, while using the interface the number of parking spots placed was not significant.

Data Collection and Visualization Platform

The game also included a data collection engine that sent data of each individual game run. At the end of each game session the interface generates and sends to an online server a data file collecting information about the identity of the user and the data of the elements placed in the virtual public space, including their location, their number, and the timestamp of the game at which the user decided to place them. This continuously updated dataset feeds an online data analytic dashboard (Fig. 6). The data collected is visualized on the map of the Superblock and on different diagrams and can be sorted based on different parameters, such as age and gender of the player, used elements and categories, areas where the elements have been placed.

The development of the data visualization and analytics platform has been an important part of this second case study as it allows for an extensive analysis of the collected data, thus allowing the designers and stakeholders to thoroughly investigate the decision-making process of the user and to potentially find user patterns. The

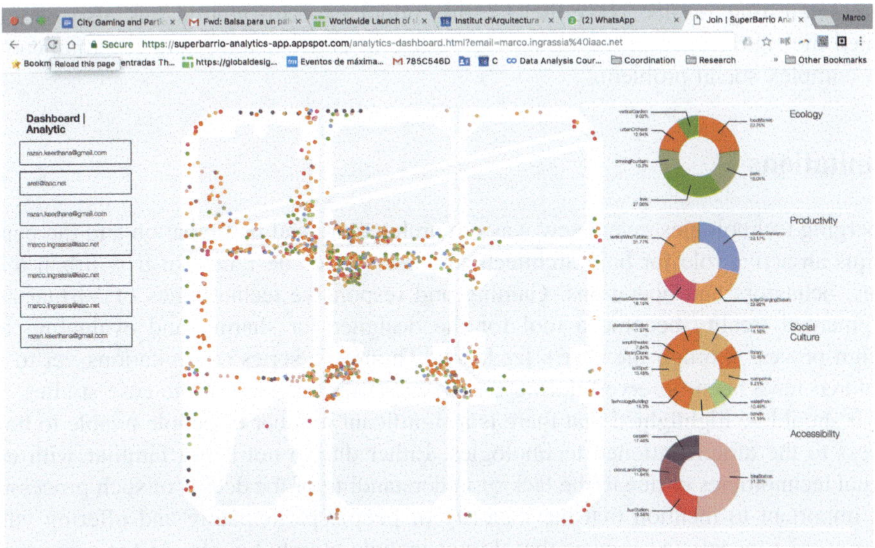

Fig. 6. Visualization platform dashboard

unstructured and feature-rich database as well as the real-time tracking and sharing platform are both open, giving accessibility to all other public or private platforms through their own API, allowing a seamless collaboration potential.

Conclusions of Case Study B

One of the most interesting points from this second case study has been the importance of the introduced metrics in the decision making of the users. A significant variation, even from the first minutes of the game was observed, which relates to the effort of keeping a balanced score. It is evident again in this case, that the design of the game is affecting the decision-making process of the user. This immediate feedback enables more informed decisions from the user while it is also educating her on her decision's impact on the complex social virtual environment, based of course on the analysis of the game's developers.

This second case study demonstrated to us and the stakeholders how specific gamification strategies can allow the users not only to visualize but also to receive information and understand the impact of certain design decisions on the district, generating awareness on complex urban dynamics. It is also valid to state that these features can allow the designer to be a mediator between the users and public entities that are promoting urban regeneration processes, allowing her to inform the public on the benefits of certain interventions through the self-fulfillment provided by the gamification of the problem. On the other hand, this approach allows a two-way communication channel, as through the user input the designers and stakeholders can identify specific use needs and areas for development, thus fulfilling the real needs of the people. Although the initial sample of users was relatively small (around 50 users) the project is currently under development with the municipality of Barcelona and a further study with a much larger user pool is being developed and expected to yield more interesting results. Nevertheless, this initial experimentation has also demonstrated the significance of the technology usage in understanding and communicating this complex social problem.

Limitations

Emerging technologies reveal new ways of architecture and participation and the paper claims an active role for both architects and users as co-designers of new urban relations, behaviors and operations. Gaming and responsive technologies of Virtual and Augmented Reality become a tool for the designer for sharing and evaluating the design process through the user's feedback. Though, a series of limitations yet to be explored have been observed during the process used on each of the case studies.

It should be highlighted that there is a significant number of people unable to have access to the aforementioned technologies. Either due to not being familiar with any digital technologies or due to the lack of understanding of the details of such processes, it's important to mention that the majority of people participating and offering valid input were concentrated in ages that do not include elderly people. At the same time, the case studies presented required a prior thorough explanation by the architects to the people participating. The projects are not yet developed at such level that can allow people to directly use these technologies by themselves. Tuning and operating of the VR and AR devices, as well as a significant time for prior explanation of the process was needed in both case studies. Finally, it is important to mention that the data collected only represented a small group of people involved in the communities, which asks for further development on refining the process to allow greater number of people participating.

Besides these limitations the case studies presented in this paper show the importance of rethinking existing processes and logics in participatory design. The use of responsive technologies opens up possibilities for bringing the user back to the center of design. The user is empowered to change, customize and adapt the environment in real time, crossing the limits of physical artefacts and traditional drawings. At the same time, such tools allow for a bidirectional raise of awareness. On one hand architects, urbanists or decision makers can educate themselves on what people need and wish and on the other hand users are being educated on various design possibilities as well as on the global impact of their decisions.

Conclusions

This paper presents two case studies on participatory design, using virtual and augmented reality technologies and game development platforms. The aim of the study is to demonstrate the potential of using these technologies to enable user participation in complex design problems of the urban environment. Although a number of limitations have been observed, relating to both the availability of technology as well as the social profile of the users, this approach has been in both cases proven to be very helpful in creating a twofold communication channel between planners, users and other stakeholders. In both case studies, it has been shown that the designer can use these technologies as a mediator, enabling user participation in the design process and allowing informed design decisions. This study is just the starting point of a further and more thorough investigation on participatory design processes and a continuation of this research is expected to yield more interesting results.

References

Anderson, C.: Makers: The New Industrial Revolution. Crown Business, New York (2012)

Beesley, P., Khan, O.: Responsive Architecture/Performing Instruments. The Architectural League of New York, New York (2009)

Carr, N.: Is google making us stupid? What the internet is doing to our brains. The Atlantic, digital edition (2008)

Hobart, S.N., Colleges, W.S.: The Fun Palace: Cedric Price's experiment in architecture and technology. Technoetic Arts J. Specul. Res. **3**(2), 73–92 (2005)

Kolarevic, B., Malkawi, A.: Performative Architecture: Beyond Instrumentality. Routledge, London (2004)

Negroponte, N.: Soft Architecture Machines. MIT Press, Cambridge (1975)

Otten, C.: Everyone is an architect. ACADIA 14: design agency. In: Proceedings of the 34th Annual Conference of the Association for Computer Aided Design in Architecture (ACADIA), pp. 81–90. Los Angeles (2014)

Sanchez, J.: Block'hood—developing an architectural simulation video game. In: Proceedings of the 33rd eCAADe Conference—vol. 1. Vienna University of Technology, Vienna, Austria, 16–18 (2015)

Sanoff, H.: Participatory Design: Theory and Techniques. Henry Sanoff, Raleigh (1990)

Toffler, A.: The Third Wave. Bantam Books, New York (1980)

Westerberg, P., Von Heland F.: Using Minecraft for Youth Participation in Urban Design and Governance, UN-Habitat (2015)

Woodbury, R.F., Shannon, S.J., Radford, A.D.: Games in early design education. Playing with metaphor. In: Proceedings of the Ninth International Conference on Computer Aided Architectural Design Futures Eindhoven, 8–11 July 2001, pp. 201–214 (2001)

The Potential of Shape Memory Alloys in Deployable Systems—A Design and Experimental Approach

Philippe Hannequart[1,2(✉)], Michael Peigney[1], Jean-François Caron[1], Olivier Baverel[1], and Emmanuel Viglino[2]

[1] Laboratoire Navier (UMR 8205), CNRS, Ecole Des Ponts ParisTech, IFSTTAR, Université Paris-Est, 77455 Marne la vallée, France
philippe.hannequart@enpc.fr
[2] Arcora, Groupe Ingérop, Rueil-Malmaison, France

Abstract. This study focuses on deployable systems actuated by shape memory alloys in the perspective of designing adaptive sun shading devices for building facades. We first set the context of smart materials for adaptive facades and underline the remarkable characteristics of shape memory alloys for mechanical actuation purposes. After outlining the constraints on the integration of this material into deployable structures, we introduce three different prototypes actuated by shape memory alloy wires. They have been fabricated and tests have been carried out on two of them. Finally, we present some perspectives on the use of these actuators for solar shading systems in façade engineering.

Keywords: Shape memory alloys · Actuator · Adaptive structures · Deployable structures

Introduction

The building envelope presents an increasing amount of mechanical devices: the building can thus adapt to changing operating conditions and meet strong technical requirements, including energy consumption specifications. However, common mechanical systems like electrical motors and rigid-link mechanisms are not fully satisfying in architecture (Fiorito et al. 2016). A whole range of new materials has been researched in the last years: the so-called "smart materials" are able to react to external stimuli and produce an output that can be exploited by engineers and designers. They can convert one form of energy (mechanical, thermal, electric, magnetic, chemical, etc. …) into another (Thill et al. 2008). In particular, those materials could lead to new kinetic elements in facades which must adapt to climatic changes. We focus on deployable solar shading systems: the large glass facades in office buildings require an efficient sun protection. But current products only offer limited geometrical freedom, and shape-changing materials could foster the emergence of new architectural components. Shape memory alloys (SMA) present some remarkable properties and are

K. De Rycke et al., *Humanizing Digital Reality*,
https://doi.org/10.1007/978-981-10-6611-5_21

being investigated through the design and analysis of prototypes resorting to different actuation principles.

Shape Memory Alloys

Morphing smart materials include SMA, shape memory polymers (Basit et al. 2013), bimetals and piezoelectric materials. Standing out among these materials, SMA can undergo large deformations (up to 8%) and return to their original shape with a high recovery force when heated, which is known as the "shape-memory effect". This feature relies on a solid-solid phase change at the crystal lattice level (from austenite to martensite and conversely) resulting in a coupling between temperature and mechanical loads. SMA are widely used as actuators or shape-changing devices in microelectronics, biomedicine and aerospace applications (Mohd Jani et al. 2014). Nickel-Titanium is the most widespread alloy because of its satisfying fatigue properties and exceptional actuation qualities. However, SMA have not been used at a large scale in architecture yet, in spite of some remarkable features like their silent actuation, an ultra-low weight, the ability to both sense and actuate, and the freedom of shape they allow.

One main characteristic of SMA actuators is their small stroke and high output force: in order to develop kinetic shading devices, an amplification of that small stroke is needed. Hysteresis is another characteristic of the thermomechanical response of SMA, and it has consequences on the behavior of SMA-based actuators. Either upon mechanical loading or upon thermal loading, the behavior of SMA is not the same during loading and during unloading. This hysteresis is related to the solid–solid phase change happening in SMA. The consequence for SMA actuators is that the position of the structure cannot be determined only by knowing the current wire temperature: it also depends on the loading history.

Integration into Morphing Structures

SMA can be manufactured in various shapes (wires, sheets, tubes, etc.…). Due to their high cost and significant environmental footprint, the use of those alloys can only be considered for a small actuating part of a façade element, transmitting efforts to a larger structure made of cheaper materials. Because of this need to minimize the SMA quantity, and because the simulation of its complex three-dimensional thermomechanical behavior is numerically challenging, we chose to focus on SMA traction wires and their one-dimensional behavior. In this case, the SMA wires need to be pre-strained in order to undergo the shape-memory effect upon heating: the wire will try to recover its initial, unstrained shape, or, if kinematically constrained, will generate high stresses which can be used for actuation. A computationally efficient, thermomechanical SMA model has been developed in a parallel research (Hannequart et al. 2017) and can be exploited to analyse the response of SMA actuators: the algorithm's one-dimensional simplification enables to simulate systems including SMA wires, which can also be embedded into a host material or fixed to a host structure. A User-Material subroutine (UMAT) has been implemented for the finite-element code ABAQUS.

For shading applications, changes of temperature due to addition of ambient temperature and absorption of the material under solar radiation could heat the Nickel-Titanium wires, resulting in a truly passive device, but an electrical heating by Joule effect has been chosen for a better control of the device. SMA wires actuated by shape-memory effect must initially find themselves in a pre-strained state. For this reason, such devices actuated by shape-memory effect need an external spring-back force in order to return to their pre-strained shape after actuation.

Various kinematic amplification strategies can be identified in engineering (Charpentier et al. 2017), in particular the eccentricity of the actuator, the bilayer effect, controlled buckling, and torsion-induced movements. They were adapted to the use of SMA as an actuator. Different prototypes of shape morphing devices for architecture could be designed and fabricated.

Prototyping

A Bilayer-Type Actuator

The first prototyped actuator exploits the bilayer effect: two thin materials are kinematically constrained together, and both materials have different thermal behaviors. A change of temperature results in different strains in both materials, which bring the whole structure to bend. In our case, the first material is a 1 mm thick wooden lamella, whose reaction to a temperature change is negligible. The second material is a 0.5 mm diameter pre-strained SMA wire provided by Ingpuls GmbH, which shortens upon heating. They have been fixed with a humidity-cured ethyl-based instant adhesive (Loctite). When the heated wire contracts due to the shape-memory effect, the wood-SMA lamella bends, and upon cooling it return to its original state (the springback force is the elastic strain energy contained in the wooden lamella). Notice that in the non-actuated state, the structure is still bent, because of the wire pre-strain, as seen in Fig. 1. This prototype is similar to the adaptive SMA-composite beam presented in Zhou and Lloyd (2009).

Actuation tests were carried out on this prototype, the SMA wire, initially at ambient temperature, being actuated by resistive heating (Joule effect). The wire temperature was measured by a thermocouple fixed on the wire with a highly thermally conductive silicon paste. Indeed, the resistivities of the martensitic and austenitic phase are different: the wire temperature cannot be inferred from the electrical input power. The radius of curvature of the lamella has been measured by digital image correlation (DIC) with the commercial software GOM Correlate (Fig. 2). It can be observed that the curvature is homogeneous along the lamella. Indeed, locally the SMA wire can be modelled as an excentric force acting on the lamella. Considering the lamella as a beam, uniform stresses in the wire lead to a uniform bending of the lamella.

The presented test has been carried out after approximately 10 previous actuation cycles. Two consecutive tests gave similar measurements, and we can thus suppose that the system's behavior is stabilised. The curvature of the lamella has then been plotted

Fig. 1. Bilayer-type actuator composed of a wooden lamella and an SMA wire: initial state at ambient temperature

Fig. 2. Curvature measurements for the bilayer-type actuator: initial state at ambient temperature (*left*), actuated state with resistive heating (*right*)

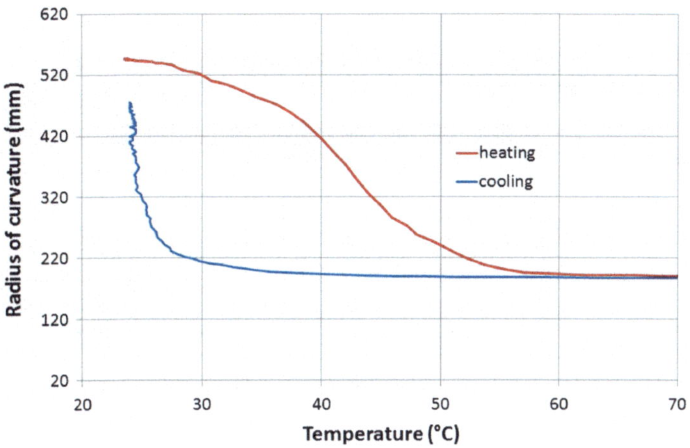

Fig. 3. Actuation test: measurements of the actuator's radius of curvature with respect to the wire temperature—heating curve and cooling curve

with respect to the wire temperature in Fig. 3. The measurements highlight the actuator's hysteresis described in Section "Introduction": there is a strong difference between the heating and the cooling curve. The cooling curve does not completely reach the initial state, maybe because of some residual strains in the wooden lamella, and probably because the test was interrupted too early. We can identify a stable "hot" state: at temperatures higher than 58 °C, the prototype always presents the same curvature. In the same way, if the difference between ambient temperature and actuation temperature were higher we could identify a "cold" state, with no geometrical variations below a certain temperature. Between those two temperatures, the hysteresis prevents from being able to determine the curvature at a given temperature: the loading history has to be known. This hysteresis is a material property which is related to the dissipative nature of the phase change, it does not vary with cycling and has to be considered in any design with SMA.

This way of embedding or fixing SMA into/onto a two-dimensional host material (ribbon, thin shell) is very promising in architecture: whereas the actuator and the moving element are often dissociated, this combination leads to visually interesting kinematics where the material itself deforms.

A Buckling Actuator

The second means of displacement amplification, illustrated in Fig. 4, relies on controlled buckling of elastic materials. We exploited the Euler buckling of two elastic steel ribbons with 1 mm thickness and 10 mm width, in a prototype similar to the one presented in Scirè Mammano and Dragoni (2015). Both ribbons contribute to realize the actuation movement and provide a spring-back force. 3D-printed hinges were manufactured in order to receive these steel ribbons as well as to crimp one central SMA wire (0.5 mm diameter). At ambient temperature (initial state), the steel ribbons

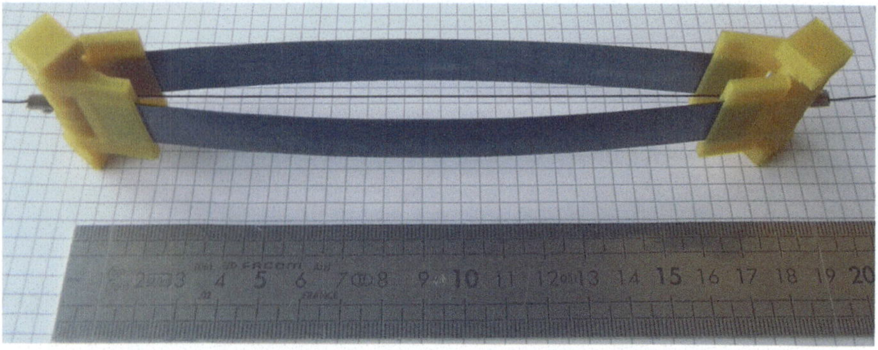

Fig. 4. Buckling actuator: initial state at ambient temperature

Fig. 5. Curvature and distance measurements for the buckling actuator: initial state at ambient temperature (*left*), actuated state with resistive heating (*right*). We define the actuator's "opening" as the length of the *green arrow*

are bent and the wire is thus pre-strained. Upon heating, the wire shortens and thus brings the ribbons to bend more. Upon cooling, the elastic strain energy contained in the ribbons provides the springback force to bring back the wire in its pre-strained state.

Actuation tests were also carried out on this prototype, with the same experimental method, starting at ambient temperature, with a heating phase followed by a cooling phase. In this experiment, the curvatures of the steel ribbons have been measured by DIC in order to determine the maximum distance between both ribbons, which we called the actuator's "opening", highlighted in Fig. 5.

The opening has thus been plotted with respect to the wire temperature in Fig. 6. In the same way as for the first prototype, the measurements highlight the actuator's hysteresis (see Section "Introduction"). We can identify a "hot" and a "cold" state: above 64 °C or below 23 °C, the actuator's geometry does not change. Between those two temperatures, the actuator's geometry depends on the loading history. The kinematic amplification described in Section "Shape Memory Alloys" is illustrated here by the actuator's opening which increases by 19 mm during actuation while the SMA wire shrinks by approximately 4%, which represents 7 mm. This system had already been

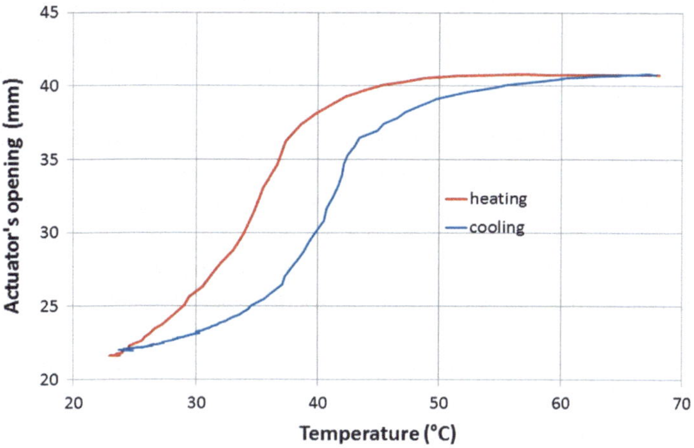

Fig. 6. Actuation test: measurements of the actuator's opening length with respect to the wire temperature—heating curve and cooling curve

actuated many times before the experiment, and we can thus consider that the studied behavior is already stabilized.

A Torsional Buckling Actuator

Torsional movements have also been investigated, as shown in Fig. 7: a slender elastic steel ribbon undergoing rotations at its supports can produce a large displacement. The supports can freely rotate around their vertical axis, and one SMA wire is stretched between each side of the support: by heating one wire, the supports rotate in two opposite directions, inducing the torsional buckling movement of the elastic ribbon. In

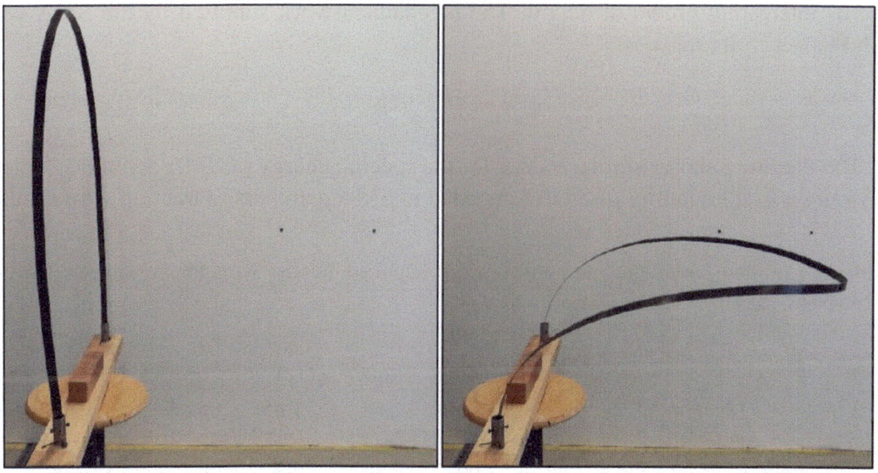

Fig. 7. Torsional buckling actuator: initial state (*left*) and deployed state upon actuation (*right*)

the specific material and geometric configuration of the manufactured prototype, the structure can remain in the deployed state without further actuation, due to gravity. Upon heating the other wire, the supports rotate back and the structure returns to its original position.

In this actuator, there is no springback force opposing to the movement of the actuators, except the elastic strain energy of the steel ribbon before it buckles. However, the two SMA wires can be actuated separately: the actuation of one wire enables the other wire to return to its pre-strained position. Thus, only one wire should be heated at one time. There is always at least one wire in a pre-strained and cold state. For this actuator, no measurements have been carried on yet.

Facade Integration

The three presented prototypes could perform the structural function in sun shading devices, supporting a large and lightweight element, opaque or translucent (membrane, fabric or composite surface) performing the occultation. They could at the same time be the actuating parts of these sun shades. Exploiting the experimental data we can calculate the mechanical work supplied by the heated wires to the system constituted by the bent lamellae, through the following formula:

$$W = \frac{1}{2} l \cdot EI \cdot \left(\chi_f^2 - \chi_i^2\right)$$

where E is the elastic modulus of the lamella, I is its moment of inertia, l is its length and χ_i and χ_f are the initial and final curvatures of the lamellae. The results are presented in Table 1.

The difference of two orders of magnitude in mechanical work mainly stems from the difference in stiffnesses of the lamellae. Probably wood is not the right material for an upscaling of the first actuator. However, the calculation for the second prototype can give us interesting orders of height. The mechanical work supplied by the SMA wire with respect to its mass is:

$$\epsilon = 980 \text{ J/kg}$$

The literature shows similar values for the specific energy of SMA actuators. This is why only a small quantity of SMA is needed in SMA actuators: a traction wire enables

Table 1. Comparison of the mechanical work supplied by the wire to the lamella for two prototypes

	E (GPa)	I (mm^4)	χ_i (m^{-1})	χ_f (m^{-1})	W (J)
First prototype (Section "A Bilayer-Type Actuator")	1.08	1.64	1.85	5.08	3.56×10^{-3}
Second prototype (Section"A Buckling Actuator")	2.06	0.83	2.33	6.12	8.23×10^{-1}

to fully exploit the potential of shape memory alloys. Even for an upscaling to a façade system, the quantity of SMA needed is expected to remain low.

Furthermore, the integration on the outer side of the façade needs the sun shades to be designed for high loads, such as wind loads and snow loads. This is why, at this step, an integration into a double-skin façade is planned, in order to avoid high design loads, to facilitate maintenance, as well as to increase service life.

Finally, the issue of the actuation temperatures inside a double skin façade needs to be addressed. Indeed, the air gap temperature can easily exceed 70 °C on summer days. For an electric control, the actuation temperature will have to be superior to the air gap temperature.

Conclusion

This preliminary study aims at proposing some actuation principles for SMA-based sun protection devices. The three manufactured prototypes present the main characteristics of SMA actuators. First, they are stable below and above two temperature limits, but between these limits they are difficult to control, because imposing a temperature does not fully characterize the state of the device. Thus, SMA actuators are better suited for applications only requiring a change from one initial to a unique final shape. Secondly, the shape memory effect in SMAs only works in one way. In order to switch between two positions, an SMA actuator needs a springback force or different SMA wires actuating the structure in two different ways, at two different times. Finally, these 3 prototypes illustrate some advantages of SMA actuators over conventional actuators, like a silent actuation and a reduced number of mechanical elements.

We can also list some remaining technical challenges in SMA actuation. One drawback of this type of actuation is the lack of control upon the system: it is difficult to obtain a feedback on the real position of the actuator. Even upon electrical activation, measuring the resistivity of the wire is not a sufficient information to determine the state of the actuator: the wire's resistivity is variable and depends on the volume fractions of martensite and austenite phase. Furthermore, the energy consumption of these devices has to be studied and compared to other commonly used solutions. Indeed, SMA can provide a high energy output with a remarkably low weight, but resistive heating is a poorly efficient energy transfer. The continuous heating of SMA wires to maintain a deployed position can be questioned.

These three prototypes could be the basis elements of adaptive shading devices for building facades. Through this critical analysis of different actuation principles, the potential of shape memory alloys for architecture and façade engineering has been clarified. This work is being conducted in strong partnership with a façade consultancy firm. Material behavior research is being carried out in parallel with the design and analysis of different prototypes, aiming at realizing SMA actuators for sun shading devices.

Acknowledgements. This work has been partially funded by Arcora & Ingérop Group and by ANRT under CIFRE Grant No. 2015/0495.

References

Basit, A., L'Hostis, G., Pac, M., Durand, B.: Thermally activated composite with two-way and multi-shape memory effects. Materials **6**(9), 4031–4045 (2013)

Charpentier, V., Hannequart, P., Adriaenssens, S., Baverel, O., Viglino, E., Eisenman, S.: Kinematic amplification strategies in plants and engineering. Smart Mater. Struct. **26**(6), 063002 (2017)

Fiorito, F., Sauchelli, M., Arroyo, D., Pesenti, M., Imperadori, M., Masera, G., Ranzi, G.: Shape morphing solar shadings: a review. Renew. Sustain. Energy Rev. **55**, 863–884 (2016)

Hannequart, P., Peigney, M., Caron, J-F.: A micromechanical model for textured polycrystalline Ni–Ti wires. In: Conference Proceedings from the International Conference on Shape Memory and Superelastic Technologies (San Diego, CA, USA, May 15–19, 2017). ASM International, Materials Park (2017)

Mohd Jani, J., Leary, M., Subic, A., Gibson, M.: A review of shape memory alloy research, applications and opportunities. Mater. Des. **56**, 1078–1113 (2014)

Scirè Mammano, G., Dragoni, E.: Modelling, simulation and characterization of a linear shape memory actuator with compliant bow-like architecture. J. Intell. Mater. Syst. Struct. **26**(6), 718–729 (2015)

Thill, C., Etches, J., Bond, I., Potter, K., Weaver, P.: Morphing skins. Aeronaut J **112**(1129), 117–139 (2008)

Zhou, G., Lloyd, P.: Design, manufacture and evaluation of bending behaviour of composite beams embedded with SMA wires. Compos. Sci. Technol. **69**(13), 2034–2041 (2009)

A Multi-scalar Approach for the Modelling and Fabrication of Free-Form Glue-Laminated Timber Structures

Tom Svilans(✉), Paul Poinet, Martin Tamke, and Mette Ramsgaard Thomsen

Centre for Information Technology and Architecture (CITA), KADK, Copenhagen, Denmark
{tsvi, paul.poinet}@kadk.dk

Abstract. This research project presents both innovative multi-scalar modelling methods and production processes aimed at facilitating the design and fabrication of free-form glue-laminated timber structures. The paper reports on a research effort that aims to elucidate and formalize the connection between material performance, multi-scalar modelling (Weinan 2011), and early-stage architectural design, in the context of free-form glue-laminated timber structures. This paper will examine how the concept of multi-scalar modelling as found in other disciplines can also be used to embed low-level material performance of glue-laminated timber into early-stage architectural design processes, thus creating opportunities for feedback across the design chain and an increased flexibility in effecting changes. The research uses physical prototypes as a means to explore and evaluate the methods presented.

Keywords: Multiscale modelling · Glue-laminated timber · Material performance

Introduction

Timber as a building material is seeing a resurgence in research and architectural and construction industries. It is sustainable, renewable, and has many properties that make it a promising alternative to common building materials such as steel and concrete. New opportunities to expand and innovate with the use of engineered timber and free-form glulams in architecture emerge with recent advances in digital simulation tools and computational design workflows. Real progress in this area is, however, inhibited by issues concerning the integration of the multiple scales, material behaviours, and constraints from fabrication processes. We present innovative design, modelling, and fabrication workflows developed in an applied architectural and industrial context. We set a focus on approaches that benefit early-stage design in multi-disciplinary architectural practice, and on material performance as a key concern for industrial timber fabricators. These approaches must remain flexible and abstract enough to be used as early as possible within the design process, and to facilitate conversations and estimates before the scope of the design is even set. The following

K. De Rycke et al., *Humanizing Digital Reality*,
https://doi.org/10.1007/978-981-10-6611-5_22

describes an effort towards this aim which integrates notions of multi-scalar modelling and material performance into the design and prototyping of an experimental free-form glue-laminated timber structure.

Multi-scalar Modelling and Integrating Material Performance

New Design-to-Fabrication Frameworks

One of the key aims of this research is to address the linearity in digital chain—from the design and development of an architectural proposal; its translation into material lengths, dimensions, and production data; and finally to its fabrication and assembly—by introducing opportunities for feedback, recursion, and bidirectionality into the process. This comes from two separate research projects, both part of the Innochain Training Network: one which looks at how the integration of material performance into early-stage architectural design can lead to new and better-informed methods of modelling and making free-form timber structures, and the other which investigates the notion of multi-scalar modelling as it applies to the design and management of complex architectural proposals. Both consider the separation and differences that exist between design and fabrication, and search for a way with which to close the gap and integrate the two ends of the architectural process. This rapprochement has been the subject of discussion in other areas, and theoretical frameworks and solutions have been proposed by other practitioners in the field. Indeed, the shift from linear and discrete modes of working between design and fabrication has also forced a reconsideration of the roles of architect and fabricator, enabled in large part by the accessibility and convergence of digital modelling tools with CAM software. The idea of the 'digital craftsman' (Scheurer 2013) has repurposed the contemporary architect/designer as a digitally-enabled designer/maker, made possible by the blurring between digital tools of design and digital tools of making. The familiarity with and integration of advanced modelling and simulation tools within contemporary design processes means that the production and manipulation of fabrication data is now more accessible than ever, and thus designers have the capacity to infuse their work with material and fabrication considerations from the very beginning. This feedback from production back to conception is also explored in the Fabrication Information Modelling (FIM) framework (Duro-Royo and Oxman 2015), which again ties the design of an architectural artefact to the processes and constraints of its making. Both frameworks seek to include fabrication parameters within the design process, though the risk is then that design decisions are driven and consumed by these low-level parameters and fabrication considerations.

Multi-scalar Modelling

Introduced within different fields—from mathematics to weather simulations—the concept of "Multi-Scalar Modelling" comes from the realization that the full behavior of a particular system cannot be represented within a single model, since important aspects of it transcend different scales and levels of resolution (Weinan 2011).

Maintaining full detailed resolution through all levels within one single environment makes the system unwieldy, computationally expensive, and difficult to manage and change. Existing architectural design research projects have already introduced Multi-scalar Modelling as a design-to-fabrication paradigm where structures and data are linked across different scales through integrative pipelines and simulation frameworks that use specific design methods—from mesh-based bidirectional information flows (Nicholas et al. 2016), to graph-based modelling and relational networks (Poinet et al. 2016)—tackling the different issues raised above.

The multi-scalar approach in this research seeks also to mitigate those potential issues by introducing bidirectional feedback between the different scales of design. That is, design decisions made at a broad level affect the low-level parameters of individual components, but changes in low-level parameters can filter back up and affect higher-level aspects of the design. The goal is to not privilege either the top-down design-led path or the bottom-up fabrication-led path, but rather to achieve a mediated middle-ground where conflicts between scales can be resolved through changes at either level. The hope is that this design methodology allows more freedom in marrying fabrication realities to architectural design intentions, and a fluidity of change even later in the design-to-fabrication process.

The mechanism by which this multi-scalar method is explored in this research is an object-oriented data ecology and graph-based organizational models. This allows low-level timber-specific material feedback—such as bending limits, springback anticipation, and fabrication constraints—to be localized to individual elements or fabrication workpieces, while also being communicated to neighbouring elements and other components through relational graphs at higher scales. Similarly, optimizations which include many individual elements can be effected at higher scales, after which the specific changes can be communicated down to the individual element models, which in turn propagate those changes through their internal material model and verify it against constraints and material limits. This sort of back-and-forth communication between an element network and its constituent parts enables design decisions that can choose to either mitigate conflicts, or else override one scale in favour of the other in general or specific cases.

Modelling and Data Management

The design and modelling of a free-form glulam structure in such a feedback-laden context required a reconsideration of standard modelling tools. Moving from generic representational data types such as surfaces, curves, and meshes to more material-specific and constrained models meant that new corresponding data structures needed to be conceived and implemented, ones which would be able to describe and keep track of things such as material limits and fabrication parameters. This effort began as a set of simple convenience functions and data structures to ease the modelling process—analyses for enforcing bending limits, for example, or quick ways to generate oriented and dimensioned extrusions that would describe a free-form glulam blank. This cascaded into the implementation of a more generic and extensible class hierarchy which started to integrate more complex arrangements of glulams, machining features, and finer control over the geometric and material properties of the model. The main goal was to encapsulate as much information as possible into discrete objects and processes that had real counterparts—glulam blanks being the main object of interest—and to separate the data from its representation—which means that geometry and visual feedback is generated on-demand, and most operations on or with these glulam data structures deal directly with the data, and not its geometrical output. These two simple but crucial considerations—encapsulation and the separation of data and representation—meant that the emerging library could remain lightweight—important, especially when handling a large number of free-form members—and modular. Variations or new forms of existing object types can be implemented through a simple class inheritance scheme, which opens up the use of the library for new and speculative glulam designs, as well as allowing it to be extended for specific project demands.

The data library mainly revolves around a generic Glulam data structure. From a multi-scalar point of view, this contains enough minimal information to keep track of individual lamellas within the Glulam, as well as lamella counts and dimensions that allow the total bounding volume and 3D representation to be generated. In such a way, the data type is kept compact and lightweight, while being capable of responding to analysis at different scales. This mentality extends to the generation and manipulation of Feature objects, which represent local relationships or properties, such as specific geometries on the glulam member or machining operations to be performed on it. The latter especially benefits from this minimalism, as toolpaths and cuts can be described more accurately and compactly by defining key workplanes and the parameters of a specific type of machining operation, for example. In this sense, it shares similarities with efforts such as the Building Transfer Language (BTL) (Stehling et al. 2014), which presents a way of describing timber machining operations in a compact, modular, and machine-agnostic way, without resorting to producing and interpreting geometrical models.

Therefore, by separating the properties of the model from its geometrical representation, it becomes trivial to link the model data to other models and inputs and thus dynamically drive the model from various scales and directions. Likewise, by encapsulating these properties within an object-oriented ecology of data types, objects retain these properties during transformations and changes to parts of the model. This is what allows the model to retain its relationships between individual members, material properties, and fabrication information during the design process. Keeping these types of operations and types discrete from each other also divides the complexity of the overall model into manageable and interchangeable chunks—an ecology of models that talk amongst themselves rather than a total whole.

Prototyping and Fabrication

Demonstrator

As a way to synthesize the multi-scalar concepts described earlier with the data management and software architecture described in the previous section, a free-form timber demonstrator was conceived and developed. This served as a case-study that allowed us to explore the proposed methods and evaluate their effectiveness as modelling and information-generating tools. The focus was on developing a modelling architecture that would be able to embody ideas of multi-scalar modelling down to the level of fabricating individual components. Improving the fabrication feasibility and precision is another on-going parallel effort. The demonstrator provided an opportunity to test the robustness of the software library and its flexibility once integrated into a multi-model and multi-resolution workflow. The criteria for success therefore focused more on the flexibility and communication between these models rather than a finished physical outcome.

As a design driver, a speculative branching glulam module was used to compose an enveloping structure (see Fig. 1). The branching module was the result of previous experiments with free-form glulams, taking advantage of timber bending limits as well as lamination techniques. It was devised as a way to create structural connections between more than two points into a single member through the process of lamination and cross-lamination, thereby creating a structural module which could be arranged into more complex and unique arrangements using simple end-to-end lap joints. With this modular branching in mind, a vaulted pavilion was conceived with a predefined branching pattern of structural members (see Fig. 2). The goal of the exercise was then to marry this predefined pattern to the material realities of the branching glulam member, while constraining it to our specific production capacities. This clash—between imposed pattern and material performance—was where the multi-scalar modelling framework was put to use. This feedback gave valuable direction for the final pavilion design: a vaulted structure between several foot conditions—a multi-legged catenary surface populated with the branching modules. At these foot conditions, where the weight of the whole pavilion was concentrated, structural members are thicker and fewer. As the structure expands from these foot conditions, the use of the branching module means that the structure divides and thins out, covering the vault surface in

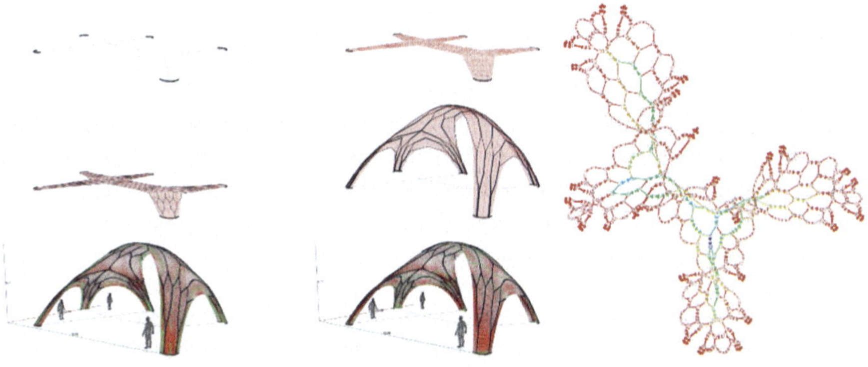

Fig. 1. Overview of the modelling process for a design iteration (*left*) and overall graph representation of the demonstrator (*right*). The *color* gradient corresponds to the degree of centrality of each component while the *line thickness* relates to the number of laminates contained within each member

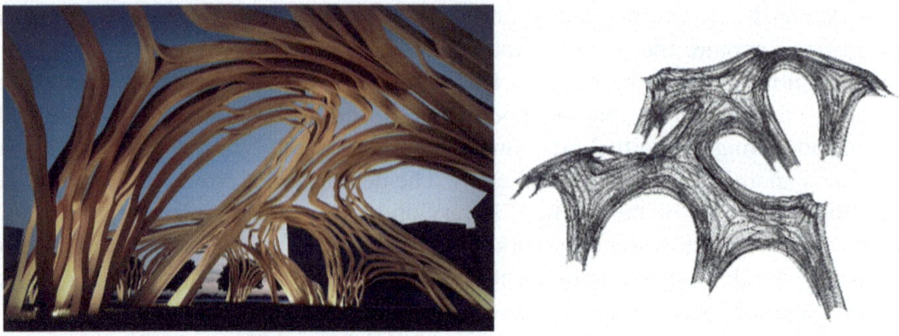

Fig. 2. Interior render of the demonstrator (*left*) and a graph representation of the whole design (*right*)

between with a network of thinner members. From this overall design, a number of individual components were brought through to physical prototyping as a way to explore the full cycle—from design to production—and explore the potential input of physical feedback into the multi-scalar model.

Multi-scalar Modelling in Practice

The architecture of the workflow was framed as a set of interconnected models operating at different scales (see Fig. 3). An undirected graph based on a mesh model at a broad level served to guide and organize the entire design—a schematic skeleton containing minimal but crucial information about the spatial location and orientation of each structural member, node, and the relationships between them. This allowed light-weight relaxation and optimization tools and structural analyses to be run over the whole network quite quickly, without delving into too much computational complexity or geometric detail. The constituent members of this network were then tied to individual beam models which began to describe each structural member in further geometric detail. Each beam model was linked to one or more material glulam models which contained information about the composition and type of glulam blank required to respond to the beam's structural and geometrical requirements. This dictated how the glulam blank would be formed: how tightly the wood fibres are aligned to the free-form beam axis influences the beams structural capacity, but has a consequent impact on the complexity of forming and constituent lamella sizes. This was then processed further into a fabrication model, which took into account fabrication limitations, specific fabrication processes and toolpath strategies, and so on. These processes were then looked at individually in terms of machine-specific workpiece fixation, accessibility of toolpaths, fabrication times, and other low-level fabrication parameters.

What this meant was that a design could be proposed and formulated into a graph, then tested through the lower-level models to yield the necessary types of glulam blanks, material dimensions, and fabrication data. More importantly, however, this allowed the aforementioned feedback between models. For example, a design iteration may cause a material model for a particular component to demand lamella sizes that are too small or impractical to use. Constraining these lamella dimensions to some other

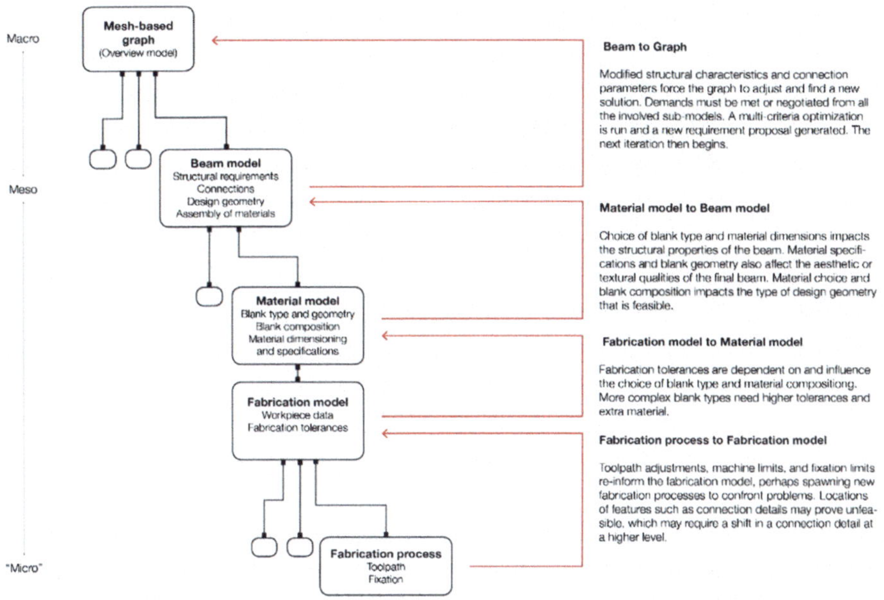

Fig. 3. Overview of the multi-scalar framework and the types of feedback between the different scales of models in this particular case

size could mean that a different type of glulam blank would have to be used for creating the module, which could have structural and geometric repercussions, and which further might force a re-evaluation of that module's role in the overall structural scheme and the roles of its neighbours (see Fig. 4). Estimation of timber quantities and dimensioning therefore became a quick byproduct of each design iteration, and the impact of design changes on fabrication time could be estimated at a glance. This also resulted in lower-level models also having an active voice in the mediation and development of the overall design scheme (Fig. 5).

3D Scanning

As a way to bring physical feedback from the production process back into the model, each fabricated prototype was also 3D scanned. The role of scanning was on several levels: firstly and most obviously, it helped us compare the fabricated result with the production model, which reiterated the live and unpredictable behaviour of laminated wood (see Fig. 6). Secondly, it allowed us to locate the geometrically complex glulam blanks in front of the cutting spindle and adjust toolpaths and fixations as required. Thirdly—and possibly most interestingly for future work—as a documentation technique, the scanning allowed us to begin to record the deviations of the physical product from the fabrication model, and use this record to inform subsequent modelling decisions. In the most common case, seeing deviations caused by springback during the glulam blank forming process allowed us to adjust subsequent material models with greater bending to compensate for this springback. This raises questions about the

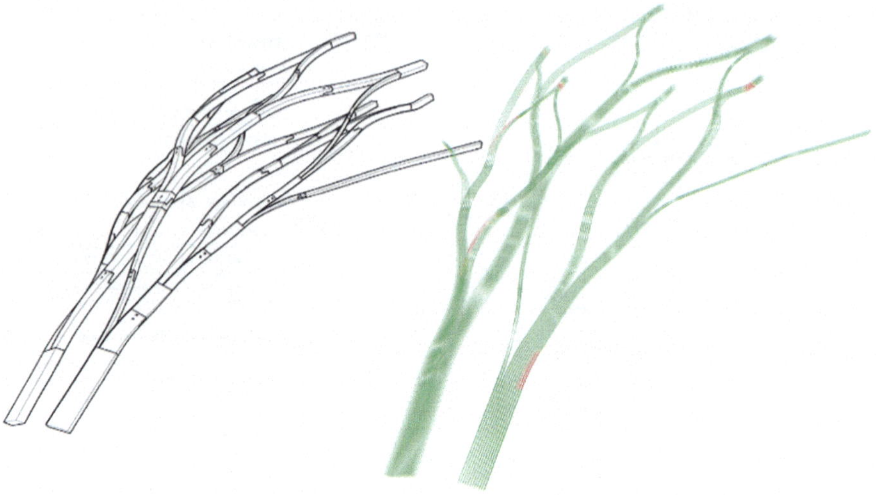

Fig. 4. The area around a foot condition extracted from the global design proposal, showing both geometric information from the beam model (*left*) and material information from the material model (*right*). In this case, the model is checked to see if the constituent lamellas of each module are within their bending limits

timeline of the design-to-production workflow and if it would be even possible to adjust larger design decisions when the first parts of the project have already been manufactured, and what the role of the accumulated scan feedback is once a particular project is over.

Conclusion and Future Work

In conclusion, as a design and information-generating framework, multi-scalar modelling is a promising alternative to other, more holistic or fragmented design methods, especially in the context of free-form timber. Designs that involve a complex and somewhat unpredictable material such as timber would greatly benefit from the feedback and low-level input that this method offers. Additionally, as a production and optimization tool, the presented workflow allowed a quick succession of design iterations, while avoiding a laborious translation from design information to fabrication information each time.

However, certain challenges remain, most notably of how this technique could better incorporate the glulam blank forming process and how it could help to tighten the tolerances of the finished products. This research is part of two larger research projects which look at both multi-scalar modelling and the role of material performance in glue-laminated timber construction. As such, future work will focus on addressing the integration of material performance in the forming and machining of glue-laminated components, and how feedback from this level could contribute and guide early-stage design decisions; as well as how physical feedback in a multi-scale design environment

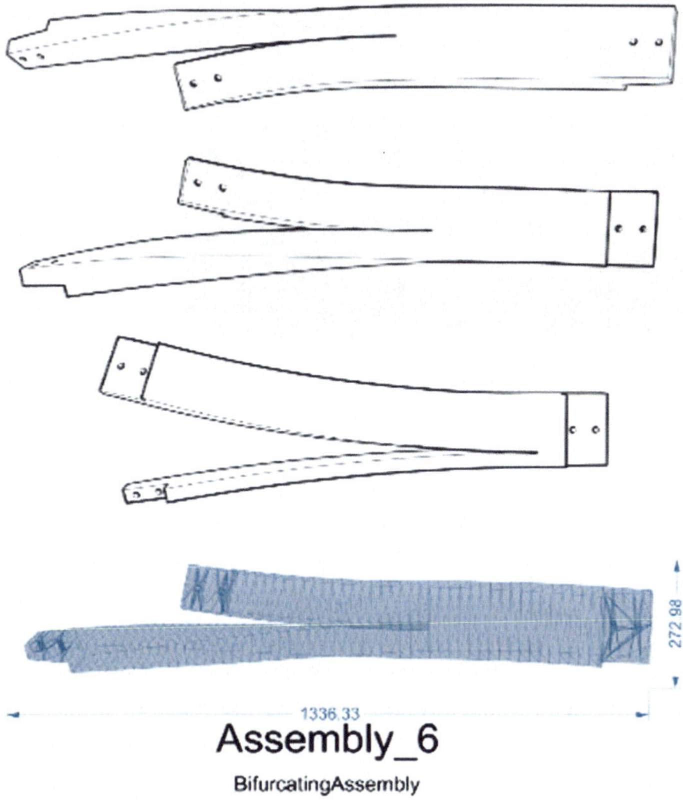

Fig. 5. The design broken down into individual component models

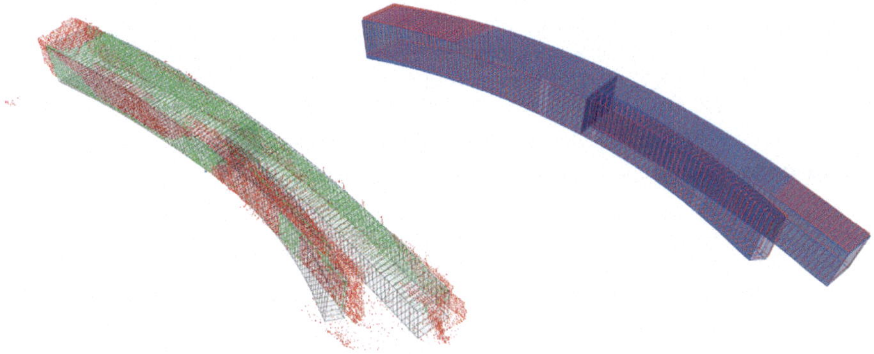

Fig. 6. Analysis of 3D scans of a bifurcating glulam blank (*left*) and its production geometry (*right*)

Fig. 7. Forming one of the bifurcating glulam components (*left*) and some finished examples (*right*)

could be recorded, accumulated, and leveraged for future use. Continued collaboration with the involved industry partners will allow this method to be tested against established workflows and methods of modelling, optimizing, and fabricating complex glue-laminated timber structures.

Most important, however, is the possibility of an informed mediation that this method enables: having multiple scales and resolutions available throughout a longer period during a project development phase allowed, in this case, a designer to make decisions faster and more frequently. By widening the scope of information available and modelling the interrelations between constraints at various scales, the method provides an increased measure of control over complexity in architectural design and points to ways in which free-form timber structures can become more feasible and accessible (Fig. 7).

Acknowledgements. This project was undertaken at the Centre for IT and Architecture, KADK, in Copenhagen, Denmark as part of the Innochain Early Training Network. This project has received funding from the European Union's Horizon 2020 research and innovation programme under the Marie Sklodowska-Curie Grant Agreement No. 642877. We would also like to express our sincere thanks to the CITAstudio Masters students that helped during the fabrication process.

References

Duro-Royo, J., Oxman, N.: Towards fabrication information modeling (FIM). In: MRS Proceedings, vol. 1800. Cambridge University Press (2015)

Nicholas, P., Zwierzycki, M., Stasiuk, D., Nørgaard, E., Thomsen, M.R.: Concepts and methodologies for multiscale modeling—a mesh-based approach for bi-directional information flows. In: ACADIA 2016—Posthuman Frontiers, pp. 110–129 (2016)

Poinet, P., Nicholas, P., Tamke, M., Thomsen, M.R.: Multi-scalar modelling for free-form timber structures. In: Proceedings of the IASS Annual Symposium, Spatial Structures in the 21st Century (2016)

Scheurer F (2013) Digital craftsmanship: from thinking to modeling to building. In: Marble S (ed) Digital Workflows in Architecture. Birkhäuser, Basel, pp. 110–129

Stehling, H., Scheurer, F., Roulier, J.: Bridging the gap from CAD to CAM: concepts, caveats and a new Grasshopper plug-in. In: Gramazio, F., Kohler, M., Langenberg, S. (eds.) Fabricate: Negotiating Design & Making, pp. 52–59 (2014)
Weinan E (2011) Principles of Multiscale Modeling. Princeton University, Princeton

Assessment of RANS Turbulence Models in Urban Environments: CFD Simulation of Airflow Around Idealized High-Rise Morphologies

Farshid Kardan[1,2(✉)], Olivier Baverel[2], and Fernando Porté Agel[1]

[1] Wind Engineering and Renewable Energy Laboratory (WIRE), Ecole Polytechnique Fédérale de Lausanne (EPFL), Lausanne, Switzerland
farshid.kardan@epfl.ch
[2] Laboratoire GSA-Géométrie Structure Architecture (Ecole Nationale Supérieure D'architecture Paris-Malaquais), University Paris-Est, Paris, France

Abstract. This paper presents the evaluation of Reynolds-averaged NavierStokes (RANS) turbulence models, including the RNG $k - \varepsilon$, SST $k - \omega$ and more recently developed SST $\gamma - Re_\theta$ models of flow past an idealized three-dimensional urban canopy. For validation purposes, the simulated vertical and spanwise profiles of mean velocity are compared with wind tunnel measurements and large eddy simulation (LES) results. These quantitative validations are essential to assess the accuracy of RANS turbulence models for the simulation of flow in built environments. Furthermore, additional CFD simulations are performed to determine the influence of three different idealized high-rise morphologies on the flow within and above the semi-idealized urban canopy. In order to assess airflow behavior, the pressure coefficient on high-rise morphologies, turbulence kinetic energy contours and vertical velocity magnitude profiles at roof level of high-rise and surrounding buildings are evaluated. The results render the SST $\gamma - Re_\theta$ model attractive and useful for the simulation of flows in real and complex urban morphologies. For the region around an idealized high-rise building, different flow patterns and strong changes in velocity magnitude and pressure coefficient are observed for different building morphologies.

Keywords: Computational fluid dynamics (CFD) · RANS models · Idealized urban canopy · Building morphologies

1 Introduction

The fast and accurate prediction of turbulent airflow in built environments is becoming increasingly essential to architectural and urban projects. Three main approaches can be used to study airflow in urban environments: full-scale measurements, reduced-scale wind tunnel experiments and computational fluid dynamics (CFD). (i) Full-scale measurement is the most accurate method, as it utilizes the physical and real scale objects. However, it is expensive and time-consuming, which makes it impractical in many situations. (ii) Reduced-scale wind tunnel experimentation is comparatively

K. De Rycke et al., *Humanizing Digital Reality*,
https://doi.org/10.1007/978-981-10-6611-5_23

faster and less expensive than the full-scale measurement method, but the analyzing process is generally limited to a few selected discrete positions in the flow field, which can make it difficult to study a large and complex environment like an urban canopy. (iii) Computational fluid dynamics (CFD) provide a time and cost effective method to study and predict turbulent flows. CFD is particularly useful to forecast urban meteorological phenomena, estimate the dispersion of air pollutants, predict fire spreading, evaluate the urban heat island effect, optimize the outdoor and indoor thermal environments, maximize the wind comfort and safety around buildings, calculate the wind loads on building structures, improve the indoor air quality by natural ventilation, and optimize the harvesting of solar and wind energy in built environments (Moonen et al. 2012; Nakayama et al. 2011; Stankovic et al. 2009).

In general, there are three main types of CFD methods (Ferziger and Peric 1997; Pope 2000): (i) direct numerical simulation (DNS), (ii) large-eddy simulation (LES) and (iii) Reynolds-averaged Navier-Stokes (RANS). The RANS simulations are highly sensitive to the choice of turbulence model, but have the advantage of having a substantially lower computational cost compared to DNS and LES. For this reason, RANS remains the CFD tool of choice for many applications, particularly those involving short-time forecasts and large-scale simulations, which can be highly computationally demanding. Considering the variety of RANS turbulence models available in the literature and the sensitivity of the results to the choice of turbulence model, a rigorous evaluation of said models is needed (Liu et al. 2013; Liu and Niu 2016; Tominaga and Stathopoulos 2013). For this purpose, a comprehensive comparison of three RANS turbulence models is conducted including the re-normalisation group (RNG) $k - \varepsilon$ (Yakhot et al. 1992), shear-stress transport (SST) $k - \omega$ (Menter 1994) and SST $\gamma - Re_\theta$ (Langtry and Menter 2009) models for flow over an idealized three-dimensional urban canopy by comparing the simulation results with the wind tunnel measurements of Brown et al. (2001) and the large-eddy simulation results of Cheng and Porté-Agel (2015). Recently, Kardan et al. (2016) conducted a detailed evaluation of RANS turbulence models for flow over a two-dimensional isolated building, and reported a relatively better performance of the SST $k - \omega$ and SST $\gamma - Re_\theta$ models in both unstructured and coarse grid resolution. However, it is also reported that further investigation into complex three-dimensional urban environments is needed to generalize the SST models' results.

High-rise buildings have become increasingly prevalent in cities as an inevitable result of population growth and lack of available land in urban areas. Since rectangular prisms, cylinders, and twisting forms are some of the most common geometries found in high-rise building designs, it is essential to assess how building morphologies affect wind behavior around such structures. In this study, following the evaluation and selection of an appropriate RANS turbulence model, additional CFD simulations are conducted to determine the effect of the aforementioned high-rise morphologies on the flow above the semi-idealized urban canopy. In order to assess the wind behavior on and around buildings, the pressure coefficient distribution, turbulent kinetic energy (TKE) and vertical profiles of velocity magnitude at roof level of high-rise and sur-rounding buildings are evaluated.

2 Numerical Setup

The simulations were performed on the Deneb cluster using the high-performance computing (HPC) unit of Swiss Federal Institute of Technology in Lausanne (EPFL). A total of 64 CPUs (Intel Xeon) were used in parallel for the simulations.

Idealized Urban Canopy Case

The CFD code Fluent (version 15.0.7) is used to perform RANS simulations of flow over an idealized urban morphology in isothermal conditions, with the aligned array consisting of 7 × 6 buildings with a height of h (=0.15 m), and between building spacing equal to h in both the streamwise and spanwise direction. Hence, the numerical setup is selected to reproduce the conditions of the wind tunnel experiment of Brown et al. (2001). The Reynolds number of the flow is 3×10^4 based on the free stream velocity and building height. The same experimental data was recently used for validation of a large-eddy simulation code (Cheng and Porté-Agel 2015). The computational domain, shown in Fig. 1, has dimensions equal to $L_x \times L_y \times L_z$ 4.8 m × 1.8 m × 2.1 m. The structured grids with grid resolution of $N_x \times N_y \times N_z = 320 \times 120 \times 210$ grid points are tested, which correspond to a number of grid points covering the building of $(n_x \times n_y \times n_z) = 10 \times 10 \times 15$, respectively.

A turbulent inflow condition with a non-uniform (logarithmic) mean velocity profile is used at the inlet, as in the reference experiment. The inlet profile of the normalized mean streamwise velocity is compared with wind tunnel experiment and LES results, and is shown in Fig. 2. Symmetry boundary conditions are specified for

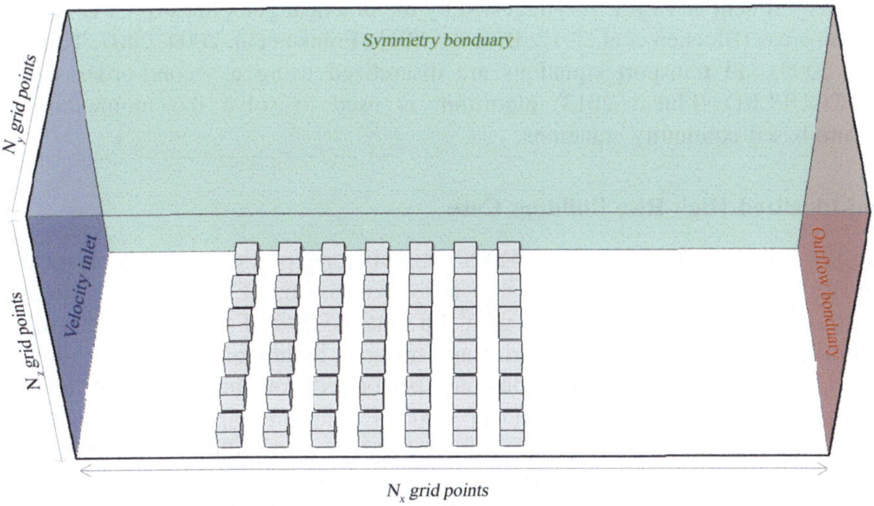

Fig. 1. The computational domain

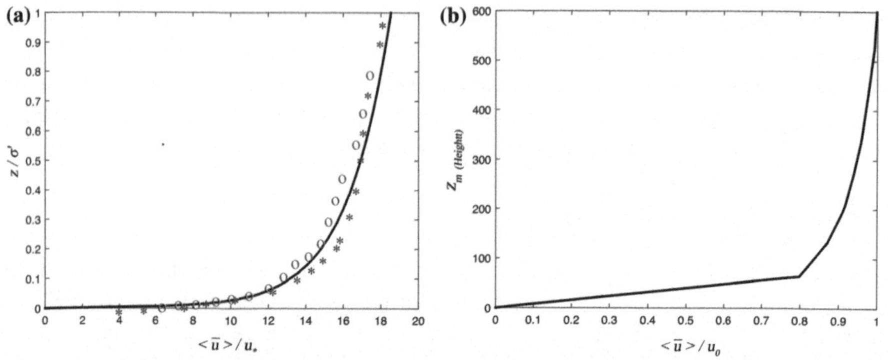

Fig. 2. Vertical inflow profiles of normalized mean streamwise velocity component **a** idealized urban canopy case and **b** semi-idealized high-rise buildings case. *Red circle symbols* wind tunnel experiment; *blue star symbols* LES; *black line* RANS

the top and sides of the computational domain (Franke et al. 2010). At the domain outlet, an outflow boundary condition is used. This implies that all streamwise derivatives of the flow variables are equal to zero at the outlet. A standard wall function based on the logarithmic law for smooth surfaces (Launder and Spalding 1974) is used at the upstream ground, downstream ground and at the surfaces of the buildings. The selected value of the sand-grain roughness height $k_s = 0.00014h$ (for the upstream ground) and $k_s = 0.014h$ (for the downstream ground and buildings surfaces) are calculated based on the relation $k_s = 9.793z_0/Cs$ (Blocken et al. 2012; Blocken 2015) where the roughness constant $C_s = 0.7$ (Tominaga 2015) and aerodynamic roughness length $z_0 = 10^{-3}h$ are used in order to reproduce the same conditions as in the wind tunnel experiment and LES. As suggested by the practical guidelines of CFD for flows in urban areas (Blocken et al. 2012; Blocken 2015; Franke et al. 2004, 2007; Tominaga et al. 2008), all transport equations are discretized using a second-order scheme. The COUPLED (Fluent 2013) algorithm is used to solve the momentum and pressure-based continuity equations.

Semi-Idealized High-Rise Building Case

As shown in Fig. 3, to perform RANS simulations of flow over semi-idealized high-rise buildings, the aligned 7 × 7 array, consisting of $l_x \times l_y$ equal to 50 m 50 m, rectangular prism buildings with differing heights of (=10 m), (=20 m), and (=50 m) for the very low-rise, low-rise and mid-rise buildings, respectively, are considered. Moreover, three simulations are performed for three different idealized high-rise morphologies with a height of (=200 m) which are implemented at the center of the semi-idealized urban canopy. The spacing between buildings is 20 m in both streamwise and spanwise directions. The computational domain has dimensions L_x $L_y \times L_z$ equal to 4800 m × 3400 m × 600 m. A turbulent inflow condition with a non-uniform (logarithmic) mean velocity profile is used at the inlet, as shown in Fig. 2.

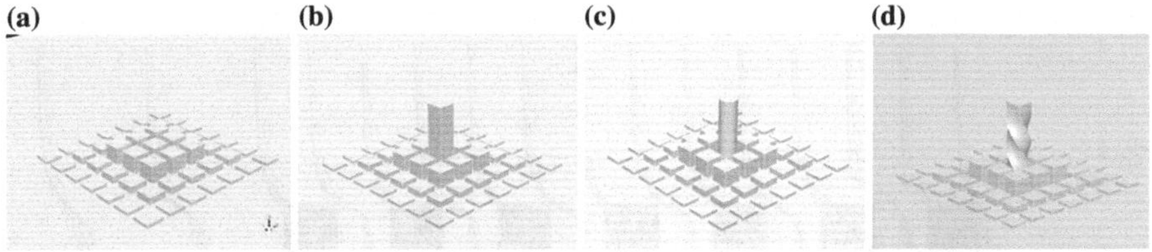

Fig. 3. Semi-idealized urban canopy **a** without a high-rise building (reference case), **b** rectangular prism high-rise morphology, **c** cylindrical high-rise morphology and **d** twisted high-rise morphology

Note that the remaining numerical setup of the simulations is similar to the one described in Section "Idealized Urban Canopy Case".

3 Results and Discussion

Vertical and Spanwise Profiles of the Idealized Urban Canopy

In this section, the normalized mean velocity simulated by the different RANS models are compared to the wind tunnel measurements of Brown et al. (2001) and the LES results of Cheng and Porté-Agel (2015) using a modulated gradient subgrid-scale (SGS) model (Lu and Porté-Agel 2010). The streamwise velocity components are denoted by (u). As shown in Fig. 4, all the RANS models reproduce the reverse flow at the top surface of the first idealized building fairly well for the mean velocity profiles. It is found that the RNG $k - \varepsilon$ model is not able to reproduce the reverse flow above and between buildings at the downstream positions from x_h = (1–4h), and a much smaller streamwise velocity is found in the aforementioned locations. In comparison, the two SST models show a slightly better prediction of the vertical profiles of streamwise velocities, which is related to the better performance of these models in adverse pressure gradient flow. In these grid conditions, the SST $\gamma - Re_\theta$ model shows the best overall prediction, followed by the SST $k - \omega$ model. The improved results of the SST $\gamma - Re_\theta$ is more obvious in spanwise velocity profiles at the downstream position of x_h = (6h). These results are consistent with the previous study of Kardan et al. (2016) that indicates that the SST models show better overall performance than the other RANS models tested for flows past an isolated building.

Vertical Profiles of Mean Velocity Around the Semi-idealized High-Rise Morphologies

In this section, three more simulations are conducted to analyze the impact of rectangular prism, cylindrical and twisting morphologies on the flow around and above the semi-idealized urban canopy, and the results are compared to a reference case where a high-rise building is absent. The SST $\gamma - Re_\theta$ turbulence model is selected based on

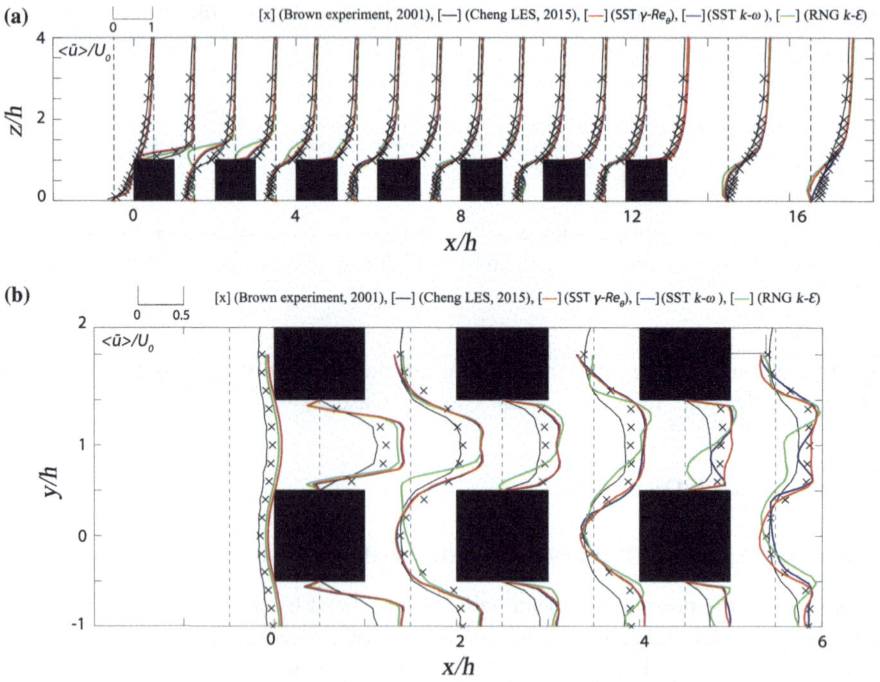

Fig. 4. **a** Vertical profiles at centerline of the idealized urban canopy and **b** spanwise profiles at $z = h/2$ of $\langle \bar{u} \rangle / U_0$. *Cross symbols* experiment; *black lines* LES; *red lines* SST $\gamma - Re_\theta$ model; *blue lines* SST $k - \omega$ model; *green lines* RNG $k - \varepsilon$ model

the results obtained in the previous section. The normalized mean velocity profiles at the center of the surrounding and high-rise buildings are projected from roof level to the domain height, and compared to that of the reference case as labeled and shown in Fig. 5.

It is found that there is symmetry between rows (a) and (c) of the results. For the upstream locations of (a, i) and (c, i), similar velocity patterns are found across the different high-rise morphologies, and that the convergence of these patterns is found at a height of (=300 m). It is worth mentioning that the velocity magnitudes upstream of the high-rise across the different morphologies are smaller than that of the reference case. It is also found that velocity magnitude is higher in the cylindrical high-rise morphology than the rectangular prism and twisted morphologies in these two locations. In the (b, i) building, all the high-rise morphologies introduce a velocity profile consisting of lower magnitudes when compared to the previously mentioned buildings. This is mainly due to the adverse pressure gradient flow that is induced by the presence of the high-rise building.

For the (a, ii) and (c, ii) buildings, similarity is found except for in the case of the twisted morphology, where some fluctuation is observed in the (a, ii) building. This is due to the irregular form of twisted geometries where opposite faces are not symmetrical. For the rectangular prism and cylindrical morphologies, the velocity

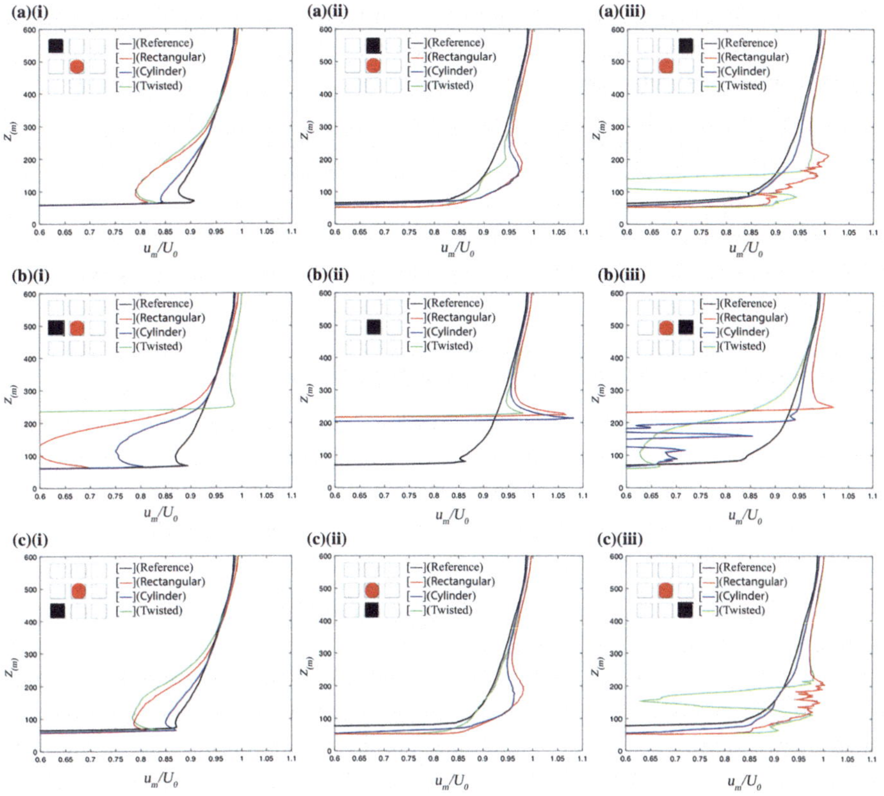

Fig. 5. Vertical profiles of the normalized mean velocity magnitude. *Black lines* reference case; *red lines* rectangular prism high-rise morphology; *blue lines* cylindrical high-rise morphology; *green lines* twisted high-rise morphology

magnitude profiles between (=50 m) and (=150 m) are significantly higher than in the reference case. The surrounding buildings downstream of the high-rise building are shown to be highly sensitive to the different types of high-rise morphologies. It is found that in (a, iii) and (c, iii) the twisted morphology velocity profile deviates the most from the reference case between (=50 m) and (=150 m), where the normalized velocity magnitude decreases to (=0.5) and (=0.62), respectively. The highest velocity magnitude across the simulations is found just above the high-rise building for the rectangular prism and cylindrical morphologies. This indicates that these two morphologies have higher potential to harvest wind energy from. However, this high velocity magnitude is not sustained, and about (=10 m) above the roof, decreases rapidly and begins to converge with the reference case.

Strong fluctuations in the vertical velocity profiles are found at (b, iii), directly downstream of the high-rise building. This indicates that the mentioned location is particularly sensitive to the morphology of the high-rise building. It is of interest to mention the differences in the vertical velocity profiles of the cylindrical and rectangular morphologies across the downstream region. In the cylindrical case, the buildings on either side of the high-rise exhibit a smooth profile similar to the reference case, whereas directly downstream of the high-rise, a large amount of fluctuation is observed. This is due to the fact that a cylindrical morphology encourages rapid re-combination of the separated flows. In the rectangular prism morphology, the velocity profile directly downstream of the high-rise building retains the shape of the profile at the high-rise itself, as it does not allow the separated stream flow to re-converge immediately. Instead, turbulent flow can be found downstream on either side of the high-rise building due to recirculation patterns found in the wake region of the building corner.

Comparison of Pressure Coefficient (C_P) and TKE Contours

Figure 6 shows the pressure coefficient distribution on the high-rise morphologies. From the results obtained, the maximum pressure coefficient is found in the case of the rectangular prism followed by the twisted and cylindrical morphologies. This indicates that the circular plan shape for tall buildings is more effective in reducing wind pressure than those of rectangular plan shapes.

In addition, the contours of the normalized turbulent kinetic energy (TKE) k/U_0^2 are also obtained along the centerline of the streamwise direction of the high-rise building morphologies, as can be seen in (Fig. 7).

In all cases, over-prediction of the TKE just above the high-rise buildings is found. Simulated patterns around the rectangular prism and twisted morphologies are more similar when compared with the cylindrical morphology, which relates to the shared rectangular plan shape of these buildings. It is worthwhile to mention that in the twisted high-rise morphology case, higher TKE is found around mid-rise buildings at upstream locations of the twisted tower, compared with those of rectangular prism and cylindrical morphologies. These results are consistent with those of the previous discussions in Section "Vertical Profiles of Mean Velocity Around the Semi-idealized High-Rise Morphologies" which indicates higher velocity fluctuation downstream of the twisted morphology.

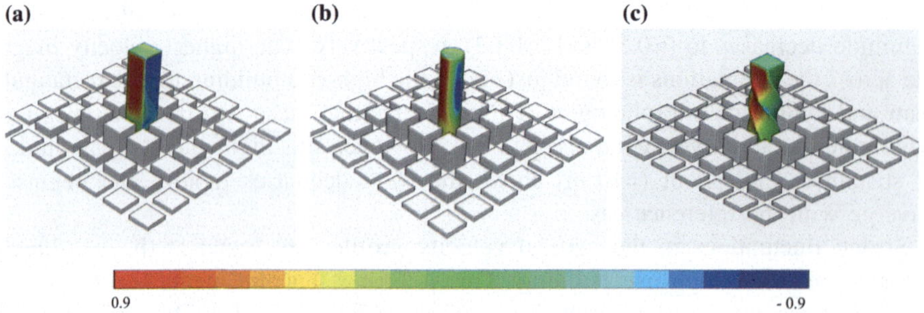

Fig. 6. Pressure coefficient distribution on **a** rectangular prism high-rise morphology, **b** cylindrical high-rise morphology and **c** twisted high-rise morphology

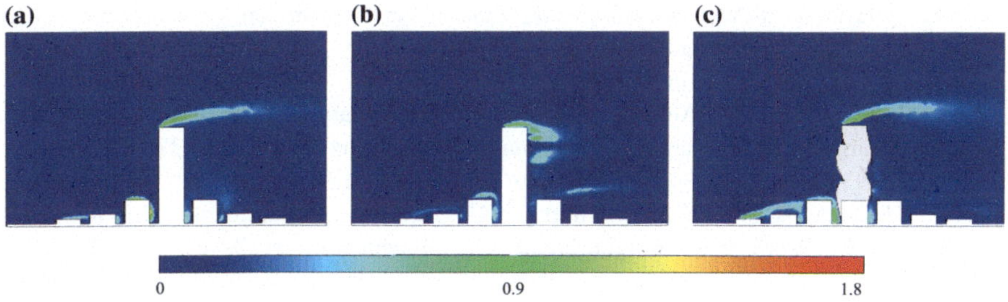

Fig. 7. Pressure coefficient distribution on **a** rectangular prism high-rise morphology, **b** cylindrical high-rise morphology and **c** twisted high-rise morphology

4 Summary

The present study assesses the performance of the RNG $k - \varepsilon$, SST $k - \omega$ and SST $\gamma - Re_\theta$ models for flow over an idealized urban canopy by comparing the simulation results with the wind tunnel measurements of Brown et al. (2001) and the large-eddy simulation results of Cheng and Porté-Agel (2015). The SST $k - \omega$ and SST $\gamma - Re_\theta$ models show improved predictions for the vertical and spanwise profiles of the mean velocity. It is found the computational costs of the SST models are significantly lower, and at least a hundred times faster, than LES. In addition, as the SST $\gamma - Re_\theta$ model is developed to simulate transitioning (from laminar to turbulent) flows, it can be particularly useful for flows around supertall buildings.

The SST $\gamma - Re_\theta$ model is used to conduct further simulations, which analyze the impact of high-rise morphologies on the flow around and above the semi-idealized urban canopy. The pressure coefficient distribution on high-rise buildings, normalized TKE contours and normalized mean velocity profiles simulated for the rectangular prism, cylindrical, and twisted morphologies are compared with the reference case, where a high-rise building is absent. It is found that different building morphologies can lead to very strong changes in velocity magnitude above the surrounding buildings. More specifically, the circular plan shape of tall building is more effective in reducing wind pressure compared with those of rectangular plan shapes. Amongst all surrounding buildings, those located directly upstream and downstream of the high-rise building were most affected by the differing morphology type. It is also found that the circular plan shape for tall buildings is more effective in reducing wind pressure compared to the rectangular plan shapes.

References

Blocken, B., Janssen, W.D., van Hooff, T.: CFD simulation for pedestrian wind comfort and wind safety in urban areas: general decision frame-work and case study for the Eindhoven University campus. Environ. Model. Softw. **30**, 15–34 (2012)

Blocken, B.: Computational fluid dynamics for urban physics: importance, scales, possibilities, limitations and ten tips and tricks towards accurate and reliable simulations. Build. Environ. **91**, 219–245 (2015)

Brown, M.J., Lawson, R.W., DeCroix, D.S., Lee, R.L.: Comparison of centerline velocity measurements obtained around 2D and 3D building arrays in a wind tunnel. Technical Report, Los Alamos National Laboratory, NM, pp. 7 (2001)

Cheng, W.C., Porté-Agel, F.: Adjustment of turbulent boundary-layer flow to idealized urban surfaces: a large-eddy simulation study. Bound. Layer Meteorol **50**, 249–270 (2015)

Fluent, A.: ANSYS fluent user's guide. Release 15.0 edn (2013)

Franke, J., Hellsten, A., Hirsch, C., Jensen, A.G., Krus, H., Schatzmann, M., Miles, P.S., Westbury, D.S., Wisse, J.A., Wright, N.G.: Recommendations on the use of CFD in wind engineering. In: van Beeck, J.E. (ed.) COST Action C14, Impact of Wind and Storm on City Life Built Environment. Von Karman Institute. Proceedings of the International Conference on Urban Wind Engineering and Building Aerodynamics, von Karman Institute, Sint-Genesius-Rode, Belgium (2004)

Franke, J., Hellsten, A., Schlu ̈nzen, H., Carissimo, B.: Best practice guideline for the CFD simulation of flows in the urban environment. COST action, pp. 51 (2007)

Franke, J., Hellsten, A., Schlu ̈nzen, H., Carissimo, B.: The best practise guideline for the CFD simulation of flows in the urban environment: an outcome of cost 732. In: Fifth International Symposium on Computational Wind Engineering, pp. 1–10 (2010)

Ferziger, J.H., Peric, M.: Computational methods for fluid dynamics, pp. 364. Springer, Berlin (1997)

Kardan, F., Cheng, W.C., Baverel, O., Porté-Agel, F.: An evaluation of recently developed RANS-based turbulence models for flow over a two-dimensional block subjected to different mesh structures and grid resolutions. E.G.U. 18 (2016)

Launder, B.E., Spalding, D.B.: The numerical computation of turbulent flows. Comput. Methods Appl. Mech. Eng. **3**, 269–289 (1974)

Langtry, R.D., Menter, F.R.: Correlation-based transition modeling for unstructured parallelized computational fluid dynamics. AIAA J. **47**, 2894–2906 (2009)

Liu, X., Niu, J., Kwok, K.C.S.: Evaluation of RANS turbulence models for simulating wind-induced mean pressures and dispersions around a complex-shaped high-rise building. Build. Simul. **6**, 151–164 (2013)

Liu, J., Niu, J.: CFD simulation of the wind environment around an isolated high-rise building: an evaluation of SRANS, LES and DES models. Build. Environ. **96**, 91–106 (2016)

Lu, H., Porté-Agel, F.: A modulated gradient model for large-eddy simulation: application to a neutral atmospheric boundary layer. Phys. Fluids **22**, 1–12 (2010)

Menter, F.R.: Two-equation eddy-viscosity turbulence models for engineering applications. AIAA J. **32**, 1598–1605 (1994)

Moonen, P., Defraeye, T., Dorer, V., Blocken, B., Carmeliet, J.: Urban physics: effect of micro-climate on comfort, health and energy demand. Front. Archit. Res. **1**(3), 197–228 (2012)

Nakayama, H., Takemi, T., Nagai, H.: LES analysis of the aerodynamic surface properties for turbulent flows over building arrays with various geometries. J. Appl. Meteorol. Climtol. **50**, 1692–1712 (2011)

Pope, S.B.: Turbulent flows, pp. 771. Cambridge University Press, Cambridge (2000)

Stankovic, S., Campbell, N., Harries, A.: Urban wind energy, pp. 200. Earthscan, Abingdon (2009)

Tominaga, Y., Mochida, A., Kataoka, H., Tsuyoshi, N., Masaru, Y., Taichi, S.: AIJ guidelines for practical applications of CFD to pedestrian wind environment around buildings. J. Wind Eng. Ind. Aerodyn. **96**, 1749–1761 (2008)

Tominaga, Y., Stathopoulos, T.: CFD simulation of near-field pollutant dispersion in the urban environment: a review of current modeling techniques. Atmos. Environ. **79**, 716–730 (2013)

Tominaga, Y.: Flow around a high-rise building using steady and un-steady RANS CFD: effect of large-scale fluctuations on the velocity statistics. J. Wind Eng. Ind. Aerodyn. **142**, 93–103 (2015)

Yakhot, Y., Orszag, S.A., Thangam, S., Gatski, T.B., Speziale, C.G.: Development of turbulence models for shear flows by a double expansion technique. Phys. Fluids A **4**, 1510–1520 (1992)

Automated Generation of Knit Patterns for Non-developable Surfaces

Mariana Popescu(✉), Matthias Rippmann, Tom Van Mele, and Philippe Block

Department of Architecture, Institute for Technology in Architecture, ETH Zürich, Stefano-Franscini-Platz 1, 8093 Zurich, Switzerland
mariana.popescu@arch.ethz.ch

Abstract. Knitting offers the possibility of creating 3D geometries, including non-developable surfaces, within a single piece of fabric without the necessity of tailoring or stitching. To create a CNC-knitted fabric, a knitting pattern is needed in the form of 2D line-by-line instructions. Currently, these knitting patterns are designed directly in 2D based on developed surfaces, primitives or rationalised schemes for non-developable geometries. Creating such patterns is time-consuming and very difficult for geometries not based on known primitives. This paper presents an approach for the automated generation of knitting patterns for a given 3D geometry. Starting from a 3D mesh, the user defines a knitting direction and the desired loop parameters corresponding to a given machine. The mesh geometry is contoured and subsequently sampled using the defined loop height. Based on the sampling of the contours the corresponding courses are generated and the so-called short-rows are included. The courses are then sampled with the defined loop width for creating the final topology. This is turned into a 2D knitting pattern in the form of squares representing loops course by course. The paper shows two examples of the approach applied to non-developable surfaces: a quarter sphere and a four-valent node.

Keywords: Knitting pattern · Non-developable surfaces · 3D knitting · Computational algorithm · Automated generation

Introduction

In architecture, textiles have been used as membranes forming tensile structures and have proven to be a feasible solution for the creation of resource-efficient formwork (Brennan et al. 2013; Veenendaal et al. 2011) or reinforcement for complex concrete geometries (Scholzen et al. 2015). Currently, most are produced as flat sheet material. Therefore, the creation of doubly curved architectural shapes with these fabrics requires extensive patterning. Knitting is a widely-used fabrication technology for textiles, which allows for the creation of fabrics with great variety in structure and the possibility of creating bespoke

The original version of this chapter was revised: For detailed information please see Erratum. The erratum to this chapter is available at https://doi.org/10.1007/978-981-10-6611-5_56.

K. De Rycke et al., *Humanizing Digital Reality*,
https://doi.org/10.1007/978-981-10-6611-5_24

geometries similar to the final shape (near-net shape) without the need for patterning, avoiding waste through offcuts (Van Vuure et al. 2003; Hong et al. 1994; Abounaim et al. 2009). Within the current standard industrial machine width of 2.5 m (Aboumain 2010), architectural-scale elements with high curvature can be fabricated to be used as skins (Thomsen et al. 2008), pavilions (Sabin 2017), or within hybrid systems of bending-active elements and knitted membranes (Ahlquist 2015a, b; Thomsen et al. 2015).

For the creation of a given piece of knit textile, a knitting pattern is necessary to define a set of instructions for the CNC-knitting machine to steer the knitting process. However, current knitting software offers only limited possibilities. Any custom, non-repetitive, non-developable knit pattern needs to be programmed by the user in a manner requiring detailed manipulation and understanding of knitting operations (ShimaSeiki 2017). Solutions allowing for easier manipulation of 2D knit patterns based on primitives, simulation of the resulting 3D shape and steering of the machine have been developed by McCann et al. (2016). For the creation of a given 3D geometry, the knit patterns contain increases (widening), decreases (narrowing), and short or incomplete rows (Fig. 1). The latter is equivalent to locally increasing the number of courses.

Strategies have been developed to generate curved and doubly curved shapes from knitted textiles for primitives such as cylinders, spheres and boxes through the use of mathematical descriptions of the shapes (Hong et al. 1994). If not automated, this process relies heavily on the proficiency of the user and it may be very difficult or impossible to achieve accurate, weft-knit fabrics for shapes beyond basic geometric primitives. Strategies for freeform surfaces, demonstrated by Thomsen et al. (2008; 2015), focus on translating a 3D shape into a 2D knitting pattern through the unrolling of developable surfaces, but do not solve non-developable surfaces.

In contrast to other industries where knitting is used for mass production (e.g. garment and shoe industry, automotive industry), the construction sector has a greater demand for non-repetitive modules using bespoke geometries to meet the requirements of contemporary architecture. The ability to create knitting patterns in a fast and flexible way for various 3D geometries is therefore especially important if knitting is to be used in construction. Possible applications demanding such flexibility are stay-in-place fabric formworks with integrated features for the guidance of reinforcement and other building elements.

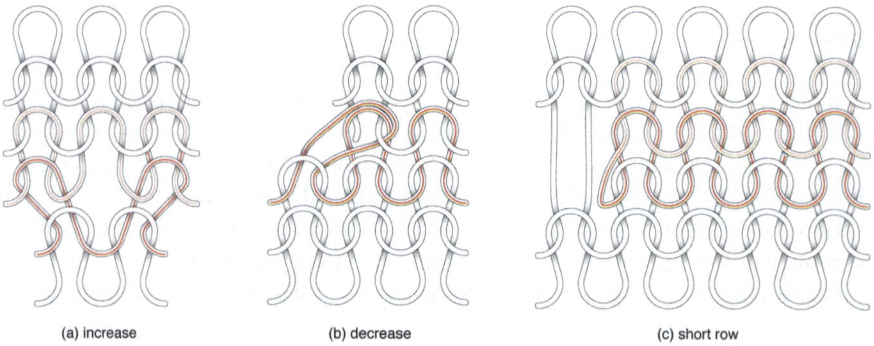

Fig. 1. Knit pattern manipulations: **a** increasing or **b** decreasing the number of loops from course to course; or **c** using short rows

This study presents an approach for directly creating knitting patterns on a given 3D geometry in an automated way without being constrained to developable surfaces. To achieve accurate results, constraints need to be taken into consideration. These are not only related to the variation in the knitting logic, but also to the accurate representation of the heterogeneous material behaviour. Similar to the work of Yuksel et al. (2012) or Igarashi et al. (2008), the presented approach relies on geometric descriptions, surface topology, loop geometry and course direction for the generation of accurate knitting patterns. In addition, the method accounts for the creation of short-rows, which follow the principles of machine knitting fabrication techniques.

In our case-studies we have applied the method among others to a quarter sphere and a four-valent node. The results for these two case studies are presented in Section "Results".

Methods

This section gives an overview of the methods used for the creation of patterns for a given 3D geometry.

As loops are of constant height and width, the knit topology is developed and represented using a quadrilateral network with only a few triangular exceptions that represent increases, decreases, or the starts/ends of short-rows. The quadrilateral network is generated with the following four steps:

1. **Patching**: Depending on the complexity of the 3D geometry, global singularities need to be defined manually to split the geometry in a number of patches.
2. **Course generation**: The surface is contoured and the courses are defined based on the given course height. The generated courses include short rows.
3. **Loop generation**: The courses found in the previous step are sampled with the loop width to define the loop topology for each course.
4. **2D knit pattern generation**: The resulting final topology is translated into a 2D knitting pattern containing each row to be knit with the corresponding number of loops.

For the case studies, we used an implementation of our methods using Rhinoceros 3D 5.0 (2017) as design environment. The implementation is developed using Python 2.7. (2001–2017) and is based on the compas framework (Van Mele 2017) using a network data-structure, which is a directed graph describing both the connectivity and the geometry of the knit (see Sections "Course Generation" and "Loop Generation").

Patching

An arbitrary quadrilateral mesh is the starting point for the generation of the knitting pattern. As a first step, depending on its complexity, the input geometry is segmented into several patches by defining singularities. These patches can coincide to the different pieces of material that need to be put together once the fabric is produced. However, as their primary role is to simplify the geometry and have better control over the pattern generation, they can be more numerous than required for the fabrication of the final piece. Each patch can therefore be generated using multiple sub-patches. The pattern generation ensures that courses are aligned to both the start and end edges of the sub-patch, in the knitting direction. This results in matching courses between different sub-patches. It also offers the user control over the pattern alignment. The sub-patching can therefore be used as a way of controlling and aligning the pattern to specific constraints or desired features.

For the case studies presented in Section "Results", the input mesh was patched manually, and a knit pattern was generated for each patch. The following sections will describe this process schematically for a single patch.

Course Generation

A minimum of two guide curves, in course direction, are defined by the user to represent the start and end course of the piece. The user also defines the resolution for contouring and the loop parameters. Figure 2a shows the chosen knitting direction and the two guide curves defining the start and end of the piece. Figure 2b depicts the loop geometry and resulting parameters: course spacing/height c in warp direction and loop width w in weft direction. The loop parameters are considered fixed user inputs as they are directly related to the chosen knitting machine and yarn parameters (tension, knit point, gauge, yarn diameter etc.). For our experiments, the loop parameters were determined as presented in Munden (1959) and Ramgulam (2011) using samples knit on a computerised Brother KH970 and ShimaSeiki SWG 091N. These parameters form

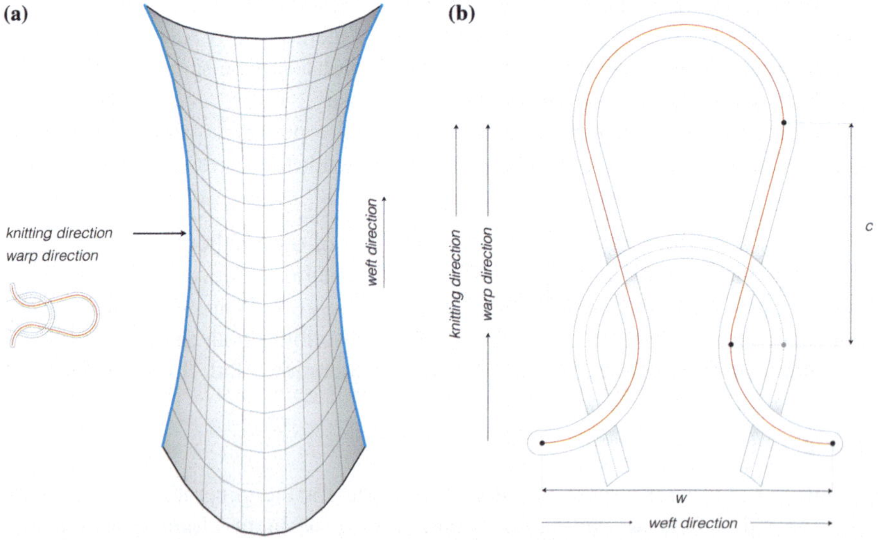

Fig. 2. **a** Mesh patch as input geometry for knitting pattern generation indicating chosen knitting direction and both start and end edges in course direction; and **b** loop geometry where c is the course spacing in warp direction (course height parameter in pattern generation) and w the loop width in weft direction (loop width parameter)

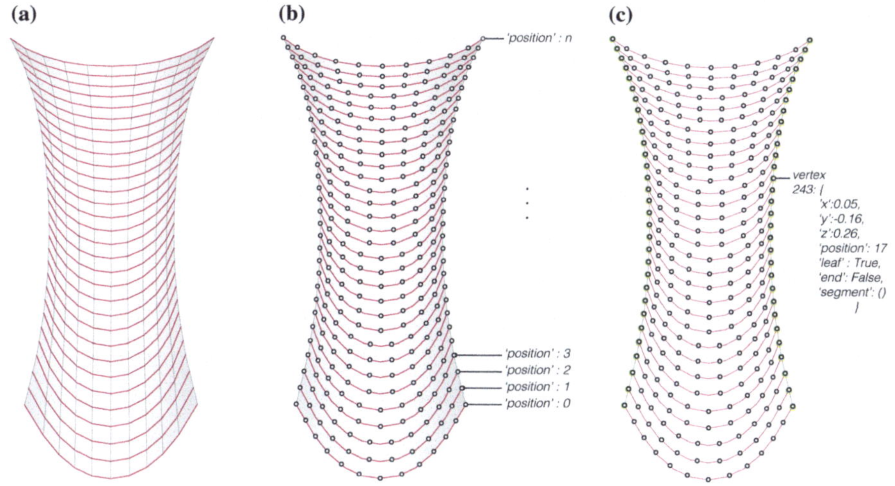

Fig. 3. **a** Contouring of the patch perpendicular to the course direction; **b** ordering and sampling of the contours with the defined course height; and **c** initial network instance with vertex attributes

the basis for generating the courses described in this section and the loop topologies described in Section "Loop Generation".

Given the user-defined parameters, the surface is covered with contours in the direction transverse to the course direction. The contours shown in Fig. 3a are created using level-sets computed on a distance field. These contours are ordered and subsequently sampled with the course height (Fig. 3b). A network of vertices and edges is initialised using the sample points.

Figure 3c depicts the initial network, where each vertex is given a '*position*' attribute corresponding to the contour line order. The '*leaf*' vertices (vertices with a single connecting edge) are also defined. At this point, the '*end*' and 'segment' attributes of the vertices have default values. Their use will be explained in Section "Loop Generation". Edges of the network store information about the directionality of the knitting. They represent the weft and warp direction of the knitting pattern and store this information in '*weft*' and '*warp*' attributes. Note that the edges in the initial network topology are neither '*warp*' nor '*weft*'. They define the contours.

First, the '*leaf*' vertices are connected to form the first course lines (Fig. 4a). Then, the vertices of the network are processed per contour, starting with the longest contour line (Fig. 4e). For each vertex of a contour, a set of potential connection vertices is defined as the four closest vertices on an adjacent contour (Fig. 4b, c). From these candidates, the vertex that creates the connection closest to perpendicular to the adjacent contour is selected (Fig. 4g).

However, when the angle formed by the two best candidates is similar, and the difference is less than 6 degrees, the candidate that forms the straightest line with the connection from the previous contour is preferred (Fig. 4h).

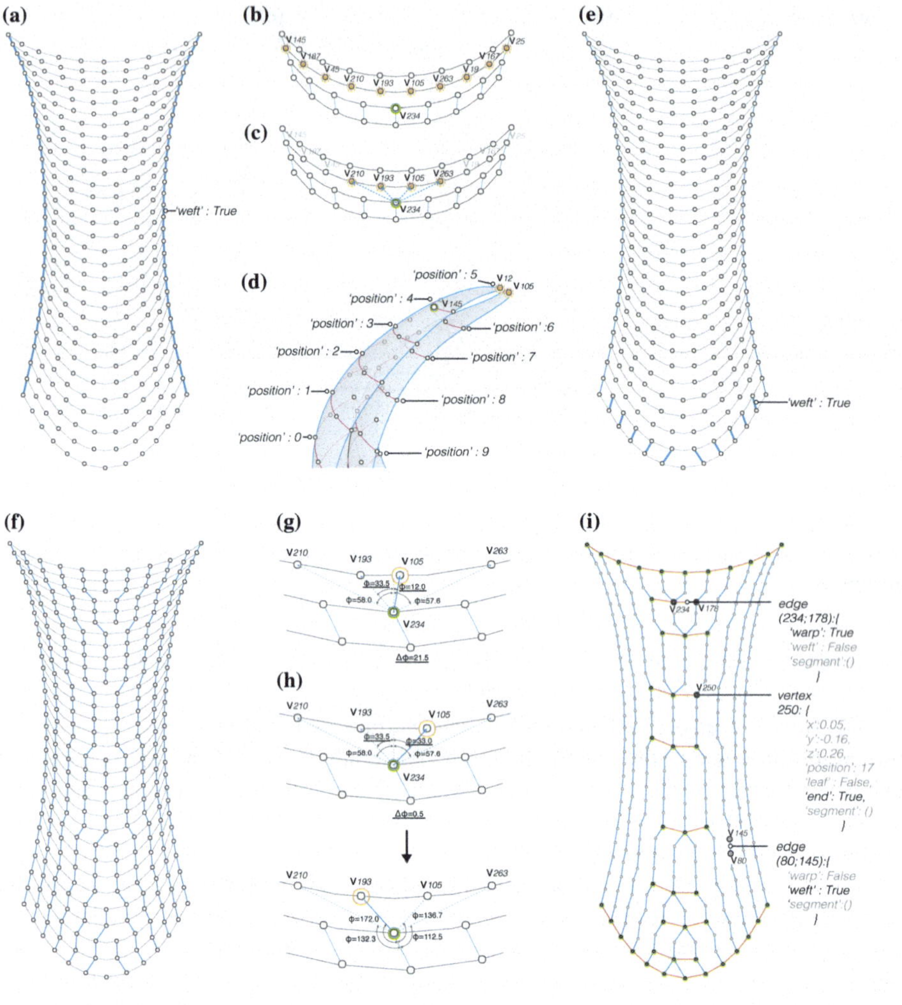

Fig. 4. **a** Connections added to all 'leaf' vertices; **b** set of connection candidates; **c** four closest connections in the set; **d** connection candidates for extreme curvature, based on 'position' attribute; **e** 'weft' connections for longest contour; **f** all 'weft' connections; **g** chosen connection based on minimum angle deviation from perpendicular to the current contour; **h** preferred connection out of similar candidates based on minimum angle change from previous 'weft' connection; and **i** resulting connections in 'weft' direction defining the courses, 'end' vertices, and first 'warp' connections

Note that the resulting '*weft*' edge is only added, if the connecting vertex has less than four already existing connections, and if it is not already connected to another vertex of the current contour.

By restricting the search for connections to vertices on the adjacent contours the approach can be applied to geometries with extreme curvature. The '*position*' attribute of the vertices ensures that vertices that may be geometrically close but not part of the adjacent contour are ignored (Fig. 4d). The resulting network is shown in Fig. 4f.

In a second pass, all vertices with fewer than four connections are revisited and connected to the closest to perpendicular target in the direction in which no '*weft*' connection exists.

Finally, all vertices with more than four connections, their immediate neighbours over a non '*weft*' edge, and ones on the first and last contour are marked as '*end*'. The non '*weft*' edges connecting two '*end*' vertices are marked as '*warp*' (Fig. 4i).

We include a pseudocode (Listing 13) snippet for generating 'weft' connections, propagating from the starting position to the last position. The same logic is used to propagate the connections in opposite direction such that connections are created in both directions starting from the longest contour.

```
if start_position < last_position:
  current_position = start_position + 1
  while current_position < last_position:
    initial_vertex_list = [all_vertices_on_current_position]
    for vertex in initial_vertex_list:
      target_position = current_position + 1
      if number of vertex neighbours > 2:
        possible_connections = [closest_neighbours_on_target_position]
        most_perpendicular = [ordered_angles_possible_connections]
        if (most_perpendicular[0] - most_perpendicular[1]) < 6:
          connect (vertex, least_angle_change_connection)
        else:
          connect (vertex, most_perpendicular_connection)
    current_position = current_position + 1
    next_vertex_list = [all_vertices_on_current_position
    for vertex in next_vertex_list:
      if current_position != last_position:
        if number of vertex neighbours < 3:
          possible_connections = [closest_neighbours_on_target_position]
          connect (vertex, most_perpendicular_connection)
```

Listing 1. Pseudocode for generating 'weft' connections starting from the longest contour and propagating towards the last contour

Loop Generation

In this step, the courses, represented by all '*weft*' connections, are sampled with the loop width.

Before the final topology of the network can be created, a mapping network is initialised that keeps track of the connected '*end*' vertices (Fig. 5a). The network is created by finding all the '*weft*' edge chains between two '*end*' vertices. These '*weft*' edge chains are sampled with the loop width. The resulting points constitute the final

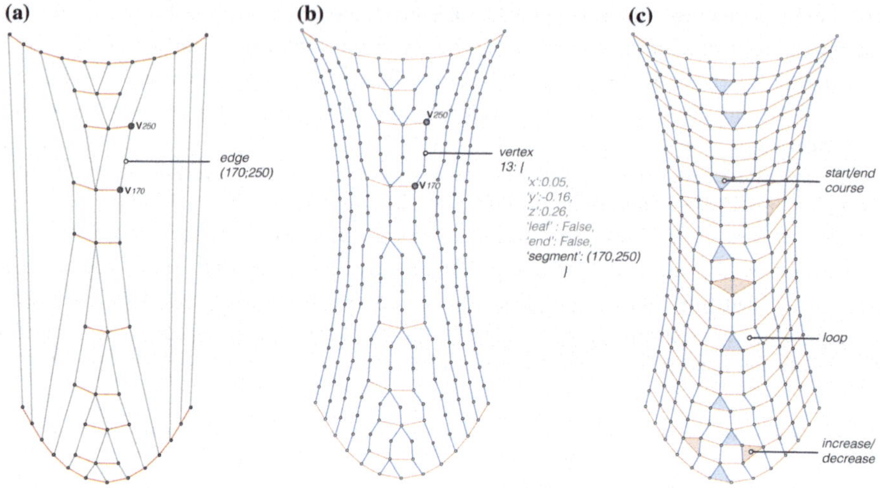

Fig. 5. **a** Mapping network—topological representation of the courses within the network; **b** final network topology based on sampling with the loop width before the generation of 'warp' connections; and **c** final network representing the topology to be knit including short rows, increases and decreases

vertices of the network while the previous vertices and edges that are neither '*weft*' nor '*warp*' are discarded (Fig. 5b). These vertices are also given a '*segment*' attribute, which identifies their position between two '*end*' vertices.

For the creation of all '*warp*' edges the same logic is applied as with the creation of the '*weft*' edges. The difference being that the restricted cloud to search for possible connections is now determined by the mapping network topology instead of the 'position' along a contour.

Figure 5c shows the final knit topology consisting of '*warp*' and '*weft*' edges. Each face of the network represents a loop. In '*weft*' direction, the triangular exceptions represent the start or end of a short-row, while the same exceptions in '*warp*' direction represent the increase or decrease in the number of loops within the next course.

2D Knitting Pattern Generation

For the translation of the topology to a 2D knitting pattern, a dual network is created where each vertex represents one loop and the edges retain the 'weft' and 'warp' directionality. Knowing the connectivity of the vertices through the edges, the topology can now be drawn as a pattern where every loop is represented by a square. Figure 6a shows the pattern representation resulting from the approach presented in this section. Figure 6b shows what the knitting pattern typically looks like in machine knitting software. Each square/pixel also represents a loop. The colour of the square gives additional information about machining operations such as transferring loops and inactivating needles (e.g. decreases), adding needles (e.g. increases), which bed the

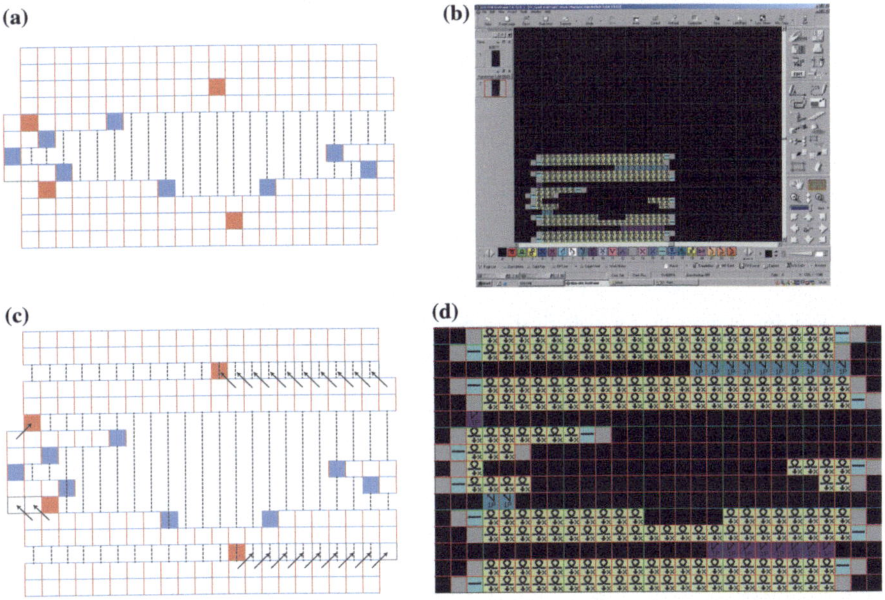

Fig. 6. **a** Resulting 2D pattern; **b** knitting pattern within SDS-one knitting machine software; **c** resulting knitting pattern showing the necessary transformations for knitting software; **d** resulting knitting pattern as represented in knitting software

needle is on etc. Figure 6c, d show the transformations applied to the pattern when translated to a knitting machine software specific pattern.

Results

The approach is tested on a series of knitted prototypes using various 3D input geometries, ranging from simple primitives to freeform surfaces. The results demonstrate that by using the presented approach, an even distribution of stitches and 3D fit of the textile can be achieved.

Reaching an accurate result is highly dependent on using accurate loop dimensions (width and height). The geometry of loops is dependent on a series of machining factors (tension settings, gauge etc.) and on the type of yarn used. With this in mind, calibration of the model can be done by determining the accurate loop dimensions through testing of plain knit samples of material.

Quarter Sphere

Primitives such as spheres and tubes are examples of non-developable surfaces for which patterns are available in most knitting software. Spherical shapes are usually described as patterns with a series of short-rows with repeating elliptical shapes. Figure 7 shows this approach as described by De Araujo et al. (2011).

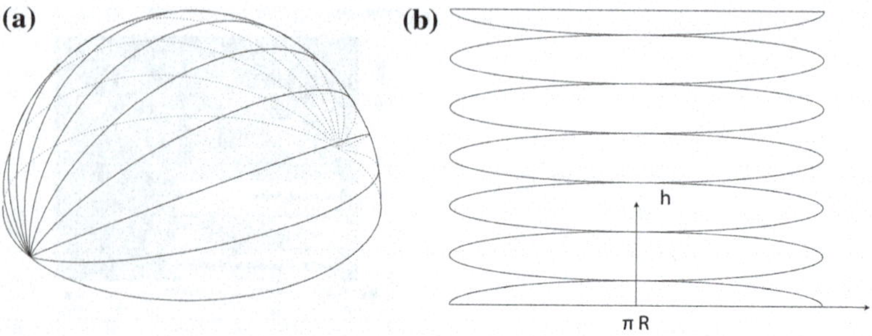

Fig. 7. Knitting approach for spherical form, after De Araujo et al. (2011, pp. 137–170): **a** theoretical 3D form; and **b** repeating knitting pattern

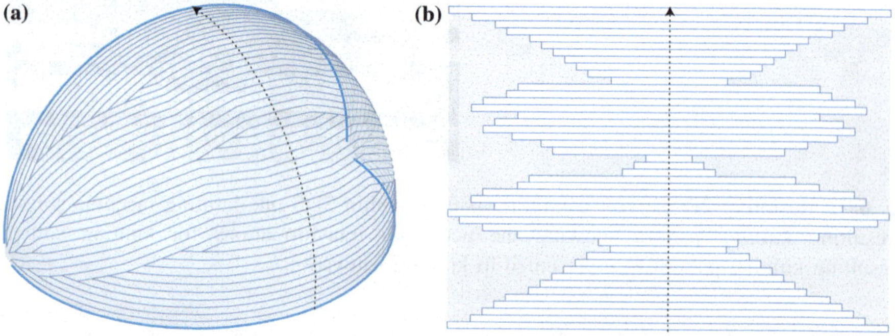

Fig. 8. Knitting pattern for a quarter sphere: **a** generated courses on the quarter sphere; and **b** 2D knitting pattern schematic

Figure 8 shows the results of the presented method applied to a quarter sphere. A large number of short-rows is necessary for creating this geometry. Noticeable is the symmetry and periodicity of the resulting pattern, which is comparable to rationalised patterns in known examples.

Four-Valent Node

This subsection presents the approach applied to a non-developable geometry of a four-valent node. This geometry is one for which knitting patterns as primitives do not readily exist. As the chosen geometry is more complex, all of the steps described in Section "Methods" are followed.

Figure 9a shows the initial patching of the geometry for the creation of the knitting pattern while Fig. 9b shows the patches that were knit as a single piece.

Figure 10a gives an overview of the resulting generated pattern, highlighting, in red, the short-rows used for the shaping of the geometry. Figure 10b shows the schematic course-by-course pattern to be knit for a piece of the node.

Figure 11 shows the resulting knit geometry using the generated pattern and highlights the short rows as they appear in the knit piece. Figure 12 shows the

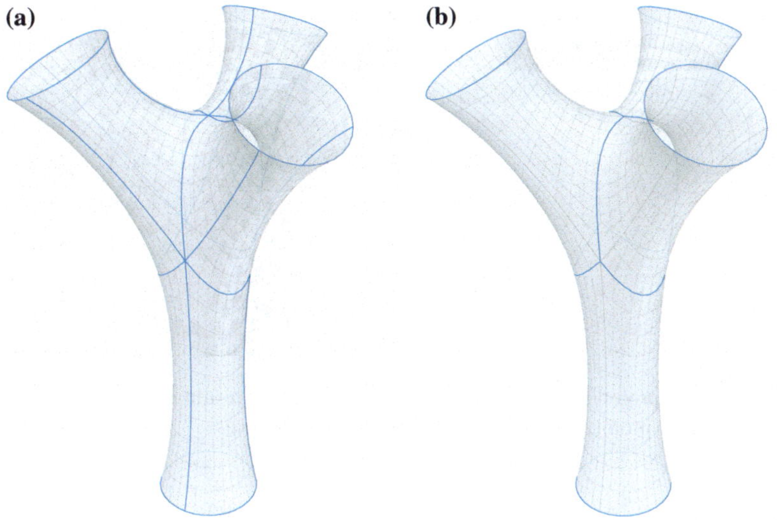

Fig. 9. **a** Patching of the surface for pattern generation; and **b** patches knit in one piece

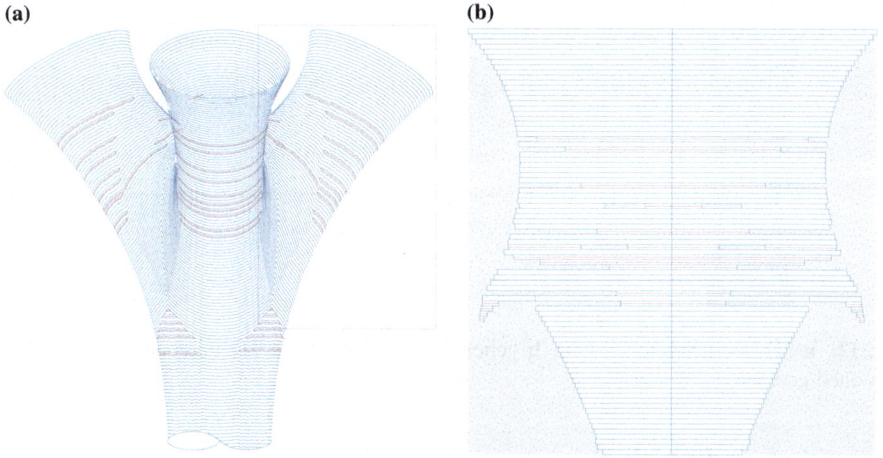

Fig. 10. **a** Courses for the four-directional minimal surface; and **b** knitting pattern for right branch (*box*) of the minimal surface

geometry as model, generated knit pattern and physical knit piece. Note that the knitted piece is tensioned into shape, which creates a discrepancy between the modelled shape and the knit shape as the modelled shape is patterned without taking into account a tensioned situation.

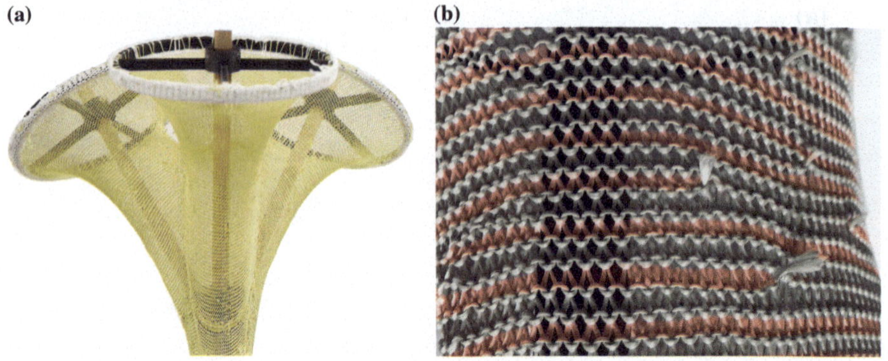

Fig. 11. **a** Physical prototype of minimal surface knit using the resulting knitting pattern; **b** short row features on knit prototype colourised for illustrative purposes

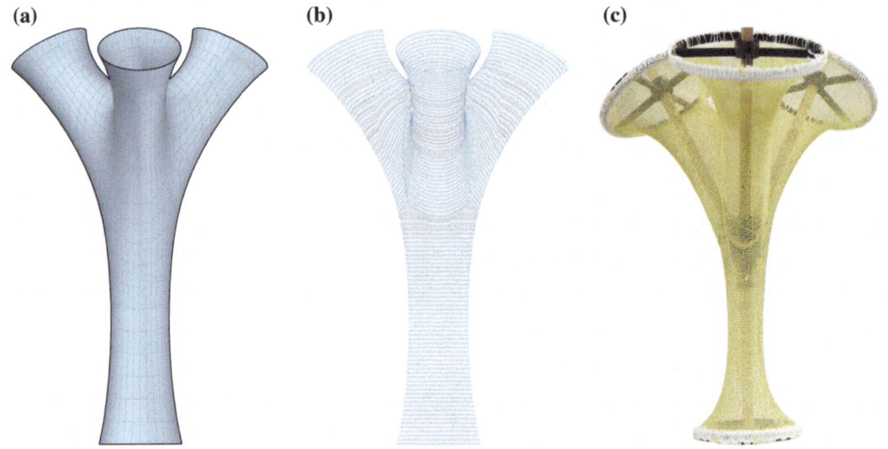

Fig. 12. **a** 3D geometry to be knit; **b** generated knit pattern on 3D geometry; **c** knitted and tensioned geometry

Discussion and Outlook

The presented approach produces knitting patterns from a given 3D geometry with accurate placement of short-rows and loops such that the input geometry can be created. The accuracy of the model in comparison to the real object is highly dependent on accurate measurements of loop geometry, which are directly correlated to the knitting machine parameters. Knowing these parameters, a good draping of the model with minimal loop distortion is achieved. Currently, all results are based on uniform loop parameters over the entire geometry. While it is possible to define varying loop dimensions and alignment for the separate (sub-)patches, loop variation is not possible within a single (sub-)patch.

Future developments will target the creation of more varied patterns with an eye on adjusting pattern densities and alignment to specified parameters and desired directions such as stress fields. The patching of the geometry plays an important role in being able to achieve the desired variations. Possible strategies for automated and optimised patching of a given geometry will be investigated in the future.

Furthermore, considering a tensioned fabric system (Fig. 11), the modelled/patterned geometry and tensioned result differ greatly. If the approach is to be applied to such systems, the inverse problem of computing an initial state given a known target tensioned state needs to be addressed.

On the fabrication side, the approach produces patterns that are in accordance to knitting machine functioning. However, transferring these patterns to the code needed for CNC knitting machines to operate is still a manual process. Specific machining instructions (carriage speed, yarn feeder position) are not part of the generated 2D patterns and need to be input in the knitting machine software. Presently, experiments have been done with ShimaSeiki's SDS-ONE (ShimaSeiki 2017) and will be further developed to create a more streamlined process.

When considering the machining process, further constraints need to be implemented on the pattern generation. For example, for the creation of short rows, needles on the machine are inactivated from one course to another. Depending on the mechanical possibilities of each machine, the recommended maximum number of needles that can be skipped, from a course to another, may be limited. Fabrication constraints such as these will be implemented in the pattern generation.

Acknowledgements. This research is supported by the NCCR Digital Fabrication, funded by the Swiss National Science Foundation. (NCCR Digital Fabrication Agreement# 51NF40-141853). Machine knitting experiments and investigation into knitting machine and knitting software working has been done in collaboration with the Institute for Textile Machinery and High Performance Materials at the Technical University of Dresden.

References

Abounaim, M.D., Hoffmann, G., Diestel, O., Cherif, C.: Development of flat knitted spacer fabrics for composites using hybrid yarns and investigation of two-dimensional mechanical properties. Text. Res. J. **79**(7), 596–610 (2009)

Aboumain, Md.: Process development for the manufacturing of flat knitted innovative 3D spacer fabrics for high performance composite applications. PhD dissertation, Technical University of Dresden (2010)

Ahlquist, S.: Integrating differentiated knit logics and pre-stress in textile hybrid structures. In: Design Modelling Symposium: Modelling Behaviour, pp. 101–111 (2015)

Ahlquist, S.: Social sensory architectures: articulating textile hybrid structures for multi-sensory responsiveness and collaborative play. In: ACADIA 2015: computational ecologies, pp. 262–273 (2015)

Brennan, J., Walker, P., Pedreschi, R., Ansell, M.: The potential of advanced textiles for fabric formwork. Proc. ICE Constr. Mater. **166**(4), 229–237 (2013)

De Araujo, M., Fangueiro, R., Hu, H.: Weft-Knitted Structures for Industrial Applications, Advances in Knitting Technology, pp. 136–170. Woodhead Publishing Limited, Sawston (2011)

Hong, H., Fangueiro, R., Araujo, M.: The development of 3D shaped knitted fabrics for technical purposes on a flat knitting machine. Indian J. Fibre Text. Res. **19**, 189–194 (1994)

Igarashi, Y., Igarashi, T., Suzuki, H.: Knitting a 3D model. Pac. Graph. **27**(7), 1737–1743 (2008)

McCann, J., Albaugh, L., Narayanan, V., Grow, A., Matusik, W., Mankoff, J., Hodgins, J.: A compiler for 3D machine knitting. In: SIGGRAPH (2016)

Munden, D.: 26—the geometry and dimensional properties of plain knit fabrics. J. Text. Inst. Trans. **50**(7), T448–T471 (1959)

Python (Copyright 2001–2017) Python Programming Language. https://www.python.org

Ramgulam, R.B.: 3—Modeling of Knitting, Advances in Knitting Technology, pp. 48–85. Woodhead Publishing, Sawston (2011)

Thomsen, MR., Hicks, T.: To knit a wall, knit as matrix for composite materials for architecture. In: *Ambience08*, pp. 107–114 (2008)

Thomsen, MR, Tamke, M., Holden Deleuran, A., Friis Tinning, I.K., Evers, H.L., Gengnagel C., Schmeck, M.: Hybrid tower, designing soft structures. In: Design Modelling Symposium: Modelling Behaviour, pp. 87–99 (2015)

Rhinoceros (Copyright 2017) Rhinoceros Modeling Tools for Designers, Version 5. https://www.rhino3d.com

Van Mele, T.: Compas: framework for computational research in architecture and structures, softwareX (in preparation) (2017)

Van Vuure, A.W., Ko, F.K., Beevers, C.: Net-shape knitting for complex composite preforms. Text. Res. J. **73**(1), 1–10 (2003)

Veenendaal, D., West, M., Block, P.: History and overview of fabric formwork: using fabrics for concrete casting. Struct. Concr. **12**(3), 164–177 (2011)

Sabin, J. http://www.jennysabin.com/1mythread-pavilion (2017)

Scholzen, A., Chudoba, R., Hegger, J.: Thin-walled shell structures made of textile-reinforced concrete. Struct. Des. Constr. **1**, 106–114 (2015)

Schimaseiki: SDS-one APEX3. http://www.shimaseiki.com/product/design/sdsone_apex/ (2017)

Yuksel, C., Kalador, M., James, D.L., Marschner, S.: Stitch meshes for modeling knitted clothing with yarn-level detail. ACM Trans. Graph. **31**(4), 1–12 (2012)

Enlisting Clustering and Graph-Traversal Methods for Cutting Pattern and Net Topology Design in Pneumatic Hybrids

Phil Ayres(✉), Petras Vestartas, and Mette Ramsgaard Thomsen

School of Architecture, Centre for Information Technology and Architecture, The Royal Danish Academy of Fine Arts, 1435 Copenhagen, Denmark
phil.ayres@kadk.dk

Abstract. Cutting patterns for architectural membranes are generally characterised by rational approaches to surface discretisation and minimisation of geometric deviation between discrete elements that comprise the membrane. In this paper, we present an alternative approach for cutting pattern generation to those described in the literature. Our method employs computational techniques of clustering and graph-traversal to operate on arbitrary design meshes. These design meshes can contain complex curvature, including anticlastic curvatures. Curvature analysis of the design mesh provides the input to the cutting pattern generation method and the net topology generation method used to produce a constraint net for a given membrane. We test our computational design approach through an iterative cycle of digital and physical prototyping before realising an air-inflated cable restrained pneumatic structural hybrid, at full-scale. Using a Lidar captured point-cloud model, we evaluate our results by comparing the geometrical deviation of the realised structure to that of the target design geometry. We argue that this work presents new potentials for membrane expression and aesthetic by allowing free-patterning of the membrane, but identify current limits of the workflow that impede the use of the design method across the breadth of current architectural membrane applications. Nevertheless, we identify possible architectural scenarios in which the current method would be suitable.

Keywords: Cutting pattern · Pneumatic membrane · Mesh segmentation

Introduction

The research project *Inflated Restraint* investigates an alternative approach to conventional methods of cutting pattern generation described in the literature. The approach favours free patterning of the membrane cutting pattern through the application of mesh segmentation approaches recently introduced to generative architectural design. This paper presents our research motivation, background and state-of-the-art, methods, results and evaluation, discussion and conclusion (Fig. 1).

K. De Rycke et al., *Humanizing Digital Reality*,
https://doi.org/10.1007/978-981-10-6611-5_25

Fig. 1. *Inflated Restraint*—one of three full-scale demonstrators in the exhibition *Complex Modelling* at KADK, fall 2016

Motivation

Our modelling approach targets the generation, analysis and fabrication of a cutting pattern for an arbitrary (non form-found) design volume with a complex surface, and the generation of cable restraint topology. Multiple models and algorithmic methods, predominantly related to mesh segmentation (Nejur and Steinfeld 2016), are employed to support different scales of design consideration at different periods in the design cycle. In short, the design cycle is organised by the defining of the global design target, the subdivision of the design mesh into curvature regions, the subdivision of regions into suitably sized patches for fabrication, the flattening of the patch geometry and the generating of a suitable topology for the restraining cable (Fig. 2). Our focus on pattern cutting is motivated by the understanding that a primary contributing factor to the performance and aesthetic quality of architectural pneumatics, and architectural membrane structures in general, is the cutting pattern. This makes it a central design concern that spans across different scales of consideration and implicates constraints of fabrication. Within the project, we identify three inter-linked scales of consideration for the membrane—the textile (micro scale), the pattern patch (meso scale) and the overall shape (macro scale). At the micro scale (textile), we find that under typical inflation pressures (in the range of 200–350 Pa) we do not activate the anisotropic characteristics of our coated textile membrane material. At the meso scale (patch), we find that naked edge relaxation of patch elements is essential for final surface quality. At the macro scale (complete membrane) we demonstrate that computationally derived cutting patterns, that appear to be arbitrary, can be fabricated efficiently (minimising waste) and approximate design targets that contain regions of synclastic and anti-clastic curvature.

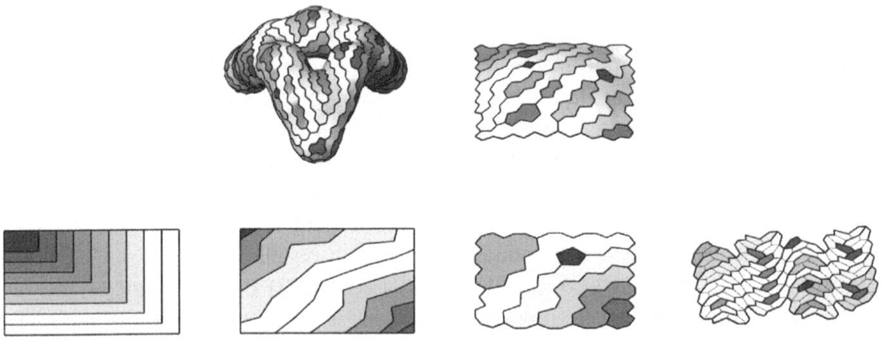

Fig. 2. Early investigations of mesh segmentation strategies demonstrating design freedom in the control of patch geometries

Background and State-of-the-Art

There is general consensus across the literature regarding design approaches for architectural membranes—both mechanically stressed and pneumatic. Design workflows tend to commence with form-finding, progress through structural analysis and conclude with the generation of a cutting pattern (Otto et al. 1982a, b; Gründig et al. 2000; Kim and Lee 2002; Philipp et al. 2015). With the cutting pattern acting as the interface between design intent and manufactured reality, the process of cutting pattern generation implicates issues of design, engineering, fabrication and, not least, aesthetics (Dent 1971; Knipper et al. 2011). In general, the aesthetic of architectural membranes follows principles of regularity and minimum deviation between membrane sub-panels. In this project, we explore alternative principles of free-patterning within the constraints of achieving pre-defined design targets.

Approaches to Cutting Pattern Generation

The principle challenge in cutting pattern generation is the sub-division of a target geometry into a set of sub-surfaces, with minimal distortion (Gründig et al. 2000). This is a trivial problem in the case of target geometries that are developable, however, in practice, mechanically stressed and pneumatic membranes generally gain structural performance through double curvature—anticlastic in the case of mechanically stressed, and synclastic in the case of pneumatics. Many approaches to the problem of distortion minimisation from doubly curved sub-surfaces to flattened pattern are found in the literature. These include target surface sub-division using geodesics (Ishii 1972), dynamic relaxation, force density method (Moncrieff and Topping 1990), finite element method using weighted least-squares minimisation (Tabarrok and Qin 1993) and a hybrid iterative flattening technique combined with equilibrium shape finding using FEM (Kim and Lee 2002). It is of note that underlying the development of methods is a change in focus in where distortion should be considered in the realisation process—initially as a problem of fabrication, then as a problem of forces encountered in construction and finally as a problem of material extension after erection.

Mesh Representation and Treatments

In all the aforementioned approaches and in computationally led methods in general, a mesh acts as the underlying data-structure for representing the membrane. However, their treatment in the cutting pattern generation process can vary significantly. In Gründig et al. (2000) the representation of individual membrane segments occurs through a mesh segmentation process that alters the mesh topology. Geodesics act as cutting geometry of the mesh with additional vertices and edges added where faces are split. The implication here is a potential loss of mesh regularity.

In Tabarrok and Qin (1993), and Kim and Lee (2002) the topology of the input mesh is pre-defined as an approximation of the design target with topology maintained through segmentation and only mesh geometry adjusted in the various relaxations. This implies that a relatively clear design geometry is already present, and may therefore act as an impediment to speculative early-design iteration.

Mesh Segmentation Methods

In the broader field of generative architectural design, relatively recent developments in methods of mesh segmentation from the field of computer graphics are finding traction with architectural design problems, particularly in the field of free-form surface rationalisation, planarisation and discretisation (Nejur and Steinfeld 2016). These approaches are predicated on the use of design meshes, or typically their dual representation, to provide a weighted graph from which mesh features can be interrogated and modified (ibid.). In this project we investigate the use of K-means clustering and graph-traversal as methods of steering mesh segmentation for generating membrane cutting patterns. By altering parameters such as face-angle and edge-traversal-distance, fine grain control of mesh segmentation can be achieved and qualitatively different results produced (Fig. 2).

Methods

In contrast to conventional approaches of pattern cutting generation that operate on form-found geometries, we begin with an unconstrained design cycle to define a macro-scale design target mesh. This target is designed using the Cocoon toolset—a plugin for Rhino/Grasshopper. To create a spatial tension we define two pneumatic bodies—one with larger volume (approx. 8.1 m^3) and areas of high anticlastic curvature, and the other with a smaller internal volume (approx. 4.7 m^3) with predominantly synclastic curvature. Both pneumatic bodies are fully closed surfaces with no open boundary connection conditions to the ground (Fig. 3, stage 1).

The k-Means Clustering and Graph-Traversal Approach

A custom mesh curvature analysis algorithm operates on the design mesh to identify areas of synclastic and anticlastic curvature. A k-means clustering algorithm is then applied to group mesh faces within the isolated synclastic curvature zones. Graph

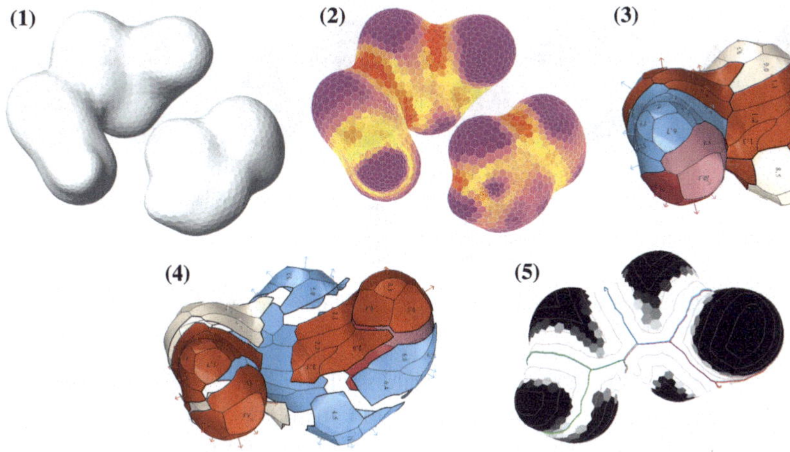

Fig. 3. Stages in the modelling workflow: **1** A defined target model; **2** curvature analysis of the design model; **3** subdivision of the surface (only the larger inflated body shown); **4** breaking into discrete models for refinement and further division into sewable patches; **5** first order cable-restraint topology generation by joining 'lowest' areas of anticlastic curvature

traversal methods (Depth First Search and Dijkstra's Shortest Path) are then used to subdivide the anti-clastic curvature zones into strips suitable for fabrication. New meshes are generated for all of these curvature regions and refined to increase fidelity as they are used downstream to define the cutting geometry of individual patches. The naked edges of all patch meshes are analysed to determine shared edges and these are relaxed to create smooth boundaries for unrolling. The Rhinocommon unroller class is then used to produce a developable surface patch with edge marking, indexing and patch ID added prior to laser cutting (Fig. 4). Without this process of relaxation along shared edges, 'pinch' artefacts would become evident during inflation as can be seen in early simulation studies (Fig. 5).

The topology of the cable restraints is determined from curvature analysis of the design target model. In this case, we firstly identify areas of anticlastic curvature, the

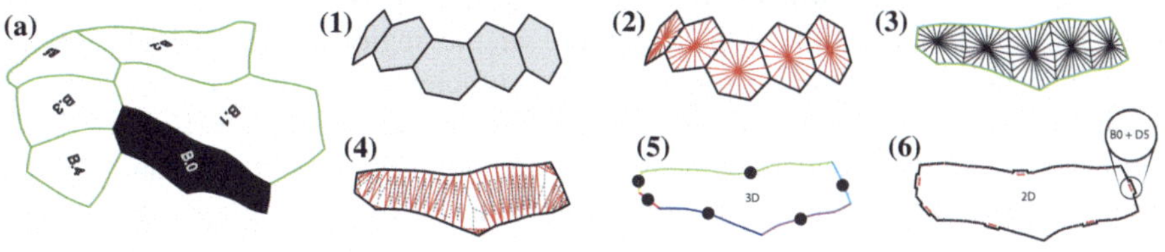

Fig. 4. Stages in the flattening of a patch (**a**): **1** Patch is defined by a collection of dual mesh polylines; **2** polylines are triangulated into a 'fan' mesh; **3** relaxation using Kangaroo 2 with weak inner springs (*black lines*) and stiff outer springs (*green lines*); **4** naked edge is flattened using the Rhinocommon function '3D polyline to mesh' resulting in minimum triangulation for unrolling; **5** edge adjacency attributes are added for the transfer from 3D to 2D; **6** Rhinocommon 'Unroller' class is used to unroll the mesh and edge adjacency labels are added

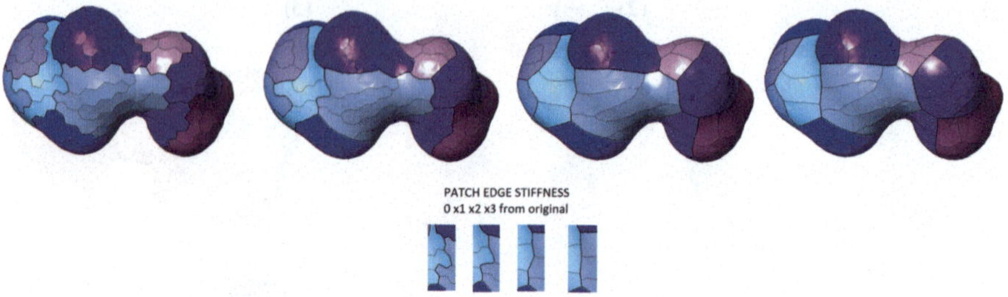

Fig. 5. Sequence showing patch naked-edge relaxation

'lowest' points of the membrane surface, to establish the primary loops of the net. This ensures that they do not exhibit 'slippage' on the membrane when under tension. Further edges are added to the net topology using search criteria that combines finding areas of lowest synclastic curvature (to relieve membrane stress) and being approximately equidistant from each other (to ensure even distribution). As such, the net topology does not have any direct geometric correspondence to the sub-division of the membrane into patches, with this being exemplified by the use of a neon coloured braided rope. In simulation, we 'inflate' the membrane model to verify its interaction with the cable restraint model. Here, the measure of success is that the net does not exhibit 'slippage' and the two systems find and maintain equilibrium across the operating pressure range.

Results and Evaluation

The full-scale demonstrator comprises two pneu. The smaller pneu is designed to have synclastic curvature which conforms to the 'natural' tendency of pneumatic form. The larger pneu includes areas of anticlastic curvature, which is not a natural pneumatic form. Exhibited together, the two pneu demonstrate the role of the membrane in steering the geometric result of the pneumatic system.

With a combined volume of approx. 12.8 m^3, inflation of the membranes takes ~7 min using a proprietary insulated duct fan (Östberg IRE 125 A1) operating at 240 V with a maximum flow of 64 l/s at a maximum pressure of 300 Pa. The pneu system reaches full inflation with a pressure in the region of 200–350 Pa. At this operating pressure, membrane stresses are not sufficient to have to consider the anisotropic characteristics of the coated textile used for the membrane (Fig. 6). This is determined using the standard equation for membrane stress that relates inflation pressure to radius of curvature, and comparing to the stress/strain graph for the textile shown in Fig. 6.

We verify the modelling workflow by comparing the design model with the physical demonstrator. Here, the measures of success are: (1) that there is a close geometric correlation between the model and the realised pneu; (2) that the inflated surface achieves the desired curvatures; (3) that the membrane has a smooth transition

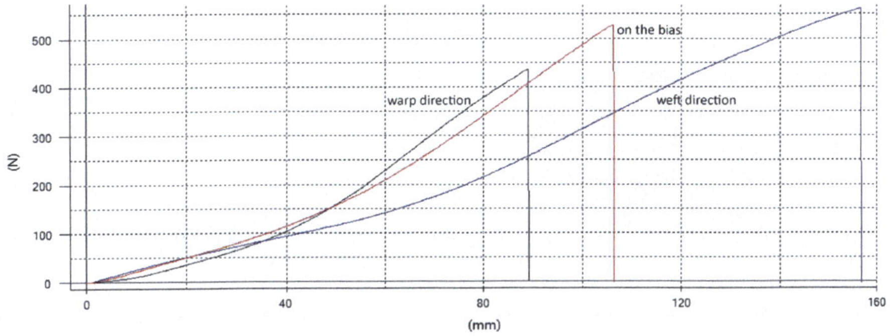

Fig. 6. Stress/strain graph for uniaxial tensile test [following Ansell and Harris (1982)] for the coated membrane textile in the direction of warp, weft and on the bias. The difference between warp and weft directions reveal the anisotropy of the textile. However, significant deviation only begins to occur after approx. 100 N/40 mm of elongation

across patches; (4) that the membrane is fully tensioned with no areas of compression resulting in unsightly and underperforming compression wrinkles.

In addition to comparison between the physical demonstrator and the design model, comparison with earlier physical prototypes (Fig. 7, left) provide evidence of refinement in the cutting pattern generation method (the addition of naked edge relaxation for patches as shown in Figs. 4 and 5) and successfully achieving measure of success 3 and 4 in the final membrane.

For the measures of success 1 and 2 we compare a Lidar captured point-cloud of the physical pneu and determine the deviation from the intended model. Surface deviation shows a general tendency towards the surface being under-inflated (Fig. 8). The sectional study (Fig. 9) reveals the deviation between the realised membrane and the design intention. This was sufficient for the generated restraint net to not be compatible with the membrane. A second iteration of net production used a mesh derived from the laser-scanning data of the physical membrane. This approach of operating from 'as-built' data to inform subsequent construction phases is described in the literature

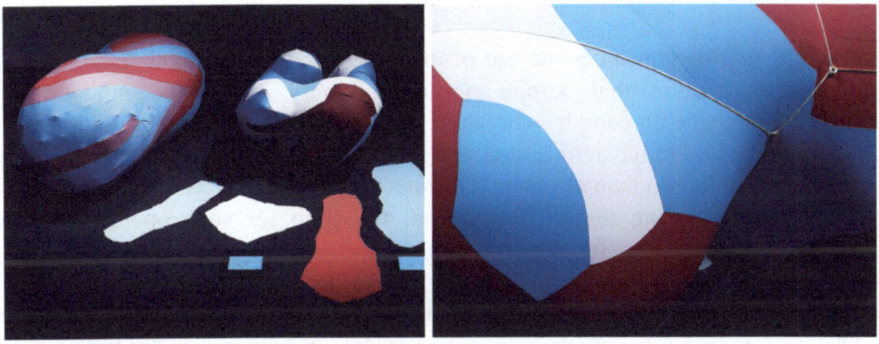

Fig. 7. Early prototypes (*left*) showing evidence of 'pinches' and wrinkling, and the final membrane (*right*) showing smooth transitions between patches and fully tensioned membrane

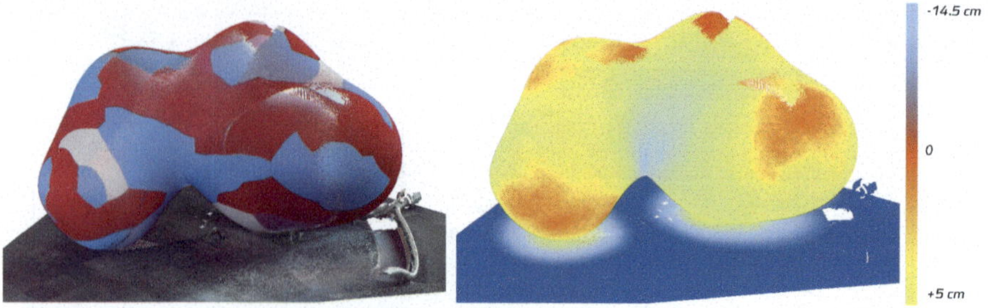

Fig. 8. 3D Lidar scan of the larger membrane (*left*) and surface deviation analysis of target model (*right*)

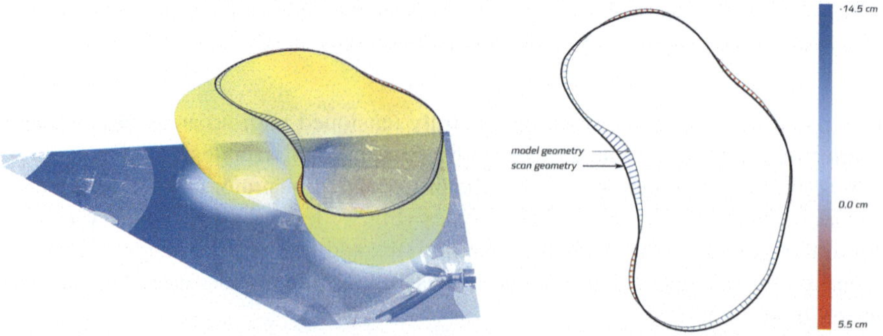

Fig. 9. Cross sectional comparison between target geometry and 3D Lidar point cloud of the larger membrane reveals larger geometric deviation in the area of pronounced anticlastic curvature

(Brandt 2012), and resulted in nets that conformed to surface for both pneumatic forms in the final demonstrator.

Discussion and Limits

Inflated Restraint demonstrates the use of mesh segmentation methods for the generation of cutting patterns, that, despite appearing arbitrary, approximate smooth pneumatic design targets with complex curvature. However, as an internal exhibition piece, the project is limited in its architectural generalisability as it has not been designed or tested against the full domain of external architectural loads such as wind and/or snow. Designing and analysing the performance of the pneu for such operational loads remains an open challenge for the workflow we have described in this paper. Where rudimentary analysis of membrane stresses can be easily calculated given inflation pressure and membrane curvature, a more involved understanding of membrane performance at the yarn scale would be a necessary area of exploration for further work. This is due to the patchwork nature of the membrane in which yarn-to-yarn relations between patches are often highly oblique and eccentric. The literature points to such conditions as being detrimental to load transfer and recommends that, given two lengths of fabric to be joined, the angle of the edges to be joined should be mirrored (Otto et al. 1982a, b). Such a constraint is a challenge to implement for the cutting pattern generation method described. Currently, fabric orientation is determined through nesting of patches to minimise waste along the length of the manufactured coated fabric. However, a multi-objective optimisation aimed at negotiating this fabrication constraint with that of minimising warp/weft orientations between neighbouring patches could be considered.

Despite the limitations outlined above, *Inflated Restraint* does demonstrate architectural relevance for circumstances in which light-weight, quick to erect, quick to demount, visually engaging and formally expressive forms of interior spatial sub-divider or screen are desired. For example, interior large span contexts such as exhibition halls, foyers, atrium spaces, and, currently under investigation, as dynamically reconfigurable architectural scenography for an immersive spatial sound and visual performance.

Conclusion and Further Work

This paper has presented a computational approach to cutting pattern generation for architectural membranes. The approach has been refined through a cycle of iterative testing that has included the production of physical prototypes. A final demonstrator comprising two cable-restrained air-inflated pneumatic membranes has been produced at full scale to evaluate the approach. The computational approach offers an alternative method of cutting-pattern generation to those described in the literature, providing designers greater freedom in the definition of target design geometry (i.e. not constrained to a form-finding process), and in steering the outcome of the cutting pattern generation through altering parameters of the mesh segmentation process. We have identified that a current trade-off to these freedoms is a more constrained domain of architectural application and have suggested what these might be.

Further work aims to address the limitations within the computational design process to allow designers more explicit structural performance feedback of derived cutting patterns. Testing of the method in the context of cutting pattern generation for mechanical stressed membranes also remains an open challenge.

Acknowledgements. *Inflated Restraint* was exhibited together with two further projects—*A Bridge Too Far* and *Lace Wall*—as part of the research exhibition *Complex Modelling* held at Meldahls Smedie, The Royal Danish Academy of Fine Arts (KADK), over the period September–December 2016.

This project was undertaken as part of the Sapere Aude Advanced Grant research project *Complex Modelling*, supported by The Danish Council for Independent Research (DFF).

The authors wish to acknowledge the collaboration of Maria Teudt.

References

Ansell, M., Harris, B.: Fabrics—characteristics and testing. In: IL15 Air Hall Handbook. Institute for Lightweight Structures, West Germany (1982)

Brandt, J.: The death of determinism. In: Ayres, P. (ed.) Persistent Modelling—Extending the Role of Architectural Representation, pp. 105–116. Routledge, London (2012)

Dent, R.: Principles of Pneumatic Architecture. Architectural Press, London (1971)

Gründig, L., Moncrieff, E., Singer, P., Ströbel, D.: High-performance cutting pattern generation of architectural textile structures. In: IASS-IACM, Fourth International Colloquium on Computation of Shell & Spatial Structures (2000)

Ishii, K.: On developing of curved surfaces of pneumatic structures. In: International Symposium on Pneumatic Structures, IAAS, Delft (1972)

Kim, J.-Y., Lee, J.-B.: A new technique for optimum cutting pattern generation of membrane structures. Eng. Struct. **24**, 745–756 (2002)

Knipper, J., Cremers, J., Gabler, M., Lienhard, J. (eds.): Construction Manual for Polymers + Membranes. Birkhäuser, Basel (2011)

Moncrieff, E., Topping, B.H.V.: Computer methods for the generation of membrane cutting patterns. Comput. Struct. **37**(4), 441–450 (1990)

Nejur, A., Steinfeld, K.: Ivy: bringing a weighted-mesh representation to bear on generative architectural design applications. In: ACADIA/2016: POSTHUMAN FRONTIERS: Data, Designers, and Cognitive Machines, Proceedings of the 36th Annual Conference of the Association for Computer Aided Design in Architecture (ACADIA), pp. 140–151 (2016)

Otto, F., Burkhardt, B., Drüsedau, H.: Manufacturing. In: IL15 Air Hall Handbook. Institute for Lightweight Structures, West Germany (1982)

Otto, F., Drüsedau, H., Hennicke, J., Schaur, E.: Architectural and Structural Design. In: IL15 Air Hall Handbook. Institute for Lightweight Structures, West Germany (1982)

Philipp, B., Breitenberger, M., Wuchner, R., Bletzinger, K.: Form-finding of architectural membranes in a CAD-environment using the AiCAD-concept. In: Ramsgaard Thomsen, M., Tamke, M., Gengnagel, C., Faircloth, B., Scheurer, F. (eds.) Modelling Behaviour: Design Modelling Symposium 2015, pp. 65–74. Springer, Berlin (2015)

Tabarrok, B., Quin, Z.: From finding and cutting pattern generation for fabric tension structures. Comput. Aided Civil Infrastruct. Eng. **8**(5), 377–384 (1993)

Simulation and Real-Time Design Tools for an Emergent Design Strategy in Architecture

Ole Klingemann(✉)

Structure and Design|Department of Design, Faculty of Architecture, University of Innsbruck, Innsbruck, Austria
Jens-Ole.Klingemann@uibk.ac.at

Abstract. Based on the example of tidal-driven ecologies, this paper investigates how the complex and unpredictable processes of dynamical systems can be applied to design solutions with the creation of new spatial structures. Such processes provide the inherent starting point for a dynamic design strategy based on the creation of new digital tools. Because this approach implicates a shift away from the static solutions of traditional planning tools and towards an ongoing process (both digital and materialized) within a continuously changing environment, this paper discusses theoretical concepts and their initial implementation. As a first step, this paper presents a customized, real-time design tool programmed in CUDA/C++, and introduces the concept of emergent parameters. Using an example of their successful application in a proto-design, it reveals the potential of a non-linear digital design strategy to generate unique structural solutions. Furthermore, it points the way towards a post-natural ecology and architecture that will overcome the traditional distinction between natural ecologies and built environment. Starting from this investigation into the emergence of proto-design patterns, an architectural informed emergent design process will be developed and simultaneously implemented in a current PhD program.

Keywords: Complex computation · Complexity · CUDA · Dynamic simulation · Emergence · Fluid dynamics · Nonlinearity · Parallel computation · Process thinking · Real-time tools · Unified-particle-physics

Emergent Design in Morphodynamic Environments

With the concept of emergence and simulation as formulated by DeLanda (2011), there arises a philosophical position that establishes the fundamental status of emergent entities and thus the irreducibility of the presented emergent structural formations. Additionally, it establishes digital simulation as a valid technique in science and philosophy for describing and understanding complex systems, while at the same opening up new possibilities for a digital design process. Stuart-Smith (2011) defined a similar design agenda with his emphasis on the organisation of matter as the primary concern of the architect and Colletti (2013) developed with Digital Poetics a related theory of design-research in architecture.

K. De Rycke et al., *Humanizing Digital Reality*,
https://doi.org/10.1007/978-981-10-6611-5_26

Although controversially discussed, DeLanda contributes to the field of architecture and closes a gap between science and design by the explanation of emergent concepts: "Simulations are partly responsible for the restoration of the legitimacy of the concept of emergence because they can stage interactions between virtual entities from which properties, tendencies, and capacities actually emerge." (DeLanda 2011, p. 6). With the development of a new tool for the design of complex structures in dialogue with nature, we are able to link this field of research with the possibilities of today's digital simulation techniques. This leads to a renaissance of process philosophy in the architectural discourse going back to Deleuze's work of Difference and Repetition (1968) and Duffy's recent defence of Deleuze (2013), with its focus on mathematical concepts.

Different/Citation

As interpreted by Duffy, Deleuze's concept of individuation and the process of production is modelled on the logic of different/citation: "The logic of the differential, as determined according to both differentiation and differenciation, designates a process of production, or genesis, which has, for Deleuze, the value of introducing a general theory of relations…" (2013, p. 45). With the process of differentiation for pattern generation and that of differentiation for the individuation of new formations, a key concept is established for an emergent design strategy that can be directly translated into the digital simulation presented here (see Fig. 1). The former is implemented in a fluid simulation by iterative solving of differential equations, together with a setup of gradients to fuel a process; the latter in the emergence of specific flow-patterns and new growth-structures. With 'Applied Virtuality' a similar concept of mathematical thinking is also used in Bühlmann's (2010) contribution to the current discourse of design theory.

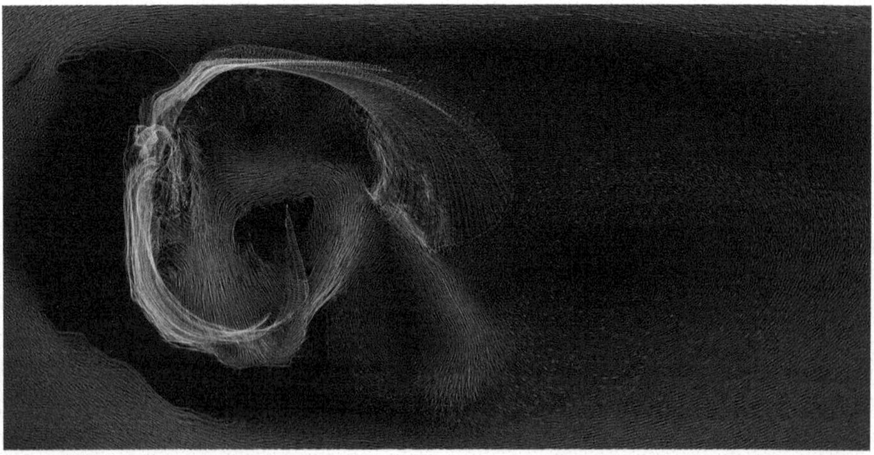

Fig. 1. Application screenshot from top with the emergence of a new land formation (represented in *white*) in the German Wadden Sea, coupled with the changed flow pattern at tidal inflow from the *left*. The current flow directions and intensities are visualized by coloured vectors

Complex Systems

The still-young field of research focusing on complex systems offers a similar introduction to pattern formation, as well as to the necessity of developing new design tools. Bar-Yam defines 'Complex Systems' as "the new approach to science studying how relationships between parts give rise to the collective behaviours of a system, and how the system interacts and forms relationships with its environment." (2009, p. 3). Where designing in complex systems is concerned, this definition gives rise to two key concepts: the part-to-whole relation in an emergent process and the interrelationship between such newly-designed systems and their dynamic environment. Using custom algorithms designed for the visualization and production of new emergent patterns, we studied the complex process of emergent pattern formation in conjunction with first proto-designs.

CUDA Real-Time Simulation Tool for Designing in Dynamical Systems

This paper describes an approach to software development from the perspective of the designer, with a view to resolving the complex tasks of design within dynamical systems and with focus on the self-organization of pattern formations in such systems. A task common to various scientific disciplines such as physics, biology, chemistry or the social sciences, studying complex systems is, according to Bar-Yam, "The idea [...] that instead of specifying each of the parts of a system we want to build, we can specify a process that will create the system that we want to make. This process would use the natural dynamics of the world to help us create what we want to make." (2001, p. 8).

Conventional CAD-software and even the use of scripting tools inside these packages is limited in terms of their problem-solving strategies and computational power. Therefore, NVIDIAs CUDA platform for parallel computation on the GPU was used to develop a new design tool that is capable of interacting with the designer in real time. Built upon the NVIDIA FLEX framework by Macklin, the concept of Unified Particle Physics (Macklin et al. 2014) is used for the overall simulation, and the model of Position Based Fluids (Macklin and Müller 2013) is used for the simulation of fluid dynamics. The tool for designing in tidal dynamics presents first steps towards an emergent design approach with an interpretation of the abstract physical models of fluid dynamics and their implementations in a digital simulation. Fluid dynamics and their different pattern formations over time are a widespread phenomenon in nature and an ideal research area for the development of precise key concepts to direct complex processes in a proper way on the one hand, and on the other hand for demonstrating the need for a co-evolutionary design strategy. While the simulation of fluids leads to the definition of emergent parameters by starting from physical models and equations and to the visualization of their emergent tendencies, the design approach extends them with the creation of new spatial strategies and parameters of growth.

Because visualization and structural algorithms are immanent to revealing the hidden virtual structures of emergent systems, they must be controllable by the designer. That is why an abstract and reduced framework for describing physical systems—such as NVIDIA FLEX—was chosen for further development. Here, new algorithms can be implemented without the traditional restrictions to expand the simulation of natural processes by various design inputs and to generate new emergent formations.

Simulation Setup and Calibration

In using this experimental approach, it is necessary to continually verify the accuracy of the pure physical simulation: in this case, the fluid simulation starting with the implementation of a fluid container with increasing inflow and valid boundary conditions presented here. The fluid container consists of 1,357,824 fluid particles, emitted in 3 vertical layers from the left (FLEX is currently limited to 2,097,152 particles when solving for 64 neighbours per particle due to its memory implementation). Particles that flow out to the right are permanently refilled at the left side. Boundary conditions are implemented with the mirroring of particles at the side-layers and a forced direction for particles in the emitting- and outflowing-layer. The successful application and calibration of parameters can then be verified with the emergence of typical flow patterns in a classical setup with a cylinder placed in the current. For the growth structures, it is essential to capture the simulation at flood (see Fig. 2).

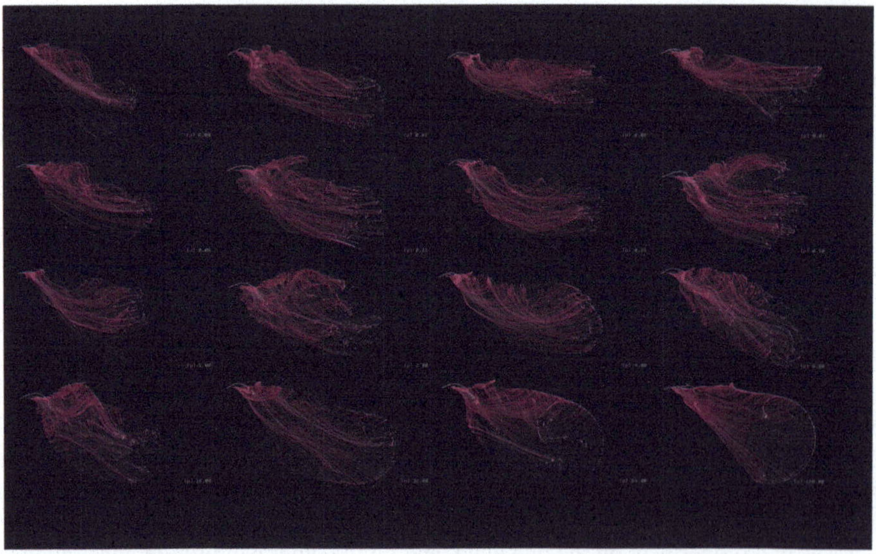

Fig. 2. Aggregation patterns with variations of the intensive parameter *[viscosity]* of the fluid simulation at inflow from the *left*. Early studies revealed the need for the simulation of an increasing inflow by starting with an empty fluid container at low tide. A constant flow would have been inaccurate for a simulation of tidal dynamics and not capable of generating new formations. These growth patterns emerge only from the recursive feedback of the grown structure without predefined obstacles. This setup already underlines a key concept for an emergent design, not to force the system by external constraints, but rather to allow new solutions to evolve dynamically

Non-linear Design and Reproducibility

In a further step, we implemented a custom growth process in CUDA/C++ with recursive feedback into the fluid simulation by utilizing passively advected particles emitted into the fluid. They form the initiators for new evolving structures, accelerating a process of aggregation according to a concept similar to that of enzymes, which can be controlled both in the initial setup as well as throughout the progressing simulation in real time. By using the same point in time and space together with the same initial values for all parameters this digital experiments are reproducible, but very sensitive to small changes due to their non-linear behaviour. Therefore, real time adjustments are crucial and must be logged in the application. At the same time the designer has to get used to a non-linear evolution after changing a parameter, but is always able to go back.

Emergent Simulation Parameters

Emergent properties describe the dynamically evolved properties of the emergent whole, which are irreducible to the properties of its parts. Similarly, Heylighen states that "the philosophy of complexity is that this (reductionism of classical science) is in general impossible: complex systems, such as organisms, societies or the Internet, have properties—emergent properties—that cannot be reduced to the mere properties of their parts." (Heylighen 2008, p. 2) They are constitutive for the dynamical system that makes them essential both for a digital simulation technique by abstraction from the physical mechanisms of its parts, and for an emergent parameterization of the design tool that provides influence over non-linear processes. To facilitate readability, these parameters are introduced in square brackets.

Identity Parameters of Emergent Wholes

Two parameterizations are already included in DeLanda's concept of defining the identity of emergent wholes or assemblages. Although they are not directly implemented as adjustable parameters, they lead to an evaluation of the inherent design potentials.

[Territorialization] specifies "the relative homogeneity or heterogeneity of the components of an assemblage […] This parameter must also specify the state of the boundaries of an assemblage: sharp and fixed or, on the contrary, fuzzy and fluctuating" [1, p. 187]. In this sense, the flow patterns and new growth structures are, on the one hand, highly de-territorialized by their soft borders and thus have a high potential for resiliency, and on the other hand they are territorialized through the homogeneity of their particles, (see Figs. 3 and 4).

[Coding] specifies the relationship of emergent wholes to the environment. "An organism, for example, may be said to be highly coded if every detail of its anatomy is rigidly determined by its genes and relatively decoded if the environment also contributes to its anatomical definition." (2011, p. 188). As a first step, the presented implementation focuses on the structural potential connected to environmental dynamics. As a result, parameters of growth are tightly connected to the parameters of

Fig. 3. Steps of a non-elastic aggregation by a single enzymatic emitter in turbulent flow at Re = 3.0 with *[A-elastic distance] == [A-maximum distance]*, forcing a scattered territorialization

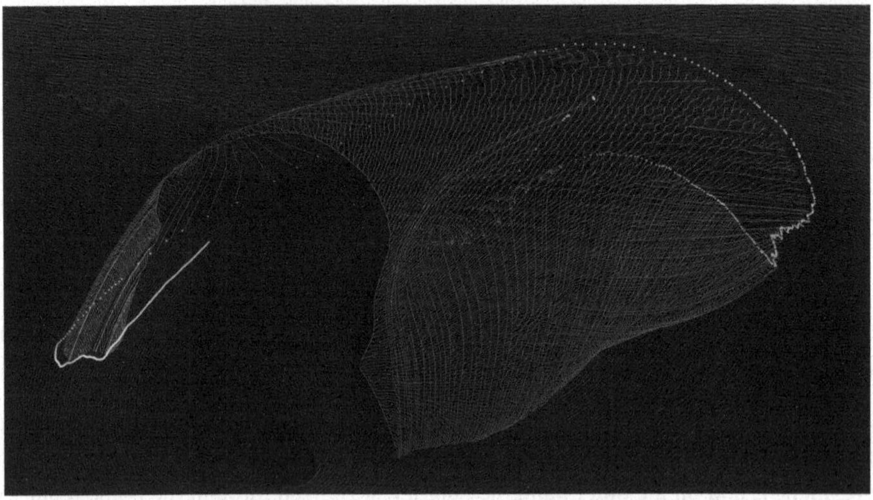

Fig. 4. Full-elastic aggregation by a single enzymatic emitter after 560 steps in turbulent flow at Re = 3.0 with *[A-elastic distance] == [A-step distance]*, generating a smooth territorialization

fluid dynamics. The result is therefore highly decoded but already leads to an evaluation of its design potential for a further development informed by architectural parameters, see Section "Evaluation and Outlook to Post-natural Ecologies".

Intensive Parameters of Fluids

Intensive properties of fluids form the basis for any parameterization in a fluid simulation and are partly variables in the nonlinear partial differential equations describing the flow of incompressible viscous fluids. The Navier–Stokes Eq. (1) add additional parameters to a non-linear parameterization that are sensitive to small changes and at the same time are primarily responsible for the emerging fluid patterns, as expressed by the dimensionless Reynolds number (2). With the Reynolds number, the transition from laminar to turbulent flow is predictable and also scalable, thus we can better adjust the intensive parameters and predict a similar behaviour in another scale.

Navier–Stokes momentum equation of incompressible fluids

$$\frac{\partial \mathbf{v}}{\partial t} = -(\mathbf{v} \cdot \nabla)\mathbf{v} - \frac{1}{\rho}\nabla p + \mu \Delta \mathbf{v} + \mathbf{f} \tag{1}$$

$\mathbf{v}$ *[flow-velocity]* of a velocity field
ρ *[density]* of the fluid
p *[pressure]*
μ *[dynamic viscosity]* of the fluid
$\mathbf{f}$ *[optional external force]* like wind or gravity

Reynolds number

$$\mathrm{Re} = \frac{\rho v L}{\mu} \tag{2}$$

v *[velocity]* of the fluid with respect to the object
ρ *[density*] of the fluid
μ *[dynamic viscosity]* of the fluid
L *[characteristic linear dimension]* of an obstacle

[Temperature] is a measure of the average kinetic energy of the particles in a fluid body and is not parameterized for now, *[flow-velocity]* is used instead.

[Pressure] is indirectly build up by a pressure gradient caused by the tidal inflow and evolving obstacles.

[Concentration] here is the concentration of sediments and will be further discussed in Section "[E] Enzymatic Parameters".

[Flow-Velocity] is adjusted to the tidal dynamics with the initial setup of the inflow and responsible for turbulent flow.

[Dynamic-Viscosity] is also adjusted to the tidal dynamics with the initial setup, but controllable during the simulation and responsible for viscous flow.

[Density] is set to 1, but can be further used for a refinement of the sedimentation process.

[Characteristic-Length] depends on the Wadden Topography and the evolving growth-structures (see Fig. 2).

Fluid Simulation Parameters

The implementation of fluid dynamics within FLEX is based on particle methods for robustness and flexibility. This introduces parameters particularly for controlling the surface behaviour of fluid bodies, which are not available in voxel based methods. Even drawbacks of this method give rise to additional parameters that are important for our design adjustments.

[Vorticity-Confinement] is used by Macklin and Müller (2013) to replace lost energy by numerical dissipation. We used this parameter for permanently controlling local turbulences and to design local features.

[Buoyancy] is an upward force exerted by a fluid and enables the continuous control over vertical velocities and thus the vertical evolution.

[Cohesion] is the attraction between particles of the same kind and produces droplet structures, not presented here.

[Surface-Tension] is the attempt of particles to minimize surface area, not presented here.

[E] Enzymatic Parameters

The natural dynamics in the Wadden Ecology are far too complex for an entire simulation. Thiede and Ahrend stated this in their numerical modelling of sediment flow around the island Sylt (2000, p. 187) for high dynamic areas with the exclusion of biological processes. We use an abstraction of their sedimentation model combined with the biological concept of enzymes. Microorganisms are influencing the physical sedimentation process and so does a dynamic design strategy by an artificial enzymatic growth process. At first, passive particles with enzymatic features to drive an aggregation of sediment particles are set up with the following parameters.

[E-Dimension] defines how many enzymes are emitted simultaneously and controls structural density, (see Figs. 3 and 4 for a single-emitter and Figs. 9, 10 and 11 for multi-emitters).

[E-Life-Time] of the enzymatic emitter, controlling the area of evolution.

[E-Time-Steps] defines the period after which a new enzyme is emitted and controls structural density.

[E-Buoyancy] of enzymes, controlling the vertical evolution.

[E-Drag] of enzymes in reaction to fluid forces, controlling structural smoothness.

[E-Damping] of the movement of enzymes, controlling the area of evolution.

[E-Ballistic] behaviour of enzymes.

[E-Active-Minimum-Life-Time], [E-Active-Maximum-Life-Time] define the life-span for emitting active enzymes, that initiate the following aggregation process.

[E-Active-Minimum-Age], [E-Active-Maximum-Age] define the life-span for a single active enzyme.

[A] Aggregation Parameters

In a second step, growth is implemented with an aggregation process by adsorption correlated to the acceleration of the active enzymes for an increase or decrease in sedimentation height.

[A-Critical-Velocity] defines the maximum velocity of enzymes for aggregating new particles.

[A-Drag] of aggregation in horizontal direction, controlling structural branching.

[A-Vertical-Drag] of aggregation in vertical direction, controlling the vertical evolution.

[A-Vertical-Damping] of aggregation, controlling the area of evolution.

[A-Vertical-Amplitude] of aggregation, controlling the sedimentation-height in relation to the tidal-height.

[A-Minimal-Distance], [A-Rigid-Distance], [A-Elastic-Distance], [A-Maximum-Distance] define the aggregation type of a new aggregated particle in relation to the distance between the enzyme and the previous aggregation, (see Figs. 3 and 4).

[A-hardening age] defines the recursive feedback time of the aggregation. It influences how fast a feedback to the fluid simulation is generated, and thus how turbulently or smoothly the flow changes are in conjunction with the ongoing growth process.

Emergent Tendencies and Tidal Patterns

Emergent Tendencies describe the finite possible states of an emergent whole, such as the tendency of a fluid to go from laminar, periodically to turbulent flow under special conditions. Over time, new patterns emerge and reveal hidden structures that are of equal importance for an emergent design approach as emergent parameters. Therefore, custom tracking techniques with *markers, trajectories, streaklines* and *celltracing* for visualizing emergent patterns in fluids are implemented in CUDA/C++, (see Figs. 5, 6, 7 and 8).

Evaluation and Outlook to Post-natural Ecologies

One successful application of this approach is the emergence of the proto-designs for the German Wadden Sea, (see Figs. 9, 10 and 11). These are early structural studies evolved from several setups and adjustments of the described emergent parameters. Nonetheless they can be evaluated by design criteria fundamental to architecture. Also during the design process, criteria that are immanent for rich spatial structures and a self-organisation have been applied. These design criteria are among others: resiliency

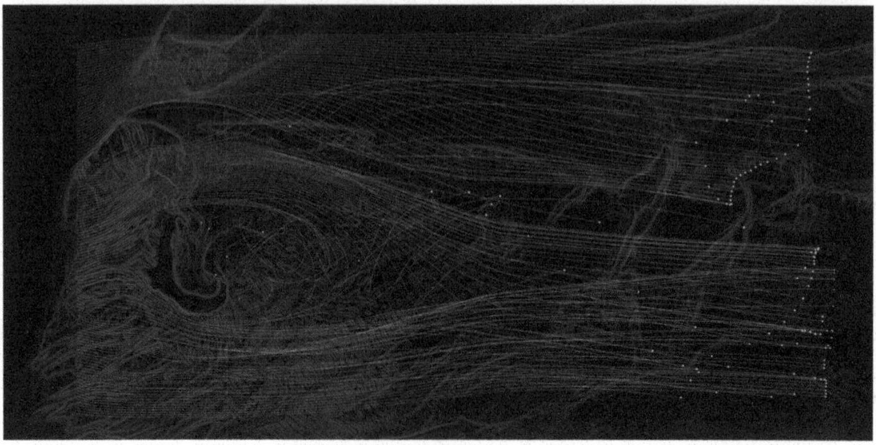

Fig. 5. Visualization of inflow patterns in the German Wadden Sea with 128 markers after 5×600 steps: *Fluid markers* (*coloured dots*) are the passively advected particles and *fluid trajectories* (*curves*) their recorded pathlines. They reveal the hidden attractors for basins of sedimentation behind the small island. This area is then used for the evolution of the following proto-designs

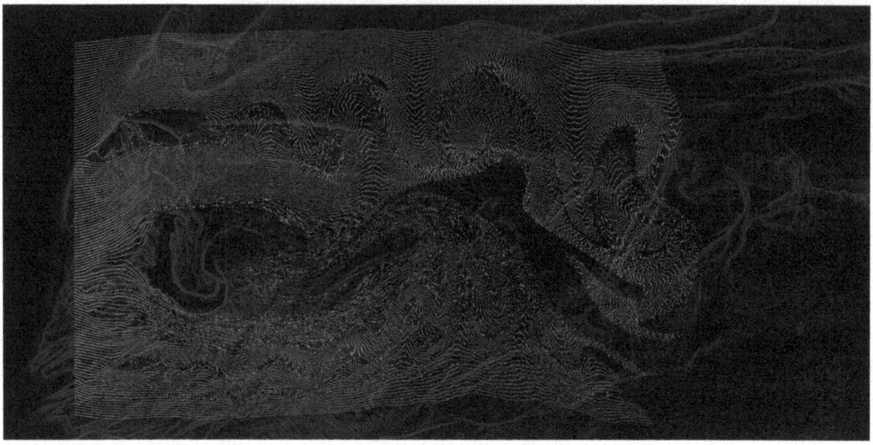

Fig. 6. 128 streakline-emitters after 5×600 steps: *Fluid streaklines* are the streaks of all markers continuously emitted from one point. They reveal turbulent areas for a differentiated design

in growth, differentiated global and local structures, density and porosity, orientation, high resolution and an overall rich ornamentation.

A conventional, scientific software is usually limited to solving one task defined at the beginning of a simulation over an extended period of time, and to showing the final results at the end of the calculation. In contrast, the real-time tool presented here allows the continuous interaction of the designer during computation, and consequently for the

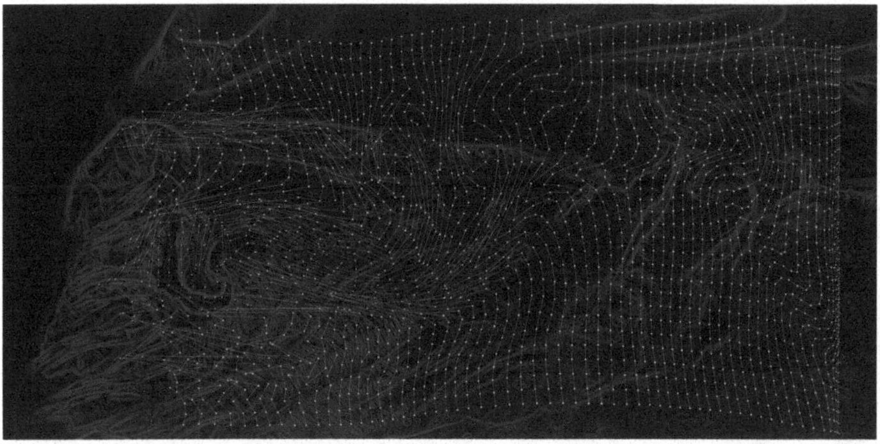

Fig. 7. Grid of 32 × 62 markers after 5 × 100 steps: *Fluid markers* (*coloured dots*) are the passively advected particles and *fluid timelines* (*connecting lines*) are the topological connections of simultaneously emitted markers. A grid of markers reveals the evolution of the whole site without blank areas and the crossings of timelines are a measure for the chaotic evolution

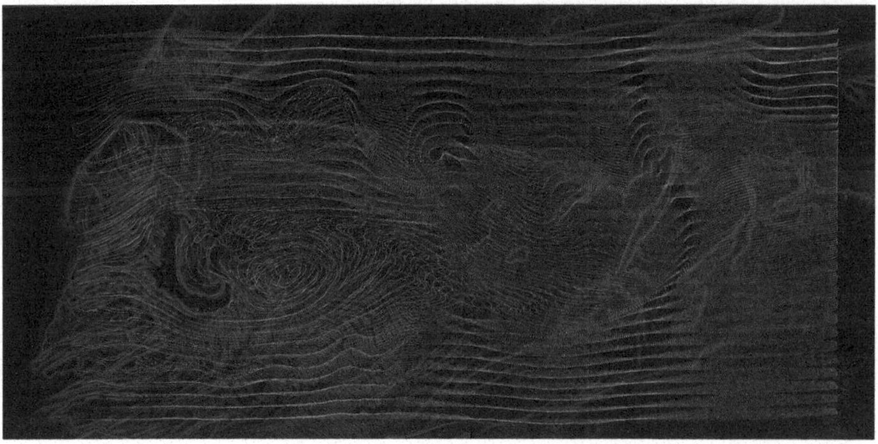

Fig. 8. Grid of 32 × 62 streakline-emitters after 5 × 100 steps: *Fluid celltracing* (*coloured dots*) are *fluid* markers emitted from a grid. They reveal the evolution of the whole site with vortices and eddies, particularly behind the small island. These areas are of special interest for the design of rich structures and give the best hint for their evolution

Fig. 9. Aggregation of 2 enzymatic emitters located at 54°44.70′N, 8°21.50′E: materialization of the new island from Fig. 1

creation of intuitive complex solutions. Beyond this verification of its design potential, the proposed tool will need to be further developed to incorporate specific architectural design features. By inventing more design-coded algorithms and architectural parameters for growth, this will automatically be the next step.

With the development of new recursive tools for a designed structural formation in dialogue with a dynamic environment, as presented here, the possibility for a simultaneous co-evolution of natural and human initiated processes finally arises. This opens up a path to realizing the full potential of sustainable design within post-natural ecologies, which will be the focus of our long-term agenda, in collaboration with other scientific disciplines working in the field of complex systems.

Fig. 10. Aggregation of 8 enzymatic emitters located at 54°44.70′N, 8°21.50′E: emergence of another new land formation in the German Wadden Sea

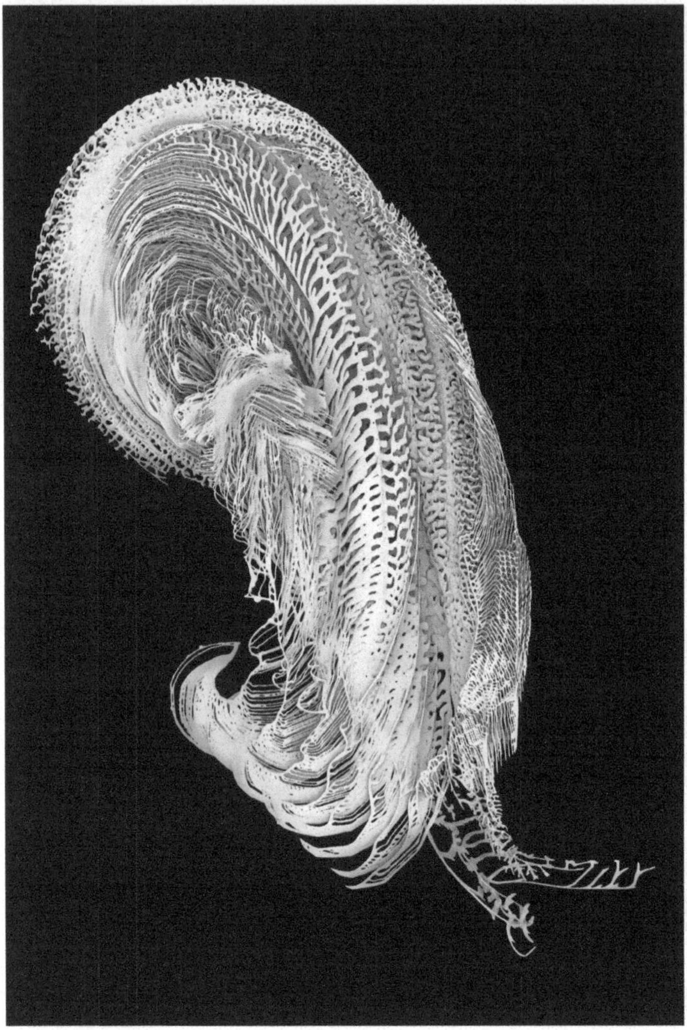

Fig. 11. Materialization of the new island from Fig. 10, digitally grown in tidal dynamics

References

Bar-Yam, Y.: General features of complex systems. In: Kiel, L. (ed.) Knowledge Management, Organizational Intelligence and Learning, and Complexity, vol I. UNESCO EOLSS Publishers, Paris (2009)

Bühlmann, V.: Applied virtuality. On the problematics around theory and design. In: Hampe, M., Lang, S. (eds.) The Design of Material, Organisms and Minds. Springer, Berlin (2010)

Colletti, M.: Digital Poetics. An Open Theory of Design-Research in Architecture. Ashgate Publishing, Farnham (2013)

DeLanda, M.: Philosophy and Simulation. The Emergence of Synthetic Reason. Continuum International Publishing Group, New York (2011)

Deleuze, G.: Différence et Repetition. Presses Universitaires de France, Paris (1968)
Duffy, S.: Deleuze and the history of mathematics. In: Defence of the 'New'. Bloomsbury Academic, New York (2013)
Heylighen, F.: Complexity and self-organization. In: Bates, M., Maack, M. (eds.) Encyclopedia of Library and Information Sciences. CRC Press, Boca Raton, FL (2008)
Macklin, M., Müller, M., Chentanez, N., Kim, T.: Unified particle physics for real-time applications. ACM Transactions on Graphics (TOG). Proceedings of ACM SIGGRAPH 2014, **33**(4). ACM, New York (2014)
Macklin, M., Müller, M.: Position based fluids. ACM Transactions on Graphics (TOG). Conference Proceedings of SIGGRAPH 2013, **32**(4). ACM, New York (2013)
Stuart-Smith, R.: Formation and Polyvalence: The Self-Organisation of Architectural Matter. Ambience'11, Borås, Sweden (2011)
Thiede, J., Ahrend, K.: Klimaänderung und Küste – Fallstudie Sylt. GEOMAR, Kiel (2000)

Robotic Fabrication Techniques for Material of Unknown Geometry

Philipp Eversmann

Eversmann Studio, München, Germany
studio@eversmann.fr

Abstract. Both natural materials such as timber and low-grade or recycled materials are extremely variable in quality and geometry in unprocessed state. Additive digital fabrication processes in robotics in combination with sensor feedback techniques offer large design freedom, high precision and material efficiency and enable a highly customized fabrication and calculation process. Separate studies have been made on scanning, efficient algorithmic arrangement and automated assembly of structures of variable timber elements. In this paper we explore a robotic fabrication process, in which we combine the techniques of scanning, digitally arranging and robotically assembling in one continuous real-time workflow. This means that the final design and appearance only emerge after a unique fabrication process, corresponding to the material used and the assembly sequence. We describe techniques for the simulation modelling and performance analysis using particle simulation, and demonstrate the feasibility through the realisation of the envelope of a robotically assembled double-story timber structure with hand-split wood plates of varying dimensions. We discuss a future use of natural, low-grade or waste material in the building industry through robotic processes. We conclude by analysing the integration of qualitative analysis, physical simulation and the degree of variability of input material and resulting complexity in the computation and fabrication process.

Keywords: Feedback process · Robotic fabrication · Particle simulation · Scanning · Real-time workflow

Introduction

Introducing natural variability in manufacturing processes rather than fabricating variable designs out of standardized products allows using local material (Stanton 2010), decreases processing operations, and can therefore save material and reduce the ecological footprint of construction. The natural variability in quality and geometry of raw timber products that is also found in low-grade or recycled materials makes their application in a controlled manufacturing process extremely challenging. Therefore, repeatable and standardized digital fabrication techniques are nowadays predominantly used to process highly engineered products. Additive robotic fabrication processes combined with sensor technology can be used for real-time feedback, e.g. in assembly processes. This opens up the possibility to integrate material variability through highly

K. De Rycke et al., *Humanizing Digital Reality*,
https://doi.org/10.1007/978-981-10-6611-5_27

customized analysis, calculation and fabrication processes. When using organic materials such as timber, the question arises at which state and format should the material ideally be integrated in additive robotic fabrication? A number of recent research projects have investigated the use of raw materials in digital design and fabrication workflows, such as the arrangement and positioning of irregular wood components through algorithmic techniques (Monier et al. 2013). Further studies show the possibility to physically scan and data process natural wood branches (Schindler et al. 2014), and connect them through robotic milling for architectural structures (Mollica and Self 2016; Self and Vercruysse 2017). Autonomous assembly of standard wood beams using feedback processes was experimented with by Jeffers (2016). The "Mine the scrap" project (http://certainmeasures.com/mts_installation.html) investigated the use of scrap material. A computational interface was created in order to scan leftover wood plates of random geometry and arrange them algorithmically to match a given design envelope most closely. This paper investigates design modelling and fabrication techniques of how material variability can be directly integrated into a continuous real-time robotic workflow and how geometrical and functional performance can be strategically simulated and evaluated. In Section "Methods", we describe the development of the design, analysis and fabrication process and the realisation of a large-scale demonstrator. In Section "Results", we evaluate results of geometric generation, analysis and fabrication. We conclude in Section "Conclusion" by discussing a future use of robotic technologies for organic, low-grade and waste materials and analyse the incorporation of qualitative variability, physical simulation and the degree of variability and resulting process complexity.

Methods

In this section, we demonstrate design modelling techniques for façades of rectangular panels of variable size, analyse their ability to adapt to a range of surface typologies and evaluate water permeability using particle simulation (Section "Design Modelling and Performance Analysis"). In Section "Real-Time Robotic Communication and Setup" we explain the robotic fabrication setup and its communication protocols and continue to describe a large-scale demonstrator in Section "Demonstrator Project".

Design Modelling and Performance Analysis

Traditional facades and roofs can be fabricated with timber panels without the need for additional waterproofing layers. When overlapped and mounted in certain angles and orientations, timber panels can last for up to 100 years depending on their wood type (http://www.holzschindel.at/holzschindeln/ratgeber). The fabrication technique plays an important role for the panel's weather endurance. The traditional method is to split the timber chunks along the fibre direction. This keeps the fibres intact and creates a highly-textured surface, which allows split panels to dry much more efficiently than sawn panels (Niemiec and Brown 1993). Their varying size (in comparison to standard sizes) also allows the use of all parts of a naturally grown tree, limiting material waste (Fig. 1).

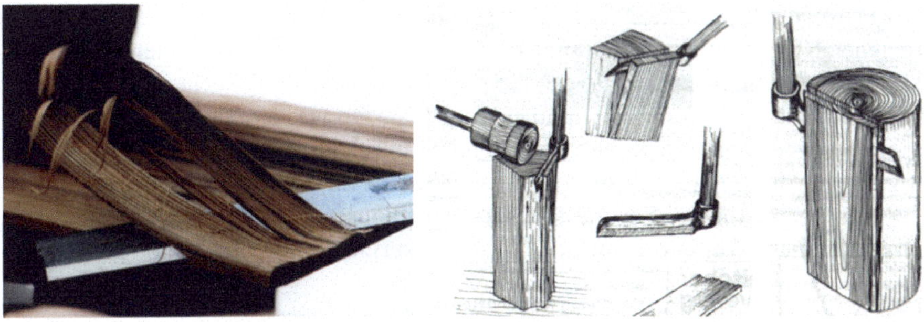

Fig. 1. *Left* Splitting of timber panels, image courtesy by Theo Ott—www.holzschindeln.de. *Right* Radial- and tangential splitting, image courtesy of Ludwig Weiss Holzschindelwerk

Panelisation Modelling. We investigated multiple geometric approaches for variable panelisation on flat and curved surface typologies with three design goals: (1) the pattern should be capable of adapting to double-curved surfaces while visual and geometric consistency remains. (2) It should be water resistant. (3) The facade should attach directly to the primary structure without additional sub-structures. The design space and its material, structural and geometric dependencies are summarized in Fig. 2. The primary structure consists of a geometric system of cuboids which are diagonally braced. Both horizontal as well as diagonal design patterns were therefore included in the algorithm. We also allowed horizontal as well as vertical overlaps between the panels. Since the algorithm had to function iteratively as well for the fabrication sequence, each panel is first placed at its ideal location and orientation along the surface curvature, and then subsequently rotated until it fits tightly to the precedent panels.

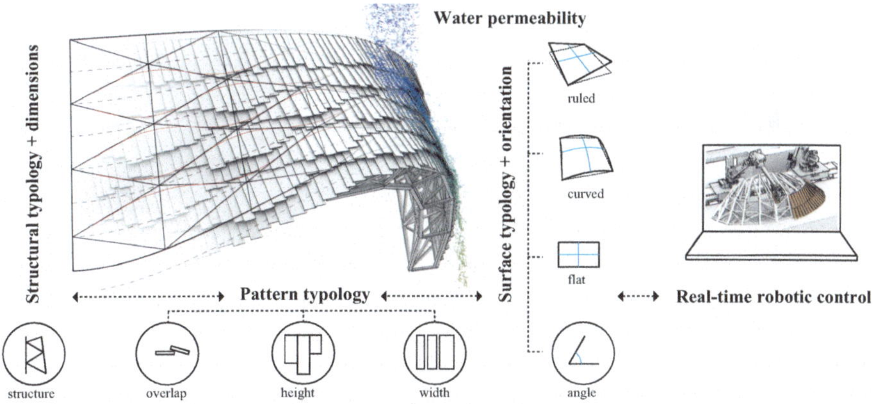

Fig. 2. Design space: overview of the relationship between the parameters and constraints of structural typology, pattern typology and variation, surface typology and angle, water permeability and fabrication system

Curvature adaptability. With these general constraints, a large number of patterns were generated. We evaluated a range of patterns for curvature and water permeability. Since the maximum distance between the panels and the driving surface already give an indication of the general surface tightness, our first step was to evaluate the variation of the parameters of height and width of the panel in relation to surface typology as well as a variation of "horizontal" or "diagonal" fixation lines (Fig. 3).

Water permeability. For the evaluation of the dependencies between the design pattern, parameter variation, surface typology, surface angle and water permeability,

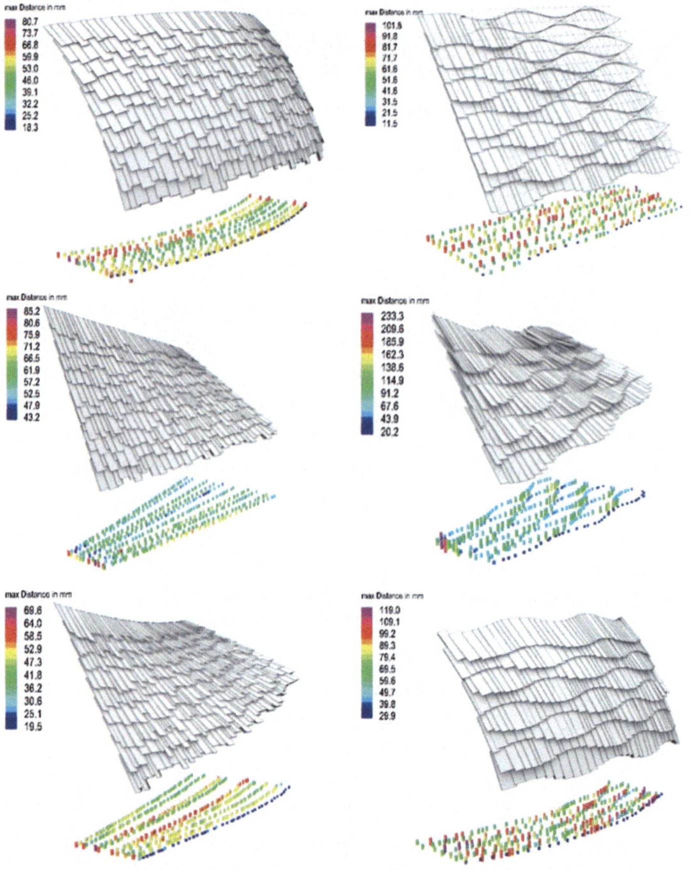

Fig. 3. Geometric adaptability: mapping of max. distance between panels and driving surface. *Left column* Horizontal rows 30 cm, width variation 8–25 cm, height variation 40–60 cm. Results for spherical curvature r = 6 m, double-ruled surface and simple-ruled surfaces, inclination 40°–80°. *Left column* Diagrid panelling pattern on flat, *ruled* and *spherical curved* surfaces, panel width randomisation 8–25 cm, angle smoothing r = 1.5 m. Results show that randomisation can cause local disturbances. Variation of vertical overlaps as shown in the diagrid panelling introduces surface ripples on flat surfaces and can cause large irregularities on curved surfaces

Surface typology + inclination

Pattern + Variability	flat 80°	flat 60°	flat 40°	flat 20°	r=6m 40°-80°	ruled 40°-80°
w 8-24cm, l 60cm, h 30cm	0.02	0.03	0.1	0.12	0.05	0.27
w 8-24cm, l 40-60cm, h 30cm	0.23	1.14	2.91	8.4	0.57	0.58
w 8-24cm, l 40-60cm, h 30cm	0.08	0.72	2.91	6.83	0.24	0.18
w 8-24cm, l 60cm, h 30cm	0.01	0.05	0.3	0.92	0.16	0.03
w 8-24cm, l 60cm, h 45cm	0.07	0.38	1.09	2.72	0.41	0.47

Fig. 4. *Right* Setup particle simulation in realflow: a rectangular dyverso-particle emitter emits a stream of particles on a panelised surface. *Left* Water permeability (in % of particles detected traversing the panels) in relation to panelisation parameters: width of butt-jointed and overlapped panels, panel height and row height in relation to drive surface geometry and inclination

we created an analysis setup using the particle simulation software Realflow (2017). Through large amounts of particles, fluid dynamics can be efficiently simulated for CG applications. The possibility to automate and trace particles through a scripting interface in Python also allows for a quantitative analysis for design and evaluation purposes (Tan et al. 2017). With a sufficiently high resolution and continuous collision detection the interaction of particles with values corresponding to the physical properties of water and a range of design patterns was simulated. Around 300.000 particles were emitted and traced for each pattern through custom Python scripts (Fig. 4). As the analysis shows, the row height can be increased by 50% without a significant increase of water permeability. Butt-jointed panels in combination with variable panel height show the highest values since a lot of joints stay open. The angle of the tested surfaces has a dramatic impact on the amount of water traversing the surface, since on lower angled surfaces, water can traverse in the opposite direction of the overlapping pattern.

Real-Time Robotic Communication and Setup

Here, the term "real-time" refers to a robotic fabrication system, which is able to react to new information acquired during the actual fabrication process rather than merely executing previously compiled machine code. Such a system relies on digital sensors and feedback processes that can react and calculate new instructions for the robotic system on run-time (Raspall et al. 2014). In order to establish a real-time digital fabrication process, it is necessary to have integrated sensors and to be able to constantly read and write on the controller in order to make geometrical calculations on an external computer. One approach is to use a custom socket connection between PC and controller (Dörfler et al. 2016). Our approach was to implement the Robotstudio PC SDK (Robotics 2015) within GH Python (Piacentino 2017). The PC SDK libraries allow developers to create applications that can directly communicate with IRC5 controllers (Csokmai and Ovidiu 2014; Dalvand and Nahavandi 2014). This has also already been demonstrated for high frequency applications as human-machine

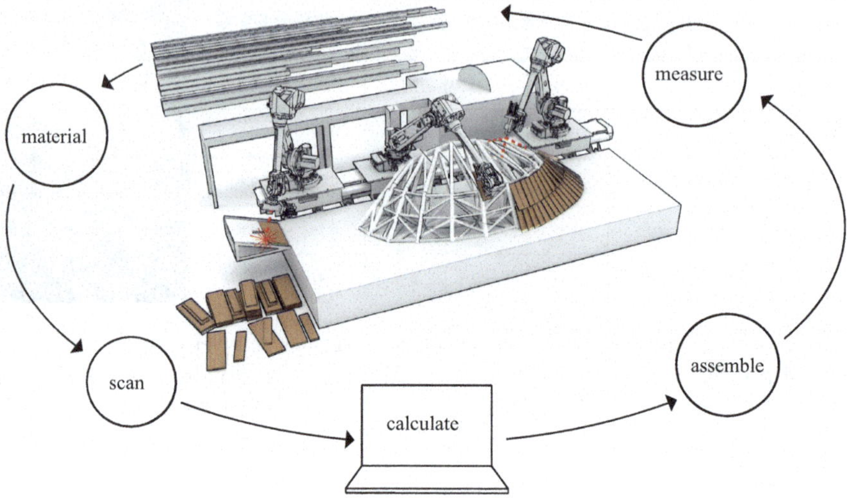

Fig. 5. Robotic fabrication workflow—Feedback process: wood panels of unknown dimensions are scanned, fabrication geometry and placement is calculated on an external PC, then written back to the robot controller in order to execute assembly. Reference points of the substructure are measured and calibrated with the numerical model in order to improve general fabrication tolerances

interaction for trajectory teaching using 3D-scanning (Landa-Hurtado et al. 2014). Our robotic setup features two robotic arms, which are able to move along a 5 m long linear axis. One robot was integrated to cut and assemble slats of variable cross-sections using an integrated CNC-circular saw, which was used for building the primary structure of building modules. The other robot was integrated to scan and mount rectangular wood panels of variable size, which was used to create the facade of the building modules. For this process, a feedback-loop of interactive operations was necessary.

In a first step, reference points of the primary structure are measured in order to calibrate the virtual model to fabrication tolerances. Then, a timber panel gets scanned and its dimensions can be read by an external PC. The parametric model described in Section "Design Modelling and Performance Analysis" was then able to calculate the position of each new element responding to architectural and fabrication constraints during fabrication. The new gripping and mounting position is then sent back to the controller while respective robotic movements are subsequently executed. Figure 5 shows the general digital fabrication workflow. We used a photoelectric sensor in combination with distance search movements of the robot to be able to determine the measurements of the plates (Fig. 6).

In order to adjust to the largely varying surface quality of the plates, we introduced a physical robotic softness through a custom end effector, featuring a vacuum gripper with elastic foam and a connection with adjustable springs. Even though all gripping and placing positions are different, the general sequence of operations is repeatable. This allows writing a general RAPID program with variables that can be changed for

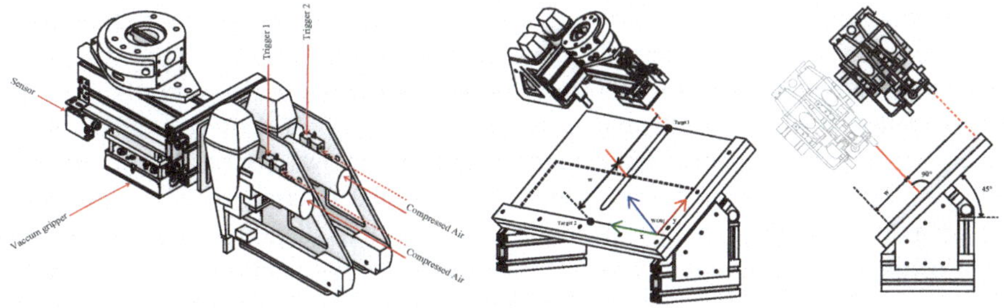

Fig. 6. *Left* Robotic endeffector design with two integrated nailguns. *Right* Feeding table for robotic scanning process

each fabrication loop. Updating of the variables occurs after the scanning process when the new positions have been calculated. We created a series of custom GH Python components which can scan the network for virtual and physical controllers, access variables and write new values. This allowed us also to simulate parts of the fabrication sequences with randomly generated panel dimensions beforehand. For movements in between the gripping and placement position, we created a standard retraction and orientation scheme that adapted to height and final positioning. The toolpath calculation remained therefore rather simple, since we could assure the building modules remained in a predefined zone, along which the robot could move freely.

Demonstrator Project

Within the course of a Master's programme in Digital Fabrication at ETH Zurich, we were able to create a double-story timber structure with an integrated timber envelope.

We investigated a design method in which the project is composed out of robotically prefabricated spatial building modules instead of a traditional fabrication method in which modules are manually assembled out of prefabricated components. The geometry consists of ruled surfaces, which creates multiple undulating ribbons that form variable vertical openings. The structure was based on a spatial truss, which was capable of adapting its geometry seamlessly between serving as wall, slab, roof and staircase. Over 4000 different wood beams were robotically cut and mounted (Fig. 7).

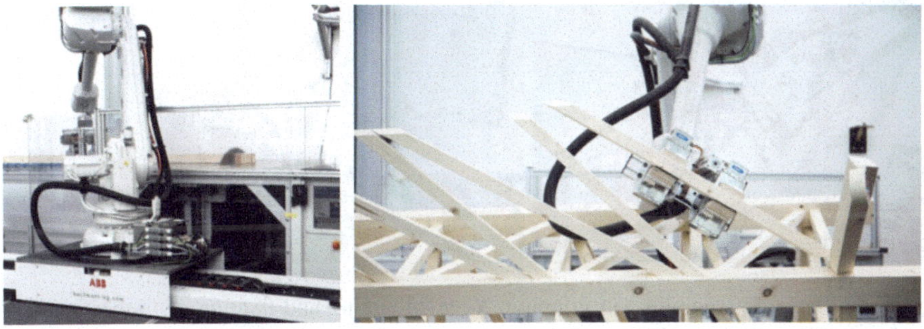

Fig. 7. *Left* Integrated saw cuts along robotic movements. *Right* Assembly of timber truss

We used four different cross sections of solid timber slats for structural differentiation and carbon-steel screws for fixation. The truss maximum vertical bracing distance on its envelope side was limited to 450 mm for surfaces between 90° and 40°, allowing fixation of the panels and sufficient water resistance (Fig. 4). Lower angled surfaces such as the roof have a denser diagonal bracing that allows double to triple vertical overlays in less inclined parts. A double and vertical overlap pattern was able to integrate horizontal (preferred by the algorithm) as well as diagonal patterns. Since the design of the primary structure was integrated with requirements of the envelope, we could directly attach the elements without additional substructures (Fig. 8). The building components could then be transported on site where they were connected to each other by steel bolts. Since each of the modules was unique and could only fit in one position, assembly errors were practically impossible (Fig. 9).

Results

We described digital modelling techniques incorporating variability in panelling dimensions and demonstrated how geometric and architectural performance can be analysed for design purposes. Patterns for irregular timber panels were analysed for geometric fitting to planar, ruled and spherical surface typologies. Water permeability was simulated through particle simulation and quantified through custom particle-tracing using Python. Results on a range of generated design patterns show that two-dimensional variability can be efficiently integrated. Overlaps in multiple directions can absorb greater geometric tolerances, but also allow for material savings through intelligently arranged overlapping patterns while keeping low values on water permeability. A robotic setup was developed for handling scanning, design calculation and mounting processes. We integrated the ABB PC SDK in a design interface of custom GH Python components, allowing the development of fabrication processes with direct interaction of geometric calculations in Rhino and robotic fabrication processes. In a large-scale demonstrator, we assembled over 2600 differently sized wood panels on a robotically-manufactured timber truss structure (Fig. 9) using a

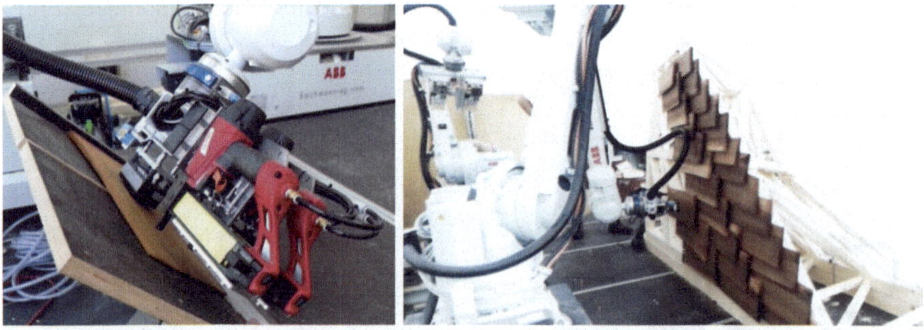

Fig. 8. *Left* Robotic scanning and gripping. *Right* Mounting and fixation of timber panel. We used hand-split timber panels of variable widths between 8 and 30 cm and a fixed height of 60 cm

Fig. 9. Interior views of the upper floor. *Right image* courtesy of Kasia Jackowska

Fig. 10. Exterior view of the double-story structure

continuous robotic feedback process responding to the provided material size in real time. Therefore, the final panel arrangement appeared only after the final production process. Since we could apply a small amount of pressure by the robot during automatic fixation, we were able to slightly elastically bend each element. This helped to approximate a geometrically complex 3-dimensional form using a simple natural material (Fig. 10).

Conclusion

Although research for integrating natural variability in architectural construction is still only at the beginning, recent advancements in robotics and control promise feasibility and efficiency at industrial scale (Vähä et al. 2013). Local materials can be directly employed, reducing the ecological footprint of construction. This can further be enabled by mobile systems for on-site digital construction (Bock 2007). While this study focusses on timber products, also other natural materials such as bamboo, straw bale, raw stones (Furrer et al. 2017), etc. and low-grade or recycled construction materials have great potentials for digital building processes. Research needs to be coupled with an investigation of functional and architectural expression for successful and profitable building applications. Furthermore, not only geometric variability, but also qualitative variability could be integrated in the design and fabrication process. Nowadays building codes account large safety values for the natural variability, which over-dimensions structural wood products. A closer computational examination might allow for a much more efficient classification and use of wood as a structural building material with large material gains. The integration of different material qualities, as found in recyclable building materials, could then be efficiently integrated. Further research needs to be done to integrate physical simulation in real-time fabrication processes. This could be used to analyse various architectural and engineering requirements also for each separate fabrication step. We deliberately excluded any sorting of material in order to simplify and streamline our fabrication process at maximum. This obviously creates constraints for the resulting geometry, which are mostly determining the visual appearance. Further studies could investigate the efficiency of a certain amount of sorting happening during the fabrication process for specific construction applications. The degree of variability of input material has a large impact on the resulting complexity of the fabrication process. In this study, the variability was constrained to two dimensions. The three-dimensional surface variation of the material (up to 20 mm on a 600 $\times$ 150 mm panel) was absorbed by the elasticity of the gripper. A similar effect could also be achieved by computationally liberating movements in certain directions of the robot (ABB Robotics 2011). Objects of unconstrained variability require three-dimensional scanning, which also leads to much more sophisticated processing techniques such as processing and mapping of large point clouds to virtual models (Tang et al. 2010). But since robotic processes are usually highly specific, our study demonstrates that functional, material and fabrication constraints can also allow the deduction of extremely simple and efficient scanning and processing techniques.

Acknowledgements. The case study project was realized in the framework of a Master class on digital fabrication with the students Jay Chenault, Alessandro Dell'Endice, Matthias Helmreich, Nicholas Hoban, Jesús Medina, Pietro Odaglia, Federico Salvalaio and Stavroula Tsafou. This study was supported by the NCCR Digital Fabrication, funded by the Swiss National Science Foundation (NCCR Digital Fabrication Agreement # 51NF40-141853). We would like to thank Philippe Fleischmann and Mike Lyrenmann for their countless efforts in helping to create our robotic setup and the companies Schilliger Holz AG, Rothoblaas, Krinner Ag, ABB and BAWO Befestigungstechnik AG for their support.

References

ABB Robotics.: Application manual-SoftMove. Robot documentation M 2004 (2011)

ABB Robotics.: Application manual PC SDK, ABB AB Robotic products. pp. 204–217 (2015)

Beyer-Holzschindel GmbH. http://www.holzschindel.at/holzschindeln/ratgeber. Accessed 30 April 2017

Bock, T.: Construction robotics. Auton. Robots **22**(3), 201–209 (2007). doi:10.1007/s10514-006-9008-5

Certain Measures, Mining the scrap. http://certainmeasures.com/mts_installation.html. Accessed 19 Feb 2017

Csokmai, L., Ovidiu, M.: Architecture of a flexible manufacturing cell control application (2014)

Dalvand, M., Nahavandi, S.: Teleoperation of ABB industrial robots. Ind. Robot Int. J. **41**(3), 286–295 (2014)

Dörfler, K., Sandy, T., Giftthaler, M., Gramazio, F., Kohler, M., Buchli, J.: Mobile Robotic Brickwork, Robotic Fabrication in Architecture, Art and Design 2016. Springer (2017). doi:10.1007/978-3-319-26378-6_15

Furrer, F., Wermelinger, M., Yoshida, H., Gramazio, F., Kohler, M., Siegwart, R., Hutter, M.: Autonomous robotic stone stacking with online next best object target pose planning, IRCA 2017. In: Proceedings of IEEE International Conference on Robotics and Automation (2017)

Jeffers, M.: Autonomous Robotic Assembly with Variable Material Properties, Robotic Fabrication in Architecture, Art and Design, pp. 48–61. Springer, New York (2016)

Landa-Hurtado, L.R., Mamani-Macaya, F.A., Fuentes-Maya, M., Mendoza-Garcia, R.F.: Kinect-based trajectory teaching for industrial robots. In: Pan-American Congress of Applied Mechanics (PACAM) (2014)

Mollica, Z., Self, M.: Tree Fork Truss. Advances in Architectural Geometry 2016, vdf Hochschulverlag AG an der ETH Zürich (2016). doi:10.3218/3778-4_9

Monier, V., Bignon, J.-C., Duchanois, G.: Use of irregular wood components to design non-standard structures. Adv. Mater. Res. **671–674**, 2337–2343 (2013). doi:10.4028/www.scientific.net/AMR.671-674.2337

Niemiec, S.S., Brown, T.D.: Care and maintenance of wood shingle and shake roofs. Oregon State University Extension Service, EC 1271 (1993)

Piacentino, G.: Grasshopper Python, McNeel & Associates. http://www.food4rhino.com/app/ghpython. Accessed 19 Feb 2017

Raspall, F., Amtsberg, F., Peters, F.: Material feedback in robotic production. In: Robotic Fabrication in Architecture, Art and Design 2014. Springer, New York, pp. 333–345 (2014)

Realflow. www.realflow.com. Last Accessed 15 May 2017

Schindler, C., Tamke, M., Tabatabai, A., Bereuter, M., Yoshida, H.: Processing branches: reactivating the performativity of natural wooden form with contemporary information technology. Int. J. Archit. Comput. **12**(2), 101–115 (2014). doi:10.1260/1475-472X.12.2.101

Self, M., Vercruysse, M.: Infinite variations, radical strategies. In: Menges, A., Sheil, B., Glynn, R., Skavara, M. (eds.) Fabricate 2017, pp. 30–35. Ucl Press, London, (2017). http://www.jstor.org/stable/j.ctt1n7qkg7.8

Stanton, C.: Digitally mediated use of localized material in architecture. In: Proceedings of the 14th Congress of the Iberoamerican Society of Digital Graphics, SIGraDi 2010, Bogotá, Colombia, November 17–19, pp. 228–231 (2010)

Tan, K.: Water simulation using realflow. Insight 03, Chapter 02, enclos. http://bit.ly/2r20wQh. Last Accessed 15 May 2017

Tang, P., Huber, D., Akinci, B., Lipman, R., Lytle, A.: Automatic reconstruction of as-built building information models from laser-scanned point clouds: a review of related techniques. Autom. Constr. **19**, 829–843 (2010)

Vähä, P., Heikkilä, T., Kilpeläinen, P., Järviluoma, M., Gambao, E.: Extending automation of building construction—survey on potential sensor technologies and robotic applications. Autom. Constr. **36**, 168–178 (2013). doi:10.1016/j.autcon.2013.08.002

Locally Varied Auxetic Structures for Doubly-Curved Shapes

Jan Friedrich[1(✉)], Sven Pfeiffer[2], and Christoph Gengnagel[1]

[1] University of the Arts, Hardenbergstrasse 33, 10623 Berlin, Germany
jan-friedrich@gmx.net
[2] Institute for Architecture, TU-Berlin, Straße des 17. Juni 152, 10623 Berlin, Germany

Abstract. In this paper we present a computerized design method which could ultimately serve to greatly simplify the production of free form reinforced concrete components. Using any desired doubly-curved shape as a starting point, we developed a digital workflow in which the spatial information of the shape is processed in such a way that it can be represented in a two-dimensional pattern. This pattern is materialized as an auxetic structure, i.e. a structure with negative transverse stretching or negative Poisson's ratio (Evans and Alderson in Adv Mater 12(9):617–628, 2000). On a macroscopic scale, auxetic behaviour is obtained by making cuts in sheet materials according to a specific regular pattern. These cuts allow the material to act as a kinematic linkage so that it can be stretched up to a certain point according to the incision pattern (Grima in J Mater Sci 41:3193–3196, 2006, J Mater Sci 43(17):5962–5971, 2008). Our innovative approach is based on the creation of auxetic structures with *locally varying* maximum extensibilities. By varying the form of the incisions, we introduce local variations in the stretching potential of the structure. Our focus resides on the fully-stretched structure: when all individual facets are maximally stretched, the auxetic structure results in *one specific* spatial shape. Based on this approach, we have created an iterative simulation process that allows us to easily identify the auxetic structure best approximating an arbitrary given surface (i.e. the target shape). Our algorithm makes it possible to transfer topological and topographical information of a given shape directly onto a specific two dimensional pattern. The expanded auxetic structure forms a matrix resembling the desired shape as closely as possible. Material specific information of the shape is further embedded in the auxetic structure by implementing an FE-analysis into the algorithm. We have thus laid the digital groundwork to produce out of this matrix, in combination with shotcrete, the desired building components as a next step.

Keywords: Free form shapes · Digital fabrication · Computational design

Introduction

Digital methods in architectural design have produced an increasing geometric complexity in recent years which in many cases requires challenging formwork for its production. Furthermore, parametric design methods tend to produce series of

K. De Rycke et al., *Humanizing Digital Reality*,
https://doi.org/10.1007/978-981-10-6611-5_28

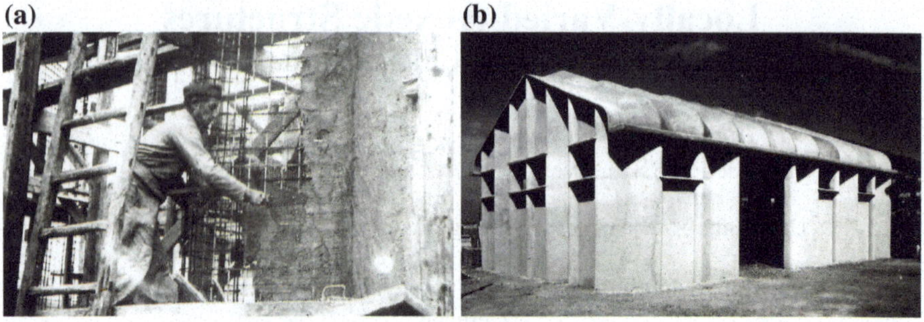

Fig. 1. Nervi, P.: Small pavilion in ferro-cement, Rome (1945). Construction site (**a**) and finished building (**b**). Images courtesy by Greco, Nervi, & Nervi, Pier Luigi (2008). Pier Luigi Nervi: Von den ersten Patenten bis zur Ausstellungshalle in Turin; 1917–1948

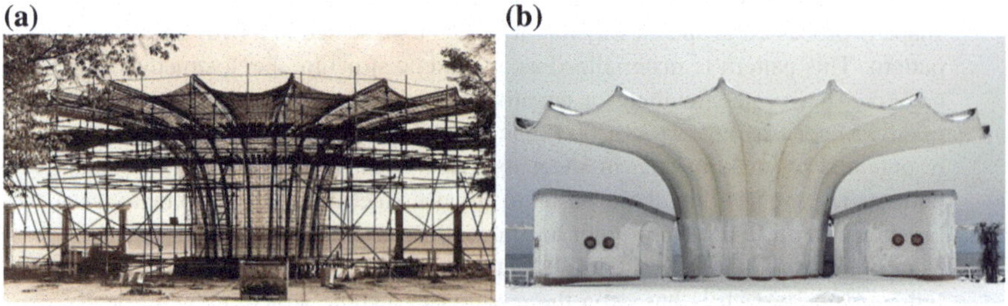

Fig. 2. Müther, U.: Musikpavillion Kurmuschel, Sassnitz (1987). Construction site (**a**) and finished building (**b**). Images courtesy by https://www.baunetzwissen.de

individual building components, each needing their own formwork, which is rendering the production process very onerous. Alternatives like 3D printing are being experimented with, yet these call for highly complex and exacting production processes, even more so with our focus on reinforced formwork (Gramazio, Kohler).

Looking a few decades back, we find pioneering shell structures that combined form setting *and* reinforcement into one structural matrix, like in Nervi's and Müther's pavilions (Figs. 1 and 2).

This approach of spraying concrete onto a matrix is inspiring, however producing such a three dimensional matrix stays cumbersome. With today's digital capabilities, might it not be possible to simplify the production by implementing a digital optimization process? Could certain aspects of this matrix (and hence, of the building's shape)—like topology, topography and materiality—be represented in a reproducible two-dimensional structure? Could this structure then be deformed in a way that would approximate the desired target shape as closely as possible?

The key to our answer lies in auxetic structures, i.e. flat materials made to exhibit auxetic behaviours through specific incisions, allowing for negative transverse stretching. As a side-effect, this provides the substantial advantage of being able to form them spatially into any bi-axially curved surface.

By a subsequent deformation process, the cuts expand to polygonal openings and each area containing incisions turns into a spatial, grid-shaped matrix, which is exactly what we are looking for. We now have a basic way of producing multiple 3D shapes from a single 2D auxetic structure. Our challenge is to figure out how to compute the specific pattern for a 2D auxetic structure that will lead to our one target 3D shape—this will be achieved through iterative variations of the auxetic structure.

Related Works

The foundational works on auxetic structures and the description of the negative Poisson's ration from Evans and Grima (2004, 2011) served as a starting point for the present study.

Investigations on computational design and fabrication with auxetic materials (Konakovic et al. 2016) showed in particular the potential for spatial deformation of auxetic structures. The structural optimization and mapping implemented in our process is based on discreet conformal mapping (Rörig et al. 2014; Springborn et al. 2008). The works of Resch (1973) and the generalizing of Resch's patterns by and Tachi (2013) served for the folding simulation. Piker's (2012) work on the digital simulation of auxetic structures and folding has given important impetus to the work. The idea of describing the auxetic structure as a kinematic linkage was inspired by Chuck Hobermans' reversibly expandable structures (Hoberman 2000). Digital processing tools were used for the structural optimization (VaryLab) and the implementation of stresses (Karamba).

Regular and Irregular Auxetic Structures

An iterative method is proposed to computationally derive from a given shape a corresponding two-dimensional auxetic structure. We start with a closer analysis of auxetic structures and their behaviour.

Global Variation: Two-Dimensional Deformation

As described by Grima (2006, 2008), auxetic behaviour on a macroscopic scale is facilitated by specific incisions in the material while keeping significant vertices connected, thus forming kinematic linkages (Fig. 1a). It is the geometry of the structure that warrants kinematic movement, according to a given set of rules: adjacent faces always stay connected through one common vertice, around which they rotate according to an alternating clockwise/counter-clockwise pattern (similar to Hoberman's "reversibly expandable structures") (Hoberman 2000). The incisions need to divide the flat material in such a way that the resulting pattern exhibits the topology of a checkerboard (Piker 2012). When the incisions follow the pattern as described and all the cuts are identical, we call this a *regular* auxetic structure.

By stretching this regular structure uniformly (in our case: transversally), the single faces begin to rotate around their common vertices and the incisions expand into

Fig. 3. Regular auxetic structure: straight cuts (**a**), partially stretched (**b**) and fully stretched (**c**)

polygonal openings. At this stage, it is the geometry of the general pattern itself that determines the maximum potential for expansion. Therefore every fully-expanded *regular* auxetic structure results in a unique corresponding flat geometry according to its specific Poisson's ratio. We chose to focus on triangular and quadrangular patterns.

Our approach puts the spotlight on that expansion process and its intermediate steps. When a material has already been stretched to a certain extent, the further potential for expansion is limited. In these partially stretched structures, the incisions are showing up as polygonal openings (Fig. 3b, c). Hence the potential for expansion is not only dependent on the geometry of the general pattern, but also determined by the type of incision pattern—straight cut or polygonal cut. To improve the control of the deformation process, our innovative approach is to use already modified patterns as our new starting point: instead of straight cuts (Fig. 3a), we are performing polygonal cuts in the flat material (Fig. 4a) resulting in an auxetic structure with a lower but clearly definable potential for maximum expansion (Fig. 4b). In other words, varying between straight and polygonal cuts gives us precise control over the deformation and thus over the size of the fully stretched structure.

This variation in incision patterns can either be applied globally to the whole structure, or more locally at the level of the individual cuts.

Local Deformation—Global Variation

We are still focusing on auxetic materials with a regular incision pattern (i.e. straight cuts as in Fig. 3). When a regular structure is evenly stretched to its maximum, it remains two-dimensional: every cut turns into a regular polygonal opening (Figs. 3c

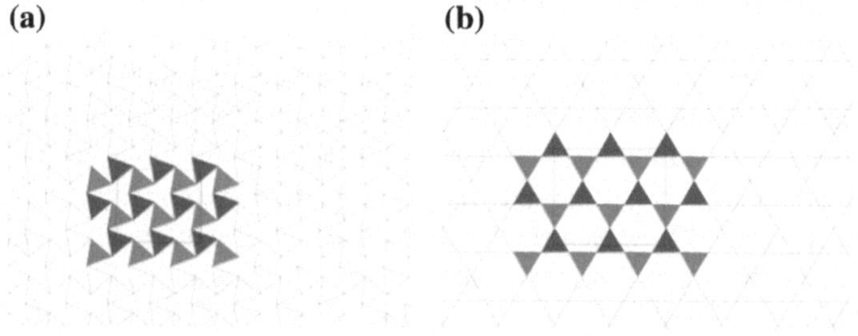

Fig. 4. Regular auxetic structure: polygonal incisions as starting point (**a**) and full expansion (**b**)

and 4b). However, if you stretch a regular structure *locally* (i.e. normally to the plane), these areas will deform spatially; the incisions in the stretched part, and only those, turn into polygonal openings. It is precisely this property of the auxetic structure to potentially transform into spatially curved shapes that we have been looking for. But there remains a problem. By only partly expanding either the whole structure or parts of it, we either obtain a variety of spatial shapes, or, one specific shape with multiple variations.

Konaković describes this very spatial deformation of auxetic structures (Konakovic et al. 2016). He goes on to develop a strategy to calculate amidst the many possible variations the one most smoothly approximating the target surface, using cone singularities to manipulate the regular auxetic structure in specific places.

In our case however, we are aiming for *one single* structure for each specific target shape and found our own way to obtain it. As an auxetic structure can only assume a clear shape in its fully stretched state, we are therefore looking for an auxetic structure that, in its fully extended state, accurately reflects the target shape. For this purpose, the type of incision must be determined individually for each facet of the general pattern. We call this an *irregular* auxetic structure.

Local Variation—Three-Dimensional Deformation

We have seen that a *regular* structure (Figs. 3 and 4) stretched at every point to its maximum results in a two-dimensional shape, every straight cut turning into a polygonal opening. Equally, when an *irregular* auxetic structure is expanded at every point, the various polygonal cuts are also transformed into maximal polygonal openings (according to the general pattern). However, since the expansion potential of each incision is locally varying, the structure will inevitably become spatial in its fully stretched state. The more the polygonal cuts in a certain area of the two-dimensional structure differ from a regular cut (i.e. the more they are looking like the final polygonal openings), the lower the three-dimensional distortion in this area of the resulting shape will be. An irregular structure (Fig. 5) stretched at every point to its maximum hence results in a precise, spatial shape.

The core of our innovation is to design the incisions for each facet individually, making it possible to work out up front which specific flat auxetic structure will, in its

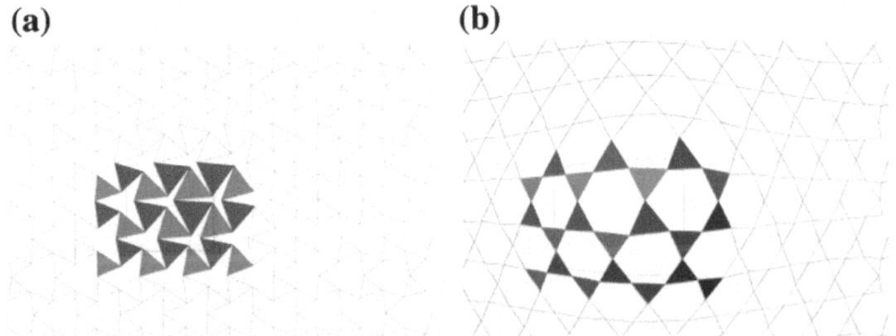

Fig. 5. Irregular auxetic structure: polygonal incisions as starting point (**a**) and full expansion (**b**)

fully expanded form, produce our target shape. In the following we present an integrated process in which the variation of the incisions is automated as a function of the target shape.

Methodology

Creating an Auxetic Structure from a Mesh

As previously described, auxetic structures have a clear topology: they are based on homogeneous triangular or quadrangular tessellations. Mathematically, such a pattern can be described as a triangular or quadrangular mesh. The computational transformation of the mesh topology is having the same effect as incisions in the two-dimensional material. The advantage is that changing mesh topology is a scriptable process that can be easily computed and iterated. Stretching can be simulated just as well. When the mesh is stretched, the topology stays the same. When the cuts expand into polygonal openings, the topology also stays the same. By adding the relevant incisions (i.e. changing the topology), the resulting mesh clearly describes a specific auxetic linkage.

Hence all that is needed in order to create an auxetic structure off the back of a given shape is to produce a regular mesh for that shape (Fig. 6). This can be done using established algorithms—we used VaryLab for this purpose, a software for discrete surface optimization and parametrization.

Our regular mesh now serves to produce the desired auxetic structure. Since it is directly based on that mesh, we can steer the incisions and polygonal openings of the auxetic structure directly through the mesh, so that each individual facet of the auxetic structure is linked to a specific mesh face.

We show in the following exactly how the size of the individual mesh faces can be used to proportionally determine the size of the incisions in the auxetic structure.

Digital Workflow

We start with an architectural model as a computationally designed shape (Fig. 8a). Here, the FE-analysis (Fig. 8b) can be used to optimize the form (Fig. 7/1). This design process results in a NURBS surface (Fig. 8c), i.e. our target shape (Fig. 7/2).

To develop the auxetic structure the surface of our doubly curved target shape needs first to be converted into a regular mesh (Fig. 9a). The smooth surface is divided up in discrete patterns. This includes a second optimization process (Fig. 7/3). There is

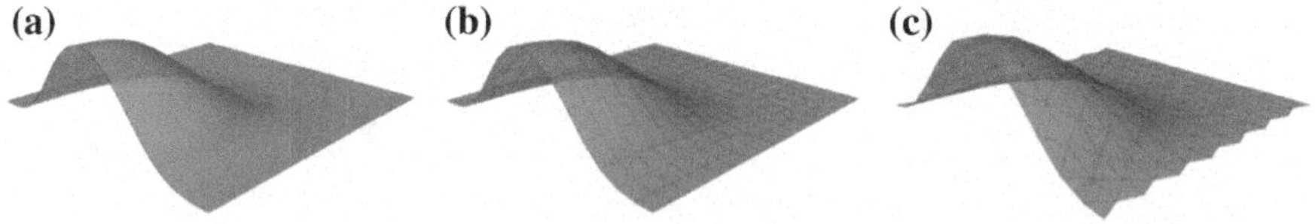

Fig. 6. Three dimensional shape (**a**), quadrangular mesh (**b**) and triangular mesh (**c**)

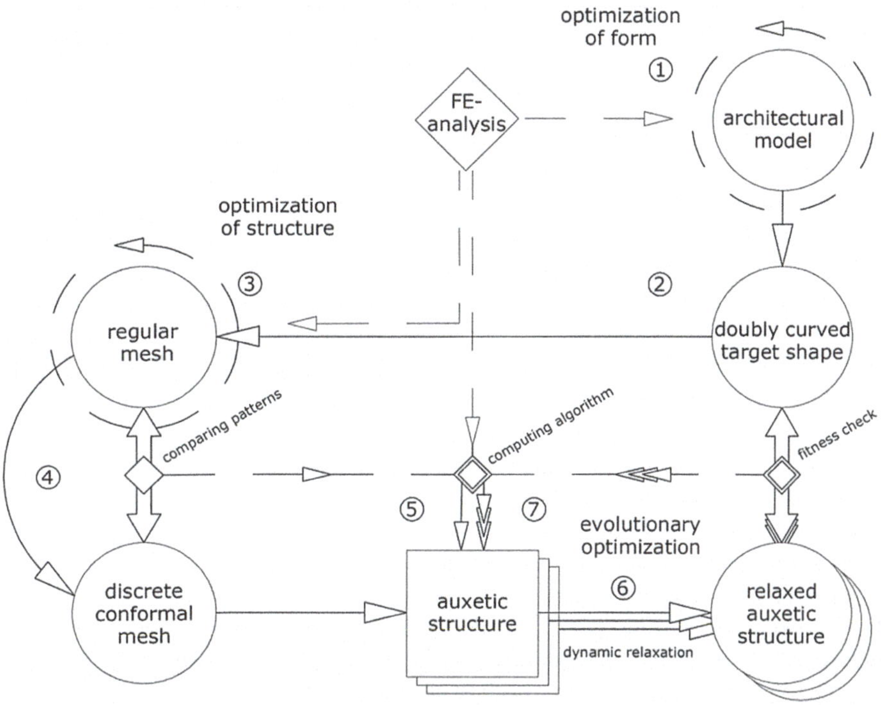

Fig. 7. Digital workflow

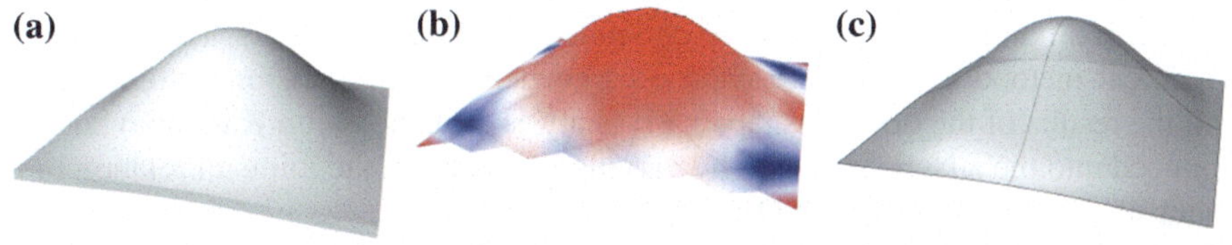

Fig. 8. Architectural model (**a**), FE-analysis (**b**) and NURBS surface (**c**)

a variety of computational tools available to do so. The FE-analysis can now be used for a structural optimization of the 3D mesh. Mesh topology and form remain the same.

At this stage, a discreet conformal 2D mesh (Fig. 9a) can be generated from the 3D mesh (Fig. 7/4) (VaryLab). Again, this one will exhibit the same mesh topology and form the basis of the auxetic structure. 2D and 3D meshes can now be compared with each other to assess the degree of deformation for each of the corresponding mesh faces (Fig. 9b). This serves as a basis for the size of the incisions making up the auxetic structure (Fig. 7/5) i.e. controlling the Poisson's ratio on each pattern. In the case of large differences between corresponding mesh-faces, that area of the auxetic structure will need to be highly extensible, i.e. incisions will have to be smaller. Where smaller differences in size appear, hardly any expansion is needed; incisions will resemble

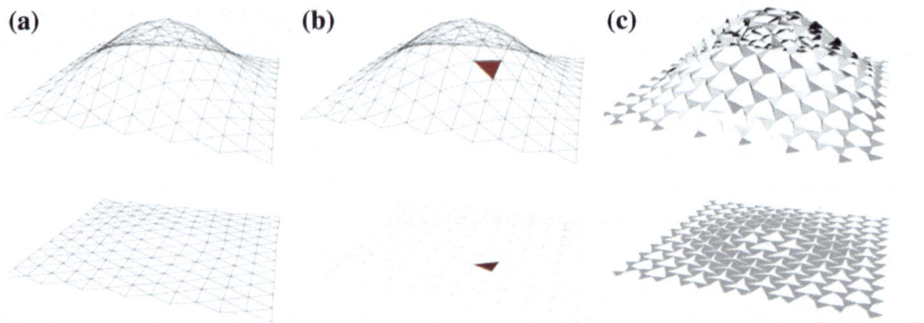

Fig. 9. Mesh and conformal map (**a**), comparison of mesh patterns (**b**), auxetic linkage, relaxed and 2D (**c**)

widely open polygons. A programmed algorithm determines the type and size of incisions needed to create a *qualitatively* correct auxetic structure corresponding to the target shape (Fig. 9c).

To improve its *quantitative* correctness, we decided to introduce a dynamic relaxation process in order to check the conformity of the generated auxetic structure with the target shape (Fig. 7/6). To allow for this, we chose to model the auxetic structure as a kinematic linkage (Fig. 10b), as previously described. We simulate materiality by assigning a specific width to each joint (Fig. 10c), according to the inner stresses of the shape (Karamba). To avoid plastic deformation in our simulation, further inner edges have to be implemented (Fig. 10d). Each of these inner edges is modelled as a linear piano hinge joint to warrant a clearly defined deformation. The resulting structure of clean simulated folds bears resemblance with Resch's patterns (1973).

The relaxation process can now be computed as a folding process. This serves as a geometrical method to check the chosen parameters for the incision pattern. At this point, the results of the original shape's FE-analysis (obtained with Karamba) could be introduced into the auxetic structure by varying the width of the joints in relation to the amount of stress (Figs. 7/7 and 11).

The two-dimensional auxetic structure we modelled in this way is now "relaxed" through a simulated dynamic relaxation process (Figs. 7/6 and 12). The result is a

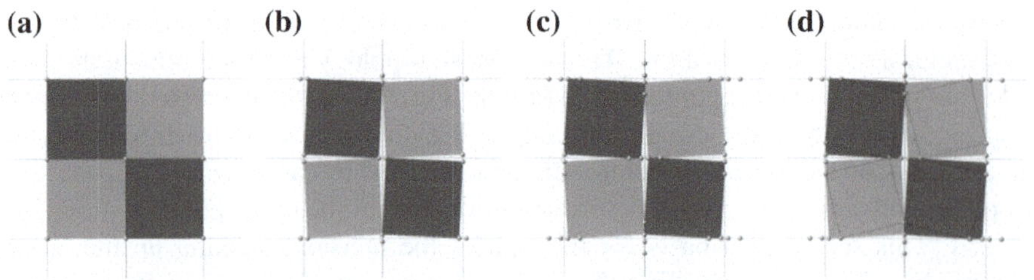

Fig. 10. Changing the topology for quadrangular mesh: checkerboard view (**a**), incisions (**b**), adding joints (**c**), adding inner edges (**d**)

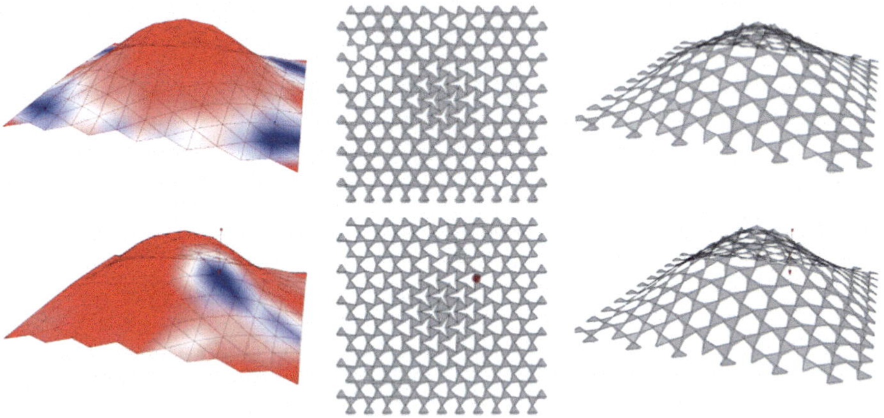

Fig. 11. Varying the width of the joints in function of the introduced stress

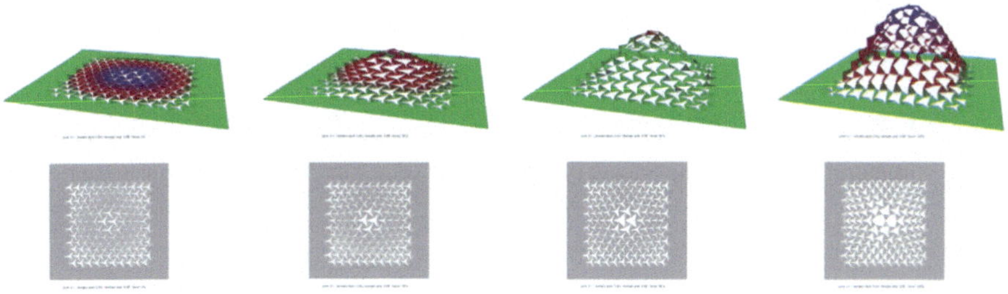

Fig. 12. Dynamic relaxation process for an irregular triangular auxetic linkage. *Green colour* indicates a good fitness, *red colour* a bad fitness

three-dimensional linkage describing a specific shape. That is now compared with the target shape and deviations are identified.

Based on these results, the auxetic linkage can now be repeatedly modified and rebuilt to compare with the target shape after dynamic relaxation. This iterative evolutionary optimization process will deliver a range of two-dimensional auxetic structures, each more closely approximating the target shape (Fig. 13).

Fabrication and Materiality

The method presented shows so far an integrated computational process for geometric analysis, form finding and algorithmic optimization. It can be applied to surfaces with both a positive or negative Gaussian curvature as well as to figures with fixed or free edges (Fig. 14).

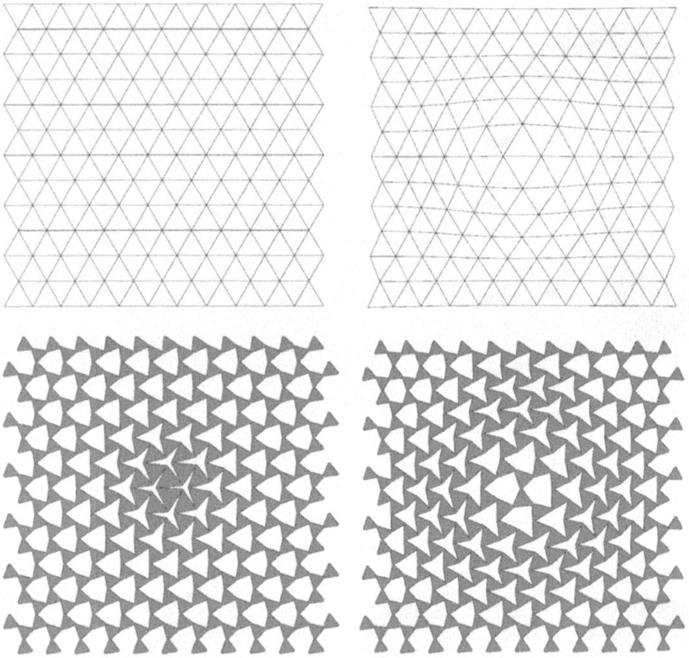

Fig. 13. Structural optimization: two very different auxetic structures based on a single target shape

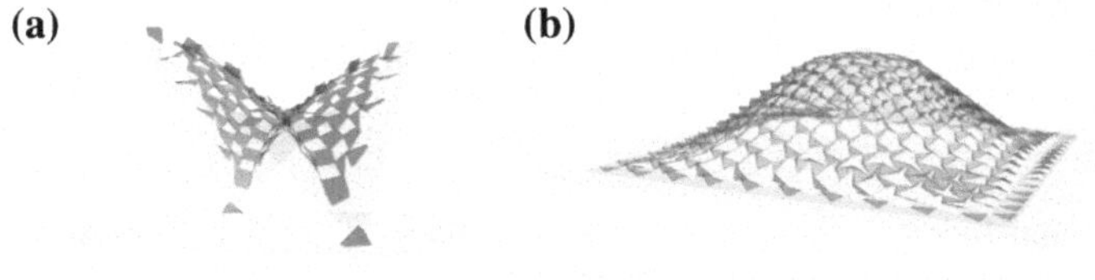

Fig. 14. Auxetic structures forming shapes with negative (**a**) and positive (**b**) Gaussian curvature

Without any further intermediate step, the auxetic structure can now be produced by means of laser-cutting or punching from flat metal sheets (Fig. 15) or textile materials.

The subsequent three-dimensional stretching of the auxetic structure can be done through a robotically controlled gradual distortion (maintaining the digital flow) or by relaxation and fixation for textiles (Fig. 16).

The absolute size of the incisions and thus also the proportions of the underlying mesh are determined by the requirements of the shotcrete, as our preliminary experimentations have shown (Fig. 17). Depending on the grain size of the chosen concrete, the hexagonal or quadrangular openings of the grid must neither exceed nor fall below

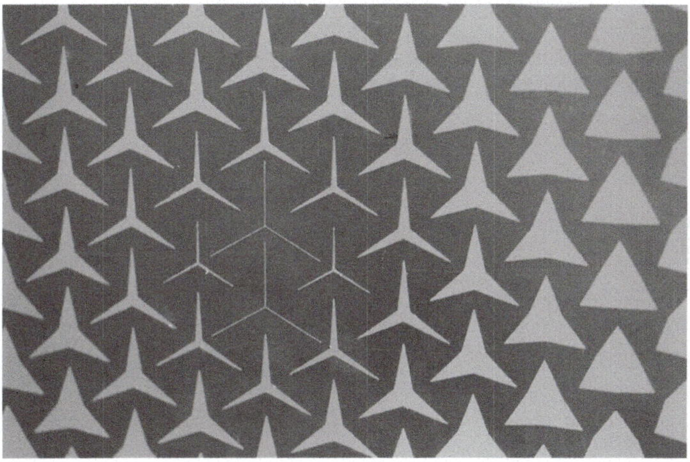

Fig. 15. Flat auxetic structure lasered on 1 mm aluminum sheet with polygonal cuts between 0.1 and 18 mm

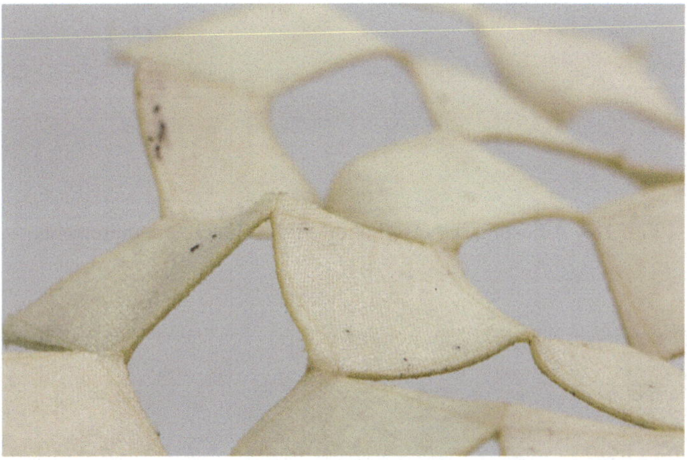

Fig. 16. Matrix made out of resin-impregnated textile based on a quadrangular auxetic structure forming a negative Gaussian curvature (maximal openings 20 mm)

a certain size—in the range of a few centimeters. These requirements are readily taken into account at the beginning of the design process by setting the mesh width (Fig. 18).

Depending on the size of the forces to be transmitted, multi-layer materials are also conceivable (Fig. 19). Overlapping layers would even allow to create a bigger, much more complex form made up of smaller individual elements.

Another imaginable application of the matrix could be as a facade element with a unique functionality and aesthetic unprecedented in traditional manufacturing processes.

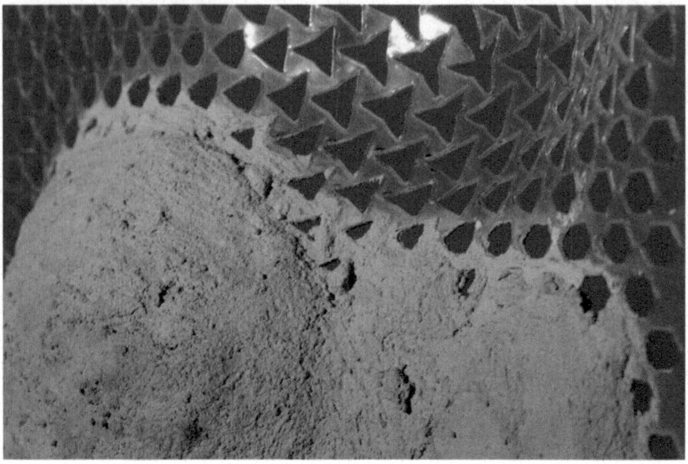

Fig. 17. Concrete on acrylic auxetic structure with maximal 7 mm hexagonal openings

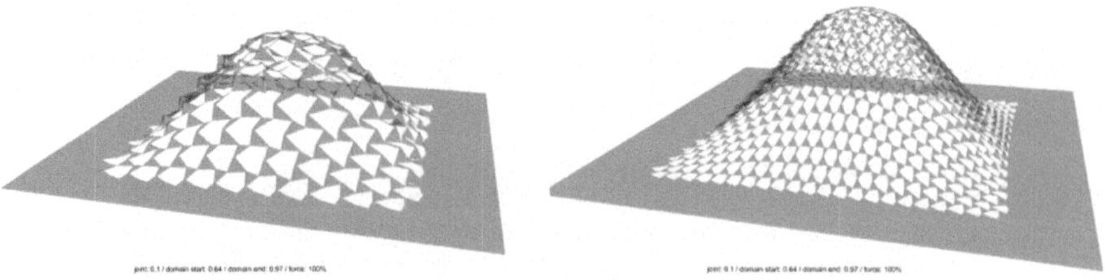

Fig. 18. Adjusting the tessellation to the shotcrete requirements

Fig. 19. Multi-layered auxetic structure made out of textile and epoxide resin forming a positive Gaussian curvature (maximum openings 22 mm)

Conclusion

The integrated digital process presented in this paper allows us to translate doubly curved shapes into an irregular two-dimensional auxetic structure—a major step towards the simplified industrial production of free form construction elements. Our design process makes it possible, based on a virtual architectural model, to introduce form, structure and material information into the auxetic structure in a single step. Further down the line in the construction process, the resulting matrix opens the possibility of doing away with complex formwork, offering reinforcement to boot.

In the end, the relaxed auxetic structure that forms our matrix is very similar to the ones used by Nervi and Müther (Figs. 1 and 2) as mentioned in the very beginning of this paper, although ours is the product of a purely digital process.

We have so far experimented with the physical implementation of our digital models on a small scale using different materials like concrete in combination with textiles, plastics and metal matrixes. These will form the basis of our further research, eventually leading to full-scale mock-ups and implementation of the production side of the process.

References

Evans, K.E., Alderson, A.: Auxetic materials: functional materials and structures from lateral thinking! Adv. Mater. **12**(9), 617–628 (2000)

Gramazio, F., Kohler, M.: Mesh mold metal. In: http://www.gramaziokohler.arch.ethz.ch. Last Accessed 10 June 2017

Grima, J.N., Evans, K.E.: Auxetic behavior from rotating triangles. J. Mater. Sci. **41**, 3193–3196 (2006)

Grima, J.N., Alderson, A., Evans, K.E.: Negative poisson's ratio from rotating rectangles. Comput. Methods Sci. Technol. **10**(2), 137–145 (2004)

Grima, J.N., Farrugia, P.S., Caruana, C., Gatt, R., Attard, D.: Auxetic behaviour from stretching connected squares. J. Mater. Sci. **43**(17), 5962–5971 (2008)

Grima, J.N., Manicaro, E., Attard, D.: Auxetic behaviour from connected different-sized squares and rectangles. Proc. R. Soc. A. **467**(2121), 439–458 (2011)

Hoberman, C.: Reversibly expandable structures having polygon links. U.S. Patent number 6,082,056 (2000)

Karamba Homepage: http://www.karamba3d.com. Last Accessed 12 May 2017

Konaković, M., Crane, K., Deng, B., Bouaziz, S., Piker, D., Pauly, M.: Beyond developable: computational design and fabrication with auxetic materials. SIGGRAPH Technical Paper, July 24–28 (2016)

Piker, D.: Variation from uniformity. https://spacesymmetrystructure.wordpress.com. Last Accessed 12 May 2017 (2012)

Resch, R.D.: The topological design of sculptural and architectural systems. In: Proceedings of the June 4–8, 1973, National Computer Conference and Exposition, AFIPS'73, pp. 643–650 (1973)

Rörig, T., Sechelmann, S., Kycia, A., Fleischmann, M.: Surface panelization using periodic conformal maps. In: Advances in Architectural Geometry, pp. 199–214 (2014)

Springborn, B., Schröder, P., Pinkall, U.: Conformal equivalence of triangle meshes. ACM Trans. Graph. **27**(3) (2008)

Tachi, T.: Freeform origami tessellations by generalizing Resch's patterns. In: Proceedings of the ASME, DETC2013-12326 (2013)

VaryLab Homepage.: http://www.varylab.com. Last Accessed 12 May 2017

This Room Is Too Dark and the Shape Is Too Long: Quantifying Architectural Design to Predict Successful Spaces

Carlo Bailey(✉), Nicole Phelan, Ann Cosgrove, and Daniel Davis

115 W 18th Street, New York, NY 10011, USA
{carlo.bailey, ann.cosgrove, daniel.davis}@wework.com,
nicole.phelan@weowrk.com

Abstract. Historically, architects have relied primarily on rules-of-thumb to layout offices. In this paper we consider whether these assumptions can be improved by using machine learning to predict the success of an office layout. We trained a support vector classifier on data from 3276 private offices from 140 buildings. 56 features of the offices were used, including whether it had a window and the office's squareness. The model was able to predict the lowest performing offices with a precision of 60–70% and a recall between 20 and 40%. This research suggests that many of the assumptions that drive architects will be able to be validated or refuted by applying machine learning to data gathered from people inhabiting the built environment.

Keywords: Machine learning · Predictive model · Office design · Building performance · Support vector classifier

Introduction

The Challenge of Workplace Design

Rising competition for talented employees are seeing companies increase the amount of time, resources, and money spent creating workplace designs that satisfy diversifying user needs. For instance, Google's New York headquarters allows employees to design their own workstations and has amenities for pets (leash rails, napping corners, etc.) (The New York Times 2013). The contemporary office must provide space for focused work, casual collaboration, events, and amenities (Harvard Business Review 2014). It is hard to know how any of these programmatic elements, or the physical attributes of the space contribute to an effective workspace. Typically designers use rules of thumb, or more recently simulation models, as a way to test and benchmark the utility of their workplace designs. Usually this takes the form of simulating the behaviour of a certain phenomena over time—e.g. daylight, crowd flows, structural, etc. But it is hard with any of these methods to understand trade-offs and interactions amongst components. For instance, is a rectangular office with a window, better than a square office without a window? Further, rules-of-thumb and simulations are often based upon anecdotal evidence gathered from a small sample of buildings, which potentially causes bias and deficiencies in these methods (Van der Voordt 2003).

K. De Rycke et al., *Humanizing Digital Reality*,
https://doi.org/10.1007/978-981-10-6611-5_29

There are decades of research into the effects that office design has on inhabitants. For instance, Frank Duffy has been researching the relationship between the physical attributes of space and organizational design since the 1970s. His early work mapped out how a company's growth pattern, group sizes, communication strategy, and view of the outside world can be either hindered or developed by the architectural qualities of the work spaces (Duffy 1977). He argued that an organization that emphasizes control via departmental division would be reinforced by buildings that are composed of separate units of space (Duffy 1977). In contrast, communication across an organization would be enabled by buildings that are composed of interconnected spaces that are not heavily subdivided. Although there has been an abundance of similar research throughout the years, studies on the reported success or failures of workplaces typologies contradict each other, and there are very few studies supported by empirically supported insights (Van der Voordt 2003). As a result, there is little consensus on how office design impacts the success of an office (Van der Voordt 2003).

At WeWork we face similar challenges to most others in the workplace industry. We design, construct, and manage workplaces around the world. Each one of our 140 locations have a unique design that reflects site, community, and building idiosyncrasies. The challenge for our in-house architects is the production of space that satisfies a plethora of professional, organisational, and functional requirements, within a single location. Any one of our buildings may include brainstorming rooms, breakout spaces, or lounges with work styles ranging from quiet, individual focused work, to casual group work. Central to this range of experiences are the private office spaces WeWork provides for its members (Fig. 1). Private offices in the context of WeWork function as design studios, showrooms, or company headquarters depending on the organisation that occupies it. At WeWork, members lease offices on a month-to-month basis, and if a particular space does not satisfy the needs of the occupant they can move out or switch to a different office the next month. Therefore, it is vital our architects get design decisions regarding offices right prior to construction, otherwise less desirable offices could sit empty for months at a time.

In this research we set out to identify the spatial factors that influence the success of offices at WeWork. Historically, these factors were identified anecdotally by designers. In this paper we consider whether these factors can be identified empirically. Using BIM data and occupancy data we developed a machine learning model using a support vector classifier (SVC) that was able to classify the least desirable offices in our portfolio with a 60–70% precision. The model can be applied to new buildings, allowing designers to get a sense for which rooms might perform badly in their layouts. In this paper we will argue that this allows our designers to move from a world where they're intuitively anticipating the quality of particular offices, to one where they are making design decisions informed by accurate statistical analysis.

Related Work

Almost since the introduction of the computer in architecture, designers have been developing algorithms that allow them to evaluate architectural designs. During the 1970s, at The Bartlett, UCL, Hillier et al. (1976) proposed a variety of analytical methods to quantify physical attributes and the configuration of buildings and urban

Fig. 1. A private office at WeWork. Private offices come in a variety of shapes and sizes that cater to entrepreneurs, freelancers, or larger businesses

layouts. More recently Penn et al. (2017) have employed these techniques to simulate occupant behaviour and predict movement flows within retail environments.

Recently, there has been a number of studies around the modelling of occupant behaviour to predict activity in buildings during the design phase of a project. Goldstein et al. (2011), proposes a method to simulate people's activity across a floor plan to predict space utilization. Using a probabilistic model that derives probabilities from distances to amenities and time of day, a model was developed that would predict meeting room usage against actual usage data. Similarly to the work done by Tabak (2009) demonstrates a method of simulating occupant behaviour in offices that uses organizational parameters as input features. The output of the model would be a predicted schedule for each one of the agents in the simulation. We extend this approach by actually calibrating our model with occupant usage information, and expanding the range of input features to include spatial attributes.

Our work differs from previous studies in the following ways:

1. Our work focusses exclusively on private offices within coworking spaces. It does not attempt to predict occupant schedules, but rather the percentage of month-to-month utilization, which is a contract structure specific to the business of WeWork.
2. Our model only uses the physical features of the space to predict utilisation. We do not factor in other non-architectural variables such as time, price, or tasks in our models. We intentionally did not include such variables since they are not directly

under the control of architects at WeWork and we wanted to give the architects a sense of how their spatial decisions were affecting the offices they were designing.
3. Our approach uses machine learning algorithms trained on 3276 observations/rooms across 140 locations. As we outlined in a previous study (Phelan et al. 2016), this vast sample is quite unique to WeWork, as we control our entire product lifecycle from design to operation.

Methodology

Data Collection

WeWork is a vertically integrated company. We have a centralized data storage system, in an Amazon Redshift database, that houses information across our entire product lifecycle. Data includes everything from harvested BIM data from our archived CAD models, to information on sales, member churn, and particular furniture in a space. Using the Python programming language, we downloaded the data and preprocessed it into a set of inputs. The population of the study consisted of 34,042 private offices located within 140 WeWork buildings. From this population we took a random sample of 3276 private offices for the study (limiting the sample by looking only at 1–8 person offices).

Input Variables

To decide which physical attributes of our spaces to train our model with, we used a combination of inherited rules of thumb about what makes a pleasant space, coupled with insights from our prior research. For example, we assumed that having a window in an office would affect its desirability (as most people enjoy views to outside), so whether or not an office has a window (true/false), was a feature we included. Similarly, we thought an office's density would be a factor—most people would prefer more open space than feeling overcrowded. We also learnt from a study trying to predict meeting room utilization (Phelan et al. 2016), that the distribution of space types in a location affects utilization. Insights, from interviews with our members suggested that a room's squareness also affects its reconfigurability and members dislike long rectangular spaces. In all, we initially collected 56 variables to feed our model. Figure 2 shows an example WeWork floor plan with program distributions across our portfolio on the right. Figure 3 shows three examples of how we took the raw geometric data from our spaces (mainly consisting of coordinate representations), and broke these down into features for the model, such as a room's squareness. Table 1 shows the list of input features we derived from our BIM data (abbreviated for conciseness).

Success Metrics

WeWork leases each office on a month-to-month basis. Good offices are almost constantly leased, whereas bad offices are hard to lease and might be empty for most of the year. To help our designers create desirable offices that are constantly leased, we

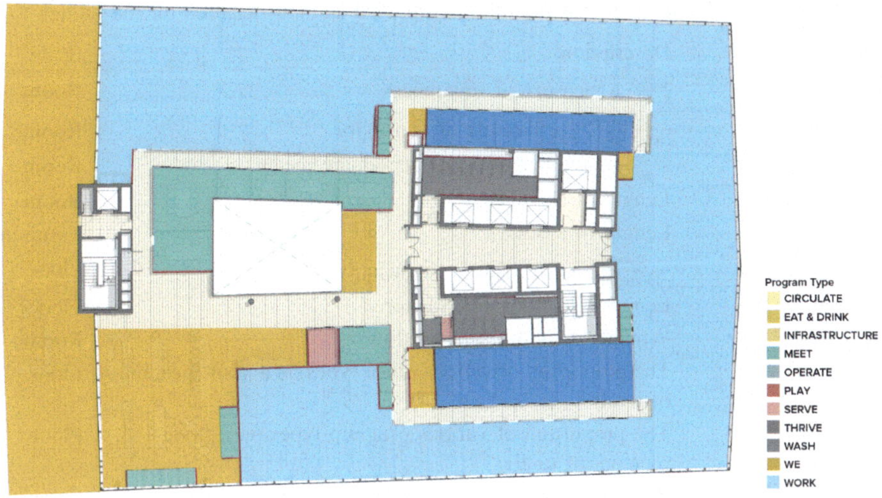

Fig. 2. Example WeWork floor plan showing a break-down of programmatic arrangement

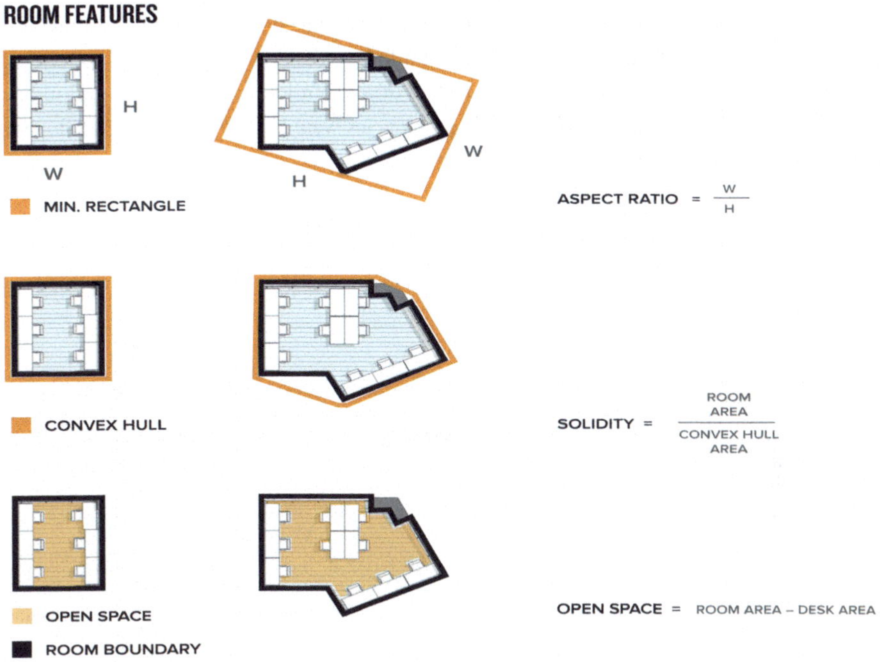

Fig. 3. Diagram showing the logic behind the solidity, aspect ratio and open space metric

wanted to predict which offices were likely to be leased and which ones were likely to remain empty. We call this success metric 'mean occupancy', which is the percentage of months a given office was leased. Figure 4 shows the distribution of mean

Table 1. A list of the physical attributes we used as input features for the model

Metric	Description	Scale
Aspect ratio	$\frac{width}{height}$ a measure of squareness	Room
Solidity	$\frac{room\ area}{cancave\ hull\ area}$ measure of concavity	Room
Density	$\frac{room\ area}{desks}$ number of desks per square meter	Room
Room depth	Length of an office	Room
Desk length	Length of a desk	Furniture
Floor solidity	$\frac{floor\ area}{concave\ hull\ area}$ measure of concavity	Floor
Perimeter length	Length of a floor's perimeter	Floor
Has window	Whether a room has a window	Room
Office type distribution	The proportion of various office types on a floor (percent of total)	Floor
Program type percentage	The proportion of various program types on a floor (percent of total)	Floor
Program adjacencies	Whether a room is adjacent to various program/room types	Floor
Floor connectivity	Whether a floor has a wework floor above or below it	Floor
Floor height	Floor to floor height	Floor
Perimeter depth	Mean depth of offices along the perimeter	Floor
Total desks	Total number of desks in a location	Building
Total area	Total usable square meters of a given location	Building

Full list abbreviated (for example, the program adjacencies were broken down by space type, but here listed as a single metric)

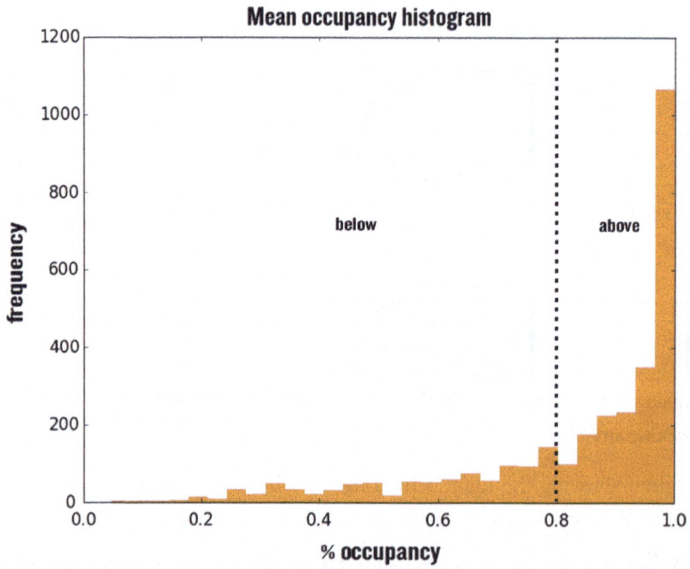

Fig. 4. Histogram of mean occupancy across our sample, with the 'above'/'below' label threshold shown

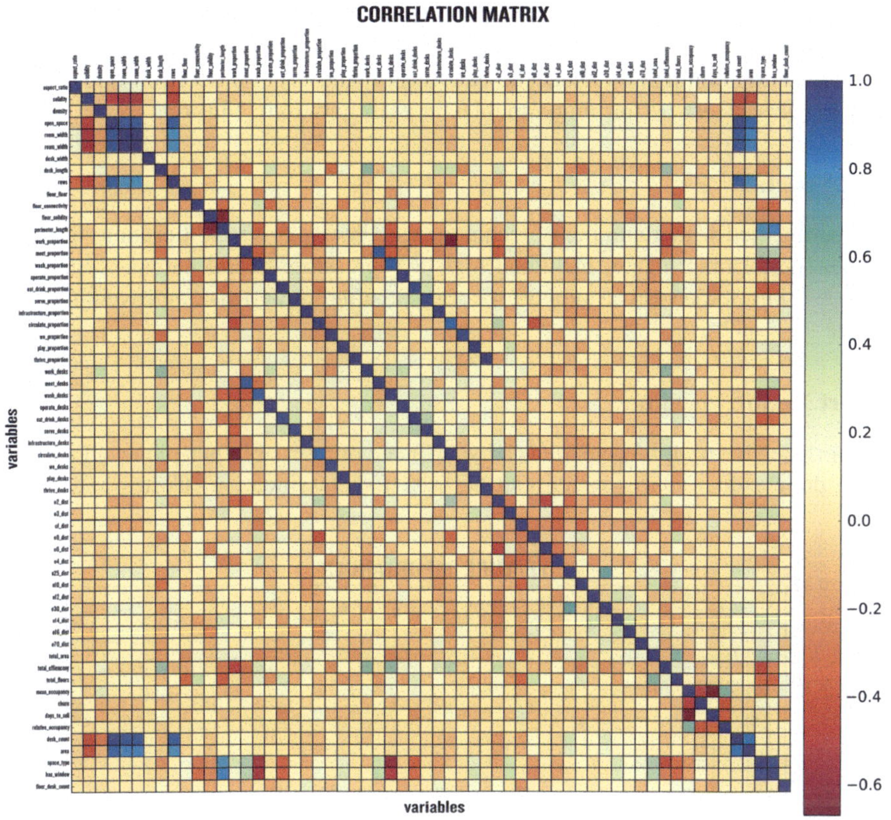

Fig. 5. Correlation matrix between all 56 variables of the dataset

occupancy within our sample. The data are extremely skewed left with most offices having 100% occupancy. Given the non-normal distribution of the data, we segmented the target variable, mean occupancy, into those that were either within or above the bottom 35th percentile (<80%) occupancy. So rather than predicting a continuous value, we attempted to build a binary classification algorithm that identified the low performing offices.

Dimensionality Reduction

In machine learning there is always a *bias-variance tradeoff* (Hastie et al. 2017) that corresponds with model complexity. Variance is when a model is overly complex (perhaps trained with too many input features), overfits the training data and does not generalize well to the test data or examples it had not seen before. In contrast, bias is the complete opposite, when a model is not complex enough, underfitting the data and does not generalize well. Given the amount of variables we considered in this study (56), there was a high chance we could create a model with high variance. We sought to mitigate this by going through various feature reduction methods to arrive at a set of the most influential variables in predicting an office's performance. These included a correlation matrix to arrive at an independent set of variables (Fig. 5), and then recursive feature elimination using the variable weights from a decision tree classifier (removing the lowest performing variables in each iteration), Fig. 6.

Model Selection

We experimented with a number of techniques to classify private offices into those that were low performing or not, including logistic regression, neural networks, and decision trees, but settled on a support vector classifier (SVC) as it proved the most accurate.

Support Vector Machines

A support Vector machine constructs a hyper-plane or a set of hyper-planes in a high or infinite dimensional space, which can be used for classification. Cortes and Vapnik

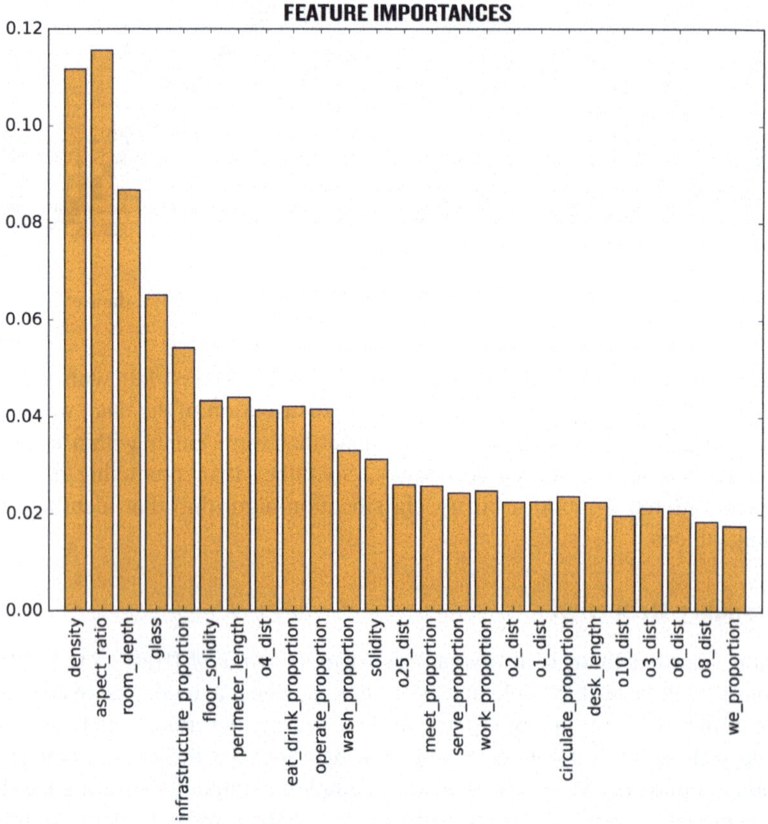

Fig. 6. Bar chart showing the magnitudes of feature importances for 25 variables, given by the decision tree classifier

(1995), demonstrates that a good separation is achieved by the hyper-plane that has the largest distance to the nearest training data points (called support vectors) of any class, since the larger the margin the lower the generalization error of the classifier. The strengths of support vector machines lie in their effectiveness in high dimensional spaces with many features, and their ability to generalize well, even when the number of dimensions is greater than the number of samples (Pedregosa et al. 2012).

Model Training

We split the data into a training and test set by building location to ensure that the model was tested on rooms within locations that it had not seen previously. The python package Scikit-Learn was used for the SVC model (Table 2). We used the algorithms radial based function (RBF) kernel for our implementation. To tune the model's parameters we used Scikit-learn's built-in *GridsearchCV* class, that searches an array of possible parameters and tests these against the training and test to arrive at the best configuration.

Results

We achieved a classification precision of 60–70% using the support vector classifier (SVC) and a recall between 20 and 40%. Precision here is the fraction of retrieved instances that were relevant (a measure of classification quality), while recall is the fraction of relevant instances that were retrieved (a measure of quantity). Figure 7 shows the model stability of the SVC's performance across 10 k-folds with an unseen validation set.

Model Predictions

From the classification model we were able to obtain not only the prediction of the class labels (above or below), but also the probability of the respective labels. This probability can be interpreted as the model's confidence on the prediction. We used this as a method for visualizing potential room performance (Fig. 8 shows model predictions heat-mapped over an existing WeWork floor plan). We serialized the model and developed a web framework with an API. The API can be queried in real-time with new data by the design team, allowing for the design exploration process to be augmented by automated design validation.

Table 2. Classification report from the predictions of the SVC model

Labels	Precision	Recall	F1-score	Support
Above	0.56	0.95	0.70	441
Below	0.73	0.14	0.24	391
Avg/total	0.64	0.57	0.48	832

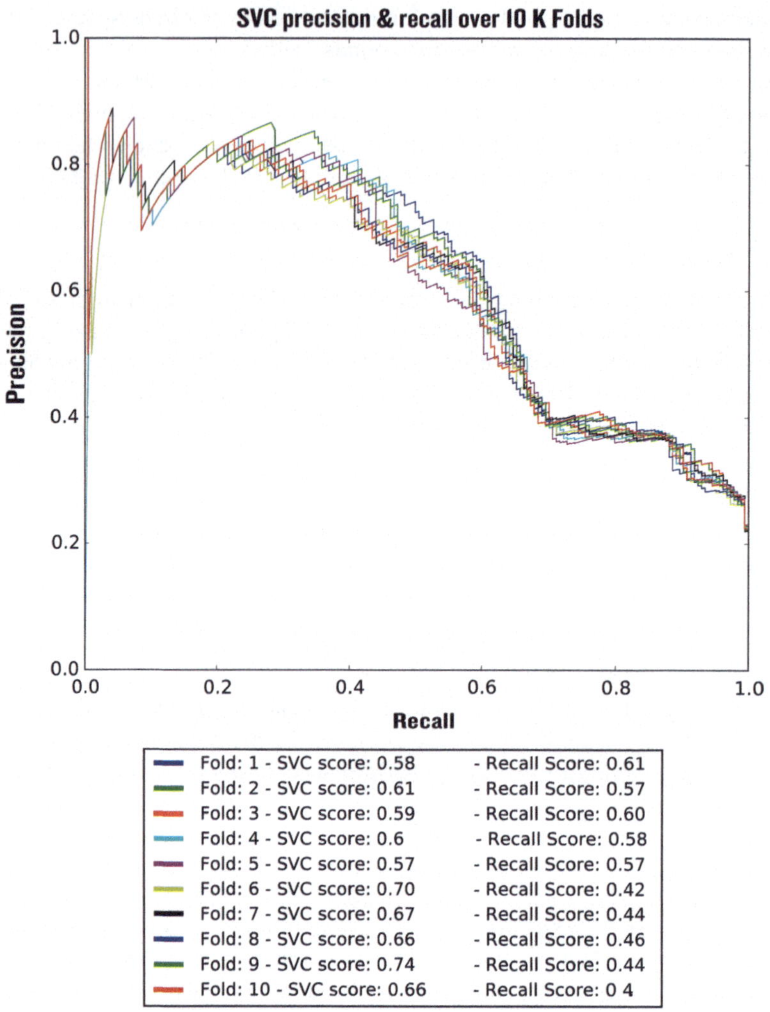

Fig. 7. Precision and recall curve for accuracy predicting 35th percentile over 10 k folds

Discussion

Architects have long sought to understand how their buildings will be used prior to construction. Usually this takes the form of simulating the behaviour of a certain phenomena over time—e.g. daylight, crowd flows, structural, etc. This research suggests that we may be able to enhance our knowledge of future building performance by training algorithms on historical usage data. We have demonstrated that predictive SVC models are well suited to learn complex interactions between design variables, which is near enough impossible to capture without machine learning algorithms. This methodology is well placed to be integrated in a generative design workflow where a

Fig. 8. SVC's prediction probabilities mapped onto an WeWork floor plan. 0.0 indicates when the model is certain the particular office will not be low performing, 1.0 indicates model certainty in low performance. These probabilities can be thought of as model confidence

predictive model could serve as the fitness function, evaluating each design with knowledge of empirical data.

Obtaining the vast amounts of empirical building data needed for further research would be a difficult task for many architects. This is because traditional contractual models in the AEC industry do not usually promote the involvement of the design team post construction. Therefore, any post-occupancy analysis would have to be carried out by the building owner. Additionally, obtaining a population size somewhere even close to that of WeWork would require a landlord with a vast portfolio. With these limitations in mind, we see opportunities for our research to be applied in the commercial real estate, medical, education, or the retail sector, which are all industries where the designs are relatively consistent, the sample size is large, and the metrics for success are clearly defined.

Although this work studied 56 design variables, from an office's boundary geometry, materiality, to room adjacencies, there were many more features that we did not include. We believe any future iteration of this method would greatly benefit from including other physical phenomena at a finer grain. Specifically, views to the outside, daylight levels, and temperature, are a few examples of factors that previous research has shown to affect inhabitants of office space (William 1997), and could improve the accuracy of a predictive model.

Conclusion

Historically, architects have relied primarily on rules-of-thumb to layout offices. Our research suggests that many of the assumptions that drive architects will be able to be validated or refuted as we apply machine learning to data gathered from people inhabiting the built environment. Already we can predict which offices people think are less desirable. As we use sensors and other technology to gather more data about how people use buildings, we anticipate a similar technique could be applied to everything from the way people gather in public spaces, to the layout of hotel rooms. Although this method is heavily reliant on digital methods, it is inherently user-focused since people's passive habitation patterns become the key drivers for the algorithm. In this way it humanizes the digital, paving the way for a more user-centric design bolstered by predictive analytics.

References

Cortes, C., Vapnik, V.: Support-vector networks. Mach Learn **20**, 273–297 (1995)

Duffy, F.: Organizational design. J. Archit. Res. **6**, 4–9 (1977)

Fisk, W.J., Rosenfield, A.H.: Estimates of improved productivity and health from better indoor environments. Indoor Air **7**, 160–172 (1997)

Goldstein, R., Tessier, A., Khan, A.: Space layout in occupant behaviour simulation. In: Proceedings of Building Simulation 2011: 12th Conference of International Building Performance Simulation Association, Sydney (2011)

Harvard Business Review: https://hbr.org/2014/10/workspaces-that-move-people (2014). Last Accessed 15 May 2017

Hastie, T., Tibshirani, R., Friedman, J.: The Elements of Statistical Learning: Data Mining, Inference, and Prediction. Springer Series in Statistics, 2nd edn. https://statweb.stanford.edu/~tibs/ElemStatLearn/printings/ESLII_print10.pdf. As of May 2017

Hillier, B., Leaman, A., Stansall, P., Bedford, M.: Space syntax. Environ. Plan. Des. **3**(2), 147–185 (1976)

Pedregosa, F., Varoquaux, G., Gramfort, A., Michel, V., Thirion, B., Grisel, O., Blondel, M., Prettenhofer, P., Weiss, R., Dubourg, V., Vanderplas, J., Passos, A., Cournapeau, D., Brucher, M., Perrot, M., Duchesnay, E.: Scikit-learn: machine learning in python. J. Mach. Learn. Res. **12**, 2825–2830 (2012)

Penn, A., Turner, A.: Space syntax based agent simulation. http://discovery.ucl.ac.uk/2027/1/penn.pdf. As of May 2017

Phelan, N., Daniel, D., Carl, A.: Evaluating architectural layouts with neural networks. In: Symposium on Simulation for Architecture and Urban Design (2016)

Tabak, V.: User simulation of space utilisation: system for office building usage simulation. Ph. D. thesis, Eindhoven University of Technology, Netherlands (2009)

The New York Times.: http://www.nytimes.com/2013/03/16/business/at-google-a-place-to-work-and-play.html (2013). Last Accessed 15 May 2017

Van der Voordt, T.J.M.: Productivity and employee satisfaction in flexible workplaces. J. Corp. Real Estate. **6**(2) (2003)

Modelling and Representing Climatic Data in the Tropics: A Web Based Pilot Project for Colombia

Roland Hudson(✉) and Rodrigo Velsaco

Universidad Piloto Colombia, Carrera 9 no. 45a-44, Bogota, Colombia

Abstract. Understanding local climate is a critical factor in the design of buildings that requires continuous research to widen its scope, deepen base information and seek data correlations. In tropical regions climate varies greatly across relatively small areas, changes in altitude and geography define conditions that add complexity to typical climate patterns. Paucity of data makes understanding tropical climates difficult. Scarcity of design strategies and computational tools designed for the tropics means bio-climatic design and basic low-energy strategies are poorly understood and underutilised. This paper describes the first phase of a collaborative project aimed at addressing these issues in Colombia.

Keywords: Tropical climate · Colombia · Bio-climatic design

Introduction

Since 2010 the eco-envolventes research project at the Universidad Piloto in Bogota has been investigating computational, bio-climatic design in Colombia (Velasco and Robles 2011; Velasco et al. 2015). In 2017 the group will make historical data from Colombian weather stations available openly online. The only data that is currently available is from a single weather station (Energyplus.net 2017) that inadequately represents the complex climate of the country.

In collaboration with Bath University and Buro Happold we present work-in-progress of a project builds on the newly available climate data for Colombia. The project is co-funded by the Univerisdad Piloto and the Newton Caldas fund via the Royal Society of Engineers. Our work starts with processing and formatting the raw climate data and producing files in formats that support a wide range of users. Beyond this we plan to make online, climate data visualisation tools that help more people understand the relationships between Colombia's climate, topography and human comfort. Ultimately through access to data and improved understanding of its implications we plan to identify how low-energy methods (Olgyay 1963; Givoni 1976) can lead to more comfortable internal environments in Colombia. We anticipate that through the applied use of low-energy strategies and computational design methods this work will have social, economic and environmental impact. Socially, improved human thermal comfort will address high levels of thermal dissatisfaction observed in the country (Rodriguez and D'Alessandro 2014; Rodriguez in review). Economically,

K. De Rycke et al., *Humanizing Digital Reality*,
https://doi.org/10.1007/978-981-10-6611-5_30

reduced energy required for heating and cooling means lower lifetime costs for a building. Environmentally, reductions in heating and cooling lowers carbon emissions. These improvements would result in better functioning buildings with a longer lifetime reducing both cost and embodied energy in modifying or replacing a building.

To move towards these bigger goals, the project proposes three objectives. First, to develop new visualization tools to communicate the relationships between human comfort, terrain and climate in the region. Second, based on this improved understanding of the impact of the region's climate develop frameworks and tools that better support the application of low-energy design strategies. Thirdly, to build a national network of stakeholders in academia and industry to provide feedback on the first two objectives and to establish future partnerships for development and application of low-energy design software and construction systems in Colombia. This paper provides a brief overview on progress to date of the first of these three objectives.

Goals for Visualisation Tools

Design and construction of buildings in Colombia is undertaken by a diverse set of individuals not all of whom have had formal design education or practical training. Our visualisation tools are therefore intended to be accessible and will implement new and existing techniques to represent the relationship between human comfort and climate. We aim to use modes of interactive representation of climatic data to help both specialist and non-specialist audience access and understand climatic data and the built environment.

Acknowledging our target audience may not be CAD/BIM/3d modelling users we chose not to develop plugins for a specific platform. This approach would favour design specialists and discriminate the builders and stakeholders who we believe also can impact the quality and comfort of Colombian buildings. We will provide standard file formats for use in specialised applications, but our focus is on using the internet to provide accessible visualisation and understanding of Colombia's diverse climate.

Methods for low-energy design tend to focus on the Northern Hemisphere, but tropical regions have very different patterns of weather largely defined by daily changes than annual changes. We sought an interface where users could select a city and see instantly the climate patterns across the year and how they differ between different parts of the world. Importantly our tools do not require knowledge of specific software applications only the ability to use a web browser. To support this goal, we have developed 2d and 3d tools that allow users to quickly select and compare climatic factors from different cities.

Intuitively we know that climate, and topography varies greatly across Colombia but this has yet to be represented in a way that is easily understandable. There is a need to clearly illustrate this and do so in a way that allows users to draw their own conclusions about these variations. Marsh (2000) suggests that ability to interact and 'play' can enhance comprehension of climate information. To embrace this approach to understanding and illustrating climatic variation we present a gridded, colour scaled mapping of Colombian climate and altitude, where the user selects what variable to visualise.

Thermal comfort is affected by variations in climate and topography. To illustrate this, we have developed two online visualisation tools. The first, an interactive psychrometric chart that responds to altitude and maps annual weather onto the chart. Users can select a city, the chart updates to reflect the city's altitude and shows number of annual hours within a comfort zone defined by user-controlled comfort parameters. The second tool is an extension of the gridded climatic/altitude tool where a grid of thermal comfort indices are calculated based on user-controlled comfort parameters. Comfort can be mapped, controlled and its spatial distribution examined.

Current climate classification techniques and associated design methods do not adequately address the Colombian context. We have developed an online, bivariate colour mapping tool to allow the climate to re-classified and viewed as a spatially continuous system (Teuling 2011). This tool allows users to explore data that includes all the climate measures, altitude and thermal comfort and seek new forms of classification.

Impact

A key innovation that our work is founded on is a new lightweight, structured, extendible data format, climaJSON, that is designed for online use but has potential for use offline.

Using open source frameworks and free data our work includes 5 new web tools which could be reconfigured for any region of the world to provide fast, simple, free and accessible climate information visualisation.

Our work could benefit the rest of the region; Ecuador, Venezuela, Guyana, Surinam, Perú, Bolivia have a paucity of free climate information (see Fig. 1) and topographical similarities to Colombia. The same situation exists in African equatorial

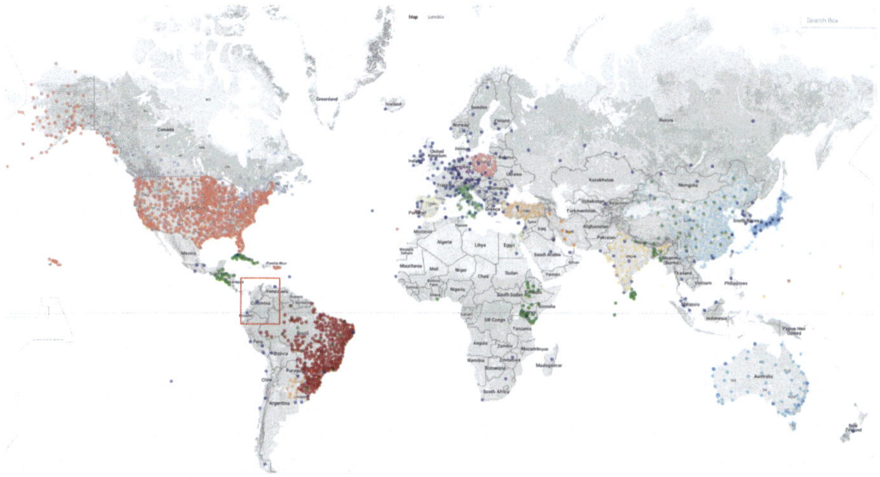

Fig. 1. Global availability of free climate data (Ladybug.tools 2017 last accessed 15 May 2017)

countries, Indonesia and Papua New Guinea suggesting our work could have global impact.

Increased economic prosperity, growth of urban centres and the need for new buildings with better human comfort indicates the work is needed within Colombia.

Background

The distribution of available free climate data on the Ladybug.tools epwmap (Ladybug.tools 2017) shows the sparsity of data in equatorial South America (see Figs. 1 and 2) and in Colombia where only data for Bogota is available.

Colombia's IDEAM manages 4483 terrain weather stations (IDEAM 2017) (see Fig. 2). Open access to this data represents a massive change in the granularity of information. Our challenge, given the combination of mountainous terrain, variable climate and quantity of data is to make this data as useful as possible to the greatest number of people to make the greatest impact.

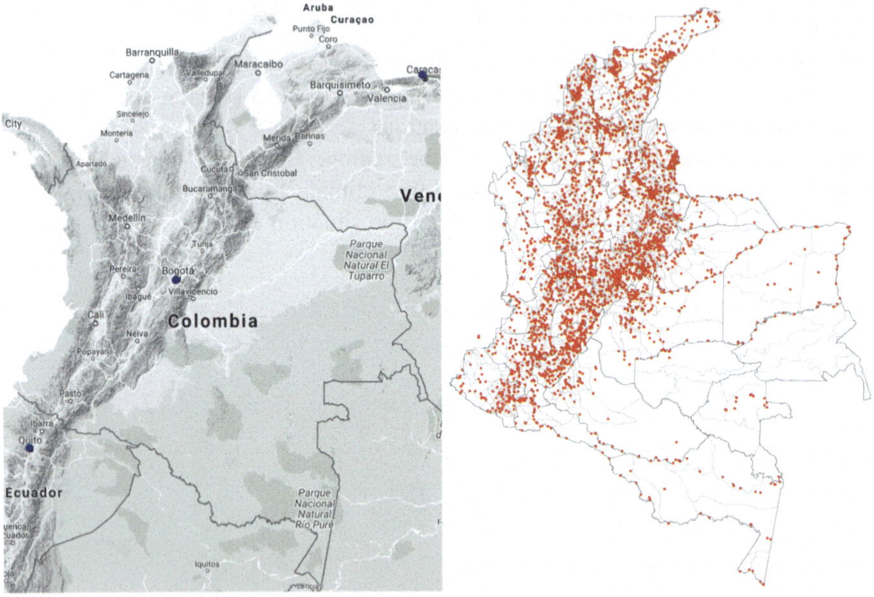

Fig. 2. Current availability of free climate data in equatorial South America (*left*) (Ladybug.-tools 2017 last accessed 15 May 2017). IDEAM's Colombian 4483 Weather Stations (*right*)

Work-in-Progress

climaJSON

Parsing the raw climate data and generating hourly values for each measured data field in each Colombia town is a key part of our work but not the focus of this paper. Using the *.epw format (EnergyPlus 2017) files will be made openly available. EnergyPlus (2017) is the US Department of Energy, Building Technology's simulation engine and the *.epw format is widely used and is compatible with many desktop applications used to perform architectural studies and energy simulations it is popular with architects and engineers.

Climate data for both *.epw and the popular *.tmy (2017) files is stored as a comma separated value (*.csv) format that maintains columns for all fields even if the values are unknown or unused, increasing file size and slowing processing. Good practice in data modelling suggests that null values should be avoided in relational models. Nulls indicate vagueness in the data either; because the value is unknown, it is known but is missing or they are not applicable (Coronel 2016). Typical data formats for climate and energy analysis are not easy to use online.

We propose a new lightweight, accessible and online format, climaJSON (2017), for storing, sharing, extending and visualising climate data using JavaScript Object Notation (JSON) (2017). The format follows the Json-schema (Json-schema.org 2017; Droettboom et al. 2017) allowing file validation and standardisation. climaJSON can be used on and off-line and parsed by open source tools such as the D3.js library (Bostock et al. 2017).

Climate Visualisation

Existing climate data visualisation tools can be classified according how they are deployed:

- Obsolete standalone applications [Weather tool in Ecotect (Marsh 2000)].
- Standalone applications [Climate consultant (Milne et al. 2009; Bhattacharya 2009)].
- Plugins to parametric extensions to a modelling applications [Ladybug for Grasshopper (Ladybug 2017), Dhour (Steinfeld and Levitt 2013)].
- Code libraries [Open Graphic Evaluative Frameworks (Steinfeld et al. 2010, 2017)].
- Online open access [online comfort tool of the Center for the Built Environment (CBE) at University of California Berkeley (Hoyt et al. 2013; Schiavon et al. 2014)].

All approaches provide sophisticated visualisation. The last class of tools is an exception, all others require download and installation on a desktop computer, licensing and technical competency. CBE's comfort tool only requires an internet connection and a web browser. Complying with published standards, providing links to further information, implemented with JavaScript and D3.js CBE's tool allows users to develop knowledge of thermal comfort via an interactive interface.

Following this accessible model our tools are built with JavaScript running on the client's computer. JavaScript is a popular web based programming language and has many powerful, fast, cross-browser libraries or API's like Three.js (org/ and May 2017) (used in the 3d climate chart) and D3.js used in all our visualisations.

2d weather mapping. Day-of the year is plotted on the x-axis and hourly data on the y-axis. Each hourly cell is colored per a standard fixed color range for easy comparison. Drop-down boxes allow users to select a city and climate data type to visualize. Hovering the mouse over an hourly cell shows the data associated with the cell.

3d weather mapping. The interactive 3d mapping tool provides additional emphasis of climate patterns at specific locations, the value of each hourly data point is used to define the z height and colour of a mesh vertex. Hourly data values are displayed the mouse hovers over a face of the mesh.

Psychrometric chart by altitude. Apsychrometric chart (PC) is a graphical representation of the composition of air (Milne and Givoni 1979). Typically, dry bulb temperature is mapped on the x-axis and humidity ratio on the y-axis (humidity ratio is the measure of grams per kilo of dry air). Curves rising from the left to right of the chart show relative humidity (RH) in 10% increments. Our PC builds on the CBE comfort tool (Hoyt et al. 2013; Schiavon et al. 2014). We use BS EN ISO7730:2005 to define a comfort zone and calculate the indices; predicted mean vote (PMV) (Fanger 1970) and predicted percentage dissatisfied (PPD). These calculations are determined by user-controlled thermal comfort parameters; wind speed, metabolic rate and level of clothing (see Fig. 5).

Users choose a location from a drop-down menu, hourly climate data from this is placed in data bins which are plotted onto the chart, darker colours indicate greater frequency. Data associated with that bin—the number of hours and ranges in temperature, relative humidity and humidity ratio, can be seen by moving the mouse over the bin. The chart also displays the number of hours for the selected location that fall within the currently defined comfort zone. The altitude of the selected location is used to define the chart (Ren 2004; Stull 2016). Humidity ratio is proportional to atmospheric pressure which is inversely proportional to altitude, this can be seen in the shape of the RH curves when a chart near sea level is compared to one at 2548 m (see Fig. 5).

Gridded Mapping of climate data. Agridded dataset is useful to determine climatic conditions between station locations (Haylock et al. 2008). The process of spatially interpolating data is complex and even more so given large variations in topography and climate. As an intermediate step before generating a full, hourly, gridded dataset for Colombia we have defined a monthly dataset to consider the possibilities offered by gridded data and visualisation methods.

CRU-TS v3.10.01 Historic Climate Database (CGIAR-CSI 2017) documents nine monthly averaged climate variables globally at a 0.5 spatial resolution for 1901–2009. A subset of this data was extracted and averaged for the years 1999–2009 in Colombia. This defined 371 location points spaced at approximately 50 km intervals, each with an averaged set of monthly climate values including daily minimum and daily maximum temperatures, vapour pressure. Data was formatted and stored in the climaJSON

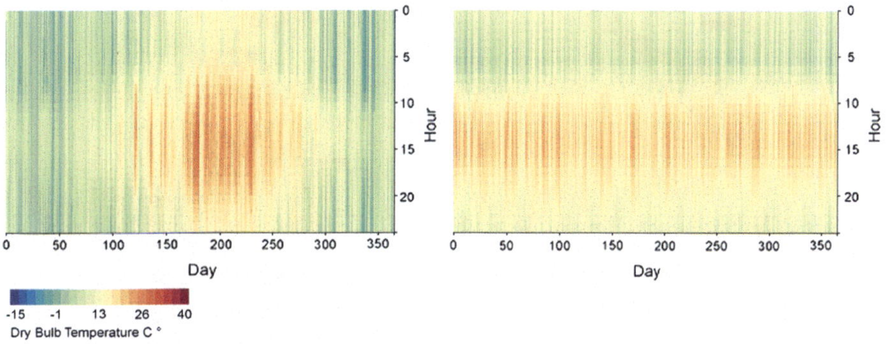

Fig. 3. 2d annual charts comparing London and Bogota

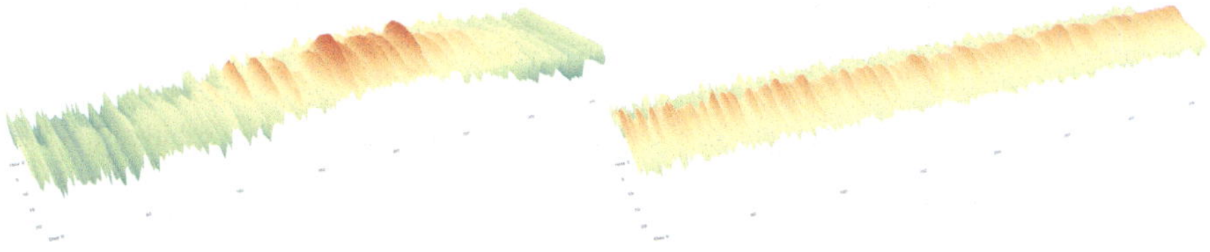

Fig. 4. 3d annual charts comparing London and Bogota

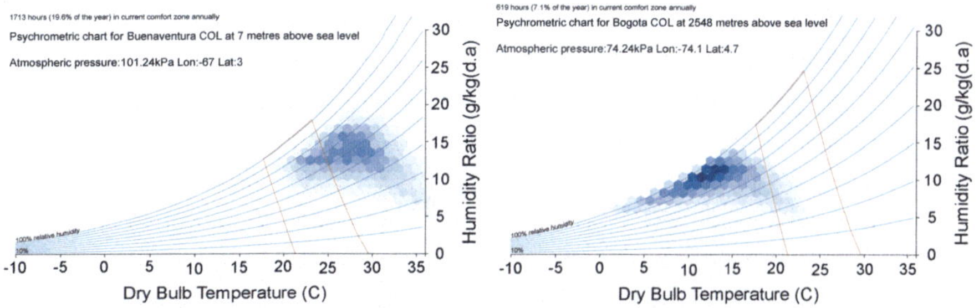

Fig. 5. Psychrometric charts at 7 and 2548 m above sea level

format. Digital elevation data from NASAs SRTM (CGIAR-CSI 2017) was sampled at each location point to add altitude to the dataset. Based on elevation, saturation pressure can be calculated for each point and combined with the measured vapour pressure to give a monthly average for relative humidity.

Our gridded dataset tool allows mapping and comparison of these monthly averaged climate values and altitude; this enables the variation of Colombia's topography and climate to be visualised and compared. Users select one of seven parameters and view the spatial distribution of the data (see Fig. 6).

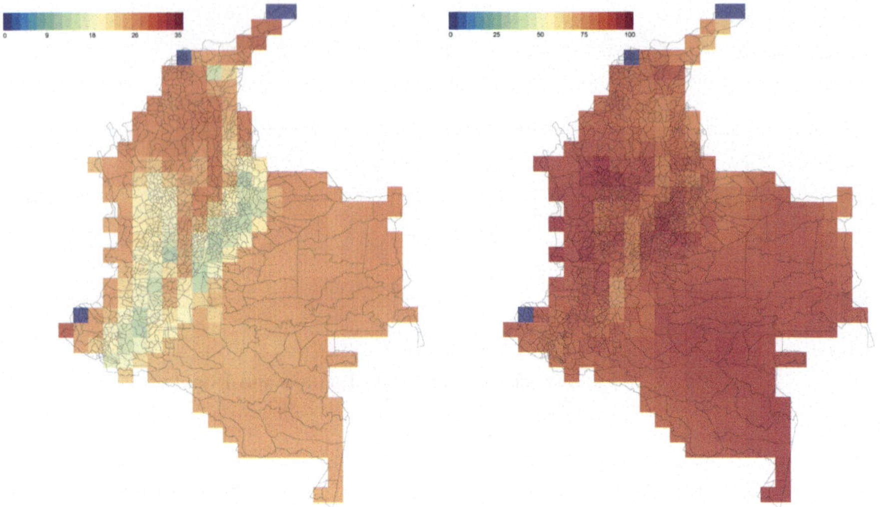

Fig. 6. Comparing monthly averages of Colombia's mean temperature and relative humidity

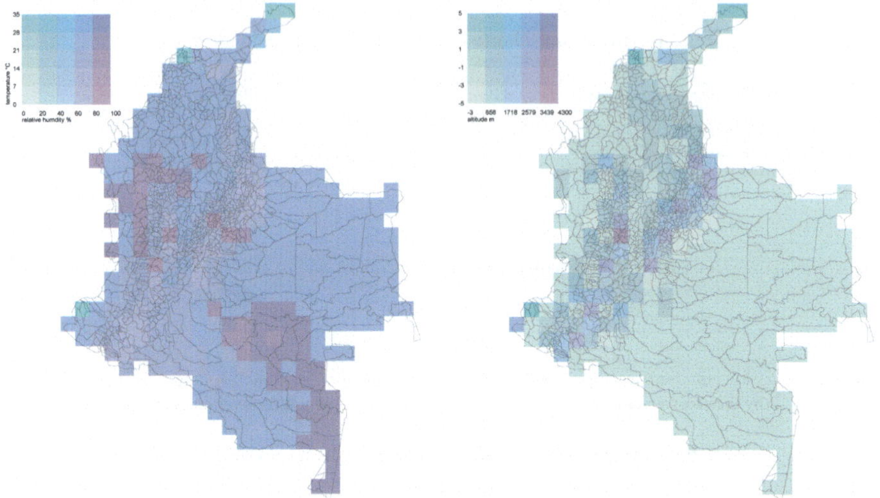

Fig. 7. Bivariate mapping of relative humidity and daily mean temperature (*left*) and altitude and daily maximum PMV (*right*)

We extended the gridded climate data tool to calculate thermal comfort indices of PMV and PPD for the average minimum and maximum daily temperatures for each cell. The calculations be dynamically fine-tuned by adjusting the user-controlled thermal comfort parameters.

Gridded Bivariate Mapping of climate data + thermal comfort. To formally capture correlations and spatial continuity, bivariate mapping allows two variables to be combined and represented with a single color from a 2d color scale. Our bivariate mapping (see Fig. 7) tool allows users to combine any two of 11 variables (six empirical and five calculated) and geographically display results overlaid on a map of Colombia.

Summary

The paper describes work-to-date of a project that aims to disseminate, raise awareness and broaden understanding of Colombia's historical climate data. Climate data visualisation is our first objective, future goals are; distributing data in standard file formats, developing design tools and strategies applicable to the region and developing a national stakeholder user-group.

Our prototype visualisation tools show:

- The function of an online JSON data format designed for climate data visualisation.
- Differences between a northern hemisphere climate and tropical climates. Figures 3 and 4 show annual variation patterns in the northern climates versus daily variation in the tropics. Design strategies exist for northern locations with annual variation but tropical design strategies are underdeveloped.
- The relationship between comfort and altitude in Colombia is not yet fully understood. Figure 6 shows how for a given dry bulb temperature humidity ratio increases with altitude, meaning higher absolute humidity's and greater sensations of wetness which intuitively we think increases thermal discomfort. Standards for high altitude thermal comfort do not yet embrace this observation. Psychrometric charts used in building design processes are typically generated at sea level or at best 2–3 standard altitudes, indicating more research into thermal comfort at altitude is needed.
- Spatial mapping of climate, terrain and comfort allows the visualisation of the uniqueness of Colombia's climatic context. Figure 7 shows the diversity and correlations in the data demonstrating how sensations of overheating and underheating can occur in short distances. Improved access to climate data and better understanding of its implications would help develop climate responsive buildings.
- Bivariate mapping of gridded data indicates the possibility of new forms of climate classification that may lead us to better understand design with thermal comfort in Colombia. Figure 7 shows correlations between mean temperatures and relative humidity (left) and altitude and thermal comfort (right).

Conclusions

We have presented 5 open access online tools for visualising newly available climate data for Colombia. The tools are designed to be as simple to use as possible to reach a wide audience as possible. and based on a new data format for online sharing and processing of climate data using popular web programming languages and open source frameworks.

Our tools and data format are prototypes and require refactoring and further testing. The latter stage of the project; building a group of stakeholders from academia and industry will enable broader user-based testing of the tools and refining requirements. We plan to implement further climate visualisation methods online in particular extending the psychrometric chart with passive analysis, design strategy zones that respond to the Colombian context. The Gridded dataset currently only involves monthly data that comes from a predefined grid. Significant work on interpolation methods is required to convert hourly station based readings to an hourly gridded data set.

As we evaluate and further develop our climate visualisation tools the potential for application in the tropics is clear. Our understanding of the region's climate is improving and we can envision the developing tools to support low-energy strategies for buildings in the tropics.

References

Bhattacharya, Y., Milne, M.: Psychrometric chart tutorial: a tool for under-standing human thermal comfort conditions. In: 38th American Solar Energy Society Conference, Buffalo. pp. 11–16 (2009)

Bostock, M.: D3.js—Data-driven documents. https://d3js.org/. Accessed 3 May 2017

CGIAR-CSI.: CRU-TS v3.10.01 Historic Climate Database for GIS|CGIAR-CSI. Cgiar-csi.org. http://www.cgiar-csi.org/data/uea-cru-ts-v3-10-01-historic-climate-database. Accessed 16 May 2017

CGIAR-CSI.: SRTM 90m Digital Elevation Database v4.1|CGIAR-CSI. Cgiar-csi.org. http://www.cgiar-csi.org/data/srtm-90m-digital-elevation-database-v4. Accessed 16 May 2017

Coronel, C., Morris, S.: Database Systems: Design, Implementation, & Management, 12th edn. Cengage Learning, Boston, p 79 (2016)

Droettboom, M.: Understanding JSON schema—understanding JSON schema 1.0 documentation. Spacetelescope.github.io. https://spacetelescope.github.io/understanding-json-schema/. Accessed 16 May 2017

Eco-envolventes.net: climaJSON. http://www.eco-envolventes.net/climaschema. Accessed 18 May 2017

ECMA: The JSON Data Interchange Format, 1st edn. ECMA. http://www.ecma-international.org . Accessed 15 May 2017

Energyplus.net: Weather Data by Region|EnergyPlus. https://energyplus.net/weather-region/south_america_wmo_region_3/COL%20%20. Accessed 8 Feb 2017

EnergyPlus: EnergyPlus™ Version 8.6 Documentation: Auxiliary Programs, 1st edn. U.S. Department of Energy, pp. 59–77. https://energyplus.net/sites/all/modules/custom/nrel_custom/pdfs/pdfs_v8.6.0/AuxiliaryPrograms.pdf. Accessed 21 Feb 2017

Fanger, P.O.: Thermal Comfort: Analysis and Applications in Environmental Engineering. McGraw-Hill, New York (1970)

Givoni, B.: Climate and Architecture, 2nd edn. Applied Science Publishers, London (1976)

Haylock, M., R., Hofstra, N., Klein, Tank, A.M.G., Klok, E.J., Jones, P.D., New, M.: A European daily high-resolution gridded dataset of surface temperature and precipitation. J. Geophys. Res (Atmos), **113** (2008)

Hoyt, T., Schiavon S., Piccioli, A., Moon, D., Steinfeld, K.: CBE Thermal Comfort Tool. Center for the Built Environment, University of California Berkeley. http://comfort.cbe.berkeley.edu/ (2013)

IDEAM.: Catálogo Nacional de Estaciones del IDEAM| Datos Abiertos Colombia. la plataforma de datos abiertos del gobierno colombiano. https://www.datos.gov.co/Ambiente-y-Desarrollo-Sostenible/Cat-logo-Nacional-de-Estaciones-del-IDEAM/hp9r-jxuu. Accessed 15 May 2017

Json-schema.org.: JSON Schema. http://json-schema.org/. Accessed 3 May 2017

Ladybug.tools.: epwmap. http://www.ladybug.tools/epwmap/. Accessed 15 May 2017

Ladybug.tools.: Ladybug Tools. http://www.ladybug.tools/. Accessed 17 May 2017

Marsh, A.: Playing around with architectural science. In: ANZAScA Conference Proceedings—Annual Conference of the Australian and New Zealand Architectural Science Association, Adelaide, South Australia (2000)

Milne, M., Liggett, R., Benson, A.: Climate consultant 4.0 develops design guidelines for each unique climate. In: Proceedings of the American Solar Energy Society, Buffalo New York (2009)

Milne, M., Givoni, B.: Architectural design based on climate. In: Watson D (ed) Energy Conservation Through Building Design. McGraw-Hill, New York (1979)

Olgyay, V.: Design With Climate: Bioclimatic Approach to Architectural Regionalism. Princeton University Press, Princeton (1963)

Rodriguez, C.M., D'Alessandro, M.: Climate and context adaptive building skins for tropical climates: a review centred on the context of Colombia. In: Proceedings of Energy Forum 2014, pp. 585–60 (2014)

Rodriguez, C.M.: Strategies to improve thermal comfort in social housing through façade retrofits: field study in Bogota, Colombia. In: Energy and Buildings (in review)

Schiavon, S., Hoyt, T., Piccioli, A.: Web application for thermal comfort visualization and calculation according to ASHRAE Standard 55. In: Building Simulation, August 2014, vol. 7, no. 4, pp. 321–334 (2014)

Steinfeld, K., Levitt, B.: Dhour: a bioclimatic information design prototyping toolkit. In: ACADIA 13: Adaptive Architecture [Proceedings of the 33rd Annual Conference of the Association for Computer Aided Design in Architecture (ACADIA), Cambridge, 24–26 Oct 2013), pp. 217–226 (2013)

Steinfeld, K., Bhiwapurkar, P., Dyson, A., Vollen, J.: Situated bioclimatic information design: a new approach to the processing and visualization of climate data. In: ACADIA 10: Life Information, New York 21–24 Oct 2010, pp. 88–96 (2010)

Steinfeld, K., Schiavon, S., Moon, D.: Open graphic evaluative frameworks: a climate analysis tool based on an open web-based weather data visualization platform. In Digital physicality—Proceedings of the 30th eCAADe Conference—Volume 1/ISBN: 978-9-4912070-2-0, Czech Technical University in Prague, Faculty of Architecture (Czech Republic), 12–14 Sept 2012, pp. 675–682 (2012)

Stull, R.: Practical Meteorology: An Algebra Based Survey of Atmospheric Science. BC Campus (2016)

Teuling, A.J.: Technical note: towards a continuous classification of climate using bivariate colour mapping. Hydrol Earth Syst Sci 15:3071–3075 (2011) doi:10.5194/hess-15-3071-2011

TMY: Users Manual for TMY3 Data Sets, 1st edn. National Renewable Energy Laboratory, pp. 3–9. http://www.nrel.gov/docs/fy08osti/43156.pdf. Accessed 21 Feb 2017

Threejs.org: three.js—Javascript 3D library. https://threejs.org/. Accessed 3 May 2017

Ren, H.S.: Construction of a generalized psychrometric chart for different pressures. Int. J. Mech. Eng. Educ. **32**(3), 212–222 (2004)

Velasco, R., Robles, D.: Eco-envolventes: a parametric design approach to generate and evaluate façade configurations for hot and humid climates. In: eCAADe 29th Conference Proceedings, pp. 539–548 (2011)

Velasco, R., Hernández, R., Marrugo, N., Díaz, C.: Notes on the design process of a responsive sun-shading system: a case study of designer and user explorations supported by computational tools. Artif Intell Eng Des Anal Manuf **29**(04), 483–502 (2015)

Thermally Informed Bending: Relating Curvature to Heat Generation Through Infrared Sensing

Dorit Aviv[1(✉)] and Eric Teitelbaum[2]

[1] School of Architecture, Princeton University, Princeton, NJ, USA
dorit@princeton.edu
[2] School of Engineering and Applied Science, Civil and Environmental Engineering, Princeton University, Princeton, NJ, USA

Abstract. The process of bending flat ductile surfaces such as metal sheets into curved surfaces that match a precise digitally predefined geometry so far has required either using a mold or a predetermined gauge and alloy-based predictive model to determine the necessary bending arc. Instead, our approach relies on the material itself to report its bending state during a simple actuation movement enabled by a robotic arm. We detect spontaneous heat generation triggered by the transition from elastic to plastic deformation using an infrared camera in real time. This information is used to determine the motion path of the robotic arm in order to reach the desired final geometry in one movement rather than incrementally. The bending process is generated through a RAPID code with a trap loop to guide the robot and a Processing script to control the thermal sensing and trap loop activation. Results for an IR sensor-guided geometric trap loop and the determination of a meaningful threshold for the onset of plastic deformation are discussed. This method's efficiency and applicability to large scales of production and to a wide array of ductile materials with varying elastic modulus, suggest a potentially significant contribution to the fabrication of curved architectural skins.

Keywords: Plastic deformation · Thermal imaging · Robotic fabrication · Feedback based fabrication

Introduction

Methods for accurately bending metal sheets into pre-defined curved surfaces have predominantly relied on the production of molds onto which flat sheets are attached and plastically deformed incrementally until they match the mold's shape. The wastefulness of mold making and the limitations of other bending methods, such as predetermined gauge- and alloy-based prediction necessary for bending a specific arc, incremental pin-based or press-brake bending methods, prompted us to develop of an alternative bending approach in which we used a robotic arm to precisely apply a large force to a metal surface and thus achieve a mold free sheet metal forming process.

The contemporary need for analytical methods to form curved surfaces for use in architectural fabrications of facades, has been highlighted by Pottmann et al. (2007).

K. De Rycke et al., *Humanizing Digital Reality*,
https://doi.org/10.1007/978-981-10-6611-5_31

Our method detects spontaneous heat generation during sheet metal deformation with an infrared camera as an indicator of the threshold of plastic deformation in real time and uses that information to determine the digital path of the robotic arm to reach the desired final geometry in one process rather than incrementally. This feedback loop allows us to permanently bend the sheet into the shape of the goal geometry. Historically, this process has been challenging because of the manner in which ductile materials spring back and can only maintain the deformation applied to them in the plastic range (Wang et al. 2008).

Once a material deforms plastically, frictional heat is generated from planes of atoms slipping over one another. During this motion, imperfections in the atomic lattices slide over one another changing their relative locations, and it is this tangling of dislocations that causes plastic deformation. The slope of the stress versus strain curve from 0 to the yield point, the elastic modulus, governs the amount of springback the material will undergo in both the elastic and plastic range. However, the value of the elastic modulus remains unchanged in the plastic regime (see Fig. 1, dashed lines C and D), so if one had an ability to precisely measure the elastic modulus for a mode of deformation for any material, the goal geometry may be generated through an informed amount of overbending. Since these properties discussed here are true for all materials, we propose a universal method for predicting over-bending by looking for the onset of plastic deformation using a thermal camera as the heat released when the material reaches its yield point. With this information, we can calibrate a bending model to determine how much to over-bend. Bending a piece of metal by the use of a robotic arm for a parametrically controlled bending path enables us to predictively overbend a strip of metal to achieve a desired goal form after the structural relaxation of the unloaded strip.

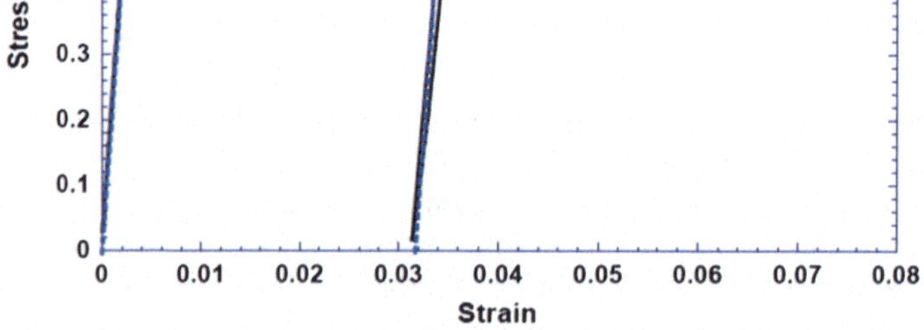

Fig. 1. Elastic and plastic deformation ranges of bent metal. *A* is the yield point, *B* is the end of creep, *C* shows the slope defined by the elastic modulus, *D* indicates the relaxation slope during unloading, identical to *C*. (Scherer 2015)

Fig. 2. Developable curved surfaces applications (Pottman et al. 2007). *Left* ARUM installation at the 2012 Venice biennale (Photo: Zaha Hadid press release). *Right* Operable roof for passive cooing fabricated by the authors. The metal sheets are bent with a single-curvature although that curvature direction or amount could be changing along the surface

In Pottmann et al. (2007), the authors deliver an overview of architectural geometry modeling and fabrication methods for curved facades. We chose to explore a case study of the third category they describe: "developable panels," i.e., surfaces that can be formed from a flat sheet of material without stretching or tearing (see Fig. 2). This allows us to begin with a proof of concept for a 2-dimensional curve as the basic generic bending process, where the plastic deformation is triggered by curvature and not by stretching or tearing of the material. As we shall explain, our method can be extrapolated into more complex types of curved panels with double curvature as well. For a 2-dimensional curve, we established precise relationship between its geometric description and a physical bending process.

We chose a NURBS curve, which allows a description of a curve with control points outside of the curve itself, that act as "weights", as they parametrically influence the curvature as attractors but are also related to the physical properties as the start and end points define the curve's tangents (Farin 2002). With this, we can incrementally increase the maximum curvature of the described curve from a straight line to a goal curve by rotation of the tangents to the curve (see Fig. 3) and simulate the bending of a piece of metal, maintaining a maximum curvature at a predictable location.

If we overstress the material sheet by a specific curvature amount past our goal geometry, equal to the stress of the metal's curvature at its yield point, we can predict the relaxed form we will receive to be our goal geometry. If we define the curvature by

$$k = 1/\mathrm{R} \tag{1}$$

we can find the curvature value at the yield point and add it to the goal geometry to over-bend the metal in order to count for its elastic relaxation kinetics. Where the final curvature is defined based on the addition of the curvature of the goal geometry and the curvature at the yield point, which is assumed to be the springback amount. The resultant curvature is then compared to the plot of possible curvatures (as seen in Fig. 4) and the curvature with the closest value is chosen as the over-bend destination.

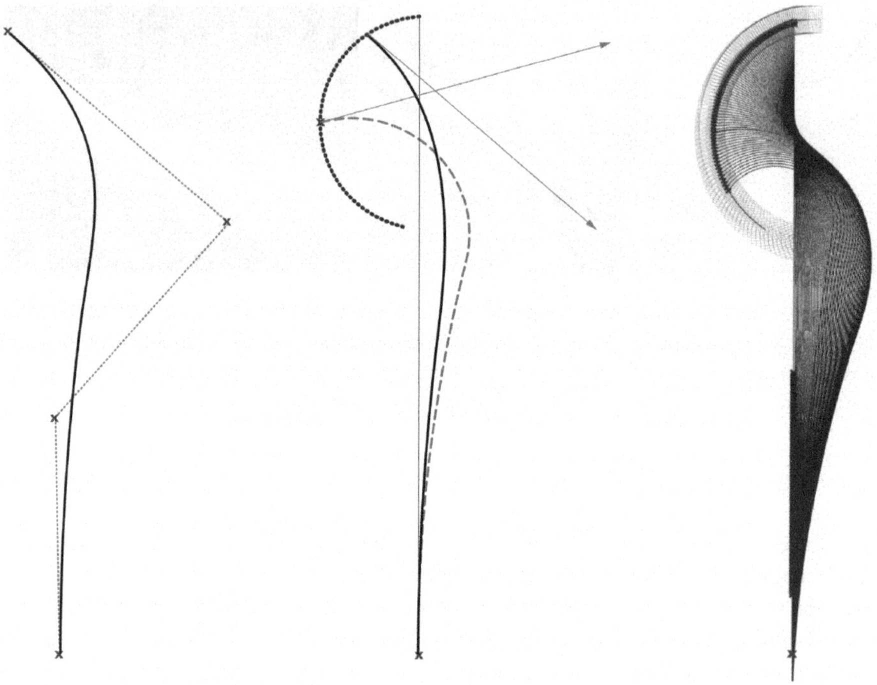

Fig. 3. Development of geometry from a goal curve into an array based on transition from straight to bent and over-bent curves. A precise gentle goal curvature is shown within an array of increasing curvatures (decreasing radii)

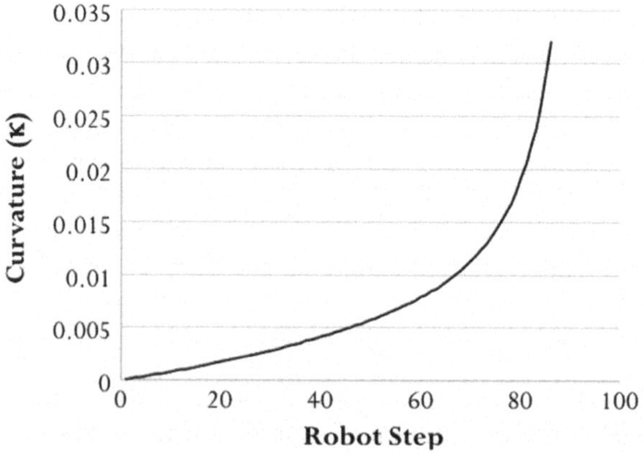

Fig. 4. Graph of change in curvature based on array position. Note that the curvature increases as the radius of the circle at that point decreases

Related Work

While controlled metal bending has been continuously developed in the automotive and aviation industries for decades, the complexity associated with predictably deforming a sheet have resulted historically in the common use of trial and error methods for achieving curved geometries. Other approaches to characterize spring back of bent metal due to residual elastic effects are typically not universally understood (Wang et al. 2008). Such approaches are also limited to highly regulated material stock, whereas recycling or less controlled operations compromise high quality feedstock. Some rigorous approaches developed initially by the automotive industry, such as incremental bending, have been applied to complex curvature using a robotic arm using sensing (Kalo and Newsum 2014). Instead of incrementally bending miniscule parts of the metal into a goal shape, we offer a process, which enables more holistic approach to control curvature. With our method of relying on the material properties for deformation detection through thermal imaging, the stock's quality or versatility is not an important parameter. Thermal imaging is used in the construction industry to perform fatigue limit evaluations, not to evaluate degree of deformation or fidelity to a curve (Luong 1998).

Thermal imaging is not currently applied to metal bending, however there are attempts at quantifying the percent temperature increase as a function of applied strain locally (Eisenlohr et al. 2012). More rigorous approaches mathematically derive heat generation as a function of local parameters, such as stress and strain tensors (Rusinek and Klepaczko 2009). Both Eisenlohr et al. (2012) and Rusinek and Klepaczko (2009) report temperature increases up to 5% or 7 °C during a plastic deformation event, often with a large spike at the onset. This is a readily observable range with modern thermal imaging cameras, and a simple proof of concept experiment we performed with a regular FLIR camera confirmed the detection of the thermal event. A thermal image from this initial experiment is shown in Fig. 5.

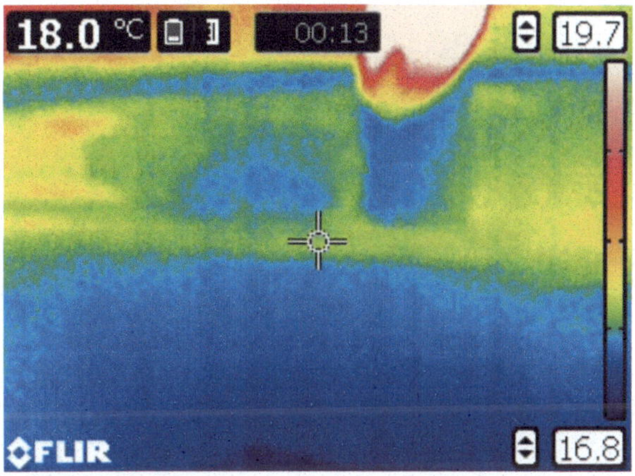

Fig. 5. Thermal image during manual bending of a thin aluminum sheet. The *yellow region* appeared due to heat generation when plastic deformation occurred and dissipated afterwards

Process Overview

In the physical setup of the experiment, we clamped the metal sheet at one end to a wooden column and at the other end to the end effector of the robotic arm. This limits the range of movement of the sheet and therefore the resultant geometry as well. This can be improved if two robotic arms at either end collaborate in a dynamic, sensor-based process. However, we chose to use this setup in order to be able to control the tangent condition at the two ends, using one robot only. When the robot is changing its position, the end of the metal attached to the wood is static and the other end changes its position and tangent direction because of the end effector's steel plate attachment (see Fig. 7). A FLIR infrared camera is positioned perpendicular to the metal's surface to be able to detect heat generated across it. The metal sheet is painted in matte black paint to allow infrared temperature sensing. A webcam is placed facing the top edge of the metal sheet to track the increasing curvature of the bent metal and compare it to the predicted curvatures from the digital interface.

As delineated in Fig. 6 above, the bending process we created involves converting a user-defined goal geometry to a robotic arm movement, informed by heat sensing from an infrared camera. Below is a step-by-step description of the process:

1. User input of a NURBS curve drawn in computer-aided-design software.
2. Curve path arrayed: the resulting curve length is re-computed using Grasshopper algorithmic modeling tools into a straight line. An array of intermediate curve is generated between the straight line and the goal geometry, maintaining a consistent

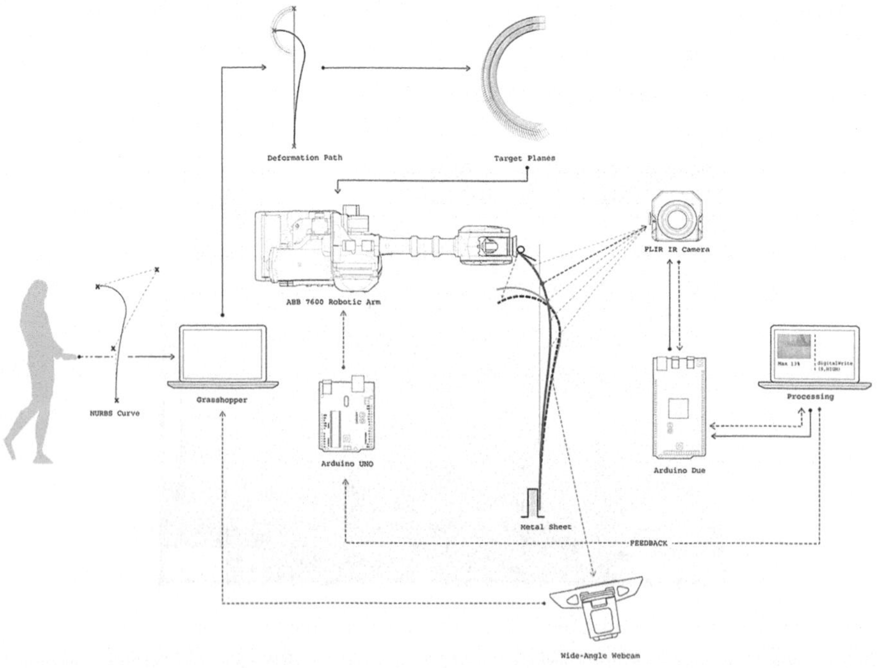

Fig. 6. Process diagram with feedback loop

Fig. 7. Physical setup of the bending procedure

point of maximum curvature (see Fig. 3). An extension of the array is created for over-bending positions generation.

3. Robot targets: The planes defined by the end point of the curve and the tangent direction at that point are translated into Robot Targets using MUSSEL (Johns et al. 2017) converting Rhino coordinates into RAPID code for an ABB 7600 robotic arm.
4. Initial movement: the RAPID code generates movement of the robotic arm to the planes specified in Grasshopper. The end effector of the robotic arm is a rigid steel plate to which the metal sheet is attached at its end, enabling control over the metal's tangent. The robotic arm follows the path of the planes defined in grasshopper, consequently increasing the bending moment of the metal sheet, until its yield stress is reached and plastic deformation generates heat in the metal surface.
5. The thermal camera used for this project is a FLIR® Lepton module which produces a 60 × 80 array of temperature pixels, sent over a serial connection to processing. This image is updated at a rate of about 2 Hz within a processing image.
6. A Processing code divides the metal into vertical strips. The average pixel color of the strip is compared to the average background pixel value. If there is a significant difference between these two values, the script sends a command to the robot to alert the robot controller to the onset of plastic deformation.
7. Trap loop: once the robot receives a signal from the Arduino, a trap loop is generated to calculate the remaining number of targets for the robot path. This calculation is based on the curvature detected when the robot was at a position

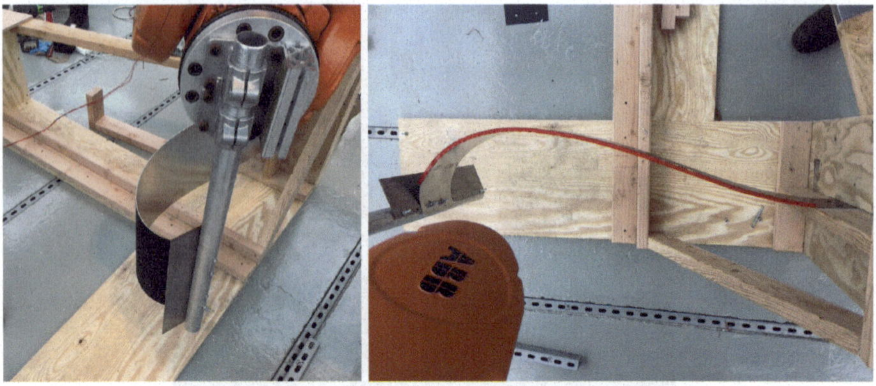

Fig. 8. Views of the robotic arm and the deforming sheet during the bending process

where plastic deformation was sensed by the FLIR camera. That curvature is added to the goal geometry curvature to over-bend the metal by the right amount.

8. Final movement: The robotic arm resumes its movement, following the final path to the calculated over-bent position and stops (Fig. 8).
9. Webcam concurrent to all movement detects the 2-dimensional projection of the metal sheet and compares the result with the predicted geometry from Grasshopper.
10. Release of metal sheet and comparison with the goal geometry.

Results and Discussion

The Processing code was successful in streaming the thermal information at rates that matched well with the robot's movement. For a faster process, the 2 Hz refresh rate would have been too slow, but the reduced speed of the robot's motion allowed this to work. The rate was also sufficient to sense a thermal event, shown as the white vertical line in the right inset of Fig. 9, before the heat equilibrated due to conduction through the material and convection to the surroundings. Finding the threshold value for the maximum reading of a vertical minus the average metal reading, divided by the average reading, was a trial and error process. The metal is examined with the sensor at different spots and with different size windows by the user before bending occurs to establish a familiarity of the range of sensor readings. As the metal bends, its 2D projection onto the temperature sensor lens decreases, which is why a temperature event trigger threshold cannot be too small, since statistically the average means less once the window shrinks.

Figure 10 shows the data threshold comparison calculation from the bend test documented in Fig. 9. It is not uncommon for readings to hover as high as 0.12, however visual confirmation confirms the value of 0.13 was indeed matched with a plastic deformation event. (Videos of the bending procedure in plan and in section can be viewed at https://vimeo.com/201171029 and https://vimeo.com/201171809, respectively.) Using a vertical segment versus individual pixels reduced noise, however

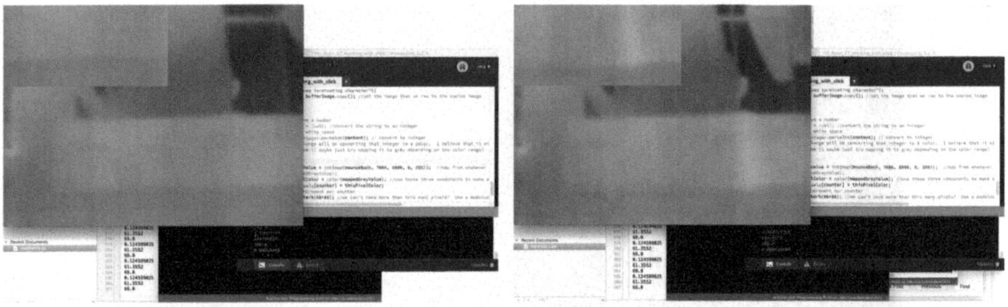

Fig. 9. Screenshots of processing code and infrared image during bending. To the *right* the image shows the *white region* appearing due to heat release during plastic deformation

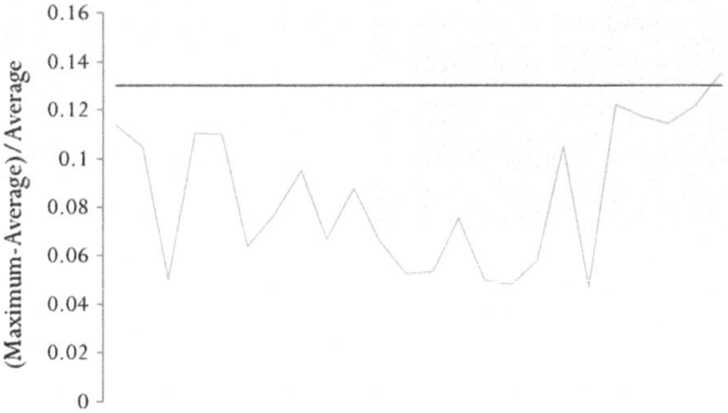

Fig. 10. Signal noise from FLIR camera and threshold for plastic deformation as detected during bending

more intelligent algorithms or different bin sizes for the vertical strip thickness could be changed in future runs to help reduce noise. Performing the sensing and bending in a room without windows and equilibrium surface temperatures would help to reduce noise as well.

Overall, dynamically updating a bending path based on a modal elastic modulus unique to each piece of material was successfully performed, and this experiment hopefully helps lay the groundwork for future experimental analysis. Images from metal bending are shown in Fig. 11.

To achieve a more dynamic process, the heat detection would be helpful in the analysis of the deformation prediction if the infrared image were calibrated with the webcam that views the metal sheet from above. This calibration will result in detection of what points on the curve yield at a given stage of the process, and allow the robotic arm to respond accordingly.

Additionally, load monitoring with a load cell on the end effector, calibrated with the thermal imaging, can be used to extract an exact elastic modulus value. Residual stresses after the robot unbends are still difficult to predict even with a calculated amount of unbending since spline curves do not entirely plastically deform. A gripper element that doesn't need an operator to unbolt could help this as well, but in terms of

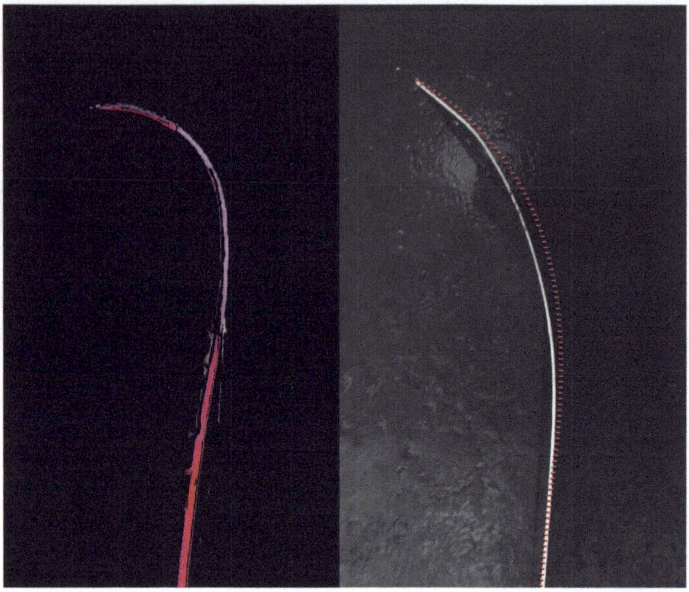

Fig. 11. Sample result for a 3 mm thick aluminum sheet. To the *right* the over-bent position as captured by the webcam. To the *left* the released metal in comparison with the goal geometry. Imprecisions are due to the imperfect release process

eliminating potentially dangerous force discharge upon unclamping material from the end effector.

Conclusion

While predictable metal stamping has been continuously developed in the automotive and aviation industries for decades, the path we have pursued here is unique. Instead of investing wasted material and effort on disposable molds or using a predetermined gauge-and alloy-based prediction, we relied on the material itself to report to us its properties during a simple actuation movement enabled by a robotic arm. We achieved this by using thermal sensing, detecting surface temperature with an infrared camera tracking the face of a deforming metal sheet. This method allows for in-depth understanding of the relationship between a defined geometry and the microstructure of the material at hand. Furthermore, this technique fits within a larger conceptual framework of embodied computation (Johns et al. 2014). In other words, the modeling of a design process is not a linear and predictable progression of intention and execution. Instead, it is a feedback loop providing live data embodied in the material itself and informing the operation of the model.

This experiment has opened up many avenues to explore regarding the potential of thermal sensing in architectural fabrication. It is a powerful yet underutilized technique for feedback control over physical processes. When calibrated with geometric properties, it offers an innovative way to think about the relationship between the physical and the digital worlds. This idea can be extended beyond the specific proposal discussed here of sensing the yield point of a material through temperature change detection. Using heat transfer as a means to investigate the response of material microstructures to changes in macro-geometry is a fruitful ground for architects to understand the physical consequences of digital design (Bechthold 2010).

Acknowledgements. We initiated this research in Ryan Luke Johns's computation seminar in Princeton University in 2016. We would like to thank him as well as Serguei Bagrianski and Axel Kilian for their help and advice.

References

Bechthold, M.: The return of the future: a second go at robotic construction. Archit. Des. **80**(4), 116–121 (2010)

Eisenlohr, A., Gutierrez-Urrutia, I., Raabe, D.: Adiabatic temperature increase associated with deformation twinning and dislocation plasticity. Acta Mater. **60**(9), 3994–4004 (2012)

Farin, G.: Chapter 1—A History of Curves and Surfaces in CAGD. In: Handbook of Computer Aided Geometric Design, pp. 1–21. Amsterdam: North-Holland (2002). www.sciencedirect.com/science/article/pii/B9780444511041500022

Johns, R.L., Kilian, A., Foley, N.: Design approaches through augmented materiality and embodied computation. In: Robotic Fabrication in Architecture, Art and Design 2014, pp. 319–332. Springer International Publishing, Cham (2014)

Johns, R.L.: MUSSEL. http://www.grasshopper3d.com/group/mussel, extracted 1/25/2017

Kalo, A., Newsum, M.J.: An investigation of robotic incremental sheet metal forming as a method for prototyping parametric architectural skins. In: McGee, W., de Leon, M.P. (eds.) Robotic Fabrication in Architecture, Art and Design 2014, pp. 33–49. Springer International Publishing, Cham (2014)

Luong, M.P.: Fatigue limit evaluation of metals using an infrared thermographic technique. Mech. Mater. **28**(1–4), 155–163 (1998). doi:10.1016/S0167-6636(97)00047-1

Pottman, H., Asperl, A., Hofer, M., Kilian, A., Bentley, D.: Architectural Geometry, vol. 724. Bentley Institute Press, Exton (2007)

Rusinek, A., Klepaczko, J.R.: Experiments on heat generated during plastic deformation and stored energy for TRIP steels. Mater Des **30**(1), 35–48 (2009)

Scherer, G.: MSE501 Introduction to Materials, Plasticity Handout. Princeton University, Princeton (2015)

Wang, J., Verma, S., Alexander, R., Gau, J.-T.: Springback control of sheet metal air bending process. J. Manuf. Process. **10**(1), 21–27 (2008)

Localised and Learnt Applications of Machine Learning for Robotic Incremental Sheet Forming

Mateusz Zwierzycki, Paul Nicholas(✉), and Mette Ramsgaard Thomsen

Centre for IT and Architecture, Royal Danish Academy of Fine Art, School of Architecture, Copenhagen, Denmark
paul.nicholas@kadk.dk

Abstract. While fabrication is becoming a well-established field for architectural robotics, new possibilities for modelling and control situate feedback, modelling methods and adaptation as key concerns. In this paper we detail two methods for implementing adaptation, in the context of Robotic Incremental Sheet Forming (ISF) and exemplified in the fabrication of a bridge structure. The methods we describe compensate for springback and improve forming tolerance by using localised in-process distance sensing to adapt tool-paths, and by using pre-process supervised machine learning to predict stringback and generate corrected fabrication models.

Keywords: Machine learning · Incremental sheet forming · Robotic fabrication

Introduction

Incremental Sheet Forming (ISF) is a formative fabrication process, in which mechanical forces are applied to a flat material so as to form it into a desired 3D shape. A particular problem associated with ISF is the springback of the metal, and the development of methods to counteract it. In this paper we counteract springback in the fabrication of architectural panels by creating two different material behaviour models—the first based on on-line distance measurement and linear regression, and the second based on a neural network. This paper particularly describes the application, architecture and accuracy improvement of the artificial neural network. The trained model is based on 102-dimensional input data, which the artificial neural network learns and interprets, so that it is able to predict the shape of the panel after the fabrication process. This work is a continuation of the research described in Nicholas. The limitations and problems encountered in the approach presented in that paper are addressed here.

Background

The structure A Bridge Too Far spans 4 m, and is made from 1 mm aluminium panels rigidized using a process of robotic incremental sheet forming (ISF). It was exhibited

K. De Rycke et al., *Humanizing Digital Reality*,
https://doi.org/10.1007/978-981-10-6611-5_32

together with two further projects—*Inflated Restraint* & *Lace Wall*—as part of the research exhibition *Complex Modelling* held at Meldahls Smedie, The Royal Danish Academy of Fine Arts (KADK), over the period September–December 2016.

In the ISF process, a rounded head tool is applied from either one or two sides of a flat sheet, and moves over the surface of that sheet to cause localized plastic deformation out of plane (Jeswiet et al. 2005). In a double-sided approach two tools, one working as a forming tool and one as a support, enable forming out of plane in opposing directions, and significant freedom and complexity in the formed geometry. The double-sided setup used to fabricate A Bridge Too Far incorporates two industrial robots working on each side of a moment frame that allows for a working area on approximately 1000 × 1000 mm.

The primary advantage of ISF is to remove the need for complex moulding and dies, which only become economically feasible with large quantities. However, this means that positional accuracy is lowered compared to other approaches, as well as more highly dependent upon a combination of material behaviour and forming parameters. Research into resultant incrementally formed geometries has shown significant deviations from planned geometries (Bambach et al. 2009), which is recognised as a key deterrent from the widespread take up of the process (Jeswiet et al. 2005). Our testing has shown that parameters including the forming velocity, the tool path, the size of the sheet and distance to supports all affect springback of the sheet and the geometric accuracy of its forming. It would be extremely difficult to develop a linear, analytical model for springback control including all of these factors, on account of their complex interdependency.

Machine Learning. As an alternative, machine learning approaches can be considered. Data mining techniques have been previously applied to prediction and minimisation of springback in sheet metal forming, including statistical regression (Behera et al. 2013) and neural networks (Khan et al. 2014). These approaches have combined Finite Element simulation with data mining techniques (Ruffini and Cao 1998). Finite Element simulation can also be combined with point cloud data, obtained via scanning of formed results, with the difference between simulation and scan taken as training data. While Khan et al. (2014) extends the prediction of springback errors to new shapes and the generation of corrective geometries, this is restricted to a simple pyramidal geometry. We go beyond state of the art by applying a deep learning based approach to significantly more complex geometries.

A second trajectory has been the use of dynamic tool path correction, in conjunction with online monitoring, to predict and correct distortion due to springback. While adaption of toolpaths, for example to change tool speed or 'jump' zones of material fracture has been proposed (Paniti and Rauschecker 2012), this trajectory has not yet been explored in ISF. A significant limit is that previous limitations on fabrication technology have required that the tool path be specified in advance, rather than as the process develops. For this reason, toolpath correction has been limited to application on simulation results, however recently developed control and communication tools for robotic control—i.e. HAL (2017), Robots (2017), KUKA|prc (2017)—together with a shift to industrial robots, rather than purpose built machines, provide new opportunities for adaptive control as a means to manage forming tolerance.

Methods

Localised Method

A Bridge Too Far (Fig. 1) relies on several hundred point connections being made precisely between upper and lower skins. In the first method, rather than rely on model-based predictions of forming, an in-process measurement-based approach was developed to achieve the required tolerances. This automated process involved measurement of formed geometry at these points, calculation of the local deviation, and automatic selection of an appropriate toolpath for a cone that iteratively corrects these deviations to make a precise connection. A single point laser distance measure mounted to the robot arm instead (Fig. 2) was used to measure, at each cone centre-point, the local deviation between actual formed depth and ideal geometry. Some knowledge of springback could also be gathered, modelled and used during this process.

Fig. 1. A Bridge Too Far exhibited at the research exhibition *Complex Modelling* KADK, 2016

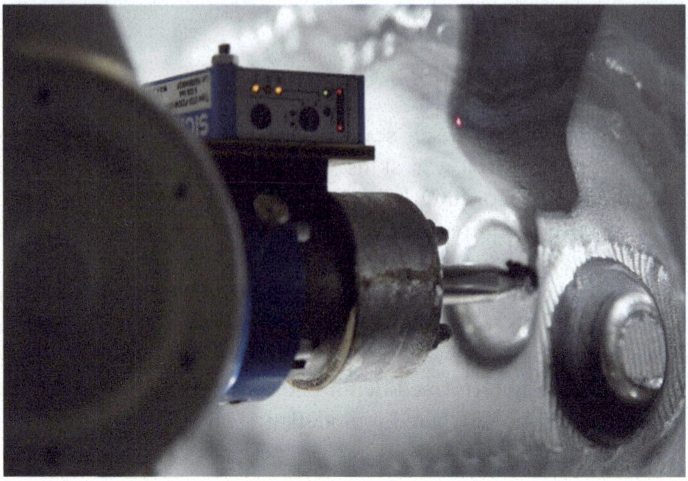

Fig. 2. Single point distance sensor mounted to the robot arm

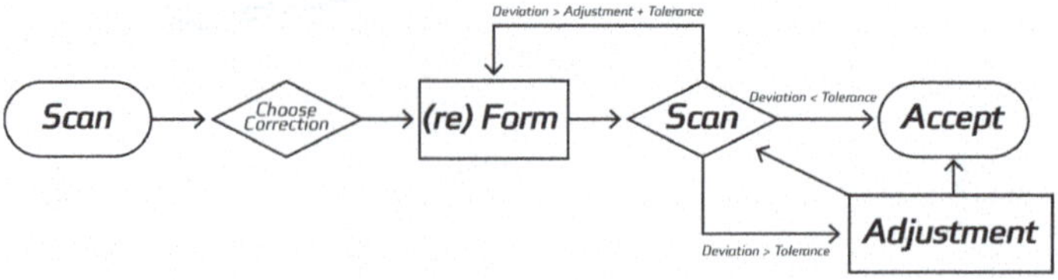

Fig. 3. Workflow diagram of the localised method

To determine the required amount of over-forming for each cone, the relationship between target depth and forming depth was modelled during the entire cone forming process using a linear regression. After forming, the resultant depth was scanned, and this data was used to refine the curve fitting model. This allowed a continued improvement in accuracy across the course of fabrication. Scanning occurred several times in the forming of each cone, and could trigger two automated correction methods. If the formed cone geometry was greater than 5 mm off the target geometry, the cone was reformed. If between 5 and 2 mm off the target geometry, the tip of the cone was extended by 2 mm; tolerances below 2 mm were considered acceptable. This procedure is illustrated as a diagram in Fig. 3.

Learnt Method

Contrary to the Localized method which acts during the fabrication process, the Learnt method is utilizing supervised machine learning to construct a material model a priori the fabrication. Therefore the adaptability competence of the overall workflow is shifted to the earlier design phase. This simplifies the fabrication process (no need for on-the-fly toolpath generation), but requires to collect the material behaviour samples

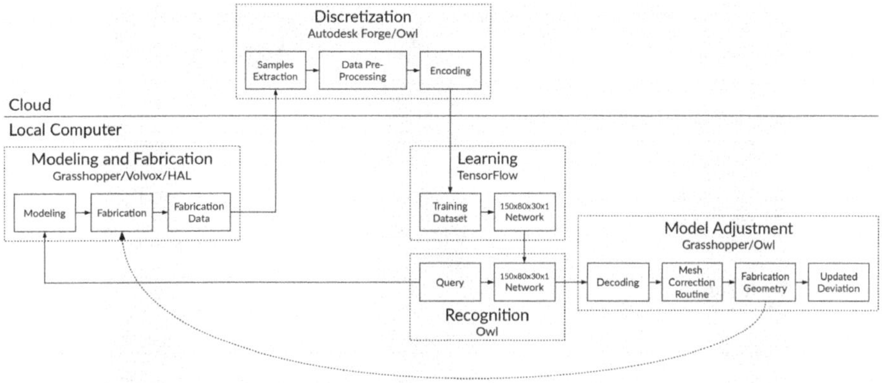

Fig. 4. Workflow diagram of the learnt method

beforehand. While this can be a time consuming venture, there is an advantage of recycling the acquired database for use in other projects or workflows. Another advantage of developing a material model compared with the toolpath adaptation seen in the Localized method, is the decoupling of the model from the fabrication technique.

In the most general terms, the Learning method is a mapping between 102-dimensional input and 1-dimensional output. The accuracy of the map is a subject of investigation, and the means required to achieve a satisfactory results are described hereafter. This section is divided into 4 subsections focusing on the different aspects of the developed workflow and the challenges encountered during the development process: *Acquisition, Clean-up, Pre-processing* and finally *Learning*. The presented workflow is a continuation and elaboration of the previously published work in Nicholas et al. (2017), which should be seen as a test bench for the methods described in this publication (Fig. 4).

Acquisition. The acquisition phase was the first problem to be addressed, as it influences directly the quality of the constructed material behaviour database. The highly reflective surface of the aluminium panels made the 3d scanning problematic because the laser-based scanner is prone to reflections, resulting with points scattered in space. While it wasn't a matter of concern with the dataset collected for the first experiment (due to its highly prototypical nature), in the presented workflow this problem was addressed to minimise the data loss or noise-induced training inaccuracy.

A chalk-based, water-soluble spray paint was used to effectively (and temporarily) cover the metal surface with a thin layer of a non-reflective coating. The scattering incidents effectively remove the points from the scope of the scan, creating "holes" in the point cloud—making the resulting outliers greatly stand-out and guide the learning process away from the most accurate state. The one issue which wasn't resolved at the hardware level of the acquisition process is the inherent noise of the 3d scanner—which is up to ±2 mm (Faro Focus3D 2017) (Fig. 5).

Clean-up. The time consuming process of 3d scanning is succeeded with manual process of further point cloud cleaning. With the toolset based on Grasshopper (2017) and Volvox (2017), the clean-up task became more intuitive and efficient, as it was

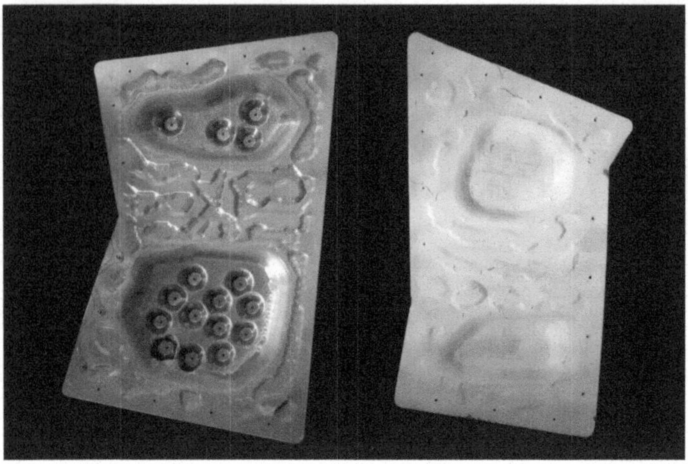

Fig. 5. The reflective surface of the panel compared with the coated surface

possible to automate the general cropping of the scan data. Removing the points is required to prepare the data for the automatized pre-processing methods.

Pre-processing. In the presented workflow context means preparation of the input Tensors for the learning process. As described above, the input is a 102 Tensor with 100 dimensions to represent the local shape of the surface as a heightfield (of size 5×5 cm), and 2 additional values: distance to the edge of the panel and distance to the edge of the fabrication rig frame. This is a slight difference introduced in the new approach compared with the one from [blinded reference], which was using a 9×9 heightfield + distance to the panel edge value. The significance of this change is further investigated in the evaluation section.

The sample extraction process, gathers up to 3000 samples per one panel. Each sample is mirrored and rotated 180° to effectively obtain 4 samples. While for this workflow the procedure takes only up to a minute for each panel, the reusability of the toolset and its scalability was a matter of concern. The Autodesk Forge platform (2017) was used to parallelize the pre-processing using a cloud-based solution. Two applications and a common library was developed to make the software setup and usage more agile. This enabled to experiment with different extraction parameters, such as the number of points per sample, the size of the sampling frame etc. In total there was over 45,000 samples extracted from 9 panels during the first training session and almost 100,000 for the reversed mapping session (Fig. 6).

Learning. The architecture of the artificial neural network is slightly expanded in comparison with the one used in the previous workflow (Nicholas et al. 2017). With the learning process based on the TensorFlow framework (2017) running on GPU, it was possible to expand the network size to 4 layers. The layers have 150, 80, 30 and 1 neurons, with 102 inputs. A 3-layered network was also a subject of tests, yet it didn't reach a comparable accuracy with the 4-layered one (Fig. 7).

The training process was performed on a single PC equipped with a GPU card, and few sessions were performed to reach the final accuracy. Each session initialized the

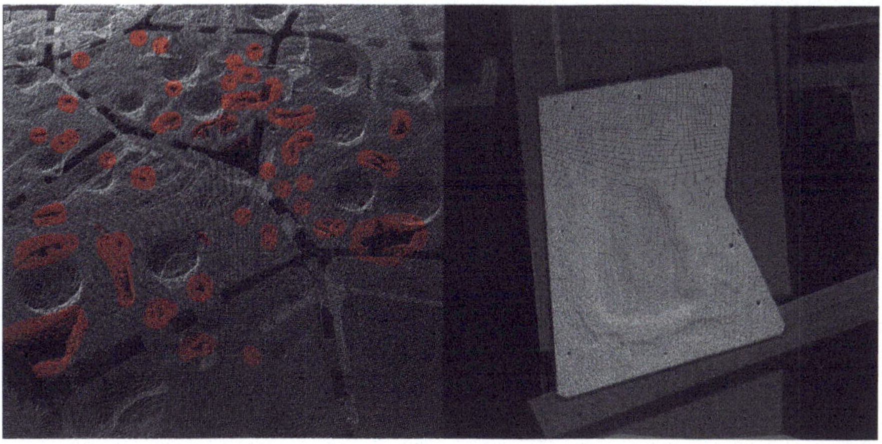

Fig. 6. From *left* an example of a point cloud with missing points (marked in *red*) which is a result of the point scattering problem; The anti-reflective coating reduces the noise and the point scattering to minimum

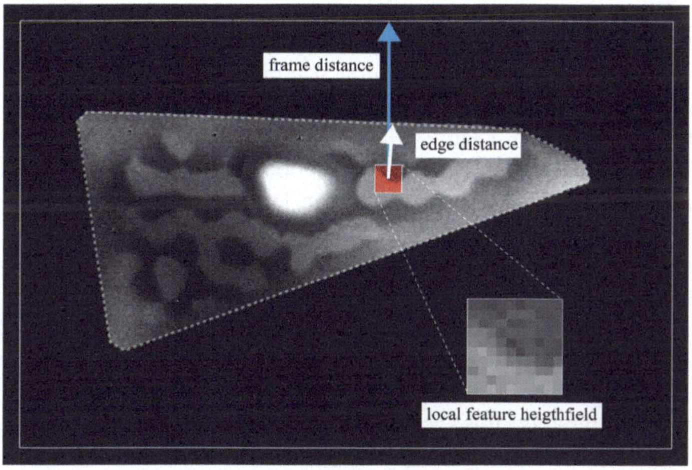

Fig. 7. Data encoded within the 102-dimensional input of each of the training samples

network with randomised values and used different training parameters (learning rate, optimizer type, batch size), taking up to 15 min to train down to the point of optimizer equilibrium.

Results

The evaluation of the neural network prediction error was conducted on ~10,000 samples from which the output mesh was reconstructed. An overlay of the input mesh (the desired geometry), the result of fabrication and the prediction are shown in Fig. 8.

Fig. 8. The prediction closely follows the fabrication results, up to the point where the scanner induced noise (up to ±2 mm) becomes an obstacle in reaching a higher prediction accuracy. The standard deviation of the prediction error is 1.075 mm

While there are areas with a larger error (of up to 7 mm), the majority of predictions (>70%) are within ±2 mm deviation (which also incorporates the hardware error of up to ±2 mm), and 90% of samples are of error less than 3 mm.

Certainly the areas with a large difference in the slope gradient yield predictions with lesser accuracy, as shown in the cross-sections in Fig. 8. Arguably the source of this problem lies in their sparsity which translates directly to the relatively small amount of the representative samples in the training dataset

Reversed prediction. The presented experiments were conveyed with a plan of predicting the fabricated geometry shape based on the input mesh model. Given the network is able to learn that relationship, it is in principle possible to train the network in the opposite direction: predicting the input mesh geometry based on the desired output. This experiment yielded a network which is able to generate the geometry with lesser accuracy. Arguably this accuracy drop is caused by the amount of the hardware-induced noise in the training set inputs. A comparison of the results obtained from the 2 networks (forward and reversed prediction) is compiled in Fig. 9.

Future Work

While for the presented fabrication method the prediction accuracy is satisfactory, there is room for improvement. An identified obstacle to achieving better results is the 3d scanning hardware used for the data acquisition. This problem will be addressed in the future by using a digitizer or a blue light laser scanner rather than a regular mid-to-high range 3d scanner.

The greatest prediction deviation in both of the models ("forward" and "backward" predicting) occurs in the areas with a steep slope. The reason of that error is considered

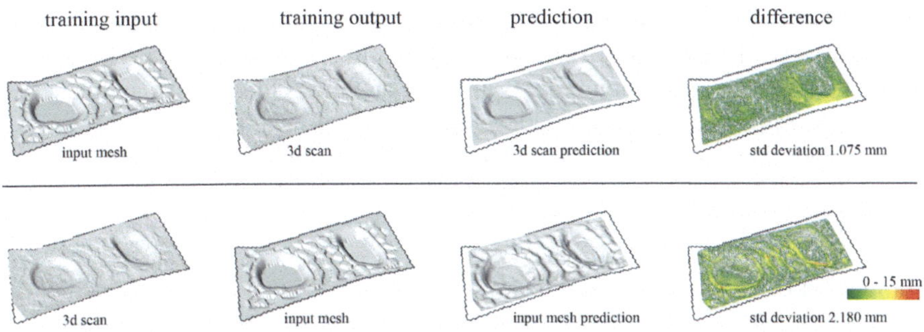

Fig. 9. The comparison of different input-output training sets and the achieved accuracy. *Top row* "forward" prediction, *bottom row* "reversed" prediction

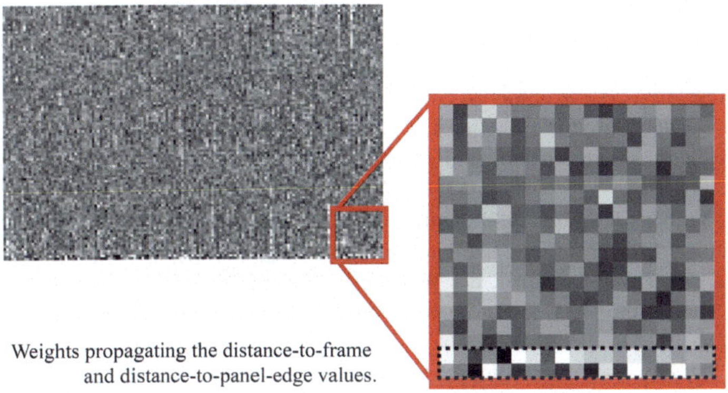

Fig. 10. A plot of the network's first layer weight values

to be the sparsity of those regions in the training dataset in comparison with the rest of the samples. This could possibly be solved by collecting more samples from a greater variety of panels.

An interesting research direction is seen in the analysis of the trained network. The plot of the first layer's weights shows a pattern formed by the last 2 rows of the weight matrix (Fig. 10). Those values are trained on the data encoding the distance to the edge of the panel and the distance to the fabrication rig frame. This pattern indicates that there is some importance of these data in the trained model, although we don't know if it is possible to quantify.

Conclusion

This paper demonstrates the use of sensing and learning to manage springback, for the purpose of reducing geometric inaccuracies within the incremental sheet forming process. Two different methods have been presented, the first based on in-process localised adaptation, and the second based on learnt prediction. Each introduces new potentials for design decision making, extending specification into an ongoing contact with the process of fabrication. However, while both methods negotiate between the design model and the fabrication process, the first method is limited to a simple process of continual correction, as it does not develop a model of material behaviour, while the second method does construct such a model. We believe that the application of machine learning in this project indicates their wider feasibility in architectural fabrication.

Acknowledgements. This project was undertaken as part of the Sapere Aude Advanced Grant research project Complex Modelling, supported by The Danish Council for Independent Research (DFF). The authors want to acknowledge the collaboration of Autodesk Forge, SICK Sensor Intelligence Denmark, Monash University Materials Science and Engineering, Bollinger + Grohmann, and the robot command and control software HAL.

References

Autodesk Forge: https://forge.autodesk.com/. Accessed 19 May 2017

Bambach, M., Taleb Araghi, B., Hirt, G.: Strategies to improve the geometric accuracy in asymmetric single point incremental forming. Prod Eng **3**, 145–156 (2009). doi:10.1007/s11740-009-0150-8

Behera, A.K., Verbert, J., Lauwers, B., Duflou, J.R.: Tool path compensation strategies for single point incremental sheet forming using multivariate adaptive regression splines. CAD Comput. Aided Des. **45**, 575–590 (2013). doi:10.1016/j.cad.2012.10.045

Faro Focus3D 120: http://www.faro.com/en-us/products/3d-surveying/faro-focus3d. Accessed 19 May 2017

Grasshopper 3D: http://www.grasshopper3d.com/. Accessed 19 May 2017

HAL: http://www.food4rhino.com/app/hal-robot-programming-control. Accessed 19 May 2017

Jeswiet, J., Bramley, A.N., Micari, F., et al.: Asymmetric single point incremental forming of sheet metal. CIRP Ann. Manuf. Technol. **54**, 623–649 (2005)

Khan, M.S., Coenen, F., Dixon, C., et al.: An intelligent process model: predicting springback in single point incremental forming. Int. J. Adv. Manuf. Technol. **76**, 2071–2082 (2014). doi:10.1007/s00170-014-6431-1

KUKA|prc: http://www.robotsinarchitecture.org/kuka-prc. Accessed 19 May 2017

Nicholas, P., Zwierzycki, M., Nørgaard, E., Stasiuk, D., Ramsgaard Thomsen, M.: Adaptive robotic fabrication for conditions of material inconsistency: increasing the geometric accuracy of incrementally formed metal panels. In: Proceedings of Fabricate 2017, pp. 114–121 (2017)

Owl Framework: https://github.com/mateuszzwierzycki/Owl. Accessed 19 May 2017

Paniti, I., Rauschecker, U.: Integration of incremental sheet forming with an adaptive control into cloud manufacturing. In: MITIP (2012)

Ruffini, R., Cao, J.: Using neural network for springback minimization in a channel forming process. SAE Tech. Pap. (1998). doi:10.4271/980082

Robots IO: https://robots.io/wp/. Accessed 19 May 2017

TensorFlow Framework: https://www.tensorflow.org/. Accessed 19 May 2017

Volvox: https://github.com/CITA-cph/Volvox. Accessed 19 May 2017

Energy Efficient Design for 3D Printed Earth Architecture

Alexandre Dubor(✉), Edouard Cabay, and Angelos Chronis

Institute for Advanced Architecture of Catalonia, Pujades 102, Barcelona, Spain
{alex, edouard.cabay, angelos.chronis}@iaac.net

Abstract. Additive manufacturing with mud has the potential to reintroduce traditional materials within our contemporary design culture, answering the current demands of sustainability, energy efficiency and cost in construction. Building upon previous research, this study proposes the design and test of real-scale wall elements that aim to take advantage of both the novel material fabrication process as well as the significant thermodynamic properties of the material to achieve a performative passive material system for bioclimatic architecture. Although this project is still at an early stage, the presented study demonstrates the potential of combining 3D printing of mud with performance analysis and simulation for the optimization of a wall prototype.

Keywords: 3D printing · Clay · Sustainability · Thermal analysis · Structural analysis · CFD · Optimization · Performance · Simulation

Introduction—3D Printing with Mud

Mud constructions, which are based on ancestral techniques that take advantage of local materials, are today a viable alternative to the contemporary challenges of energy efficiency. Their close to zero ecological footprint is a great advantage in meeting current energy consumption and emission goals. Almost all mineral ground that contains clay can be used for construction. At all stages of its use, clay requires very little embodied energy (e.g. no baking energy and little processing). Its maintenance and repair are easy; it can be recycled and it does not generate waste. Mud constructions allow substantial savings in winter heating and summer cooling thanks to their thermal inertia. Finally, thanks to their ability to absorb and evaporate, clays regulate humidity, promoting a healthy indoor climate. Although mud as a construction material is freely and abundantly available, traditional mud constructions require labor-intensive processes that can't compete with contemporary materials in terms of cost and quality.

As additive manufacturing (AM) is continuously gaining momentum in the construction industry today, new applications are constantly being discovered. While AM for architecture is yet at an early stage, its clear potential has been identified by industries, contractors and architects. The advancements of CAAD have already led to the design of complex and optimized building geometries and AM is enabling advanced material fabrication and distribution processes, setting the ground for more efficient buildings. AM has the potential to quickly and precisely create these complex

K. De Rycke et al., *Humanizing Digital Reality*,
https://doi.org/10.1007/978-981-10-6611-5_33

Fig. 1. 3D printing with clay

geometries previously too expensive to make, while permitting a drastic reduction of production waste.

AM with mud has the potential to reintroduce this traditional material within our contemporary culture, answering the current exigencies of quality, cost and efficiency. Although the technology is still in its infantry, examples of additive manufacturing with earth materials, such as clay and mud, already exist in the literature. Friedman et al. (2014) used robotic fabrication and clay deposition to create 3D weaving patterns for table top objects (albeit with no real architectural scale application). Kawiti and Schnabel (2016) also used 3D clay printing but within the context of an indigenous project development and with local materials, closer to the interest of this paper (Fig. 1). The main differences though in our approach are the use of robotic fabrication, aiming at full real-scale 3D printing and our focus on not just the structural but also the environmental performance of clay.

Research Context

The project Pylos developed by Giannakopoulos, has proven the possibility to use additive manufacturing with mud at an architectural scale. For this purpose, the project developed a new material mix, naturally sourced and biodegradable, with a measured strength three times stronger than typical unbaked clay material. A custom extruder mounted on a robotic arm has been also developed along with specific CAD/CAM software that allows the 3D printing of complex geometries (Giannakopoulos 2017). Building upon the potential of this previous project, this study proposes the design and test of real-scale wall elements that aim to take advantage of both the novel material fabrication process as well as the significant thermodynamic properties of the material to achieve a performative passive material system for bioclimatic architecture.

Mud, as a building solution can be useful in any climate. This study has been carried out with an aim to address design solutions for different locations, using local climatic information as a driver for the design of site specific components. The case study presented here has been developed based on the specific climatic conditions of the context of the Valldaura natural park in the province of Barcelona, where the final full-scale wall prototype is constructed. Local information for solar incidences, temperature and humidity have been used for the simulations and the other design drivers of the wall, to optimize its passive performance capacities. Although the design, analysis and fabrication processes described here are site-specific, the intention of the study is to evaluate the potential of the material and the additive manufacturing techniques to produce performative building components in any climate or site. Since the simulation and performance monitoring methods are not site-specific and the design and fabrication processes are fully customizable, our approach can be replicated for different climates and different performance objectives.

Simulation and Performance Monitoring

The overall development of the project presented here involved a thorough research on various aspects of the problem, such as ancestral precedents of earth constructions and their bioclimatic performance, thermodynamic properties of walls and historical use of earthen materials, which despite their significance in the project are not discussed in this paper. This paper, on the other hand, is focusing on the performance evaluation methods used in the project and their impact on the design and fabrication of the prototype wall. To assess and optimize the different aspects of the performance of the wall, i.e. its structural, thermal and ventilation performance in conjunction always with the properties of the material a series of simulation and physical tests were carried out. The diverse and often conflicting performance objectives of the project required a multifaceted performance evaluation, which informed the design and fabrication at different stages. It is not the aim of this paper to provide an exhaustive analysis of each part of the performance assessment of the developed prototype, but rather to give an overview of the process and assess its results at this initial step of the research.

Physical tests and digital simulations have been performed on wall prototypes, using solar radiation, daylight, thermal conductivity, thermal convection, thermal mass and structural behavior analyses. Simulation models were developed in a range of software platforms, such as RhinoCFD, a computational fluid dynamics (CFD) plugin for McNeel Rhinoceros 3D (Rhino), Ladybug and Karamba, two add-ons for the Grasshopper plugin of Rhino to assess the airflow, solar and structural performance of the proposed prototype walls. In parallel, a series of open source performance monitoring machines, such as a hygrothermal monitoring apparatus and a load machine were also developed to validate the design assumptions on physical wall prototypes (Fig. 2). A light visualization exercise was also performed as part of the simulation and design evaluation process by recreating the sun path with the help of an industrial robot. A very important aspect of this study was that the different performance objectives of the wall were constantly assessed in parallel to each other, with each one informing the others throughout the design and development of the project. Due to the complexity of the problem, the performance analysis and monitoring methods used in this study were

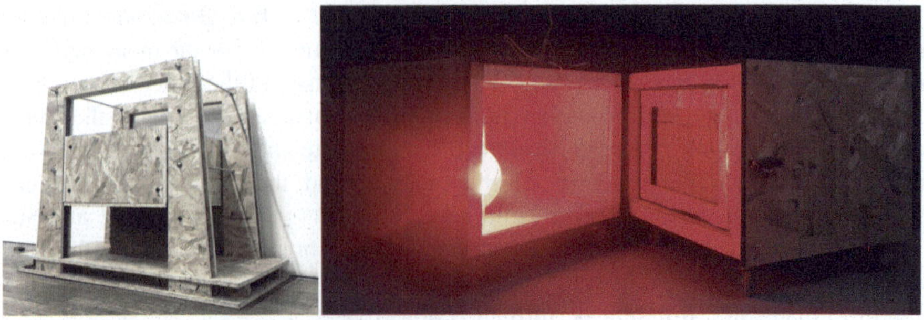

Fig. 2. Monitoring and performance testing apparati

not directed towards achieving the greatest possible accuracy but primarily to adequately inform the design and fabrication of the project, enabling the research team to identify the best compromise for the performance objectives at each iteration of the project.

Performance Results

Structural Performance

The structural performance of the 3D printed wall was evidently a fundamental aspect of the project. Although, the performative objectives of the study were mainly focused towards the thermodynamic properties of the material, the structural stability of the wall is nevertheless imperative. The structural analysis and optimization was twofold, with on one hand a focus on the material composition and its consequent structural performance and, on the other, with a focus on the structural optimization of the 3D printed geometry.

In terms of the material composition, the control of the composition's parameters, such as the amount of water, the addition of other ingredients, such as fibers, sand and proteins were investigated for their significant effect on the material's viscosity, shrinkage and brittleness, with consequent effects on its structural rigidity. Issues like material cracks, elimination of gravitational shrinkage and speeding up of the drying process were dealt with a meticulous experimentation with the material composition. The main issue of the material decomposition related to the non-uniform shrinkage of the material (up to 7%) and the low strength of the uncured clay while printing (compressive strength <0.2 MPA within the first 6 h of drying). The extensive material tests and physical prototypes helped define an optimal robotic toolpath and profile for the 3D printed geometries that maximizes their structural performance.

By comparing the structural resistance under compression of a wide range of different geometries and topologies, the researchers also identified the most important design parameters of the wall's inner and outer parts. Working with a curvilinear pattern without straight line segments and using a radius larger than 20 mm and pattern that proved to be more resistant to cracks and overhang fails and had a much better shrinkage behavior. The continuous periodic decomposition pattern without self-intersections also proved to create stronger bonds between the printing paths while offering a fast and efficient printing process by avoiding the need to stop the extrusion for every intersection. Physical tests using a 3-point flexural strength test were also conducted to assess the structural strength at each stage of the material development process.

Further to a thorough material experimentation, the structural optimization of the modular geometry was also a key aspect of the structural performance of the wall. Digital simulations using Karamba for Grasshopper were done in parallel with physical prototypes and strength tests to find a structurally sound strategy for the 3D printed modules. Issues such as buckling, toolpath joinery and accumulation of vertical load were analyzed in a simplified simulation model, using a uniform material in finite element analysis (FEA). A more thorough structural simulation approach, which would take into account the anisotropic properties of the material over space and time would allow for a more accurate optimization process, but the simplification of the structural analysis undertaken and its seamless integration in the existing design workflow, which was based on the Grasshopper platform, provided valuable feedback to the design process, driving important design decision on the topology and the geometry of the wall. Over a period of 3 months of research, more than 50 different geometry and typology variations of the wall were simulated and evaluated in conjunction with the other performance objectives and a total of 5 different solutions were then physically tested using a load bearing test to validate the performance of the digital simulations. Despite the simplification of the simulation model, our tests verified our initial assumptions and confirmed the informed design decisions taken. The outcome geometry was a conical diagrid structure that was informed by all the aforementioned structural and material issues (Fig. 3).

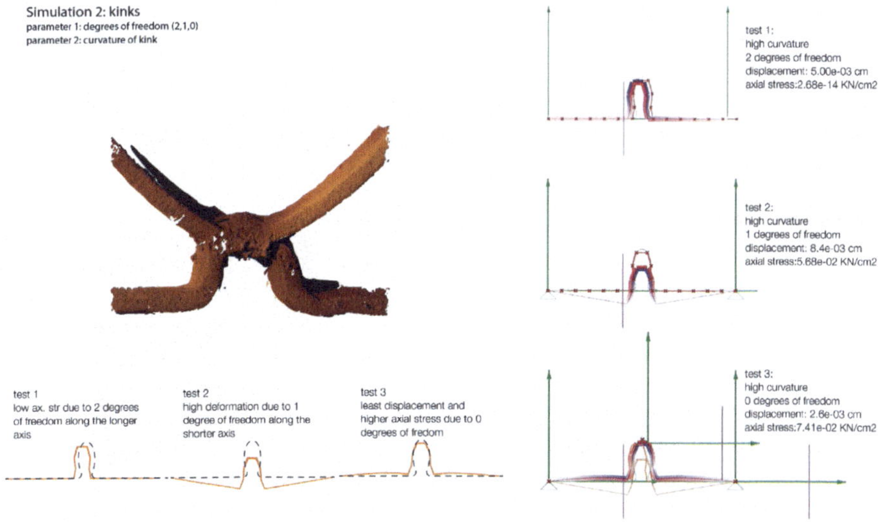

Fig. 3. Structural performance simulation and physical tests—shrinkage

Thermal Performance

The thermodynamic properties of clay have been of the most significant interest in this study. Further to optimizing the thermal inertia of the material itself through experimenting with the material composition, a great effort was also put in using the inner and outer geometry of the wall to receive, store and appropriately dissipate heat. Again, several different topological and geometrical strategies were implemented to assess their different effect on the wall's performance. The focus on the wall's infill geometry was towards creating a heat sink, through a modular branching logic that would allow the gradual flux of heat from the outer to the inner part of the wall, to optimize the storage of thermal energy during the day and its dissipation to the inside during the night. The outer surface of the wall's geometry, on the other hand, was optimized for solar radiation performance during the winter and summer periods. The two strategies combined create a performative wall, whose inner and outer geometries are optimized to receive, store and dissipate internally the heat of the environment, according the specific needs of the site's climate.

To optimize the outer surface of the wall for solar radiation, simulations were run at all stages of development using Ladybug, an add-on that interfaces Radiance to Grasshopper. At each development stage, the different geometry and topology strategies of the outer surface that were informed by the structural analysis were assessed through solar radiation tests and their geometrical and topological features were used to minimize and maximize the summer and winter radiation, respectively. The final conical diagrid geometry of the outer surface of the wall was informed by these simulations, by finding a compromise for the diagrid' s offset inclination that supports both the structural overhang and the optimum solar incident angle. Using Ladybug for the solar radiation simulations was quite useful, as it was both accurate enough for the purpose of the study as well as integrated in our computational design platform. The solar studies performed allowed us to maximize the total incident radiation by more than 4 times in the winter and minimize it by 70% in the summer. Further to the surface modulation, a comprehensive solar radiation study was also done to optimize the overall geometry of the final wall prototype for the specific site of construction. The final shape of the wall was strategically positioned on the north-south axis to enable greater control of the surface modulation (Fig. 4).

Ta assess and optimize the thermal performance of the infill of the wall, thermal conduction simulations as well as physical experiments with a custom developed

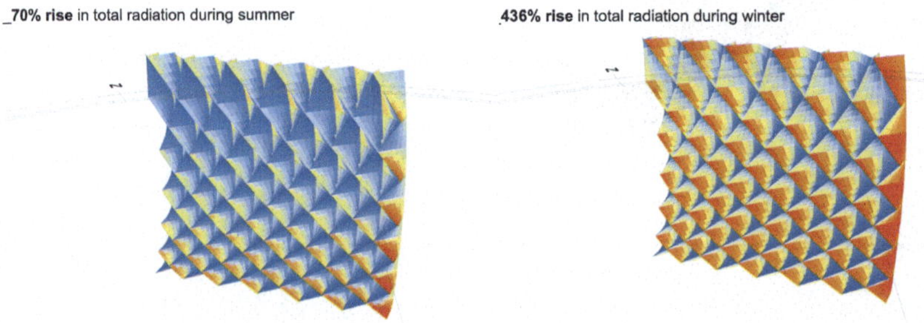

Fig. 4. Solar radiation analysis

apparatus were performed. Initially, heat transfer simulations were done using Energy2D, a visual multi-physics simulation tool, which helped establish basic principles for the infill geometry. However, as the material mix used has not been characterized and thus its heat transfer coefficients are unknown and also due to the overall complexity of accurately simulating the heat transfer of the wall, physical tests were used to measure the performance of different wall modules physically. A custom apparatus of two compartments with a heat lamp was used and the temperature flux was recorded with two sensors. A comparison with a standard wall geometry of the same amount of material showed a significant performance benefit of the optimized infill geometry driving the project to further investigation of its thermal properties. The heat sink geometry strategy demonstrated both a significant temperature drop (10 °C) as well as, most importantly, a significant reduction on the heat transfer time, which is in our case, most useful for the day–night heat cycle objective. Finally, a thermal camera was also used to assess both the solar radiation and self-shading performance of both the inner and outer surfaces of the wall at all stages of the wall development (Figs. 5 and 6).

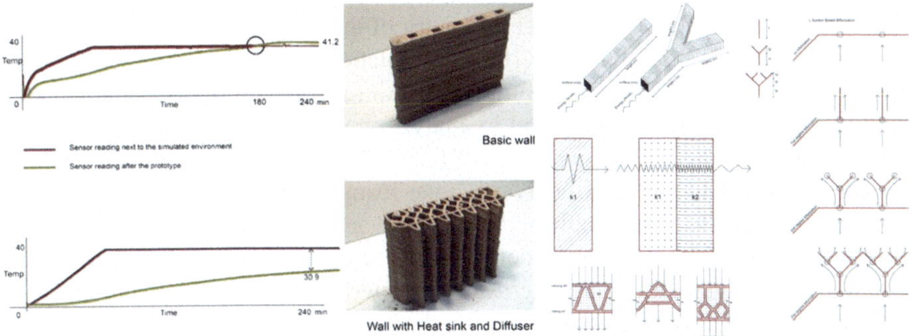

Fig. 5. Thermal strategy. Comparison with standard wall (*left*) and branching strategy (*right*)

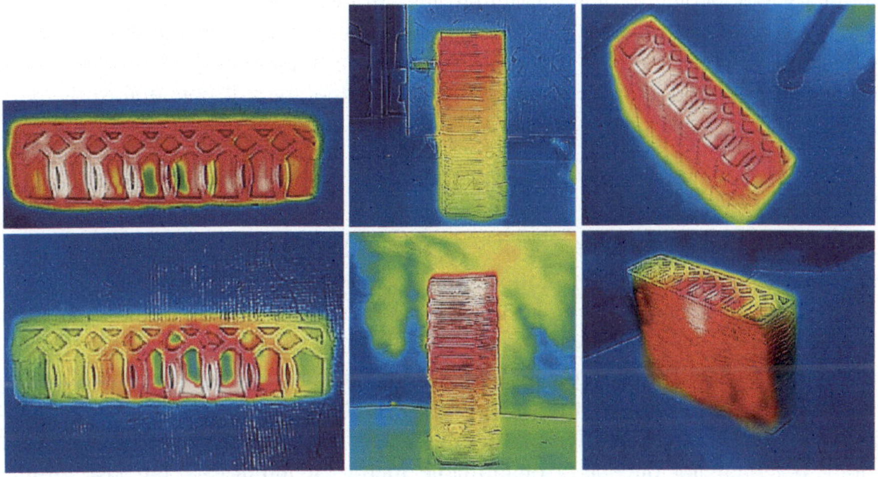

Fig. 6. Thermal imaging of the wall infill

While the precision of the digital and physical thermal analysis tools used has not always been as accurate as intended, due to both resources and complexity of the project, they nevertheless were proven to be very valuable in the performative design workflow, by giving comparative data between each design iteration and allowing the team to get insight on the performance solution space.

Ventilation Performance

Equally important aspect of the environmental performance of the wall has been its ventilation capacity. The ventilation goal was twofold, as it was first and foremost an important environmental driver but also important in the natural drying process of clay. In terms of its environmental performance, the ventilation strategies focused on creating both channeled openings and micro perforations throughout the inner geometry of the wall, which would allow a natural ventilation of the enclosed space. Taking advantage of both the Bernoulli effect but also the inherent thermal properties of the material, airflow channels were designed to drag the air through diminishing openings through the wall and channeling it through the colder parts of the infill to cool the air.

To assess the potential of the inner geometry of the wall in driving and channeling the air, computational fluid dynamics (CFD) simulations were performed on a number of different geometries. The CFD simulations were done to firstly understand the feasibility of the proposed strategy and to analyze the effect of the wall's geometry on the pressure field of the air on the surface of the wall and identify optimal positions for the openings. The CFD simulations were done using RhinoCFD, a plugin that integrates PHOENICS CFD, an established CFD software with Rhino. The initial studies, despite their simplifications, clearly demonstrated a differentiation effect on the pressure field caused by the different patterns of the external surface of the wall and thus a significant potential for air channeling through the wall. CFD simulations with small openings on the wall also demonstrated strong drafts within the air channels, although these need further work to be validated (Fig. 7). The ventilation studies performed were preliminary and inconclusive at this initial stage of the research, due to the complexity of the given task, and thus the final prototype that was erected on site did not include any openings, as the goal is to first assess the thermal properties of the material itself. However, further work is currently underway to explore the full potential of the wall's cavities

Daylight Performance

The daylight performance of the wall was also addressed, both through openings as well as again through micro perforations on the external surface of the wall. A strategy of interior light channeling through the wall was taken to minimize the solar gains but keep the light levels high. To assess the daylighting performance of the different strategies, a series of digital simulations were done using Ladybug for Grasshopper, as well as a physical test, using the robotic arm coupled with a dimmable spot light to emulate the different sun positions. The robotic arm test allowed the assessment of the daylight potential not only as a quantifiable metric of luminosity but also from an aesthetic point of view of the ambience of the light created. Generally, although the

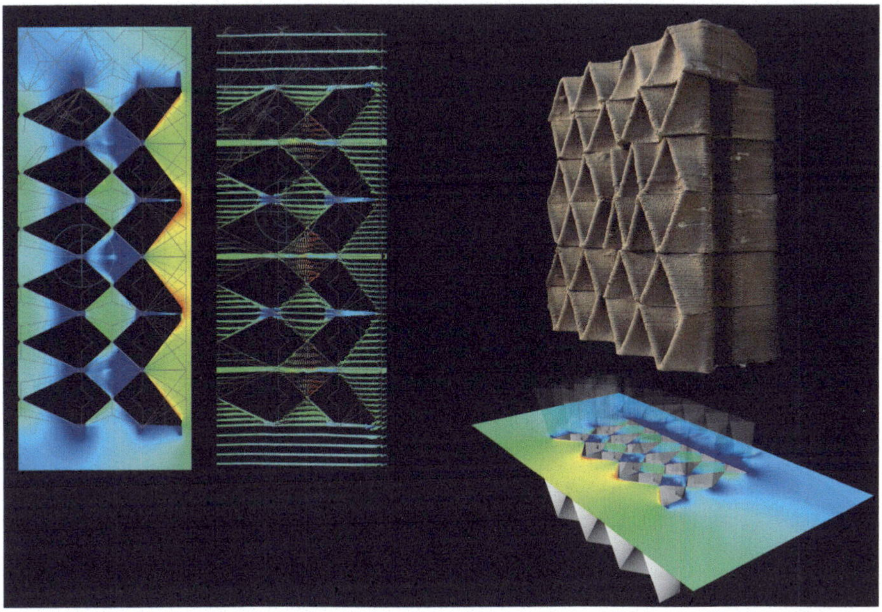

Fig. 7. CFD analysis of potential wall openings

daylight studies were an important part of the overall study they are not considered significant for the thermodynamic performance of the wall and there were also not included in the final erected prototype, thus they are outside the scope of this paper.

Full Scale Wall Prototype

As a conclusion for this first phase of the research, a prototype of a 2.85 m long, 0.35 m thick wall has been fabricated and assembled at the Valldaura campus of the Institute for Advanced Architecture of Catalonia. It is composed of 55 modules, parametrically designed to optimize the solar radiation, airflow and for structural performance. The wall's geometry has been conceived as a gradient in both horizontal and vertical directions, with various levels of self-shading to optimize according to the easterly and westerly sun. A number of sensors have also been put in place to measure the thermal performance and the ventilation potential of the wall. This first prototype has been envisioned as an initial stepping stone of this research and further work is currently being done to further investigate the potential of this approach. Nevertheless, this first prototype has been an important driver for research and the study associated with it has drawn important conclusions on the material composition, the geometry optimization and the environmental potential of the approach (Fig. 8).

Fig. 8. Full scale prototype erected on site

Conclusions

While this investigation is still on-going, the initial results presented in this paper demonstrate both the potential as well as the inherent limitations of our approach. More specifically, key design limitations, which are dictated by the layered additive manufacturing process of a slow curing material have been identified and used as a driver for further research. Design workflow limitations have also been identified in the currently available simulation tools integrated in computational design software, urging the need to develop new simulation approaches. The issue of resolution and accuracy of the simulation models in adequately informing design decisions for a geometrically complex additive manufacturing process is also in need for further investigation. Furthermore, the potential use of the simulation models for design optimization, which relies on the agility of these models is also an important point.

Such optimization potential lies, for example, on the design of the geometry of the exterior, its heat gain performance on the infill pattern for its thermal conductivity and storage and on the placement of openings for thermal convection and ventilation. Another important aspect of the optimization process proposed is the combination of digital simulation and fast physical prototyping with performance testing of 1:10 wall element prototypes using simple testing apparati and low-cost sensors.

Although this project is still at an early stage, the present study demonstrates the potential of combining 3D printing of mud with performance analysis and simulation for the optimization of a 3D printed wall. The study identifies important pitfalls in the design and performance assessment workflow and serves as a fruitful first step towards the further development aiming to reintroduce mud as a sustainable material for contemporary digital fabrication techniques.

Acknowledgements. This project has received funding from the European Union's Horizon 2020 research and innovation programme under the Marie Sklodowska-Curie Grant Agreement No. 642877.

TerraPerforma is a project of IAAC, Institute of Advanced Architecture Catalonia, developed within the Open Thesis Fabrication program in 2016–2017.

Program Directors: Edouard Cabay, Alexandre Dubor.

Research Advisors: Areti Markopoulou, Angelos Chronis, Sofoklis Giannakopoulos, Manja Van De Warp, Mathilde Marengo, Grégoire Durrens, Djordje Stanojevic, Rodrigo Aguirre, Kunaljit Singh Chadha, Ji Won Jun, Ángel Muñoz, Wilfredo Carazas Aedo, Josep Perelló, Pierre-Elie Herve, Jean-Baptiste Izard, Jonathan Minchi.

Researchers: Sameera Chukkappali, Iason Giraud, Abdullah Ibrahim, Raaghav Chentur Naagendran, Lidia Ratoi, Lili Tayefi, Tanuj Thomas.

References

Friedman, J., Kim, H. & Mesa, O., Woven Clay. In: ACADIA 14, Design Agency—Projects of the 34th Annual Conference of the Association for Computer Aided Design in Architecture, pp. 223–226. Los Angeles. (2014)

Giannakopoulos, S.: Pylos. http://pylos.iaac.net/main.html. Accessed 7 July 2017

Jackson, A.: Makers: the new industrial revolution. J. Des. Hist. **27**(3), 311–312 (2014)

Kawiti, D., Schnabel, M.A.: Indigenous parametricism—material computation. Living systems and micro-utopias: towards continuous designing. In: Proceedings of the 21st International Conference on Computer-Aided Architectural Design Research in Asia (CAADRIA 2016), pp. 63–72. Melbourne (2016)

Kwon, H.: Experimentation and analysis of contour crafting (cc) process using uncured ceramic materials. University of Southern California, Los Angeles (2002)

Ice Formwork for Ultra-High Performance Concrete: Simulation of Ice Melting Deformations

Vasily Sitnikov(✉)

KTH Royal Institute of Technology, 100 44, Stockholm, Sweden
vasily.sitnikov@arch.kth.se

Abstract. This research project asks how can we efficiently solve structural problems with much smaller masses of concrete while maintaining the same level of cement consumption, through the use of Ultra-High Performance Concrete (UHPC), degradable ice formwork and through an active involvement of digital simulation. A survey undertaken in the recent advancements of concrete-related technologies has revealed an opportunity to propose a new concept of casting. The new method of fabrication of UHPC elements employs ice as the main material of the temporary formwork construction, involves automation of the fabrication process and solves the problems of waste material and manual labor at the stage of demolding. This casting method is being developed for cases, where unique and customized concrete elements of complex geometry are needed, e.g. topology-optimized structural elements. To impose a desirable geometry onto ice, the concept considers the use of 2.5D CNC milling operations. To minimize machining time, milling is thought to be combined with controllable melting of ice that allows to achieve high-quality surface finish on a large scale in a very simple way. The main hypothesis of the research is, that the control over melting could be possible through digital simulation, that is the melting deformations can be pre-calculated if parameters of the material system are known.

Keywords: Ultra-high performance concrete · Formwork · Simulation

Context—Fabrication of Complex Geometry in Concrete

A Strategy for Efficient Use of Ultra-High Performance Concrete in Architectural Constructions

The contemporary industry of prefabricated structural concrete elements aims at minimizing its environmental and economic cost. For this purpose, the industry employs a strategy of reducing the share of cement in concrete, decreasing it down to the minimum of approx. 10% (Fig. 1a). This approach results in a relatively weak concrete, leading to thicker profiles of floor slabs, beams and columns. The increasing mass of these concrete elements negatively affects the total embedded energy caused by transportation and handling, increasing the final CO_2 footprint. The current research assesses alternative strategies of improving the overall performance of the industry.

K. De Rycke et al., *Humanizing Digital Reality*,
https://doi.org/10.1007/978-981-10-6611-5_34

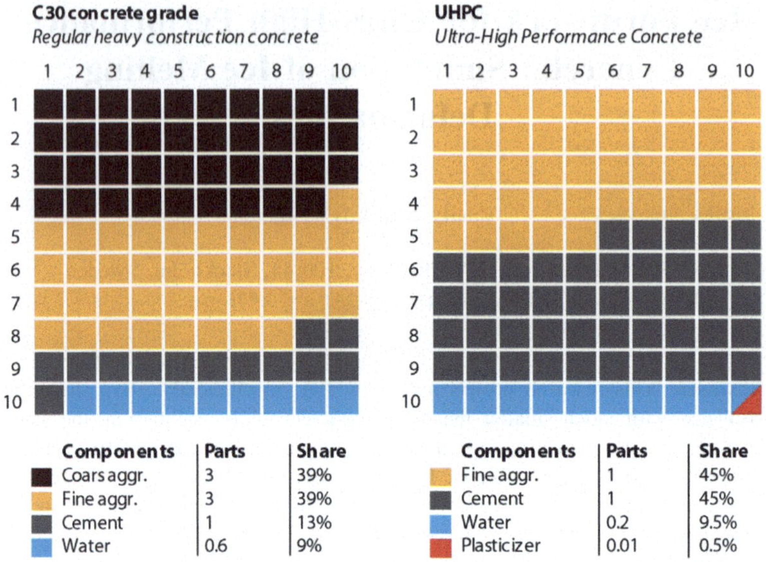

Fig. 1. Comparison of mix design of C30 concrete grade and UHPC

Counter-intuitively, a possible solution is to boost mechanical properties of concrete by increasing the content of cement. Higher material performance allows us to solve structural problems with much smaller masses of concrete, maintaining the same level of cement consumption per structure. This strategy is based on the use of Ultra-High Performance Concrete (UHPC), where cement constitutes approximately 40% of the bulk mass (Fig. 1b). While the chemical composition of UHPC is almost identical with regular concrete, the mechanical properties are simply incomparable. For instance, some types of UHPC gain compressive strength of 100 MPa in twenty-four hours of hardening (Fig. 1c), while regular concrete grades can barely reach half of this strength in a month time (Sitnikov and Sitnikov 2017).

The radically improved material properties of UHPC entail new concepts in the field of structural design. For instance, Phillipe Block claims that the new material can save up to 70% of the mass of concrete constructions, if structural geometry is redefined according to the improved material properties (López et al. 2014). However, contemporary computational methods of structural optimization generate complex doubly-curved geometry which is very uneconomical to produce using existing industrial methods. In many cases, the only fabrication method is based on CNC-milled EPS formwork. Considering the high amount of non-degradable waste caused by this type of production, and the shortage of opportunities to reuse these molds, the specificity of design needed for UHPC concrete arguably needs another system for mold production.

A Strategy for Minimizing Formwork-Related Material Waste

Driven by the intention to eliminate the material waste in the process of production of concrete structures, several researchers have proposed constructing formwork out of fully reusable materials, adopting the format of "closed loop recycling". Following this trace of thought, Mainka et al. (2016) and Oesterle et al. (2012) have developed strategies for producing concrete constructions using formwork made from industrial wax. Considering that wax can be CNC-milled and can be fully recycled at a relatively low cost, the casting technology was assumed to have a straightforward implementation in the building industry. However, wax is an extremely sensitive material with variable softness that reacts dynamically to changes in temperature. This lack of stability makes it difficult to use for on-site concrete casting.

But the employment of an easily melting, reusable material for construction of a temporary formwork is evidently a good choice for sustainable and economic fabrication strategy of bespoke concrete elements. To use such a process actively involves a new technological parameter—temperature—that so far has been used rather passively by the concrete industry. In these circumstances, a viable way to reduce risks of fabrication is to conduct concrete casting procedures in a controlled environment, that is prefabrication of concrete elements in an equipped factory, followed by transportation of this elements on construction site. Historically, the form of prefabrication has proved its efficiency in countries where climate conditions does not allow in situ casting during the winter season. However, when talking about 70% lighter elements of UHPC, the expenses on transportation are getting substantially reduced, offering new cases for efficient implementation of prefabrication.

Arriving to the Concept of Ice-Formwork

Control over temperature throughout the casting process enables even broader range of materials that can constitute a temporary container for concrete of a specific geometry. In this case even ice can be thought of as a main material for the formwork, providing not only a wasteless, but an ultimately clean process of fabrication. Given that the energy consumed by the refrigeration system can come from renewable sources, the concept has a potential to be fully sustainable.

To anticipate results of the implementation of this proposal in practice, the current research seeks to look into all the effective changes that fundamentally reorganize design and fabrication workflows, and reconfigure the economy of the process, starting with the re-reading of industrial refrigerator as a new kind of kiln.

Research in the Material Science of Concrete

The concept of the ice formwork for concrete builds upon the assumption that, apart from the increased mechanical properties, the chemical design of UHPC entails other qualities, which distinguish it from conventional concrete grades. Therefore, the research attempts to exploit the side effects caused by the extremely low water-cement ratio possessed by UHPC, which offers drastically different ways of processing it.

(a)

Results of Initial Testing

Compressive and Tensile Strength Gain (MPa)

	Days:	1d.	3d.	7d.	*8d.*	14d.
Reference samples	Compr.	69.2	85.2	105.5	-	116.3
	Tension	5.4	9.1	10.7	-	12.1
Examined samples	Compr.	5.8	11.7	17.5	52.4	79.8
	Tension	0.8	1.9	2.9	9.1	11.3

(b)

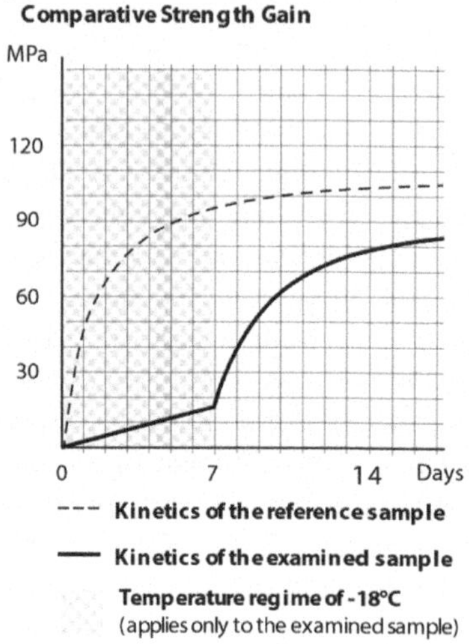

Fig. 2. Kinetics of sub-freezing hydration of UHPC (**a** *left*, strength tests results of reference and examined samples; **b** *right*, graphs illustrating kinetics of hydration)

First of all, the almost complete absence of unbounded water in UHPC mixes provides possibilities for hardening in sub-freezing temperatures. This aspect was investigated and proved with a series of material tests (Fig. 2a) during the first stage of the current research. With an addition of a very basic antifreeze admixture it was possible to conduct concrete setting and hardening in the temperature of −18 °C.

Second of all, experiments carried out in this research show that the temperature increase as UHPC hardens can be kept below 0 °C. The material studies identified that the amount of heat released due to the exothermal reaction of cement hydration is directly dependent on the initial temperature of concrete mix (Fig. 2b).

Above mentioned means that, if UHPC mix of a negative temperature is poured in a formwork made of ice, it will harden without melting the contact surface. This fact is illustrated in the Fig. 3, where very minor details that are present on ice surface of formwork remain on the surface of the negative concrete cast. These findings constitute the initial argumentation of the concept of ice formwork.

The Ice Machining Techniques—Imposing Geometry onto Ice

The Basic Types of Automated Methods of Processing Ice Shape

From the design perspective, the most important question is how to impose a desirable geometry onto ice. Assessment of this question has revealed a range of already established computer numerical controlled methods for processing ice.

The most established method of ice processing is the three-axe CNC-milling. Machines, designed specifically for ice milling, has been developed by many companies and are now available on the market in different sizes. Yet, they have no

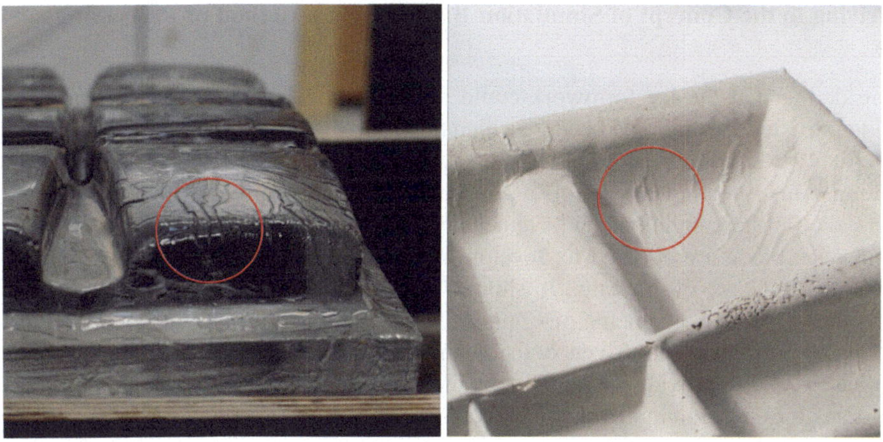

Fig. 3. Accuracy of the local geometry transfer and the surface finish quality. *Left* ice formwork, *right* negative cast in concrete

practical implementation, and are used only for advertisement and entertainment. This type of machining allows to make 2.5D reliefs of several decimeters in depth, that is reliefs without undercuts.

Next, abrasive wire-sawing has been used by Søndergaard and Feringa (2017) to produce large-scale elements of ice of a ruled-surface geometry for a temporary pavilion in Kiruna, Sweden. The wire saw actuated by an industrial robot exemplifies a possibility of machining ice on an architectural scale, producing elements that are directly implemented in architectural structures. However, this concept is limited by merely ruled geometry and therefore can't fully respond to the demand of structurally-optimized geometry.

Finally, several concepts for ice 3D printing have been proposed by different authors. Perhaps, the most well-established has been developed at McGill Center for Intelligent Machines. The PhD dissertation of Barnett (2012) delivers a robot-assisted rapid prototyping system for producing ice object of almost any geometry by incremental deposition and crystallization of water. While reaching the tolerance of 0.5 mm, the method requires parallel deposition of a supporting material (shortening methyl ester) that should be removed manually after the printing is finished. It is hard to overcome these obstacles while scaling the fabrication up for the intended architectural application. However, an alternative strategy has been proposed by Peter Novikov from Asmbld, the architectural technology consultancy. A project of an Ice 3D Printer offers a method of aggregating ice through direct deposition of overcooled purified water, utilizing the effect of immediate crystallization. While being a very promising tool, it is still on a very early stage of development.

Due to the objectives of this research it was considered that CNC-milling is the best fitting method in terms of speed, precision, scale range and formal freedom.

Arriving to the Concept of Simulation-Based Hybrid Method of Processing Ice

Considering that the ice formwork could be most efficient in cases when producing unique structurally optimized concrete elements, and therefore to be for single use only, it is highly desirable to minimized as much as possible the time-consuming process of milling. This intention has led to a concept of a simulation-based method of hybrid fabrication, where minimal and elementary milling interventions in a volume of a regular ice block are allowed to develop into a more complex geometry through the following process of melting. It should be mentioned that the intention to link computer-generated geometry of optimized structures to the natural phenomenon of melting deformations is driven both by the necessity to minimize the effort during fabrication and to provide a natural materialization strategy, that does not solely rely on the transitory machining equipment, but aspires to involve new perpetual physical processes in the culture of fabrication.

Conducted laboratory experiments (Figs. 4, 5 and 6) helped to identify conditions that reduce complexity and provide homogenous, incremental melting that is quite

Fig. 4. Melting test #1—a negative cast of the surface developed by projecting a heated water jet onto the surface of ice. Full documentation of the fabrication process: goo.gl/FPIMW8

Fig. 5. Melting probe #2—a negative cast of the surface molten by static contact with warm water

suitable for geometrical simulation without excessive challenges. The chosen condition is illustrated with the Fig. 6 and consist of a static interaction between ice and saline water solution of equal negative temperature.

Underlying Physical Principles of Ice Melting in Saline Water Solution

The physical phenomenon that underlies the above-mentioned formation process happens on the molecular level and depends on quite a few factors. When a block of ice is submerged in a saline water of the same negative temperature, the randomly walking molecules of liquid start to collide with the static molecules of ice, held by its crystalline lattice. At every collision of a salt molecule with ice, one of the water molecules leave the crystalline lattice, changing its state from solid to liquid. Therefore, the interface between ice and water is constantly retreating. Important to notice that the molecules located on the flat areas are open only from 180°, and those located on edges of the block are open from 270°. That is free-swimming molecules will collide with the edges more frequently than with molecules located on the flat sides, resulting in disproportional deformation of the initial volume and development of doubly-curved

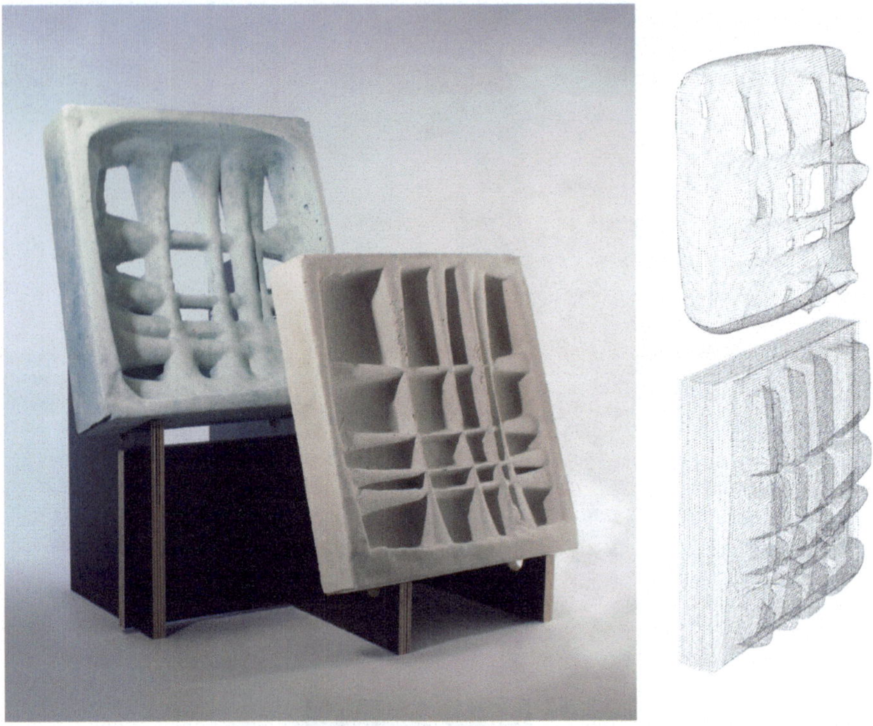

Fig. 6. Melting probes #3—a negative casts of the geometry of the ice surface created through submerging ice in a saline solution of sub-zero temperature. When the geometry of ice was imprinted in concrete it was digitally recovered through 3D scan of the negative concrete casts for further analysis

surface. Depending only on the concentration of salt and on the level of constant temperature, the same deformations will progress faster or slower.

Even though this physical condition of melting largely avoids ambiguity of bulk flow caused by thermal convection and turbulence, it still involves aspects of diffusion and flux due to the dilution of saline solution with fresh water that separates from ice. In fact, in the process of constantly decreasing concentration of salt, the self-contained system of saline solution, ice and constant negative temperature can reach a state of equilibrium, so that deformation to the interface would no longer occur (Kim and Yethiraj 2008). Therefore, it is necessary to consider changes in concentration that are conditioned by the diffusion coefficient of salt in water at a given temperature, and by the flux caused by difference in local densities.

A Preliminary Framework for Iterative Digital Simulation of Melting Deformations

In the recent decades, there has been proposed a number of strategies for computer simulations of ice melting. It was oceanology that has first raised the mathematical problem of the moving boundary (Stefan problem) for numerical simulation of sea ice melting. On the other hand, molecular simulations used in chemistry allow precise modelling of almost any molecular reaction, including water phase change. However, the only discipline that has been interested in melting of life-sized lump of ice is the industry of visual animation. Since the scientific credibility is not the objective of this discipline, the epistemological value of melting animations is quite small. Nevertheless, contemporary methods of animation encompass aspects of thermal transfer and fluid dynamics on a quite advanced level, and can contribute to the intention of this research.

In terms of geometrical modelling, animations either use level-set or particle-based algorithms. The implementation of particle-based method is advisable when the intention is to capture the deformations caused by fluid dynamics, such as deformations of water flowing down the surface of ice (Iwasaki et al. 2010). However, in cases, when melting is caused not by mechanical, but rather chemical factors, the level-set algorithms seem to be a better choice (Wojtan et al. 2007).

The intended simulation, therefore, is based on the representation of polygon mesh with marching cubes algorithm (Lorensen and Cline 1987). The initial surface condition that separates liquid and solid phases (that is the initial geometry of ice) is used to assign the initial decimal values between 0 and 1 to corners that belong to ice or saline solution correspondingly. These values will be recalculated at every iteration according to the diffusion level of salt particles at a given negative temperature. The flux caused by the difference in local densities should also be taken into account. It is obvious, that the total sum of all the corners' values in this closed system should be constant throughout the simulated period. At every iteration, the polygon mesh that represents the interface will be transformed according to the new distribution of values, using linear interpolation to separate vertices with values <0.5 (ice) from those of >0.5 (saline solution). Due to the fact that changes of corner values are conditioned by changes in the neighboring corners, in principal it can be computed using cellular automata algorithm.

To evaluate the accuracy of simulation and estimate its tolerance, the additional laboratory tests will be conducted. By comparing the spatial data calculated through simulation of melting with the data gained through 3D-scanning of actual material samples, the level of correspondence between digital model and physical process can be identified. A simple way to scan deformations of ice melting is to imprint its geometry in concrete, that is to make a negative cast. The feasibility of this method has been already tested. As a well-functioning methodological tool, it can either serve to improve and calibrate the simulation.

Conclusion

The current stage of the research does not allow to draw an overall conclusion, since a lot of aspects lack certified documentation and need to be further examined. However, the crucial aspects have been validated with principal empirical tests. It is, therefore, possible to claim that concrete casting in ice, as well as fabrication of ice formwork through melting deformations, is viable and potentially efficient (Figs. 7, 8 and 9).

Fig. 7. Melting test #3—the ice formwork

Fig. 8. Melting test #1—the process of producing ice formwork (full documentation: goo.gl/FPIMW8)

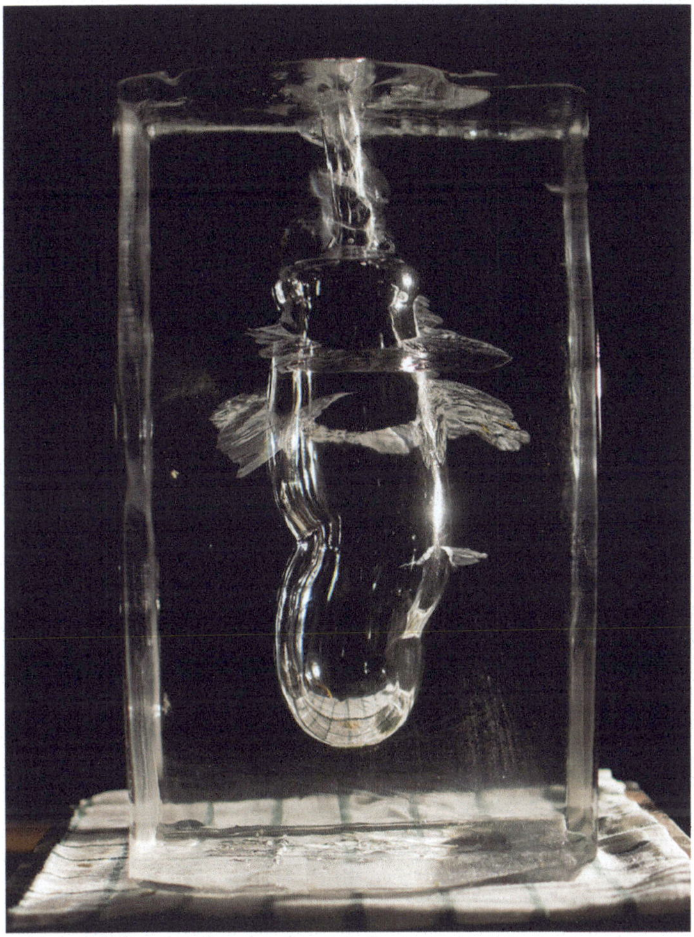

Fig. 9. Melting test #1—the ready formwork (full documentation: goo.gl/FPIMW8)

The research targets on fabrication of a full-scale structural element out of UHPC. The process of its fabrication is thought to be the major experiment of this research and the main epistemological source for evaluation of the concept, revealing all the hidden aspects that can't be anticipated beforehand. However, the studies that are separately undertaken in material science of concrete and physics of phase transfer can, to a certain extent, ensure a positive result and reduce risks related to the increased scale of technological operations.

Acknowledgements. This project is a part of the Innochain Research Training Network. It has received funding from the European Union's Horizon 2020 research and innovation programme under the Marie Sklodowska-Curie Grant Agreement No 642877.

References

Barnett, E.: The Design of an Integrated System for Rapid Prototyping with Ice. McGill University, Montreal (2012)

Iwasaki, K., Uchida, H., Dobashi, Y., Nishita, Y.: Fast particle-based visual simulation of ice melting. J. Compil. Pac. Graph. **29**(7), 2215–2223 (2010)

Kim, J.S., Yethiraj, A.: The effect of salt on the melting of ice: a molecular dynamics simulation study. J. Chem. Phys. **129**(12), 124504 (2008)

López, D.L., Veenendaal, D., Akbarzadeh, M., Block, P.: Prototype of an ultra-thin, concrete vaulted floor system. In: Proceedings of the IASS-SLTE Symposium, Brasilia (2014)

Lorensen, W.E., Cline, H.E.: Marching cubes: a high resolution 3D surface construction algorithm. In: Proceedings of the Annual Conference of the ACM (1987)

Mainka, J., Kloft, H., Baron, S., Hoffmeister, H.-W., Dröder, K.: Non-Waste-Wachsschalungen: Neuartige Präyisionsschalungen aus Recycelbaren Inustriewachsen. Beton- und Stahlbetonbau **111**(12), 784–793 (2016)

Oesterle, S., Vansteenkiste, A., Mirjan, A.: Zero waste free-form formwork. In: Proceedings of Second International Conference on Flexible Formwork, pp. 258–267. Bath (2012)

Sitnikov, I., Sitnikov, V.: Production of thin-walled unreinforced elements out of UHPC. Concr. Plant Int. **17**(2), 46–50 (2017)

Søndergaard, A., Feringa J.: Scaling architectural robotics: construction of the Kirk Kapital headquarters. In: Proceedings of Fabricate 2017, pp. 264–271. UCL Press, London (2017)

Wojtan, C., Carlson, M., Mucha, P.J., Turk G.: Animating corrosion and erosion. In: Ebert, D., Mérillou, S. (eds.) Eurographics Workshop on Natural Phenomena 2017, Prague (2007)

Toward a Spatial Model for Outdoor Thermal Comfort

Evan Greenberg[1(✉)], Anna Mavrogianni[1], and Sean Hanna[2]

[1] Bartlett School of Architecture, UCL, 22 Gordon Street, London WC1H 0QB, UK
{evan.greenberg,a.mavrogianni}@ucl.ac.uk
[2] UCL IEDE, Central House, 14 Upper Woburn Place, London WC1H 0NN, UK
s.hanna@ucl.ac.uk

Abstract. This paper presents the methods used and initial data analysis implemented toward the development of a spatial-architectural model for outdoor thermal comfort. Variables have been identified from existing thermal comfort models and qualified relative to their spatiality while data from a subset of these has been collected for preliminary statistical modelling, used to develop an initial understanding of the significance of spatial variables on thermal comfort. Although the specific analyses undertaken thus far are only introductory and do not provide definitive conclusions on their significance, the current study begins to bring to light the possibility that some explicitly spatial variables may explain enough variation in outdoor thermal comfort to justify the creation of a thermal comfort model for architectural design. The benefits of this approach on urban design and urban analysis is discussed.

Keywords: Thermal comfort · Spatial analysis · Regression modelling

Introduction

It has been suggested that microclimatic variation in the urban environment can promote the increased use and creation of programmed space which contributes to the social and cultural life of the city (Greenberg and Jeronimidis 2013) while increasing the thermal comfort range (Steemers et al. 2004). However, determining comfort in a dynamic urban environment is not a trivial task. Over the past 100 years or so, hundreds of indices have been developed in order to arrive at the calculation of a single value of thermal comfort, focusing primarily on the six basic human thermal comfort parameters. From initial field studies using a kata thermometer (Hill et al. 1916) undertaken in the early part of the 20th century to the most contemporary computationally-driven indices, such as the Universal Thermal Climate Index (Jendritzky et al. 2001), this quantification of thermal comfort is largely driven through the understanding of the environment, human physiology and behavior. Both the Modified Temperature Index (Adamenko and Khairullin 1972) and the Actual Sensation Vote (Nikolopoulou 2004) address the relationship between a model of thermal comfort and space—and specifically, the outdoor urban environment—but ultimately consider these

K. De Rycke et al., *Humanizing Digital Reality*,
https://doi.org/10.1007/978-981-10-6611-5_35

same environmental physiological and behavioral parameters which cannot necessarily be controlled through architectural design. If architects can only design space, rather than the climate of a space or the physiology of a human within a space, it is crucial to develop an understanding of the spatial variables affecting thermal comfort. This paper presents the methods used and initial data analysis implemented toward the development of a spatial-architectural model for outdoor thermal comfort and discusses the implications of such a model on architectural design. The methods and preliminary analysis presented in this paper are part of a larger research investigating the extent of significance of these spatial variables on thermal comfort.

Methods

Identification of Parameters and Parameter Classes

While this research ultimately aims to propose a model for outdoor thermal comfort, a review of 64 indoor and outdoor thermal comfort indices has been conducted in order to gain a broad view of the variables included within the models and to identify those which are spatial in character. Previous cataloguing of thermal comfort indices undertaken by de Freitas and Grigorieva (2015), Blazejczyk et al. (2012), Auliciems and Szokolay (1997) and Parsons (2014) were used as primary sources of identification.

In order to categorize variables considered in each index, six variable classes have been identified as physiological, behavioral, environmental, psychological, positional and demographical. Physiological variables are those that relate to the body functions of the individual human body. Behavioral variables are those actions which are carried out by the individual. Environmental variables are associated with the physical conditions and surroundings. Psychological variables[1] are those that are personal, subjective or related to history, memory or thought processes of the individual. Positional variables concern the location or climatic context at which the index was calculated or aimed. Finally, demographical variables belong to those characteristics of a specific population type for which the index was calculated or aimed.

Data Recording

The study of variables within the thermal comfort indices explored led to the identification of 96 distinct variables. In order to determine the significance of the variables on thermal comfort, an analysis of all independent variables was required. Therefore, literature on each of the 64 thermal comfort indices identified has been reviewed fully

[1] Psychological variables included here, aside from the variable decision, are based on the definitions put forward by Nikolopoulou and Steemers (Nikolopoulou and Steemers 2003). Naturalness is defined as the existence of naturally-produced conditions; time is defined as the length of time exposed to conditions in a space; environmental stimulation is considered as the presence of variable climatic conditions; perceived control is the perceived ability to change location; short-term experience is the memory or history of exposure to immediately previous conditions; decision is considered as the determined selection of a specific environment, as defined by the authors.

in order to appropriate values for each variable across all indices. A number of assumptions have been made within this method in order to reduce the number of independent variables. For instance, any indoor environment is assumed to be without direct solar radiation. A school environment, for example, will likely have students sitting at their desks which would be abstracted through an appropriate metabolic rate. In this regard, it would therefore be acceptable to eliminate an occupational variable. Similarly, the individual country which has been identified as a specific variable has been omitted; this information can be accounted for through psychological parameters, due to the fact that the overarching cultural significance to thermal comfort is acclimatization, and acclimatization can be considered within variables such as time (time spent within a space) and long-term experience (experience built over numerous visits to a space or over long periods of time). Furthermore, in an attempt to include as many data points as possible within the global set, assumptions have been made for those data points that were missing based on logical deduction. For example, where air humidity or air pressure values have been omitted within a data set in the literature, a logical assumption has been made using published weather data sets (TMY3 or IWEC, for example) for those specific geographical locations in order to arrive at an approximate value for these variables.

After recording data collected during original field experiments described in the literature and logically assumed, the matrix of indices and variables was edited in order to arrive at a data set with a large portion of recorded data points, resulting in a set of 753 data points across 27 variables and 14 indices.

Preliminary Analysis of Thermal Comfort Data

It is beneficial to perform preliminary analysis on the data set, in order to develop an understanding of patterns and relationships between each independent variable and the dependent variable. In doing so, a scatter plot was generated for each of the 24 continuous independent variables related to thermal comfort. While three ordinal variables have been identified, they have been omitted from this study at present due to the complexity in their analysis, but will be included in future studies. The Pearson's Correlation Coefficient, or r, was calculated for each independent variable relative to the dependent variable, and ranked (Fig. 1) for the purposes of comparison. Due to the large variation in data sets from different thermal comfort indices, a specific judgement on how to interpret the correlation coefficient has not been made. However, it can be noted that five variables have a correlation coefficient above ± 0.6, four variables between ± 0.4 and ± 0.59, while the remaining 15 variables showed low to no correlation.

Single Variable Linear Regression and Best-Fit Curves

Further to calculating the Pearson's Correlation Coefficient for each continuous independent variable and thermal comfort, the coefficient of determination or R^2, was calculated, and the best-fit curve and its equation was found for those variables with R^2 values greater than 0.50 (Fig. 2). While r gives the correlation and positive-negative

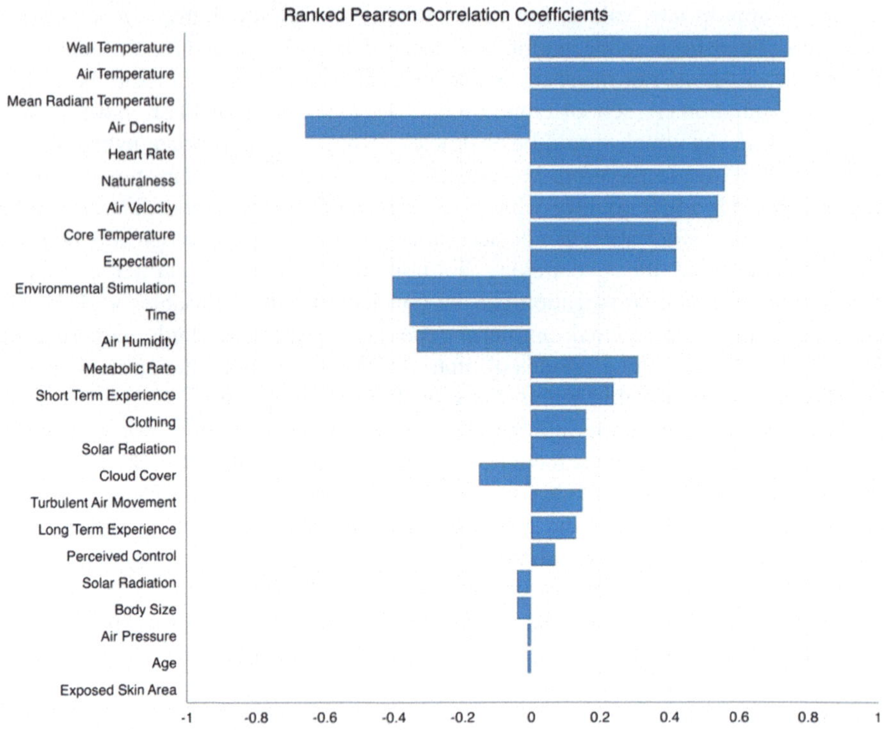

Fig. 1. Ranked Pearson correlation coefficients of independent variables effecting thermal comfort. Image courtesy of authors

relationship between an independent and its dependent variable, R^2 returns the percentage of data that can be explained by the best-fit curve (Moore and McCabe 2006).

Only three variables (wall temperature, air temperature and mean radiant temperature) have R^2 values over 0.5, meaning that approximately 50% of the variation in the predicted thermal comfort levels can be explained by the variation in one of these independent variables alone. Four independent variables have R^2 values between 0.2 and 0.5, and five additional variables have R^2 values between 0.1 and 0.2. Twelve remaining variables have R^2 values below 0.1. All continuous independent variables have been ranked (Fig. 3) for the purposes of comparison.

There is no doubt that the above quantitative analysis holds great value; at the same time, however, it is important to view the data qualitatively in order to discern patterns and potential non-linear relationships that may not accurately explain the data (Anscombe 1973; Matejka and Fitzmaurice 2017). For this reason, each scatter plot has been assessed visually. For a number of scatter plots, it has been found that specific indices caused outliers to change the coefficient of determination and the resulting best-fit curve. In these instances, data from these indices have been removed and the coefficient of determination recalculated. An altered ranking of continuous independent variables following this visual analysis has been developed (Fig. 4) for the purposes of comparison.

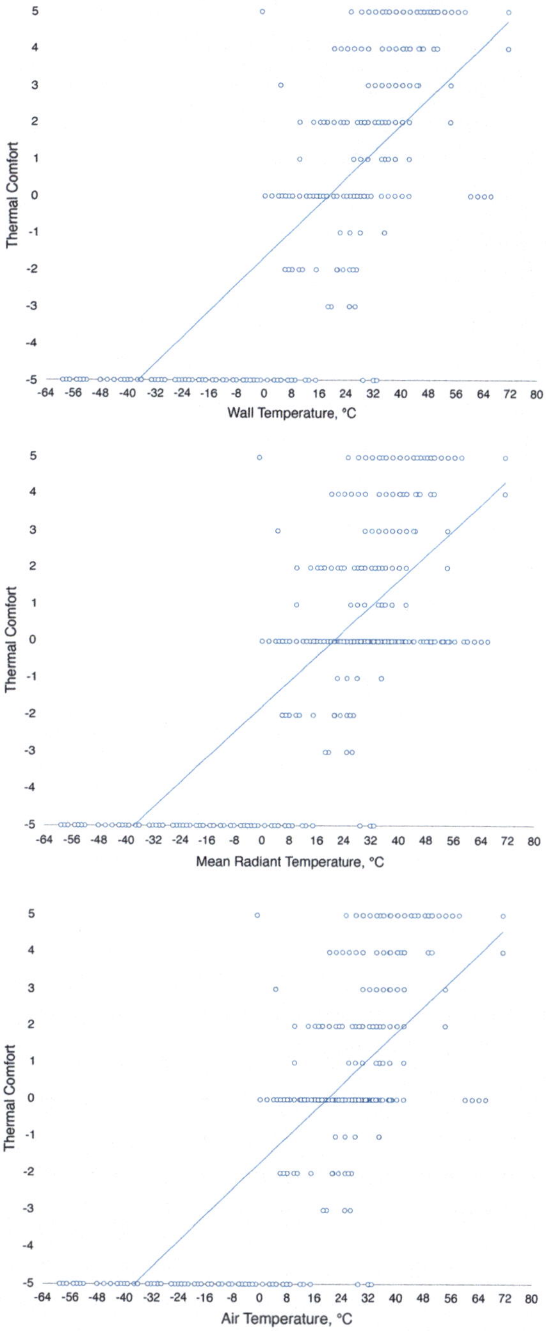

Fig. 2. Scatter plots for three continuous independent variables plotted against thermal comfort, with best-fit curves with R^2 values greater than 0.50. Image courtesy of authors

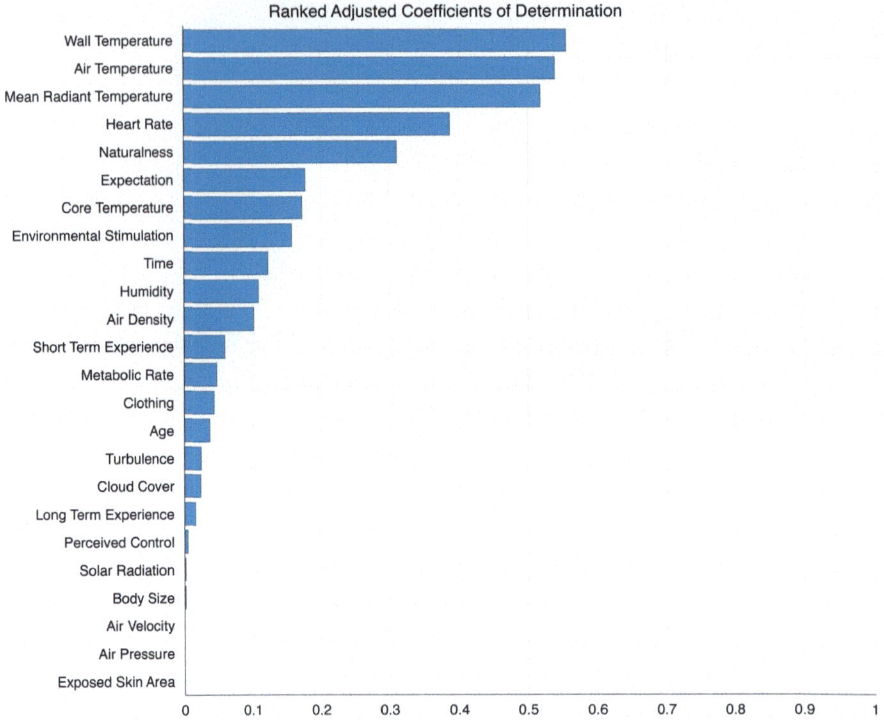

Fig. 3. Ranked coefficients of determination of independent variables effecting thermal comfort. Image courtesy of authors

Analysis of Parameter Spatiality

In making the distinction between those parameters that are spatial from those that are not, parameters that are at a scale larger than the space under examination, such as weather patterns, as well as those that are controlled by or linked to the individual have not been included. Because all indices identified do not take into account the same parameters, it is not possible to compare them through absolute terms. There fore, a system of relative spatiality (Fig. 5), as defined by orders of magnitude, has been defined in order to determine those parameters which are to be included in a proposed spatial model for thermal comfort. Those parameters which are inherently spatial are considered of the zeroth order; those which are directly affected by space are of the first order; those which are affected by first-order parameters are of the second order, and so on. Those parameters which are in no way affected by space are of an infinite order. Including all parameters which are in some way spatial (while omitting those which are of an infinite order) would result in a highly complicated model with numerous variables which could not be controlled architecturally. Therefore only those that are of a zeroth or first order will be considered as spatial within a new model for thermal comfort as they provide a real opportunity for spatial impact in the design process. It is important to mention that those parameters which have been identified previously as

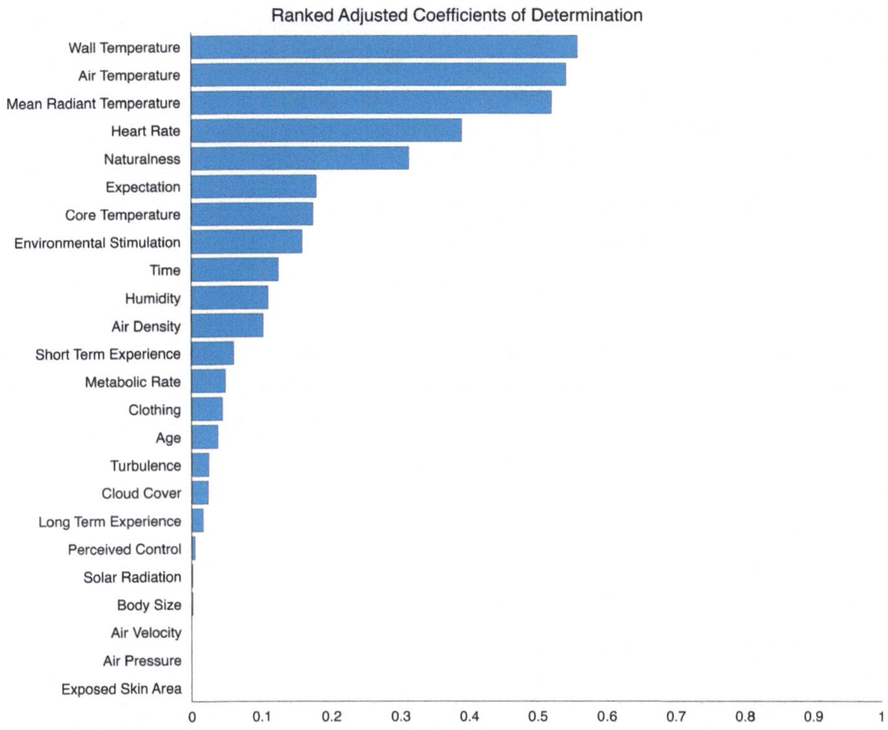

Fig. 4. Ranked adjusted coefficients of determination of independent variables effecting thermal comfort. Image courtesy of authors

discreet but are in fact specific instances of more generic parameters have been considered only in the generic. For example, metabolic rate, or activity, has been identified in various specific quantities such as light activities or moderate exercise, but in evaluating their spatiality are considered together as metabolic rate.

Determining the order of spatiality of all parameters results in the establishment of 14 spatial parameters whose significance toward thermal comfort can be considered. These parameters include: location (indoors or outdoors), naturalness, time, environmental stimulation, perceived control, short-term experience, decision, mean radiant temperature, wall temperature, solar radiation, air velocity, turbulent air movement, and metabolic rate of three parameter classes: environment, psychology, and behavior.

Discussion

Understanding the relationships between all independent variables and their effect on thermal comfort is crucial in developing a new model for thermal comfort. However, it is paramount to understand the particular significance of those variables that are spatial so that they may be leveraged by architects in the design of architectural space. The above statistical analysis has shown that the regression equation for the highest

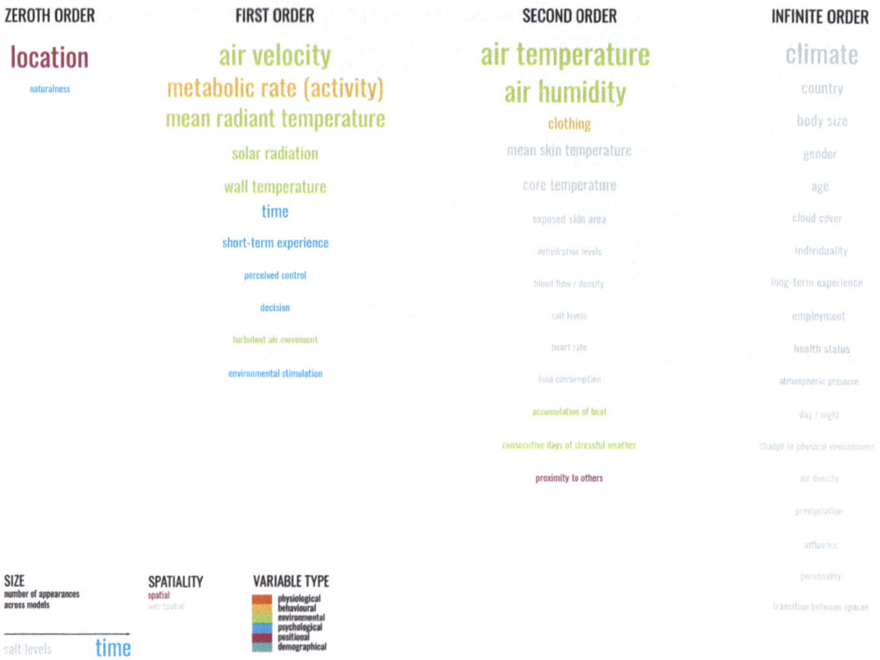

Fig. 5. Qualitative analysis of variable spatial order, outlining variable type, its relative occurrence across thermal comfort indices and its spatial order. Image courtesy of authors

significance of an independent variable (wall temperature) can account for only approximately 50% of all variation in the dependent variable (thermal comfort).

Although this may at first glance appear to be a very low significance, this is difficult to ascertain from this study alone. The collected data set includes data points from different models and therefore likely possesses hierarchical relationships that should be statistically considered in future.

While it is beneficial to approach each individual independent variable analytically, it is clear that thermal comfort cannot be explained by one independent variable alone, and the relationship between independent variables may better explain their significance on thermal comfort. A regression model will be developed in the next phase of this research in order to determine the independent variables which affect thermal comfort, and distinctly, the significance of those independent variables that are spatial. Particular emphasis will be made on exploring hierarchical statistical models in order to account for the numerous existing indices studied and the varied data that has been recorded.

This presents a noteworthy and consequential opportunity for designers to modify the conventional approach to design, where an independent variable, such as time or naturalness, for example, may hold similar or greater importance than traditional design variables such as morphology or orientation. While a definitive conclusion cannot be made from the specific analyses undertaken thus far, the current study begins to bring to light the possibility that some explicitly spatial variables may explain enough variation in outdoor thermal comfort to justify the creation of a thermal comfort model for architectural design.

Future Outlook

Thermal Comfort as an Urban Design Tool

This new spatial-architectural model for thermal comfort has the potential to be used as an urban design tool, capable of aiding in the design of outdoor spaces, their adjacencies and ultimately their programs. This presents a significant shift from current thermal comfort models, often used as verification tools or as a mechanism to inform users of how behavior, such as the increase or decrease of clothing layers, must be amended. In contrast, if it is assumed that spatial parameters such as wall and mean radiant temperatures as well as naturalness, for example, are statistically significant in determining thermal comfort, this provides architects an opportunity to leverage design decisions around these parameters. For example, the materiality and color of external surfaces of buildings and ground may hold particular importance, as the surface albedo would contribute directly to both wall temperature and mean radiant temperature. The materiality of a building system has a direct relationship to structure; in this case, it is not difficult to imagine that structural systems, intrinsically tied to architectural quality and morphology, could thus begin to connect closely to outdoor thermal comfort. Designers too could begin to develop novel approaches to solar access or wind flow, for instance, so that artificial lighting or ventilation is decreased and naturalness is amplified.

Thermal Comfort as an Urban Analysis Tool

Furthermore, this thermal comfort model can be compared to existing spatial analysis methods to develop an understanding of the ability for thermal comfort to predict spatial use. While the Space Syntax method, for example, employs topological and geometric analyses to determine spatial use (Al Sayed et al. 2014), the thermal comfort model sought in this research provides an opportunity to explore spatial analysis through architectural design. This alteration to an approach toward urban analysis may allow for an enhancement in empathic design in a field where increased quantitative design is often seen as lacking in human quality.

Conclusions

The research presented here demonstrates the methods and preliminary statistical analysis of a set of independent variables in determining thermal comfort. Pearson correlation coefficient values have been calculated and ranked in order to develop an understanding of the relative correlations of independent variables and thermal comfort. Coefficients of determination have also been calculated; while the greatest significance has only shown a 50% description for output data, a high proportion of independent variables have R^2 values well above the mean value; these findings begin to suggest that the relative significance of independent variables may hold importance, rather than specific absolute values. Only a multilevel regression analysis of all independent variables will allow for a real understanding of the significance of independent variables on thermal comfort. Nonetheless, a new approach to design may emerge from the utilization of a spatial model for thermal comfort, where design parameters that affect those variables have great importance and potentially begin to transform analysis-driven architectural design toward a humanistic perspective.

References

Adamenko, V., Khairullin, K.: Evaluation of conditions under which unprotected parts of the human body may freeze in urban air during winter. Bound-Layer Meteorol. **2**, 510–518 (1972)

Al Sayed, K., Turner, A., Hillier, B., Iida, S., Penn, A.: Space syntax methodology. Bartlett School of Architecture, UCL, London (2014)

Anscombe, F.J.: Graphs in statistical analysis. Am. Stat. **27**, 17–21 (1973)

Auliciems, A., Szokolay, S.V.: Thermal comfort. PLEA (1997)

Blazejczyk, K., Epstein, Y., Jendritzky, G., Staiger, H., Tinz, B.: Comparison of UTCI to selected thermal indices. Int. J. Biometeorol. **56**, 515–535 (2012)

De Freitas, C.R., Grigorieva, E.A.: A comprehensive catalogue and classification of human thermal climate indices. Int. J. Biometeorol. **59**, 109–120 (2015)

Greenberg, E., Jeronimidis, G.: Variation and distribution: forest patterns as a model for urban morphologies. Archit. Des. **83**(4) (2013). (London: Wiley)

Hill, L., Griffith, O., Flack, M.: The measurement of the rate of heat-loss at body temperature by convection, radiation and evaporation. Philos. Trans. R. Soc. Lond. Ser. B Contain. Pap. Biol. Charact. **207**, 183–220 (1916)

Jendritzky, G., Maarouf, A., Staiger, H. Looking for a Universal Thermal Climate Index (UTCI) for outdoor applications. In: Proceedings of the Windsor Conference on Thermal Standards, pp. 5–8 (2001)

Matejka, J., Fitzmaurice, G.: Same stats, different graphs: generating datasets with varied appearance and identical statistics through simulated annealing. In: Proceedings of the 2017 Chi Conference on Human Factors in Computing Systems. ACM, Denver, Colorado, USA (2017)

Moore, D.S., McCabe, G.P.: Introduction to the Practice of Statistics. W. H. Freeman and Company, New York (2006)

Nikolopoulou, M. Designing Open Spaces in the Urban Environment: A Bioclimatic Approach. Centre for Renewable Energy Sources, EESD, FP5 (2004)

Nikolopoulou, M., Steemers, K.: Thermal comfort and psychological adaptation as a guide for designing urban spaces. Energy Build. **35**, 95–101 (2003)

Parsons, K.: Human Thermal Environments: The Effects of Hot, Moderate, and Cold Environments on Human Health, Comfort, and Performance. CRC Press, Boca Raton (2014)

Steemers, K., Ramos, M., Sinou, M.: Urban diversity. In: Steemers, K., Steane, M.A. (eds.) Environmental Diversity in Architecture. Spon Press, Abingdon (2004)

Survey-Based Simulation of User Satisfaction for Generative Design in Architecture

Lorenzo Villaggi, James Stoddart, Danil Nagy(✉), and David Benjamin

The Living, an Autodesk Studio, New York, NY, USA
danil.nagy@autodesk.com

Abstract. This paper describes a novel humanist approach to generative design through the measurement and simulation of user satisfaction in an office environment. This technique involves surveying hundreds of employees for their adjacency preference and work style preference, and then calculating how well different floor plan layouts match these preferences. This approach offers an example of how design goals which are typically considered to be "qualitative"—and beyond measurement—might become part of a "quantitative" generative design workflow for architectural design and space planning. The methodology is demonstrated through an application in the design of a 49,000 square foot office space for 270 people.

Keywords: Modelling and design of data · Generative design · Simulation · User satisfaction · Survey methods

Introduction

Generative design integrates artificial intelligence into the design process by using metaheuristic search algorithms to discover novel and high-performing results within a given design system. Its framework is dependent on three main components: (1) a generative geometry model that defines a 'design space' of possible design solutions; (2) a series of measures or metrics that describe the objectives or goals of the design problem; (3) a metaheuristic search algorithm such as a genetic algorithm which can search through the design space to find a variety of high-performing design options based on the stated objectives. This paper is an extension and in-depth overview of the second component of the generative design workflow. In this work we focus on the development and use of a comprehensive system to automatically evaluate qualitative properties of office space designs based on user survey data.

A Humanist Design Approach

While architectural design is an extremely complex task composed of many competing small-scale requirements, the traditional design process avoids complexity and specificity through generalization. In fact, the standardization of anthropometric needs into idealized canons has a long history: from Da Vinci's *Vitruvian Man* and Le Corbusier's *Modulor* to *Dreyfuss*' Joe and Josephine ergonomic charts. More recently, Nicholas

K. De Rycke et al., *Humanizing Digital Reality*,
https://doi.org/10.1007/978-981-10-6611-5_36

Negroponte discussed the human designer's limitations in accommodating particular occupant-level requirements such as "single individual or family needs" (Negroponte 1969). To address these limitations, he argued for a humanist approach based on a collaboration between human designers and intelligent machines. This approach would allow designers to navigate complex problems by "treating pieces of information individually and in detail" (Negroponte 1969). Building upon this possibility, our work extends the capabilities of generative design to the inclusion of human desires and work pattern requirements with the objective of discovering design solutions that satisfy nuanced user preferences.

Measuring Performance

The application of generative design to engineering applications is well known. In this context, however, performance measures tend to be straight forward, dealing primarily with goals such as maximizing structural performance (by minimizing stress, structural displacement, etc.) while minimizing material use. In architectural design, however, measuring performance can be more difficult. Although many software tools exist for measuring a building's overall performance (such as structural analysis, air circulation and daylighting), these engineering-based measures are rarely the only or even the primary goals of a design project.

Even more important are occupant-level concerns such as how the space will be used, how the space feels, and how the layout of the space matches the needs of the program. Such measures are typically considered "qualitative" rather than "quantitative", and thus can be difficult to measure deterministically from a design model (Fig. 1). Because of this difficulty, there are few existing software tools or guidelines for measuring such occupant-level goals.

Fig. 1. Measuring occupant-level preference data in an office environment

Office Design and Productivity

The open-plan office has gained popularity among designers and enterprises worldwide (Konnikova 2017; Waber et al. 2014), as it is assumed to inspire collaboration and foster informal interaction within the office. While unplanned exchanges have been proven to boost productivity (Pentland 2012), many employees find the noise and lack of privacy caused by open layouts to negatively impact their personal performance (Kim and de Dear 2013). In fact, several studies have shown that open workspaces actually decrease overall user satisfaction and do not lead directly to productivity gains (Gensler: US Workplace Survey 2016; Brennan et al. 2002).

In a recent survey of UK and US workplaces, Gensler urges businesses to invest in environments that match work processes and are attentive to user needs (Gensler: US Workplace Survey 2016). Haynes and Duffy further stress the importance of designing layouts from the occupants' perspective (Haynes 2007a; Duffy 1992). Despite the potential benefits of such an approach, it can be difficult for designers to manage the complexity involved with satisfying many users' needs simultaneously. This is especially true in contemporary workplace settings where not only human desires but also work patterns can vary dramatically across different job roles (Fig. 2).

Quantification of Preference

To incorporate occupant-level measures in the generative design process for office space planning, we developed a new method that simulates future occupancy based on surveys of future occupants. These surveys ask each user about their preferences within the office space, both in terms of *spatial* and *work style* preference.

Spatial preference involves asking each occupant to select and prioritize all the people, teams, equipment, and amenities that they want to be near. Work style preference involves asking each occupant to indicate their preferred ambient conditions (quite/loud space, bright/dark space, etc.) (Fig. 2). Based on the survey results we encode this data in our model through a series of metrics which become objectives or constraints to guide the generative design process. These occupant-level metrics are:

1. *Adjacency*, which directly measures the travel distance from each employee to their preferred neighbors and amenities.
2. *Work style*, which measures the suitability of an office area's ambient environment and distraction levels to the assigned team's surveyed preferences.

To derive the work style metric, we also need calculate four additional measures:

- *Distraction*, which measures the amount and distribution of high-activity zones to determine noise and distraction levels throughout the office.
- *Productivity*, which measures concentration levels at individual desks based on sight lines to other desks and other noise sources.
- *Daylight*, which measures the amount of natural daylight at each work space throughout the year.
- *Views to outside*, which measures each work space's visual access to an exterior window.

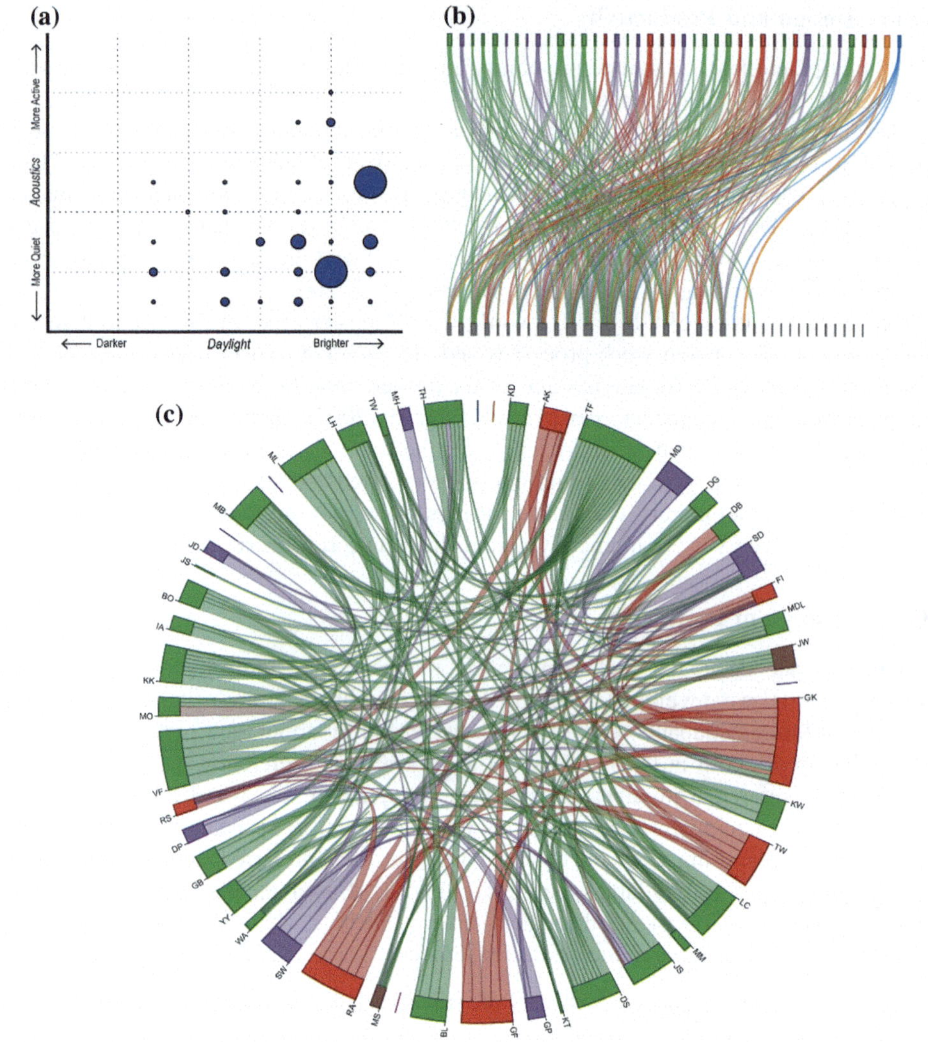

Fig. 2. Visualization of user work style preference based on survey data: **a** ambient preference, **b** team to amenity adjacency, **c** team to team adjacency

By integrating these two human-level metrics into a generative design workflow, we can automate the process of discovering high-performing office layouts that maximize its suitability to the individual preferences of its future occupants.

Related Work

A comprehensive review by Calixto and Celani of 22 years of applications of evolutionary algorithms to problems in space planning reveals a general focus on the efficacy and efficiency of the optimization algorithms in generating satisfactory results with little or no discussion about the metrics by which the designs should be evaluated (Calixto and Celani 2015). Researchers have tended towards known computationally difficult problems such as facility assignment without considering human-level metrics such as preference or feeling which are crucial to the success of a spatial layout.

To broaden the applications of generative design tools to real-world space planning problems, our work extends this method to occupant-level measures that automatically simulate the qualitative needs of the future occupants of the space. To ground the work, this section discusses related research in the quantification of occupant-level measures and their impact on workspace occupants' satisfaction, well-being, and mood.

Adjacency

Adjacency requirements have been widely used as an indispensable evaluation criteria in space allocation problems. Krejcirik (1969) uses adjacency requirements between rooms in a semi-automated space planning method. Muther (1973) defines "relationship charts" composed of weighted proximity relationships between rooms to inform layout designs. Krarup and Pruzan (1978) advance a model for layout adjacency evaluation. In a similar fashion, Liggett discusses the use of graph techniques to generate layouts to meet adjacency requirements (Liggett 2000), while Arvin and House (2002) and more recently Helme et al. (2014) use adjacency as a topological constraint in computerized space planning tools.

Work Style Preference

Recent work in user satisfaction in office spaces reveals that the lack of correspondence between workspace design and worker desires is one of the major causes in drops of employee productivity (Gensler: US Workplace Survey 2016; Haynes 2007a; Duffy 1992; Haynes 2007c; Stallworth and Kleiner 1996; Becker and Steele 1995). Though the importance of such measures for workspace design is well understood, examples of its actual quantification and implementation in design methods have been limited.

To address the need for such methods we propose a novel work style metric which calculates the correspondence between work place environment and stated preferences in terms of daylight, views, and distraction levels. We chose these criteria based on literature about the factors which most often contribute to workplace satisfaction.

Productivity and Distraction. Research about the relationship between workspace layouts and employee satisfaction is primarily focused on the measure of user productivity (Leblebici 2012; Haynes 2008; Uzee 1999; Leaman and Bordass 1993). Haynes, however, explains how productivity remains ill-defined and its quantification lacks standard guidelines (Leblebici 2012; Haynes 2007b). Nevertheless, the literature shows that there is sufficient evidence to claim that certain properties of the office environment do affect user productivity: the room layout and environmental comfort

(Leblebici 2012; Mak and Lui 2012; Mawson 2002; Oldham and Fried 1987), employee proximity (Clements-Croome and Baizhan 2000) and correspondence between work patterns and spatial configuration (Gensler: US Workplace Survey 2016; Haynes 2007a; Stallworth and Kleiner 1996; Mawson 2002). A major cause of productivity loss is distraction (Leblebici 2012; Haynes 2008), defined by Mawson as "anything that takes attention away from the task to be performed (like) noise and visual disturbance" (Mawson 2002).

Daylight and Views to Outside. While there is no consensus about the necessary amount of light for user's well-being (Veitch et al. 2004), the relationship between access to natural daylight and user satisfaction and health has been widely studied and the literature shows enough evidence to claim correlations between the two (Clements-Croome and Baizhan 2000; An et al. 2016; Mahbob et al. 2011; Veitch 2006; Wilkins 1993). Similarly, access to windows and views to the outdoors and its positive effects on user well-being has been extensively studied (Mahbob et al. 2011; Kim and Wineman 2005; Menzies and Wherrett 2005; Sims 2002).

Methodology

Our workflow for simulating user-based desires is divided into two primary steps: (1) the collection and structuring of data for use within the design model and for design validation, and (2) the design of a series of quantitative performance measures which can directly measure the performance of each design.

Data Gathering

The data acquisition process is structured to move from broad, human-interpretable data which is used to shape the design problem to specific, machine-readable data that is integrated directly into the generative design workflow. The result is a dataset with enough specificity to capture the complexity of user desires while being general enough to ensure completeness and offer some degree of flexibility in the final architecture.

Our method uses three stages of surveying to arrive at the final dataset: (1) early exploratory sessions to determine broad survey goals & constraints, (2) individual questionnaires of all 270 employees to identify the domain of user preferences, and (3) team-based questionnaires structured for direct operability with the design model and measures. While the initial two stages don't produce an actionable dataset for direct use within the design model, they were critical in ensuring that the final dataset and design model were correctly addressing user needs.

Early Survey: Focus Group. Early sessions were carried out within small focus groups to collect responses to a series of prompts regarding spatial concerns and workspace ambitions. The responses from these sessions were collected and distilled into broad goals that were used to explore initial geometric strategies and measurement methods.

Intermediate Survey: Individual Questionnaires. A follow-up anonymized individual questionnaire identified the domain of user needs, including typical daily usage of spaces, individual workspace needs, and ambient condition preferences along

					OPEN		ENCLOSED		PRIVATE			
Division	Team Manager	Team Size	Answered	% Answered	Hot Desk	Cubicle	Collab. Team Room	Quiet Team Room	Semi-Private Office	Mini-Private Office	Typ. Private Office	Pref. Strength
[illegible]	[illegible]	46	26	56.52%	5.96	3.56	4.62	4.15	3.15	3.35	3.19	
[illegible]	[illegible]	19	13	68.42%	6.15	4.00	3.92	3.85	3.23	3.62	3.23	41.76%
[illegible]	[illegible]	12	7	58.33%	5.86	3.29	5.43	4.57	3.43	2.71	2.71	44.90%
[illegible]	[illegible]	54	18	33.33%	5.72	3.56	4.61	4.00	3.67	2.78	2.67	
[illegible]	[illegible]	3	1	33.33%	7.00	3.00	4.00	1.00	6.00	2.00	5.00	85.71%
[illegible]	[illegible]	8	1	12.50%	7.00	1.00	3.00	4.00	6.00	5.00	2.00	85.71%
[illegible]	[illegible]	4	2	50.00%	6.00	2.50	6.50	5.50	3.50	2.00	2.00	57.14%
[illegible]	[illegible]	3	0	0.00%	0.00	0.00	0.00	0.00	0.00	0.00	0.00	N/A
[illegible]	[illegible]	5	2	40.00%	7.00	5.50	4.50	3.50	2.50	4.00	1.00	85.71%
[illegible]	[illegible]	6	0	0.00%	0.00	0.00	0.00	0.00	0.00	0.00	0.00	N/A
[illegible]	[illegible]	4	2	50.00%	7.00	3.00	6.00	4.00	3.00	2.00	3.00	71.43%
[illegible]	[illegible]	7	3	42.86%	7.00	5.67	3.00	3.00	4.00	3.00	2.33	66.67%
[illegible]	[illegible]	78	31	39.74%	6.13	3.87	3.23	3.39	3.77	3.81	4.00	
[illegible]	[illegible]	12	4	33.33%	7.00	4.25	2.75	3.25	3.25	3.50	4.00	60.71%
[illegible]	[illegible]	9	5	55.56%	5.80	2.20	3.80	3.60	4.00	4.20	4.40	51.43%
[illegible]	[illegible]	13	5	38.46%	6.00	4.80	3.40	4.40	3.60	3.20	2.60	48.57%
[illegible]	[illegible]	29	11	37.93%	5.73	5.09	3.55	2.91	3.55	3.18	4.00	40.26%
[illegible]	[illegible]	5	0	0.00%	0.00	0.00	0.00	0.00	0.00	0.00	0.00	N/A
[illegible]	[illegible]	6	4	66.67%	7.00	2.25	2.50	3.75	3.75	4.50	4.25	67.86%
[illegible]	[illegible]	22	10	45.45%	6.04	2.00	4.24	4.60	3.37	3.36	4.42	
[illegible]	[illegible]	5	2	40.00%	7.00	2.00	4.00	4.00	3.00	3.00	5.00	71.43%
[illegible]	[illegible]	2	0	0.00%	0.00	0.00	0.00	0.00	0.00	0.00	0.00	N/A
[illegible]	[illegible]	4	3	75.00%	5.33	2.00	4.33	5.00	3.00	3.67	4.67	47.62%
[illegible]	[illegible]	1	0	0.00%	0.00	0.00	0.00	0.00	0.00	0.00	0.00	N/A
[illegible]	[illegible]	1	0	0.00%	0.00	0.00	0.00	0.00	0.00	0.00	0.00	N/A
[illegible]	[illegible]	9	5	55.56%	5.80	2.00	4.40	4.80	4.00	3.40	3.60	54.29%
AVERAGE SCORE:					4.72	2.41	3.17	2.94	2.81	2.28	2.41	
PREFERENCE 1:					0 0.00%	7 35.00%	1 5.00%	2 10.00%	1 5.00%	6 30.00%	3 15.00%	
PREFERENCE 2:					0 0.00%	2 10.53%	1 5.26%	2 10.53%	3 15.79%	5 26.32%	6 31.58%	
PREFERENCE 3:					0 0.00%	4 21.05%	4 21.05%	2 10.53%	3 15.79%	3 15.79%	3 15.79%	
TOTALS:					0	13	6	6	7	14	12	

Fig. 3. Statistical analysis of surveyed data was used to understand general trends and ranges in user preference

with basic demographic data (Fig. 3). With a 70% response rate, this survey allowed us to identify areas of high variability across employees (in terms of amenity usage, ambient preferences and inter-team collaboration), as well as areas of high agreement (Fig. 2). These finding were used to refine the design model by constraining areas of low-variability and calibrating its combinatorial capabilities to capture the breadth of user needs.

Final Survey: Team-Based Data. For the final survey, an aggregation at the team level was deemed sufficient to reflect the diversity of user preferences while providing flexibility for future growth and restructuring. A questionnaire was supplied to each team leader and non-team-based individual to collect: (1) identification data, (2) environmental preferences for ambient light and activity levels, (3) amenity adjacency preferences, and (4) team adjacency preferences ((2), (3) and (4) were complemented by user weighting) (see Fig. 4). By taking this approach and reaching out directly to team leaders we were also able to obtain 100% participation in the survey. Data collected from the final survey was translated into an object-oriented structure, using standard JSON conventions, to allow for linkage to data specific geometric elements in the design model (see Fig. 1).

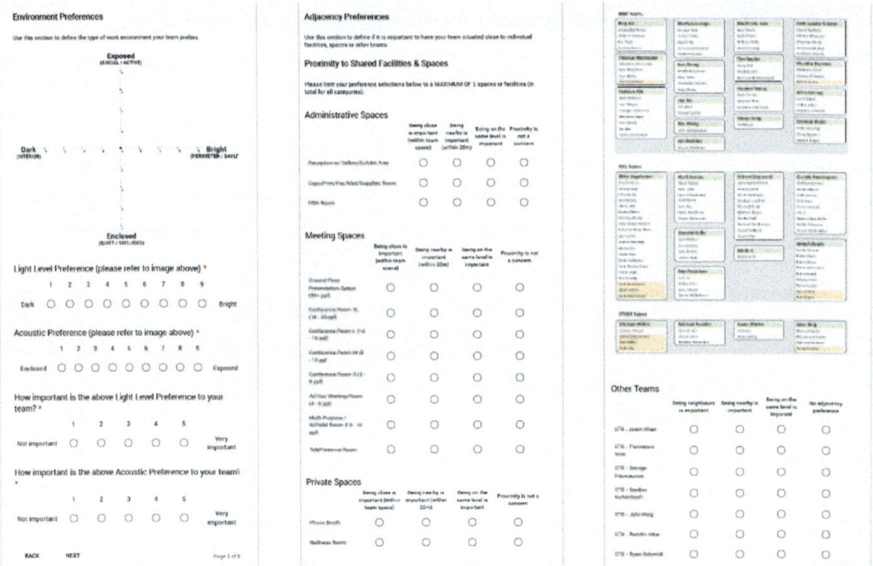

Fig. 4. Web-based questionnaire issued to users with descriptions and diagrams of answer ranges

Designing Measures

Based on the results of the survey, we can calculate two human-level metrics that evaluate the performance of a given office layout according to future occupant preferences (see Figs. 5 and 6).

Adjacency. The adjacency metric is directly calculated from the geometry of the plan by measuring the travel distance from each employee to their preferred neighbors and amenities. For each design iteration, a mesh-based traversal graph—represented as nodes and edges—is generated to describe all possible movement pathways within the

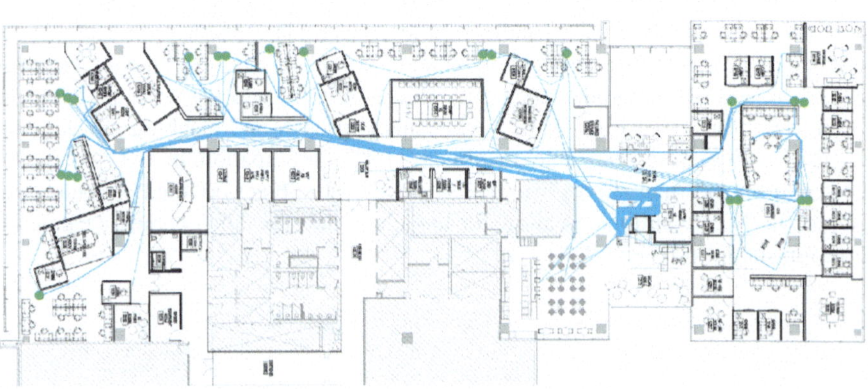

Fig. 5. Visualization of *adjacency metric* in a sample office design

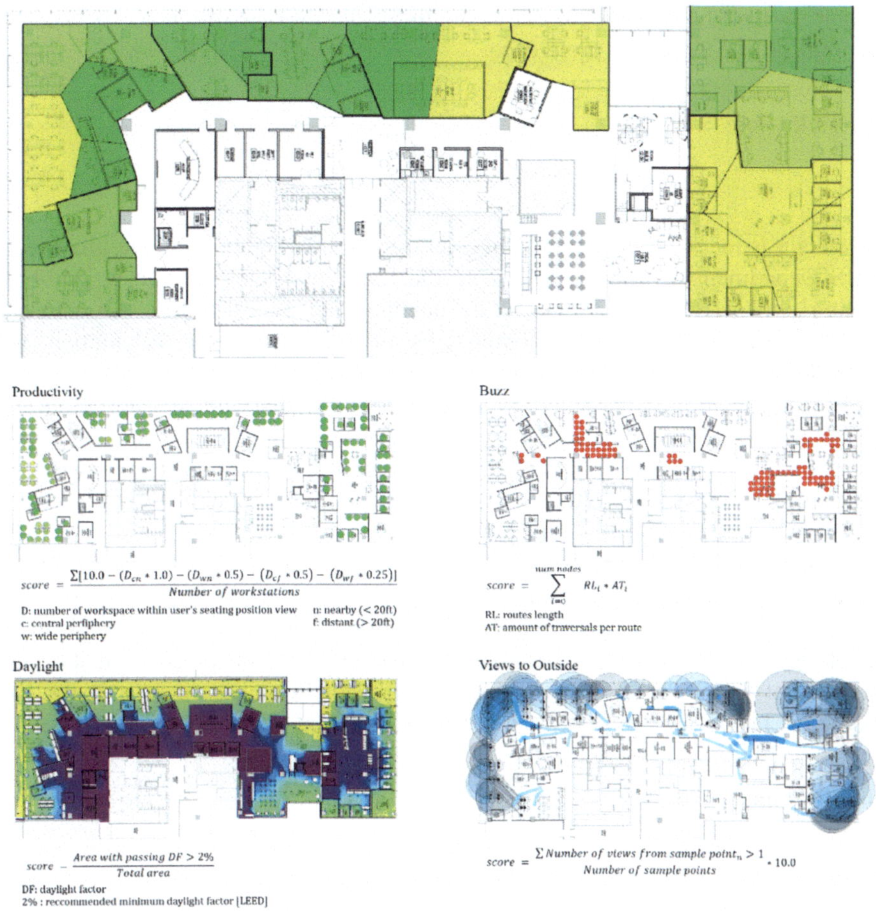

Fig. 6. Visualizations of *work style metric* and related sub-metrics

plan. Team desk and amenity positions are then assigned to nodes within the graph. The system iterates through each assigned team desk, calculates the shortest path to nodes associated with preferences, and sums a total distance weighted by designated importance. The objective of the optimization is to minimize this total travel distance.

$$\text{Adjacency score} = \frac{\sum [SPL * (1 + \Delta Floors * VM)]}{NSP} * 10.0 \tag{1}$$

SPL shortest path length
VM vertical multiplier
NSP number of shortest paths

Work Style. The work style metric calculates the suitability of daylight, view, and distraction measurements at each desk to the assigned team's surveyed preferences.

The objective of the optimization is to minimize the average difference between the measurement and the stated preference.

$$\text{Work style score} = \frac{Err_{daylight} * W_{daylight} + Err_{activity} * W_{activity} + Err_{views} * W_{views}}{W_{daylight} + W_{activity} + W_{views}} * 10.0 \quad (2)$$

Daylight is measured using Radiance, an open-source lighting engine provided by the US Department of Energy, following the Daylight Factor method at grid-based sampling points (Reinhart 2011). While daylight can be calculated using existing analysis tools, the other three components of the work style calculation were custom-designed and built directly into the generative design model.

Views to outside are measured by finding non-occluded lines-of-sight from each desk to any exterior window. *Buzz* measures the amount and distribution of high-activity zones by taking the routes calculated for the adjacency metric and aggregating them by nodes. This node activity is used to calculate the likely noise and distraction at each desk.

Productivity measures the concentration levels at individual desks based on sight lines to other desks and proximity to noise sources. Each desk is scored with a cumulative penalty of visual and auditory distractions. Nearby desks with unobstructed lines-of-sight are tabulated with graduated penalties based on position in the visual cone (central vs. periphery) and distance. Nearby high-traffic nodes from the Buzz measure are tabulated with graduated penalties based on distance.

Validation

To validate our metrics, we had to compare the results of our quantitative measures with the perception of actual occupants in a real office space. To do this we modelled the users' existing office space and scored it according to our six measures. For comparison, several volunteers from the user group were given high-level descriptions of the measures and asked to score the existing office based on their experience and judgment.

In the first analysis, five of the six measures had high similarity between the computed and perceived scores. However, we noticed a substantial discrepancy in the distraction measure. After further analysis we determined that the discrepancy resulted from the fact that our computed distraction measure was not accounting for the way that visible circulation zones and areas of congestion contribute to distraction. After revising our distraction measure to include these congestion areas, all six measures were found to match the intuitions of the users in their current office space.

Results

To use these metrics in the final office design, we first developed a generative space planning model which could create a large variety of valid floor plan layouts for the office. We then used a variant of the NSGA-II genetic algorithm (Deb et al. 2002) to

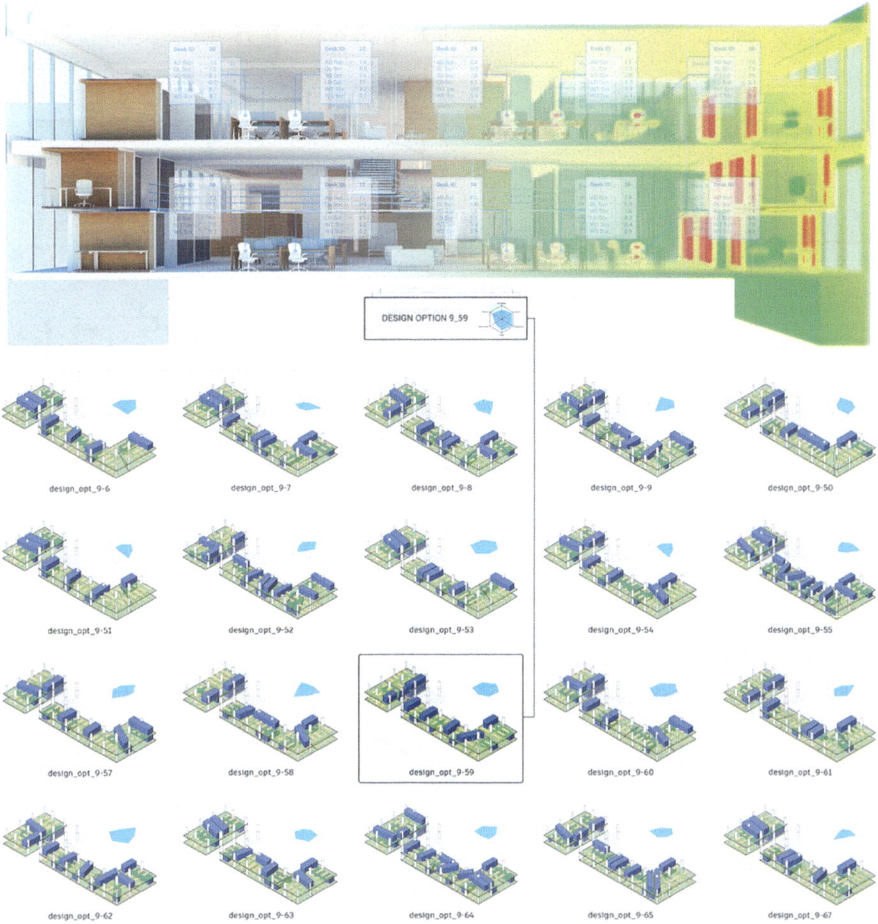

Fig. 7. Selection of high performing designs based on user-defined preference metrics and overlaying of preference data on 3D model

optimize the model and find a set of optimal designs according to the goals of the project.

Although a full description of the optimization process is beyond the scope of this paper, its result was a collection of high-performing schematic design solutions which could be further analyzed and used as a basis for the final design of the office space. To accommodate all the goals of the project the optimization also included several higher-level design goals beyond the occupant-level satisfaction metrics described here. However, by including these metrics we discovered some interesting design solutions which met the varying needs of the future inhabitants in some non-intuitive ways (see Fig. 7). For example, some of the designs featured "back-alley" connections between group work zones that increased adjacency between them. Many of the designs also used non-typical wall and room alignments to obscure sources of

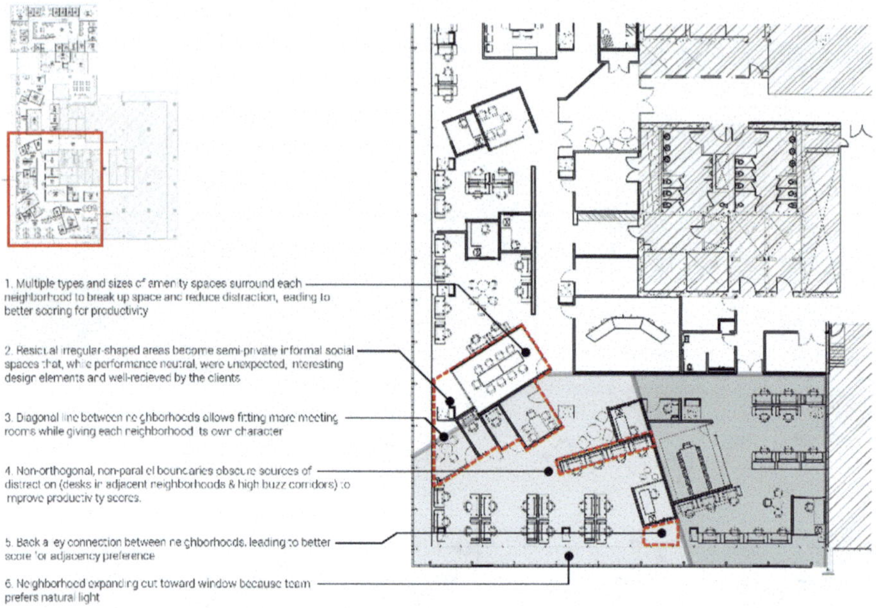

Fig. 8. Non-intuitive physical features of the selected design reflect high performing metrics and high correspondence between local spatial configurations and individual needs

distraction and create more daylight for groups which preferred a more open space and more isolation for groups that preferred more privacy (Fig. 8).

Conclusions

This paper demonstrates an extension of the capabilities of generative design to a more humanist design approach that can satisfy occupant desires in terms of spatial quality and organization. The paper focuses on the evaluative component of generative design for architectural space planning, and describes a series of novel metrics for automatically evaluating user satisfaction within workspaces. The paper also demonstrates an application of this method to the design of a real office space.

To further validate the measures described in this paper we plan to conduct post-occupancy evaluations of the office space once it is built. Through integrated sensors and ongoing digital surveys, data about user satisfaction will be continuously collected and used to further calibrate our measures. We also plan to re-apply the generative design system at later stages throughout the life of the office space to produce new solutions that accommodate changes in both user satisfaction and corporate organizational structure.

References

An, M., Colarelli, S.M., O'Brien, K., Boyajian, M.E.: Why we need more nature at work: effects of natural elements and sunlight on employee mental health and work attitudes. PLoS ONE **11**(5), e0155614 (2016)

Arvin, S.A., House, D.H.: Modeling architectural design objectives in physically based space planning. Autom. Constr. **11**(2), 213–225 (2002)

Becker, F.D., Steele, F.: Workplace by Design: Mapping the High-Performance Workscape. Jossey-Bass, Hoboken (1995)

Brennan, A., Chugh, J.S., Kline, T.: Traditional versus open office design: a longitudinal field study. Environ. Behav. **34**(3), 279–299 (2002)

Calixto, V., Celani, G.: A literature review for space planning optimization using an evolutionary algorithm approach: 1992-2014. Blucher Des. Proc. **2**(3), 662–671 (2015)

Clements-Croome, D. and Baizhan, L.: Productivity and indoor environment. In: Proceedings of Healthy Buildings, vol. 1, pp. 629–634 (2000)

Deb, K., Pratap, A., Agarwal, S., Meyarivan, T.A.M.T.: A fast and elitist multiobjective genetic algorithm: NSGA-II. IEEE Trans. Evol. Comput. **6**(2), 182–197 (2002)

Duffy, F.: The Changing Workplace. Phaidon Press, London (1992)

Gensler: US Workplace Survey 2016. https://www.gensler.com/research-insight/workplace-surveys/us. Last accessed 10 May 2017

Haynes, B.: Office productivity: a shift from cost reduction to human contribution. Facilities **25** (11/12), 452–462 (2007a)

Haynes, B.: An evaluation of office productivity measurement. J. Corp. Real Estate **9**(3), 144–155 (2007b)

Haynes, B.: Workplace Connectivity: A Study of its Impacts on Self-Assessed Productivity (No. eres2007-297). European Real Estate Society (ERES) (2007c)

Haynes, B.: The impact of office layout on productivity. J. Facil. Manage. **6**(3), 189–201 (2008)

Helme, L., Derix, C., Izaki, Å.: Spatial configuration: semi-automatic methods for layout generation in practice. J. Space Syntax **5**(1), 35–49 (2014)

Kim, J., de Dear, R.: Workspace satisfaction: the privacy-communication trade-off in open-plan offices. J. Environ. Psychol. **36**, 18–26 (2013)

Kim, J., Wineman, J.: Are windows and views really better. A quantitative analysis of the economic and psychological value of views. Lighting Research Center, Rensselaer Polytechnic Institute, New York (2005)

Konnikova, M.: The open office trap. The New Yorker. http://www.newyorker.com/business/currency/the-open-office-trap. Last accessed 20 Apr 2017

Krarup, J., Pruzan, P.M.: Computer-aided layout design. Math. Program. Use, 75–94 (1978)

Krejcirik, M.: Computer-aided plant layout. Comput. Aided Des. **2**(1), 7–19 (1969)

Leaman, A., Bordass, B.: Building design, complexity and manageability. Facilities **11**(9), 16–27 (1993)

Leblebici, D.: Impact of workplace quality on employee's productivity: case study of a bank in Turkey. J. Bus. Econ. Finance **1**(1), 38–49 (2012)

Liggett, R.S.: Automated facilities layout: past, present and future. Autom. Constr. **9**(2), 197–215 (2000)

Mahbob, N.S., Kamaruzzaman, S.N., Salleh, N., Sulaiman, R.: A correlation studies of indoor environmental quality (IEQ) towards productive workplace (2011)

Mak, C.M., Lui, Y.P.: The effect of sound on office productivity. Build. Serv. Eng. Res. Technol. **33**(3), 339–345 (2012)

Mawson, A.: The Workplace and Its Impact on Productivity. Advanced Workplace Associates, London (2002)

Menzies, G.F., Wherrett, J.R.: Windows in the workplace: examining issues of environmental sustainability and occupant comfort in the selection of multi-glazed windows. Energy Build. **37**(6), 623–630 (2005)

Muther, R.: Systematic layout planning. In: MIRP Books (Chapter 5), 4th edn., p. 14 (1973)

Negroponte, N.: Towards a humanism through machines. Technol. Rev. **71**(6), 44 (1969)

Oldham, G.R., Fried, Y.: Employee reactions to workspace characteristics. J. Appl. Psychol. **72** (1), 75 (1987)

Pentland, A.: The new science of building great teams. Harv. Bus. Rev. **90**(4), 60–69 (2012)

Reinhart, C.F.: Simulation based daylight performance predictions. In: Hensen, J., Lamberts, R. (eds.) Building Performance Simulation for Design and Operation. Taylor & Francis, Abingdon (2011)

Sims, L.: Enhance User Satisfaction, Performance with Daylighting. Research and development, pp. 109–110 (2002)

Stallworth Jr., O.E., Kleiner, B.H.: Recent developments in office design. Facilities **14**(1/2), 34–42 (1996)

Uzee, J.: The inclusive approach: creating a place where people want to work. Facil. Manage. J. Int. Facil. Manage. Assoc. **1999**, 26–30 (1999)

Veitch, J.: Lighting for high-quality workplaces. Creat. Product. Workplace **3**, 7 (2006)

Veitch, J., Van den Beld, G., Brainard, G., Roberts, J.E.: Ocular Lighting Effects on Human Physiology and Behaviour, p. 158. CIE Publication, Cambridge (2004)

Waber, B., Magnolfi, J., Lindsay, G.: Workspaces that move people. Harv. Bus. Rev. **92**(10), 68–77 (2014)

Wilkins, A.J.: Health and efficiency in lighting practice. Energy **18**(2), 123–129 (1993)

Navigating the Intangible Spatial-Data-Driven Design Modelling in Architecture

Jens Pedersen[1], Ryan Hughes[1(✉)], and Corneel Cannaerts[2]

[1] Aarhus School of Architecture, Nørreport 20, 8000 Aarhus, Denmark
rhu@aarch.dk
[2] KU Leuven Faculty of Architecture, Hoogstraat 51, 9000 Ghent, Belgium

Abstract. Over the past few decades, digital technologies have played an increasingly substantial role in how we relate to our environment. Through mobile devices, sensors, big data and ubiquitous computing amongst other technologies, our physical surroundings can be augmented with intangible data, suggesting that architects of the future could start to view the increasingly digitally saturated world around them as an information-rich environment (McCullough in Ambient commons: attention in the age of embodied information, The MIT Press, Cambridge, 2013). The adoption of computational design and digital fabrication processes in has given architects and designers the opportunity to design and fabricate architecture with unseen material performance and precision. However, attempts to combine these tools with methods for navigating and integrating the vast amount of available data into the design process have appeared slow and complicated. This research proposes a method for data capture and visualization which, despite its infancy, displays potentials for use in projects ranging in scale from the urban to the interior evaluation of existing buildings. The working research question is as follows: "How can we develop a near real-time data capture and visualization method, which can be used across multiple scales in various architectural design processes?"

Keywords: Data capture · Data visualization · Data modelling

Context

Contemporary design tools have become increasingly accessible and open, resulting in a highly flexible and customisable design environment (Davis and Peters 2013), enabling architects to interface directly with a wide variety of technologies, from simulation and analysis to robotics, microelectronics and sensing. This research project builds upon the Arduino open source electronics platform and employs the Grasshopper algorithmic modelling environment for Rhino, contributing to the democratisation of spatial sensing and data visualisation for architects. There have been several precedents in using sensors and data visualisation to map intangible layers within design environments with a focus towards implementation in landscape architecture (Fraguada et al. 2013; Melsom et al. 2012) or urban design (Offenhuber and Ratti 2014). Our project focuses on architectural design and integrates sensing and

K. De Rycke et al., *Humanizing Digital Reality*,
https://doi.org/10.1007/978-981-10-6611-5_37

mapping of intangible data from the scale of the site through to the building and the interior.

This research develops bespoke spatial data gathering hardware and explores computational design workflows that permit the integration of this data into the architectural design process.

The project has two primary goals:

- To develop low-cost, mobile sensing and data capturing units that are wearable and can be mounted to various manned or autonomous vehicles;
- To develop a computational toolset and workflows for the integration and digital representation of various data streams within an architectural design environment, allowing architects to navigate and manipulate these representations to:
  - Get a better understanding of the intangible spatial aspects and;
  - Inform computational design models with more complex datasets.

First Prototype

The first prototype of the sensing unit was developed during the Sense, Adapt, Create Field Station summer-school in Berlin in September 2016. Starting from the site and buildings of Teufelsberg, the workshop challenged students to reflect on our environment, consisting of several tangible and intangible layers and question how architecture relates to those fields. As part of this overall brief the students were introduced to electronics prototyping and several sensors. The summer school resulted in several field stations or probes that interacted with chosen fields of the site (Fig. 1).

The mobile sensing unit employed in the summer school was developed by the tutors of the sensing workshop, and logged soil humidity, carbon dioxide, ambient light, air humidity, temperature and wind speed levels as well as the orientation of the device. While this was useful within the framework of the workshop, it had some shortcomings: the geolocation was not integrated in the sensing unit, but consisted of an external smartphone, resulting in a difference in granularity between location and sensor data. The representation of the sensor data was thus limited to standard graphs represented in a spreadsheet (Ballestrem 2016).

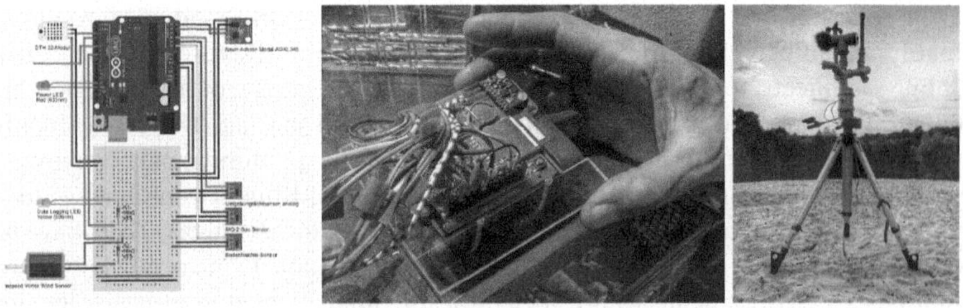

Fig. 1. Images and diagrams of the prototype Arduino-based mobile data logger for the field station project

The research was further developed into a second prototype at the Aarhus School of Architecture in 2017. Limitations of the first prototype were resolved, and several data capturing and visualization methods were further explored, as discussed in detail below.

Data Capture

The specific spatial conditions and needs of architectural design tasks place certain requirements on data capture methods that can be difficult to meet when dealing with imperceptible data and static measurement devices. Inconsistencies between the dimensionality of the data being recorded and its representation space can result in abstract re-interpretations and lead to false readings and ultimately, useless data. To model intangible information in a volumetric manner that is of value in an architectural design process it is necessary to move beyond static, point-based data collection methods and manoeuvre the hardware spatially whilst simultaneously linking the streamed data with the physical location.

Our solution was to develop an open-source, low-cost hardware framework for adaptive-resolution data capture at an architectural scale on the Arduino platform, providing easy access to a wide variety of sensor datasets and offering great potential for expansion and modular task assignment (Melsom et al. 2012). The base system was developed on a battery-powered microcontroller to which we connected additional hardware for geolocation and data recording, permitting us to simultaneously log incoming generic data alongside its coordinates, altitude and a timestamp. These data-packets were then appended to with project-specific sensor data, here humidity, temperature, sound and ambient light levels as well as a gaseous composition reading were collected (Fig. 2).

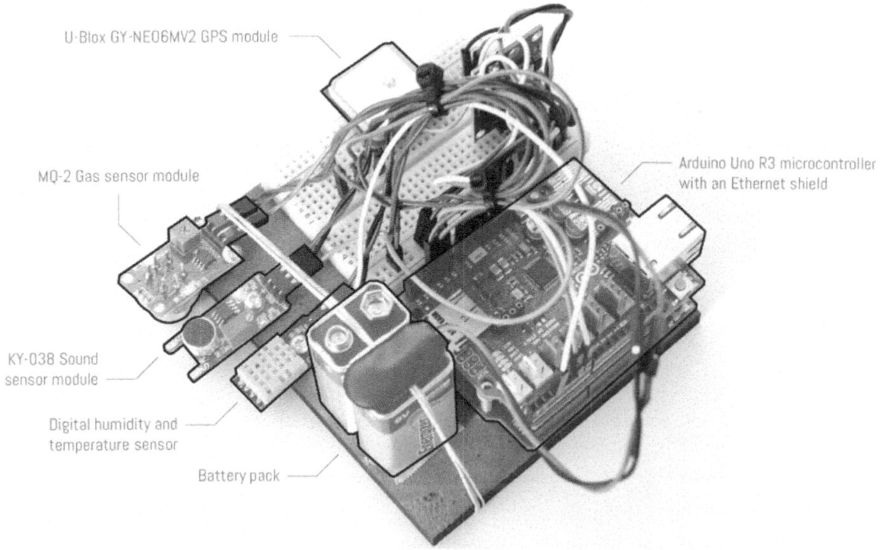

Fig. 2. Data collection device developed on the open-source Arduino microelectronics platform for this project

One of the problems with using low-cost and open-source hardware is the fidelity of the readings, in that it is expected that we will have spikes in the data unless it is smoothed. To compensate for this, all of the data we have gathered was at first logged at short intervals (5–15 ms) in groups of 100 reading. These groups are then averaged, leaving one value which is logged, affecting the data as shown in Figs. 3, 4 and 5.

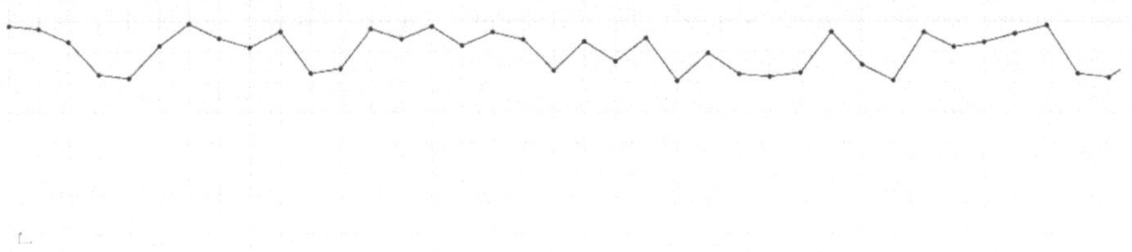

Fig. 3. Unsmoothed data

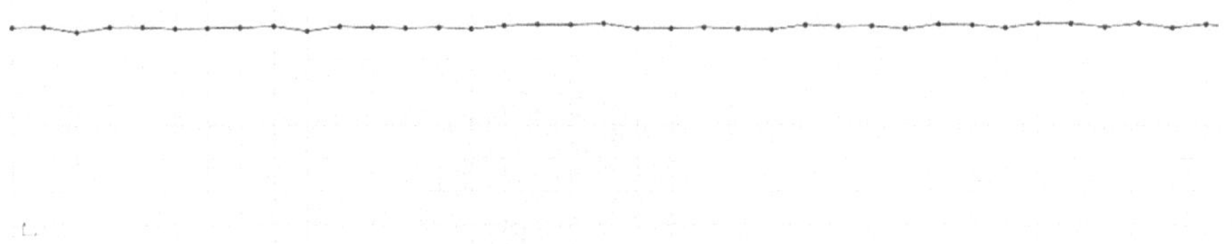

Fig. 4. Smoothed data

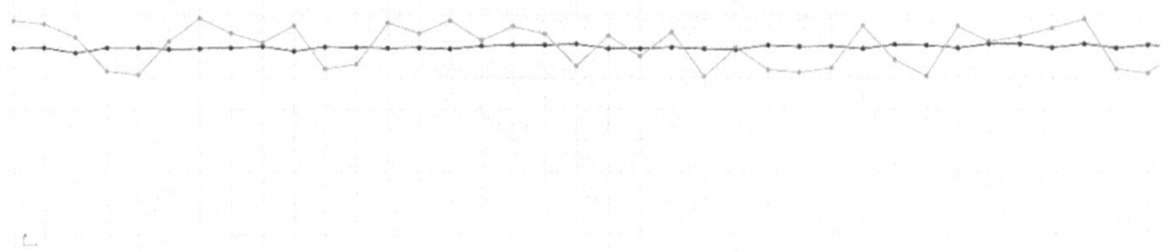

Fig. 5. Comparison

With these basic tools, this project expands the previous work in this field by making it possible to assign the data to real-world coordinates which are then mapped to points in a digital 3D environment through a specific data structure which will be further detailed in the following section.

The link between the captured data and both its physical and digital position is non-trivial as it permits a bi-directional workflow between the data capture process and the visualisation and control model. This live interface, coupled with an autonomous physical motion platform, allows the iterative re-evaluation and sampling of the datascape at points where it is deemed important by the designer to have more information, resulting in an adaptive-resolution data capture process. Additionally, the developed workflow is motion-system agnostic, resulting in a higher diversity of possible landscapes for sampling, depending on the manoeuvrability, precision and range of the motion platform.

Data Visualization and Processing

As described in previous research (Fraguada et al. 2013; Melsom et al. 2012), visualizing and processing collected data is a difficult task, and has therefore received a lot of attention in this project. It is our ambition to enable designers to capture and visualize data in a CAD environment native to designers and architects, allowing them to stream/write data in a time efficient way.

We interpret data through a modelling environment well known to the practice of architecture and design: Rhinoceros 3D, for which we have developed a series of custom tools and data interpretation methods in Grasshopper. This paper describes the tools related to the act of reading and writing data to the digital environment and visualizing the data. The methods of visualization enable the production of color gradient maps, volumetric models or simply permit the designer to query and alter the data. We believe that these functionalities can become important tools for architects in the future.

The methods revolve around a bespoke voxel modeler that we can read and write information from and to. Voxels are three dimensional pixels consisting of a volumetric grid of points where each pixel can have a scalar assigned. In the case of this research each grid position has an implicit knowledge of its neighboring environment, structured in a grid of cubes, as described in Fig. 6.

While there are many simpler methods for storing data, the structure described above is a necessity since we are building a marching cube algorithm on top of the voxels (Bourke 1994), not by building the elements of Fig. 6 as explicit geometry but as a data structure, meaning we build arrays [x, y] with ID pointers for the specified information. Classically, marching cube algorithms are used to make implicit surfaces (Bourke 1997), using various different mathematical formulas to wrap geometries in a mesh. However, in this case we will not use this technique to generate the surfaces, but volumetric representations of the gathered data from the sensor module.

It is these properties, and the structure of voxel grids that allow them to excel at the representation of regularly-sampled but non-homogenous data that we are interested in. We have adopted this voxel data structure and developed a method for converting a 3D

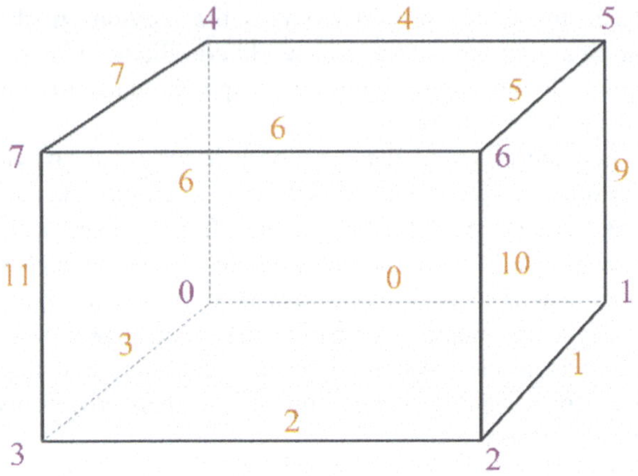

Fig. 6. The structure of a voxel. The *violet numbers* refer to vertex ID's while the *orange numbers* refer to edge ID's

point position to an ID in the data structure, making it possible to read and write data very efficiently. The formula below uses the following information: '*p*'—a point in space, *pMin*—the first data point of the voxels, *deltaX, deltaY* and *deltaZ*—the spacing between the points in their respective directions and '*nX*', '*nY*' and '*nZ*'—the number of points in their respective directions. The formula is:

$$\begin{aligned} \text{Id}{=}(\text{int})(|(\text{p.X} - pMin.\text{X})/\text{deltaX}\ | + |(\text{p.Y} - pMin.\text{Y})/\text{deltaY}| \\ *\text{nX} + |(\text{p.Z} - pMin.\text{Z})/\text{deltaZ}\ | * \text{nX} * \text{nY}) \end{aligned} \quad (1)$$

This means that in order to use our geolocation data it should first be converted into an XYZ co-ordinate, so that it can be converted to an ID using the formula (1). We do so using cylindrical equidistant projection (Weisstein 2017). This method of conversion has issues with increasingly distorted points the further you get away from equator, an issue we chose to overlook for two reasons: firstly, it is computationally efficient since the factor $\cos(\theta^1)$ only needs to be computed once; secondly, the distortion is deemed negligible since the data points need to be remapped to the voxel field, where the precision of this depend on the voxel spacing. In formula (2), λ is longitude, θ is latitude and θ^1 is a latitude in the middle of the sample data.

$$x = (\lambda - \lambda 0) * \cos\left(\theta^1\right) \quad (2)$$

$$y = \theta \quad (3)$$

This makes it possible to read and write data with a high degree of precision to- and from—the system in near real-time. Due to the infancy of the project we have tested the workflow through logged data for a small architectural installation in Vennelystparken, Aarhus, Denmark.

Using the remapped geolocations, it was possible to map the data to a three-metre spacing voxel grid. This made it possible to produce numerous ways of representing data through colour gradients and coloured points (Figs. 7 and 8). This capability is not new, but we were able to extend the use of the data by converting it to volumetric representations, exemplified by Figs. 9 and 10, which shows how sound data starts to

Fig. 7. From *left* to *right*: the site; sampled points mapped to the environment; a simple visualization of gas sensor data

Fig. 8. Gradient of sound measured on site, visualizing the difference between the quieter area of the park to the *left* and the busier pathway in the *middle*. The area to the far right reads *blue*, as no readings were taken on the road

Fig. 9. Perspective view of volumetric representation of sound levels on site

Fig. 10. Perspective view of volumetric representation of sound levels on site

take an abstract shape. The functionality and potentials of this method are presented in the following section.

Conclusion

The workflow of data capture and interpretation has been applied here to a simple design exercise, where the data has been used to visualize combinatorial data maps and create volumetric representations for the associated data. The project aims to build on previous work by offering data collection and representation methods at both the urban and building scale, including renovation projects and interior installations. One of the shortcomings of the workflow is the inability of the geolocation system to work indoors, posing a challenge for interior data collection on a smaller scale. While there are several systems available for indoor spatial triangulation, we chose to work with a GPS based setup to provide initial data for the development of volumetric representations, where we then focused. These volumetric representations can serve many purposes, for instance they can be a new way of presenting intangible information around us, but they could also be used as abstract zoning models. Meaning they could inform certain architectural decisions. This idea opens up new avenues of architectural thinking similarly to that of draftsmen in the early 1900s that explored the New York zoning laws (Koolhaas 1994).

While the research is still in its early stages and needs to be developed further in larger case studies, the preliminary findings demonstrate that the developed method could be useful in the early stages of the design process, as it allows for the construction of abstract spatial data models within the design environment familiar to architects, and permits the integration of these models into a computational design workflow.

We plan to continue the development of the project via large scale testing through implementation in architectural projects and teaching workshops with students, where different modes of data capturing, geolocation and representation will be tested.

References

Ballestrem, M.: Sense Adapt Create. Fieldstations, TU Berlin, 10 Sept. 2016. Web. 6 June 2017

Bourke, P.: Polygonising a scalar field. Retrieved 21 Feb 2017, from http://paulbourke.net/geometry/polygonise/ (1994)

Bourke, P.: Implicit surfaces. Retrieved 7 June 2017, from http://paulbourke.net/geometry/implicitsurf/ (1997)

Davis, D., Peters, B.: Designing ecosystems, customising the architectural design environment with software plug-ins. In: Peters, B., De Kestelier, X. (eds.) Computation Works: The Building of Algorithmic Thought, pp. 124–131 (2013)

Fraguada, L., Girot, C., Melsom, J.: Ambient terrain. In: Stouffs, R., Sariyildiz, S. (eds.) Computation and Performance—Proceedings of the 31st eCAADe Conference, Faculty of Architecture, Delft University of Technology, Delft, The Netherlands, 18–20 Sept 2013, pp. 433–438 (2013)

Koolhaas R (1994) Delirious New York. The Monacelli Press, New York

McCullough M (2013) Ambient Commons: Attention in the Age of Embodied Information. The MIT Press, Cambridge

Melsom, J., Fraguada, L., Girot, C.: Synchronous horizons: redefining spatial design in landscape architecture through ambient data collection and volumetric manipulation. In: ACADIA 12: Synthetic Digital Ecologies [Proceedings of the 32nd Annual Conference of the Association for Computer Aided Design in Architecture (ACADIA)] San Francisco 18–21 Oct 2012, pp. 355–361 (2012)

Offenhuber D, Ratti C (eds) (2014) Decoding the City: Urbanism in the Age of Big Data. Birkhauser Verlag, Basel

Weisstein, E.W.: Cylindrical equidistant projection. From MathWorld—A Wolfram Web Resource. http://mathworld.wolfram.com/CylindricalEquidistantProjection.html (2017)

Tailoring the Bending Behaviour of Material Patterns for the Induction of Double Curvature

Riccardo La Magna(✉) and Jan Knippers

ITKE—University of Stuttgart, Keplerstr, 70174 Stuttgart, Germany
ric.lamagna@gmail.com

Abstract. Cellular structures derive their macro mechanical properties from the specific arrangement of material and voids, therefore geometry is the main responsible for the peculiar mechanical behaviour of these classes of metamaterials. By varying the geometry and topology of the pattern, the mechanical response changes accordingly, allowing to achieve a wide range of properties which can be specifically tailored and adapted to the needs of the designer. In this paper, the bending behaviour of certain classes of cellular structures will be discussed, with the specific aim of achieving non-standard deformations. The main objective of the research focuses on the ability of inducing double-curvature in flat produced patterns as a by-product of elastic bending.

Keywords: Bending · Curvature · Metamaterials · Material design

Introduction

It is well-known that attempting to wrap a piece of paper around a sphere will inevitably lead to crumpling and wrinkling of the sheet. This everyday life fact has deep consequences on many aspects of the world that surrounds us, from why we fold a slice of pizza to the way the globe is represented on maps. This effect was perfectly captured by Gauss in his remarkable *Theorema Egregium*, and it is also manifest in the mechanical behaviour of plates and shells (Williams 2014). This means that the design of plate-based deformable structures is severely limited by the geometry of the elements. To overcome such limitations alternatives must be devised.

Strategies to induce double curvature in structural elements can vary in scope and nature, ranging from the choice of material, such as membranes, to the plastic deformation of metal sheets. The current paper specifically focuses on the elastic deformation of designed metamaterials which present exceptional and counterintuitive features in terms of bending deformation. A metamaterial is a material engineered to have a property that is not found in nature (Kshetrimayum 2004). Metamaterials don't derive their behaviour directly from the properties of the base materials, but rather from their designed structure. Taking advantage of the enhanced mechanical properties of cellular structures, functionally graded materials with specific deformation characteristics can be created. In the current paper two classes of cellular structures will be taken into considerations, namely the regular honeycomb and its auxetic dual (Gibson and Ashby 1999).

K. De Rycke et al., *Humanizing Digital Reality*,
https://doi.org/10.1007/978-981-10-6611-5_38

Poisson's Ratio and Its Influence on Curvature

The bending behaviour of functionally graded cellular structures heavily depends on the macroscopic Poisson's ratio of the pattern. We first introduce the definition of Poisson's ratio for linear, isotropic materials to focus on the main aspects that derive from it. Poisson's ratio v is an elastic constant defined as the ratio between the lateral contraction and the elongation in the infinitesimal uniaxial extension of a homogenous isotropic body:

$$v = -\frac{\varepsilon_{transversal}}{\varepsilon_{longitudinal}} \tag{1}$$

Tensile deformations are considered positive and compressive deformations are considered negative. The definition of Poisson's ratio contains a minus sign so that normal materials have a positive ratio. It is a pure, dimensionless number as it is the ratio of two quantities which are dimensionless themselves. The theoretical limit of Poisson's ratio for isotropic materials is $-1 < v < 0.5$. This derives from the fundamental assumption that the function of energy density per unit volume for a linear elastic solid ψ must be strictly increasing for all conceivable deformations, i.e. $\psi \geq 0$ (Mott and Roland 2013). This quantity is a function of the elastic constants and deformation. For an isotropic material the elastic constants (Young's modulus E, Poisson's ratio v, the shear modulus G and the bulk modulus K) are linearly dependent on each other, only two of the four being linearly independent. By formulating Poisson's ratio in terms of shear bulk modulus we have:

$$\begin{aligned} K(G, v) &= \frac{2G(1+v)}{3(1-2v)} > 0 \\ G(K, v) &= \frac{3K(1-2v)}{2(1+v)} > 0 \end{aligned} \tag{2}$$

To satisfy both of these constraints, and to ensure that the shear modulus G and the bulk modulus K are bounded, the restriction $-1 < v < 0.5$ must be placed on the values of Poisson's ratio. Almost all common materials, especially those typically employed in the construction industry, have a positive Poisson's ratio, i.e. they become narrower in cross section when they are stretched (Fig. 1 top). The reason of this behaviour is that most materials resist a change in volume, determined by the bulk modulus K, more than they resist a change in shape, as determined by the shear modulus G.

Due to Poisson's effect, when a material is compressed in one direction it will expand in the other two directions perpendicular to the one of compression. How much it will expand in the transversal direction is captured by the value of Poisson's ratio. On the other hand, if the material is stretched rather than compressed, it will contract in the directions transverse to the direction of stretching (Fig. 1 top).

As a consequence, bending will directly be influenced by the dual existence of Poisson's effect for stretching and compressive actions. Considering a homogenous beam made of isotropic material undergoing bending, the cross section will present a

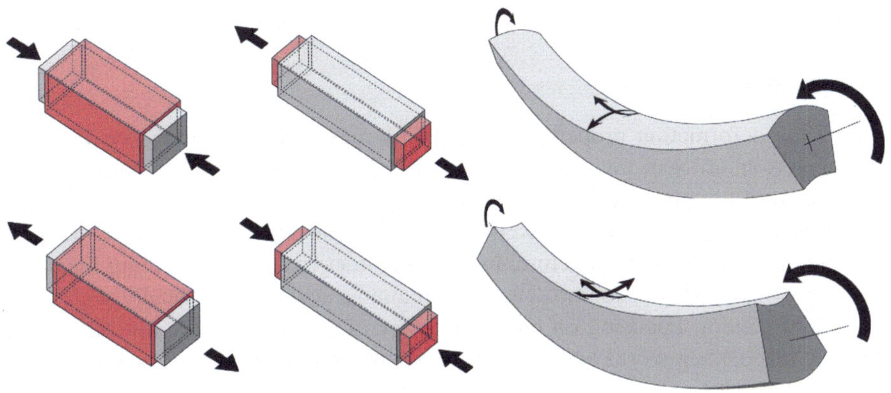

Fig. 1. *Top* Poisson's effect for conventional materials. *Bottom* auxetic materials with negative Poisson's ratio display the opposite effect of conventional materials

typical butterfly distribution of the normal stresses. For the top beam in Fig. 1, the top half of the cross section will be under compression and the bottom half under tension, with the neutral surface running exactly along the centre of mass of the beam. As the top half is subject to compressive forces, it will expand laterally to compensate the change of volume following the described Poisson's effect. The same is true for the bottom half which will shrink laterally because under tension. This differential change of length over the beam's cross section is responsible for inducing curvature in the transversal direction of bending deformation. In summary, Poisson's ratio governs the curvature in a direction perpendicular to the direction of bending. This effect can be easily visualised in the anticlastic curvature that emerges from the bending of a rubber eraser.

Common materials have a positive Poisson's ratio. Although materials with negative Poisson's ratio are theoretically prescribed by the relationships seen above, in reality they are extremely rare. In nature only certain rocks and minerals have been found to display auxetic behaviour. Whilst a positive Poisson's ratio material will contract laterally when stretched, an auxetic material will expand its transversal cross section (Fig. 1 bottom). The opposite is true for a compressive axial force acting on the material, which will have the effect of shrinking the specimen in the perpendicular direction. By inverting the sense of lateral expansion (or shrinking), also has a direct effect on the response to bending. Looking back at the example analysed previously, we now consider a beam made of homogeneous and isotropic material, but this time we assume that the material possesses auxetic behaviour (Fig. 1 bottom). By applying a bending load, we will still have the top half of the beam under compression and the bottom half under tension. We will now have the top half shrinking under compression and the bottom half expanding under tension. The sense of deformation is exactly mirrored with respect to the neutral surface of the beam. As a consequence of this inversion between expanding and shrinking areas, the hypothetical auxetic beam will present a synclastic curvature, as opposed to the anticlastic curvature of the conventional beam discussed previously.

Cellular Patterns

Auxetic Patterns

To study the deformation properties of auxetic honeycombs, we focus on Fig. 2 left which depicts an array of inverted honeycomb cells packed together to form a cellular solid structure. Being interested in assessing the Poisson's ratio of this structural object, we apply a controlled support displacement to one of the edges and measure the degree of lateral deformation of the structure. Indeed, the sides expand, confirming the auxetic nature of the inverted honeycomb. The pattern almost scales isometrically with respect to the initial pattern. Focusing on a detailed view of a cell, it can be noticed how the expanding mechanism works and where it is deriving from. Therefore, at a macro level the inverted honeycomb exactly behaves auxetically.

Having assessed the sign of Poisson's ratio for inverted honeycombs, we need to determine the value of the ratio itself. Comparing the axial dilation Δl_a of the structure with the transversal expansion Δl_t it can be seen that the two values are almost identical. As seen from Eq. (1), Poisson's ratio is the ratio between these two quantities, which in this case are identical. This means that the value of Poisson's ratio is practically -1. This implies that inverted honeycombs are placed exactly at the lower boundary of the theoretical limits prescribed for Poisson's ratio.

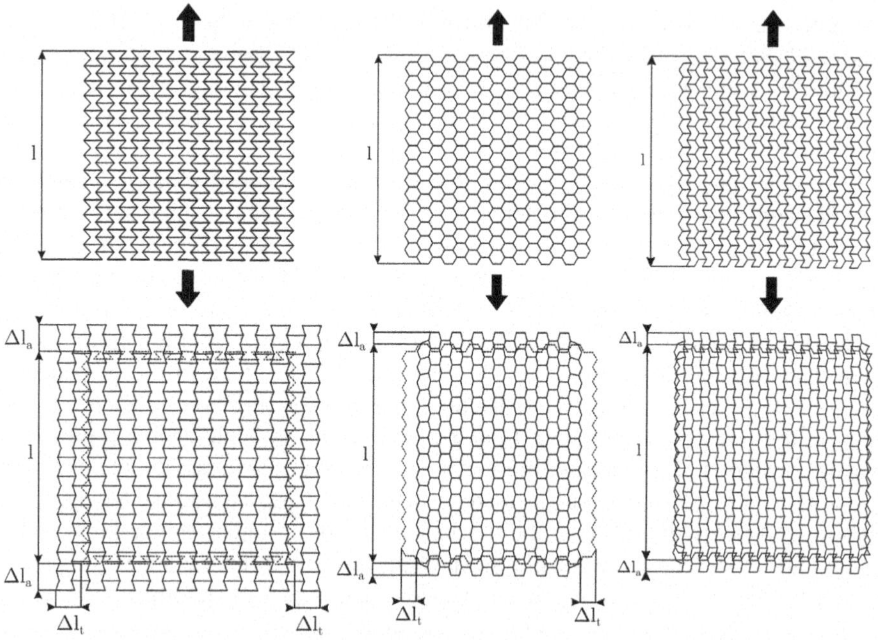

Fig. 2. *Left* Deformation of an auxetic pattern. *Centre* deformation of a regular honeycomb. *Right* zero Poisson's ratio pattern under axial displacement

The Regular Honeycomb

The upper limit of Poisson's ratio exists for isotropic materials. This derives from the interdependency of the elasticity constants for isotropic materials seen in Eq. (2). Theoretically, anisotropic materials can assume any value (Ting and Chen 2005), as the limit of incompressibility is no longer as restrictive as for their isotropic counterpart. The regular hexagonal honeycomb belongs to that class of anisotropic structures which present a positive Poisson's ratio almost equal to one. This is shown in Fig. 2 centre, where a regular honeycomb cellular structure is subject to an axial support displacement. Opposite to what has been seen for the inverted honeycomb, in this case the lateral contraction is indeed positive, as it stretches towards the interior of the honeycomb. Moreover, the values Δl_t and Δl_a are the same, confirming that regular honeycombs do indeed approach a value of Poisson's ratio equal to +1, which is way beyond the incompressible limit of 0.5 for conventional, isotropic materials.

The Zero Poisson Pattern

In Fig. 3 the geometries of the regular and inverted honeycombs are shown with the transitioning pattern in between. By adopting an internal angle of 120° for the inverted honeycomb cell it is possible to completely tile the plane with the same number of regular hexagons. As the top and bottom edges of the auxetic cells are coincident with those of the regular hexagon, the only edges that need to be adjusted are the internal convex ones. This can be achieved by shifting the internal vertices one at the time towards the exterior, until the internal angle of the regular honeycomb cell is reached. In this way, all the transitions between these two states are totally valid and do not break the regular tessellation of the plane. Here, only one of the two vertices is shifted until the 'fish scale' pattern is reached, after which the second vertex is allowed to transition from its concave configuration to the convex one.

Lying at the transition between the positive and the negative pattern, the 'fish scale' pattern (Fig. 2 right) is supposed to possess a 0 Poisson's ratio. For this reason it will often be referred to as the zero Poisson ratio pattern or the neutral pattern to distinguish it from the regular and auxetic honeycombs. From Fig. 2 right, which again shows the amount of lateral contraction/expansion with respect to the axial stretching, it is clear that no transversal effect occurs in the deformation process, confirming that indeed the pattern possesses a vanishing Poisson's ratio.

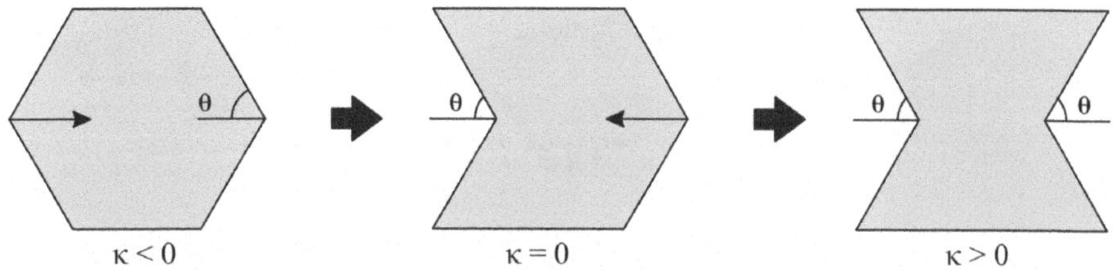

Fig. 3. Pattern transition between the regular, zero Poisson's ratio and auxetic honeycomb

Bending and Its Effects on Gaussian Curvature

To conclude this section, we focus on the deformation effects that bending has on the different pattern typologies. We do this with the aid of prototypical test specimens. The specimens were created using a polyjet 3D-printing method. The three main patterns were printed in the size of 10 cm × 10 cm × 1 cm with a cell wall thickness of 1 mm. Figure 4 shows the three specimens in their bent state. Comparing the different patterns, it can be seen that for the negative Poisson's ratio a spherical curvature is achieved, in the case of the regular honeycomb the Gaussian curvature is negative and finally for the 0 Poisson pattern the bending simply returns a cylindrical surface with vanishing Gaussian curvature. This indeed confirms the assumptions that have been made so far regarding the bending behaviour of the patterns in relationship to the values of Poisson's ratio. For the three test cases, the bending process consists of a support displacement applied to the centre of the specimens, similar to the displacement applied in the test cases of Fig. 2 but in the opposite direction. As the transversal strain is almost the same magnitude as the induced axial strain, also the amount of bending will be comparable in both directions. Furthermore, the intrinsic deformation mechanisms of the individual patterns allow to stretch these cellular structures way beyond the stretching percentages of typical materials. Depending on the geometry of the cells and the internal angle between the plates, the maximum stretching can be controlled and tweaked to reach very high values.

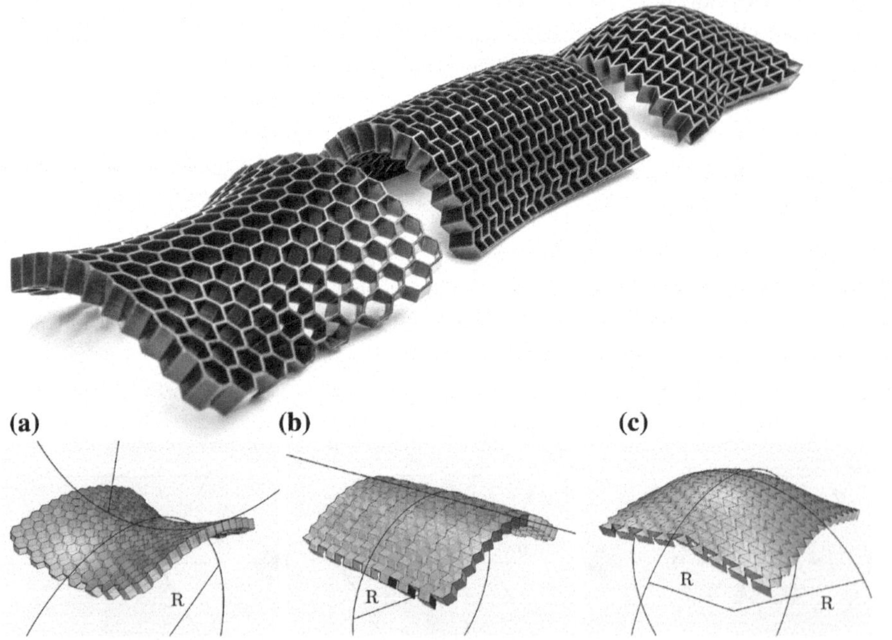

Fig. 4. Specimens showing the effects of bending on the Gaussian curvature. **a** Anticlastic curvature, **b** cylindrical surface, **c** synclastic curvature

Methodology

Given a double curved surface, what is the best flat pattern design which returns the target surface under bending? The problem can be equivalently stated in mechanical terms, as the sought design is that which minimises the strain energy of the system, the strain energy being a measure of how much the deformed pattern deviates from the target geometry (Audoly and Pomeau 2010). In practice, we seek for patterns which will induce only bending and torsional actions in the cell walls of the honeycombs, as the strain energy under bending and torsion is much lower compared to membrane strain (La Magna et al. 2016). The problem needs to be broken down in several steps. These can be summarised as following:

1. Analyse the Gaussian curvature of the base surface.
2. Flatten the base surface keeping track of the position of the original curvatures.
3. Create a corresponding pattern which matches the curvature values.
4. Map the pattern to the flattened domain.
5. Define the actuation mode and place the actuators on the flattened pattern.
6. Bend.

The flattening of a double curved surface is a complex challenge which has been providing topics of research for generations of mathematicians and geometers. Probably the best-known field of application is map projections. A map projection is a systematic transformation of the latitudes and longitudes of locations on the surface of a sphere into locations on a plane (Snyder 1989). As seen in Gauss's *Theorema Egregium*, a sphere's surface cannot be represented on a plane without distortion. The flattened pattern has to maintain the correspondence between one the principal curvature directions, which will correspond with the main bending axis, on the three-dimensional surface and its planarised version. In the case of the torus in Fig. 5, the radial directions, being the ones of main bending, are the ones preserved in length during the flattening operation.

The pattern generation makes use of a transformation of the plane known as Schwarz-Christoffel mapping. In complex analysis, a Schwarz-Christoffel mapping is a conformal transformation of the upper half-plane onto the interior of a simple polygon. Being a conformal mapping between two domains, it means that the angles between elements contained in the first domain will be exactly preserved under the transformation. For the implementation, a specific tool for Matlab was employed. The tool is called Schwarz-Christoffel Toolbox for Matlab and is being actively developed by Driscoll and Trefethen (2002) and available for free at the time of this writing.

Figure 5 shows the complete pattern generation process for the embedded approach method in the case of a torus. The surface of the torus is first flattened and mapped to a rectangular domain, the so-called canonical domain. The canonical domain is then populated with the elements of the pattern (Fig. 5a, b) and interpolated depending on their curvature value (Fig. 5c). The pattern is then mapped back onto the flattened domain of the torus (Fig. 5d). Finally, for the Finite Element simulation that will follow, the contracting cables are placed in the correspondent position as shown in Fig. 5e.

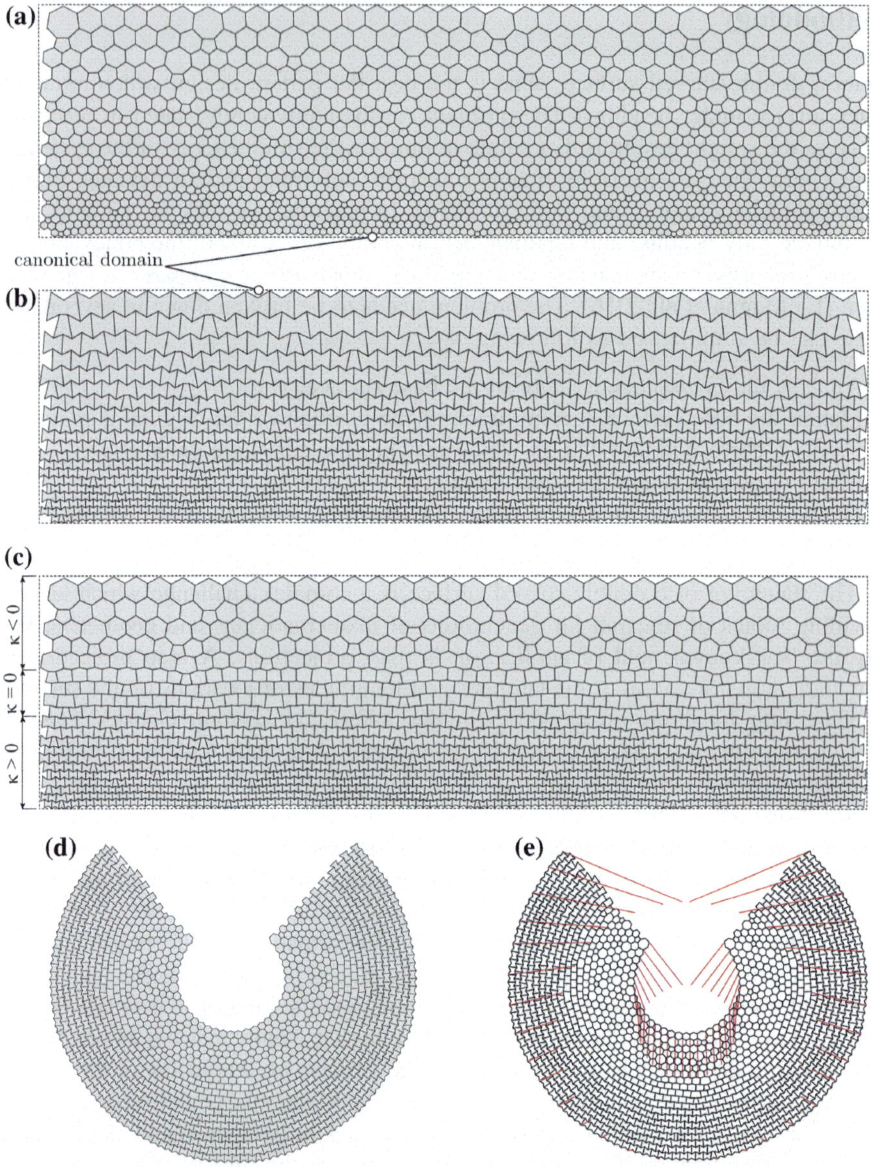

Fig. 5. Generation process of the bending pattern. **a** Regular honeycomb pattern embedded in the canonical domain, **b** dual auxetic honeycomb pattern, **c** interpolation of the pattern, **d** conformal mapping of the pattern from the canonical domain to the physical domain through the Schwarz-Christoffel mapping, **e** view of the arrangement of the contracting cables

Test Cases

The method presented was tested numerically and physically. Due to the manufacturing complexity, 3D printing technology was employed to fabricate the test pieces. The employed Polyjet 3D printing machine was an Objet Connex 500 by Stratasys. The material used for the prototype is defined by its Shore hardness A95.

Physical Prototyping

For the physical prototyping, the chosen geometry was a half torus sectioned along the horizontal plane with an external diameter of 323 mm and an internal radius of 117 mm. The height of the printed pattern was 10 mm, which is roughly comparable to the size of the individual cells. This specific geometry was chosen as the most representative as its smooth Gaussian curvature transition between the external synclastic part and the internal anticlastic part poses a challenging geometrical problem.

Figure 6 shows the final printed prototype. The torus is presented in its flat state, intermediate deformation state and fully bent state. From the images, the bent cellular pattern is clearly assuming the shape of the torus. Both negative and positive Gaussian

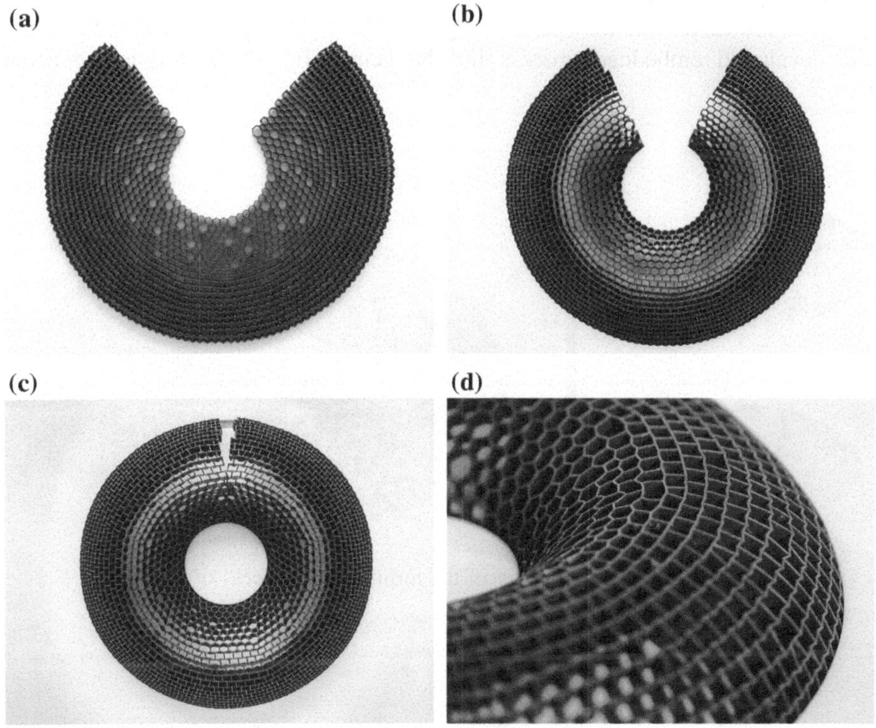

Fig. 6. **a** 3D-printed pattern of the flattened torus, **b** intermediate bending stage, **c** torus in its fully bent configuration, **d** detail of the fully bent torus

curvatures of the internal and external portions of the donut are visible, as well as the zero curvature transition between the two zones.

Numerical Simulation

Besides assessing the embedded approach through physical prototyping, the results were tested also numerically with a full Finite Element simulation of the torus test case and a second test case similar to the well-known Downland gridshell (Harris et al. 2003).

The simulation was run using the contracting cable approach by Lienhard et al. (2014). Under the applied prestress load, the cables shrink and bring the supports of the model into their correct and final position. The material properties are the same of the printed physical prototype. The results of the simulation are reported in Figs. 7 and 8. As in the case of the physical prototype, also the form-found geometry well approximates the toroidal surface used as a base for testing the embedded approach. The von Mises plots show an even and constant stress distribution over the elements of the structure. An area of higher stress concentration is recognisable in the transition zone where the Gaussian curvature of the surfaces is supposed to vanish. The area of transition is slightly stiffer than the interior and the exterior of the torus and the sides of the Downland geometry. This is partially due to the higher buckling resistance of this particular pattern geometry. Nonetheless, the simulation results show agreement with the physical prototype, and the initial base geometry, besides demonstrating the validity of the developed embedded process for the generation of bending-active cellular patterns.

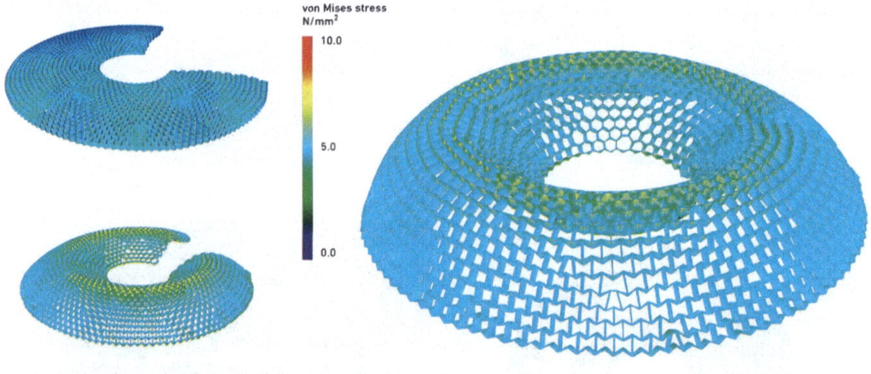

Fig. 7. Von Mises stress plot of the form-finding process of the torus

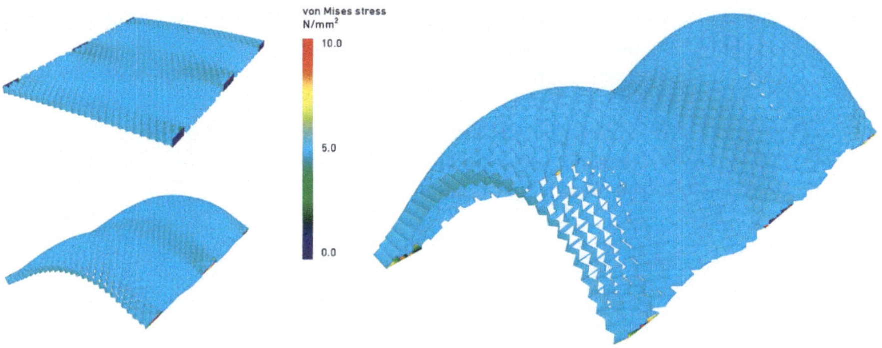

Fig. 8. Von Mises stress plot of the form-finding of the Downland test case

Conclusions

The approach presented here was tested for two ideal study cases which present a challenging geometry due to the smooth Gaussian curvature transition between different areas of the surface. The simulation methods employed for the analysis of large elastic deformations of structures has been proven to be adequate to capture the geometrical changes in the elements and the stress levels induced in them. Future developments of the process will include further material prototyping to test different kinds of geometries and the generation strategy of the cellular pattern. In terms of fabrication, ongoing research is currently focusing on improving the scale of the systems to test the structural and mechanical behaviour of the structures at larger scales. The design explorations and method development conducted here served as a starting point for future improvements which will hopefully lead to new and exciting applications.

References

Audoly, B., Pomeau, Y.: Elasticity and Geometry: From Hair Curls to the Non-Linear Response of Shells. Oxford University Press, Oxford (2010)

Driscoll, T.A., Trefethen, L.N.: Schwarz-Christoffel Mapping. Cambridge Monographs on Applied and Computational Mathematics. Cambridge University Press, Cambridge (2002)

Gibson, L.J., Ashby, M.F.: Cellular Solids: Structure and Properties. Cambridge University Press, Cambridge (1999)

Harris, R., Romer, J., Kelly, O., Johnson, S.: Design and construction of the Downland Gridshell. Build. Res. Inf. **31**(6), 427–454 (2003)

Kshetrimayum, R.S.: A brief intro to metamaterials. IEEE Potentials **23**(5), 44–46 (2004)

La Magna, R., Schleicher, S., Knippers, K.: Bending-active plates: form and structure. Adv. Archit. Geom. **2016**, 170–186 (2016)

Lienhard, J., La Magna, R., Knippers, J.: Form-finding bending-active structures with temporary ultra-elastic contraction elements. Mob. Rapidly Assem. Struct. IV (2014). doi:10.2495/mar140091

Mott, P.H., Roland, C.M.: Limits of Poisson's ratio in isotropic materials—general result for arbitrary deformation. Phys. Scr. **87** (2013)

Snyder, J.P.: Album of Map Projections, United States Geological Survey Professional Paper. United States Government Printing Office, Washington (1989)
Ting, T.C.T., Chen, T.: Poisson's ratio for anisotropic elastic materials can have no bounds. Q. J. Mech. Appl. Math. **58**(1), 73–82 (2005)
Williams, C.: Appendix B—differential geometry and shell theory. In: Adriaenssens, S., Block, P., Veenendaal, D., Williams, C. (eds.) Shell Structures for Architecture: Form Finding and Optimization, pp. 281–289. Routledge, London (2014)

Design of Space Truss Based Insulating Walls for Robotic Fabrication in Concrete

Romain Duballet[1,3](✉), Olivier Baverel[1], and Justin Dirrenberger[2,3]

[1] Laboratoire Navier, UMR 8205, Ecole des Ponts, IFSTTAR, CNRS, UPE, Champs-sur-Marne, France
romain.duballet@gmail.com

[2] Laboratoire PIMM, Ensam, CNRS, Cnam, 151 bd de l'Hôpital, 75013 Paris, France

[3] XtreeE, Immeuble Le Cargo, 157 bd Macdonald, 75019 Paris, France

Abstract. This work focuses on the design of ultra-light concrete walls for individual or collective housing, the normative context being constrained masonry. It is stated that current block work building is very inefficient in terms of quantity of concrete used for cinderblocks and mortar joints, and with regards to thermal insulation. Here is proposed a robotic manufacturing technique based on mortar extrusion that allows producing more efficient walls. First we present the fabrication concept, then design criteria for such objects. In the last section we show a comparative study on different geometries. We conclude with a discussion on the performances of this proposed building system.

Keywords: Concrete printing · Robotic construction · Space truss

Introduction

Cement consumption is one of the major environmental issues of our century. Concrete is now the most used manufactured material in the world, with 6 billion cubic meters produced every year. Considering current and future needs in housing, it is not meant to decrease. Between 2011 and 2013, China has produced more concrete (6.6 Gigatons) than the US did during the whole twentieth century (4.5 Gigatons).[1] Given its polluting impact, it is now crucial to learn how to build with less cement.

In the past two years, numerous studies have flourished on the topic of robotic mortar extrusion, mainly oriented toward what is denoted by concrete 3D printing. This term refers to a family of constructive approaches consisting in progressively stacking relatively small quantities of specifically formulated mortar to fabricate a given form. The first publication on such strategies go back to Pegna (1997) and its first explicit mention to Khoshnevis (2004). If one can find reviewed work on the associated fabrication processes (Lim et al. 2011, 2012; Gosselin et al. 2016), concrete formulation (Le et al. 2012a, b; Feng et al. 2015) or robotic control (Bosscher et al. 2007), few has been done to prove the true interest of concrete printing in construction. The two main aspects often mentioned about these techniques are the speed and ease offered to

[1] Source: USGS Cement Statistics 1900–2012; USGS, Mineral industry of China 1990–2013.

K. De Rycke et al., *Humanizing Digital Reality*,
https://doi.org/10.1007/978-981-10-6611-5_39

building, on one hand, and the new attainable geometrical freedom, on the other hand. If Pegna's original paper focused on the concept of "free-form construction", emphasizing the interest of geometrical freedom, most today's application of concrete printing exhibits nearly traditional constructive elements. If the fabrication process is in itself innovative, and very promising for the future of construction work, its impact on the fabricated object still remains marginal. Our position on such matter is that such technologies should be strongly linked with their final usage from the beginning. The problem that we address here is the following: how this technology can allow great reduction of overall concrete consumption in the building industry?

In this paper we present a novel constructive approach based on robotic extrusion of mortar. The final goal is the production of structural-insulating walls of new performances. Some of the co-authors presented a work of similar purpose in DMS 2015 (Duballet et al. 2015), where the studied process involved extrusion of cement paste mixed with polystyrene beads. Here the mortar is extruded only for its structural performances, but we take advantage of a novel assembly strategy of insulating blocks, that also serve as support for the printing.

Masonry with Robotic Mortar Extrusion

Constrained masonry, the assembly technique of breeze blocks and mortar restrained in a reinforced concrete frame, is a very popular building system, especially for individual and collective housing, for it is at once cheap, fast and easily implemented. From a purely mechanical point of view, it is however quite inefficient. In the case of a one or two-storey house, the need in mechanical resistance for the wall itself, considering the presence of the reinforced concrete frame, is indeed far lower than the breeze blocks/mortar system can provide. The main role of this staking is in fact to allow solid continuity between the concrete frames, for bracing purpose and to act as separating wall. Up to a limit, it could be said that the mortar between the blocks is the only needed element to provide resistance. Such considerations leads us to the idea of assembling insulating blocks instead of breeze blocks, leaving the mechanical role to the mortar in between, and getting thermal performances in addition.

The new system for the wall is now a generalization of the previous one: a continuous spatial structure in mortar, and thermic insulation in the negative space. The three questions to answer are then (a) which shape for the mortar structure, (b) can it be easily fabricated, and (c) are the overall performances meeting the expectations?

The key aspect of the technique is to assemble specifically shaped insulating blocks by printing a mortar joint at the edges location (Fig. 1). The mortar is extruded through a nozzle controlled by a robotic arm, as described in Gosselin et al. (2016). The mortar acts as joint for the insulating blocks, while they act as printing support for the mortar. It can be decomposed into the following steps:

1. Insulating blocks fabrication
   For prototyping in our laboratory we will work with extruded polystyrene panels from which the blocks will be extracted, thanks to a hot-wire robotic cutting process. Other ways of making the blocks can be thought of, like molding of mineral

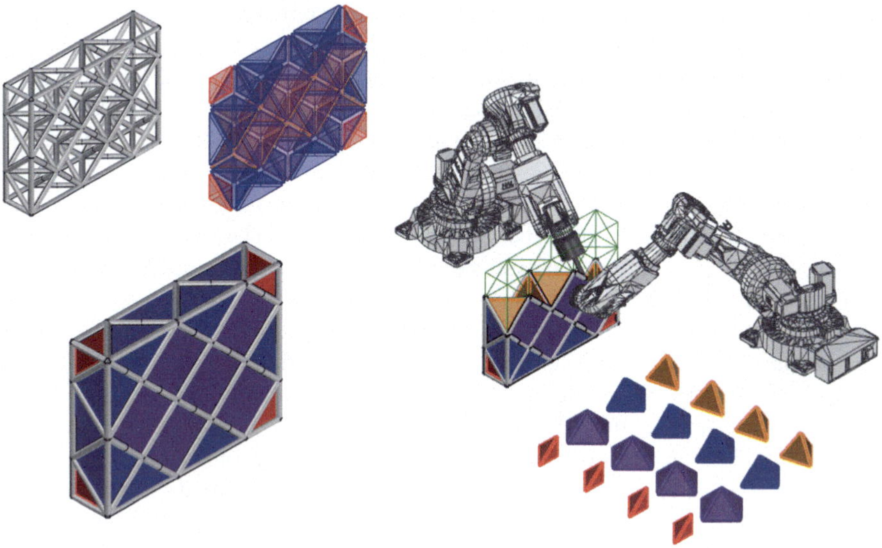

Fig. 1. Concept of fabrication

foam for instance. The blocks are shaped into polyhedra, the edges of which are replaced by a channel to be filled with mortar.

2. Insulating blocks assembly
   The blocks are assembled by a pick and place robot. They form a layer of printing support for the mortar to be extruded on. This step is repeated for each layer of blocks, and alternating with step 3.
3. Mortar extrusion
   The mortar is extruded by a printing system (Gosselin et al. 2016). The already deposited blocks channels are filled by the printed truss members, on which will be brought the next insulating blocks (back to step 2).
4. Confining elements
   Tie columns and ring beams that will ensure masonry confinement.

This system brings some shape constraints for the blocks, they must form a space tessellation, the edges of which will form a mortar space truss. Since this truss is printed on the blocks, it must not be done too close to verticality. In addition, for assembly purpose, the number of crossing elements at each nodes must be limited. We have conducted a comparison of different space tessellations. In the next section we present our design strategy, and the last section deals with the results.

Design Criteria

Thanks to this manufacturing technique it is realistic to hope for ultra-light space truss walls that drastically reduce concrete consumption while reaching current needs in thermal insulation. Three constructive solutions are used as a comparison, traditional block work wall with an additional insulating layer, a contemporary solution for walls based on cellular concrete blocks and a pre-wall system where concrete is cast between layers of insulating material. The overall size of the wall is fixed (2.5 m × 3 m), as well as a goal U-value for thermal efficiency of 0.09 W m^{-2} K, corresponding to the lower side of 2020 Thermal Regulations (imposing a U-value for the wall between 0.15 and 0.1 W m^{-2} K).

We have conducted a parametric study on such walls performances, taking into account both mechanical and thermal performances. The study is made on Grasshopper, and the structural analysis performed with Karamba 3D. We work with a mortar of relatively high compressive value, as it can be advised for such light structures. Material performances are to be taken with care for dimensioning, because the extrusion process can limit its final quality. Our normative context is constrained masonry, governed by Eurocode 6, which does not specifically give a value for calculated maximum tensile stress. We choose to work with the Eurocode 2 value $f_t = 0.3\ f_{ck}^{2/3}$ taking an additional security factor into account. Our hypothesis for calculus is a C90/105 concrete, and since it is supposed to be non-reinforced mortar, we limit tension resistance to 0.3 MPa, which is a strongly conservative hypothesis in our case. The thermal performances are calculated with a geometrical mean, taking into account a security factor corresponding to member thickness irregularities that can be expected at the nodes when extruding such a structure.

The parameters are then (1) the type of space truss grid (see next section), (2) the truss thickness and (3) the bars diameters. The compared objectives are (a) additional insulation need (if needed, to reach target U-value of 0.09 W m^{-2} K), (b) mechanical efficiency and (c) surface weight of the wall.

Grid Comparison

Different grid topologies are investigated for the concrete space truss in terms of structural efficiency and compatibility with the proposed manufacturing method. These topologies are taken from the edges of a space tessellation of the bounding box of the wall, so that a geometric duality with the polyhedral insulating blocks can be obtained. Considering that fresh mortar will be printed on the blocks, some geometrical configurations are to be avoided. We have retained five potential topologies that does not present internal vertical members that would be hard to print. They are listed in Table 1 and shown on Fig. 2.

Table 1. Grid types

Name	Description	Maximum node valence
TriPr	Triangle prisms	7
SemiOcTe	Semi-octahedra/tetrahedra	8
OcTe	Octahedra/tetrahedra	9
CnTri	Counter running triangle (tetrahedra)	10
Hexa	Hexagonal pyramids	12

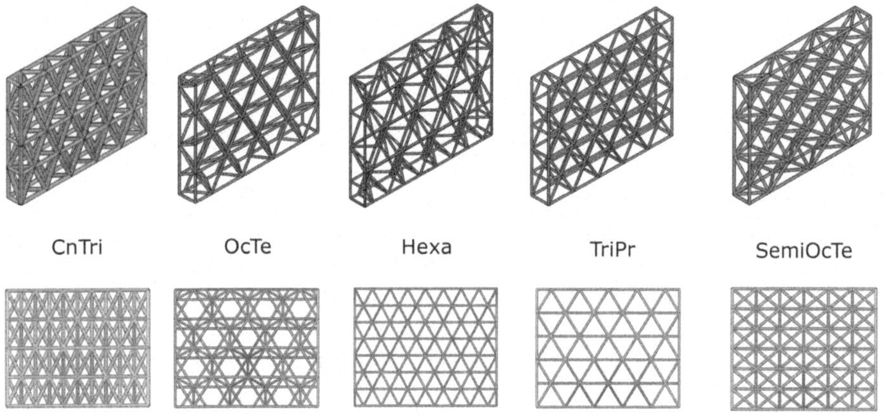

Fig. 2. Grid topologies

We suppose that the lateral and horizontal frontier are supported, as for constrained masonry, and impose three load cases corresponding respectively to self-weight, overall lateral pressure (wind) and a specific horizontal point load of 2 kN. This last case has been taken from design methods for guardrails. The need for such verification comes from the attainable lightness for the wall (around 50 kg m^{-2}). Traditional block work construction is indeed very often far more resistant that one could need, in non-seismic areas. In our case, this additional load case is critical. From possible structural failures—stresses in material, local and global buckling—in every case the tensile stress in members is the one critical. We conduct a heuristic calculation on every topology with Grasshopper plugin Octopus, plotting performances in a three dimensional space corresponding to the following quantities:

X axis Additional insulating thickness to reach target U-value (cm)
Y axis Maximum tensile stress in members (MPa)
Z axis Weight (kg m^{-2})

On Fig. 3 are plotted the Pareto fronts of each topology. The colors map additional information on them.

On all the left graphs is represented a color scaled mapping of the overall thickness of the wall (truss + additional insulation), typically varying from about 40 cm (green) to 90 cm (dark blue). The middle graphs show the points on the Pareto fronts below the maximum tensile stress of 0.3 MPa in green that are all the acceptable solutions, the others are in red. Finally the graphs on the right isolate the ten lightest (weight criteria on axis z) of such feasible individuals. The hexagonal pyramids solution would not allow an acceptable tensile stress for decent weight so the results are not plotted and the topology rejected.

On Table 2 are listed the performances of the best solution for each topology.

On Fig. 4 are plotted the weight variations with overall thickness for the ten best solutions of each configuration. We observe that the Triangle Prisms topology is the

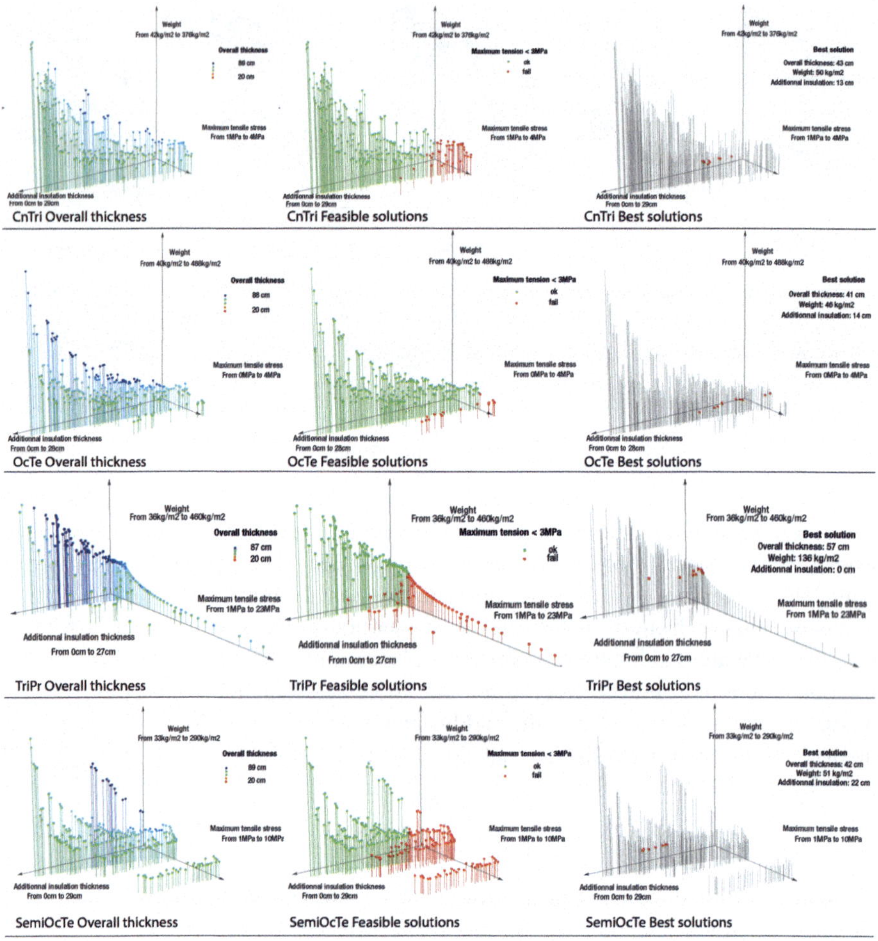

Fig. 3. Pareto fronts comparison

Table 2. Optimal grids comparison

Name	Thickness (cm)	Weight (kg m^{-2})	Add. insulation (cm)	Valence max
TriPr	76	136	0	7
SemiOcTe	42	51	22	8
OcTe	41	46	14	9
CnTri	43	50	13	10

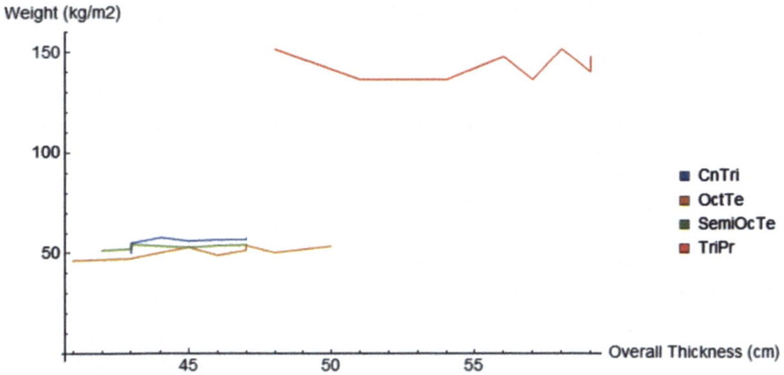

Fig. 4. Ten best individuals' weight/overall thickness

worst configuration. The three other solutions are of similar results for our criteria, therefore the best choice is the Semi Octahedra/Tetrahedra configuration that exhibits the lowest node valence, useful for fabrication purpose.

Results and Discussion

We have presented here a new fabrication concept for light weight masonry walls of novel structural and thermal performances. The key aspect of it is to replace the traditional rectilinear breezeblock and mortar system by the printing of a mortar space truss structure, supported by polyhedrically shaped insulating blocks. This system takes part in the global effort toward automation in construction, to build better and more efficiently, to face current needs in housing and to reduce environmental impact. This generic problem can be addressed at three main levels: a change in building materials themselves, a better geometrical control of build objects, and/or a transformation of building systems. Our proposition is a renewal of traditional constrained masonry system, by allowing it to handle more efficient geometries such as space trusses. The result is a relatively simple assembly system that allow great performances of produced objects, as shown in Table 3.

Table 3. Performance comparison with other systems

Wall system	Overall thickness (cm)	Weight (kg m^{-2})	U-value (W m^{-2} K)
Breeze blocks	40	180	0.1
Cellular concrete	56	150	0.09
Pre-wall	42	220	0.15
Printed truss	42	50	0.09

We conclude with a comparison with three other systems engaging concrete:

1. A traditional breeze block and mortar system, with 20 cm thick blocks (25 × 50 cm^2), and 1 cm mortar joint. With a performance insulator, like graphite polystyrene ($\lambda = 0.031$ W m^{-1} K^{-1}), a 20 cm layer is needed to reach target thermal performance.
2. A cellular concrete solution, 36 cm thick with 20 cm graphite polystyrene.
3. A pre-wall system consisting in a insulating mold of one 5 cm and one 25 cm insulation layer connected with steel bars, for casting a 12 cm thick concrete wall. Minimum U value is of 0.15 W m^{-2} K.

The presented building system can not only be seen as a renewal of traditional masonry, but also as a generalization of "concrete printing", when this term denotes only the stacking of extruded mortar laces to form an object. In fact, this technique, although studied under the assumption that it is an effort toward "free-form construction" (Pegna 1997), depending on its explicit materialization, allows a certain family of shapes to be build, in a certain way, and with a given technological mean. We have proposed in another paper (Duballet et al. 2017) a classification of generalized robotic extrusion building methods, taking into account not only the extrusion itself, but also different types of support and assembly strategies, as well as scale considerations. From that perspective, the present system is a generalized mortar printing approach, mixing assembly of external elements during the process, and making use of printing support left in place. Such considerations help us understand more precisely what those robotic techniques can become for today's construction.

Acknowledgements. The authors would like to thank V. Esnault at LafargeHolcim Research and Development for fruitful discussions. Support from LafargeHolcim is gratefully acknowledged.

References

Bosscher, P., Williams, R.L., Bryson, L.S., Castro-Lacouture, D.: Cable-suspended robotic contour crafting system. Autom. Constr. **17**(1), 45–55 (2007)

Duballet, R., Baverel, O., Dirrenberger, J.: A proposed classification for building systems based on concrete 3D printing. Autom. Constr. (2017)

Duballet, R., Gosselin, C., Roux, P.: Additive manufacturing and multi-objective optimization of graded polystyrene aggregate concrete structures. In: Design Modelling Symposium (2015)

Feng, P., Meng, X., Chen, J.-F., Ye, L.: Mechanical properties of structure 3D printed with cementitious powders. Constr. Build. Mater. **93**, 486–497 (2015). http://www.sciencedirect.com/science/article/pii/S095006181500690X

Gosselin, C., Duballet, R., Roux, P., Gaudillière, N., Dirrenberger, J., Morel, P.: Large-scale 3D printing of ultra-high performance concrete—a new processing route for architects and builders. Mater. Des. **100**, 102–109 (2016). http://www.sciencedirect.com/science/article/pii/S0264127516303811

Khoshnevis, B.: Automated construction by contour crafting—related robotics and information technologies. Autom. Constr. **13**(1), 5–19 (2004)

Le, T.T., Austin, S.A., Lim, S., Buswell, R.A., Gibb, A., Thorpe, T.: Mix design and fresh properties for high-performance printing concrete. Mater. Struct. **45**, 1221–1232 (2012a). doi:10.1617/s11527-012-9828-z

Le, T.T., Austin, S.A., Lim, S., Buswell, R.A., Law, R., Gibb, A.G.F., Thorpe, T.: Hardened properties of highperformance printing concrete. Cem. Concr. Res. **42**(3), 558–566 (2012b)

Lim, S., Buswell, R., Le, T., Wackrow, R., Austin, S., Gibb, A., Thorpe, T.: Development of a viable concrete printing process. In: Proceedings of the 28th International Symposium on Automation and Robotics in Construction (ISARC2011), pp. 665–670 (2011)

Lim, S., Buswell, R.A., Le, T.T., Austin, S.A., Gibb, A.G.F., Thorpe, T.: Developments in construction-scale additive manufacturing processes. Autom. Constr. **21**(1), 262–268 (2012)

Pegna, J.: Exploratory investigation of solid freeform construction. Autom. Constr. **5**(5), 427–437 (1997)

Collaborative Models for Design Computation and Form Finding—New Workflows in Versioning Design Processes

Jonas Runberger[1(✉)] and Julian Lienhard[2]

[1] Chalmers Department of Architecture and Civil Engineering, Gothenburg, Sweden
jonas.Runberger@white.se
[2] Structure GmbH, HCU Hamburg, Hamburg, Germany

Abstract. This paper explores the role of frameworks and conventions in design computation workflows for collaboration on the development of design, structure and detailed fabrication within a visual scripting environment. The frameworks were developed in the computational team Dsearch of the architectural practice White Arkitekter and the engineering practice of structure, and was used in the part-time 12-month design and build workshop Textile Hybrids at the HafenCity Universität Hamburg. During the workshop, the framework was further developed to facilitate iterative design/analysis studies between design models in a visual scripting environment and a FEM simulation environment through cloud-based data exchange protocols. The authors regard this continuous re-development of workflow frameworks during design development as emergent, and regard this as a valuable and potential mode of development also for architecture and engineering practice. This is shown in the individual practices of the authors, where additional layers have been added.

Keywords: Computational design workflows · Modelling of data · Collaborative design

Introduction

Generic models for design collaboration enabled by digital technology has been discussed as re-usable schema for interactions between designers and software (Oxman 2006), or as compartmentalized strategies for the capture of design intent and the rationalization of geometries (Hudson 2010). Both approaches seek to establish generic traits that can be classified and categorized. In this paper project-specific models for workflow are presented as emergent during the development itself, which in turn makes them specific to the conditions of the very particular material and structural set-up.

Given that visual scripting environments such as Grasshopper are node-based, with a functionality depending only on the association and topology of the contained definition elements, there is technically no need to use any other standards. In a collaborative environment however, it is already known that the legibility of visual scripts can be imperative in order to clarify intent and functionality (Davies et al. 2011). In this case, this was addressed through the use of frameworks for computational design

K. De Rycke et al., *Humanizing Digital Reality*,
https://doi.org/10.1007/978-981-10-6611-5_40

workflows developed within Dsearch at White Arkitekter. Initially set up as a pure graphic standard employed to cluster and encapsulate and colour code key parts of visual scripting elements (Runberger and Magnusson 2015), this standard has been expanded to include clustered functions in the form of User Objects (Fig. 5). The collected framework is based on several plugins developed elsewhere, but have contextualized and integrated into the conventions.[1]

The next level of organization within the development at Dsearch workflow is the design system; the assemblage of linked methods and tools, but also physical and computational models and drawings; the human actors of the project and the groups they form—design team, R&D team, specialists, clients and external consultants. In a professional setting the design system is set up to handle all design issues, but also policies and contracts—the complete framework that facilitate and condition design development through computational means (Magnusson and Runberger 2017a) (Fig. 2).

Background and Experiment

The workshop assignment focused on hybrid structures, i.e. the combining of two load-bearing systems with different mechanics in order to achieve more efficient structures. The design focused specifically on textile hybrids, which are the result of the combination of form- and bending-active structures, where the two systems of mechanically pre-stressed textile membranes and bending-active beam elements become interdependent (Lienhard and Knippers 2015).

Based on this approach, a group of 20 architectural and structural engineering students developed designs from first concepts to design refinement and final fabrication/assembly. An initial design competition saw five competing teams develop site specific concepts for a foyer space at HafenCity Universität Hamburg, combining material experiments with computational design and form-finding. After a collective process of evaluation, one concept was selected (Fig. 1), and three teams were given this design concept, the set of computational design and analysis tools, and the collaborative workflow framework as starting points for further development. Guided by experienced tutors with many years of design, engineering and management experience from computationally developed projects, the students took on shifting roles of designers, structural engineers, project managers and fabricators. The particular challenge was found in the nature of a textile hybrid system as a closed force equilibrium, providing challenges that would not be present in pure bending-active structures (Schleicher et al. 2015). Therefore, it was not possible to split up the model into independent units for individual design teams, instead a distributed but connected design system needed to be set up, enabling evaluation at different stages of form finding development (Fig. 2).

[1] Key plugins used within the framework include Conduit, Elefront, Flux, Honeybee, Human, Human UI, Kangaroo 2, Ladybug, Lunchbox, MeshEdit, Metahopper, Python and Weaverbird.

Fig. 1. Render of winning design concept from first stage of Textile Hybrids workshop at the HCU Hamburg

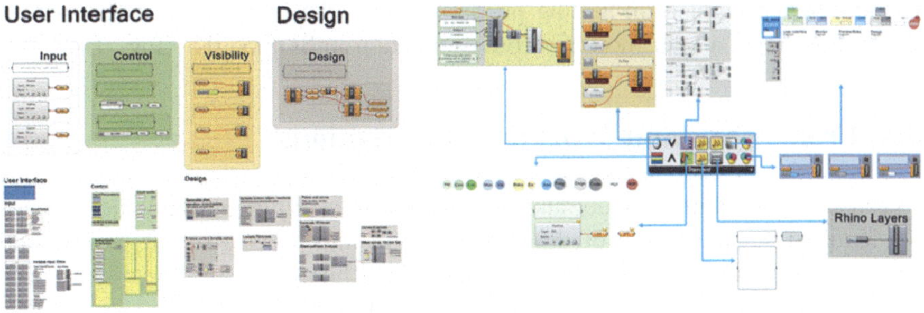

Fig. 2. Colour coding and clustering in the conventions developed by Dsearch. User Objects included in the given framework, including streamlined functions for Baking, use of Flux, accessing graphic standards etc

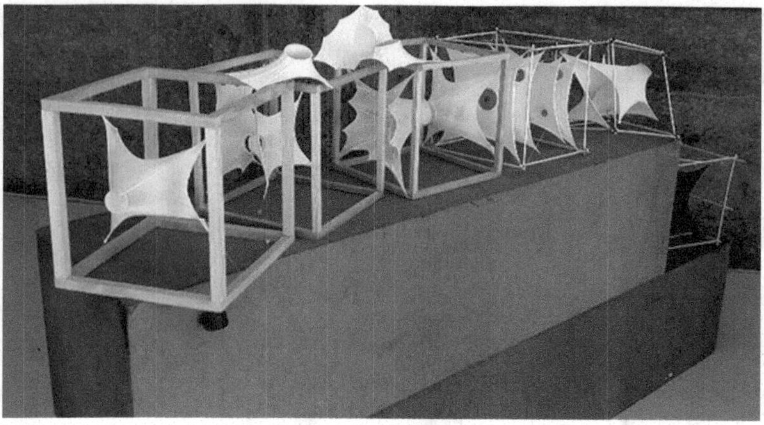

Fig. 3. Physical models in textile and glass-fibre reinforced plastic rods, within frames

Beyond the early conceptual physical models (Fig. 3), the collaborative work was primarily based on design iterations explored in digital design models and Finite Element analysis tools. A fully digital process is quite well established for the simulation and form-finding of membrane structures. Similar approaches for bending-active structures is less developed however, and the combination of the two in textile hybrids currently being developed by a few specialized research groups (Adriaenssens 2008; Ahlquist et al. 2013; Deleuran et al. 2016). The workshop employed Finite Element Analysis as a central part of the design process, handling both the form-finding and structural design phase as well as generating cutting patterns in the production phase. Thus, the engineering tools became an integral part of the design process rather than a mirrored instrument where FEM is used to check feasibility. In turn this required the setting up of bidirectional information flows that enabled fast iterations for design generation and evaluation. One important factor in this case was to include both an early more light-weight form-finding process through the integration of Rhino Membrane,[2] followed by a later more rigorous simulation in SOFiSTiK[3] (Fig. 4).

Initial Framework and Workflow Conventions

The need for a distributed model and the integration of different simulation environments required an integration of different platform into a design system. The initial design system for the workshop was set up using Rhinoceros and Grasshopper as a main technical platform, with associated simulation software requiring external data exchanges as well as an online management and communications platform. This set up also defined the initial roles of the teams, including Object Planning (design and

[2] Rhino Membrane is a Rhino plugin for form-finding tensile structures, with an associated plugin for Grasshopper. http://www.food4rhino.com/app/rhinomembrane-v20.

[3] SOFiSTiK is a software for finite element analysis. http://www.sofistik.com.

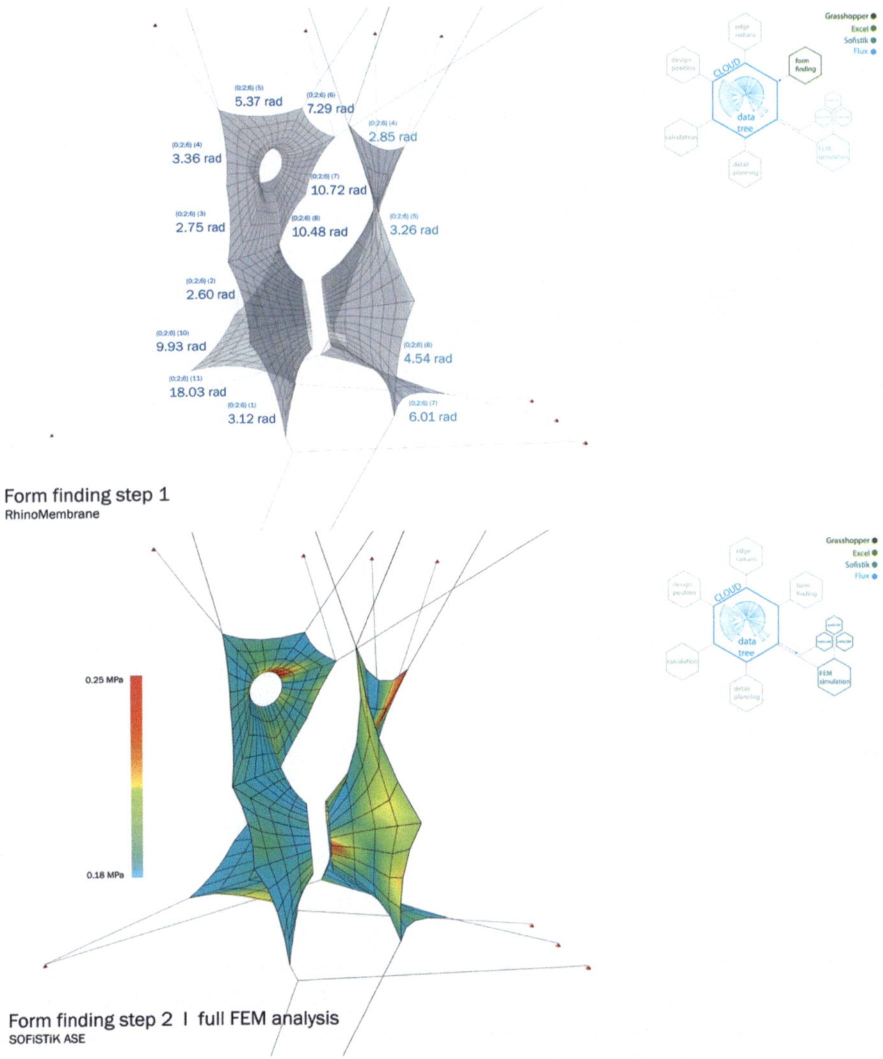

Fig. 4. Initial form-finding in Rhino Membrane and subsequent FEM analysis in SOFiSTiK

detailing), Project Management (material purchase and general time management) and Structure Planning (form-finding and structural simulations) (Fig. 5). With an expectation that data exchange between platforms would be required, this framework also

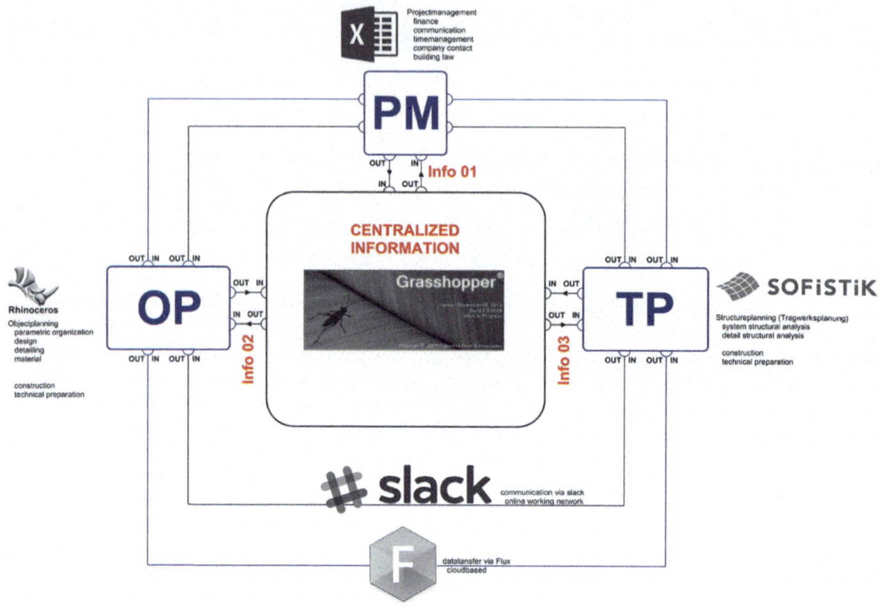

Fig. 5. The initially assumed setup of software and data flows that was used for the outset of the design system, also including the organization of the (human) design team resources

integrated the cloud based data exchange platform Flux[4] and the project communication platform Slack.[5]

Process Development

The academic setting and the clear objective in terms of structural and material applications allowed the development of a comprehensive model for the general workflow, as well as the establishment of a design system integrating several different design and simulation environments. Given these conditions, the student teams could, over time, establish a workflow and a design system that enabled several design iterations, and which also could facilitate late changes to parts of the proposal. The controlled process also gave the opportunity operate efficiently in the parallel teams with different roles. The general workflow (Fig. 6) allowed an initial design phase in which early form finding could be conducted within Rhino and Grasshopper using analytical equations to calculate force directions, and within Rhino Membrane for form finding iterations, with simplified models using straight non-bending members. In this first stage the cable forces were optimized for their resultant force directions to point in

[4] Flux is a cloud based platform for data exchange between different Grasshopper files as well as between different platforms such as Rhino, Grasshopper, Dynamo, Revit and Excel. http://Flux.io.

[5] Slack is an online communications platform for teams, that allows both open threads for all members to exchange information and direct person to person chat functions. http://Slack.com.

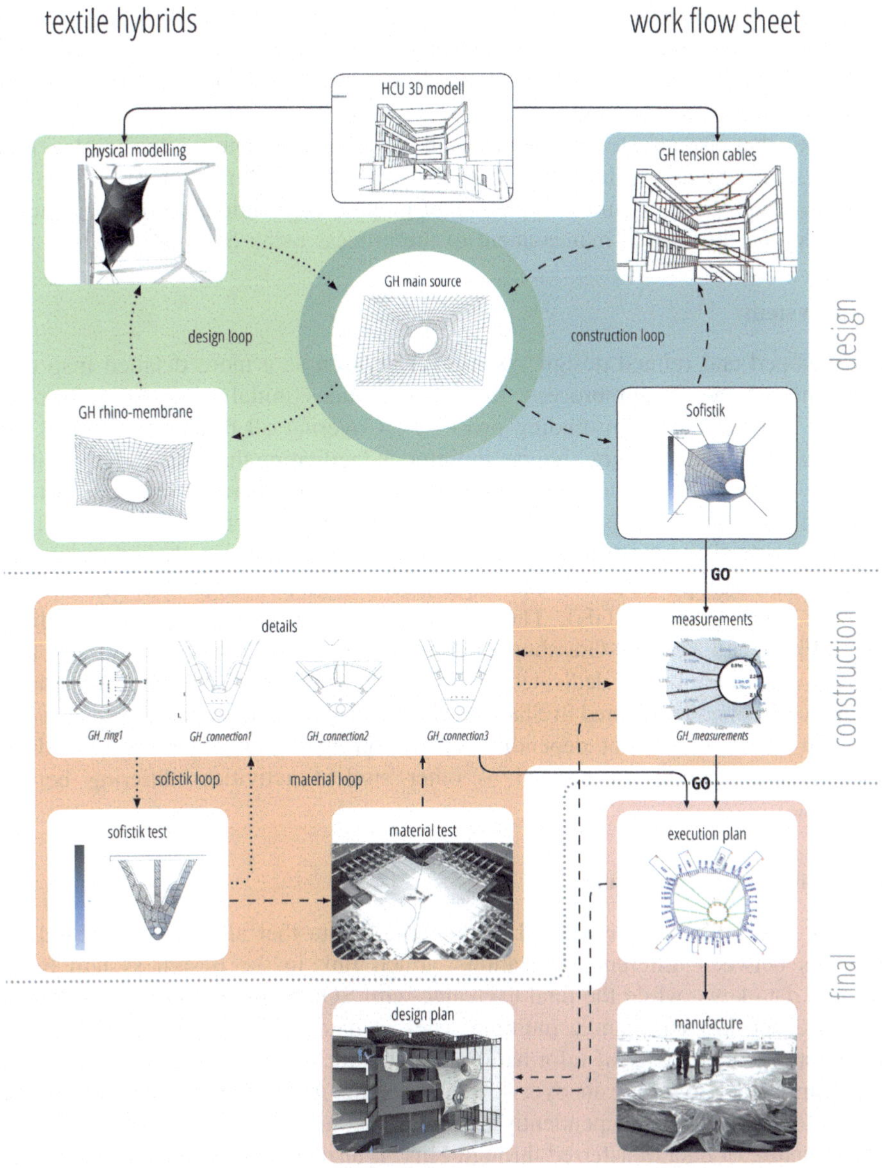

Fig. 6. Overview over general workflow indicating iterations of design and key thresholds between design, construction and final manufacture and assembly, conceptualized in the beginning of the design

the directions of supports and GFRP beams. At a second stage key elements (nodes, cables, beams and quad elements) where extracted from the model using the Rhino STiKbug plugin,[6] to be imported into Sofistik for static FEM calculations. During the development process, there were primarily four emerging aspects of workflow and conventions that also affected the process and the project outcome: a further developed design system organizing information exchange and workflow, the specific use of cloud-based data transfers in Flux, the use of graphic conventions in the Grasshopper definitions, and the overall management of the process using Slack.

Design System

The developed and refined design system can be seen as a more detailed map of all constituents of the design process, that also includes initial trials for information exchange and the dual form-finding processes. Covering all different technical platforms as well as dedicated grasshopper definitions facilitating the data exchange, it also informally depicts the relation between different teams for parallel development. The central information model is linked to six different environments, some more closely associated with direct Grasshopper access (such as rhMembrane.gh), others requires dedicated Grasshopper files for exchange (such as GHtoSofi.gh and SofiToGH.gh providing links to SOFiSTiK). The exchanges of data were generally facilitated through Flux, with a central Grasshopper definition as a main information model, and a set if individual definitions dedicated to information exchange. All other documented communication was conducted in Slack, enabling distant tutors to participate as well as inter-team collaboration not depending on synchronizing time (during periods the workshop was running in parallel to other student activities, differing between participants).

Cloud-Based Data Transfers

The Flux data keys are pre-defined containers of data that allows cloud-based data exchanges between different applications.[7] Each link in the design system has its dedicated data keys, while the final exchange with SOFiSTiK was conducted through text files. Each key can handle multiple types of data, which in this case allowed a simple yet precise transfer of information that also included important meta-data such as a central key providing an overview of all other keys. This feature allowed the different teams to work independently over periods, and still be aware of any changes in the structure of data transferred through Flux. It also contained an overview of what keys and what data was to be exchanged to and from all different files, such as the Central Information Model definition, the rhMembrane definition and the definitions providing connections to and from Sofistik (Fig. 7). The online interface could also

[6] STiKbug is a plugin at development stage, and not yet released, courtesy of R. La Magna.

[7] http://flux.io.

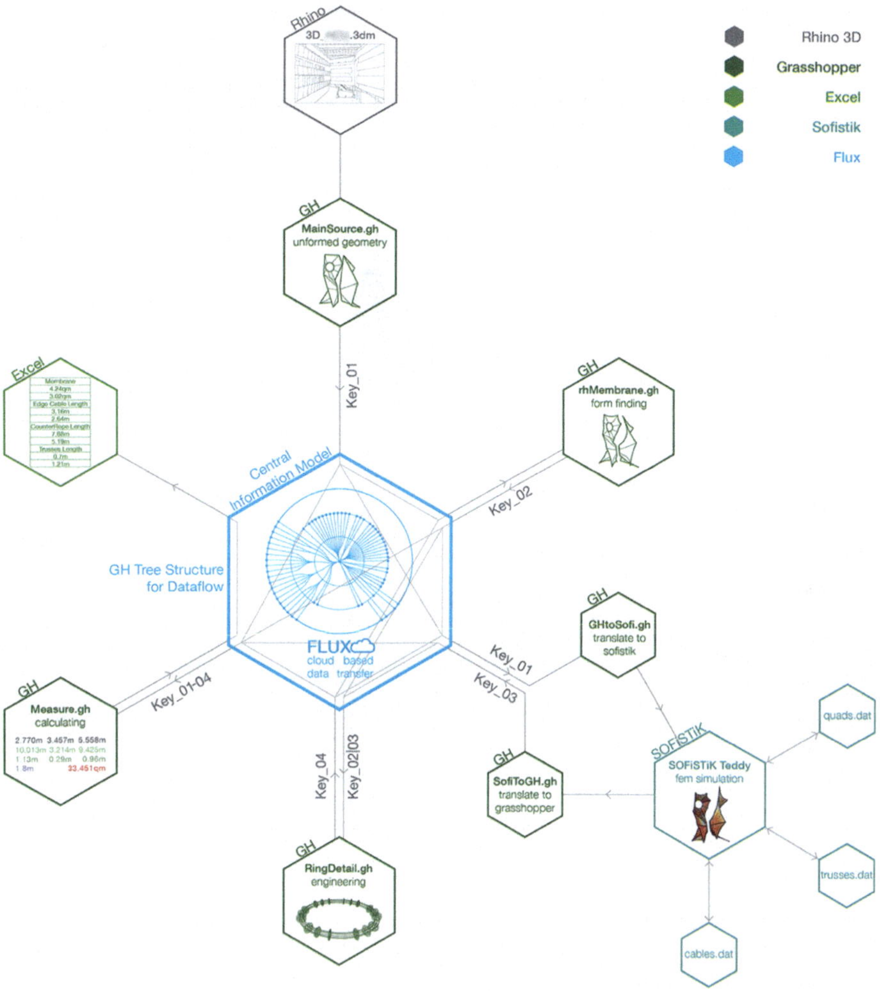

Fig. 7. The refined setup of software and data flows that constituted the design system

during the process provide visual cues to the status of geometrical data (Fig. 8). The inherent capacity for Flux to retain uploaded data furthermore allowed parallel work independent in time.

Graphic Conventions

The introduction of graphic conventions initially provided an improved legibility of all graphic definitions used. Further development adapted the conventions to include project specific aspects such as clear identification of geometry control mechanisms within the central file, specific modules for streaming data to different locations within the files, the monitoring of data key content within the Grasshopper definition, and

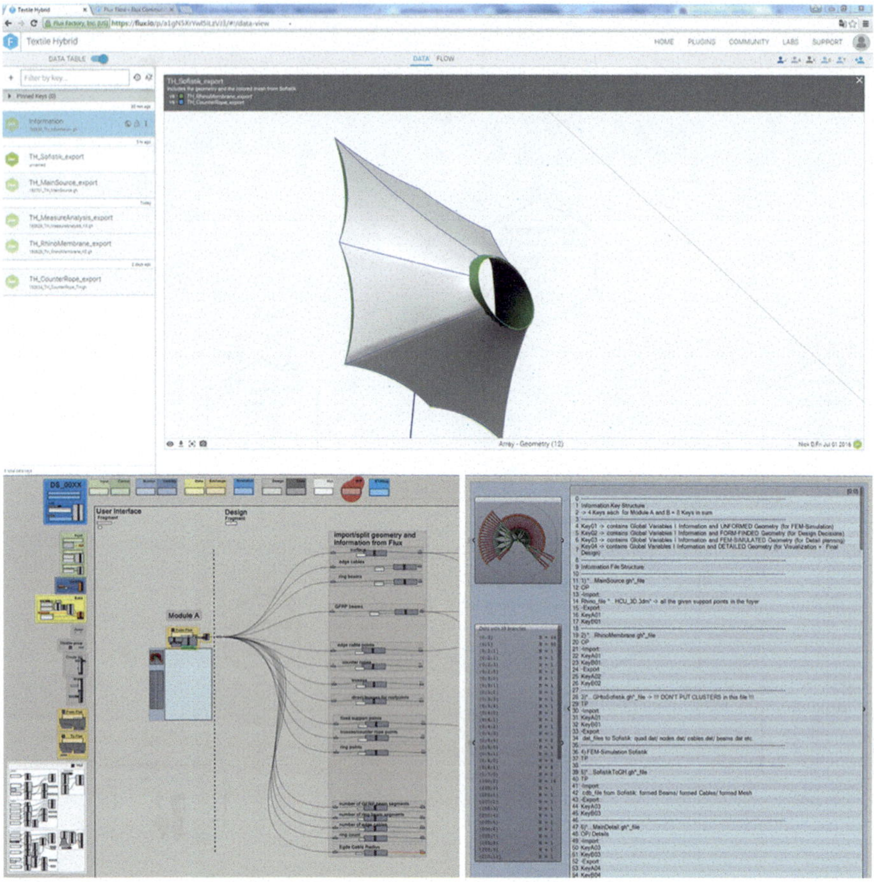

Fig. 8. Flux on-line preview of geometry uploaded from Grasshopper, extract of Grasshopper definition dedicated to receiving data from Flux for further processing, and a multi data key containing data structure, annotation and description of all data keys used

principles for the dedicated exchange definitions (Fig. 9). The use of clear naming conventions for data and different modules were an additional asset in the collaborative work. The overall initial conventions also included principles for dividing the Grasshopper canvas into different fields such as Control, Design and Export were further refined given that each definition included a complex setup of export functions to Flux.

Process Management

Slack was introduced as a general platform for communication, allowing tutors and supervisor to be engaged in the development from a distance.[8] Over time it also became

[8] http://Slack.com.

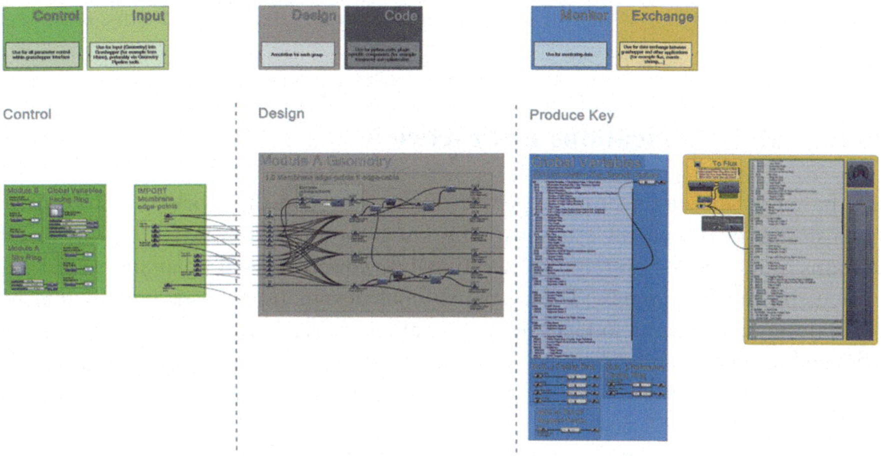

Fig. 9. Adapted graphical conventions

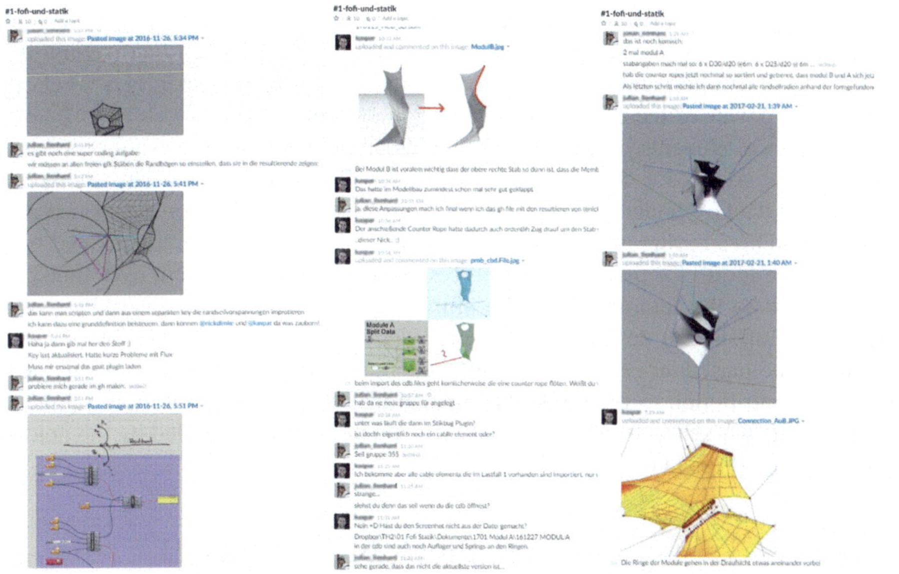

Fig. 10. Examples from the slack management in the thread specific to form-finding and statics

the platform for all exchange within the team, which in turn made the project development narrative documented in real time, including key decisions as well as general conversations and even arguments and disagreements. The platform includes the possibility to post images and files, further allowing direct discussions and evaluations over specific annotated model files or images. At times the communication also addressed specific items in the data streams in Flux, allowing direct feedback from tutors or team members, which in effect integrated the different platforms seamlessly (Fig. 10).

Findings and Applications in Practice

While based on the engineering expertise of structure and the workflows for computational design development established by Dsearch at White Arkitekter, the specific conditions of the workshop required additional development and adaptation conducted by the participating students. The computational design framework was in this way adapted during the process to facilitate new challenges as the process commenced, such as the need for version history management and archiving during design iterations. The process faced critical conditions at several points in time. Several issues such as fire hazard led to the use of a pure glass-fibre fabric with a stretch factor of less than 0.2%, something that posed a major challenge to the manufacturing process which was entirely carried out by students and became visible in the final outcome in terms of wrinkling (Fig. 13). At a late stage an initially included mirror had to be removed since an appropriate product could not be found. This first led to a heated and documented debate on Slack, but once the decision was made, the design system and the frameworks set up allowed for design adjustments and the generation of a new equilibrium system, detail solution and associated production data in the timespan of 1 week, thus proving the power of the comprehensive design system.

The team of students here acted not only as participants in the design process, but also as focus groups for the assessment of the employed framework. In analogy, the cloud based data exchange protocol not only operated as a vessel for data transfer, but also as an archive during iterative design development. This was complemented by the Slack on-line forum for managing the process, where the development over time could be monitored by the authors. The mode of investigating principles for workflow by the engagement of student design team has precedence (Davis et al. 2011), but the experiences of the particular set-up for design and engineering collaboration is regarded to be a new contribution to the field of computational design. The workflow framework applied in the workshop is under constant development, as exemplified by the practices of the authors (Fig. 11).

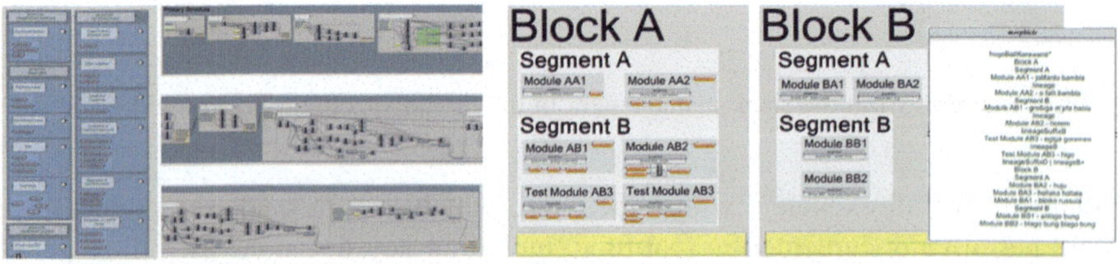

Fig. 11. Monitoring devices providing controls of previews of each definition module, and the assembly of morphological notes in an autolog, as part of the Dsearch adaptation of the framework after the workshop

Dsearch has in parallel further developed the framework in project applications. In several respects the lessons already learned from the Textile Hybrids workshop has informed the progress of development of standards and conventions, such as different trials for better monitoring of the step-by-step functionality of individual definitions. This is also partly based on additional sources where several layers of definition representations are proposed for enhanced legibility as well as preliminary specifications of definition functionality similar to pseudo-code (Zboinska 2015). First steps are also taken towards automatic documentation within definitions in terms of definition development as well as the morphological functions of definitions (the process of generating architectural form) in an autolog, which in turn could be linked to documentation and communication platforms such as Slack (Magnusson et al. 2017).

Within the practice of structure the workflows have also been adopted. So far parametric models were mostly used in the development of structures or parts of projects that can be distinctively and independently defined such as the ETFE roof in Imst, Austria (Fig. 12). In this case, the roof consists of pneumatically pre-stressed membranes, which are combined with a cable net to control the shape and maximum membrane forces in the ETFE film. For the design of the roof and the inner beam structure, several form-finding routines were integrated into the design system. The parametric model was used throughout the early design phases to define the shape of the roof and perform structural analysis with a direct linked to the FEM environment. The developments made during the Textile Hybrids workshop highlighted possibilities in working on a central and cloud based information model with separate expert groups by storing topology information in a versioned data tree. This approach now renders the possibility to introduce parametric workflows also to more complex types of structures with involvement of separate groups of experts (Fig. 13).

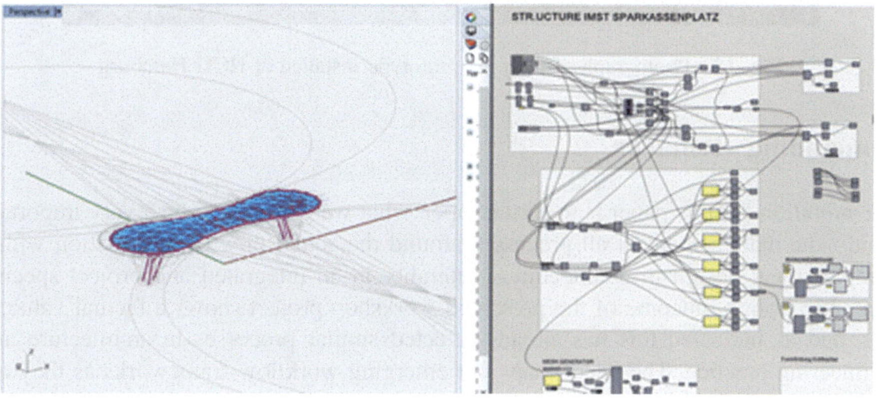

Fig. 12. The ETFE roof structure of Imst, employing aspects of the workflow framework

Fig. 13. Photograph of the final prototype installed at HCU Hamburg

Concluding Remarks

The ambition for this paper is to further open what we consider to be a very important topic—the management of all processes around the actual flow of information within computational design in architecture, preferably in an integrated and project specific way. The design outcome of the presented workshop project shows a factual value of this, and as indicated this has already affected similar processes in architecture and engineering practice. The paper maps out emerging workflow frameworks as they are continuously formed and reformed during computational design development in visual scripting environments. In terms of the experiment conducted with students this includes principles that allow iterative development for initial and final form finding, detail design and fabrication planning, as well as material testing across different computational design platforms and general project communication specific to computational design processes.

Experimental design workshops involving students always include unknown factors beyond design issues, such as skill level, ambition, engagement and general scheduling. In the case of the Hybrid Structures workshop, several teams achieved very advanced results, and not only carried through workflow-related conventions, but also contributed to new findings. The dynamic conditions of practice provide other challenges, such as changing project conditions, lessons learned from prototypes, or new fabrication limitations post tendering. Practice conditions rarely allow the establishment of comprehensive design systems, instead computational tools are often used to automate parts of the overall design process or fabrication deliveries for more complex situations of projects, merged with manually produced CAD information.

To establish analogous emergent workflow models in practice is challenging, but shows potential since it ensures correct processing and delivery of data, proper documentation of process in order to learn from past mistakes, systematizing re-occurring functions into re-usable scripts, as well as creating empirical data for future research. The employed workflow frameworks and the integration not only of design and simulation tools, but also communication platforms, are targeting these issues. Not only does this add another layer to best practice within computational design, it can also challenge conventional design practice and standardized quality systems.

Acknowledgements. Student Team: Marcelo Acevedo Pardo, Fabrizio M. Amoruso, Vasileios Angelakoudis, Nick Dimke, Kaspar Ehrhardt, Luisa Höltig, Timo van der Horst, Tobias Hövermann, Michel Kokorus, Dennis Mönkemeyer, Anies Ruhani-Shishevan, Gert Salzer, Anton Samoruko, Dino Tomasello, Zeliha Atabek, Björn Bahnsen, Niklas Dürr, Maria Fritzenschaft, Matthis Gericke, Mahmoud Ghazala Einieh, Can-Peter Grothmann, Moritz Seifert, Timo Volkmann

Surveying: Erik Jensen, Dominik Trau]

Supervision: Lehrstuhl Tragwerksentwurf Prof. Michael Staffa, Wiebke Brahms

Membrane Manufacturing workshop: Jakob Frick

HCU Labs: Thomas Kniephoff, Jens Ohlendieck, Kai Schramme

Fire Simulation: Ingenieurgesellschaft Stürzl mbH

Sponsors: MAX HOFFMANN, SOFISTIK, Serge Ferrari, Textilbau GmbH

References

Adriaenssens, S.: Feasibility study of medium span spliced spline stressed membranes. Int. J. Space Struct. **23**(4), 243–251 (2008)

Ahlquist, S., Lienhard, J., Knippers, J., Menges, A.: Physical and numerical prototyping for integrated bending and form-active textile hybrid structures. In: Gengnagel, C., Kilian, A., Nembrini, J., Scheurer, F. (eds.) Rethinking Prototyping: Proceeding of the Design Modelling Symposium, Berlin, pp. 1–14 (2013)

Davis, D., Burry, J., Burry, M.: Understanding visual scripts: improving collaboration through modular programming. Int. J. Archit. Comput. **9**(4), 361–375 (2011)

Deleuran, A.H., Pauly, M., Tamke, M., Tinning, I.F., Thomsen, M.R.: Exploratory topology modelling of form-active hybrid structures. Procedia Eng. 155, 71–80 (2016)

Hudson, R.: Strategies for parametric design in architecture: an application of practice led research. Ph.D. Thesis, University of Bath (2010)

Lienhard, J., Knippers, J.: Bending-active textile hybrids. J. Int. Assoc. Shell Spat. Struct. **56**(1), 37 (2015)

Magnusson, F., Runberger, J.: Design system assemblages: the continuous curation of design computation processes in architectural practice. In: Hay, R., Samuel, F. (eds.) Professional Practices in the Built Environment, pp. 197–204. University of Reading, Reading (2017)

Magnusson, F., Runberger, J., Zboinska, M., Ondejcik, V.: Morphology & development—knowledge management in architectural design computation practice. (eCAADe 35, forthcoming) (2017)

Oxman, R.: Theory and design in the first digital age. Des. Stud. **27**(3), 229–265 (2006)

Runberger, J., Magnusson, F.: Harnessing the informal processes around the computational design model. In: Thomsen, M.R., Tamke, M., Gengnagel, C., Faircloth, B., Scheurer, F. (eds.) Modelling Behaviour: Design Modelling Symposium 2015, pp. 329–339. Springer, Berlin (2015)

Schleicher, S., La Magna, R.: Bending-active plates: form-finding and form-conversion. In: Thomsen, M.R., Tamke, M., Gengnagel, C., Faircloth, B., Scheurer, F. (eds.) Modelling Behaviour: Design Modelling Symposium 2015, pp. 53–64. Springer, Berlin (2015)

Zboinska, M.: Boosting the efficiency of architectural visual scripts: a visual script structuring strategy based on a set of paradigms from programming and software engineering. In: Thomsen, M.R, Tamke, M., Gengnagel, C., Faircloth, B., Scheurer, F.(eds.) Modelling Behaviour: Design Modelling Symposium 2015, pp. 479–490. Springer, Berlin (2015)

Modelling Workflow Data, Collaboration and Dynamic Modelling Practice

Kåre Stokholm Poulsgaard[1(✉)] and Kenn Clausen[2]

[1] Institute for Science, Innovation and Society, University of Oxford, Oxford, UK
lncs@springer.com
[2] 3XN Architects and GXN Innovation, Kanonbådsvej 8, 1437 Copenhagen, Denmark

Abstract. This paper details the development of a digital model for a complex, large-scale façade design and presents an analytic framework for approaching design modelling as a case of distributed creative practice. Drawing on digital design theory, philosophy of mind and anthropology, we introduce the framework of cognitive ecology to analyse modelling practice around the façade design for the new IOC headquarters and show how this practice entails the simultaneous development of computational processes and collaborative workflows. Situated across 3XN architects and GXN innovation's internal R&D and design departments, we discuss how dynamic and coupled workflows add value to new forms of collaborative practice that are vitally engaged with extending capacity for computational and creative design thinking.

Keywords: Design modelling · Workflow · Computational design · Cognitive science

Introduction

As architectural modelling matures the scope and implications of its practice evolve. Increasingly, advanced computational tools play decisive roles in architectural design, leading to a profound questioning of relations between creativity and technics, geometry, data, tectonics, and materials (e.g. Oxman and Oxman 2011; Menges 2012; Picon 2010; Terzidis 2006; Kolarevic 2003). While these discussions often centre around formal and technical speculation, the transformative impact of design modelling holds the power to fundamentally restructure all stages in the design and delivery of complex buildings. In this paper, we discuss how computational modelling at 3XN architects and GXN innovation is expanding to entail the design of distributed collaborative processes that transform internal and external workflows at the studio.

K. De Rycke et al., *Humanizing Digital Reality*,
https://doi.org/10.1007/978-981-10-6611-5_41

Modelling Workflow

This paper presents the modelling of a large-scale façade design for the new International Olympic Committee (IOC) headquarters in Lausanne (Fig. 1). We discuss how the model developed from a tool for extensive shape research informed by design intent to an environment for integrating complex formal concerns with structural and fabrication constraints as the project developed from a single to a double skinned facade. This data-driven integration of different areas of expertise from designers and collaborators lead to an increasing significance of modelling workflow as well as form.

While form is naturally visible in completed buildings, the workflow and calculations underlying formal expression in finished structures remain hidden—as noted by architect Richard Garber: "[b]uildings alone, especially complex ones, cannot convey the collaborative activities that design teams have developed in the service of construction execution." (Garber 2017, p. 10) On projects like the IOC headquarters, where complex form puts high demands on all partners during both design and construction phases, workflow modelling emerges as a manifest corollary of design and an essential part of project delivery. To perceive the full significance of these complementary sides of modelling practice it is necessary to expand the scope of analysis. Here, we forward an analytic framework integrating elements from computational design theory, philosophy of mind and anthropology to advance the notion that models and modelers not only serve as tools for computational design thinking (Menges and Ahlquist 2011), but also come to act as vital environments for distributed cognitive processes mixing geometry and data in ways that can transform the social and technical institutions of architecture. Shifting the unit of analysis from isolated individuals to dynamic patterns of interaction amongst designers, collaborators, design models and digital workflows will allow for an exploration of cognitive interdependencies within the design studio while facilitating a discussion of the linked modelling of form and workflow (Fig. 2).

Fig. 1. Rendered output of the IOC façade (*left*). Mock-up of the final façade design (*right*)

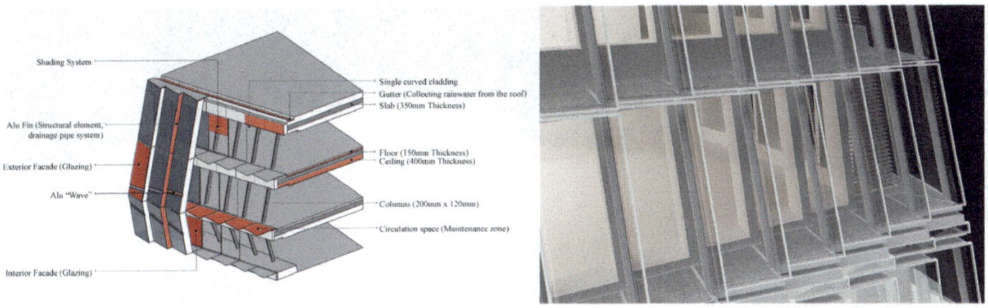

Fig. 2. Some of the complex interdependencies informing design of façade elements (*left*) and section from the façade model (*right*)

Cognitive Ecology

What are the theoretical, methodological and empirical commitments of taking workflow modelling serious? How might we begin to make sense of collaborative processes that are at once computational, aesthetic and material? How do these processes shape architectural imagination and practice? Proponents of digital modelling in architecture have been dealing vigorously with these questions for some time (e.g. Thomsen et al. 2006; Burry 2011; Davis 2013; Oxman 2008), but tend to stop short of analysing modelling as a distributed cognitive process bridging architectural design and construction. However, as design modelling matures, there is much to gain from applying a wider systemic perspective to its analysis (cf. Hight and Perri 2006; Garber 2009, 2017).

Architecture is a deeply distributed practice, intimately bound up with specific worldviews, technologies, materials and institutions, all of which shape creative collaboration. In this paper, we introduce the notion of *cognitive ecology* as an analytic framework for understanding design and workflow modelling across its various dimensions. Cognitive ecology has been advanced in anthropology to explain how the cognitive properties of synthetic or biological systems come to differ from the properties of individuals within these systems (Hutchins 2010, 2005 cf. Bateson 1972). It entails the study of cognition in context; a cognitive ecology describes a bounded system advancing and anchoring collective human thought. In this vein, computational design models and workflows can be analysed as co-constitutive of cognitive ecologies encompassing designers, engineers, software, scripts, and screens as well as organisation and memory of projects and the wider project team. This analytical framework is committed to a systemic perspective and shifts the unit of analysis from the inherent properties of elements within the system (designer, script, software) to the emergent properties of dynamic patterns of interaction between these elements (Varela et al. 1991; Thompson 2007; Malafouris 2013).

The sheer complexity of contemporary modelling practice seems to lend itself well to this type of analysis as computation, mathematics, and simulation become ever more integrated into design thinking. Computational tools bring together diverse teams around the creation of advanced 3D forms by integrating the vast calculations that often

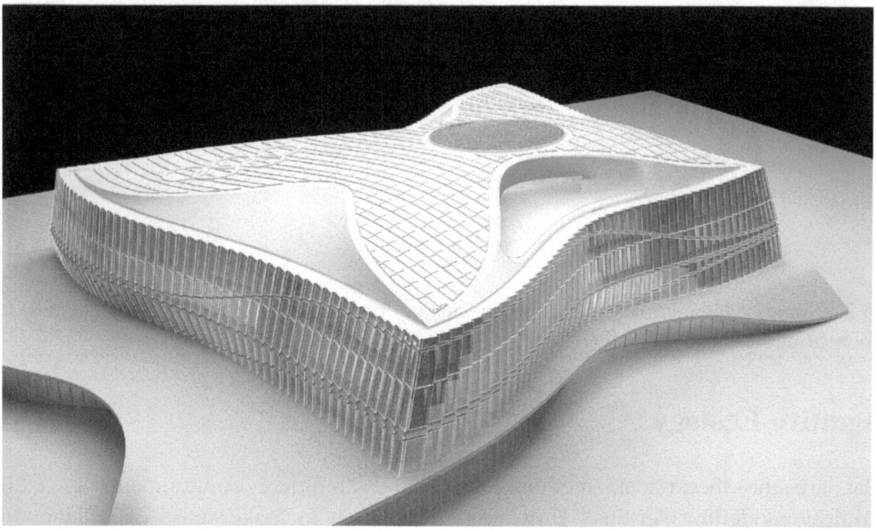

Fig. 3. Full façade model

underlie these into computational systems where geometry translates to data, and data to geometry. This shared informational basis permit functional integration across a range of software environments allowing for modular manipulation, combination and re-use of input and outputs from models across all phases of architectural design and fabrication. Data exchange and dynamic interaction between people, tools, expertise and models lead to emergent designs whose formal and computational properties exceed anything that could be achieved in isolation. In line with an ecological perspective on cognition, it seems the designing mind is vitally collaborative, distributed and emergent in contemporary design modelling (Poulsgaard and Malafouris 2017). Design thinking is empowered by the dynamics of specific modelling environments, models and workflows and creative agency becomes and emergent property, located not in the brain of the individual designer but in distributed transactions within larger cognitive ecologies. This perspective permits extending the analysis of design modelling, to also encompass collaborative workflows, through empirical investigations of the modelling and integration of diverse software, data streams, and areas of expertise during design and construction of complex buildings (Fig. 3).

Case: IOC Façade Model

In 2013, the International Olympic Committee celebrated the 100th anniversary of its establishment in Lausanne, confirming the Swiss town as its base for another 100 years to come by holding an architectural competition for a new headquarters. 3XN architects developed the winning entry as an embodiment of Olympic values with a Scandinavian twist: a context aware dynamic design seeking to merge aesthetics, functionality, and performance. In recognition of the symbolism of the Games and needs of the organization, the new IOC headquarters was designed around three key elements: movement, flexibility and sustainability. The 3XN architects and GXN innovation design and modelling teams has been faced with developing and maintaining this design language while steering the project through design development and construction documentation with a team of expert collaborators.

Modelling Ecology

By using three freeform curves as basis for subsequent façade detailing, the modelling team established and explored parametric links between curves and surfaces in the façade geometry (Fig. 4). Within a short time, the team had constructed a flexible parametric model that formalised design concept in a computational environment; this allowed them to iteratively test geometric impacts of design interventions and fabrication constraints while fleshing out design concept from competition phases during early design development.

This initial exploration sought to establish a workflow that could combine flexibility and speed with high precision in order to meet tight deadlines with collaborators. The dynamic design of the building and façade meant that any iteration could have unpredictable consequences as it scaled through the parametric model while affecting geometric integration, structural integrity, aesthetics, comfort and building performance. To understand and manage these complex relations, the team sought a solution that would tie the specialisation and expertise of design modelling and design development into an efficiently working whole. To do so, the modelling team established a live data structure linking façade model across Rhino and Revit software environments using Grasshopper, Dynamo, Flux and Python to create a two-way data link (Fig. 5). The initial data structure was set up around a comprehensive building grid for locating each individual façade module; additionally, individual modules were assigned four corner points, a perpendicular vector describing its angle and orientation, and a unique id number. This established an explicit data structure allowing for connecting and

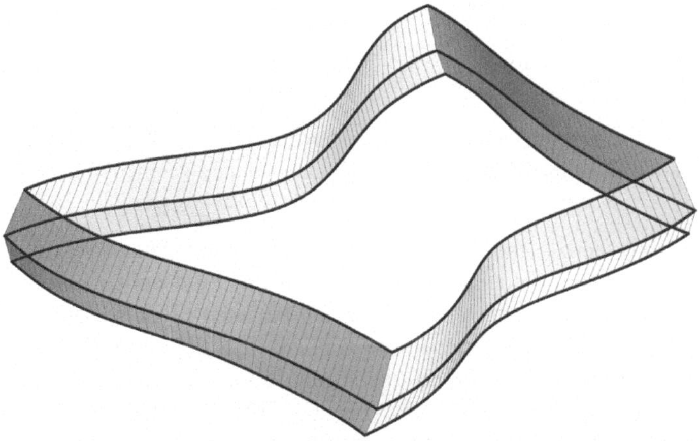

Fig. 4. Three free-form curves guide the global design of the façade model

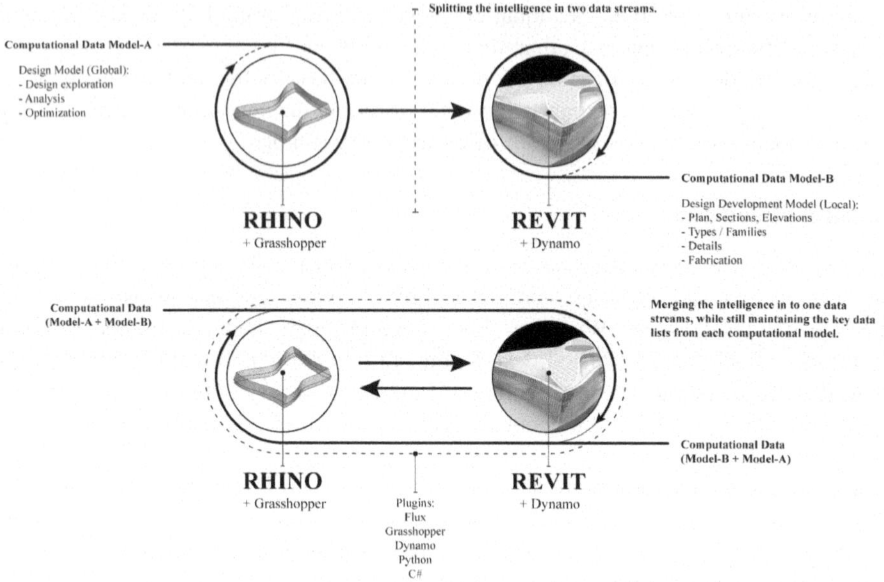

Fig. 5. Modelling environment establishing two-way data feed between Rhino and Revit to keep building intelligence live and able to inform local and global design development

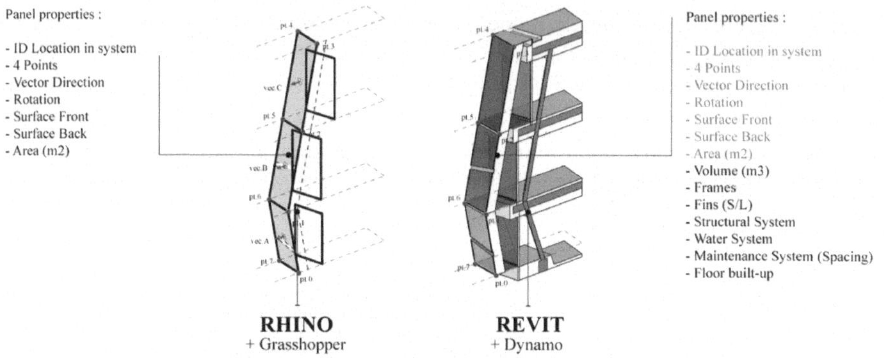

Fig. 6. Predefined data structure linking elements across modelling environments enables iterative integration of element data in Revit while maintaining geometric flexibility in Rhino

updating elements across the Revit–Rhino environment. This live data connection meant that any iteration, calculation or change in one environment would inform the other. The link established continuous feedback between models, connecting different resolutions of design and detailing; Rhino was continuously used for global and local surface level analysis and optimisation, while Revit was used for local detailing, documentation and type based data flow (Fig. 6). This permitted efficient collaboration with partners and consultants around specific structural problems using Revit while the design team maintained control of overall design expression and modelling in Rhino.

Through this dialectic between local problem solving and global design modelling, the model and workflow integrated aesthetic concerns with technical solutions to problems posed and explored in with a range of collaborators during design development phases.

Iterative Computation

Feedback from façade engineers and structural engineers led to a specific focus on geometric optimisation of façade elements and the introduction of a double-skinned façade to strike a balance between structural performance, maintenance criteria, and aesthetics (Fig. 7). The double-skinned façade placed high demands on both design team and collaborators for developing design and construction documentation while maintaining the overall dynamic expression. As each section of the façade is unique and all elements are computationally connected, this required continuous iteration and control of a wide variety of parameters and their integration; as complexity grew, the model and data structure developed during early design phases proved an essential environment for collaboratively solving the geometry of the complex façade under tight deadlines.

Utilising the linked modelling environment and data structure, 2D sections detailing construction principles for the double-skinned façade were developed in association with façade and structural engineers. These principles initially focused on defining a viable angle domain for the loadbearing steel columns supporting façade and floors (Fig. 8), as well as minimum and maximum spacing between inner and outer skin required for maintenance (Fig. 9).

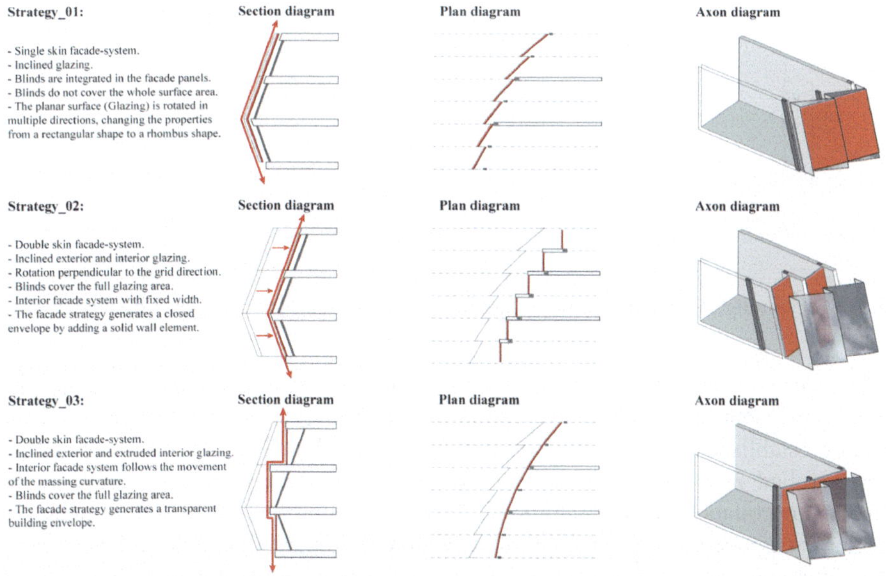

Fig. 7. Design research for geometric implications and optimization around introduction of double skinned façade

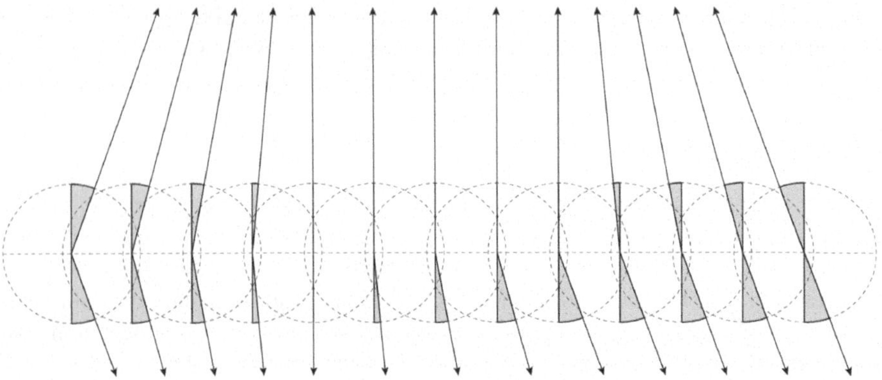

Fig. 8. Exploration of inclination and angle domain for outer façade

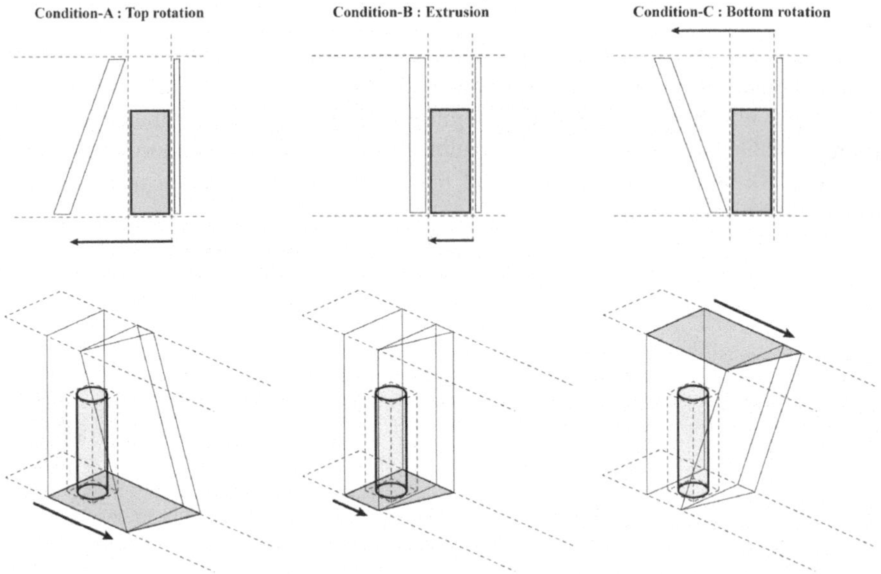

Fig. 9. Computing 3D spacing between interior and exterior skin to meet performance demands

The design team applied these principles across the full 3D model to generate relations between each individual section of the façade following the grid and data structure established during initial modelling phases (Fig. 10). Once these principles where applied at a surface level, the model was used to iteratively circulate data packages to collaborators for additional design detailing and structural computation. Structural engineers received continuously updated angle domains and vectors for calculating structural performance of the façade, while façade engineers received the latest updated surface model, plan outlines and updated area calculations for façade elements and glazing; all partners had access to the full design model as it developed

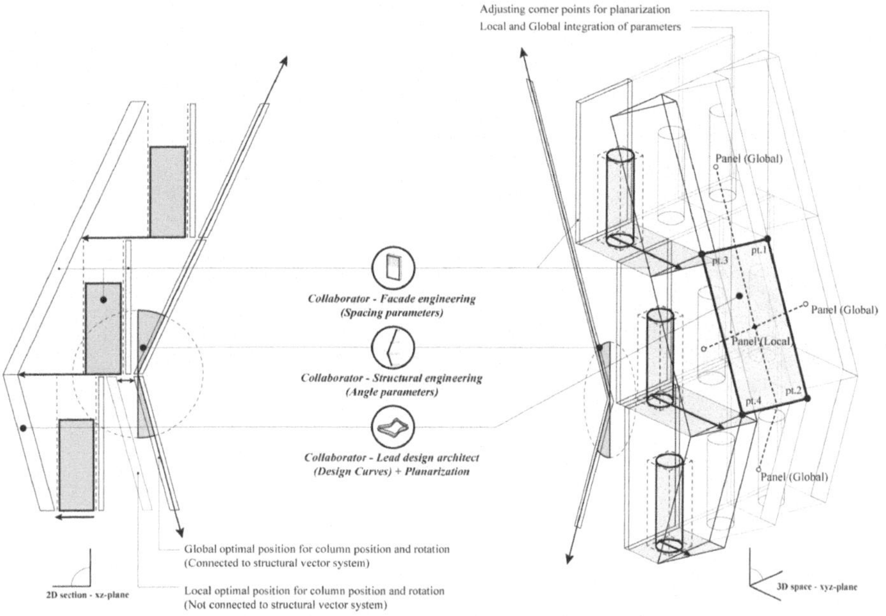

Fig. 10. Local and global integration of structural principles and design parameters across 2D and 3D environments

with time, allowing them to follow the evolving façade as additional parameters and data were added. For each successive iteration, the design team used the modelling environment for comparative analysis and manipulation across form, angle and spacing domains to ensure that all sections remained within the established baseline parameters while also maintaining the global design expression (Fig. 11). In turn, sections and construction details were continuously updated in collaboration with the different engineering teams as more and more detailed information was added to the model: window glazing and frames, dimensions of columns, joints, materials, insulation and so forth. During these stages, design exploration expanded from surface level analysis of the effects of applying specific construction principles to also incorporating actual building dimensions and construction details (Fig. 13). The data structure linking elements across Rhino and Revit environments served as an essential backbone for quick and iterative analysis and exchange of building data during these stages; this proved crucial for enabling both centralised and de-central exchange within the wider team of collaborators by establishing a shared and live frame of reference as the project evolved (Fig. 12).

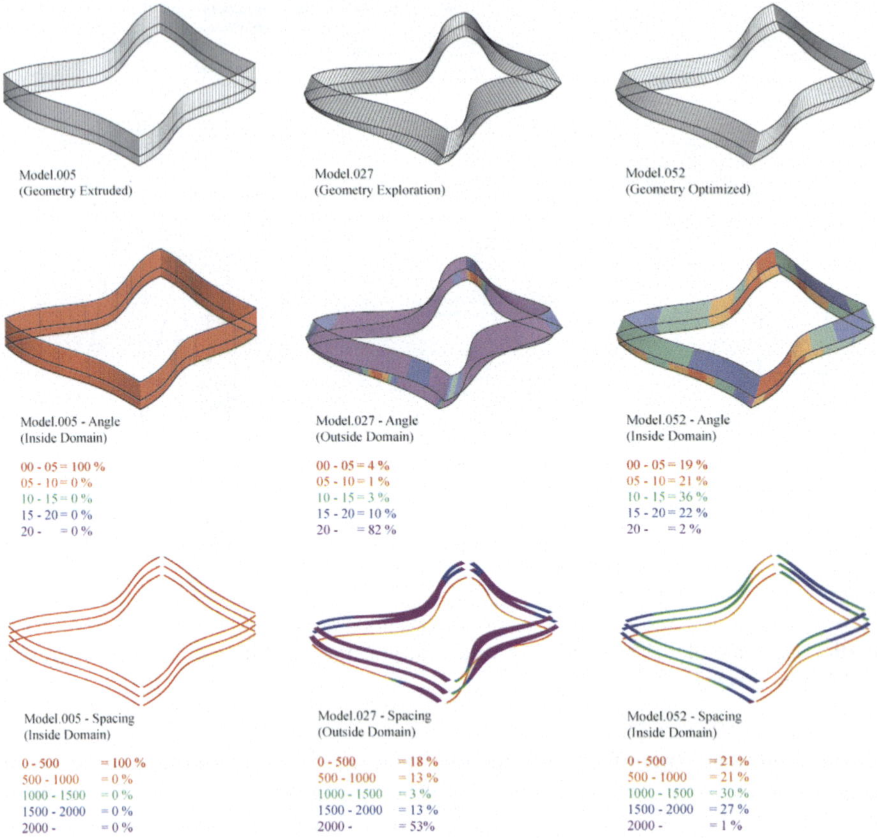

Fig. 11. Iterative design research into form (*top row*), angle (*middle*) and spacing (*bottom*) domains. Comparative analysis across these domains allowed the modelling team to evaluate performance and aesthetics across a large number of design iterations to find the best integration meeting structural criteria while maintaining a dynamic design expression

Discussion: Collaborative Workflow and Cognitive Ecology

Architects Menges and Ahlquist (2011, p. 16) propose that computational design thinking requires a deep understanding of how design models operate as form and as mathematical ordering constructs. In this, the position of the designer is changing as work moves from relative free-form design exploration to rule-based discovery within computational environments (cf. Simon 1996, p. 124; Cross 2006, p. 32). This raises interesting questions about the structuring of creative work within these highly rule bound environments. Combining modelling practice and the framework of cognitive ecology brings to light the co-constitutive relationship between design thinking, model and workflow as they co-evolve, increasing capacity for creative collaboration. We argue that the IOC model and workflow established a cognitive ecology for technical and aesthetic design exploration by coupling the distinct areas of expertise of designers,

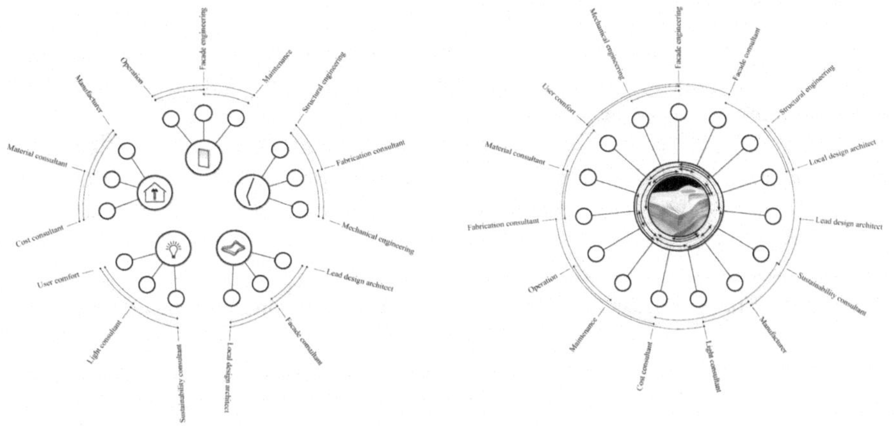

Fig. 12. Contrary to decoupled workflows freezing geometry to distribute subsets of the model (*left*), modelling workflow on the IOC façade, created a coupled system where all elements and data remained online for collaborators via access to a central model during design development and construction documentation (*right*)

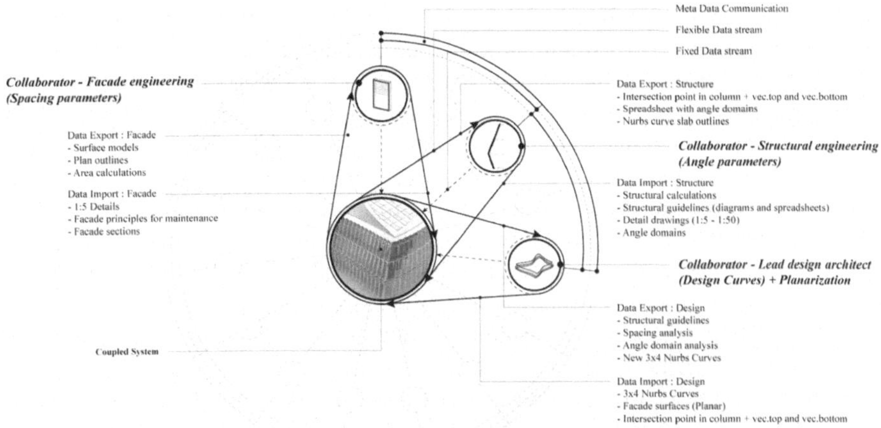

Fig. 13. Iterative computational modelling and evaluation of construction principles around inside and outside angling domain for the interior and exterior skin and spacing between them. This was one area amongst many that was managed and solved via the coupled workflow established by the model

façade engineers, and structural engineers (amongst others) and leveraging the different affordances and possibilities of Rhino and Revit software environments (Fig. 12). This dynamic coupling is integral to computational design thinking and deeply reliant on the simultaneous modelling of form and workflow.

Anthropologist Hutchins (2005, 1995) has shown how the solution of complex problems involving input and expertise from a variety of fields often come to rely on external environments for selective combination and manipulation of these fields. These environments serve as anchors; two or more input spaces are selectively projected into a separate, blended space anchored in materials or technics, which thereby develops emergent cognitive properties not available from the individual inputs. These cognitive ecologies permit collaborators to selectively access, combine and process specific aspects within them while ignoring others and this greatly enhances the emergent possibilities for creative problem solving (cf. Vallée-Tourangeau and Villejoubert 2013). These process often come to rely on material and technical environments that are: "sufficiently immutable to hold the conceptual relationships fixed while other operations are performed," but supple enough to allow continuous manipulation (Hutchins 2005, p. 1562). Modelling and workflow on the IOC project established such a cognitive ecology, allowing for the projection and selective blending of expertise, principles and calculations from designers, structural engineers, and façade engineers amongst others. This relied in part on the establishment of a clear and explicit data structure that acted as a notation system; it set out rules, classes and relations between them, and became integral to structuring collaboration on the complex design project by allowing several collaborators to work within well-established and 'sufficiently immutable' boundaries. However, the linked modelling environment also allowed the manipulation of data as geometry and a more open-ended exploration of how different computational problems and solutions would affect the overall design expression. This helped recast complex mathematical optimisation problems into visually coded on-screen geometry, that could be compared and manipulated on the fly by the design team who thereby maintained control of form and design expression throughout the complex project. The simultaneous rigidity of data driven notation systems and fluidity inherent in the linked modelling environment meant that design exploration and collaboration on the IOC project could move between two poles: at one end was aesthetic interpretation with rich scope for dynamic iteration and change, at the other, the data structure serving as notation system in which the structural options were comprehensively coded. The efficacy of the resulting model and workflow lay in its possibilities for functional simplification, the way it allowed both design team and collaborators to selectively zoom in on, mix and process specific aspects of local and global design development while ignoring others. This permitted research into highly complex and dynamic systems at a human scale allowing the design team to continuously manipulate the model and input of collaborators to find the best fit between optimal performance and design aesthetics.

Properly linked, digital models centralise and decentralise at once. Drawing many diverse actors and expertise into a common infrastructure, they allow for decentralised communication and some autonomy at the edges while also standardising conditions of communication between them. The distributed perspective on cognition emphasise that the emergent properties of a cognitive ecology arise from the dynamic and continuous coupling of elements within it (Hutchins 2010; Varela et al. 1991); in emergent systems there is significant added value in keeping and managing a linked and dynamic ecology rather than a fully distributed one-way setup which lessens feedback flows. At the IOC project, recursive cycles of design exploration with additional data from collaborators

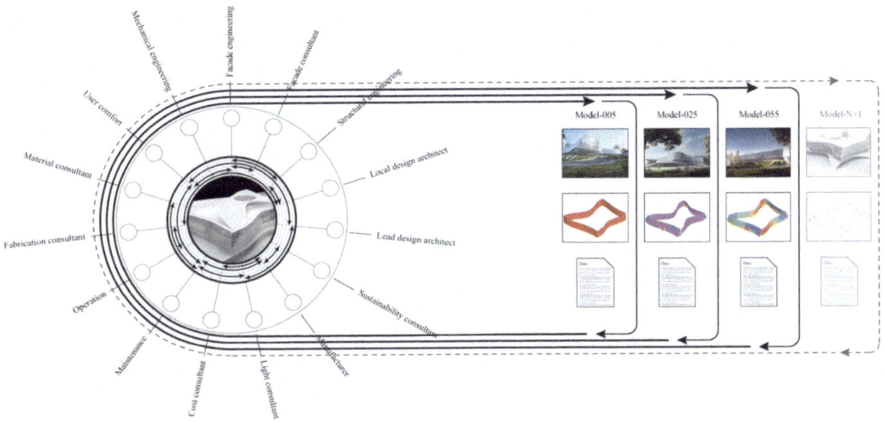

Fig. 14. The IOC model and data structure served as a robust environment for integrating design solutions while creating project documentation at different resolutions; each iteration permitting detailed visualisations, ongoing structural and aesthetic analysis, and updated data packages for servicing individual collaborators

informing each cycle, allowed the full team of architects, engineers and specialists to generate a wider number of explorative solutions for evaluation and selection as the design matured towards construction (Fig. 14). In this way, the dynamic flexibility of the original design intent found its counterpart in the iterative development of design model, workflow and collaborative processes.

Conclusion

As large-scale models mature and incorporates increasing levels of information, the scope and reach of modelling necessarily expands. The IOC model developed from a design tool for extensive shape research, to include additional factors and data as the project developed, including structural behaviour, maintenance conditions and fabrication constraints—all while facilitating creative design thinking and collaborative workflow on the project.

At 3XN architects and GXN innovation, models increasingly become essential nodes in studio practice around complex projects by creating a dynamic environment for ongoing computation of internal and external inputs while also performing as a medium for aesthetic collaboration and communication. To explore the meaning of this shift, we have introduced the framework of cognitive ecology, analysing the co-constitutive relationship between design thinking, modelling and workflow. In this perspective, modelling becomes a hybrid evolving system comprising computational, aesthetic, and structural factors dynamically interacting with each other. This combined approach opens new venues for the study of digital technology and practice within computational design modelling, and for thinking about the relationship between practice, innovation and change within the architectural professional community.

Acknowledgements. The authors would like to thank our colleagues, collaborators and client in the wider project.

References

Bateson, G.: Steps to an Ecology of Mind: Collected Essays in Anthropology, Psychiatry, Evolution, and Epistemology. Chandler Pub, San Francisco (1972)
Burry, M.: Scripting Cultures: Architectural Design and Programming. Wiley, Chichester (2011)
Cross, N.: Designerly Ways of Knowing. Springer, London (2006)
Davis, D.: Modelled on software engineering: flexible parametric models in the practice of architecture. Ph.D. Dissertation. School of Architecture and Design, RMIT University (2013)
Garber, R.: Closing the Gap: Information Models in Contemporary Design Practice. Wiley, Chichester (2009)
Garber, R.: Workflows: Expanding Architecture's Territory in the Design and Delivery of Buildings. Wiley, Chichester (2017)
Hight, C., Perry, C.: Collective Intelligence in Design. Wiley, Chichester (2006)
Hutchins, E.: Cognition in the Wild. MIT Press, Cambridge (1995)
Hutchins, E.: Material Anchors for Conceptual Blends. J. Pragmat. **37**(10), 1555–1557 (2005)
Hutchins, E.: Cognitive ecology. Top. Cognit. Sci. **2**(4), 705–715 (2010)
Kolarevic, B.: Architecture in the Digital Age: Design and Manufacturing. Spon, New York (2003)
Malafouris, L.: How Things Shape the Mind: A Theory of Material Engagement. MIT Press, Cambridge (2013)
Menges, A.: Material Computation: Higher Integration in Morphogenetic Design. Wiley, Chichester (2012)
Menges, A., Ahlquist, S.: Computational Design Thinking. Wiley, Chichester (2011)
Oxman, R.: Digital architecture as a challenge for design pedagogy: theory, knowledge, models and medium. Des. Stud. **29**, 99–120 (2008)
Oxman, R., Oxman, R.: Theories of the Digital in Architecture. Routledge, Abingdon (2011)
Picon, A.: Digital Culture in Architecture: An Introduction for the Design Profession. Birkhäuser, Basel (2010)
Poulsgaard, K., Malafouris, L.: Models, mathematics and materials in digital architecture. In: Cowley, S., Vallée-Tourangeau, F. (eds.) Cognition Beyond the Brain: Computation, Interactivity and Human Artifice, 2nd edn. Springer, London (2017)
Simon, H.: The Sciences of the Artificial. MIT Press, Cambridge (1996)
Terzidis, K.: Algorithmic Architecture. Architectural, Oxford (2006)
Thomsen, M., Ayres, P., Nicholas, P.: CITA Works. Riverside Architectural Press, Oxford (2006)
Thompson, E.: Mind in Life: Biology, Phenomenology, and the Sciences of Mind. Harvard University Press, Boston (2007)
Vallée-Tourangeau, F., Villejoubert, G.: Naturalising problem-solving. In: Cowley, S., Vallée-Tourangeau, F. (eds.) Cognition Beyond the Brain: Computation, Interactivity and Human Artifice. Springer, London (2013)
Varela, F., Tompson, E., Rosch, E.: The Embodied Mind: Cognitive Science and Human Experience. MIT Press, Cambridge (1991)

Equilibrium-Aware Shape Design for Concrete Printing

Shajay Bhooshan(✉), Tom Van Mele, and Philippe Block

ETH Zurich, Institute of Technology in Architecture, Block Research Group, Stefano-Franscini-Platz 5, HIL H 46.2, 8093 Zurich, Switzerland
bhooshan@arch.ethz.ch

Abstract. This paper proposes an Interactive Design Environment and outlines the underlying computational framework to adapt equilibrium modelling techniques from rigid-block masonry to the multi-phase material typically used in large-scale additive manufacture. It focuses on enabling the synthesis of geometries that are structurally and materially feasible vis-a-vis 3D printing with compression-dominant materials. The premise of the proposed research is that the material printed in layers can be viewed as micro-scale bricks that are initially soft, and harden over time—a kind of micro-stereotomy. This insight, and the observation about the necessity of design exploration yields the principal contributions of the paper: A computational framework that allows for geometric reasoning about shape and exploration of associated design-space, and early, proof-of-concept results.

Keywords: 3D printing · Equilibrium modelling · Shape design · Unreinforced masonry · Concrete and clay printing

Motivation

The shape-design and computational framework that is outlined in this paper is motivated by the rapid growth in desktop and large scale 3D printing and the lack of exploratory design tools thereof. The framework is based on a novel insight regarding the applicability of design and analysis methods used in unreinforced masonry to large scale, layered 3D printing with compression dominant materials such as concrete and clay.

Large Scale 3D Printing and Lack of Tools for Design Exploration

Khoshnevis et al. (2006) of Contour-crafting®, pioneers in large-scale 3D printing, deliberate two particular, shape-design related working points for the further development of the technology: (1) expressiveness of geometric modelling for designers and (2) constructive guidance about the feasibility constraints imposed by the fabrication process. Similar design and data related issues were also outlined and nominally addressed, by the other pioneering

The original version of this chapter was revised: For detailed information please see Erratum. The erratum to this chapter is available at https://doi.org/10.1007/978-981-10-6611-5_56.

K. De Rycke et al., *Humanizing Digital Reality*,
https://doi.org/10.1007/978-981-10-6611-5_42

effort at Loughborough University (Buswell et al. 2008). Further, necessity and opportunity for exploring the very design related aspects that have hitherto been side-lined, is highlighted in the maturation of the concrete printing technology in the last decade, along with the advent of several new competitors (Khoshnevis 2004; Buswell et al. 2007; Dini 2009; CyBe-Construction 2016; XtreeE 2016).

Compatibility of Funicular Structures and Concrete Printing

The dominant structural properties of printed concrete are high density (~2.2 Tons/m^3), high compressive strength (~100 MPa), relative weak flexural strength (10% of compressive strength), and a unique orthotropic behaviour induced by the layered printing process, yielding a weak tensile bond between layers (~0.5–3 MPa) (Le et al. 2012a; Perrot et al. 2016; Wangler et al. 2016) (Fig. 1). These properties make it well suited for the adaptation of the design methods of unreinforced masonry structures, including the historic methods of graphics statics (Culmann 1875), their application to the design of concrete structures (Zastavni 2008; Fivet and Zastavni 2012), and their recent computational extensions (Block and Ochsendorf 2007; De Goes et al. 2013; Vouga et al. 2012; Tang et al. 2012; Liu et al. 2013; Van Mele et al. 2014). In addition, the structural aspects of the practice of stereotomy (Sakarovitch 2003; Fallacara 2009; Rippmann 2016), which align material layout with expected force flow through the structure, are particularly relevant. The layered filaments of concrete printing, especially when they can be aligned along orthonormal directions of a surface (Gosselin et al. 2016), would require similar careful alignment. The premise of the proposed research is that the material printed in layers can be viewed as micro-scale bricks that are initially soft, and harden into rigid filaments (blocks) over time—a kind of micro-stereotomy (Fig. 1 right).

Objectives and Contributions

The primary objective of the research presented, is the synthesis of geometries that are structurally and materially feasible vis-a-vis 3D printing with compression-dominant materials. In other words, equilibrium aware, shape-design for concrete printing.

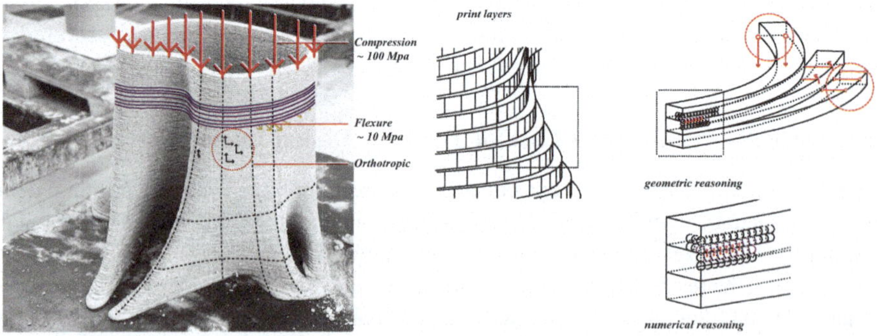

Fig. 1. *Left* dominant material properties of printed concrete. *Right* intention to discretise the print layers into micro-blocks, to enable geometric reasoning about equilibrium and print feasibility

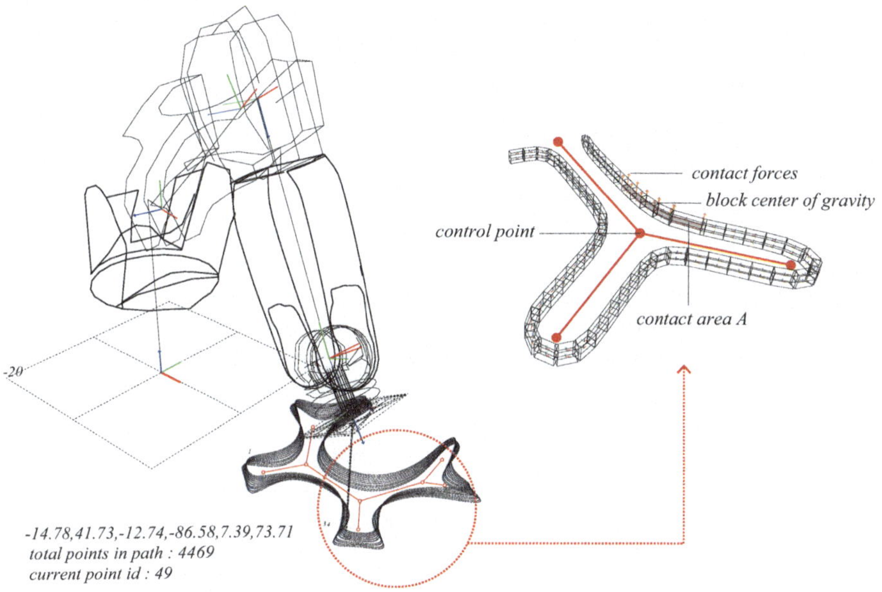

Fig. 2. Interactive design explorer (IDE) for concrete printing

The insight relating unreinforced masonry to 3D Printing and the preceding observation about the necessity of design exploration for the further development of 3D printing as an architectural tool yields the principal contributions of the paper: A computational framework that allows for a geometric reasoning about shape and exploration of the associated design-space of novel and feasible shapes. As a specific contribution, the proposed Interactive Design Environment (IDE) (Fig. 2) and the underlying computational framework (CF) adapt equilibrium modelling techniques from rigid-block masonry to the multi-phase material typically used in large-scale additive manufacturing. Lastly, proof-of-concept physical outcomes are an additional contribution.

Related Work

Design Paradigm Aspects

The importance of geometric methods of structural design, and the benefits of integrating structural and construction-aware computational methods in the early-design phase is well argued in Lachauer (2015) and Rippmann (2016). Recent developments in the field of Architectural Geometry, with its explicit aim of "incorporation of essential aspects of function, fabrication and statics into the shape modelling process" (Jiang et al. 2015), extend this so-called Computer Aided Geometric Design (CAGD) paradigm to architectural design. In the realm of large-scale 3D printing however, these design aspects have not received much attention, barring a few recent efforts (De Kestelier and Buswell 2009; Chien et al. 2016; Wolfs and Salet 2015).

Computational Aspects

Computational modelling of 3D printing should consider both structural stability of complex geometries made of printed concrete and the feasibility of printing them (Fig. 3). The first is a problem of developing geometries aligned with expected loading and force-flow. The second is additionally, dependent on time, material rheology etc. Materially, the hardened state of the concrete can be considered orthotropic (Fig. 1) and the fresh state is often considered a so-called visco-plastic Bingham material (Mechtcherine et al. 2014; Tanigawa and Mori 1989).

Equilibrium of Soft-Rigid Material

The proposed computational framework (CF) intends to unify the representation of both the aspects above. The idea here is to scale-down and adapt equilibrium modelling methods based on rigid-body mechanics, to include semi-rigid and time-dependent behaviour. Thus the two classes of methods that encode rigid-body mechanics—rigid-body dynamics (RBD) and rigid-block equilibrium (RBE)—are important precedents. The research presented here is based on the established methods of RBD (Hahn 1988; Baraff 1989; Hart et al. 1988) and RBE (Livesley 1978; Whiting 2012; Heyman 1966; Livesley 1992). Specifically, the proposed method extends the RBE method of Frick et al. (2015) to accommodate the soft-rigid aspects of 3D printing of concrete.

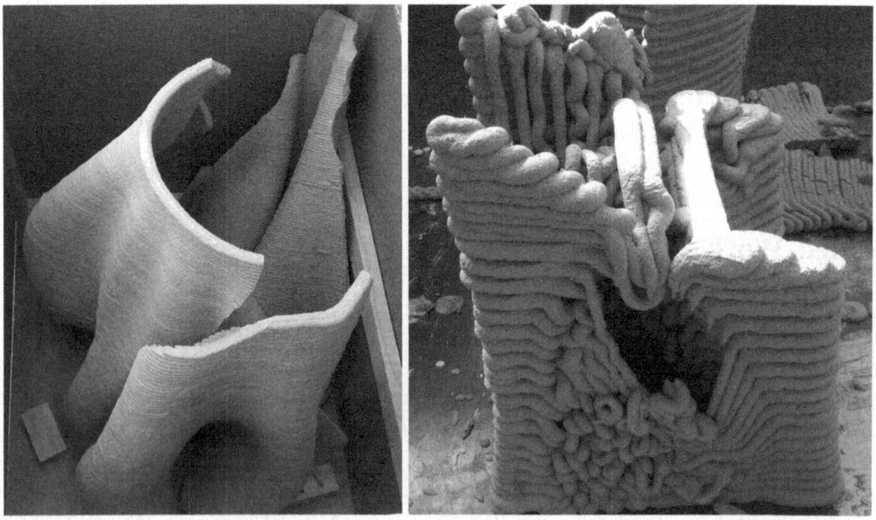

Fig. 3. Two particular phenomena to be addressed: stability of the printed shapes (*left*) and the feasibility of printing them (*right*)

It can be noted that several other state-of-the-art methods were considered including the ubiquitous Finite Element Method (Tanigawa and Mori 1989; Kwon 2002), Material-point methods (Sulsky et al. 1995; Stomakhin et al. 2013), and Discrete Element Methods (Cundall and Strack 1979; Harada 2007; Shyshko and Mechtcherine 2008). The reasons for the choice stem from the motivations mentioned previously: operating in the CAGD paradigm and therefore, extending methods that use geometry-based approaches to solve the equilibrium problem.

Computational Framework

The proposed research considers design of geometries for concrete printing as micro-stereotomy, where the micro-scale material layout is a function of structural and fabrication constraints. Further, the objective of the proposed computational framework (CF) is to directly and procedurally generate the trajectories of the print head of the 3D Printer—the so-called task graphs (Khoshnevis et al. 2006)—utilizing the various degrees of freedom of the printer, whilst ensuring stability of the printed shape and its print feasibility. These two aspects represent the principal components of the IDE (Fig. 4) and are outlined next.

Synthesis of Task Graphs

Task graphs can be produced by processing either explicit, mesh representations (Fig. 5 left and middle) or implicit, distance field representations of geometry. The proposed method for procedural, force-aligned stereotomy is to utilise the latter. In other words, task graphs are a function of the level curves of user-defined scalar functions (Fig. 6). Thus, the IDE helps design appropriate scalar functions, whose contours represent task graphs. This is expanded in Sect. 4.

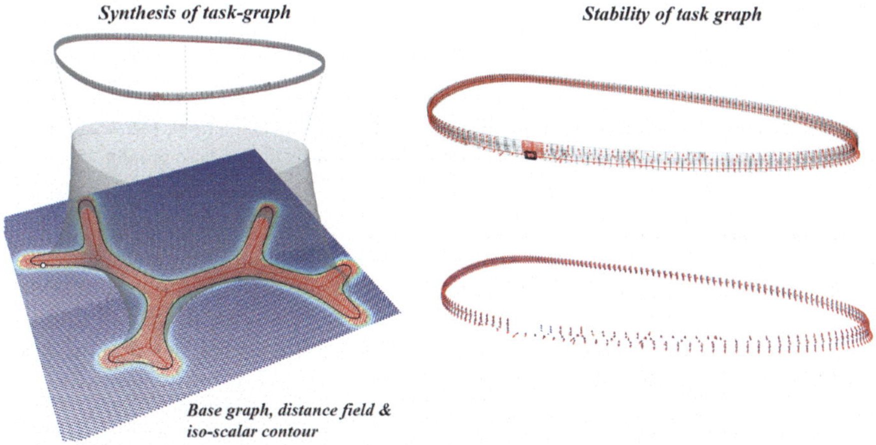

Fig. 4. Synthesis of task graphs from scalar fields (*right*), and the evaluation of their static equilibrium (*left*)

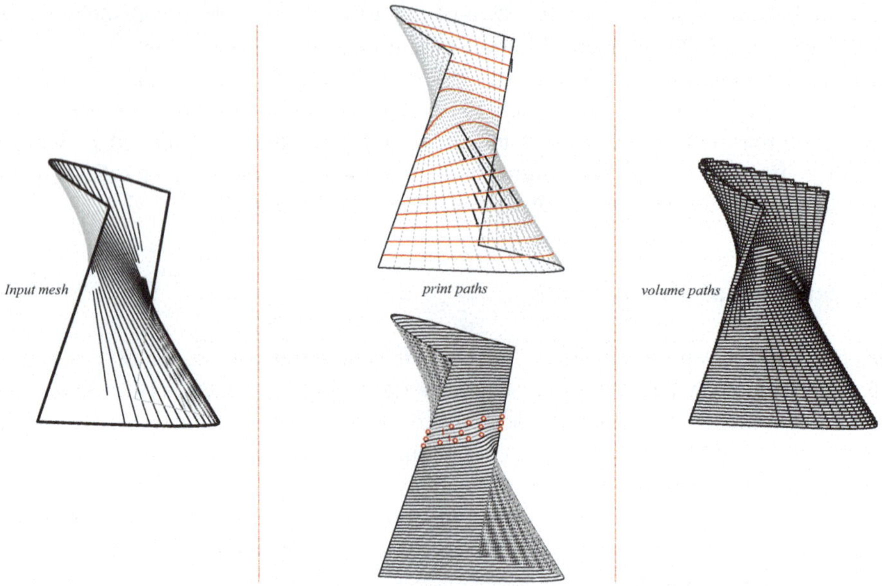

Fig. 5. Processing of mesh representation (*left*) to generate horizontal or field-aligned task graphs (*middle*) and subsequent attaching of a micro-block model to evaluate their stability (*right*)

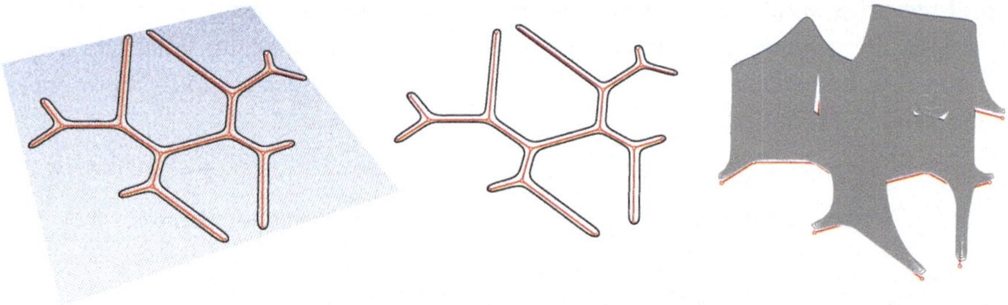

Fig. 6. (*Left*) a user-specified skeleton graphs (*red*) and related distance field. (*Middle*) Extracted level curve. (*Right*) 2.5D evolution of level curve (*blue*) and resulting task graphs

Stability of Task Graphs

The task graphs, once extracted from either implicit or explicit representations, are populated with micro-scale blocks (Fig. 5 right). These blocks are needed in evaluating the task graphs for their structural and material feasibility using the proposed static equilibrium model, as described in Sect. 5.

Procedural Shape Design

To aid the user-controlled synthesis of the task graphs, the IDE utilises a combination of implicit scalar field representation and resulting level curves as the data structure to represent the geometries to be printed. Additionally, the level curve can be evolved as it is extruded in the z-direction, resulting in a so-called 2.5D extrusion (Fig. 6 right). A skeletal graph is employed to manipulate the scalar field and thus the resulting task graphs (Fig. 6 middle). In combination, it bears similarity to the so-called skeleton-mesh co-representation (SMC) used in popular digital sculpting software such as Autodesk MudBox™ and Pixelogic ZBrush™ (Ji et al. 2010; Bærentzen et al. 2014). The choice for this data structure is a consequent of the design and fabrication related aspects outlined below.

A Forward Approach to Synthesize Feasible Geometries

A common method to produce materially feasible geometries is to start with a reasonable guess of such geometry and perturb it to a near-by state that satisfies structural and fabrication constraints. Such a process could be termed as the *inverse* approach to *modelling* of geometry. For a summary of the vast amount of methods and literature available in mesh segmentation and other techniques of stereotomic design, the reader is referred Sect. 6.3 of Rippmann (2016). Conversely, a *forward* approach aims to procedurally synthesize geometries that satisfy the same constraints. The likelihood of discovering novel shapes and possibilities is higher with the forward approach and this is an important motivation as previously mentioned.

Fabrication Constraints

Discretisation of geometry into layers and layers further into micro-blocks must allow for the newest layer of material to remain accessible to the print head. This is achieved by formulating the scalar field representation as a distance field defined in relation to the skeleton graph. In other words, the skeletal graph is in fact the medial axis of the scalar field, and medial axis guarantees the outward radiating property (Fig. 7). Further, it would be beneficial to align the print layers in orthonormal directions to the expected force flow. Level curves of scalar fields naturally achieve this.

Interactivity

To keep with the edit-and-observe contemporary paradigm of CAGD, it is important to provide appropriate, non-numeric methods of user-control of the critical geometric parameters of the computational model (Fig. 8).

Static Equilibrium Model

As mentioned previously, this second aspect of the IDE, extends the RBE method to accommodate the soft-rigid aspects of 3D printing of concrete.

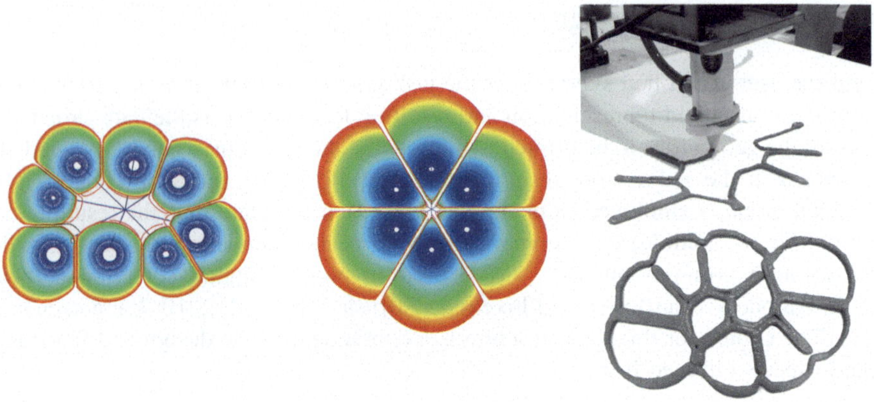

Fig. 7. Outward radiating task graph as a result of a radial distance function

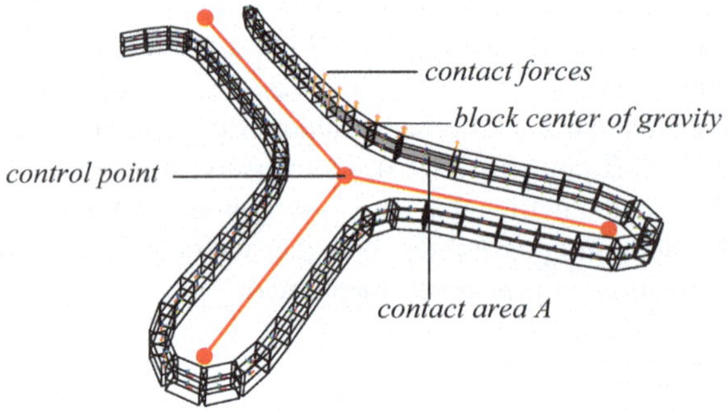

Fig. 8. A user-specified skeletal graphs and a micro-block model attached to the resulting task graph. The intermediate representation of the level curve/task graph is hidden in this case. The skeletal graph serves as a manipulator whilst satisfying fabrication constraints

Rigid-Block Equilibrium

Popular implementations of RBE and RBD mentioned previously, share a common mathematical formulation: computations at a single time instance of RBD calculates the linear and angular accelerations of each block, whilst RBE inverse computes the forces and torques on each block by setting their accelerations to zero. Further, the two algorithms converge to be exactly the same when RBD computes the so-called *resting contact* forces—forces needed to keep two static rigid bodies from inter-penetrating (Baraff 1997). In this case, both algorithms find the necessary equilibrating forces (shown in orange in Fig. 9) that satisfy linear constraints using quadratic programming.

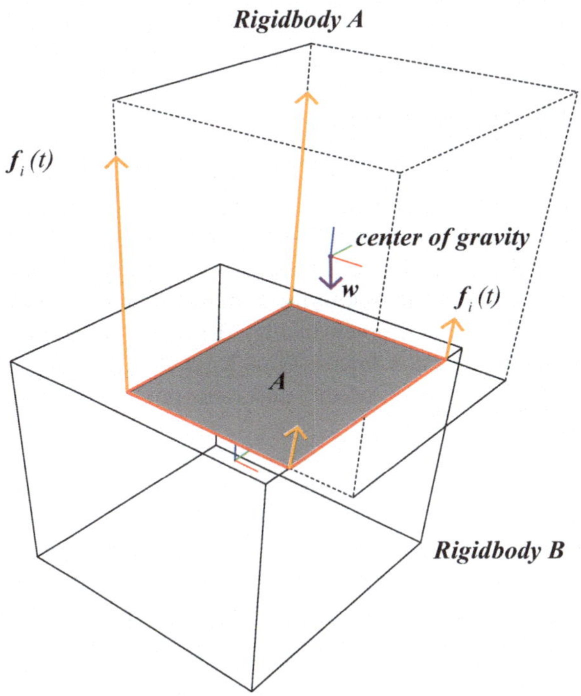

Fig. 9. Contact interface and forces when two rigid-bodies are in resting contact

Equilibrium Formulation

The formulation for resting contact forces in RBD (Baraff 1997) (Fig. 9) is

$$\boldsymbol{A}_{\mathrm{eq}}\boldsymbol{f} + \boldsymbol{b} = \boldsymbol{a} \tag{1}$$

where $\boldsymbol{a}$ is a column vector that encodes accelerations of the blocks, $\boldsymbol{A}_{\mathrm{eq}}$ is the so-called $n \times n$ equilibrium matrix, $\boldsymbol{b}$ is a column of constants, $\boldsymbol{f}$ a column vector of unknown force magnitudes, and n is the number of vertices in the contact interface.

When computing *resting contact* forces, $\boldsymbol{b}$ is a column vector of weights of the rigid blocks, $\boldsymbol{a}$ is 0. Thus (1) becomes,

$$\boldsymbol{A}_{\mathrm{eq}}\boldsymbol{f} + \boldsymbol{w} = 0 \tag{2}$$

This is the equilibrium formulation used in RBE to solve for $\boldsymbol{f}$ in a least-squares sense

$$\min g(\boldsymbol{f}) = \boldsymbol{f}^T \boldsymbol{H} \boldsymbol{f} + \boldsymbol{b}^T$$

$$s.t \quad A_{\mathrm{eq}}\boldsymbol{f} + \boldsymbol{w} = 0 \tag{3}$$

For derivation, detailed explanation and pseudo code, see (Livesley 1978; Frick et al. 2015; Baraff 1997).

Soft-Rigid Body Equilibrium (SRBE)

The proposed CF extends the RBE method above, to semi-rigid and rigidifying materials such as fresh concrete and clay. The insight here is that, the necessary forces at the contact interface, $f_i(t)$ and highlighted orange in Fig. 10, that ensure static equilibrium of the rigidified blocks is a function of the configuration of the blocks. On the other hand, the capacity of the rigidifying material to be able to offer the necessary reaction forces $g_i(t)$ (highlighted green in Fig. 10), are a function of time. In other words, the time-dependent reaction forces (Wangler et al. 2016; Le et al. 2012b) can be added as an additional constraint to the quadratic program (3) that computes the equilibrating forces.

Implementation

The current implementation of the CF shown in Fig. 11, utilises extensions to the RBE method as stated in Frick et al. (2015), and resting contact formulations and quadratic programming (QP) as stated in Baraff (1997). The ALGLIB library is used to numerically solve the QP.

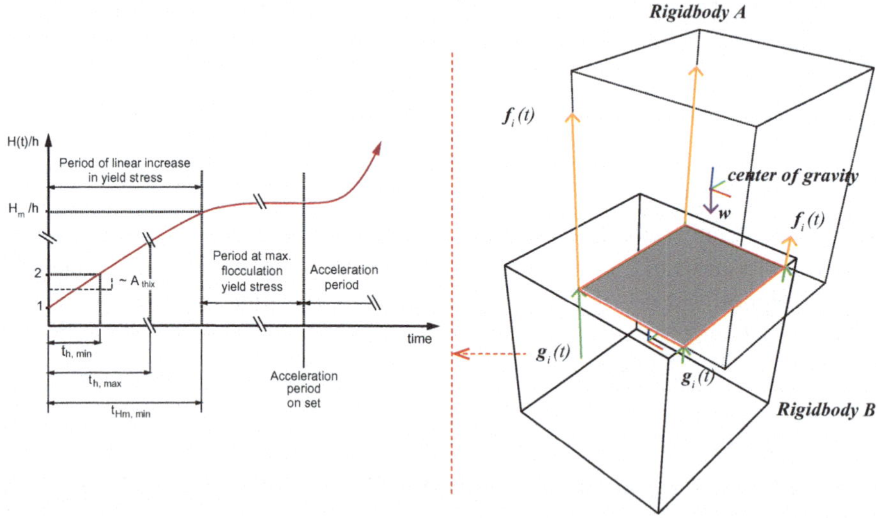

Fig. 10. Soft-rigid body equilibrium forces (*right*) and (*left*) time-dependent evolution of the bearing capacity of the material. Graph from Wangler et al. (2016)

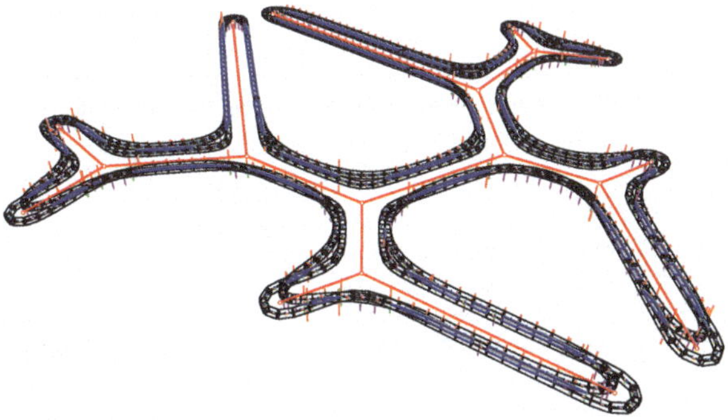

Fig. 11. Application of method to test print feasibility. In the current implementation, areas where a non-trivial contact (*blue*) interface exists, the sRBE method is executed to evaluate feasibility. If such contact is not found, the infeasibility is indicated by *pink arrows*. The skeletal graph (*red*) is manipulated based on this visual feedback

Contact Interfaces and Collisions

In order to compute the forces or accelerations, both RBD and RBE methods require efficient routines to calculate the so-called contact interface between two colliding or touching rigid bodies (Cundall 1988; Frick et al. 2016). This paper implements only the *face-to-face* contact interface as described in Frick et al. (2016), since it was found to be sufficient to provide interactive feedback to designers. The other interfaces including *face-edge*, are likely to be a result of un-printable geometry.

Results

The synthesized and evaluated task graphs are subsequently processed for printer related parameters such as accessibility of the print-head, speed etc (Fig. 12). The exported print-file is subsequently processed by the manufacturer for additional parameters before being printed (Fig. 13).

The 3D printed outcomes of the shapes designed using the proposed IDE, demonstrate areas of local overhang and under-cuts (Figs. 13 right and 14 left and middle). The overall stability of the prints and their feasibility were able to be reasoned geometrically as exemplified by the inverted printing direction of the shape shown in Fig. 13. The results also demonstrate the usefulness of the 2.5D extrusion method to generate shapes of considerable complexity, which can nonetheless be controlled with nominal parameters. Furthermore, the potential exploitation of non-horizontal print layers is also indicated (Fig. 14 right) and is discussed next.

Discussion and Future Work

The early exploratory design outcomes and physical realisations also demonstrate the potential avenues for refinement and extension of the proposed method. Significant of these are discussed below.

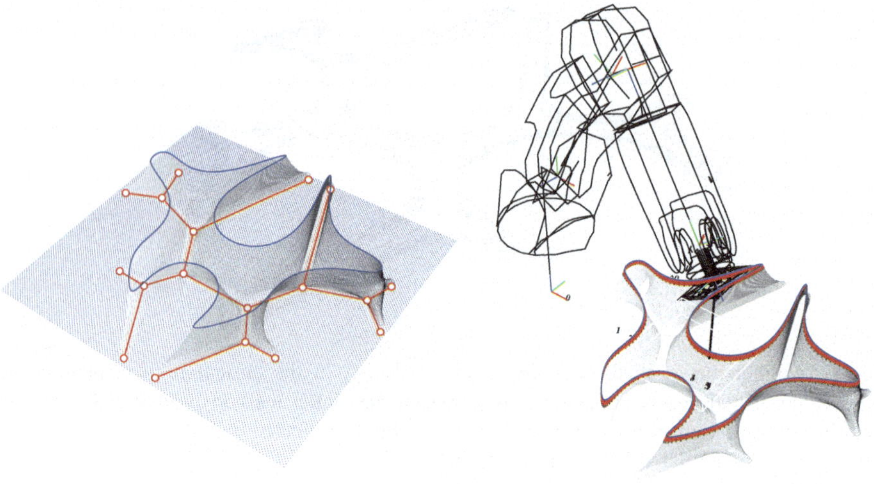

Fig. 12. Post-processing of task graphs for printed related parameters within the IDE

Fig. 13. *Left* processing of an outcome produced by proposed IDE by the manufacturer. *Right* prototype demonstrating the application of the proposed method. Printed at the RexLab, Institute of Experimental Architecture, at Innsbruck. Images courtesy Johannes Ladinig

Fig. 14. (*Left* and *Middle*) Results from the 2.5D extrusion. (*Right*) Non-horizontal extrusion. Printed at the RexLab, Institute of Experimental Architecture, at Innsbruck. Images courtesy Johannes Ladinig

Extension to 3D

The proposed 2.5D synthesis method can be extended to fully 3D spatial skeletons (Figs. 15 and 14 right) as seen in the work of Antonio Gaudi, particularly in the *Sagrada Familia* (Hernandez 2006; Monreal 2012). However, this involves the non-trivial aspect of processing of nodes (Fig. 15 middle) including assigning an appropriate print direction to the nodes, the relative sizes of the edges and nodes (Fig. 15 right) and assigning a sequence for printing. This is currently being implemented.

Exploration of Design Space and Unification of Synthesis and Evaluation

The current linear work-flow between the two main design aspects—synthesis and evaluation, and the subsequent evaluation of the printer-related parameters is already proving useful in exploring the design-space of structurally and materially feasible geometries (Figs. 14 and 16). The unification of the structural and printer-related parameters as constraints within the synthesis of the task graphs will be beneficial to

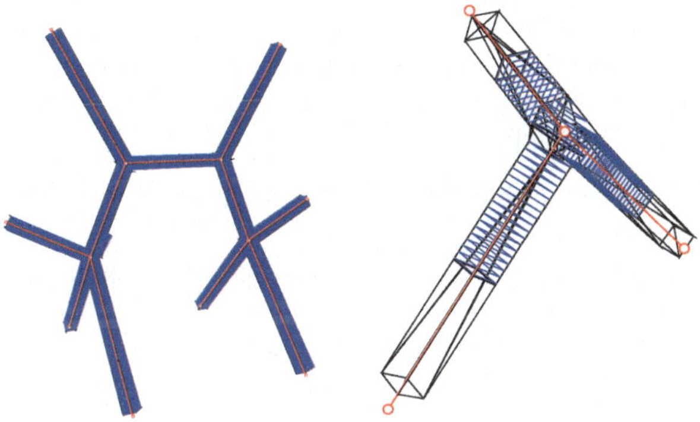

Fig. 15. Extending the 2.5D evolution to multiple, non-vertical, normal directions

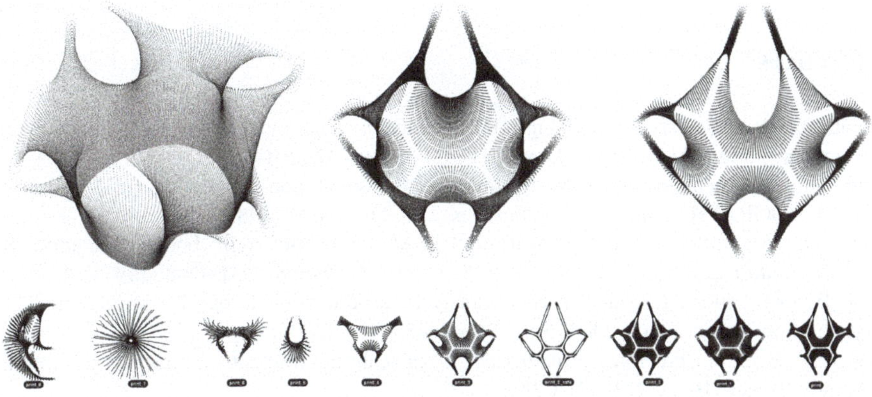

Fig. 16. Exploration of structurally and materially feasible design-space

fully support the contemporary edit-and-observe design paradigm, as also for the automated exploration of the parameter-space.

Extension to Other Materials

The proposed sRBE method, given its foundations in design methods for unreinforced rigid-block masonry, could allow for its use in masonry applications that require the modelling of mortar. Current RBE methods do not fully take into account presence of mortar. Since the time-dependent properties of rigidifying materials are encapsulated within a graph, the sRBE method could potentially be extended to other materials that can similarly be encoded—clay, certain food materials such as chocolate and pasta etc.

Conclusions

An Interactive Design Environment (IDE) for the shape design and production of clay printing was outlined, along with an underlying computational framework. Early proof-of-concept physical outcomes were also shown. The novel insight of extending methods from design and analysis of unreinforced masonry to large-scale 3D printing and the proposed computational framework bears promise in the aspects listed below. It can be remembered that these were previously earmarked to be addressed (Khoshnevis et al. 2006) but had hitherto been side-lined:

- abstraction of numerical analysis of feasibility into geometric reasoning,
- developing of an atlas of feasible and novel geometries,
- the applicability of historic techniques of masonry construction and construction, to layered printing,
- data structures amenable to interactive design exploration and
- unifying the equilibrium of printed geometries, along with the print feasibility of the parts.

All aspects mentioned above, though showing promise, require further careful design of data structures, mathematical formulations, and algorithms to fully realise the objectives.

References

Bærentzen, J.A., Abdrashitov, R., Singh, K.: Interactive shape modeling using a skeleton-mesh co-representation. ACM Trans. Graph. (TOG) **33**(4), 132 (2014)

Baraff, D.: Analytical methods for dynamic simulation of non-penetrating rigid bodies. In: ACM SIGGRAPH Computer Graphics, pp. 223–232. ACM (1989)

Baraff, D.: An introduction to physically based modeling: rigid body simulation II—nonpenetration constraints. In: SIGGRAPH Course Notes, pp. D31–D68 (1997)

Block, P., Ochsendorf, J.: Thrust network analysis: a new methodology for three-dimensional equilibrium. J. Int. Assoc. Shell Spat. Struct. **155**, 167 (2007)

Buswell, R.A., et al.: Freeform construction: mega-scale rapid manufacturing for construction. Autom. Constr. **16**(2), 224–231 (2007)

Buswell, R.A., et al.: Design, data and process issues for mega-scale rapid manufacturing machines used for construction. Autom. Constr. **17**(8), 923–929 (2008)

Chien, S., et al.: Parametric customisation of a 3D concrete printed pavilion. In: Proceedings of the 21st International Conference of the Association for Computer-Aided Architectural Design Research in Asia CAADRIA 2016, pp. 549–558. The Association for Computer-Aided Architectural Design Research in Asia (CAADRIA), Hong Kong (2016)

Culmann, K.: Die graphische statik. Meyer & Zeller (A. Reimann), Zurich (1875)

Cundall, P.A.: Formulation of a three-dimensional distinct element model—part I. A scheme to detect and represent contacts in a system composed of many polyhedral blocks. In: International Journal of Rock Mechanics and Mining Sciences & Geomechanics Abstracts, pp. 107–116. Elsevier (1988)

Cundall, P.A., Strack, O.D.L.: A discrete numerical model for granular assemblies. Geotechnique **29**(1), 47–65 (1979)

CyBe-Construction: CyBe Construction—Redefining Construction. Available at: https://www.cybe.en/ (2016)

De Goes, F., et al.: On the equilibrium of simplicial masonry structures. ACM Trans. Graph. (TOG) **32**(4), 93 (2013)

De Kestelier, X., Buswell, R.A.: A digital design environment for large-scale rapid manufacturing. In: Proceedings of the 29th Annual Conference of the Association for Computer Aided Design in Architecture, pp. 201–208 (2009)

Dini, E.: D-Shape. Monolite UK Ltd. http://www.d-shape.com/cose.htm (2009)

Fallacara, G.: Toward a stereotomic design: experimental constructions and didactic experiences. In: Proceedings of the Third International Congress on Construction History, p. 553 (2009)

Fivet, C., Zastavni, D.: Robert Maillart's key methods from the Salginatobel Bridge design process (1928). J. Int. Assoc. Shell Spat. Struct. J. IASS **53**(171), 39–47 (2012)

Frick, U., Van Mele, T., Block, P.: Data management and modelling of complex interfaces in imperfect discrete-element assemblies. In: Proceedings of the International Association for Shell and Spatial Structures (IASS) Symposium, Tokyo, Japan (2016)

Frick, U., Van Mele, T., Block, P.: Decomposing three-dimensional shapes into self-supporting, discrete-element assemblies. In: Modelling Behaviour, pp. 187–201. Springer (2015)

Gosselin, C., et al.: Large-scale 3D printing of ultra-high performance concrete—a new processing route for architects and builders. Mater. Des. **100**, 102–109 (2016)

Hahn, J.K.: Realistic animation of rigid bodies. In: ACM SIGGRAPH Computer Graphics, pp. 299–308. ACM (1988)

Harada, T.: Real-time rigid body simulation on GPUs. GPU Gems **3**, 123–148 (2007)

Hart, R., Cundall, P.A., Lemos, J.: Formulation of a three-dimensional distinct element model—part II. Mechanical calculations for motion and interaction of a system composed of many polyhedral blocks. In: International Journal of Rock Mechanics and Mining Sciences & Geomechanics Abstracts, pp. 117–125. Elsevier (1988)

Hernandez, C.R.B.: Thinking parametric design: introducing parametric Gaudi. Des. Stud. **27**(3), 309–324 (2006)

Heyman, J.: The stone skeleton. Int. J. Solids Struct. **2**(2), 249–279 (1966)

Ji, Z., Liu, L., Wang, Y.: B-Mesh: a modeling system for base meshes of 3D articulated shapes. In: Computer Graphics Forum, pp. 2169–2177. Wiley Online Library (2010)

Jiang, C., et al.: Interactive modeling of architectural freeform structures: combining geometry with fabrication and statics. In: Advances in Architectural Geometry 2014, pp. 95–108. Springer (2015)

Khoshnevis, B.: Automated construction by contour crafting—related robotics and information technologies. Autom. Constr. **13**(1), 5–19 (2004)

Khoshnevis, B., et al.: Mega-scale fabrication by contour crafting. Int. J. Ind. Syst. Eng. **1**(3), 301–320 (2006)

Kwon, H.: Experimentation and analysis of contour crafting (CC) process using uncured ceramic materials, thesis, Faculty of the graduate school University of Southern California (2002)

Lachauer, L.: Interactive Equilibrium Modelling—A New Approach to the Computer-Aided Exploration of Structures in Architecture. Ph.D. thesis. ETH Zurich, Department of Architecture, Zurich (2015)

Le, T.T., Austin, S.A., Lim, S., Buswell, R.A., Law, R., et al.: Hardened properties of high-performance printing concrete. Cem. Concr. Res. **42**(3), 558–566 (2012a)

Le, T.T., Austin, S.A., Lim, S., Buswell, R.A., Gibb, A.G.F., et al.: Mix design and fresh properties for high-performance printing concrete. Mater. Struct. **45**(8), 1221–1232 (2012b)

Liu, Y., et al.: Computing self-supporting surfaces by regular triangulation. ACM Trans. Graph. (TOG) **32**(4), 92 (2013)

Livesley, R.K.: Limit analysis of structures formed from rigid blocks. Int. J. Numer. Methods Eng. **12**(12), 1853–1871 (1978)

Livesley, R.K.: A computational model for the limit analysis of three-dimensional masonry structures. Meccanica **27**(3), 161–172 (1992)

Mechtcherine, V., et al.: Simulation of fresh concrete flow using Discrete Element Method (DEM): theory and applications. Mater. Struct. **47**(4), 615–630 (2014)

Monreal, A.: T-norms, T-conorms, aggregation operators and Gaudí's columns. In: Soft Computing in Humanities and Social Sciences, pp. 497–515. Springer (2012)

Perrot, A., Rangeard, D., Pierre, A.: Structural built-up of cement-based materials used for 3D-printing extrusion techniques. Mater. Struct. **49**(4), 1213–1220 (2016)

Rippmann, M.: Funicular Shell Design: Geometric Approaches to Form Finding and Fabrication of Discrete Funicular Structures. Ph.D. Thesis, ETH Zurich, Department of Architecture, Zurich (2016)

Sakarovitch, J.: Stereotomy, a multifaceted technique. In: Huerta, S. (ed.) Proceedings of the 1st International Congress on Construction History, Madrid, pp. 69–79 (2003)

Shyshko, S., Mechtcherine, V.: Simulating the workability of fresh concrete. In: Proceedings of the International RILEM Symposium Of Concrete Modelling—CONMOD, pp. 173–181 (2008)

Stomakhin, A., et al.: A material point method for snow simulation. ACM Trans. Graph. (TOG) **32**(4), 102 (2013)

Sulsky, D., Zhou, S.-J., Schreyer, H.L.: Application of a particle-in-cell method to solid mechanics. Comput. Phys. Commun. **87**(1), 236–252 (1995)

Tang, C., et al.: Form-Finding with Polyhedral Meshes Made Simple. http://www.geometrie.tugraz.at/wallner/formfinding2.pdf (2012)

Tanigawa, Y., Mori, H.: Analytical study on deformation of fresh concrete. J. Eng. Mech. **115**(3), 493–508 (1989)

Van Mele, T., et al.: Best-fit thrust network analysis. In: Shell Structures for Architecture-Form Finding and Optimization, pp. 157–170. Routledge (2014)

Vouga, E., et al.: Design of self-supporting surfaces. ACM Trans. Graph. **31**(4) (2012)

Wangler, T., et al.: Digital concrete: opportunities and challenges. RILEM Tech. Lett. **1**, 67–75 (2016)

Whiting, E.J.W.: Design of structurally-sound masonry buildings using 3D static analysis, Diss, Massachusetts Institute of Technology (2012)

Wolfs, R., Salet, T.A.M.: 3D printing of concrete structures, graduation thesis, Technische Universiteit Eindhoven (2015)

XtreeE: XtreeE| The large-scale 3D (2016)

Zastavni, D.: The structural design of Maillart's Chiasso Shed (1924): a graphic procedure. Struct. Eng. Int. **18**(3), 247–252 (2008)

Monolithic Earthen Shells Digital Fabrication: Hybrid Workflow

Maite Bravo[1(✉)] and Stephanie Chaltiel[2]

[1] Institute of Advanced Architecture of Catalunya IAAC, Barcelona, Spain
maite.bravo@iaac.net
[2] Institute of Advanced Architecture of Catalunya IAAC, University Polytechnic of Catalunya UPC, Barcelona, Spain
stephanie.chaltiel@iaac.net

Abstract. The digital fabrication process of monolithic shell structures is presenting some challenges related to the interface between computational design and fabrication techniques, such as the translation from the digital to the physical when a definite materiality appears during the final phase of the fabrication process, which is typically approached as a linear and predetermined sequence, most often not allowing any kind of adjustments or rectifications. This stage proves to be critical for monolithic shells, because their integrity relies on the continuity of the construction phases for its stability and durability. This paper exposes the implementation process required for new computational and physical methods for monolithic shells construction using digital fabrication techniques combined with skilled/non-monotonous manual craft. It proposes an additive manufacturing deposition process for paste clay mixes with the potential of embedding a certain level of interactivity between the fabricator and the materialized model during the fabrication process, to allow for real time adjustments or corrections. Two case studies are featured, detailing the steps required to take advantage of the flow involving manual craft, digital modeling, and digital fabrication tools. These methods aim at renewing shells construction processes, in terms of the time it takes to build them, the optimization of materials and labor, and the feasibility of the dweller to be involved during different construction phases.

Keywords: Monolithic shells · Robotic fabrication · Digital craft · Earth architecture

Mud Shell Construction

Earth construction for monolithic shells can be found in ancient traditions (Stulz and Mukerji 1993), with some strong references in the Nubian vaults (Sudan), the Harran Beehive Houses (Turkey), and the Musgum mud huts (Cameroon), among others (Niroumand et al. 2013). These modest but relevant examples are usually constructed with or without formwork with small lumps of sticky clay thrown by hand in layers. A similar and singular earth construction method derived from the "Wattle and daub" or "Torchi" (Rael 2009) composed of vertical wattles, woven branches forming a support surface, and daubed with clay or mud to make-up an assembly of interlocked

K. De Rycke et al., *Humanizing Digital Reality*,
https://doi.org/10.1007/978-981-10-6611-5_43

materials. Through the use of digital design and robotic fabrication, these traditional techniques could be systematized to respond to local conditions, and to embed additional optimization criteria.

The stability for the construction of monolithic shell structures is typically provided by a temporary formwork until the material reaches sufficient strength, which exhibits several challenges, such as the extensive use of labor and materials. As a result, fabric formwork has been explored (Veenendaal et al. 2011) because of its ability to provide a temporary support that can be adjusted and offers greater freedom for forms, a controlled smooth finish, easy removal techniques, and an overall inexpensive solution. Fabric formwork in concrete include projects such as Mark West (C.A.S.T) (1998), Matsys (2017), and Cloud 9 STGILAT pavilion (2015), among many others.

Additive manufacturing is being currently implemented at diverse scales and techniques in earth construction, where robotic fabrication is taking a relevant place with two main strategies: extrusion and spraying. Extrusion is conducted through a nozzle by which the paste like material is deposited, achieving complex geometries, but heavily restricted to the size and characteristics of the deposition apparatus. Examples include *Co-de-it, inFORMed clay matter* (2017); *G Code Clay* (2017); *Wasp* (2017); and *Mataerial/Pylos/On-Site-Robotics* (2017). The spraying technique deposits material through a nozzle under pressure piped to a mix of cement or clay that is simultaneous drying while printing in the air. Spraying has a long history using shotcrete (both wet and dry concrete mix) (Gilbert Gedeon 1993), and it's widely used in a variety of construction applications including civil works, pools, and buildings.

This paper explores the implementation of digital fabrication for monolithic shells using mud spray techniques.

Hybrid Workflow for Mud Shell Additive Manufacturing

The proposed fabrication method for mud monolithic shells are based on a collaborative framework that merges manual actions carefully coordinated with digital and robotic dynamics. The organization of a critical sequence must be properly formulated, so an iterative workflow in calibrated loops that are detailed below.

Placement, Consolidation, and Finishing Loops

Loop 1- PLACEMENT: The initial manual formwork set-up is composed of a wooden baseboard with holes supporting groups of bending rods to create supporting arches (typically 3–4 outer arches and 1–3 middle nerves), from where a stretched fabric is secured. The scan #1 of the fabric formwork uses high-resolution photography and Agisoft image processing software, then exported into a CAD environment (Rhino 3d) to generate a 3d mesh that is simplified by reducing the number of divisions or redrawing of main curves. Afterwards, the optimization #1 includes different structural simulations using Karamba, a parametric structural engineering tool, to display analysis of tension or compression areas, or stress lines for the potential placement of additional reinforcement. Robotic trajectories are designed in Kuka PRC (digital processes) with settings based on previous tests, but require constant readjustment on the trajectory and velocity (distance and angle to the structure) to achieve a homogeneous deposition and a proper mix adhesion to the fabric formwork. The material spray #1 starts with the manual clay preparation that includes soil test, mixing, and feeding the sprayer. The first material layer is a liquid mix "barbotine" (clay and water), that is sprayed on top of the fabric to facilitate an optimum adhesion for the following layers.

Loop 2—CONSOLIDATION: This stage includes the material placement in several layers, therefore repeated iteratively. The scan #2 is crucial, because it allows keeping records of the morphology in progress. The optimization #2 seeks a minimum use of material (thickness) for maximum strength, simulations show the shell regions defined by displacement criteria, or some expected deformations of the shell in progress such as the sagging on both sides of the ribs. However, these optimizations are only approximations, because they are based in Karamba on the properties of concrete, as clay is not yet available in the software. Robotic trajectories must be constantly recalibrated and reinserted into the Kuka PRC software (trajectory, thickness, etc.). The middle layers contain clay, sand and natural fibers (that add significant shell thickness while being light in weight), a mix prepared in a large bucket and stirred by hand. A custom-made spray tool nozzle allows different perforations, and must be calibrated in terms of the angle, distance, and speed of each deposition.

Loop 3—FINISHES: Includes the embedment of special finishes (additives), curing time, thickness checkup, final coating, and the removal of temporary formwork.

This precise protocol needs to be respected to ensure the solidity of the resulting structure.

Case Studies

Recent academic experiments investigate new methods and provide a more realistic set up towards patented fabrication techniques. Two case studies presented in this paper propose a complete strategy for the digital fabrication of mud shells reaching a height up to 2.0 m.

Case Study 1:

May 2016. 25 h. seminar. First year master students. IAAC, Barcelona.

Seminar: "*Phriends for Shells*" ("Phriends" was defined as the safe interaction between people and robots during the fabrication progress).

Participants: 23.

Tutors: Author 1, D. Stanejovic (robotic expert), Y. Mendez (assistant).

Five earthen shells of 1.5 m × 1.5 m × 1 m (height) with perforations were built during this 4 weeks seminar.

Loop 1—PLACEMENT: Bending rods in clusters of 2–3 members bundled together with a rope (Fig. 1) were used as supporting arches where the stretched lycra was secured. Openings with perforations were defined with laser cut rings and triangles that were used to create a temporary formwork that was removed afterwards.

Fig. 1. Cluster of 3 bending reeds bundled together. [Photos by the participants]. 2016

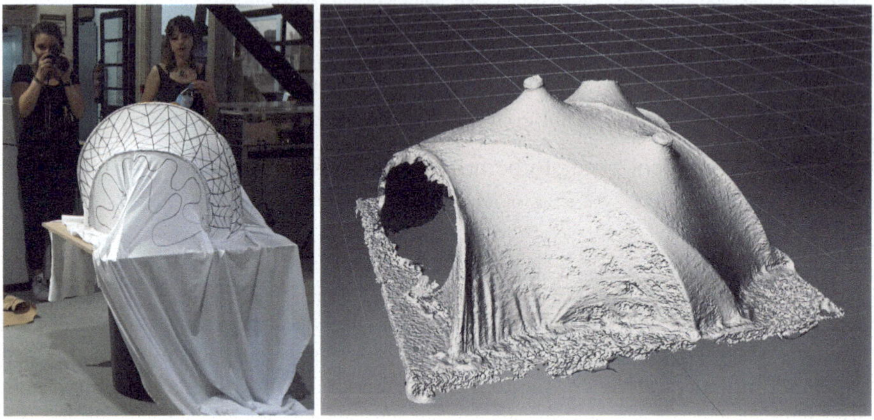

Fig. 2. 3D scanning #1 and rebuild mesh on Rhino 3D. Earthen Shells workshop. 2017

The scan #1 was done with a Sony camera on manual mode, and to facilitate the accuracy of the scan, the lycra fabric required contrasting marks done manually (Fig. 2).

The optimization #1 in Karamba simulations analysis (Fig. 3) revealed that the inclusion of perforations (triangles or circles), position, density and proximity to the main nerves could be parametrized within the shell's geometry and could minimize the amount of tension areas. The study of the perforations tested in both digital and physical models appeared to lead to the most successful designs.

The material spray #1 includes a first layer containing 1U clay + 2U water. The deposition tool was a Wagner heavy paint sprayer Flexio 590 HVLP with external air

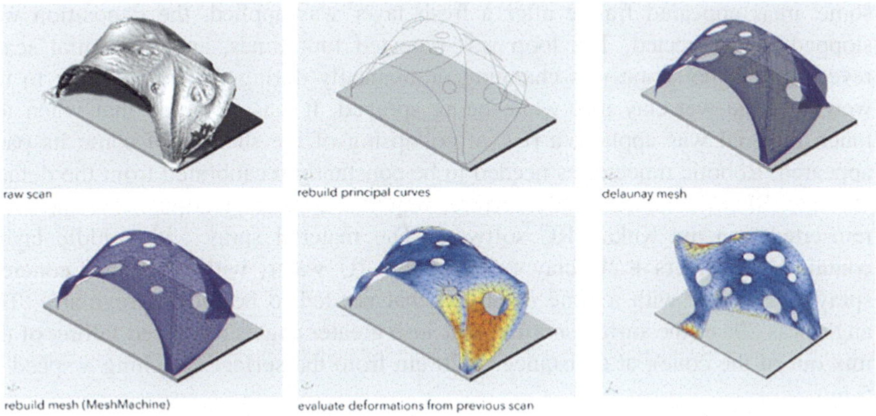

Fig. 3. Initial 3d scan, rebuild mesh, and deformations evaluation in Karamba simulation [Group 1, Iaac seminar, May 2016]

Fig. 4. Kuka robot with end effector: Wagner heavy paint sprayer applying the clay mix of the first layers on the earthen shell. Iaac seminar May 2016. [Photos by the seminar's participants]

compressor (kit part). The deposition was at a 45° angle, at 30 cm from the surface, and at a speed of 2 m/s.

Loop 2 to 6—CONSOLIDATION A: Five layers were sprayed (Fig. 4) in iterative loops. The scanning process was time consuming because it required recalibrating the trajectory of the applied spray each time. The optimization #2 in Karamba revealed some considerable deformations around the perforations areas while being applied, because of the considerable weight of the sprayed clay mixes. A reevaluation of the initial optimization considered a variation in thickness; therefore if some areas appeared fragile after a fresh layer was applied, the deposition was stopped and corrected. The loop was repeated four times, and sequential scans revealed that the shape was changing dramatically during the process due to the weight of the wet clay mix while being sprayed. It was observed that when too much material was applied, a risk of collapsing of the shell or of some its parts appeared. Robotic trajectories needed to be constantly recalibrated from the default 45° and 30 cm distance to the surface, required constant readjustment and were reinserted into the Kuka PRC software. The material spray with middle layers contained 3U fibers + 2U clay + 2U sand + 2U water, with a manual concrete sprayer "Sablon" with a cone of 30 cm that needed to be refilled regularly. The angle was 20° to the surface normal (because greater angles provoked falling of the mix out of the cone), at a distance of 20 cm from the surface, reaching a speed of 5 m/s.

Loop 7—FINISHES: A last layer was robotically sprayed with a mix that included 1U clay + 1U sand (water was not needed) + 2U stabilizers to prevent cracks and to provide better waterproofing (highly viscous cactus solution by boiling thin slices of cactus and stirring until it turned into a highly viscous mix). The deposition tool was set at a speed of 2 m/s; a 45° angle; and a distance of 30 cm from the shell. The shells reached at least a total 2 cm in overall thickness. The curing time was a total of 3 full days to complete 5 coatings (working indoors during summer at 25 °C). Drying time was at least around 4–5 h in between each spray layer, and the lycra fabric and bending rods were removed manually.

Case Study 2:

May 2017; 25 h seminar, first year master students. IAAC, Barcelona.

Seminar: "*Earthen Shells Digital and Manual Fabrication.*"

Participants: 13

Tutors: Author 1, A. Ibrahim, N. Elgewely, A.Chronis, K. Dhanabalan, K. Singh Chadha.

This seminar proposes the design and fabrication of three earthen shells with some 3d texturing and perforations with embedded natural resin translucent modules, reaching 1.5 m × 1.5 m × 1.0 m in size (Fig. 5).

Loop 1- PLACEMENT: The scan #1 used two Parrot AR Drones, 2.0 Elite Edition (about 70 pictures), controlled from smart phones and tablets by open codes that can be modified by non-experts. Optimization #1 was run with Karamba simulations with tension and compression regions of the shell paired with stress lines. The material spray #1 included a first layer only contains 2U water + 1U clay. The robotic spray was set at a constant distance of 3 cm from the shell to the end of the nozzle, at a fixed angle of 45° to the surface shell normal at all points.

Loop 2—CONSOLIDATION A: The scan #2 was performed, and the optimization in Karamba highlighted tension and compression areas, revealing a considerable shell deformation during the fabrication process, which lead to evaluate the inclusion of a real time recalibration tool with preliminary augmented reality experiments (Fig. 6), that were tested to explore the physical structure projecting the virtual model simulating the thickness variations, or by changing the location of the perforations. The material spray #2 includes applying strips of jute fabric in

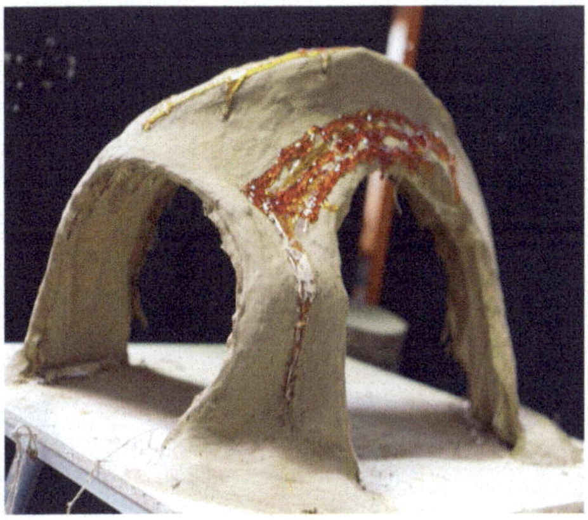

Fig. 5. Earthen shells robotic fabrication in clay mix and resin. Earthen Shells workshop 2017

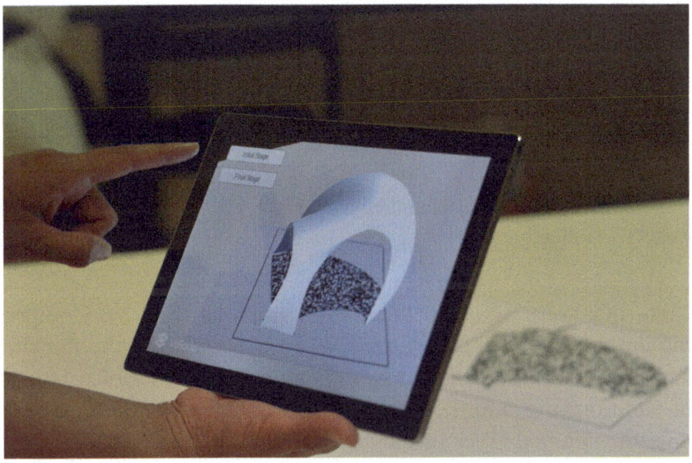

Fig. 6. Augmented reality tests. 2017. Earthen Shells' workshop

between robotically sprayed clay mix to reinforce the structure, and to allow a subtle placement of non-clay modules, or the translucent embedded openings. Once the shell surfaces are modeled in Rhino 3d, the geometries are unrolled to facilitate the design of a laser cut cutting pattern on the jute fabric. The middle layers contain clay, sand and fibers, deposited with a concrete hand sprayer placed as the "end effector" of the robotic arm. The trajectory is set to adapt to the shells form by spraying at a velocity of 2 m/s, in horizontal straight lines (contours), at an average 20 cm from the surface (Fig. 7).

Loop 3—CONSOLIDATION B: The scan #3 as the structure was already fixed on the robotic turning table some photos taken from the robot. From this scan, the structure appeared to have sagged on each part of the supporting arches. The optimization #3 revealed shell areas working primarily under tension or under

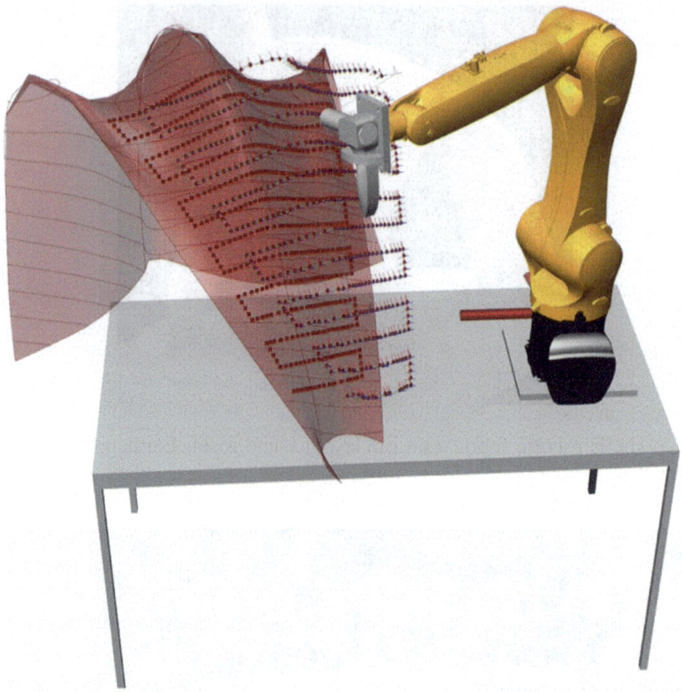

Fig. 7. Continuous robotic spraying simulation in Kuka prc. Smart Geometry Cluster "Mud, Textile and Robots", Gothenburg 2016

compression, that were assigned thickness variations (Fig. 8). Compression areas were given the thickness (T), while tension areas were given the thickness (T − 1). The thickness variation is achieved by changing the number of robotically sprayed layers. In addition, the middle layers of sprayed fibrous mix had a thickness of 1 cm, while the sprayed layer of clay sand and water was 2 mm.

Loop 4—FINISH A: The material Spray #3 for the upper layer included 1U clay + 1U sand mix + 1U water + 1U olive oil (a natural stabilizer for waterproofing and to avoid cracks). The deposition Tool was a Wagner spray paint was used by hand or attached to the robot for the upper layers application, 25 cm from the surface angle of 45° to the surface normal, and velocity of deposition 2 m/s. Total thickness was 3 cm, and curing time was only 2 h (outdoor hot and dry weather at 30 °C). Curing time checkup included knocking in different areas to verify that the structure was stable, allowing the manual removal of the temporary formwork.

Loop 5—FINISH B: This layer includes some natural resin modules robotically poured while hot (Fig. 9). The readjustment of the velocity, pouring angle, and material temperature, allowed different textures and effects to be explored.

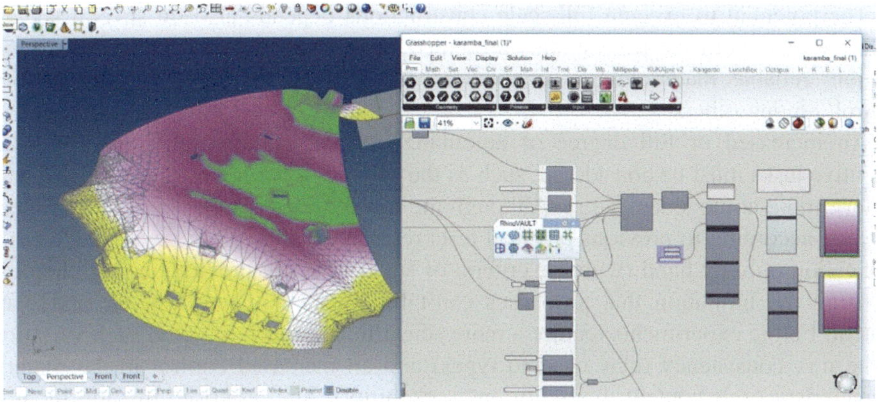

Fig. 8. Karamba simulation highlighting areas working in compression or in tension "Earthen Shells Workshop". 2017

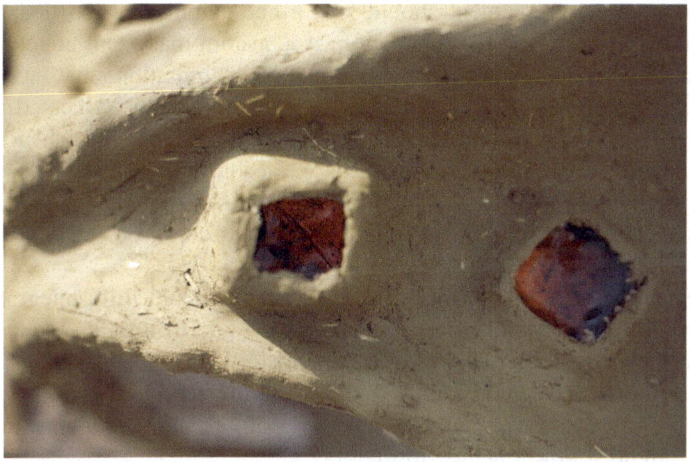

Fig. 9. Natural resin robotic pouring. Earthen Shells´workshop 2017

Next Steps for Mud Shell Digital Fabrication

This paper proposed a framework for additive manufacturing spray protocol for the construction of earthen monolithic shells that integrates manual with automated processes for novel collaborative techniques. These collaborations are not just conceived as isolated chain of tasks, but a process where manual manufacturing responds to and creates the demand for specific robotic actions in an integrated feedback loop, happening according to a precise phasing. These protocols have been tested in several instances, and are still being continuously improved and enhanced, and hope to be implemented at full scale in the near future.

The potential to execute full-scale constructions depends also on several specific factors drawn from the case studies presented in this paper, such as greater automatization, reliable material tests, continuous adjustment of reinforcement, real time control, among others.

An increased or full degree of automatization for certain rather laborious and repetitive tasks must be considered, such as the stirring of the heavy clay mix so that to achieve more homogeneous mix. Although the material proportions in the experiments were set according to previous physical tests results, the mixes were evaluated at each application only by hand (with the criteria of sticking to the previous layer), therefore exposing the limitation that properties can't be constant in terms of air and water content. Future experiments require a more scientific approach to avoid such variations in the mix consistency (clay or sand types) and an increased control in the material characteristics (air content, humidity, or workability). The proposed method includes two main families of manual actions collaborating with robotic spraying: deposition of the reinforcement strips, and application of the middle and last coating by hand. Some aspects of this fabrication stage are done by hand, such as the reinforcement strips placement according to strength and weakness of the overall structure in progress, but the location of the reinforcement should be further rationalized, so that it could be layered robotically involving the fabrication of a custom-made end effector.

Full-scale implementation must resolve some other limitations that include: testing for changing material properties (as material can be locally collected); structural engineering analysis of different shell morphologies (load, torsion, shear) with load tests according to the ratio height/thickness or according to variations on material thicknesses; the incidence of local conditions with live sensors to collect environmental data (weather, humidity, temperature, etc.), and optimization protocols to respond to local conditions (such as wind).

Varying the location and density of reinforcements could fully integrate the optimization analysis stage by identifying stress lines (Fig. 10) to add localized strength according to simulations and physical real-time distortion prone areas (Fig. 11). Stress lines-reinforcement can include adding thickness in the clay poor, or incorporating an additional material such as the applied single layer of jute strips, bamboo strands, sugar cane, reeds, amongst others as lost formworks.

Finishing by hand certain inaccessible and limited areas might be necessary to increase the viability of the shell and its resilience against the rain (i.e. around the perforations), and for ongoing maintenance reasons.

Fig. 10. Stress lines optimization in Karamba. Earthen Shells workshop. 2017

The implementation of real time control using augmented reality could provide an increased decision-making power for the builder, revealing an unexpected degree of freedom and creativity such processes could bring to monolithic shell construction. Real time 3d scanning available in other industries will improve the technique by allowing manual workers and robotic actions controller to adapt constantly to the structure's deformation and non-predictable changes in shape in specific regions of the shell (such in case study 2). Or the possibility of implementing remote control in real time could lead to new horizons in construction protocols, if external expert feedback can be implemented for remote locations while under construction.

These explorations show how the successive and simultaneous actions performed collaboratively by robots and humans would not only engender new workflows for mud shell construction, but propose a fruitful dialogue between computational techniques and revisited traditional crafts, that when implemented at full-scale could open an unexplored and innovative agenda for additive manufacturing.

Acknowledgements. This research forms part of the Marie Skłodowska-Curie Actions, ITN Innochain, Grant Number #642877.

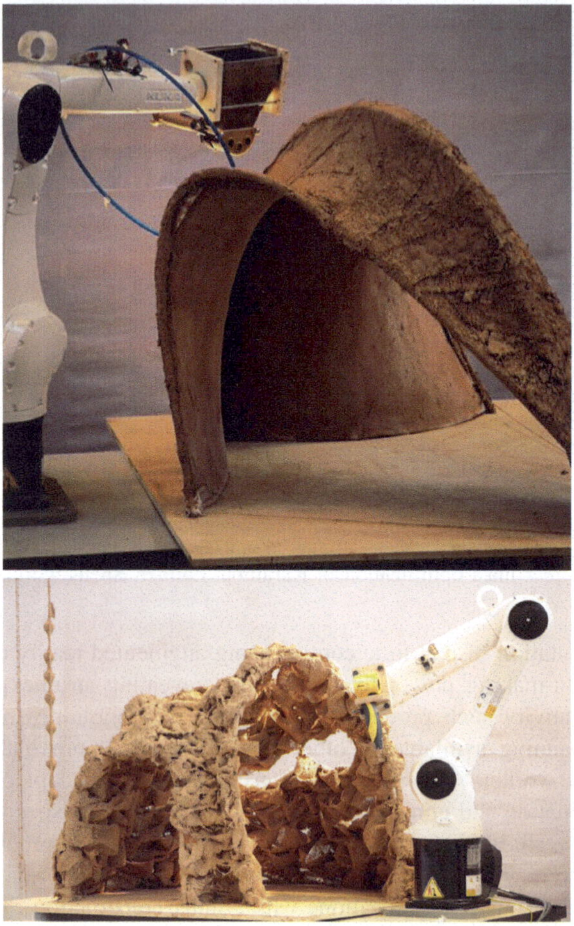

Fig. 11. Robotically sprayed mud shell with temporary fabric formwork (*top*) and weaved jute fabric as lost formwork (*bottom*). Smart Geometry 2016. Gothenburg. Photo Andrea Quetara

References

Cloud 9, STGILAT pavilion (2015). http://arcaute-zaragoza.com/stgilat.html. Last Accessed 19 May 2017

3D Printers Manufacturer, WASP. http://www.wasproject.it/w/en/. Accessed 11 June 2017

"GCODE.Clay|Emerging Objects." http://www.emergingobjects.com/project/gcode-clay/. Accessed 11 June 2017

Gilbert Gedeon, P.E.: Introduction to Shotcrete Applications. 31 Jan 1993

Iaac, Large Scale 3d printing: Mataerial, Pylos, On-Site-Robotics. https://iaac.net/research-projects/. Accessed 11 June 2017

"Informed Matter." http://www.co-de-it.com/wordpress/category/informed-matter. Last Accessed 19 May 2017

Matsys: flexible formwork, http://matsysdesign.com/tag/flexible-formwork/XXX. Last Accessed 19 May 2017

Niroumand, H., Zain, M.F.M., Jamil, M.: Assessing of critical parameters on earth architecture and earth buildings as a vernacular and sustainable architecture in various countries. Procedia Soc. Behav. Sci. **89**, 248–260 (2013). doi:10.1016/j.sbspro.2013.08.843

Rael R (2009) Earth Architecture. Princeton Architectural Press, Princeton

Stulz, R., Mukerji, K.: Schweizerische Kontaktstelle für Angepasste Technik am ILE. Appropriate Building Materials: A Catalogue of Potential Solutions. SKAT (1993)

Veenendaal, D., West, M., Block, P.: History and overview of fabric formwork: using fabrics for concrete casting (2011). doi:10.1002/suco.201100014

West, M.: C.A.S.T. The Centre for Architectural Structures and Technology (CAST), Faculty of Architecture, University of Manitoba (1998)

Application of Machine Learning Within the Integrative Design and Fabrication of Robotic Rod Bending Processes

Maria Smigielska(✉)

Zurich, Switzerland
maria@mariasni.com

Abstract. This paper presents the results of independent research that aims to investigate the potential and methodology of using **Machine Learning** (**ML**) algorithms for precision control of material deformation and increased geometrical and structural performances in robotic **rod bending technology** (**RBT**). Another focus lies in integrative methods where design, material properties analysis, structural analysis, optimization and fabrication of robotically rod bended space-frames are merged into one coherent data model and allows for bi-directional information flows, shifting from absolute dimensional architectural descriptions towards the definition of relational systems. The working methodology thus combines robotic RBT and ML with integrated fabrication methods as an alternative to over-specialized and enclosed industrial processes. A design project for the front desk of a gallery in Paris serves as a proof of concept of this research and becomes the starting point for future developments of this methodology.

Keywords: Robotic rod bending · Machine learning · Adaptive processes

Background

The bending operation is most generally a cold forming process, in which metal is stressed beyond its yield strength. Until 1991 advancement in **rod bending technology** (**RBT**) remained insignificant and the process was still operating with traditional table benders, with a few minor innovations in electronics and safety controls (KRB Machinery Co. 1993). The first conceptual schemes of **computer integration** for the design, fabrication, delivery and placement of steel reinforcement bars (Miltenberger et al. 1991) appeared in 1991 and shifted this technology from a labour-intensive and hazardous one to a safer, faster and almost fully automated process. Yet, the problem of dependency on human expertise was not solved, because those top-down systems could not intelligently compensate for variable springback and other changing conditions. The first truly adaptive system appeared in 2000 (Dunston and Bernold 2000), when machine learning techniques were applied in this field of RBT. The first definition of **Machine learning** (**ML**) is that it gives computers the ability to learn without being explicitly programmed (Samuel 1959). It finds out the highly non-linear relationships between the given parameters and as such, provides the understanding of the analyzed

K. De Rycke et al., *Humanizing Digital Reality*,
https://doi.org/10.1007/978-981-10-6611-5_44

processes by mechanization and enhancement of human knowledge. The paper mentioned above (Dunston and Bernold 2000) presents a comparison of the following prediction methods: (1) multiple linear regression; (2) in-process relaxation and (3) neural networks applied for RBT. All of them result with similar error performance exceeding ±2.5°. Such error is acceptable for standard applications of reinforcement bars, however, it might not be enough for more complex architectural applications involving metal bars as finished elements or different disciplines that demand higher precision (medical applications such as orthodontic archwires or spinal fusion surgery elements). This field requires further investigation.

ML techniques have been widely tested in the **metal sheet forming**—technology with similar constraints to rod bending. The Review of Artificial Neural Network applications (Kashid and Kumar 2012) from 2012 compares over 60 scientific papers on different approaches to this topic. It proves that ML techniques can not only achieve zero spring-back angle, but are also used for other applications like process control, process planning, quality control, finite element simulation, feature recognition, tool design, and cutting tool selection for sheet metal forming. As such, it reduces complexity, minimizes the dependency on human expertise and time taken to design the equipment. One of the recent example projects, titled *A Bridge Too Far* utilised robotic incremental sheet forming (Nicholas et al. 2016), employs Neural Networks to gain deeper understanding of the metal forming process and the resulting imprecision.

Research outcomes of such extent are missing in the field of ML applied to robotic RBT, even if the applications of these technologies have developed significantly during recent years. The project developed by *supermanoeuvre* for 2012 Venice Architecture Biennale employs robotic rod bending processes and is followed by the research *Adaptive Part Variation* (Vasey et al. 2014), where two robotic arms are capable of bending, cutting, placing, and welding steel rods into a larger assembly with the use of computer vision systems. Another ongoing research project *Mesh Mould Metal* (Hack et al. 2015) is developing robotic wire bending and welding manipulators attached to robotic arm to build formworks and reinforcements in a single continuous operation.

Introduction

This paper presents the development of the robotic RBT with incorporation of ML techniques for the precise control of metal deformation. Another focus lies in integrative methods where design, material properties analysis, structural analysis, optimization and fabrication data of robotically rod bended space-frames are merged into one coherent data model to allow for bi-directional information flows.

A design project for a gallery front desk is taken as a case study.Commissioned by AA[n + 1] Art and Architecture gallery in Paris, the front desk will serve as a piece of functional furniture (front desk and step to access an elevated office door), as well as being an exhibited piece in the gallery itself (Figs. 2 and 6). It is a space frame structure (5.5 m long, 1.4 m wide and 0.75 m high) comprising of 6 mm cold-drawn S235JR steel rods, differentiated through a process of non-standard robotic rod bending and welded on site. The front desk consists of 800 unique elements of varied length (from 0.12 to 1 m), varied number of 2- and 3-dimensional bends (between 2 and 7 bends) and variedangles (2400 unique values between 1° and 144°). The elements are of 341 m length and weight 80 kg in total.

In parallel to the front desk project, smaller product design objects are prototyped to both test new design strategies and verify fabrication processes. In addition, design and fabrication workshops have been conducted to further test these methods. A four days students' workshop at ENSA Paris-Malaquais, Digital Knowledge in February 2017 focused on incorporating FEM analysis on two levels: (1) as a design driving factor- by rationalizing stress lines resulting from specific structural loads/support conditions and then (2) structurally analyzing it as a second step. Lastly, the prototypes of 4 mm steel rods were fabricated with the robotic RBT (Fig. 14).

Methods

Design Process

The digital workflow for the front desk project is entirely embedded in the Grasshopper environment to provide continuity of the information flow between different steps of design, analysis and production data (Fig. 1). It begins with two parametric guide curves generating a proto-structural envelope, to be then optimized with a voxel-based structural analysis (GH Millipede) and evolutionary solver (Galapagos) (Fig. 3). Secondly, the surface is rationalized with varied morphostructural rod grids and then analysed with a finite element method (GH Karamba), optimized with another evolutionary solver in order to provide a better understanding of the structural conditions (cantilevering, anchorage, u-shape beam, 3d truss elements) (Fig. 5). For the FEM analysis the space-frame geometry was simplified to polylines, ignoring the arcs of the bent parts. Since the structural system differs from the typical space-frame model, a custom definition for the point welds was created (Figs. 2, 4 and 6).

The final structural pattern divided the shape into 110 rigid segments, each of which includes the following elements: left and right principal bars, left and right diagonal bar of type A and B, which fill the side spaces and two types of zig-zags pattern that fill the top part of spacing of the frame (Fig. 4).

The digital workflow provides not only a geometrical model and structural calculations, but also includes the custom made tools written in GH Python for deconstruction of the curves to prepare data for fabrication and another tool simulating the

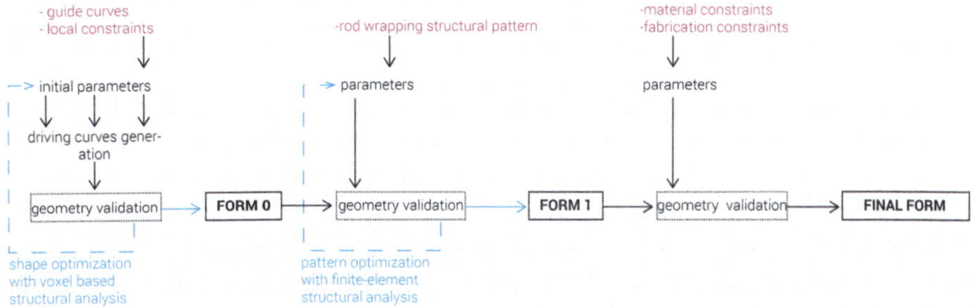

Fig. 1. Data informed design scheme for the front desk project

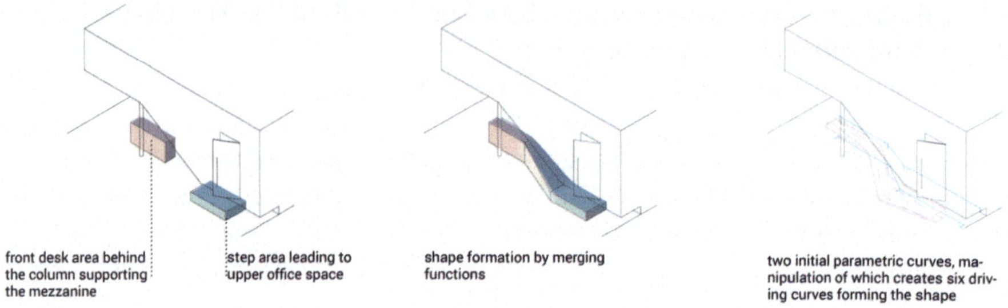

Fig. 2. Functional furniture (front desk and step to access a surelevated office door)

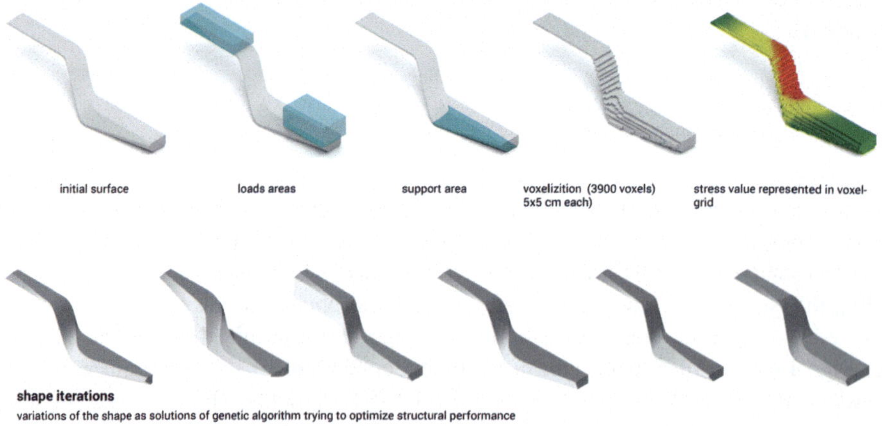

Fig. 3. Generation of the surface by voxel-based structural analysis (Millipede) and evolutionary solver

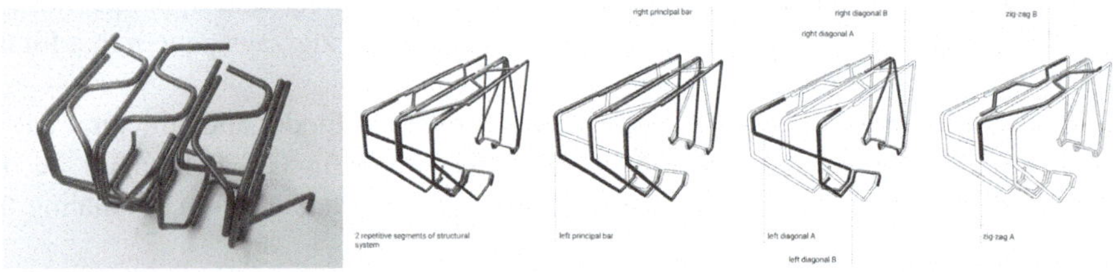

Fig. 4. Final 1:1 mockup and structural pattern of two neighboring segments of the front desk

bending process in order to predict potential collisions. Such setup stands out from conventional execution planning since there is no need for intermediate conversion to machine code. The rods are organized with an indexing system and their geometry is translated as data set containing the information about its angles, orientation (positive-clockwise, negative-counterclockwise bends) and distances between consecutive bends to be sent to the text file. Data is fed directly to RobotStudio, where another custom made tool for bending and translation operations of the robotic arm was written (again without any intermediate robot control plugins).

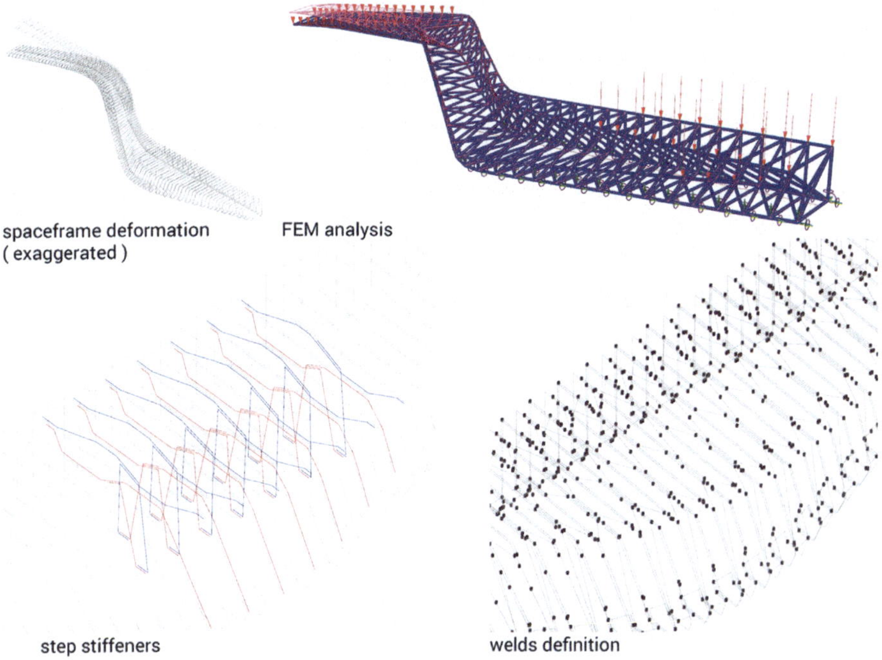

Fig. 5. *From Left Top*. Space-frame deformation (exaggerated representation); FEM analysis (GH Karamba); structural stiffeners in the step area; welds definition

Fig. 6. The final view of the front desk

Fabrication Process (Toolheads and Software)

In order to simplify the industrial process for RBT, a single robotic arm with no additional numerically controlled machines was chosen for the fabrication process. Numerous robotic toolheads were iteratively designed, prototyped and tested to find a simple, yet versatile solution. The rotary station is attached to the robotic arm (as opposed to standard CNC machines with table benders) and serves as both bender and gripper to precisely move the rods along its axis (Fig. 7). The bending

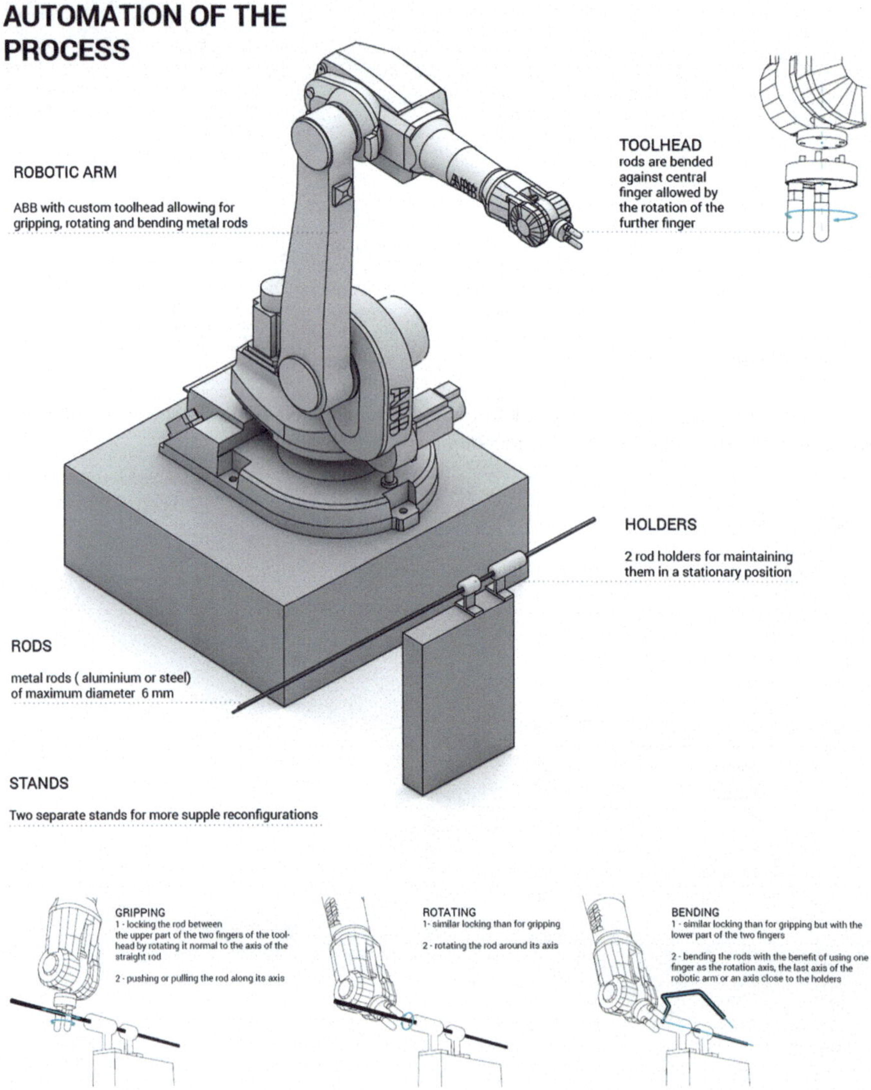

Fig. 7. Automation of the process of rod bending and axial translation with custom toolhead attached to the robotic arm

operation happens around the central finger parallel to Z axis of the 6th axis of the robotic arm.

First iteration of the toolhead was dedicated to an ABB4600 robotic arm. The next iteration of the toolhead is dedicated to an ABB1200 robotic arm and is equipped with replaceable cylinders for customization of bending radii (Fig. 8) that are installed as rollers, which reduce the friction between the toolhead and the rod while bending. Due to the limited weight payload of the ABB1200, the shape of the toolhead is optimized to its smallest weight.

The rod holding station itself is designed to be paired with different cylinders fitting to rods of different size and topology (round, square, hexagon, from 3 to 6 mm diameter) as well as with the mechanism preventing the rod from axial rotation under its own weight (Fig. 8). The station is designed as stand-alone with the single support to avoid the potential collision with rods while bending (first iteration of the holding station needed to be attached to the table covered from the side which caused many rod-table collisions during the process of 3D bending).

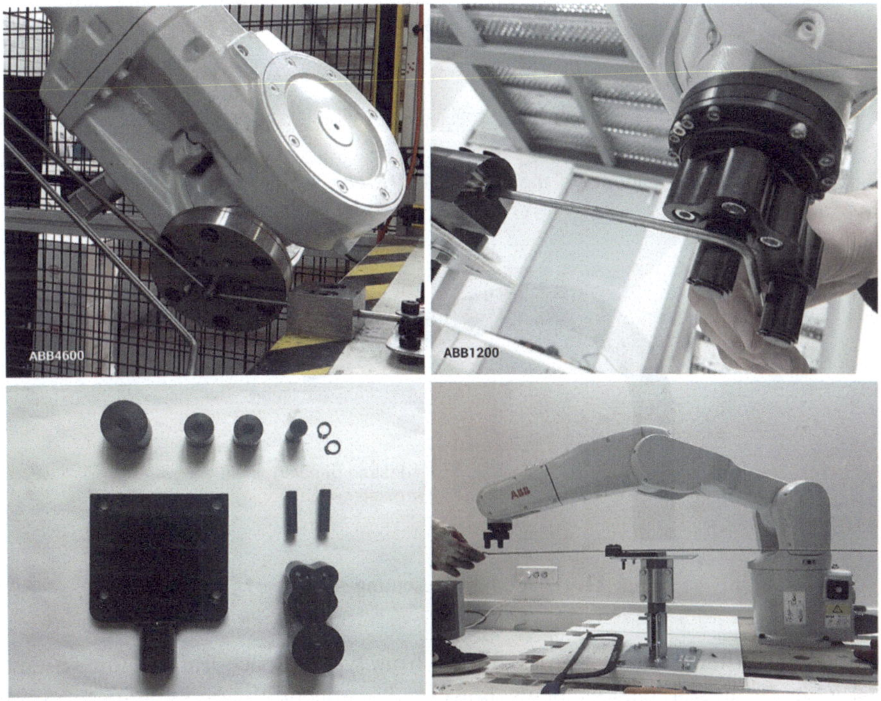

Fig. 8. *From Left Top*: Evolution of the toolhead and customization of equipment: Toolhead 1.0 for robotic arm ABB 4600; Toolhead 2.0 for the robotic arm ABB 1200 which is customized according to different bending radii, different rod size and its topology; Overview of the toolhead elements; stand-alone and single-support rod holding station

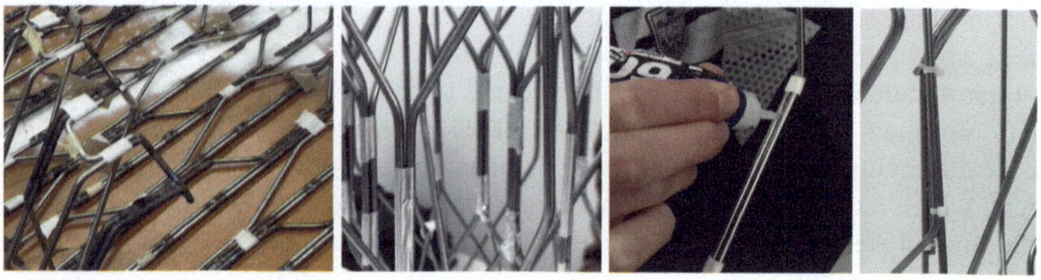

Fig. 9. TIG welding, tape, glue, zippers as alternative joinery for preassemble and final assemble phase

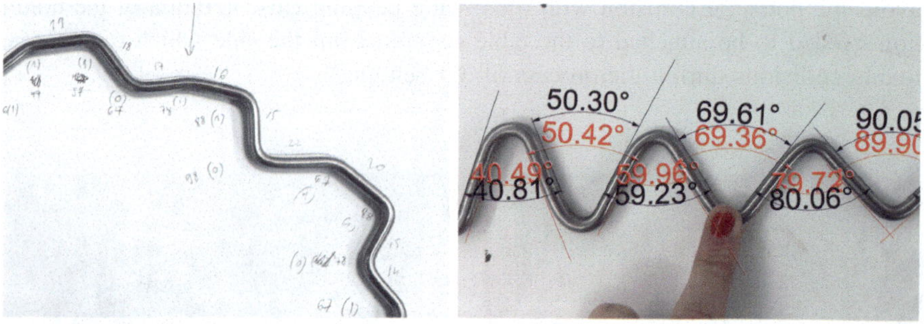

Fig. 10. *Left*: manual readouts of the bent angles; *Right*: averaged photo readouts method

Fig. 11. Diagram representing springback

Structural system of the front desk differs from typical space-frame setup, based on nodal joints of several axial elements, however it opens up the opportunity to explore and recombine different ways of connecting metal elements that preserve its structural continuity (welding, mechanical splicing, lap splicing). Although the initial exploration of the joinery consists of non standard solutions like structural glues, resin, 3d printed customized connectors, the final choice led to TIG welding on site (Fig. 9). Each of

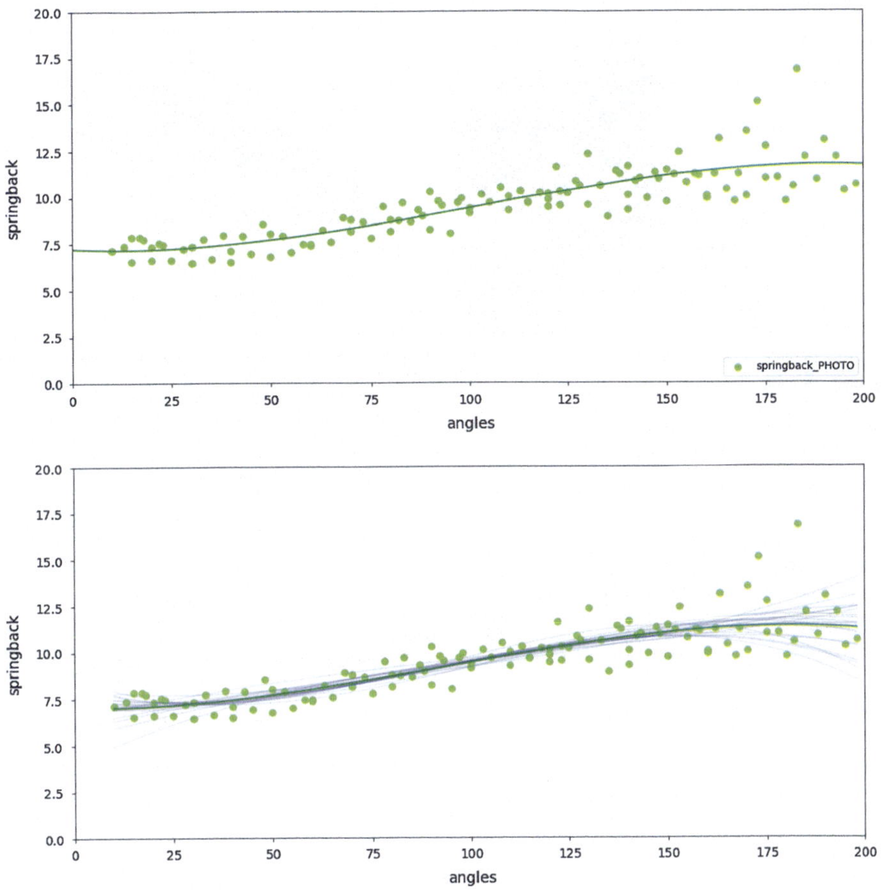

Fig. 12. Polynomial regression and bagging of polynomial regression as examples of machine learning applied for material deformation control (X—bent angles, Y—sprinback)

110 segments of the front desk is preassembled and welded separately and then sequentially welded together to create bigger parts. To provide easy transportation and assembly on site- the final piece consists of 3 chunks connected with screws.

The bending process is relatively fast- all the 800 elements were bent with one robot operator within 2 weeks time, but the manual assembly and welding is labour-intensive and highly error-generating, therefore this is one of the directions of the research further development (Fig. 13).

Fig. 13. Preassemble, welding, 3 final pieces and its joinery

Machine Learning for Material Control

In order to produce over 2400 different bending angles, it was necessary to build the continuous-valued predictive model of the metal behavior. Out of many ML methods- regression was chosen as it can model a functional relationship between two quantities [in this case- desired bending angles as X value and its corresponding springback value Y, calculated as a subtraction of real bent angles from desired bending angles (Fig. 11). To create the exhaustive training data, the set of physical bending tests of various angles between 5° and 200° was done and its springback was measured with different methods. The first trial involved manual readouts, resulted in imprecisions, especially for low and high angles (below 30° and over 150°) (Fig. 10, left). Another method-the averaged photo readouts (averaging several manual readouts from the photographed samples), resulted in higher precision, allowed to retrieve a n-degree polynomial (3° was finally chosen) predicting the deformations of any desired angle with the error of ±1° (Fig. 10, right). The final dataset consisted of 115 points and was found to be exhaustive. The automation of precise readouts (with computer vision systems), as well as increasing the number of input training data for multidimensional ML analysis (material properties, environmental data) are some of the future development goals. ML methods, including polynomial regression and bagging of polynomial regression were written with an open source scikit-learn library in Python (Figs. 12 and 13).

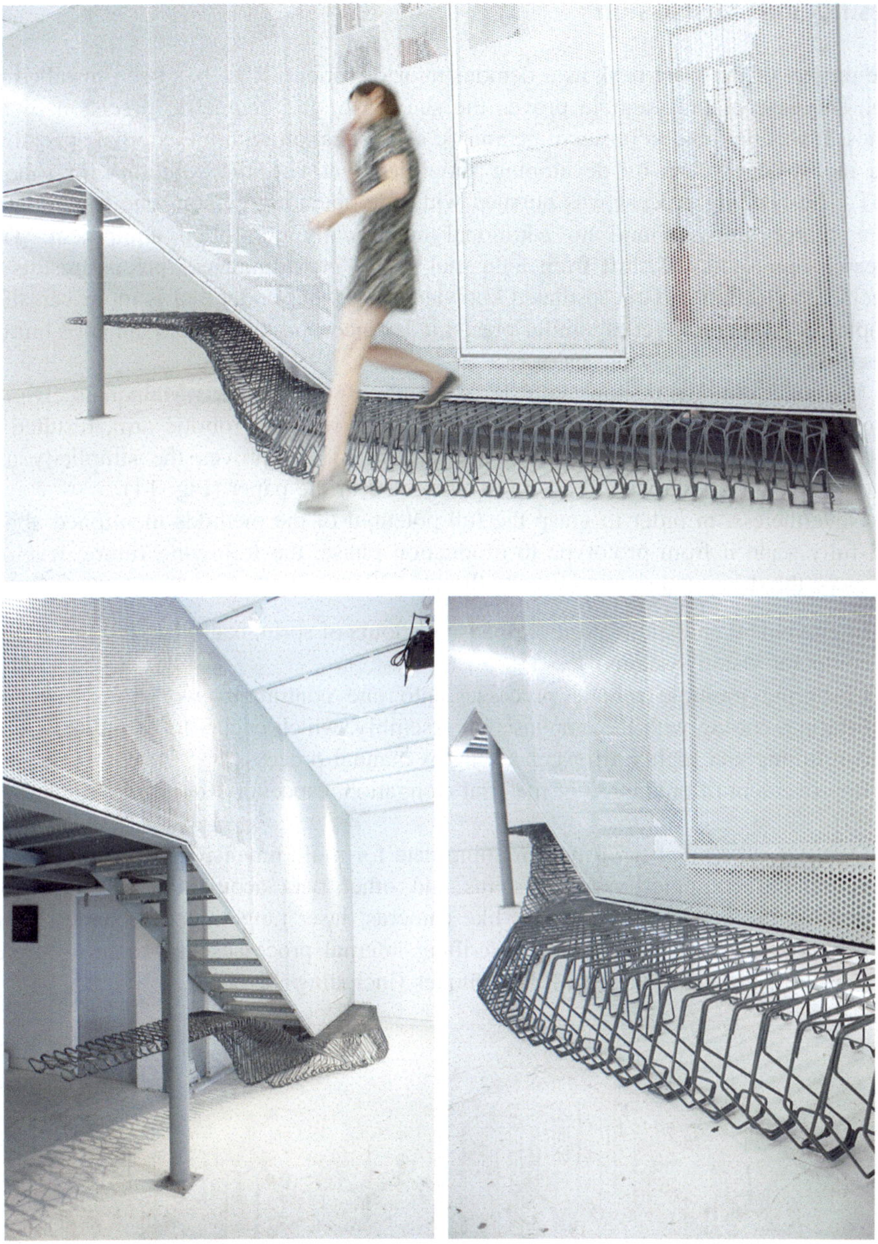

Results and Discussion

The project of the front desk as a demonstrator of robotic RBT has been installed on site. The ease of preassemble proves the success of the geometric precision of the bends. It was possible to preserve geometric sophistication without external specialists and industrial partners by developing simplified, yet versatile solutions for robotic RBT. The bending process was pursued with one operating person, one robotic arm, open source software and no additional numerically controlled equipment. This research represents the shift from industrial setup- providing very precise results of specific application and encapsulated knowledge into the model that is more versatile, simpler in mechanics, yet of similar precision by incorporating ML to enhance human expertise.

Moreover, the workshop with 12 students of ENSA Paris-Malaquais, Digital Knowledge, who had no prior experience to work with the robotic arm, resulted in prototyping mid size models within just 2 days, which proves the simplicity and accessibility of the fabrication solutions presented in the paper (Fig. 14).

Nevertheless, in order to grasp the full potential of the methods mentioned above and fully scale it from prototype to production phase, the following future developments of the design/fabrication can be drawn:

- Improvement and automation of precise readouts of springback data (by integrating computer vision systems)
- Clustering multiple robotic processes into one continuous workflow: gripping, bending, cutting with the emphasis on assembly, which occurs to be the most time consuming and highly error-generating as manual process
- Improvement of prediction of material elongation, especially for angles higher than 90°
- Increasing the number of input training data for multidimensional ML analysis by employing computer vision systems and other data acquisition tools (sensors probing the outside environment like cameras, laser rangefinders, scanners, temperature sensors or other tools describing internal processes like force or stretch sensors) and testing varied ML techniques (including ANN)

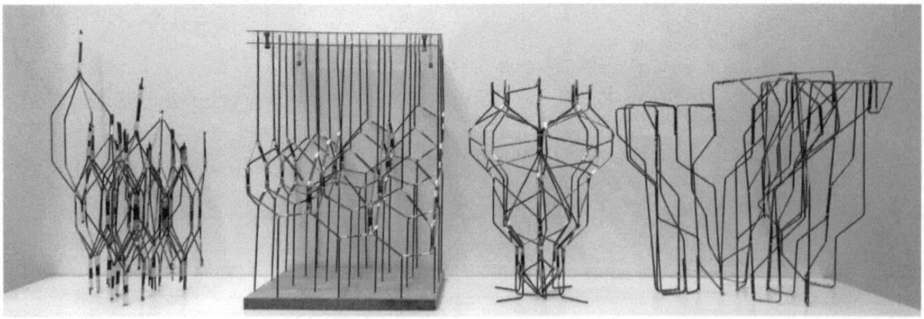

Fig. 14. Results of the 4 days workshop with ENSA Paris-Malaquais, Digital Knowledge, February 2017

- Attaching the rotary bending station to robotic arm could open up the possibility to merge bending and positioning in space processes, what could result in significant reduction of production and assembly time, while keeping the construction precision on high level.

Final Note

Machine learning has the potential to liberate design-to-construction processes of overly constrained closed industrial systems, in order to finally develop simplified, but versatile robotic automation for the prototypical building industry. It speculates on architectural discourse about two notions of precision, control and its various trade-offs.

Merging back design and making sounds like a very promising scenario, yet it is not without difficulties to face. The most common upon ML techniques mentioned in the Review ANN (Kashid and Kumar 2012), is that experienced persons are required to operate the developed ANN system. It might seem that ANN does not provide liberty, but only shifts the control centre from black-boxed industrial knowledge to encapsulated ML expertise. Yet, architecture as a multifarious discipline, has always been embracing different fields of knowledge and technologies of its time and that is also the case for now. It is necessary to gain the digital literacy by embracing ML techniques in order to achieve full reciprocity between architecture and construction, which in the end allow a escape from both routine manual work and over constrained industrial systems, towards solutions which are simpler, smarter and more versatile. Such an approachstands for creating new possibilities by augmenting human expertise with computational data, knowledge acquisition and knowledge inference within design, fabrication and assembly processes.

Acknowledgements. The author would like to thank the AREA Institute (Applied Research & Entrepreneurship for Architecture), Paris and ABB Robotics France for providing the access to the robotic facility and for the support during the design and production processes of the front desk.

References

Dunston, PhS, Bernold, L.E.: Adaptive control for safe and quality rebar fabrication. J. Constr. Eng. Manage. **126**(2), 122–129 (2000)

Hack, N., Lauer, W., Gramazio, F., Kohler, M.: Mesh mould: robotically fabricated metal meshes as concrete formwork and reinforcement. In: FERRO-11: Proceedings of the 11th International Symposium on Ferrocement and 3rd ICTRC International Conference on Textile Reinforced Concrete (2015)

KRB Machinery Co.: DMI standard bender operations and maintenance instruction manual, York, Pa (1993)

Kashid, S., Kumar, Sh: Applications of artificial neural network to sheet metal work—a review. Am. J. Intell. Syst. **2**(7), 168–176 (2012)

Miltenberger, M.A., Bernold, L.E.: CAD-integrated rebar bending. In: Cohn, L.F., Rasdorf, W. (eds) Proceedings of 7th Conference, Computing in Civil Engineering and Symposium on Data Bases, pp. 819–828. ASCE, New York (1991)

Nicholas, P., Zwierzycki, M., Stasiuk, D., Norgaard, E., Leinweber, S., Thomsen, M.: Adaptive meshing for bi-directional information flows: a multi-scale approach to integrating feedback between design, simulation, and fabrication. Adv Archit Geom 260–273 (2016)

Samuel, A.L.: Some studies in machine learning using the game of checkers. IBM J Res Dev (1959)

Vasey, L., Maxwell, I., Pigram, D.: A near real-time approach to construction tolerances, adaptive part variation. Robot Fabr Archit Art Design **2014**, 291–304 (2014)

Data and Design: Using Knowledge Generation/Visual Analytic Paradigms to Understand Mobile Social Media in Urban Design

Eric Sauda[1(✉)], Ginette Wessel[2], and Alireza Karduni[1]

[1] UNC Charlotte, Charlotte, NC, USA
ericsauda@uncc.edu
[2] Roger Williams University, Bristol, RI, USA

Abstract. Architects and designers have recently become interested in the use of "big data". The most common paradigm guiding this work is the optimization of a limited number of factors, e.g. façade designs maximizing light distribution. For most design problems, however, such optimization is oversimplified and reductive; the goal of design is the discovery of possibilities in conditions of complexity and uncertainty. This paper studies the use of Twitter as an extended case study for uncovering different methods for the analysis of urban social data, concluding that a visual analytic system that uses a knowledge generation approach is the best option to flexibly and effectively explore and understand large multidimensional urban datasets.

Keywords: Visual analytics · Knowledge generation · Social media · Urbanism

Introduction

As smart phones have become part of everyday life, communication technology and its role in shaping cities and social life became an essential topical area in architecture and urban design. The enormous volume of data being generated (estimated to reach 438 billion messages per day in 2019) (Ericcson 2016; Arce 2015) now provides new information about the social realities of cities. This development has expanded research and teaching the areas of visual analytics, urban modeling, and communication studies (Chang 2007; Offenhuber 2014; Waddel 2002), which have become important resources in environmental design disciplines. Our work expands upon this discourse, addressing the intersection of the built environment and virtual communication.

Mobile social media (Twitter, Four Square, Instagram, etc. with mobile devices such as tablets and smart phones) plays a particular role in our research as it combines many important geospatial, temporal and semantic elements as a data source. These elements have been considered important in conducting geospatial analysis (Andrienko 2010). Twitter's online social networking service is particularly interesting both for its widespread use and the API (Application Programmer Interface) that allows access to data for software development and research purposes (gnip 2016). Twitter is also well

K. De Rycke et al., *Humanizing Digital Reality*,
https://doi.org/10.1007/978-981-10-6611-5_45

studied (for a 10 year old service) and has been used toward understanding a variety of urban scale phenomenon from locations of traffic accidents or other emergencies and unplanned events such as Tahrir Square, to understanding spatial movement of food trucks (MacEachren 2011; Tufekci 2012; Wessel 2015). We recognize the limits of Twitter data; it can be demographically skewed toward the younger and the wealthier, and it is only reliable in large, aggregated data sets. Our results are subject to verification by other sources, but the methods would remain applicable.

Our interest in mobile social media investigates the relationship between the flow of information generated by mobile social media and the urban form of the city. One extreme position suggests information and communication technologies (ICT) have rendered location meaningless; such a position was taken by Webber (1964) as early as the 1960s based on his analysis of the extreme sprawl of American cities such as Los Angeles. At the other extreme are theorists who believe that the essential character of cities has remained more or less unchanged; new urbanism (Leccese 2000) is the latest such belief in the "timelessness" of urban form.

We have tried to make both positions sound credible, but both the rearguard and the avant-garde ignore current urban conditions in favor of longing for the past or promoting a bright future of uncertain prospect. Our effort is focused on finding specific on-going evidence of ICT as an indicator of the spatial form in the city, and conversely how the form of the city influences the uses of ICT. Mobile social media data is an ideal empirical source for such an analysis, combining spatial, temporal and topical information and providing large data sets for analysis. It is important (and difficult) because it is unstructured data, mostly unfiltered by existing categorization.

In this paper, we present three models of how to approach the analysis and use of Twitter data. The first two began as focused attempts to accumulate and explore large-scale Twitter data. The third is a model that incorporates lessons learned from the first two models into a novel visual analytic interface. This work is aimed at developing epistemological alternatives; we are continuing to use our insights to study both the nature of ICT in the city and on the usefulness of this approach for practitioners and researchers. Our team includes professors, researchers and advanced students from architecture, urban design, computer science and visualization.

First Approach: Problem Formation and Selective Data Acquisition

We began by collecting data using the Twitter API, which allows anyone to collect tweets by incorporating a number of delimiting variables such as time, location, hashtag, etc. Using this method, approximately 10,000 tweets were collected. We formulated specific problems with an important spatial component that Twitter data might help us to understand, gathering information that was either geographically limited or constrained by a keyword. This avoided a potential flaw with Twitter data; only about 2% was geo-tagged by the user. The use of a tightly defined search captures a selection of all users in that area.

An example will illustrate this approach We guessed that the spatial and temporal pattern of tweets during travel might vary from everyday patterns. Based on the

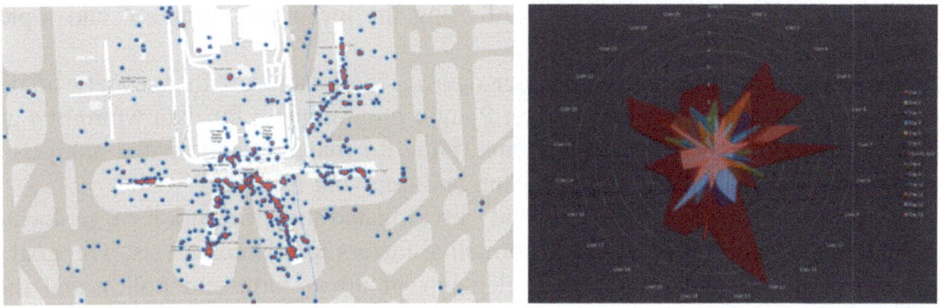

Fig. 1. Heat map of geolocated tweets at Charlotte Douglas Airport (*left*), Radial graph of tweet traffic on day of travel in red and days on either side of departure for 12 individual travellers (*right*)

collection of geolocated tweets at an international airport over a 2-week period, we were able to generate heat maps of the tweet locations within the airport. This was a fairly predictable pattern highlighting the concourses, where people have free time as they wait, in addition to last minute or just arriving tweets on the taxiway.

We were then able to identify individual users (without breaching their privacy) and their tweet history of a 2-week period centered on their tweets at the airport. We generated a graph that compared the frequency of their tweets in the airport versus the frequency of their tweets on the days before and after the airport sample. We found a clear pattern of accelerated tweets during their time at the airport, shown in the radial graph by the large size of the red graph area (Fig. 1).

Based on our work, we identified both strengths and weaknesses to this problem formulation approach. It allowed us to focus our inquiry and deal with the limits of the public API (limits to the number of tweets we can receive by a query and the number of times we can query the Twitter database). At the same time, the specificity of our search forestalled any wider exploration of large data sets without a heavy load of assumption.

Second Approach: Large Scale Data and Selective Query

In the spring of 2016, we joined a campus-wide Data Science Initiative, which afforded us access to large archives of Twitter data not accessible from the public API. We acquired a collection of about 2 million tweets of New York City and an equal number of Los Angeles metropolitan area over a 2-month period. Unlike our previous use of the Twitter public API that used specific keywords to search, we used a large geographic boundary to get access to a comprehensive set of geolocated tweets for those two cities. Our datasets consisted of only the tweets that originated from the specified geographic boundaries. Our goal was to create a permanent archive that we could use for the study of a variety of issues.

Our aim was to introduce data visualization and analytics to analyze social media data for the purposes of urban studies. One research group focused on the locations of the tweet data relative to morphological and spatial characteristics of the city and the other focused on social characteristics of using the locational, temporal and topical dimensions of the data.

For each group, we used the Twitter data in combination with a variety of data analysis methods such as custom written Python programming for data wrangling and analysis, Geographic Information Systems (GIS) for mapping and spatial analysis of the data, MongoDB database for accessing and querying the data.

The *urban form and information* study investigated the distribution of geolocated tweets and their relationship to other normative factors in urban analysis. The goal of this group was to study and analyze elements of urban form and policy and their relationship with tweet data; for example focusing on the analysis of transit oriented development (TOD) areas in Los Angeles or the connectivity of streets.

We started by creating a map of the locations of each tweet (1,000,000 total) in Los Angeles County and New York City. We then proceeded to develop a map that represented a cumulative tweet count for each census track within the county, with a single dot sized to represent the total count for each track. Next, we overlaid the census tract tweet count index with population, age and ethnicity data derived from American Community Survey 2014 (Bureau 2016).

In addition to the visual evidence provided by these overlaid maps, we also used linear regression analysis to find relationships between different variables. The results of this analysis showed no significant correlation between any of the factors obtained from the U.S. Census. The results of these regression analyses showed us that the locations of geolocated tweets were not directly and linearly affected by population demographic information.

We suspected that land use might be a more important predictor of tweet location than demographic data. We derived land use and zoning data from each city's respective planning governments (Planning 2016) and mapped the tweet data against different land use types (Fig. 2):

- Residential density
- Institutional density
- Mixed use zoning
- Parks and open space
- Commercial development.

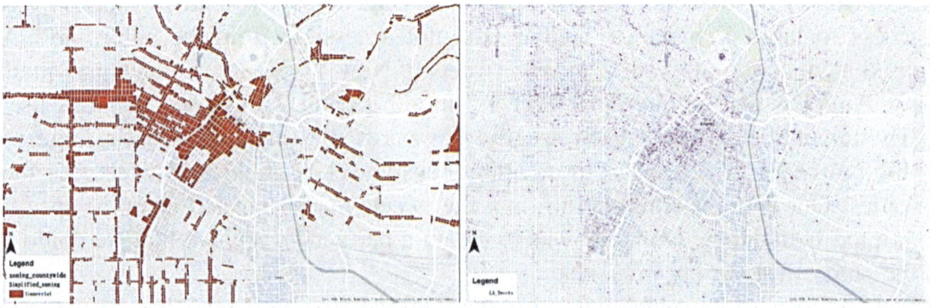

Fig. 2. Maps showing commercial distribution and tweet distribution in an area in Los Angeles. Tweet data density is closely aligned with commercial development

We performed a regression analysis on these findings to determine the correlation between these factors. The only clear correlation was between commercial development and tweet density.

We were also interested in the relationship between mass transit and tweet density. We located each Metro station in Los Angeles and every subway station in lower Manhattan, measuring the tweet density at each within a one-mile radius. We discovered influence on tweet activity but it was not decisive. We also calculated tweet density for 1, 0.5 and 0.25 mile radius from each station to try to test the influence of the station on the tweet activity. We found both a general rise in tweet activity surrounding stations and a significant number of anomalies.

Finally, we were interested in analyzing and visualizing the movement of Twitter users. Our first attempt connected the locations of every user who tweeted more than two times as a straight line. However, due to the nature and scale of the dataset, this type of visualization created an unusable spaghetti like map identified by other researchers (von landesberger 2016). To solve the issue, we developed a unique method that identified Twitter users with multiple tweets over a day and plotted their locations onto the street map using Dijkstra's shortest path algorithm (Dijkstra 1959). To calculate the shortest path we used ArcGIS Network Analyst (esri 2016). We collected the street data from OpenStreetMaps (2016) and processed our network files with GISF2E (Karduni 2016) to develop a representation of the use of the street system. We found considerable asymmetry of the Twitter users on the street grid using on the already identified location of commercial activity.

The *urban form and society study* tackled a much more complex problem. They studied how urban space is related to social factors, including both the identification of preferred locations for groups of tweets and sub groups.

This group began with an analysis of the demographic factors including population density, race, language, age, income and gender. Each of these factors were analyzed using ordinary least square regressions and displayed in a scatter plot matrix. No significant correlation was found between any of these factors and the location of tweets.

We next turned our attention to a comparison of English and Spanish language tweets. We compared the distribution of geolocated tweets with Hispanic information from the U. S. Census. We also studied the distribution of English and Spanish language tweets over the period of a month, identifying the top five spikes by location in each language.

Seeking to discover the relationship between events and tweet data, we identified major events from archives of the Los Angeles Times newspaper for a 1-month period. For each event, we identified the volume of tweets and plotted geolocation information.

Using the Dijkstra's shortest path algorithm mentioned above, we constructed tweet trails for all English and Spanish language tweets. We were then able to compare the similarities and differences of movement between the two groups (Fig. 3). We were then able to construct heat maps highlighting the concentration of each group's movement that showed areas where different groups of people tweeting in different languages were frequently traveling.

While our studies yielded interesting insights into the use of social media, the amount of special knowledge and the effort entailed by this approach limited its general adoption and the ability to use it in design settings for rapid feedback and scenario testing.

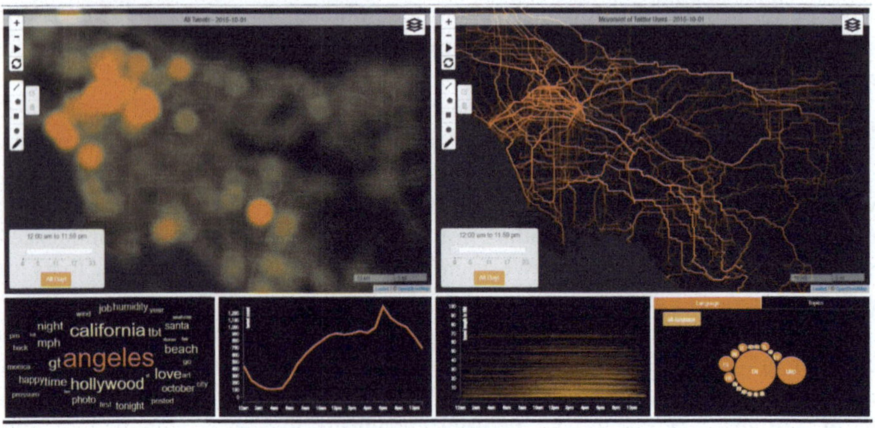

Fig. 3. Urban Activity Explorer, a visual analytic interface showing Los Angeles County. Heat map of tweets location (*top left*), flow map (*top right*), word cloud, timelines and language/topic views (*bottom*)

Third Approach: Interactive Visual Analytics and Large Data Sets

Based on the limits that we identified using selective queries, we partnered with the Charlotte Visualization Center at UNC Cahrlotte to design a visual analytic interface using New York City and Los Angeles data sets.

Visual analytics is the emerging field of analytical reasoning facilitated by interactive visual interfaces. The field's important features include the ability to deal with high-dimensional data sets, to present information visually, and to allow researchers to interact with this information. Visual analytics' fundamental premise is that analysis is better undertaken as a symbiosis of the computational power of computers with the pattern and sense-making capacity of human users (Ribarsky 1994; Chen 2009).

In this phase, we created Urban Activity Explorer, a new visual analytic interface, initially using the 2 million tweets we have collected in Los Angeles County as a sample data set. The complete description of this system has been published separately (Karduni 2017).

A central and unique feature of Urban Activity Explorer is the map views that afford interactive exploration of the geolocations of the tweet data set. One view includes a heat map showing the intensity of tweets that allows the identification of clusters of activity. A second view includes a flow map, which identifies the paths of all individuals in the data set that have multiple tweets for a single day. Built upon the results of our second approach, we mapped these multiple locations onto the street grid, which provided a view that in the aggregate represents the flow of tweets mapped onto the urban infrastructure.

The power of visual analytic interfaces is largely derived from the connection between data representation. In our case, we connect the map views to two forms of topic representation and two forms of timelines.

One topic view includes a simple word cloud (a count of the frequency of the occurrence of a word within a body of text). This view provides a quick overview of topical concerns, especially as the size of the text archive is smaller. The other view includes a topic model in Latent Dirichet Allocation (LDA) (Hoffman 2010) that uses a more complex analysis of the distribution of connections between terms. LDA is more useful with large archives, especially when they may have more complex information.

The first timeline illustrates a simple graph of occurrences of any selected item over time. The user can select any time period to investigate peaks and valleys of tweets. The second timeline accompanies the flow analysis and allows users to simultaneously select both a time interval and a limit for travel distances.

Interaction is a key aspect of our interface that was designed to allow for free investigation of a very large data set, rather than providing answers to one specific question. This allows a diverse range of users to investigate the data the without knowing ahead of time what factors might be important or what they might find. This ability to generate insight and knowledge-building is a key aspect of visual analytic interfaces (Chang 2009).

Future Work and Conclusions

Over 3 years in our research group we began with spatially constrained, modestly sized data sets, progressed to large scale data sets combined with custom scripted queries and finally to the design of a visual analytic interface that will be used as a framework for urban design. We conclude that a visual analytic system offers the most opportunity for integration into design issues for designers. For a system to be useful for urban design it must provide quick feedback, allow rapid iterative search and combine computational data analysis with human insight. Methodologically the next research initiatives within our group will include the explicit incorporation of knowledge generation paradigms and user testing our interface with urban design (Sacha 2014).

Insight and knowledge are key aspects of visual analytic systems. We plan to incorporate the knowledge generation paradigm research of Daniel Keim into our work with Urban Activity Explorer, but more generally we propose such systems as a particularly useful and appropriate approach for design. Keim articulates three levels of reasoning that form a framework for human computation. An important feature of these "loops" is that they represent not steps in a linear sequence, but rather three activity loops with distinct goals and processes that are engaged iteratively. The exploration loop is the exploration of a large data set to test preliminary ideas about the organization of the data. The verification loop is the creation of a hypothesis that is testable within the world of the visual analytic system. The knowledge generation loop tests and extends the interpretative power of a hypothesis.

Our work to this point has focused on a study of epistemological methods for ICT in urban settings; our next research goal is to conduct a user study of our Urban Activity Explorer with professionals and scholars. We believe that our interface provides a valuable tool to incorporate principles and techniques of information visualization about cities that are both qualitative (tweet topics) and quantitative (temporal information and counts). It also allows users to generate new interpretations and insights about cities and human activity that were previously indiscernible with traditional mapping techniques or unilateral data. We will test our approach with designers at all levels of experience and skill.

Acknowledgements. The authors wish to acknowledge our on-going collaboration with the Charlotte Visualization Center at UNC Charlotte, particularly Dr. Isaac Cho, Dr. Wenwen Dou and the late Dr. William Ribarsky. Our work would not have been possible with them. We also acknowledge the insights and work of student in graduate seminars during academic years 2014–17 in the School of Architecture at UNC Charlotte and the School of Architecture, Art and Historic Preservation at Roger Williams University. The collaborative work in these classes was vital to the development of this research.

References

Andrienko, G., et al.: Space, time and visual analytics. Int. J. Geogr. Inf. Sci. **24**(10), 1577–1600 (2010)

Arce, N.: 438 billion: daily volume of mobile and online messages by 2019. TechTimes 2015. http://www.techtimes.com/articles/67352/20150709/438-billion-daily-volume-of-mobile-and-online-messages-by-2019.htm (2015)

Bureau, U.S.C.: Data Releases (2016)

Chang, R., et al.: Defining insight for visual analytics. IEEE Comput. Graph. Appl. **29**(2), 14–17 (2009)

Chang, R., et al.: Legible cities: focus-dependent multi-resolution visualization of urban relationships. IEEE Trans. Vis. Comput. Graph. **13**(6), 1169–1175 (2007)

Chen, M., et al. Data, information, and knowledge in visualization. IEEE Comput. Graph. Appl. **29**(1), 12–19 (2009)

Dijkstra, E.W.: A note on two problems in connexion with graphs. Numer. Math. **1**(1), 269–271 (1959)

Erricson: Ericsson Mobility Report (2016)

esri.com. Route analysis (2016). http://desktop.arcgis.com/en/arcmap/latest/extensions/network-analyst/route.htm

gnip. Gnip—About (2016). https://gnip.com/about/

Hoffman, M., Bach, F.R., Blei, D.M.: Online learning for latent Dirichlet allocation. In: Advances in Neural Information Processing Systems (2010)

Karduni, A., Cho, I., Wessel, G., Ribarsky, W., Sauda, E.: Urban space explorer: a visual analytics system for understanding urban social media activities. IEEE Computer Graphics and Applications, accepted for publication, special issue on Geographic Data Science (2017)

Karduni, A., Kermanshah, A., Derrible, S.: A Protocol to Convert Spatial Polyline Data to Network Formats and Applications to World Urban Road Networks. Scientific Data, London (2016)

Leccese, M., McCormick, K.: Charter of the New Urbanism. McGraw-Hill Professional, New York (2000)

MacEachren, A.M., et al.: Senseplace2: Geotwitter analytics support for situational awareness. In: IEEE Conference on Visual Analytics Science and Technology (VAST). IEEE (2011)

Offenhuber, D., Ratti, C.: Decoding the City: Urbanism in the Age of Big Data. Birkhäuser, Basel (2014)

OpenStreetMaps. About OpenStreetMaps. (2016). https://www.openstreetmap.org/about

Planning, D.o.R.: Maps & GIS|DRP (2016)

Ribarsky, W., Foley, J.: Next-generation data visualization tools. In: Scientific visualization, Academic Press Ltd (1994)
Sacha, D., et al.: Knowledge generation model for visual analytics. IEEE Trans. Vis. Comput. Graph. **20**(12), 1604–1613 (2014)
Tufekci, Z., Wilson, C.: Social media and the decision to participate in political protest: observations from Tahrir Square. J. Commun. **62**(2), 363–379 (2012)
von Landesberger, T., et al.: Mobilitygraphs: visual analysis of mass mobility dynamics via spatio-temporal graphs and clustering. IEEE Trans. Vis. Comput. Graph. **22**(1), 11–20 (2016)
Waddell, P.: UrbanSim: modeling urban development for land use, transportation, and environmental planning. J. Am. Plan. Assoc. **68**(3), 297–314 (2002)
Webber, M.M.: The urban place and the nonplace urban realm (1964)
Wessel, G., Ziemkiewicz, C., Sauda, E.: Revaluating Urban Space Through Tweets: an Analysis of Twitter-Based Mobile Food Vendors and Online Communication. New Media & Society, p. 1461444814567987 (2015)

Simulating Pedestrian Movement

Renee Puusepp[1(✉)], Taavi Lõoke[1], Damiano Cerrone[2], and Kristjan Männigo[2]

[1] Estonian Academy of Arts, Tallinn, Estonia
renee.puusepp@artun.ee
[2] Spin Unit, Tallinn, Estonia

Abstract. This paper discusses the results of an extended pedestrian movement project carried out in the center of Tallinn, Estonia. Having gone through several rounds of in situ studies and software prototyping, we present a synthetic method for modelling data of urban qualities as perceived by individual pedestrians. The proposed method includes capturing the walkability of city elements as a landscape of explicit values. This landscape is turned into proximity fields that in turn are fed into an agent-based model for simulating the route selection mechanism of individual pedestrians. The described model is used to demonstrate how each simulated pedestrian is influenced by the perceived qualities of urban environment. Based on our observations and computational modelling experiments, we argue that pedestrian traffic in the city can be successfully simulated with agent-based models. Furthermore, such simulation models, when finely calibrated in the field observations, allow urban planners, city authorities and decision makers to evaluate the impact of future urban design scenarios to pedestrian flows in the city.

Keywords: Pedestrian movement · Simulation · Agent-based modelling · Synthetic modelling

Introduction

Pedestrian ecology in a busy city is a complex subject. There is a multitude of origin-destination (OD) points, a large number of different motivators and de-motivators in the urban environment as well as internal factors that affect the route selection mechanism of individuals. Unlike with cars and public transportation, it is difficult to reduce pedestrian movement to a network of paths and intersections. Graph-based approach presumes that all possible movement tracks are predefined while in reality pedestrians can take such surprising shortcuts in the city that any graph to contain all possible routes becomes quickly unmanageable. Instead, we argue that the urban streetscape from the point of view of pedestrian movement could be seen rather like a multi-layered landscape that offers walking opportunities of different quality. In respect to walkability and paying homage to Gibson's (1979) theory of visual perception, let's call this landscape the *landscape of affordances*.

Navigational decisions through such a landscape are dependent on the geometric layout of urban space as well as on behavioural characteristics of individual pedestrians. The perception of affordances depends on the pedestrian's age, gender, cultural

K. De Rycke et al., *Humanizing Digital Reality*,
https://doi.org/10.1007/978-981-10-6611-5_46

background, personal preferences, physical abilities and risk-adversity, but also on the temporary variables such as intensity of traffic, weather and lighting conditions that vary strongly during the course of a day, a week, and a year. Thus, one can say that all pedestrians have their own *landscape of affordances* that depend on their persona as well as on local and temporal conditions.

By following synthetic modelling methods (Puusepp et al. 2016) we can model the *landscape of affordances* and use it in order to explore the phenomenon of pedestrian movement in cities. The key is to combine field observations with modelling and running computational simulations. Although perfectly reasonable, it seems that not too many researchers in the field use empirical knowledge gained through in situ studies for supporting the functional logic of their simulation models. Of the few examples, creators of STREET (Haklay et al. 2001) and PEDFLOW (Kukla et al. 2001) provide some observational data to support their modelling decisions. Batty (2003) has managed to evidence his simulation models with one in situ study, although the study did not take place in the urban environment but in Tate Britain art gallery.

It seems that the majority of pedestrian simulation modellers have the aspiration of creating a universally applicable model. We believe that the route selection process of pedestrians varies significantly from location to location, from city to city, and from country to country due to differences in socio-cultural context, climate and spatial conditions. While we cannot confidently prove that yet, since it would require carrying out studies of similar scale in many different cities, we can at least show how it works in the centre of Tallinn, Estonia.

There are a number of possible approaches to simulating pedestrian movement from micro-scale simulations using computational physics to macro-scale network models for transportation planning (Haklay et al. 2001). In our study of meso-scale movements we have borrowed concepts from graph theory, environmental psychology, psychogeography and complexity modelling, in order to propose a hybrid model that combines agent-based movement simulation with computationally generated raster maps of cognitive distances.

Observing and Tracking Movement

There are many methods for tracking and mapping pedestrian movement in the city. Methods range from simply counting the number of passing people to high-tech solutions such as mobile positioning, purpose made sensory networks, and video capture combined with artificial image recognition software. We have excluded those methods partly for financial reasons and partly for following our core principles—we wanted to get the first-hand experience of the movement. On one hand we wanted to quantify pedestrians flow and its trajectories. On the other hand and inspired by experimental studies in the field of psychogeography we wanted to learn more about the *appealing or repelling character of certain places* as described in Debord's seminal essay (http://library.nothingness.org/articles/SI/en/display/2). To achieve both objectives at the same time, we decided to follow a randomly chosen but fairly evenly distributed sample of pedestrians starting from dedicated gates—street corners, mall entrances and transportation stops. The task was to record the movement track using smart phones and GPS tracking software, and take notes of pedestrian behaviour in respect to their route selection decisions.

There appeared to be a clear educational aspect to this practice of tracking. Namely, we started to realise the full complexity of pedestrian ecology. This realisation of complexity reminded us that any simulation done to date including our own simulations are going to remain very rough generalisations of reality for years to come. Thus, acknowledging the full complexity of the real world phenomenon helped us to take the outcome of any simulation model with a healthy skepsis.

Some of the early field observations turned out quite surprising. We observed several 'illegal' paths taken by pedestrians crossing roads at the most unbelievable and dangerous locations dominated by cars, trams and buses. We also saw that people are attracted and repelled by different features and properties of the urban environment depending on their age, gender and fitness. We observed some people taking long detours in order to avoid a pedestrian tunnel, while others—mostly young males—chose the shortest path through the tunnel. We saw that streets were frequented by different types of pedestrians depending on the time of the day and day of the week. Besides morning and evening rush hours when the majority of pedestrians were either going to work or back home, there were hours during the day, when streets were occupied by schoolchildren, students, moms with prams and elderly people. But perhaps the most important observation was that not many people changed their mode of operation—those who came by public transport also tended to leave by bus or tram; those who walked into the area also walked away, and those who came from the mall entered another one.

During two isolated field studies we managed to collect over 700 individual pedestrian tracks by following randomly selected pedestrians in the chosen area and recording movement with GPS-enabled smart phones. Figure 1 depicts a map of GPS tracks collected during one of these studies in October 2016. The number of collected

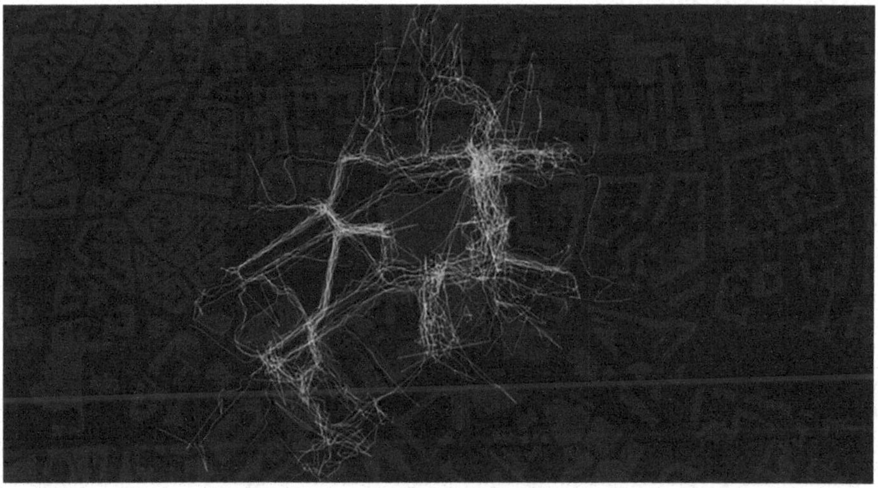

Fig. 1. Collected GPS tracks of pedestrians

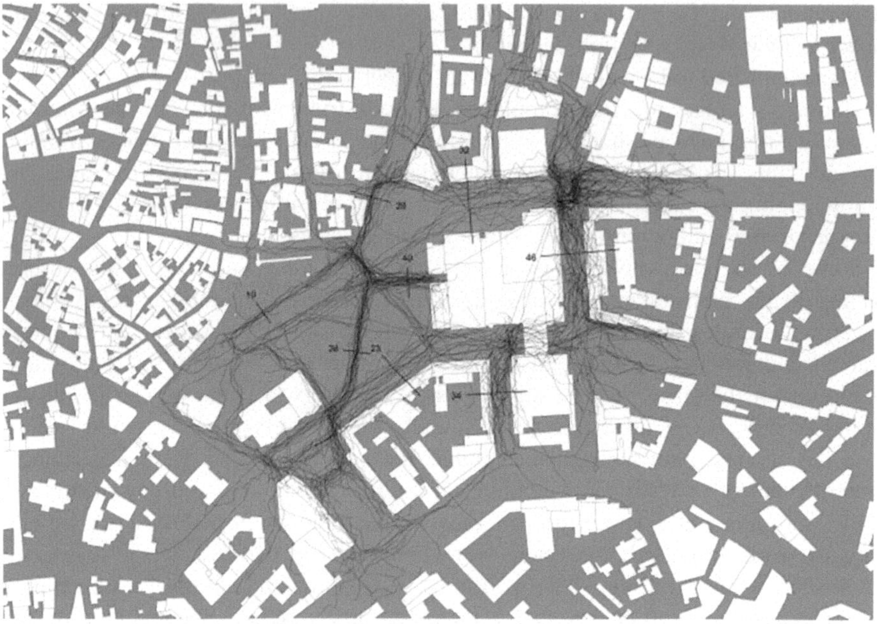

Fig. 2. Collected movement paths counted at street sections

tracks sufficed for producing a matrix of origin-destination combinations showing not only where from and where to people go but also the exact selection of their route and the time of their journey. Figure 2 shows a patch of 506 movement tracks collected and the count of tracks passing selected street sections.

Analysing Movement

Data collection and making qualitative observations in the city was logically followed by quantitative analysis that helped us to validate initial assumptions and generate comparative material for calibrating the simulation model. But before carrying out the analysis, messy real-world data had to be cleaned up. Tracking pedestrian movement with GPS-enabled devices works quite well in low-rise suburban environments and open areas but generates lots of unwanted artefacts when used in narrow and high street canyons where the GPS signal is blocked and reflected by surrounding buildings.

GPS tracks had to be adjusted to reflect the actual movement of people. This was done in GIS software (Q-GIS) where tracks were redrawn to follow the network of footpaths and sidewalks. Also, whenever a pedestrian had taken a shortcut by crossing a road or any other area not intended for walking, a new 'illegal' path was added to the network. This network was later used for analysing the use of shortest paths. But quite unexpectedly this network also served as a visual analysis tool and a source of information for preparing the *landscape of affordances*. Figure 3 shows the network that helps to identify character of different streets—those streets that are crossed in several places are clearly more pedestrian friendly than those crossed in fewer places.

Once all collected tracks were rationalised and merged into a single network it became possible to analyse each individual pedestrian track and compare it to the shortest possible route within the network. The entire operation was carried out in Rhinoceros 3D and it's plugin Grasshopper and involved the following steps:

- Building a network of unique line segments connected in crossings or points where line segments were overlapping (see Fig. 3).
- Calculating the shortest paths between every unique pair of endpoints with Dijkstra's (1959) algorithm based on the network's metric distances.
- Measuring the metric length of each pedestrian track and comparing it to its shortest path distance (see Fig. 4). For every unique pair of origin-destination combination, the percentage value of using the shortest path was calculated.
- Calculating the overall percentage of how many pedestrians use the shortest path. The used track was considered the shortest path if its length was no more than 10 m longer than the actual shortest path.

As a result of the above calculations we can conclude that the use of shortest paths by pedestrians in the Tallinn city centre is quite low. Only about 35% of recorded

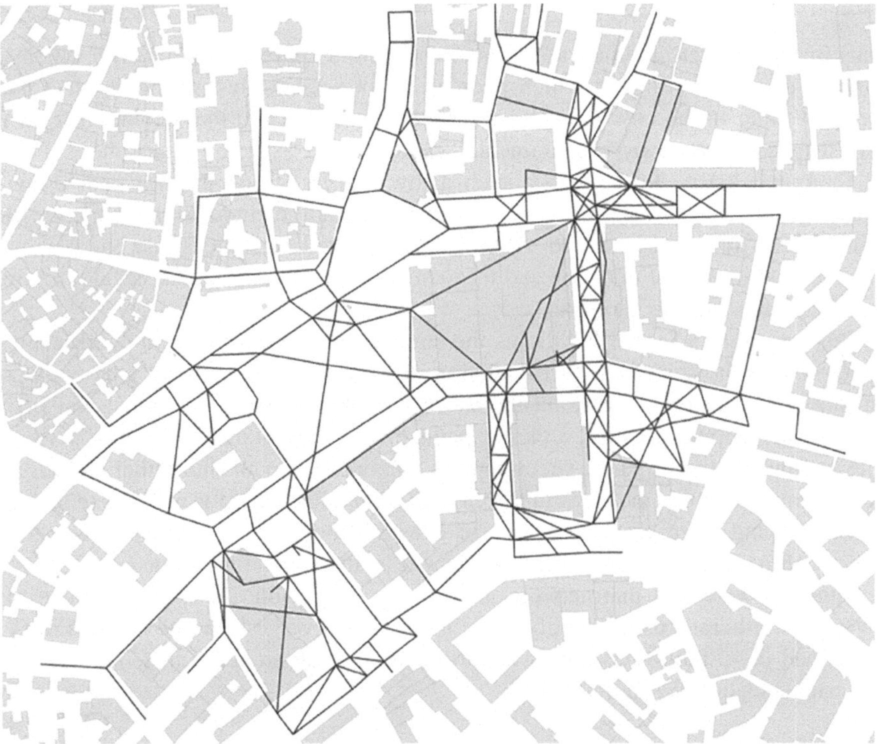

Fig. 3. Rationalised network of pedestrian tracks

Fig. 4. A recorded path (length: 587 m) shown in *black* compared to the shortest path (length: 541 m) shown in *red*

tracks in our dataset qualified as the shortest path. This result aligns with other scientific research according to which people do not necessarily use the shortest paths when travelling in the city (e.g. Zhu and Levinson 2015) and that deviation from the shortest path is caused by environmental variable such as urban atmosphere (Foltete and Piombini 2010). Nevertheless, such a low percentage of shortest path usage is surprising and can perhaps be partly attributed to our method of network creation against which all of the recorded tracks were compared. It suffices just one person to cross the road in 'illegal' location and this can make all tracks that use 'legal' crossings nearby count as not the shortest.

This conundrum further highlights the complexity of performing network analysis and especially the shortest path calculation on pedestrian movement. One could say that most of pedestrians are using the shortest paths, but each individual walker has their own perception of which walkable paths the network is made of. Thus, it seemed futile to consider the walkable streetscape as a network of paths but rather a *landscape of affordances* where some areas in the city are more affordable for walking (e.g. parks, pavements), and others are less so (e.g. roads with heavy traffic). Hereinafter we refer to such an affordance of an urban layout element as its walkability value.

Although we found that network analysis is not suitable for analysing pedestrian movements, it helped us to identify those tracks that diverge from the shortest ones. It also helped us to build hypotheses about qualities of urban space that influence people's path selection decisions.

Simulating the Movement of Individual Pedestrians

We have adopted a two-part process for preparing and running simulations. Firstly, we developed a plugin for CAD software (Rhinoceros) for drawing the simulation map. Secondly, we programmed an application for running agent-based simulations in NetLogo. Such a separation of map preparation and movement simulation uses the strongest features of both applications used—the ease of drafting urban layouts in Rhinoceros; and the ease of controlling agents' movement in NetLogo.

In the simulation model, the mobile agent is assigned an origin and a destination to reach. The agent navigates the environment by hill-climbing a field of values generated in the map preparation phase. Figure 7 shows the bodily architecture of the mobile agent and the navigational principles. Conceptually, the agent possesses overall knowledge of the urban layout and cannot get lost, but it can potentially react to purely local cues. Such an abstraction of pedestrians is obviously very crude and ignores some types of pedestrians such as aimlessly wandering tourists, strollers and dog-walkers. However, our field observations seem to suggest that these types of walkers are fairly rare in the area.

In the map preparation phase the *landscape of affordances* is drafted as two-dimensional urban elements with a walkability value. While the walkability value is assigned by the author of the map and is clearly subjective, there are some very obvious principles that can be followed. For example, traversing a pedestrian path or a sidewalk is more affordable than a lawn, while a road would be much more expensive. A building or a fence cannot be traversed at all. Once the vector map of closed curves is drafted and walkability values assigned, the map is converted into a raster grid with every grid cell inheriting the walkability value from its parent urban element.

The next step in the process is to define origins and destinations (OD-s) on the map and calculate proximity maps for each of these. The calculation of proximities is performed on the raster grid using cellular automata (CA) based value propagation rules where the transition of proximity values is affected by the walkability values of each cell. The resultant proximity map as shown in Fig. 5 can be considered as a cognitive map that represents knowledge about the configuration of the environment (Clayton 2012). Such a map is obviously dependent on the map-drawer's perception, but it is a useful tool for visualising the newly generated data.

The raster data of cognitive distances—the proximity map—is imported to Netlogo and attached to Netlogo's patches. Then an agent is placed on an origin, given a destination and the simulation can be started. The agent moves closer to its assigned destination by hill climbing the pre-calculated proximity map. The bodily architecture of agent is simple. As shown in Fig. 6 the agent has a notional body with three sensors—one in front and two symmetrically on sides at a specific angle from the front sensor—a notional field-of-vision (FoV). These sensors read the proximity values from the environment, and the agent moves in the direction of the 'winning' sensor—one with the highest proximity value. We have chosen this architecture since it is one of the simplest functional 'body plans' of for a mobile agent (Puusepp 2011).

Early trials have shown that 1 patch per 1 m^2 (1 ppm) of urban space is a good enough resolution for running the simulation. This is somewhat specific to the

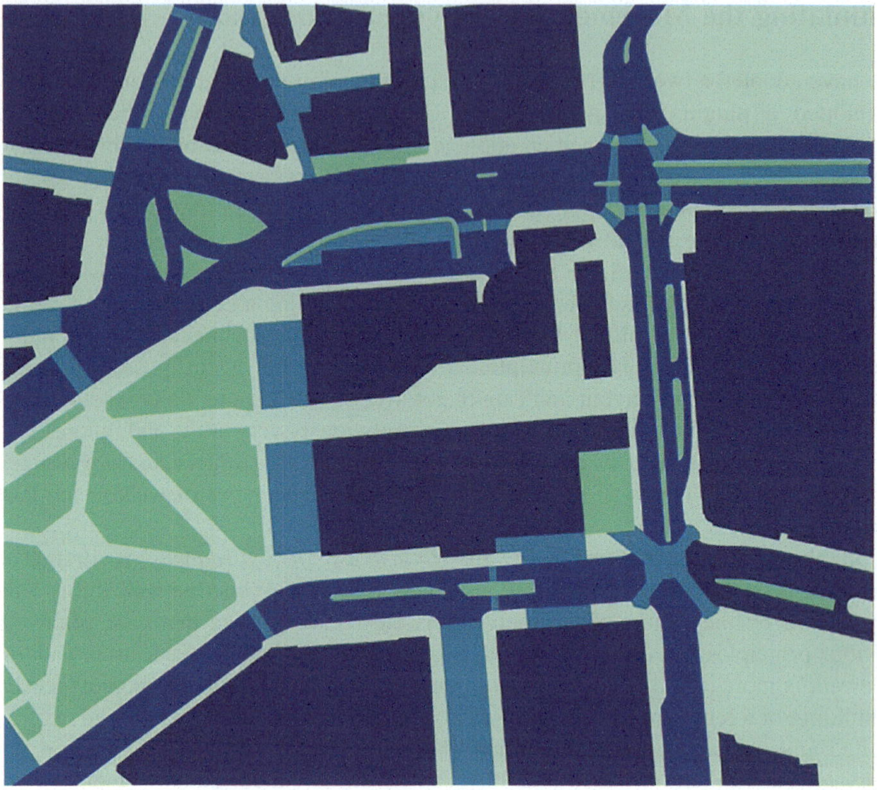

Fig. 5. Walkability landscape—less walkable areas are *darker*

environment that we are studying—the downtown of Tallinn—as pavements and pedestrian passages in this area are rarely narrower than a meter (Fig. 7).

Modifying walkability values of urban elements can make simulated walkers to choose drastically different paths. Figure 8 depicts two slightly different landscapes of affordances that lead to different route selections—in the first instance the agent takes a longer trip in order to avoid the pedestrian tunnel (left) while in the other it goes through the tunnel (right). Let us point out here that both of these behaviours were observed in real pedestrians during the field works as well. Indeed, one should compare agent movement paths to the observed paths of pedestrians and fine-tune the *landscape of affordances* accordingly. The aim is to get agents using similar routes as real-world pedestrians. It is essential to do so in order to close the loop of synthetic modelling and achieve realistic simulation results. Getting acceptable results may take several cycles depending on the complexity and size of the area.

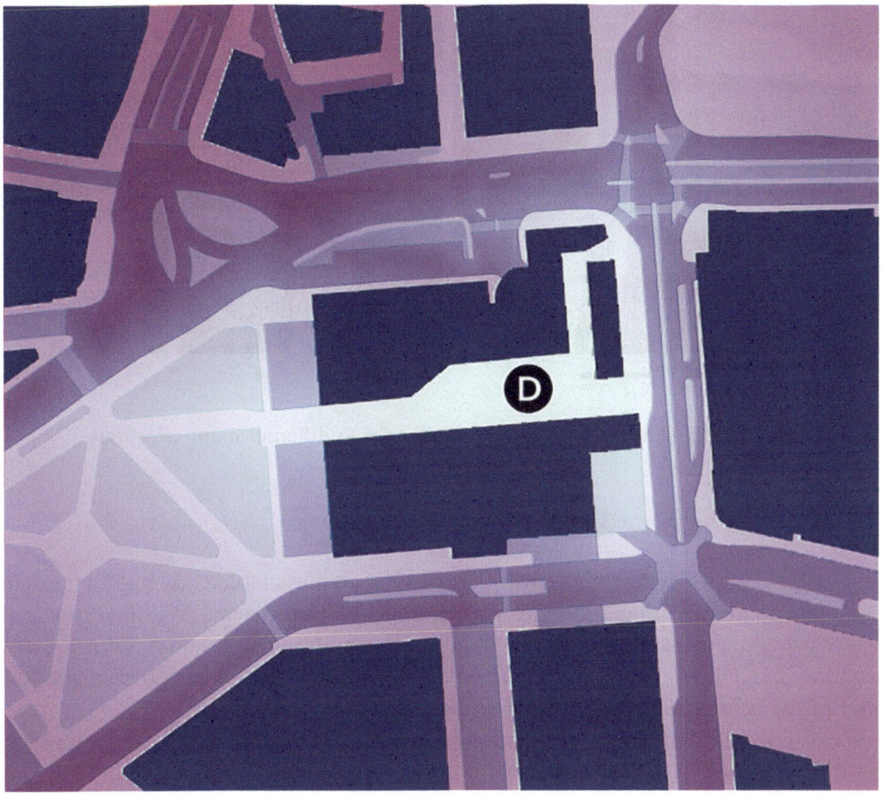

Fig. 6. The *whiter* the closer—the proximity map to a destination (D)

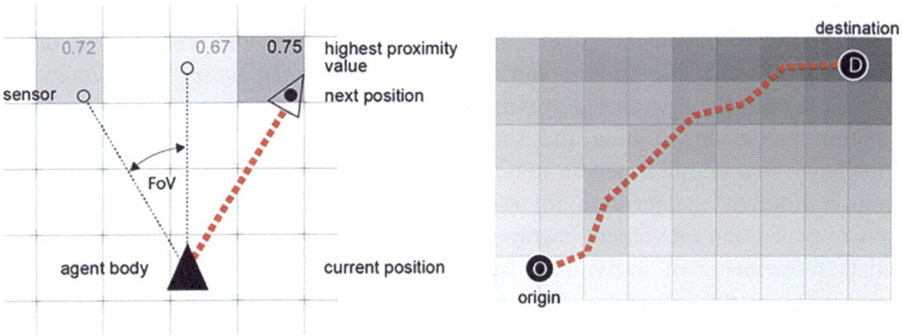

Fig. 7. Agent's architecture and route selection mechanism

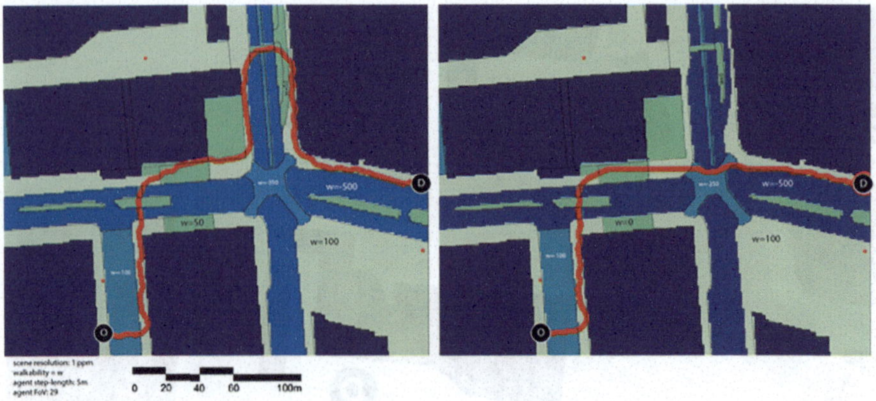

Fig. 8. Different walkability values (w) lead to different route selections

Conclusions

Simulating the entire pedestrian ecology in a city is an intractable task, but it is possible to simulate route selection mechanism of individual pedestrians. We have made the case for using agent-based modelling paradigm for building simulations of meso-scale pedestrian movements in busy urban environments. We have tested our proposed model in the centre of Tallinn within an area of about 1 km^2.

We discovered that unlike with vehicular traffic in the city, graph-based approaches are not the best suited for movement simulations of such scale since pedestrians do not always move along designated pathways and they have a significant freedom of selecting their exact walking trajectory. Instead of using graph-based calculations, we have opted for computationally more expensive model where all areas accessible to pedestrians make up a landscape of values that define how affordable the urban environment is for walking.

Walkability of pavements, roads, grass and other traversable areas can be defined on the basis of field observations and captured as the *landscape of affordances*. This landscape is used for calculating proximity maps that serve as input to the agent based simulation model for finding the shortest cognitive paths in the landscape. We acknowledge that one real person may perceive affordances in the urban environment in quite a different manner than any other person. While it is difficult to create simulations of pedestrian movement encompassing all pedestrian types, it is manageable to simulate movements of individuals by using synthetic modelling loop. The entire process involves the following steps: (1) observing pedestrian movement in the city, (2) preparing the landscape of affordances as a CAD drawing in order to capture the walkability of urban elements, (3) feeding this data into the agent based simulation model, (4) running the simulation and comparing the results to the real-world pedestrian routes, and (5) fine-tuning the *landscape of affordances*. In order to close the synthetic modelling loop and generate meaningful and useful simulation data, steps 2–5 are repeated until the simulated paths match the observed ones.

We have shown that, if affordances of all urban elements in respect to walking are defined well enough, our suggested method is a feasible way to predict the movement of individuals. We believe that this could be a useful method for assessing the impact of new street and urban design proposals to pedestrian traffic. Creating and fine-tuning a vector map of urban elements with attached walkability values is a feasible method for data modelling—turning qualities of urban environment into quantitative data.

We see three main areas where we can improve the suggested concept, methods and algorithms. Firstly, we want to feed a realistic OD flow demand matrix to the simulation model. This would help us to assess the pedestrian movement in bulk and make some quantitative predictions of pedestrian flows in the city. Secondly, we want to create different types of agents and diversify their route selection and movement algorithms. Thirdly, we want to add a layer to the simulation in order to represent temporal urban qualities and thus strive towards richer ways of quantifying urban qualities.

References

Batty, M.: Agent-based pedestrian modelling. In: Longley,P., Batty, M. (eds.) Advanced Spatial Analysis: The CASA Book of GIS, pp 81–108. ESRI, New York (2003)

Clayton, S.D.: The Oxford Handbook of Environmental and Conservation Psychology. University Press, Oxford (2012)

Debord, G.: Introduction to a critique of urban geography. http://library.nothingness.org/articles/SI/en/display/2. Last accessed 6 July 2017

Dijkstra, E.W.: A note on two problems in connection with graphs. Numer. Math. **1**, 269–271 (1959)

Foltete, J.C., Piombini, A.: Deviations in pedestrian itineraries in urban areas: a method to assess the role of environmental factors. Environ. Plan. B Plan. Des. **37**, 723–739 (2010)

Gibson, J.: The Ecological Approach to VISUAL Perception. Houghton Mifflin, Boston (1979)

Haklay, M., O'Sullivan, D., Thurstain-Goodwin, M., Schelhorn, T.: "So go downtown": simulating pedestrian movement in town centres. Environ. Plan. B Plan. Des. **28**, 343–359 (2001)

Kukla, R., Kerridge, J., Willis, A., Hine, J.:PEDFLOW: development of an autonomous agent model of pedestrian flow. Transp. Res. Rec. **1774**, 11–17 (2001)

Puusepp, R.: Generating Circulation Diagrams for Architecture and Urban Design Using Multi-agent Systems. University of East London (2011)

Puusepp, R., Cerrone, D., Melioranski, M.: Synthetic modelling of pedestrian movement—Tallinn case study report. In: Herneoja, A., Österlund, T., Markkanen, P. (eds.) 34th eCAADe Conference, pp. 473–481. Oulu (2016)

Zhu, S., Levinson, D.: Do people use the shortest path? An empirical test of Wardrop's first principle. PLoS ONE **10**, 1–18 (2015)

Structural Patterning of Surfaces

Renaud Danhaive(✉) and Caitlin Mueller

Massachusetts Institute of Technology, 77 Massachusetts Avenue, Cambridge, MA 02139, USA
danhaive@mit.edu

Abstract. This research introduces new computational workflows to design structural patterns on surfaces for architecture. Developed for design but rooted in the rigor of structural optimization, the strategies are aimed at maximizing structural performance, as measured by weight or stiffness, while offering a diversity of solutions responding to non-structural priorities. Structural patterns are important because they create meaningful difference on an otherwise homogeneous architectural surface; by introducing thickness, ribs, or voids, patterns accumulate material where it is most useful for structural or other purposes. Furthermore, structural patterning with greater complexity is enabled by the recent advancements in digital fabrication. However, designers currently lack systematic, rigorous approaches to explore structural patterns in a generalized manner. The workflows presented here address this issue with a method built on recent developments in computational design modelling, including isogeometric analysis, a NURBS-based finite element method. Most previous work simply maps patterns onto a surface based on stress magnitudes in its continuous, unpatterned state, thus neglecting how the introduction of holes interrupts the flow of forces in the structure. In contrast, this work uses a new trimmed surface analysis algorithm developed by the authors to model force flow in patterned surfaces with greater fidelity. Specifically, the proposed framework includes two interrelated computational modelling schemes. The first one operates on a human-designed topology pre-populated with an input cell shape and position. An optimization algorithm then improves the design through simple geometric operations in the parametric space applied independently on each cell or globally using control surfaces. The second modelling scheme starts with an untrimmed surface design and successively optimizes its shape and introduces new cells. Both optimization systems are flexible and can incorporate constraints and new geometric rules.

Keywords: Structural optimization · Structural pattern · Isogeometric analysis

Introduction

This paper proposes new computational methods for the design of patterned structural surfaces. Surfaces, often referred to as plates or shells when used in architecture and structural engineering, are ubiquitous both in the built environment and our daily life. Used as structures in their own right or as elements in complex plate assemblies, structural surfaces are members whose thickness is much smaller than their other dimensions. If well-designed, surface structures are extremely efficient, and are

K. De Rycke et al., *Humanizing Digital Reality*,
https://doi.org/10.1007/978-981-10-6611-5_47

frequently used in engineering design. Their potential for structural efficiency can be further enhanced through the introduction of apertures, which in turn can add a meaningful layer of tectonic hierarchy. With the intention of empowering architects and engineers with new ways to explore such structures, this research proposes new computation modelling systems integrating performance and creativity for the conceptual design of patterned structural surfaces.

Background and Motivation

"...However strange it appears, despite the diversity of our pursuits and the variety of objects, our main preoccupation was, in a way, making holes...", said Robert Le Ricolais (Motro 2007). A variety of methods have been developed in the last 30 years to help structural designers satiate this preoccupation. In fact, an entire research field, topology optimization, was born for that purpose, and has been extensively studied since its inception at the end of the 1980s (Rozvany 2009). Initially used by engineers in industries with stringent structural performance requirements, topology optimization is now integrated in web-based tools, apps (Aage et al. 2013), and add-ons for CAD environments (Panagiotis and Sawako n.d.), and has slowly found its place in the toolbox of architects and engineers as a means for reducing the amount of material used in architectural structures.

Topology optimization does not care for art, and it has not been developed for architectural design; rather, its best methods are mathematically defined optimization frameworks, and they leave little room for authorship beyond the definition of a design domain and volume requirements. Indeed, no control is possible over the shapes of the apertures, their locations, their numbers, etc. As a result, topology optimization merely is a source of inspiration for structural designers, rather than a rich design methodology.

Recent work in the computer graphics community has started addressing this issue. In particular, their work has focused on developing methods mapping regular patterns on meshes as to produce 3D-printed artefacts with desirable aesthetic and structural properties (Dumas et al. 2015). In particular, Schumacher et al. (2016) proposed Stenciling as an approach to map decorative patterns on mesh surfaces. Employing a fill ratio-based simulation, and inspired by the density-based methods of topology optimization, the method uses optimization methods to scale stencils on the surface as to reduce the compliance of a shell structure. While the results are promising and the method developed is effective, some of the drawbacks of traditional density-based methods, such as a lack of design control, remain. In a different field, Paulino and Gain (2015) devised finite elements with arbitrary shapes to integrate pattern design and topology optimization, but their method is too reminiscent of traditional optimization techniques to result in a general design-oriented method. Comparatively, the proposed frameworks are more general since the definition of pattern used here is a broad one. In the context of this paper, patterns are defined as a collection of cells whose shapes and positions can be altered, and mapped onto a surface to become voids or thickness depressions. Operations over the pattern are limitless, and the system offers unprecedented design freedom.

Designing with Patterns

Patterns in architecture have long been used for symbolic meaning or ornamental purposes. Often associated with deep vernacular meaning and symbolism, as in Gothic and Islamic architecture, patterns can be both ornamental and functional. Their use in architecture and structural works is uninterrupted throughout the history of construction, and the twenty-first century is no exception. In particular, there has been a recent surge in the performative use of patterns, particularly in structural engineering as contemporary structural engineers and architects cleverly use them to reduce the use of structural materials and to provide aesthetic and experiential qualities to their designs.

Design Heuristics and Force Flow Interruption

A simple heuristic to design patterned structures is to analyse the stresses in a base structure and nucleate apertures or scale pre-populated cells accordingly—the higher the stresses the smaller the aperture, and vice versa. The simplicity and ease of application of this heuristic rule makes it particularly adapted to conceptual design. Unfortunately, this naïve approach may be misleading as holes disrupt the flow of forces in a structure, rendering the initial stress configuration obsolete. Compared to density-based methods with similar heuristics (Querin 1997), it is harder to integrate such rules in iterative systems because, once the hole created, the stress can no longer be estimated. Despite the flaws of such methods, their underlying assumption is fundamentally correct. The structural hierarchy that naturally occurs in continuous media can indeed be leveraged to save material, but there is a need for systematic and creative ways of exploring such hierarchy.

Outline

The paper is organized as follows. Section "Pattern Mapping and Isogeometric" details technical aspects on which the research relies. Section "Structural Patterning and Optimization" introduces the workflows proposed for the design and optimization of patterned surfaces. Section "Conceptual Design Examples" demonstrates the applicability of the modelling system for the conceptual design of surface structures through two design examples. Finally, Section "Conclusions" summarizes the contributions, offers directions for future work, and reflects on the impact of research.

Pattern Mapping and Isogeometric Analysis

Contemporary CAD software employs Non-Uniform Rational B-Splines (NURBS) as their main geometry format for curve and surface modelling. The work presented here is similarly underpinned by the use of NURBS, which makes it generally applicable in most CAD environments. The following describes how the computational schemes developed rely on parametrized surfaces for pattern mapping and isogeometric analysis, with an emphasis on NURBS.

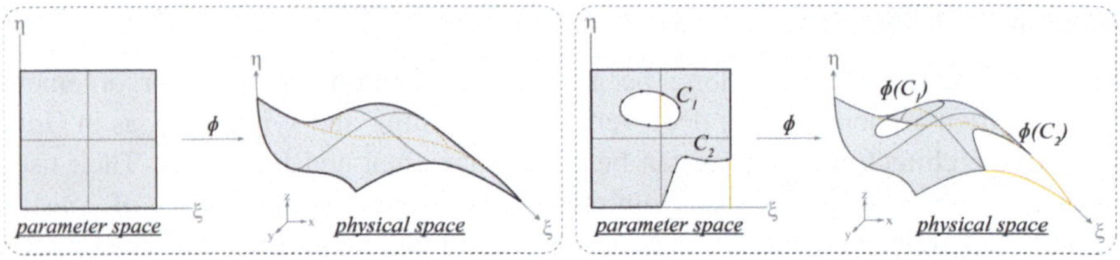

Fig. 1. Mapping from parameter to physical spaces: by definition, NURBS surfaces are parametrized surfaces that map a 2D domain to a 3D Euclidean space (*left rectangle*); a trimmed surface is defined by its parametrization/mapping and trimming curves defined in the parameter space

Mapping Patterns

Parametrized surfaces, including NURBS, map a 2D parametric space onto a surface in dimension 3, as shown on Fig. 1. Any two-dimensional object defined in the parameter space can thus undergo the same mapping, which makes for a convenient way to describe mapped patterns. Specifically, physical trimming curves are defined as mappings of curves which differentiate solid from void in the parameter domain (see Fig. 1).

Manipulating patterns directly on a non-planar surface is hard, because it entails manipulating 3D curves while enforcing that they lie locally in the tangent plane of the surface. But, doing so in the parameter space automatically ensures it, because the pattern curves undergo the same mapping as any point in the parameter space. Computationally, this is certainly less intensive. From a modelling standpoint, operating on curves indirectly in the parameter space may be frustrating; a solution is to define trimming curves directly in the physical space, through projections or Boolean operations, and subsequently pull them back in the parameter space. Once pulled back in the parameter space, the curves may be modified in the parameter space.

Isogeometric Analysis of Patterned Structures

In 2005, Hughes et al. (2005) introduced the concept of isogeometric analysis. The method shares common features with the finite element method and meshless methods and its main underlying objective is to unify CAD and finite element analysis. Specifically, they propose to match the exact NURBS geometry of a problem with the analysis model by using the NURBS basis functions to approximate the solution. The method does not require any form of meshing, and it alleviates some of the drawbacks of classical finite element analysis; more importantly, it allows for streamlined modelling workflows between parameter and physical space, which justifies its use for the present research.

An initial drawback of isogeometric analysis was its inability to deal with trimmed geometries, which stems from the absence of a mathematical relationship between parametrized surfaces and their trimming curves. As a result, in the early days of the method, geometries with complex topologies had to be reconstructed into multiple NURBS patches to become analysis-ready, which defeated the initial intent of the method itself.

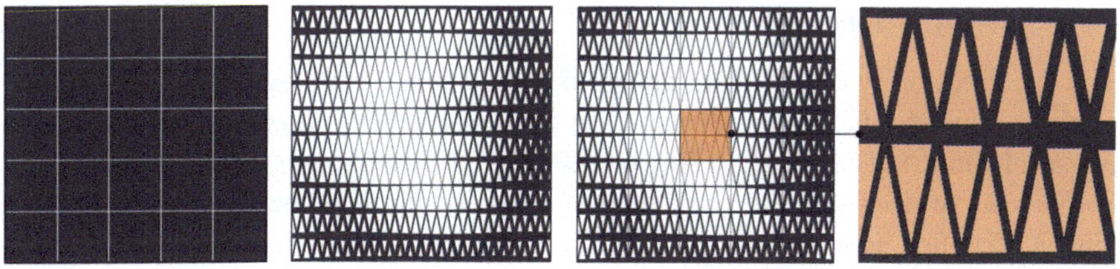

Fig. 2. Integration process in parameter space for trimmed isogeometric analysis: the parameter space as in untrimmed analysis is divided into rectangular cells (*left*), its intersections with the trimming pattern (*centre left*) are found and the latter is split to reconstruct new integration cells described with NURBS (*centre right*) and that use positive and negative integration as highlighted *on the right* for a specific cell. Note that this method is illustrated here with polyline cells but works with any form of closed trimming curve

Fortunately, multiple approaches have been put forth since then: surface refinement (Kang and Youn 2016), trimming curve approximation (Nagy and Benson 2015), and surface reconstruction (Schmidt et al. 2012). The authors have developed a novel method for trimmed isogeometric analysis that differs from existing work in that it requires neither refinement, nor simplification, nor surface reconstruction. Similarly to classical finite element analysis, isogeometric analysis requires integration over the domain in order to build the stiffness matrix used in analysis. In isogeometric analysis, integration is performed over the parameter space, itself divided in quadrilateral cells. When the parameter space is made up of alternative regions of void and solid, there is a need for an adapted integration scheme. Our method identifies trimmed integration cells, and builds cells with arbitrary NURBS boundaries. Technical details on the method are beyond the scope of this paper, but Fig. 2 illustrates its principle. Integration is then performed on each arbitrary domain using an approach similar to Sommariva and Vianello's (2009). This method is appropriate for our purposes because, with it, the numerical legwork is performed in the parameter space. Therefore, we can fully describe the patterns in the latter, while the 3D trimming curves never need to be accessed nor manipulated by our algorithms, except for design visualization. Our method is particularly appropriate for the workflows introduced in Section "Structural Patterning and Optimization". In contrast, with regular finite element methods, a surface needs to be physically trimmed and meshed to become analysis-ready, a computationally intensive process that is unsuitable for iterative workflows that require re-trimming and re-meshing at each step.

Structural Patterning and Optimization

In this section, two workflows for structural pattern design are introduced. First, the different geometric operations included in both workflows are introduced in Section "Pattern Operations and Workflow Integration". The workflows are implemented in Rhino (Robert McNeel and Associates 2017) but are extensible to other CAD platforms.

Pattern Operations and Workflow Integration

This work relies on two workflows integrating geometric operations on a set of pattern operations applied on a human-designed initial cell topology (FixTop workflow) or on interactively inserted holes (EvoTop workflow). The operations include scaling, grid distortion, curve insertion and modification. The operation sequences are diagrammed in Fig. 3. The first scheme, FIXed TOPology (FixTop), starts with an input topology and an associated connectivity that remains unchanged. This topology is populated with pre-defined shapes. These cells are then scaled using a scaling control surface, with a process detailed in Fig. 4. Compared to an individual scaling of each curve, the use of a scaling control surface has two advantages: firstly, the designer gains control through the type of surface chosen, smooth or step-like for examples, and, secondly, the control surface greatly decreases the number of design variables for dense patterns, as illustrated by the examples in Section "Conceptual Design Examples". Similarly, the designer has the possibility to incorporate distortion of the initial topology in the process, again using control surfaces (see Fig. 5). The resulting surface can be optimized by controlling the variables affecting the scaling, and distortion, which are most often described by control points, and the system can accommodate an infinite variety of patterns.

The second scheme, EVOlutive TOPology (EvoTop), is a simple cell insertion scheme. Starting from a solid configuration, cells are inserted and improved successively as many times as prescribed. The evolution of each cell is controlled through its control points by an optimization algorithm. Both modelling workflows are open, flexible, and can be combined or augmented with additional geometric rules.

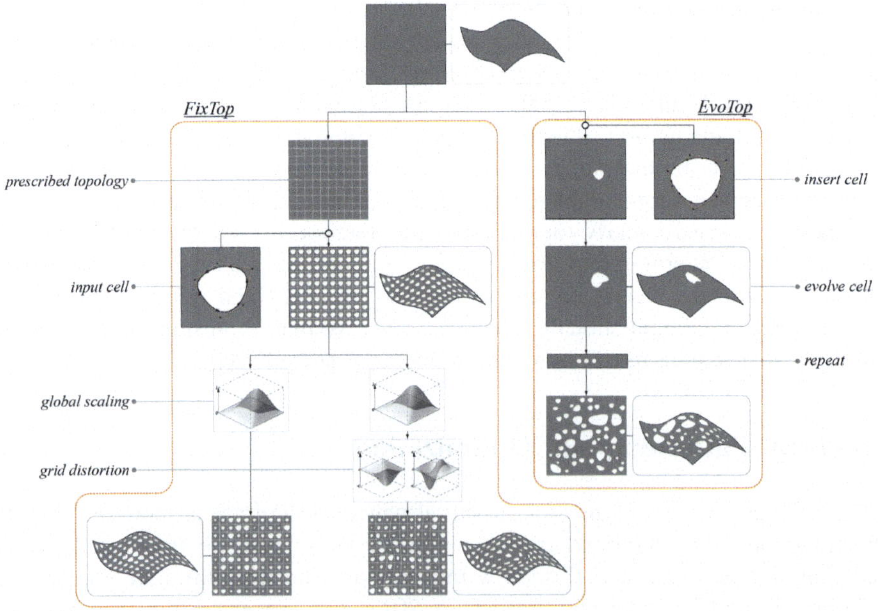

Fig. 3. FixTop and EvoTop modelling schemes

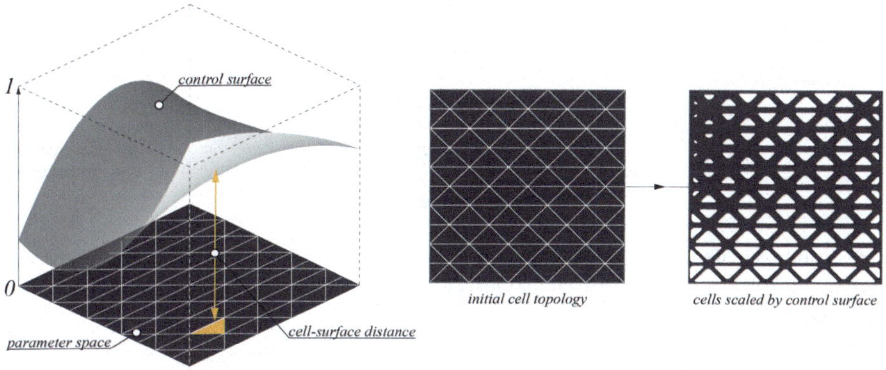

Fig. 4. Global pattern scaling in parameter space with control surface: the vertical distance between a 2D cell (from its centroid) and a surface defined over the parameter space informs its scaling (*left*), input topology in parameter space (*centre*), cells scaled by control surface in parameter space (*right*)

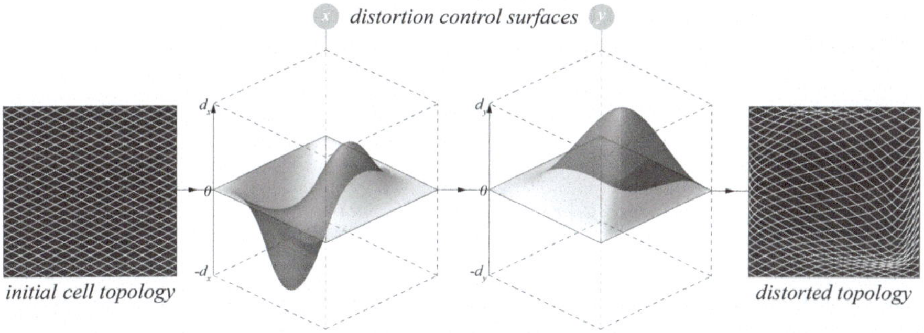

Fig. 5. Topology distortion in parameter space with control surfaces: using the same strategy as for scaling, an initial cell topology (*left*) is distorted through the use of two control surfaces. The x- and y-control surfaces move the topology points in the x- and y-direction, respectively, based on vertical distance. Connectivity is maintained throughout

Design Parameters and Automatic Geometry Parametrization

Given the operations at play in the pattern generation strategies, the design parameters are of two distinct but related types, geometric and numerical. Numerical design variables are easily handled. On the other hand, the geometric entities manipulated, curves and surfaces, are defined by a set of control points in plan or in space. If a computational system is to manipulate those automatically, it needs to turn a geometry into a set of numerical variables. Instead of a tedious manual surface or curve parametrization by the designer in the visual scripting environment, we developed an automatic parametrization system, which, given a geometry, registers each of its control point as a set of three variables to which it assigns bounds set by the designer (see Fig. 6). Control points may also be fixed in any direction in order to remain unchanged during optimization.

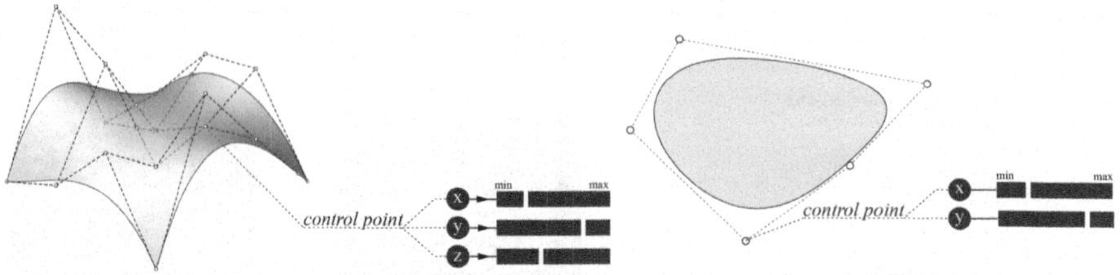

Fig. 6. Automatic geometric parametrization: the control points of the surface (*left*) or the curve (*right*) are automatically parametrized and turned into numerical variables for pattern generation and optimization

Optimization

The workflows introduced in Section "Pattern Operations and Workflow Integration" can be used to generate pattern geometries at will, but they become powerful when integrated with structural feedback and optimization to guide their geometric operations. In the following, the optimization framework and its implementations are presented generally and related to the specifics of structural design.

A general constrained optimization problem can be formulated as

$$\begin{gathered} minimize_{x \in S} f(x) \\ with\ S = \left\{(x_1, \ldots, x_n) \in \mathbb{R}^n | x_i^{min} \leq x_i \leq x_i^{max} (i = 1, \ldots, n\right\} \\ submitted\ to: \\ g_i(x) \leq 0 (i = 1, \ldots, m) \end{gathered}$$

where $f(x)$ is the objective function, typically weight or compliance in structural design, x is the design vector (a collection of scalar design variables describing the geometry), x_i^{min} and x_i^{max} are the bounds imposed on the design variables, and $g_i(x)$ are the constraints imposed, displacement or material use limits for examples. If one is able to formulate a design objective as a function and design variables as scalars, this formulation is apt to represent a design problem.

Convexity is a desirable property for an optimization problem, because it guarantees that any local minimum is a global minimum as well. However, in architectural and structural design, most problems are non-convex, and in fact contain many local optima. This is a hurdle for most optimization algorithms, as they get stuck in local wells instead of finding the global minimum. Unfortunately, the examples in the following are no exception. Thus, they will require the use of a combination of global stochastic and local derivative-free constrained optimization algorithms: a first pass by a global algorithm seeks to converge to a design located in the same well as the global optimum, and the local algorithm then refines that solution to converge to what one hopes to be the global optimum. Despite this two-pass system, because of the stochastic processes involved, there is no global optimality guarantee, but the initial design is nonetheless improved.

The research in this paper uses the C#/.Net wrapper of the NLOpt library (Johnson 2017) by King (2017) and integrated it in Rhino/Grasshopper via a custom plug-in written by the authors. This work uses ISRES (Runarsson and Yao 2005) and COBYLA (Powell 1998) as the global and local derivative-free constrained optimization algorithms, respectively. ISRES is an algorithm that internally employs penalty functions to handle constraints but does not guarantee that the final design will be in the feasible region of the optimization problem; COBYLA, when converged, offers that guarantee.

Conceptual Design Examples

The modelling system developed is used on two design case studies presented in the following.

Perforated Beam

A steel prismatic beam is optimized for stiffness by introducing 500 perforations generated by the FixTop workflow. Compliance, measured in terms of strain energy in a plane stress regime, is minimized, and the problem is constrained by a minimum material saving target of 50%. Table 1 summarizes the material properties of the beam, and Fig. 7 summarizes the problem setup and the results. The first design is generated based on a diamond-like pattern with global scaling and distortion. The last two designs are variations of a triangular pattern obtained with global scaling. The advantage of using a surface to control the individual cells is obvious here. Indeed, it reduces the problem complexity from 500 design variables to 64.

Doubly Curved Shell

The power of the ideas presented here is best illustrated when applied on curved shells. Indeed, as mentioned in Section "Pattern Mapping and Isogeometric Analysis", the methods do not require physical surface trimming, and all geometric operations applied on the pattern are two-dimensional. As a result, the optimization process is less computationally expensive and geometric operations remain tractable. A doubly curved shell (see Fig. 8), whose properties are summarized in Table 2, is optimized for strain energy with a saving target of minimum 30%, using both FixTop and EvoTop. The total vertical load applied on the shell is kept constant by multiplying the specific weight in Table 2 inversely proportionally to the material savings. Structural analysis is performed using a linear Kirchhoff–Love model (Kiendl et al. 2009), and the results are shown in Fig. 9.

Conclusions

This paper has formalized two workflows for performance-driven structural patterning, integrating geometric rules, and optimization. The resulting modelling system can be extended and enriched by designers at will. The design examples illustrate its value for

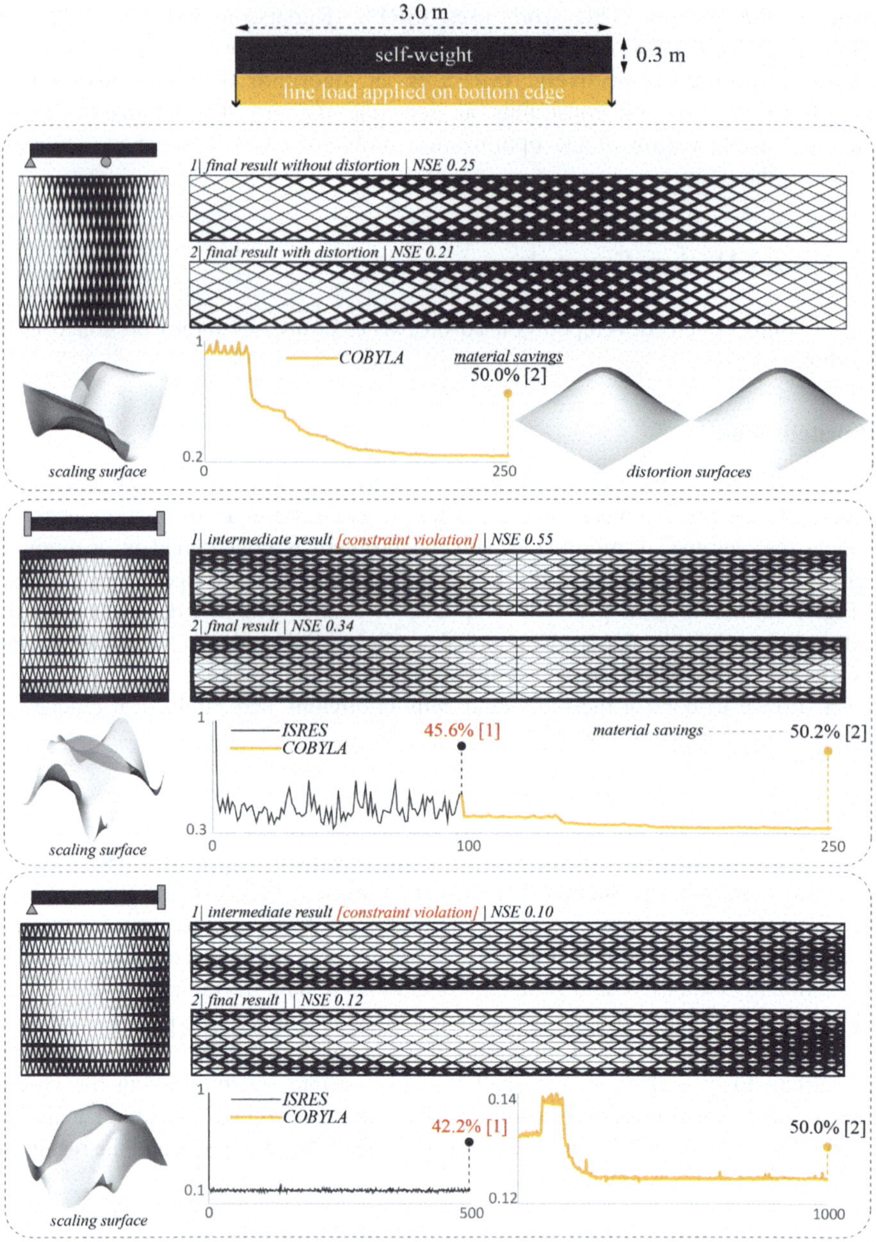
3.0 m
self-weight
0.3 m
line load applied on bottom edge
1| final result without distortion | NSE 0.25
2| final result with distortion | NSE 0.21
COBYLA
material savings
50.0% [2]
scaling surface
distortion surfaces
1| intermediate result [constraint violation] | NSE 0.55
2| final result | NSE 0.34
ISRES
COBYLA
45.6% [1]
material savings
50.2% [2]
scaling surface
1| intermediate result [constraint violation] | NSE 0.10
2| final result | | NSE 0.12
ISRES
COBYLA
42.2% [1]
50.0% [2]
scaling surface

◀ **Fig. 7.** Perforated beam optimized using FixTop: Three solutions based on different boundary conditions are found. The beam is loaded by its own self-weight and line load applied on its bottom edge, with the dimensions indicated on *top* of the figure. For each set of boundary condition, two designs are shown. For the top condition, the first design corresponds to the application of the FixTop scheme with global scaling, and the second design is obtained using FixTop with scaling and distortion. For the two other conditions, the first design corresponds to the solution found using ISRES, and the second one is the COBYLA-refined solution. Each result is accompanied by the resulting parameter space cell distribution, its corresponding control surface(s), and a graph of the optimization history for the strain energy normalized (NSE) against the starting condition (homogeneous perforations with 80% material savings)

Table 1. Problem setup—perforated beam design example

Young's modulus E (GPa)	Poisson ratio ν	Thickness (m)	Specific weight (N/m)	Line load (N/m)
210×10^9	0.3	0.02	80×10^4	−2000

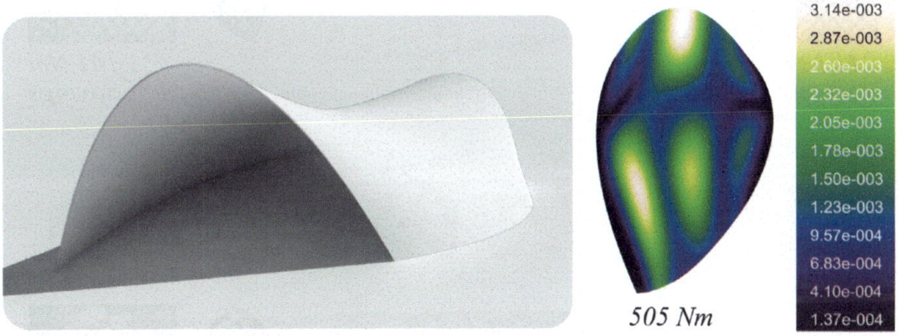

Fig. 8. Doubly curved shell: the shell is supported in all directions along its base edges (*left*) and loaded by its own-self-weight, which results in a total displacement map (*right*) with an associated strain energy of 505 Nm

Table 2. Material properties—doubly curved shell design example

Young's modulus E (GPa)	Poisson ratio ν	Thickness (m)	Specific weight (N/m)	Max. clear span (m)
45×10^9	0.2	0.08	25×10^4	18

creative structural design. In addition, it is not restricted to structural analysis and can be leveraged to optimize for other architectural objectives.

Future Perspectives

This work is paving the way for rich future research. Evidently, the modelling paradigm is open to new geometric rules and would benefit from a broader set of geometric operations. While we have focused our attention on optimizing patterned structures

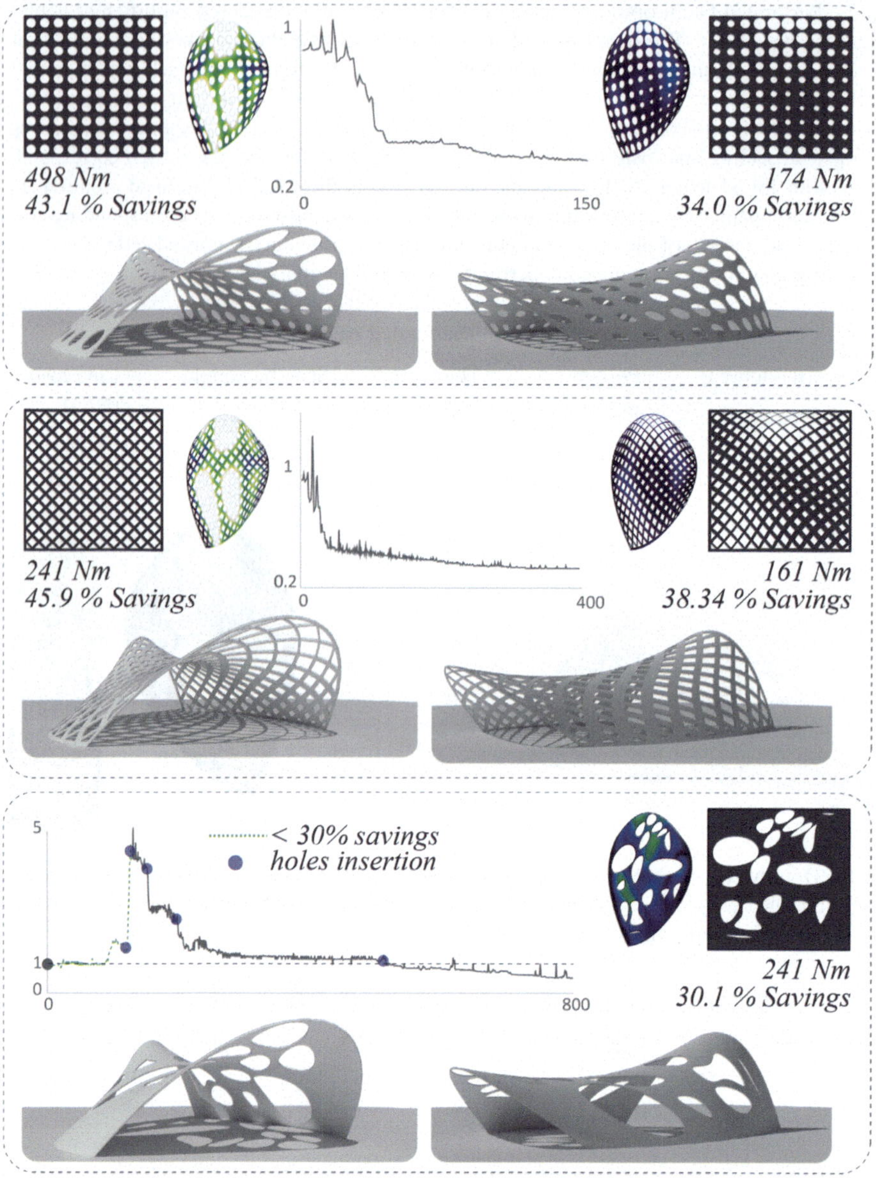

Fig. 9. Shell optimized using FixTop (*top two*) and EvoTop (*bottom*). The two first designs are optimized using FixTop similarly as the perforated beam example. On the *left side* is the starting conditions in the parameter space and its associated displacement map. The *centre* shows the optimization history of the strain energy normalized against the solid configuration, and the final result is shown on the *right side*, as well as below with two perspective views. The last design is found using EvoTop. The start condition is the solid, unpatterned configuration. Successively, holes with arbitrary shapes are added and subsequently optimized. Only the final configuration is represented on the figure. This example shows the usefulness of EvoTop to handle arbitrary designs. In particular, even with foolishly drawn apertures, the system was able to find a solution performing better than the solid one

based on their linear elastic response, including buckling and vibration considerations as objectives or constraints is an important future step. More case studies need to be developed and tested with diverse objective functions and constraints, related to stress or fabrication for example. The influence of the algorithms on the obtained results deserve consideration as well. On a practical note, implementing the system with T-Splines support (Sederberg et al. 2003), and other new parametrized surface formats, will increase its power and versatility.

Closing Remark

Designing patterned structures requires significant expertise and complex computational systems. The introduced modelling system provides simple ways to integrate creativity and rigor in the conceptual design of such structures through the combination of geometric rules and optimization techniques. Because the method is seamlessly integrated into the CAD environment, it is possible to interweave pattern modelling and structural optimization in novel, design-oriented ways. It offers designers creative and systematic ways to explore uncharted territories in high-performance structural design for architecture.

Acknowledgements. The authors would like to thank Daniel Åkesson and Vedad Alic for providing isogeometric analysis code that helped accelerate the early stages of this research project.

References

Aage, N., et al.: Interactive topology optimization on hand-held devices. Struct. Multidiscip. Optim. **47**(1), 1–6 (2013). doi:10.1007/s00158-012-0827-z

Dumas, J., et al.: By-example synthesis of structurally sound patterns. ACM Trans. Graph. **34**(4), 137:1–137:12 (2015). doi:10.1145/2766984

Hughes, T.J.R., Cottrell, J.A., Bazilevs, Y.: Isogeometric analysis: CAD, finite elements, NURBS, exact geometry and mesh refinement. Comput. Methods Appl. Mech. Eng. **194**(39–41):4135–4195. Available at: http://linkinghub.elsevier.com/retrieve/pii/S0045782504005171 (2005)

Johnson, S.G.: The NLopt nonlinear-optimization package. Available at: http://ab-initio.mit.edu/nlopt (2017). Accessed 16 May 2017

Kang, P., Youn, S.-K.: Isogeometric topology optimization of shell structures using trimmed NURBS surfaces. In: Finite Elements in Analysis and Design, vol. 120, pp. 18–40 (2016)

Kiendl, J., et al.: Isogeometric shell analysis with Kirchhoff-Love elements. Comput. Methods Appl. Mech. Eng. **198**(49–52), 3902–3914 (2009). doi:10.1016/j.cma.2009.08.013

King, B.: NLopt C# Wrapper. Available at: https://github.com/BrannonKing/NLoptNet (2017). Accessed 16 May 2017

Motro, R.: Robert Le Ricolais (1894–1977). Father of Spatial Structures. In: International Journal of Space Structures pp. 233–238 (2007)

Nagy, A.P., Benson, D.J.: On the numerical integration of trimmed isogeometric elements. Comput. Methods. Appl. Mech. Eng. **284**, 165–185 (2015)

Panagiotis, M., Sawako, K.: Millipede. Available at: http://www.sawapan.eu/. Accessed 30 April 2017

Paulino, G.H., Gain, A.L.: Bridging art and engineering using Escher-based virtual elements. Struct. Multidiscip. Optim. **51**(4), 867–883 (2015). doi:10.1007/s00158-014-1179-7

Powell, M.J.D.: Direct search algorithms for optimization. Acta Numer. **7**:287–336. Available at: http://www.damtp.cam.ac.uk/user/na/reports.html (1998)

Querin, O.M.: Evolutionary Structural Optimisation: Stress Based Formulation and Implementation, thesis, Department of Aeronautical Engineering, University of Sydney, Australia (1997)

Robert McNeel and Associates: Rhinoceros. Available at: https://www.rhino3d.com/ (2017). Accessed 12 May 2017

Rozvany, G.I.N.: A critical review of established methods of structural topology optimization. Struct. Multidiscip. Optim. **37**(3), 217–237 (2009)

Runarsson, T.P., Yao, X.: Search biases in constrained evolutionary optimization. IEEE Trans. Syst. Man. Cybern. Part C Appl. Rev. **35**(2), 233–243 (2005)

Schmidt, R., Wüchner, R., Bletzinger, K.-U.: Isogeometric analysis of trimmed NURBS geometries. Comput. Methods Appl. Mech. Eng. **241**, 93–111 (2012)

Schumacher, C., Thomaszewski, B., Gross, M.: Stenciling: designing structurally-sound surfaces with decorative patterns. Comput. Graph. Forum **35**(5):101–110. Available at: http://doi.wiley.com/10.1111/cgf.12967 (2016). Accessed 1 May 2017

Sederberg, T., Zheng, J., Bakenov, A.L.: T-splines and T-NURCCs. ACM Trans.:477–484. Available at: http://dl.acm.org/citation.cfm?id=882295 (2003)

Sommariva, A., Vianello, M.: Gauss-Green cubature and moment computation over arbitrary geometries. J. Comput. Appl. Math. **231**(2), 886–896 (2009)

Free-Form Wooden Structures: Parametric Optimization of Double-Curved Lattice Structures

Klaas De Rycke[1,2], Louis Bergis[1(✉)], and Ewa Jankowska-Kus[1]

[1] B+G Ingénierie Bollinger + Grohmann S.a.R.L, Paris, France
lbergis@bollinger-grohmann.fr
[2] ENSA-Versailles, France

Abstract. This paper discusses methodologies of design, simulation and fabrication of free-form lattice timber pavilions, based on experience gained from three such projects developed in cooperation with RDAI Architects. Due to multi-level interdependencies between structural feasibility, fabrication technology and form-shaping, informed digital models taking into consideration specialist knowledge from different professions were developed. The authors' experience is used to explain complex workflows covering all stages of design, study and fabrication.

Keywords: Parametric design · Digital production · Wood structures · Karamba

Introduction

A Subsection Sample

Late 2009, RDAI architects entrusted Bollinger + Grohmann to collaborate as structural engineers on a challenging design proposal with very short deadlines; the Hermès Rive Gauche flagship store. In the continuity of that assignment two similar projects were prepared, each with different parameters depending on the location and function.

The first, GUM, is a wooden cladding and base for a stairway. The stairs are 5.5 m high, lead from the ground level to the first floor and consist of two flights wrapped around a core. The single-layer cladding is slanted and each of 140 elements is a unique double-curved piece. The other is a pavilion placed in a private garden in the Hunting Valley, Ohio, USA, serving as a picnic place. It is composed of 90 wooden laths arranged on two layers, wreathing a central space. Each one is a unique double-curved piece of rectangular section (70 × 50 mm). The structure measures 7.3 m × 10.3 m on the ground level and is 4.6 m high. At the top, elements are placed closer to each other and form a circular opening.

All the structures are comprised of numerous timber double-curved unique laths, manufactured with highest precision and quality. They are autonomous self-supporting pieces without additional bracing. At the moment of development, few structures in comparable scale with such a high level of aesthetical requirements were built,

K. De Rycke et al., *Humanizing Digital Reality*,
https://doi.org/10.1007/978-981-10-6611-5_48

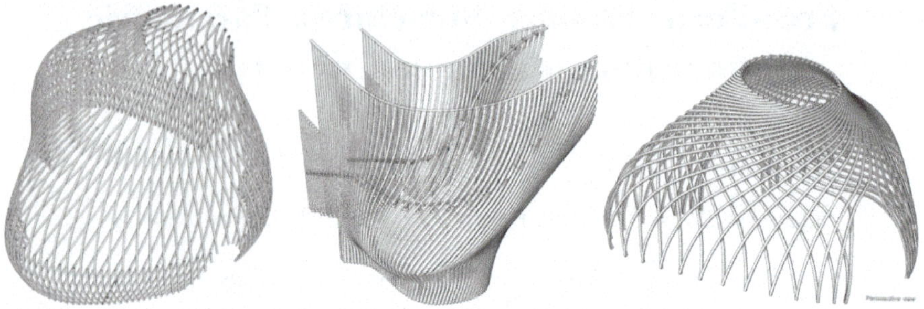

Fig. 1. From the *left* Hermès Rive Gauche (HRG), Paris, France, 2010; GUM, Moscow, Russia, 2015; Hunting Valley Pavilion (HVP), Cleveland, USA, 2016

Fig. 2. Photo of the Hermès Rive Gauche flagship store (copyrights: Michel Denancé). According to the privacy agreements, only this photo is published as representative for the series of projects

therefore innovative fabrication and workflow technologies had to be invented. This paper presents the necessary design parameters for development of the project; from fabrication techniques to geometrical and structural parameters. It also demonstrates implementation of a horizontal, iterative workflow integrating work of designers, engineers and manufacturers (Figs. 1, 2; Table 1).

Table 1. Projects' data

	Hermès Rive Gauche (HRG)[a]	GUM	Hunting Valley Pavilion (HVP)
Date	2009–2010	2014–2015	2014–2016
Pavilion dimensions			
Plan (m)	12 × 8	5.8 × 4.8	10.3 × 7.3
Height (m)	8.5	5.5	4.6
Laths production			
Type of wood	Ash	Ash	Pinus radiata (acetylated)
Process of fabrication	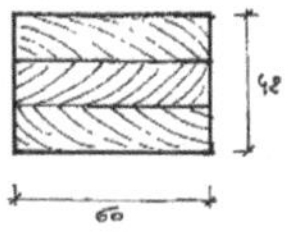	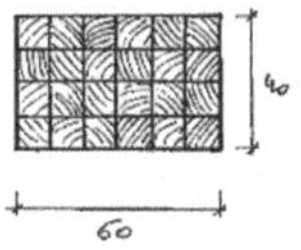	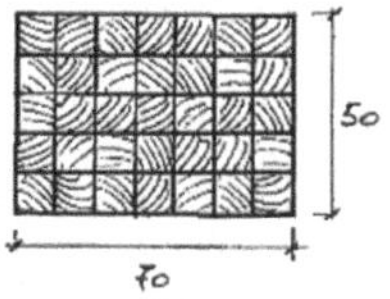
Number of laths	270	140	90
Section (mm)	60 × 42	60 × 40	70 × 50
Linear lath length (m)	1300	632	657
Lamella dimension (mm)	60 × 14	10 × 10	10 × 10
Quantity of timber (m^3)	3.3	1.5	2.3
Number of connections	2134	474	1264

[a]Average value per structure

Design Development

Geometry

The three designs base input for the script was a double-curved surface, later referred to as a base surface. Its shape was an effect of design arrangements with architects, analyzed, processed and, in result, described by one or two sets of curves smoothly enclosing the internal volume. As imagined by the architects, the curves were starting at the top and flow uninterruptedly to the bottom of the structure. Therefore, due to complexity of pavilion topologies, their local distribution varied. The main challenge was to find equilibrium between the laths shape, density and sections, since the influences of form-defining parameters were sometimes contradictory. As an example; although application of bigger sections provides more structural rigidity and larger holes would be possible, bigger sections amplify the self-weight, weakening the structural efficiency and are more difficult for bending. Using slimmer elements, although solving the weight problem and proving easier to bend, show problems for finding enough space for bolting at the intersections.

As described in Lattice spaces, from optimization through customization of new developable wooden surfaces in order to control the topology of the wooden slats, the base surfaces are reverse-engineered by re-parameterization and extraction of control

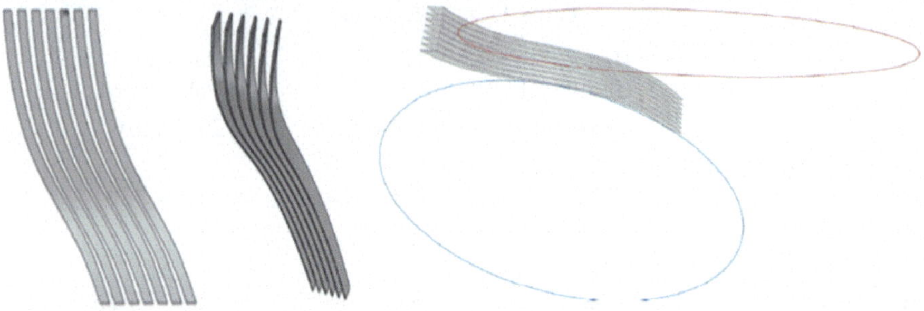

Fig. 3. Curvature of lath analysis and decomposition into lamellas of one single lath

points. Afterwards, it was possible to optimize its geometry so that it fulfilled design requirements e.g. concerning minimum radius of curvature and avoiding collision with existing objects. Next group of parameters defines the surface's tessellation and generates two sets of curves which are to become centerlines of wooden laths. Then, volumetric information was applied, defined by the distribution, rotation, orientation, sections, tolerances, collision control etc. In all described projects laths were oriented in accordance to the surface normal vectors. Therefore, the production process requires a distinction between two directions of curvatures (Fig. 3).

To unroll the lamellas, we firstly considered that the double curve is the sum of two single curves. Taken into account the rectangular cross section, additional torsion corresponding to the twist has been taken into account. In the case of HRG, the primary curvature of the initial curves comes from the base surface shape and rules the laths' bending, the secondary curvature comes from the way the lath is draped on the surface and rules the laths cutting out of a flat panel. The limit of the minimum radius of curvature is defined by the primary curve. Each type of wood has its own elastic limit and minimum bending radius which further depends on the moisture degree and off course on the thickness of the piece. For HRG, structural calculations for a leafed wood showed that a section of minimum 48 mm was necessary. This could not be a solid bent wood so it consists of lamellae. We need for structural reasons at least three lamellae, the thickness of each lamella was thus set at 18 mm and herewith the radius of the primary curvature is fixed. The secondary curvature is derived from the tessellation of the generator surface. As each piece is milled out of a flat timber panel, during the design process it has to be unrolled and oriented properly along the grains. Therefore, the secondary curvature is directly dependent on the fiber orientation and has to be limited for aesthetic and structural reasons.

It is important to mention that the change in lamella orientation generates additional torsion which should be limited to 15°/m for structural and fabrication reasons. The process of manufacturing executed for the GUM and the HVP reduced those issues by taking into account the curvature of the centerline, directly linked to the lamella thickness (limited, respectively, to 80 and 100 cm).

The geometry of the base surface and applied tessellation which are firstly determined by esthetical considerations has direct influence on the structural feasibility and manufacturing process. These characteristics had to be taken into consideration during

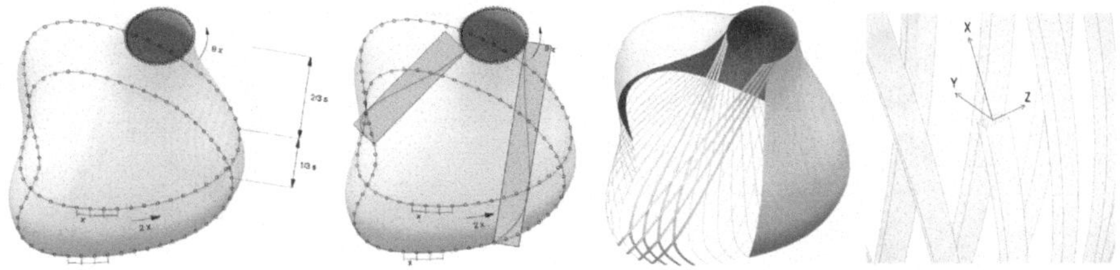

Fig. 4. Re-parameterization of the surface, centerlines and lath generations on an HRG hut—data extracted from the Rhino3D model

the optimization process, as described in Principles of Wood Science and Technology/Solid Wood (Kollmann and Wilfred 1968).

In all these projects, sets of curves are created from a division of the base surface into horizontal sections. Each horizontal curve is then regularly divided so as to generate control points of the centerlines, as illustrated on Fig. 4. In HRG and HVP, those centerlines are created by selecting control points with a regular shift, generating a lozenge pattern. This selection is ruled by different parameters allowing a control of the tessellation as illustrated on Fig. 6. Such tessellation process creates distortion and size variation of the cells between different horizontal sections.

In the case of the HRG, structural and esthetical reasons called for densification of the grid. Additional stiffening of the lower part was obtained by introducing more control points and thus vertical subdivision by three (Fig. 7). For the HVP, on the other hand, the differentiation of cells distribution is exaggerated with the use of an irregular control points' selection, creating transparency in the lower part and densification of the upper one but also adding more diverse curves (Figs. 8, 10).

For each control point, a vector normal to the surface's closest point is generated, allowing for an offset of the centerline for each lath and lamellae layer and generation of the lath volume oriented parallel to the surface. Although the algorithm generates a general rule for all the laths, in order to provide smooth continuity, for the HVP also local distribution—variation of lath density and orientation—was taken into consideration (Figs. 4, 5, 9)

Structural Feasibility

Due to the mentioned circumstances, structural and fabrication feasibility studies were incorporated in the design process from the earliest stages. For HVP the geometry is developed in Rhinoceros3D and Rhinoscript and transferred to Dlubal-RSTAB for structural analysis. For GUM and HVP the geometry is generated directly by a unique grasshopper algorithm and Karamba3d is used for structural analysis (Fig. 11). Curves representing central lines of the laths are discretized into segments of maximum length of 15 cm. Each segment is modelled taking into consideration parts orientation, grain distortion and subsequent lowering of structural capacity, lamellas superposition, layers' superposition, overlap between layers and subsequent rigidity of each individual node. Additional stresses due to the bending and torsion process are also considered.

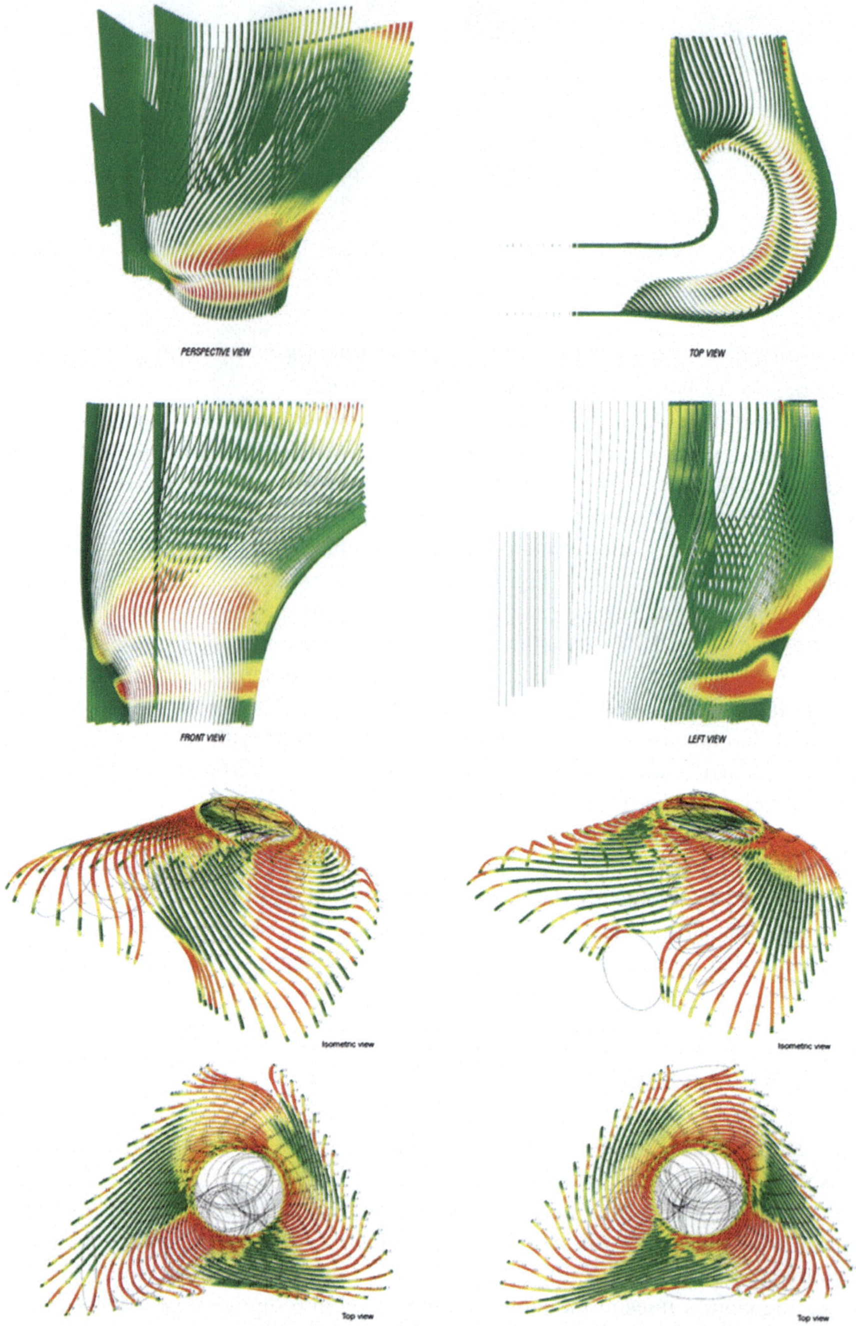

Fig. 5. Surface and laths curvatures, laths torsion analysis from the GUM (*upper*) and HVP (*lower*). Data extracted from the Grasshopper3D model

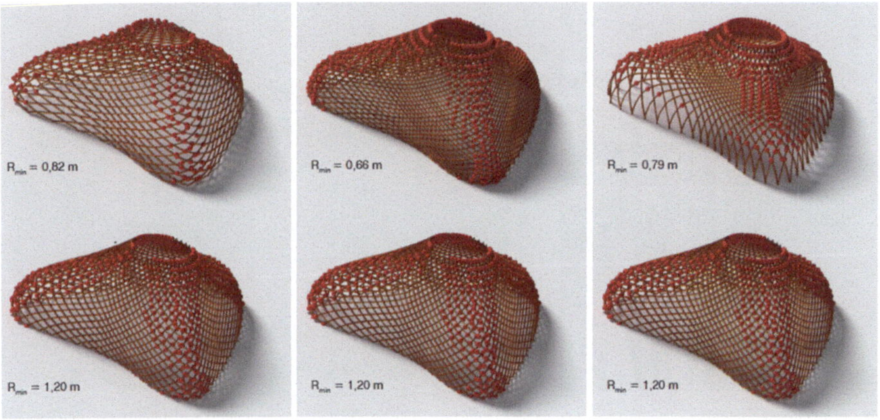

Fig. 6. Variation on the HVP tessellation and curvature analysis—data extracted from the Rhinoceros3D model

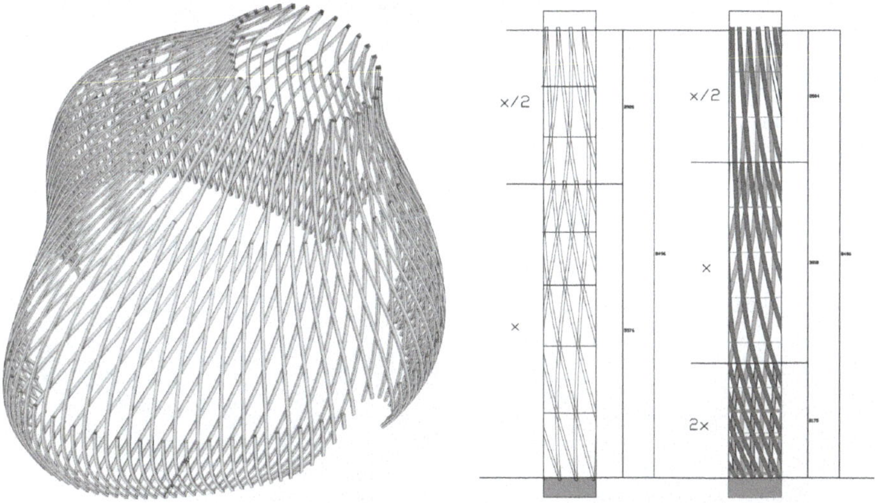

Fig. 7. Differentiation of cells distribution on HRG. Data extracted from the Rhino3D model

Fig. 8. Differentiation of cells distribution on HVP (*lower*). Data extracted from the Rhino3D model

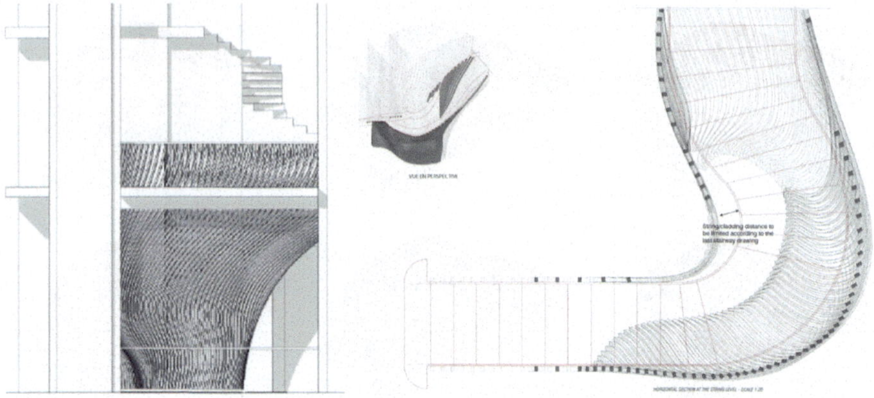

Fig. 9. GUM cladding—from the *left side* elevation, isometric view and horizontal section

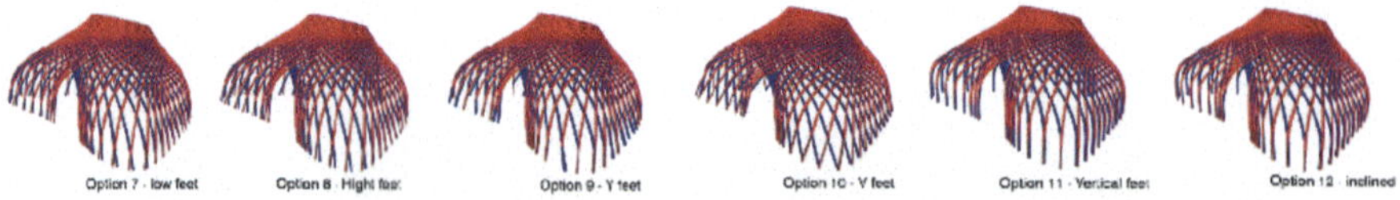

Fig. 10. Different propositions of feet geometry (HVP). Data extracted from the Rhinoceros3D model

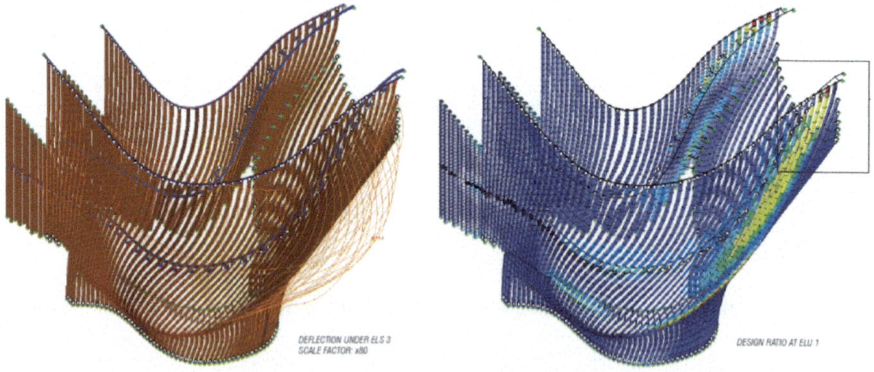

Fig. 11. Isometric views of the GUM model: amplified deflection (*left*) and strains (*right*)

As laths' lengths are larger than the maximum available dimension of lamellas, lamellas are glued with a shift along the length and structural resistance reduced.

Joints between layers are modelled with stiff connectors with correct eccentricities and to control rotational stiffness. Angles between the laths vary and so do distances between screws, therefore structural models are directly informed through the algorithm about the rotational stiffness of each singular joint based on the corresponding

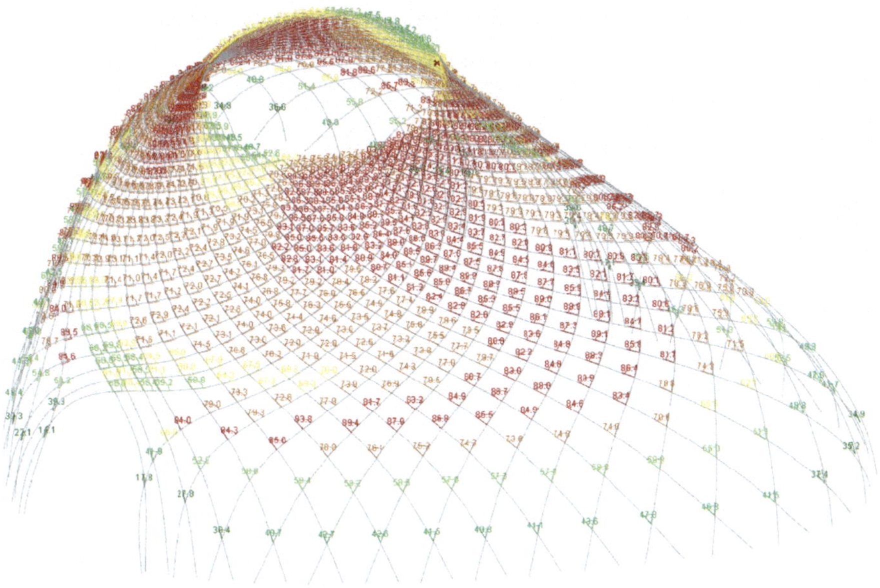

Fig. 12. Cross batten connection angles (HVP, isometric view)

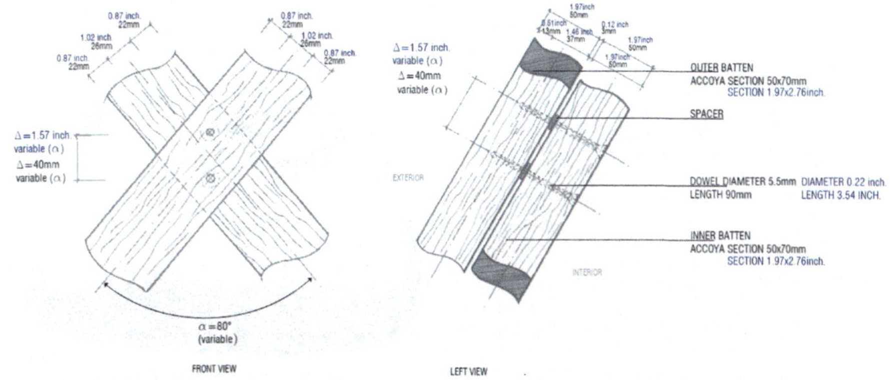

Fig. 13. Laths intersection connection detail (HVP). Front view (*left*), side view (*right*)

specific geometrical value. Instead of a pure hinged connection the relative stiffness of the connection controls the torsion of the laths and helps to reduce bracing; in HVP (which is an outside pavilion) the stiffness and density of connections in the upper part was made deliberately and made it possible to avoid any additional stiffeners (Figs. 12, 13; Table 2).

In order to obtain a complete simulation understanding, various inputs are continuously extracted from the geometrical algorithm such as cells' dimensions—to quantify snow accretion—and slope angles. These can then be used again to change the global repartition of laths on the global surface (Fig. 14).

Table 2. Connection member rigidity in function of the batten

	α	Δ (mm)	Iy (mm^4)	It (mm^4)
	80 < α < 90°	37	511	747
	70 < α < 80°	40.5	511	890
	60 < α < 70°	45.33	511	1121
	50 < α < 60°	52	511	1475
	40 < α < 50°	62	511	2063
	0 < α < 40°	76	511	3151

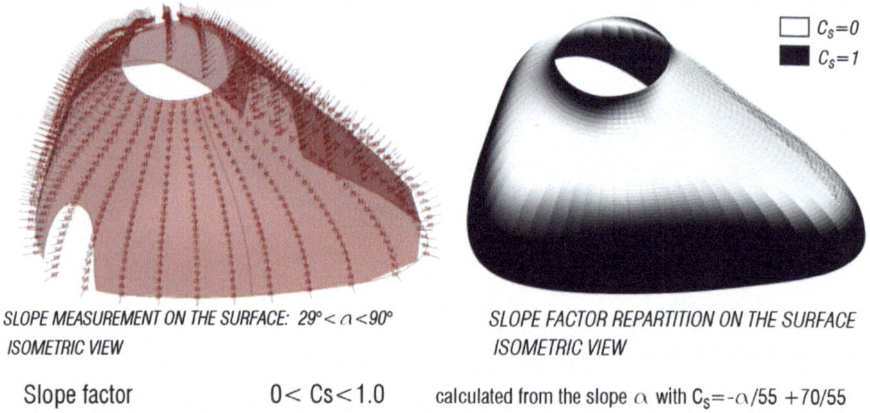

Fig. 14. Surface repartition in accordance to the local slope measurement. Isometric views of the HVP: the slope measurement (*left*), slope factor repartition (*right*). Data extracted from the Grasshopper3D model

Fabrication

Process Selection

Since the mid 60s, bent wood is mainly achieved through lamination. Long before that, the first techniques were to carve out the curvature in solid wood. With ship building a new technique with steam bending came along and was perfected by the Thonet brothers The main challenge is to find a balance between obtaining the smallest possible radius of curvature, rationalizing the use of material and still maintaining the structural qualities. Five following methods were considered, tested and discussed, as described by Mangerin in *Le courrier du bois* (2011):

- Steam bending of solid wood: time- and labor-consuming, high level of precision considering the unicity and length of pieces difficult to reach. Each lath would need a single mold.
- Mechanical simple-curved laths: requiring too much geometrical constraints (very thin slats for bending to produce satisfying aesthetic results.
- Steam bending of solid wood continuous laths segments with high radius of curvature and milling of those with small curvature radius: lath and grain continuity not satisfactory, significant material loss, time and labor consuming.
- Laminated from thin strips
- Laminated from thicker strips
- Milled from solid wood
- Gluing of solid wood pieces: laths would were divided into simple-curved parts. Although a double-curvature effect could be obtained by adapting orientation of subsequent elements, that solution wouldn't be acceptable for aesthetic reasons.
- Composition of milled layers glued together: primary direction curvature obtained by the shape of the cut-out, secondary—by gluing on a CNC processed scaffolding. This process was selected as the most structural-wise and material-wise efficient and respecting desired architectural qualities (Fig. 15).

For the project of Hermès Rive Gauche, each lath is comprised of 3 milled pieces with dimensions of 14 × 60 mm. The initial and global algorithm is fed into the robots and automatically generates the forms and holes for connections. Laths are glued and bent to the desired form by attaching them to underlying scaffolding at the height of the horizontal control point line while the glue hardens. The process was conducted in a workshop, in a controlled environment. Eventually, the laths are submitted to multiple sanding to ensure a perfect finishing.

Other approaches, for the lamella consist off course of having thinner lamellas (aesthetically less interesting) but also in dissecting the basic section in two directions. In Fig. 16, the third element can respond relatively easy to bending in each direction without needing any cut from a flat panel and thus grain and structural rigidity loss. It responds aesthetically less well and is a bit more labor intense since more has to be glued. It can however be bent more easily in two directions.

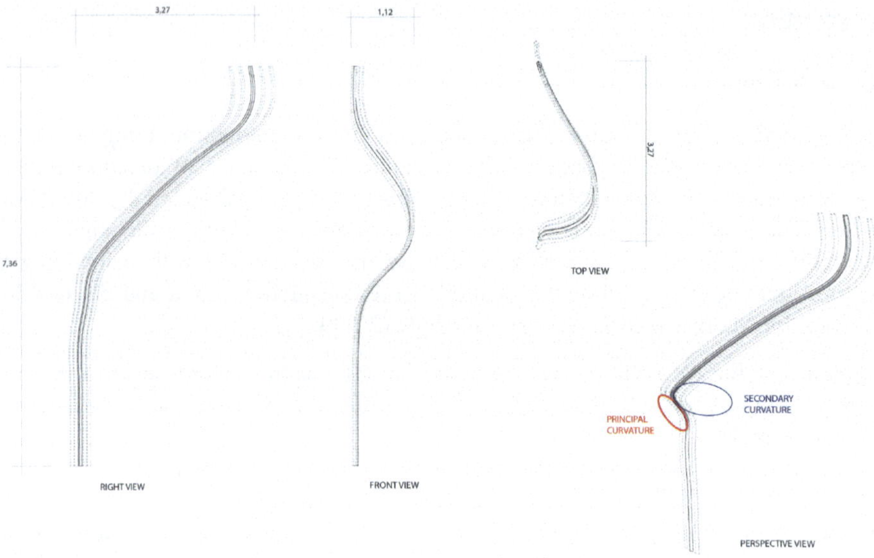

Fig. 15. Lath description from the GUM cladding—data extracted from Rhinoceros3D

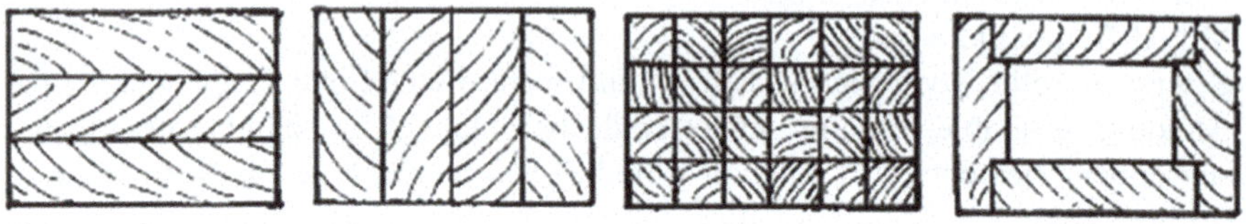

Fig. 16. Cross sections resulting from the different fabrication processes (*left* HRV)

Grain Orientation

In considering double curved laths, the grain orientation is one of the major issues not only for aesthetic reasons (in order to obtain an appearance of a solid wooden piece, fibers' orientation differentiation and also too many lamellae should be avoided), but also for structural concerns. Timber is an anisotropic material with structural characteristics vary depending on the internal structure e.g. for C30 timber, the compression strain of 27 N/mm^2 when following the grains' direction, changes to 2.7 N/mm^2 when measured perpendicularly to them. Therefore, the process used for Hermès Rive Gauche required the subdivision of each lamella segment in order to limit the angle between the lath centerlines and the fibers' orientation to 30°. Although parts were assembled considering a small shift so as to obtain sufficient overlaying, the process partially interrupts the structural continuity; only two thirds of the section could be taken into account for calculation.

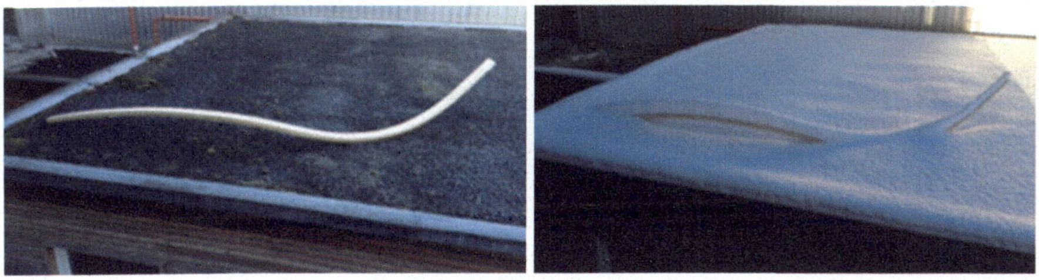

Fig. 17. Pictures of the tested acetylated double-curved wooden lath after fabrication (*left*) and exposed to temperature and hygrometry variation (*right*)

Material Selection

Pre-selection of wood type is very important and should off course be made according to its mechanical characteristics Such as strength and bending but also according to the durability class, the resistance, the bending stiffness, the grain length and direction, the fabrication, the esthetical qualities, the humidity, etc.

The HRG for example was initially planned to be fabricated in oak, after analysis of the mock-up however ash was chosen for its superior aesthetic reasons. The same material was used for the GUM. As the HVP is placed outside and uncovered, submitted to variation of temperature and humidity (Eurocode—service class 3), it required improved durability criteria. Therefore, acetylated solid wood was chosen (acetylation is a chemical modification of the wood that alters the cell structure in order to improve the material's durability). As this process has not been implemented for double curvature after treatment before, test were executed.

Further to improve weather resistance, timber defects and screws' heads were covered with wooden stoppers in order to improve parts' quality and avoid humidity penetration (Fig. 17).

Design Workflow

Approach

With increasing specialization of CAE industry, more and more data is being encapsulated within specialized engineering and planning agencies, gradually weakening the power of the architectural profession. However, informed approach has the ability of empowering designers with specialist knowledge at an early design stage, making it possible for them to change former constraints into form-shaping information. Thanks to a diversity of available computational tools, the design process may now be enhanced in various ways. Depending on chosen algorithms, not only the speed and precision of calculation can be increased, but the research spectrum is broadened, resulting in discovering solutions which otherwise might not have been taken into consideration. Moreover, advancements in computer power, modern simulation tools and numerically controlled fabrication let us question the existing process model. By

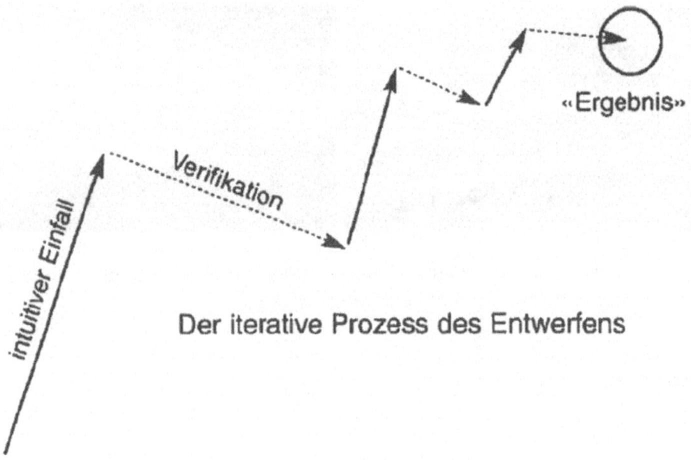

Fig. 18. The iterative process of the design (Hossdorf 2003)

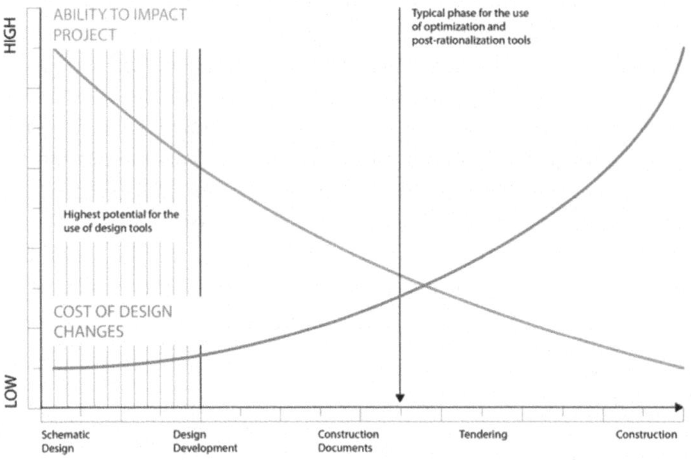

Fig. 19. Usage of optimization tools in the design process (MacLeamy 2004)

incorporating multiple information inputs from the beginning, the linear, hierarchical design may be transformed into a much more efficient horizontal, parallel investigation.

According to Patrick MacLeamy, the earlier on in the process the intervention, the more beneficial it is to the final outcome. The period of the highest potential spreads between the Schematic Design and Design Development phases; taking action at this stage leads to opening new possibilities by creating a constant feedback loop between architects and consulting specialists. Such a horizontal process with real-time evaluation of parallel design options creates space for new structural and fabrication solutions to emerge (Figs. 18, 19).

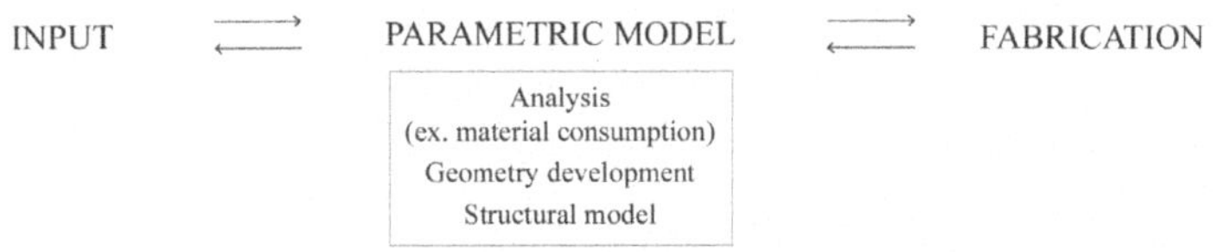

Fig. 20. Integrated workflow scheme

Implementation

All described pavilions are developed in 3d vector software (Rhinoceros3D) extended with parametric plug-ins (Rhinoscript, Grasshopper and Karamba). The goal was to integrate inputs from structural calculation and fabrication technology as early as possible. Described in a form of a shape-generating algorithm, the structural feasibility became part of the essence of the design process. Part of structural engineering work was thus scripting tools to enable to shape the projects within the boundaries of structural and fabrication feasibility and simultaneously within esthetical and architectural boundaries (Fig. 20).

For every parameter (structural, fab-technical or architectural) the starting point was a 3d sketch illustrating desired volumes and patterns. Next, parametric models linked to the structural solvers were developed. With each structure, the efforts were gradually being shifted from the execution to the design phase. In consequence, for the Hunting Valley Pavilion a complete informed model was created, encompassing geometrical information, structural model and fabrication outputs. It was written as a complete process, encompassing all steps from geometrical analysis to the structural model definition. Moreover, it contained geometry not only of the final pavilion but also of the underlying scaffolding. Therefore, whenever a change in the input geometry had occurred, the whole model would regenerate and reanalyze within minutes, resulting in significant labor and cost savings. It furthermore puts on a par without specific weighting structural, esthetical and fab-technical parameters.

Conclusion

The fabrication of free-form high precision and high finishing wooden structure is possible by the integration of material, structural, fabrication and geometric knowledge into the architectural, form. This is achievable within singular single data-model that all partners are using since the early stage. From the shown projects the development of specific tools, workflow and design approach has emerged which allows each member of the team to have a full control of different parameters (structural, esthetical, manufacturing) for each member of the team. Integrated, informed design approach has the ability to empower architects with specialist knowledge at an early design stage, making it possible to use former constraints as assets, not obstacles. Moreover, it transforms the workflow from a conventional, hierarchical work into a horizontal, iterative, creative process, broadening the spectrum of possible solutions.

Described projects illustrate different semi-digital fabrication possibilities for complex wooden structures without laborious processes as the Thonet brother's technique, limiting the loss of material, not constraining the overall shape, at the same time staying close to massive wood without using too much glue and lamellae. The described approach to geometry, material and topology is a process with attention to structural efficiency and fabrication requirements; only one or two layers of complex geometry timber laths, density and angle optimization for parameters like connections stiffness.

References

Bohnenberger, S., De Rycke, K., Weilandt, A.: Lattice spaces, form optimization through customization of non-developable wooden surfaces. In: eCAADe (2011)

Hossdorf, H.: Des Erlebnis Ingenieur zu sein. Springer, Basel (2003)

Kollmann, F.F.P., Wilfred, A.: Principles of Wood Science and Technology/Solid Wood. Springer, New York (1968)

MacLeamy, P.: MacLeamy Curves. In Collaboration, integrated information, and the project lifecycle in building design and construction and operation (WP-1202, August, 2004)

Mangerin, P.: Le courrier du bois, vol. 172, pp. 10–12. Decon, Carrot-Bijgaande (2011)

The Mere Exposure Effect in Architecture

Michael-Paul "Jack" James(✉), Christopher Beorkrem,
and Jefferson Ellinger

University of North Carolina at Charlotte, Charlotte, NC 28223, USA
mjames@UNCC.edu

Abstract. Per Schelling's model of Segregation, the population will innately segregate itself based on preferences, often leading to organization by race and class. This subdividing of communities through segregation increases social tension, discourages communication, and isolates those who are different. In 1968, Robert Zajonc proved that subjects rated a familiar stimulus more positively than similar yet unfamiliar stimuli. The mere exposure effect is a phenomenon by which people develop a preference for things solely because they are familiar with them. Architecture can diminish the impact of social segregation through mere exposure by examining the effects of architectural interventions and programs. Through mere exposure designers can create new connections between members of society by rethinking circulation paths, carefully considering the geolocation of program, and creating more effective public space. By incorporating modern social behavioral analytics into design logics, social spaces can facilitate more productive engagements between occupants. Examining the effect of unit location on circulation and noting the most effective locations for public goods, developers and city planners will increase communication between community members. Increasing communication as a primary goal of design will facilitate the development of stronger communities. Although the tool specifically targets residential complexes, the concept is scalable. Providing a designer an automated method for evaluation and data collection based on the mere exposure effect in urban design and architecture can create more informed design of public space. The goal is to create a more diverse and sustainable community through an informed understanding of how space and program influence behavior.

Keywords: Schelling's model of segregation · Mere exposure effect · Architecture · Revit · Dynamo · Zajonc · Multifamily · Community

Theory

Introduction

Architectural Design has ignored the natural processes of social segregation, letting other factors dominate the occupational organization of the built environment, ultimately isolating communities and their members. By incorporating modern social behavioral analytics into design logics, social spaces can facilitate more productive engagements between occupants. Armed with tools based on current social theory,

K. De Rycke et al., *Humanizing Digital Reality*,
https://doi.org/10.1007/978-981-10-6611-5_49

designers can create a more diverse and sustainable community through an informed understanding of how space and program influence behavior.

The Problem: Schelling's Model of Segregation

Per Schelling's model of Segregation, the population will innately segregate itself based on preferences, often leading to organization by race, religion, and class. As Schelling states, the process is not solely a result of prejudice but primarily of preference, making solutions to undesirable segregation elusive even when the population applauds diversity (Schelling 1969).

The mathematical model begins with a random grid of two types of objects that have a defining characteristic such as a color in conjunction with a set number of vacant spaces (Fig. 1). An overarching criterion is established to determine when an object is satisfied or dissatisfied. For instance, each color wants to be adjacent to at least 33% of like colors. If only one of the eight neighbors (12.5%) are of the same color, then the object is dissatisfied (Fig. 2). Each dissatisfied object moves to a vacant space (Fig. 3). Criteria are once again applied to the object to determine its state. If it meets the criteria, in this case 37.5% of the same color, it is satisfied and remains in place (Fig. 4).

After the initial random distribution (Fig. 5), 20 iterations were applied to the model. With only a 33% preference, the segregation of colors becomes apparent (Fig. 6). This model did not stabilize and contained spaces that continued to change indefinitely. This allegorical social model suggests that segregation will continue even with the removal of all prejudice and dislikes, at times creating locations of unrest. Segregation will occur motivated by preference alone without the aid of negative repulsors. Regardless of motivation, this subdividing of communities through segregation increases social tension, discourages communication, and isolates those who are

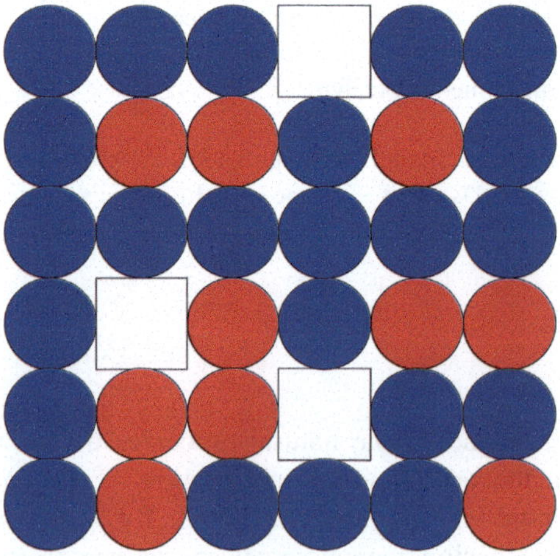

Fig. 1. Random grid

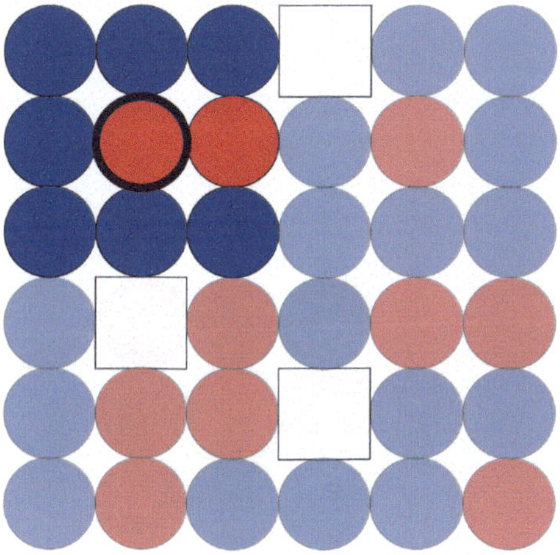

Fig. 2. Unmet criteria

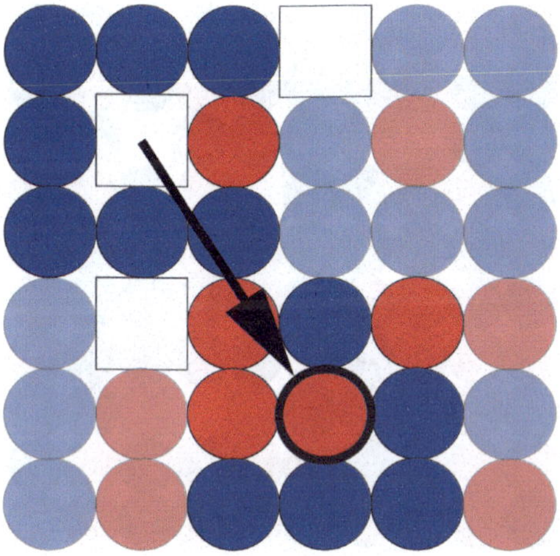

Fig. 3. Migration

different. The effects of isolation have been called an 'epidemic of loneliness', shortening lifespans and impacting the quality of life (Khullar 2016) (Fig. 7).

The Solution: Mere Exposure Effect

In the 1960s, Robert Zajonc experimented with the role of preference for familiar stimuli over novel ones. By controlling how many times a subject was exposed to a particular stimulus, Zajonc concluded that preference increases with higher exposure frequencies to a certain point and then stagnates or at times declines (Zajonc 1968). Specifically, Zajonc demonstrated that subjects rated an unknown symbol with more "goodness" if viewed more often (Fig. 8). In 1989, Robert Borstein presented twenty

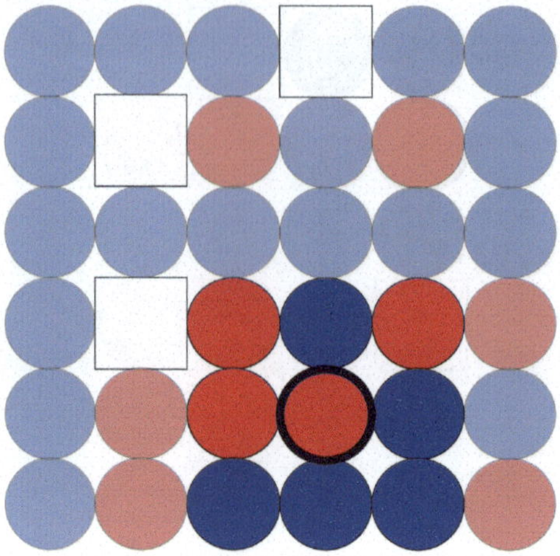

Fig. 4. Met criteria (James 2017)

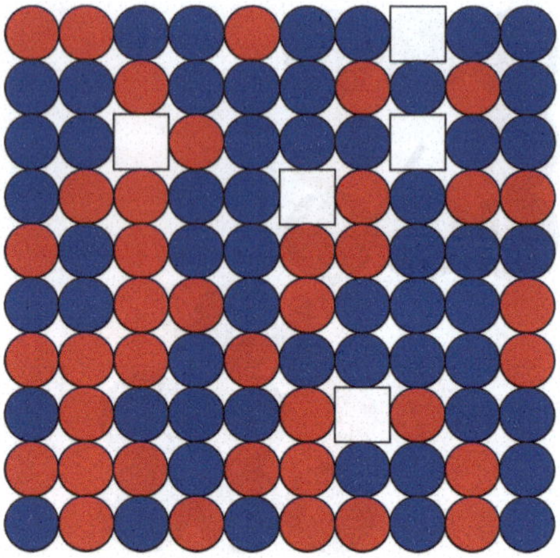

Fig. 5. Initial distribution

years of research reporting essential conditions and factors when mere exposure effects are weaker or stronger. Those factors include the number of exposures, time between exposures, personality types, as well as differences between simple and complex stimuli (Bornstein 1989). Extensive research and rigorous experimentation has been conducted to quantify the effects of exposure. This research is one of the driving principles used in marketing (Grimes and Kitchen 2007). Simply stated, the mere exposure effect is a phenomenon by which people develop a preference for things solely because they are familiar with them.

Even with 50 years of mere exposure research, architectural design has yet to implement or even introduce techniques to utilize the mere exposure effect in

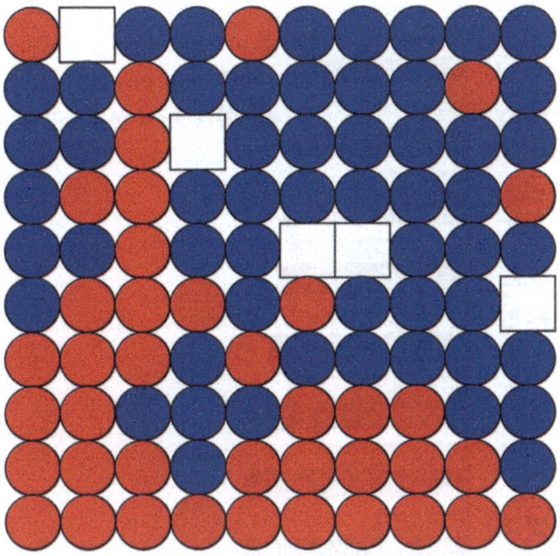

Fig. 6. After 20 iterations

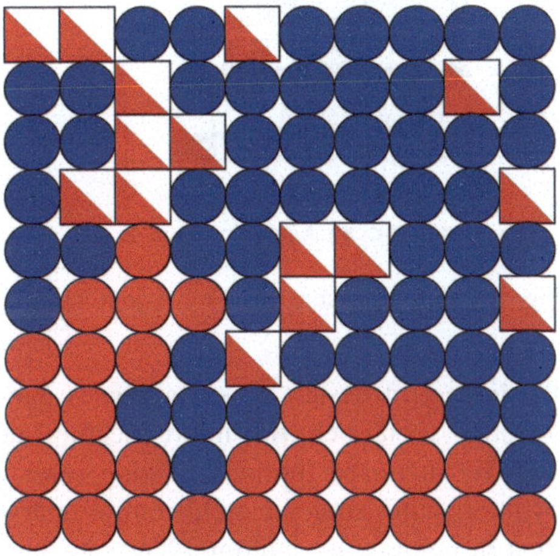

Fig. 7. Constant transfer (James 2017)

minimizing the impact of negative social segregation. This is in part because of the complexity of the social condition, and in part because of the lack of funding for research on post occupancy. Innovations in GIS offer opportunities to aggregate relevant data to determine the exposure effects of various arrangements within the urban environment. However, understanding and organizing the complex matrix of city organizations is still driven primarily by anecdotal evidence. The multifamily model offers a unique microcosm with numerous iterations of organization paradigms and programs. However, the development community is reluctant to release information concerning their highly crafted building programs and forms duplicated throughout the world. Although hindered with a lack of relevant data, the potential social returns of

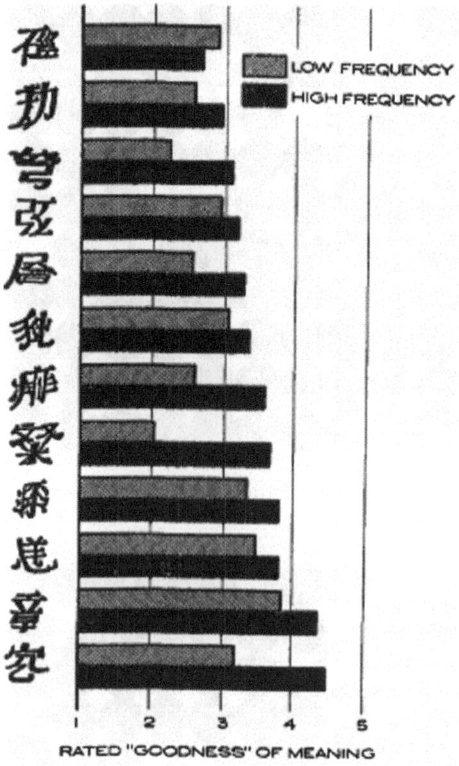

Fig. 8. Mere exposure chart by Zajonc (1968)

carefully considering the impact of design on social issues garner the need to create tools for modeling and testing social behavior in architectural design.

Considering what can be supposed through existing research in mere exposure, examining the effects of interjecting strategic architectural interventions and programs offer great potential gains in community building. Through mere exposure designers can create new connections between members of society by rethinking circulation paths, carefully considering the geolocation of program, and creating more effective public space.

Many undesirable segregations occur due to a designer's lack of awareness of how their design contributes to this phenomenon. Applying the theory of mere exposure to early schematic design could reveal the impacts of various designs on community and create novel solutions for public spaces. These solutions offer new possibilities to integrate community members at variable scales of the built environment, from urban design to multifamily housing. These core principles create the social platform for a tool which integrates the mere exposure effect to evaluate the organization of public spaces in multi-family housing.

Understanding the Impact of Circulation

Assuming that the ideal amounts of exposure will increase the preference for one's neighbor, one first must determine how much time occupants would see each other. Modeling a simplistic statistical model for circulation in a rectangular multifamily apartment wrapping a parking deck (often referred to as a Texas doughnut) reveals symbiotic relationship between occupants, public space, and circulation. The model assumed the apartment contained one gym utilized by guest for 30 min 4 times a week and one restaurant utilized by guest for 60 min 3 times a week. Each occupant would both drive to work and walk to a nearby destination once a day. The configuration contained a total number of 25 residential units (Fig. 9).

Considering staggered start times and the typical times of occupation, the model reveals that walking to all destinations combined created a total amount of exposure of less than 10 min between all residents per day. Participating for 30 min 4 times a week between the times of 4 and 7 pm will result in an average agglomeration of 980 total minutes of exposure time between residents. Participating for 60 min 3 times a week between the times of 5 and 9 pm will result in an average agglomeration of 1080 min of exposure time between residents. The staggering difference in exposure between public participation and public circulation advocates a particular role of the architect in aiding social engagement.

To understand the math behind the model, the individual exposure time to the public takes the average time per week a resident participates in a public activity such as walking to the car, eating at a restaurant, or working out in a gym. The average number of uses per week is then divided by 7 days. That number is multiplied by the average duration of the activity. For instance, if someone uses an amenity twice a week for 70 min then they would be in the amenity an average of (70 * 2/7=) 20 min a day.

To determine the total exposure time to other residents, the individual exposure time to the public is divided by the total exposure opportunity. For instance, if the

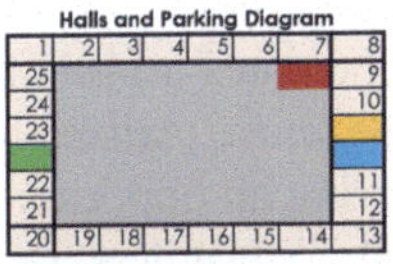

Walk Speed (FPS)	4.5	
Activity Time #1 (Min)	30.0	
Times Per Week	3.0	5.0
Act#1 Time	4.0	7.0
Location	A8	
Activity Time #2 (Min)	60.0	
Times Per Week	2.0	4.0
Act#2 Times	5.0	9.0
Location	H7	
Work Start	6.5	9.0
Work Return	4.5	6.0
Location	G5	
Pedestrian Start	5.0	8.0
Pedestrian Return	6.0	10.0
Location	H8	

Individuals	Activity # 1			Activity Start + Return			
Apartment #	Exposure Time (Min)	Exposure Ratio	TotalExposure of Unit (Min)	Distance	Exposure Time (Sec)	Exposure Ratio	TotalExposure of Unit (Sec)
A4	17.14	0.095	39.18	120	26.7	0.002	2.519
B4	17.14	0.095	39.18	150	33.3	0.003	3.128
C4	17.14	0.095	39.18	180	40.0	0.004	3.728
D4	17.14	0.095	39.18	210	46.7	0.004	4.321
E4	17.14	0.095	39.18	240	53.3	0.005	4.905
F4	17.14	0.095	39.18	270	60.0	0.006	5.481
G4	17.14	0.095	39.18	300	66.7	0.006	6.049
H4	17.14	0.095	39.18	330	73.3	0.007	6.609
H5	17.14	0.095	39.18	300	66.7	0.006	6.049
H6	17.14	0.095	39.18	270	60.0	0.006	5.481
H9	17.14	0.095	39.18	240	53.3	0.005	4.905
H10	17.14	0.095	39.18	270	60.0	0.006	5.481
H11	17.14	0.095	39.18	300	66.7	0.006	6.049
G11	17.14	0.095	39.18	270	60.0	0.006	5.481
F11	17.14	0.095	39.18	240	53.3	0.005	4.905
E11	17.14	0.095	39.18	210	46.7	0.004	4.321
D11	17.14	0.095	39.18	180	40.0	0.004	3.728
C11	17.14	0.095	39.18	150	33.3	0.003	3.128
B11	17.14	0.095	39.18	120	26.7	0.002	2.519
A11	17.14	0.095	39.18	90	20.0	0.002	1.901
A10	17.14	0.095	39.18	60	13.3	0.001	1.276
A9	17.14	0.095	39.18	30	6.7	0.001	0.642
A7	17.14	0.095	39.18	30	6.7	0.001	0.642
A6	17.14	0.095	39.18	60	13.3	0.001	1.276
A5	17.14	0.095	39.18	90	20.0	0.002	1.901
Totals	Exp to Space 428.57	Exp to Others	979.59	Exp to Space	1046.7	Exp to Others	96.428
	Avg Min of Exposure to Others		979.59	Avg Min of Exposure to Others			1.607

Fig. 9. A partial section of the Multifamily Housing Exposure Model which lists the distance between destinations; amenity use times, durations, and frequencies; each residents average time of exposure to neighbors; and total exposure times for all residents (James 2017)

restaurant serves dinner from 5:00 to 9:00 pm, that would be 4 h of opportunity. This forms the exposure ratio (20 min/240 min=) of 0.0833. All the exposure times of other residents are added together then multiplied by the exposure ratio. If 25 residents lived in the apartment who used the amenity with the same assumptions that would total 24 other residents at 20 min each or 480 min of average time in the amenity each day. Finally, the exposure ratio (0.0833) is multiplied by the aggregate exposure time by all others (480 min) to find how much time each resident could see other residents (40 min). Roughly interpreted this means that on average, the resident is in the room with two other people. The math is overly simplistic but provides a road map of how organizational strategies affect exposure times.

The model offers three insights into the residential social dynamic. First, circulation has little impact on exposure to other residents. Even creative endeavors cannot statically justify community building as a primary goal. Second, circulation does however impact the amount of exposure to activities. The residents will pass by public spaces multiple times a day, increasing exposure to public programmatic elements thus increasing preference for those elements. Finally, Community members accumulate significantly higher exposure times to each other when they occupy public space and participate in local activities. The participation in public space results in nearly 10,000% more exposure than public circulation. In conclusion, if architects can create circulation patterns that increase exposure to public space, or position public spaces at locations that maximize their exposure through circulation, preference for that space will encourage occupation. The occupation of public spaces increases exposure to other community members, increasing preference for those members. Thus, design vicariously can increase the residents' exposure to each other by increasing their exposure to public space.

The Foundation of the Design Modeling Tool

Three assumptions based on the Schelling's model of segregation, the mere exposure effect, and the multifamily housing exposure model form the foundation for the design modeling tool described here. First, one must accept that preferences influence how communities are formed. The uniformity of innocuous preferences creates a basis for collaboration and conflict resolution of larger issues. Second, one must realize that preferences are influenced by exposure and continually evolve, offering an opportunity to align local communities through common experiences. Finally, designers must create circulation patterns that encourage the observation and occupation of public space, knowing that the public participation provides the best opportunity for creating preference for other community members. To sum up, utilizing diverse types of public amenities while increasing exposure to them through circulation; design can increase preference for public spaces which increases exposure between community members, encouraging communication towards the creation of a better sense of community.

Development of a Computational Tool

Analyzing the Texas Donut, a Multifamily Model

The multifamily model offers the most reasonable resource for analyzing the mere exposure effect in design due to its scale and diversity in program (Fig. 10). Because of the innate necessity of the automobile in southern culture, the 'Texas doughnut' has become one of the dominate multifamily types (Fig. 11) in the area. The use of the tool on the Texas doughnut aims to consider the impact on a range of socioeconomic groups.

The computational tool, built in Autodesk Revit BIM modeler Dynamo, uses an existing Revit model. In the rooms category, a shared parameter called circulation is created to label public spaces and corridors. Another shared parameter in the category doors is created to create destinations. The inputs to the script are a Revit floor plan for a multifamily apartment; the rooms designated as Circulation; doors labeled with the appropriate tags; and the frequency of use for each type of path. The script outputs a heat map of high traffic and low traffic areas; and all circulation patterns from apartments to each specified destination.

The script converts all the rooms into a single surface and divides it into a grid for pathway analysis. Each labeled door is divided into types and organized to create destinations from the apartments to each destination such as an amenity, utility, main entrance, parking entrance, and pedestrian exits. The Lunchbox node Curves ShortestWalk determines the shortest path from every door to every destination using the length of each grid line. For multiple pedestrian exits or multiple entrances to an amenity, the tool uses a minimum distance function to pick the most likely path. The tool draws each path in the Dynamo script visualizer and has an option to draw it directly on the Revit model.

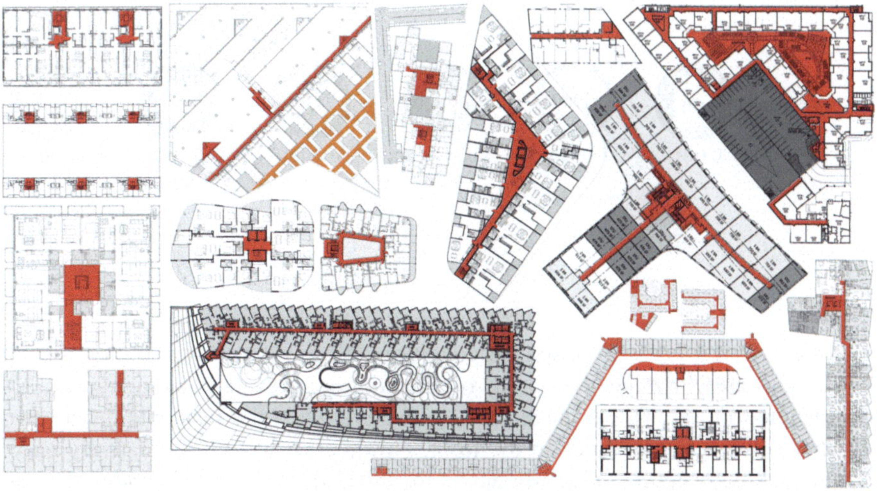

Fig. 10. Circulation diagrams of multifamily plans (James 2017)

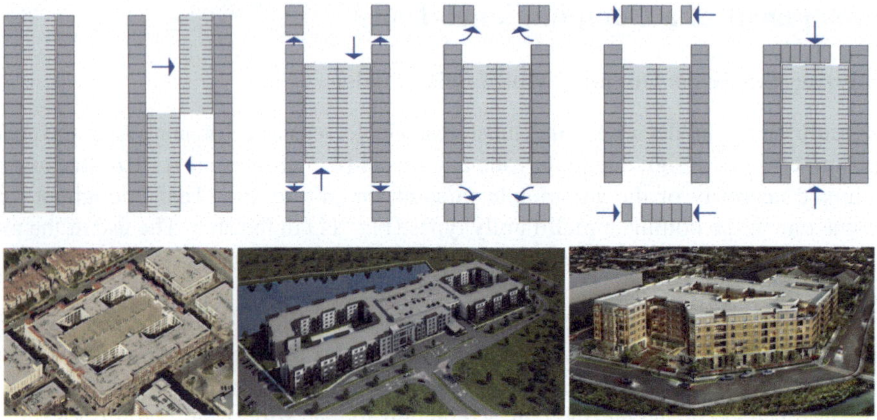

Fig. 11. Diagram of formation of Texas doughnut (*above*) (James 2017); images of typical Texas doughnut formations (*below*). Images from unknown authors, web references listed from left to right ([Texas Donut] [image] n.d)

When considering how space is traversed in the multifamily model, some assumptions were made. On average, the typical resident would use the pedestrian exit twice a day in order to go on a walk, visit a local pub, or go to the store. The resident would take out the trash every other day, go to the gym every other day, and go to the main entrance every other day to receive guest or food delivery. Finally, the car would be used to go to and from work once a day and to run an errand every other day. These assumptions are based on conjecture and can be modified in the script via user controlled sliders. After determining all circulation possibilities, the tool assigns the user score for each path to calculate the approximate exposure time at each point in the circulation path. The weighted score of each point is graphically represented by a heat map. The heat map uses the colors red for low frequency, green for average frequency, and blue for high frequency.

Using the Tool

In Fig. 12, layout 1 has a front-loaded condition. When the parking, utility, amenity, and front exit are close together, areas of isolation are prevalent throughout the corridors, denoted in red. This arrangement is effective when the residents do not avoid each other and use the facilities in the front. However, it would be relatively easy to avoid interaction by getting an apartment on the left corridor or right corridor. The right diagram of Fig. 12, shows that moving even one of the major destinations results in an

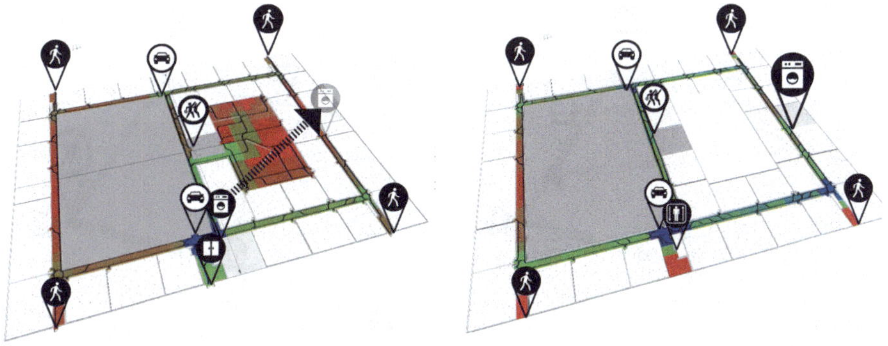

Fig. 12. Layout analysis 1 and 2 of a Texas Doughnut (James 2017)

immediate change in movement as well as a leveling out of corridor exposure creating more opportunity for the architecture to engage its users. Blue spaces now show up at four junctions, indicating high levels of exposure. Two of the hallways f increased level of circulation based exposure. However, the far-left corridor still has a low level of exposure.

In Fig. 13, moving the amenity to the exterior in layout 3 has little positive impact. The area of isolation moves to the center but suggests mild improvement over the left side of layout 2. The area around the amenity has a hint of red deteriorating the previous circulation through the center of layout 3. When considering amenities, this is by far the least effective arrangement evident by the most amount of red. In Fig. 14, layout 4 takes a more dramatic approach by moving utilities and amenities to the corners and separating the parking entrance from the main entrance. This arrangement has circulation throughout all hallways. It would be difficult for a resident to avoid other residents in this configuration. The areas of isolation are minimal and high traffic areas (blue) creeps into every corner (Fig. 15).

In Fig. 16, layout 5 tests more significant changes to investigate alternative opportunities. The heavy red area on the left corridor returns to the heat map, signaling

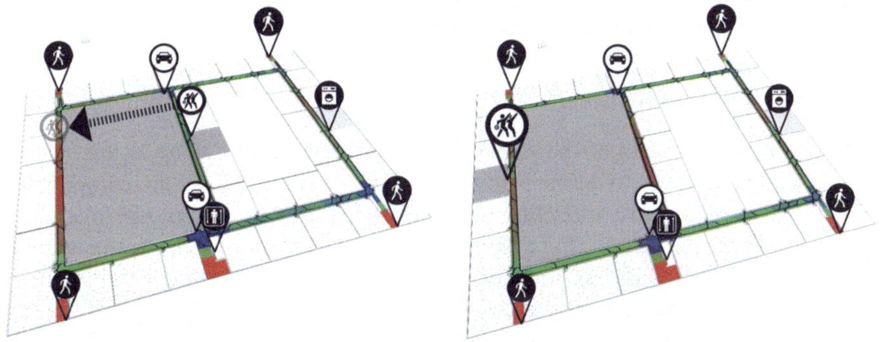

Fig. 13. Layout analysis 2 and 3 of a Texas doughnut (James 2017)

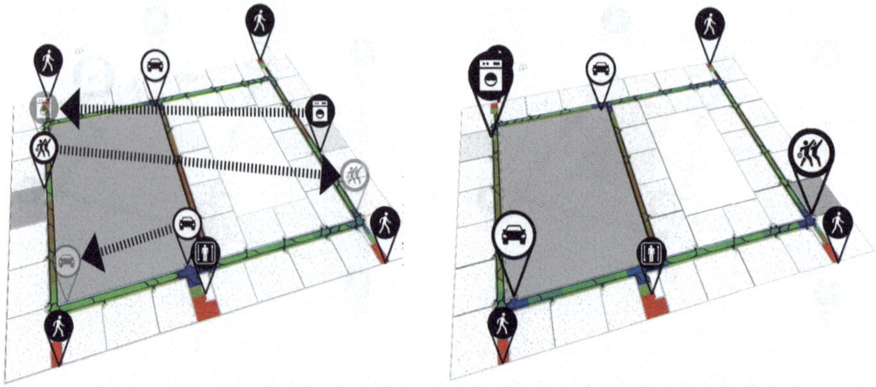

Fig. 14. Layout analysis 3 and 4 of a Texas doughnut (James 2017)

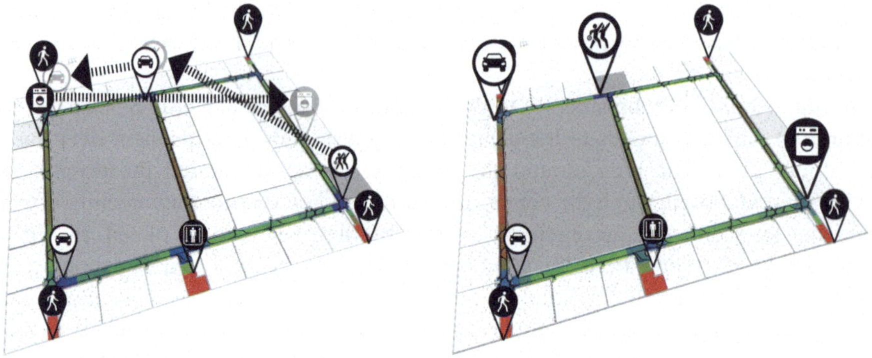

Fig. 15. Layout analysis 4 and 5 of a Texas doughnut (James 2017)

the left corridor has little to no exposure. Moving the parking to one side isn't a reasonable option because it will force residents of the right corridor to have unnecessarily long walks to their cars. The increased exposure is falsely inflated which doesn't create a realistic solution.

Models from Other Architects

To assess the effectiveness of the tool, other design styles need to be tested. Creating unique shared parameters to define pathways and destinations allows the computational tool to be adapted to most modeling styles in Revit design. The model to the right designed by FMK Architects (Fig. 16) has four apartments across from the amenities and main office. Each floor has 8 apartments with an open-air corridor. The script fails because it relies on exit doors as destinations and did not consider open air configurations. Doors and walls were inserted into the model for this analysis. In service of a more effective tool, a door family that consists of a two-dimensional rectangle placed at

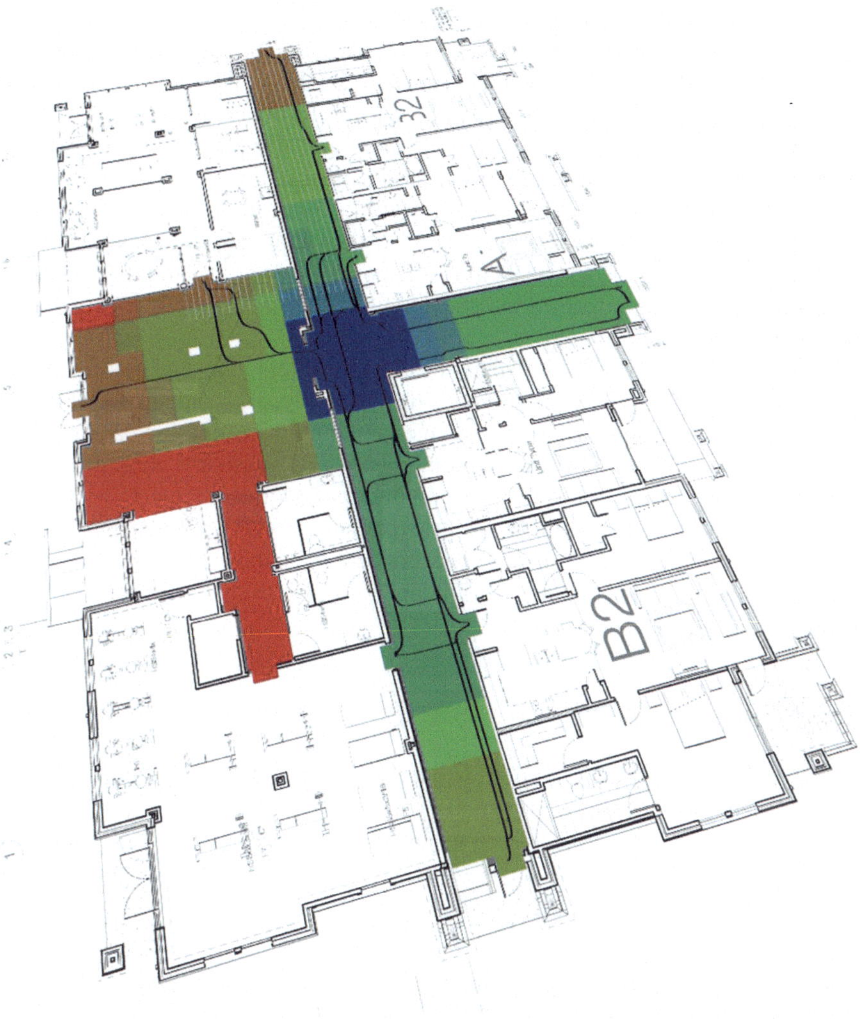

Fig. 16. FMK architect model analyzed (James 2017)

exterior transitions allows for reasonable modifications that do not interfere with the built-in Revit visualization, making the script more adaptable. This model doesn't challenge the analytical possibilities of the tool only attempts to demonstrate its adaptability.

Conclusion

The tool described in this paper aims to foster the beginning of a critical dialogue about social connections and community building. The tool offers a form of analysis to begin gathering data to compare hypothetical results with the actual results in post occupancy analysis. The multifamily model warrants the same consideration in the architectural dialogue as museums, parks, and universities which are so eloquently discussed and carefully considered in projects such as Daniel Libeskind's Jüdisches Museum Berlin, Bernard Tschumi's Parc de la Villette, and Rem Koolhaus' McCormick Tribune Campus Center. The potential returns on effective amenity placement and circulation patterns within the multifamily residential design field could expand the influence of architecture to impact the usability and comfort of the typically overlooked components of these types of projects. As designers expand their craft beyond materials utilizing social behavioral analytics, the scalability of the model would naturally expand into urban design through an informed understanding of how architecture affects the social character of its occupants.

Considering both the effect of unit location on circulation and noting the most effective locations for public goods; designers, developers, and city planners will increase the overall communication between community members. Increasing exposure and therefore the likelihood of direct communication as a primary goal of design will aid the development of stronger social communities. Although the tool specifically targets residential complexes, the concept explained here is scalable. Further development through experimentation and data collection of the mere exposure effect in urban design and architecture could aid the informed design of public space and pathways.

References

Bornstein, R.F.: Exposure and affect—overview and meta-analysis of research, 1968–1987. Psychol. Bull. **106**(2), 265–289 (1989)

Grimes, A., Kitchen, P.J.: Researching mere exposure effects to advertising. Int. J. Mark. Res. **49** (2), 191–219 (2007)

James, M.P.: Thesis Final 2017. Tiff (2017)

Khullar, D.: Loneliness is a Health Hazard, but There Are Remedies. New York Times, New York, 22 Dec, 2016, p. A3 (2016)

Schelling, T.C.: Models of segregation. Am. Econ. Rev. **59**(2), 488–493 (1969)

[Texas Donut] [image] (n.d.) [Photograph]. Retrieved from http://oldurbanist.blogspot.de/2013/06/common-garage-parking-in-practice-part.html

[Texas Donut] [image] (n.d.) [Photograph]. Retrieved from http://www.dailyherald.com/article/20170117/business/170119057/

[Texas Donut] [image] (2002) [Photograph]. Retrieved from http://buildingwithpurpose.us/clt/wp-content/uploads/2013/07/CrescentDilworth.jpg

Zajonc, R.B.: Attitudinal effects of mere exposure. J. Pers. Soc. Psychol. **9**(2, Pt.2), 15 (1968)

[Kak-Tos]: A Tool for Optimizing Conceptual Mass Design and Orientation for Rainwater Harvesting Facades

Christopher Beorkrem(✉) and Ashley Damiano

University of North Carolina at Charlotte, Charlotte, NC 28223, USA
cbeorkrem@uncc.edu

Abstract. Water is a valuable, reusable resource that we tend to take for granted. While rooftop rainwater collection systems are beneficial, the facades of high-rise buildings in particular have the potential to increase collection amounts due to the greater amount of surface area. Increasing the amount of rainwater collected and reused is beneficial environmentally, economically, as well as socially. With limited precedence for collection of wind-driven rain on the vertical façade of a building, Kaktos is intended as an aid for designers during the schematic massing studies of a project with the hopes of challenging the way we integrate rainwater collection into the architecture of our buildings. Using imported weather data, Kaktos enables the designer to calculate collection potentials for wind-driven rain, optimize the orientation of the form on the site, and analyze the flow of water over the form's surface.

Keywords: Rainwater · Facade · Design

Introduction

MIT researchers "expect 5 billion (52%) of the world's projected 9.7 billion people to live in water-stressed areas by 2050" (Roberts 2014). Each year globally water becomes a more valuable resource. This is as much a social problem as it is an environmental problem. Impervious surfaces can make up a significant portion of the surface area in highly developed areas, causing water to quickly shed away from where it is most needed. As the urban runoff flows over the impervious surfaces, it collects pollutants—dirt, oils, fertilizers, bacteria, etc. These pollutants end up in our storm water systems, or worse—our soil, ponds, streams, and oceans. This kind of pollution can kill wildlife, vegetation, and contaminate drinking water. The hardscape of urban areas also causes storm runoff, which travels too rapidly to absorb into soils. As a result, wells are drying up and needing to be lowered to access new lower water levels.

With the expected increase in urban density, improvements in global construction methods and design standards, high-rise buildings are becoming much more ubiquitous. Currently, the best practices for rainwater harvesting consider only the rainfall resource available to the horizontal surfaces of a site: the roof and surrounding land. However, these buildings have the potential to collect and use rainwater not just on their roofs or from excesses of their mechanical systems, but also with their facades.

K. De Rycke et al., *Humanizing Digital Reality*,
https://doi.org/10.1007/978-981-10-6611-5_50

Typical gray water consumption in a high-rise office building is well beyond that which can be collected through conventional means. For example, a toilet uses approximately 3 gallons of water per flush. If an average employee flushes the toilet 2–3 times a day, conservatively, he or she uses 7.8 gallons per day in just toilet flushing. In a 500,000 sf office high-rise with 4000 employees, and 200 work days in a year, this would equate to 6.2 M gallons of gray water consumption annually. The collection of rainwater from high-rise facades could help mitigate the amount of water that is being flushed down the drain, allowing users to be more conscientious about the sources of our gray water.

Wind-Driven Rain (WDR) is rain that is propelled horizontally by wind, therefore falling at an oblique angle. There is an extensive amount of previous research into the impact of WDR in a myriad of fields. In their paper "A review of wind-driven rain research in building science", Blockena and Carmelieta note "The large number of parameters and their variability make the quantification of WDR a highly complex problem. It is not surprising that despite research efforts spanning over almost a century, WDR is still an active research subject in building science and a lot of work remains to be done." (Blockena and Carmelieta 2004). While current Computational Fluid Dynamics (CFD) research is incredibly complex, ongoing real world testing by a variety of methods still proves that computational solutions about WDR are at best estimations.

Given this we can presume that while still an estimation, the changes that could occur in those estimates with better models are simply proportional, therefore we have based our models on a simple formula based on Blockena's and Carmelieta's conclusions:

> Measurements of both free WDR and WDR on buildings have indicated that the intensity of WDR increases approximately proportionally with wind speed and horizontal rainfall intensity. Measurements of WDR on buildings have revealed part of the complex wetting pattern of a facade: top corners, top and side edges are most exposed to WDR.

As far back as 1955, WDR research has presumed this proportional relationship. Hoppestad established a WDR index based on proportional calculations and the weather data that is collected in each location: Wind direction, wind velocity and rainfall amounts. Based on this he determined:

> The annual mean WDR index gives, it is believed, a reasonably precise method of comparing different sites with respect to total amounts of WDR on walls. It enables a designer to compare the exposure of a place with that at another with which he is already familiar. (Hoppestad and Slagregn 1955)

We can therefore presume for this study that Hoppestad's semi-empirical method and equation provides a proportional method for understanding the potential changes of WDR on building facades:

$$R_{Wdr} = R_h \cdot \frac{U}{V_t}$$

Equation 1: *Hoppestad's equation where* R_{wdr} *is total WDR and* R_h *is the rainfall rate, U is the wind speed (m/s) and* V_t *is the raindrop terminal velocity of fall (m/s).*

This *semi-empirical* equation serves as a basis for our evaluation and though other factors clearly contribute and can change the impact of those changes in the equation would be relatively uniform given the evaluation criteria established for the tool. Each of the variables would consistently alter the evaluation of WDR amounts that could potentially be collected.

In this paper, we will examine the impacts and potential for WDR to be harvested on the facades of buildings, and will describe a plug-in tool for schematic design for *Autodesk Revit*, which can assist in the definition of optimal geometries and orientation for a building to harvest water in its particular location and site. The tool described here is intended to create comparisons between building masses and their orientation based on conditions in particular locations and sites. These base calculations presumably would adjust uniformly as the science around WDR adjusts or as other factors complicate models.

Tool Logic

Based on our calculations, we approximate that the roof of a 100′ × 200′ vertical high rise 400′ in height would collect 12,400 gallons with 1 in. of precipitation. However, when we add in the quantities of rainwater that can be collected on the facades of the same tower the total collection increases significantly. We calculate that a 5 mph wind drives rain to fall at a 21° angle, and 10 mph wind drives it at a 36° angle onto the facade. The rainfall that can be collected on the four facades of the building we approximate to be 28,800 gallons per year or 152% more than the roof alone if the wind is blowing on average at 5 mph, and 46,628 gallons per year or 276% more than the roof alone if blown at an average of 10 mph during rainfall. According to the National Oceanic and Atmospheric Administration, the City of Charlotte, North Carolina receives an average of 43 in. of annual precipitation. Based on our calculations, we approximate that the roof of the same high rise would collect 71,667 cubic feet, or 536,104 gallons per year on its 20,000 SF roof surface. With 5 mph winds, we can calculate that approximately 814,878 gallons per year to be collected on the four facades, and 1,479,647 gallons with an average of 10 mph winds annually (Fig. 1).

The tool outlined here works by importing the weather data for a given location from the National Oceanic and Atmospheric Administration (in our initial studies we used a 5-year window). This data is first simplified to include only days where measurable rainfall occurred. The data is then further stripped to only include the measurement of precipitation, wind direction and velocity, presuming a median rain drop size. The tool functions by projecting rays onto the building mass along the angle calculated that the WDR would fall during that given day, and calculates how much of the WDR would hit each of the sides of the mass (Fig. 2).

Based on these calculations the tool can then provide a total annual calculation for WDR on each facade or can rotate the mass based on a user specification to test for the optimal orientation of the given shape. Predictably, the optimal orientation being that in which the greatest surface area of the mass is facing the median windward direction on rainy days, which will allow the most amount of WDR to hit the largest surface area.

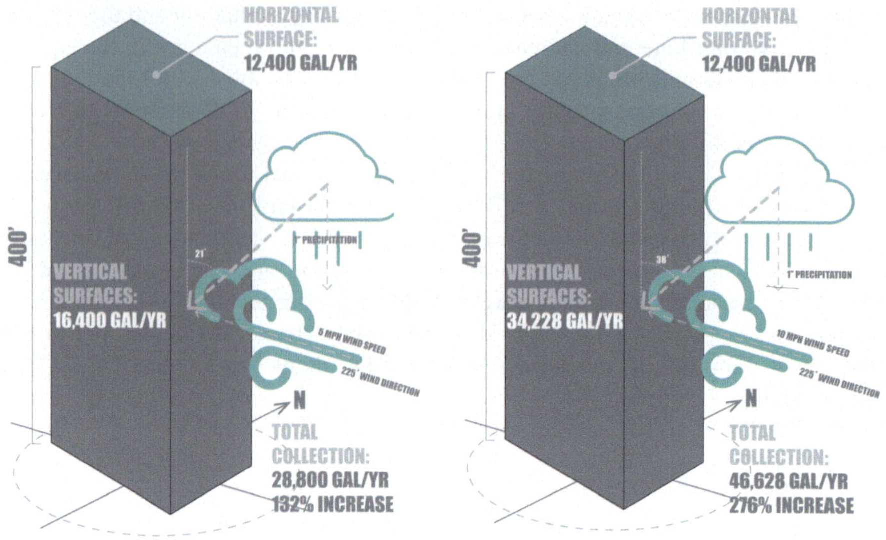

Fig. 1. Kaktos tool diagram

The user can specify the amount of variation in the rotations to determine how computationally intensive the testing is (Fig. 3).

Tool Analysis

As seen in Table 1, both form and orientation play significant roles in the amount of WDR that hits each facade. By adding the facades of the mass into the calculations for rainwater harvesting, the percentage of WDR that interacts with the form at least doubles for each instance, except for forms 5 and 6—both of which contain larger horizontal surfaces and shorter facades than forms 1–4. The increase in WDR on the facades through optimal orientation on the site differs for each form. Re-orienting forms 1 and 3 from their original test position once again doubles the amount of WDR on the facades, whereas re-orienting forms 2 and 6 has minimal effect on the amount of WDR. The intent of this tool is to aid designers in massing and orientation studies. Using *Revit*'s conceptual mass tool allows designers to study iterations quickly, analyze them using Kaktos, and make changes accordingly.

Lastly, the tool can be used to calculate the optimal locations for rainwater collection to minimize the frequency of large quantities flowing along the surface of the building. This aspect of the tool works by tracing the gravitational flow of the water along the facade using a sphere-mapping method (Beorkrem et al. 2013). Sphere-mapping uses a technique of intersecting spheres with a complex mass to find the approximate circular shaped intersection between the mass and the sphere. The shortest line between the center of the sphere and this circular intersection represents the likely path for the water. Using the endpoint of the shortest line, another sphere is placed and this process continues iteratively while a total water amount is tallied to

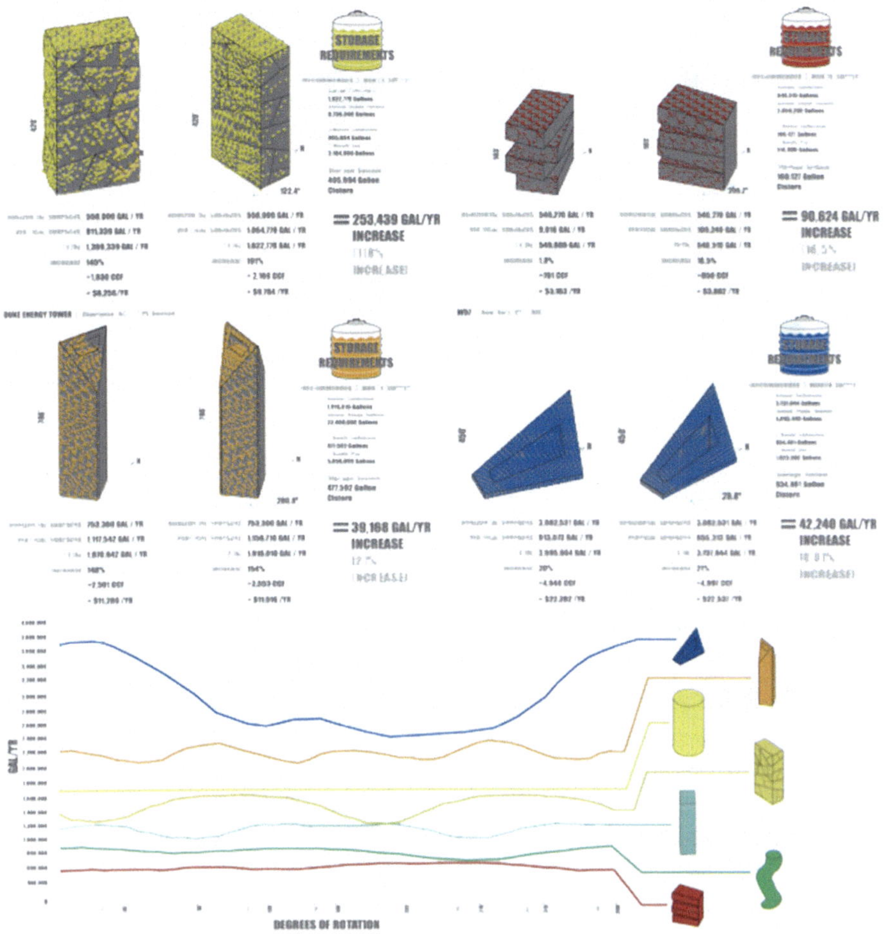

Fig. 2. Testing a variety of building masses

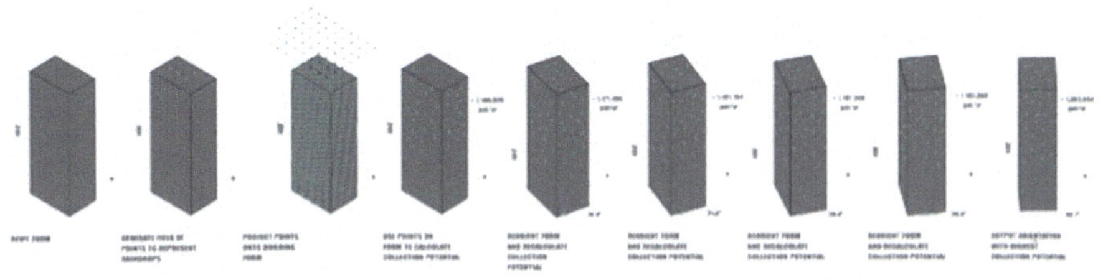

Fig. 3. Mass orientation testing

Table 1. Mass comparison/orientation comparison

	1	2	3	4	5	6
	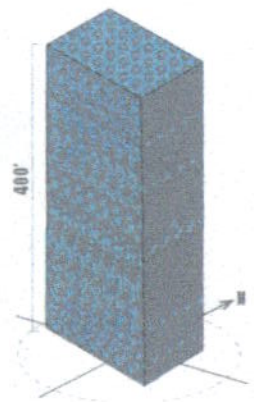	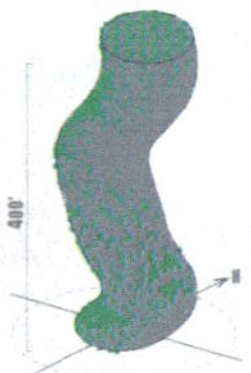	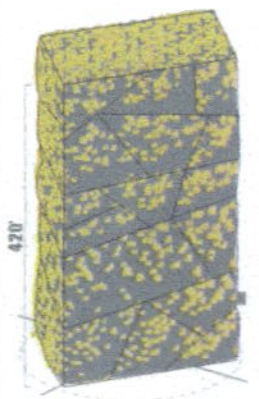	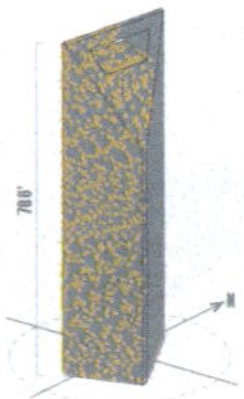	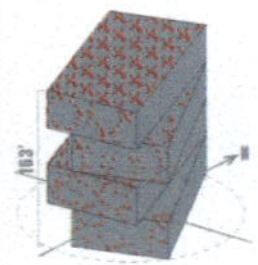	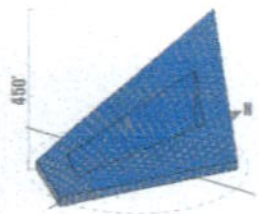
Horizontal surfaces (gal/year)	558,000	219,015	558,000	753,300	540,270	3,082,531
Vertical Surfaces (gal/year)	558,620	636,534	811,339	1,117,542	9616	613,073
Total	1,146,620	855,549	1,369,339	1,870,842	549,886	3,695,604
Increase (%)	100	290	145	148	1.8	20
	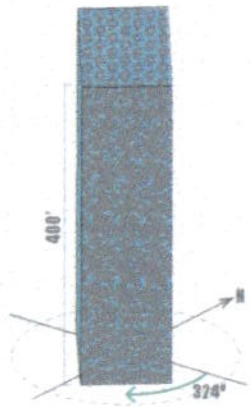	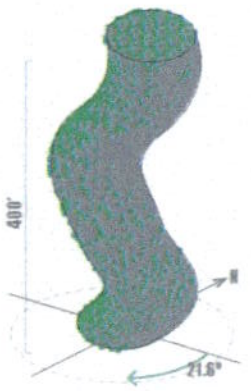	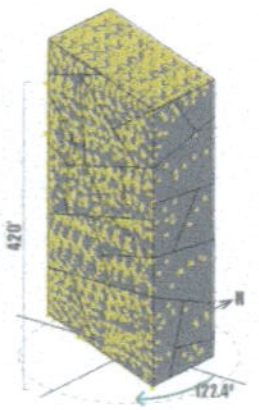	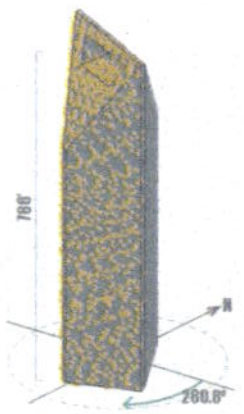	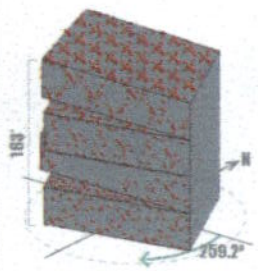	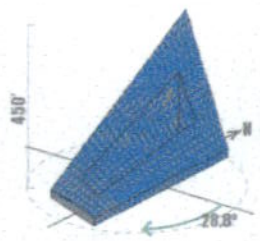
Rotation angle	324°	21.6°	122.4°	280.8°	259.2°	28.8°
Horizontal surfaces (gal/year)	558,000	219,015	558,000	753,300	540,270	3,082,531

(*continued*)

Table 1. *(continued)*

Vertical surfaces (gal/year)	644,684	638,070	1,064,778	1,156,710	100,240	655,313
Total	1,202,684	857,085	1,622,778	1,910,010	640,510	3,737,844
Increase (%)	115	291	191	154	18.5	21
Increase (gal/year)	56,064	1536	253,439	39,168	90,624	42,240
Increase (%)	115	0.2	118	21	16.5	0.01

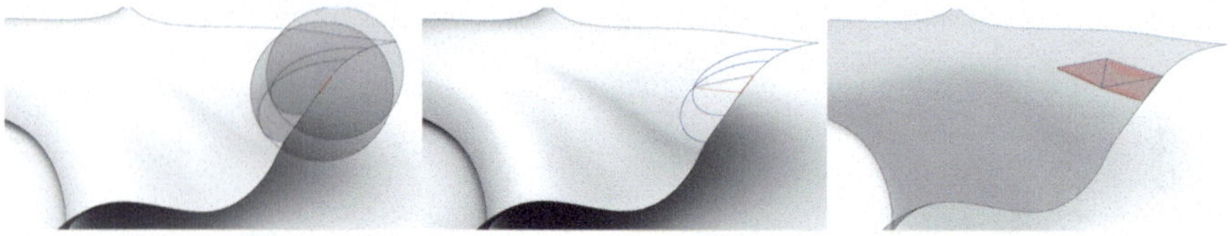

Fig. 4. Sphere-mapping process. Sphere intersecting with surface (*left*). Resulting lines projected on surface (*center*). Resulting mapped triangles (*right*)

indicate when a total reaches a user-defined amount at which point the water can be collected. Placing intermediate collection points on the facade of a building removes the water before it reaches the ground, reducing the amount of urban runoff and preventing the unnecessary use of energy to pump the water back up through the building (Fig. 4).

This part of the tool also challenges designers to factor the way water flows over the facade during early schematic phases of building design, whether that be for aesthetic reasons (using the rainwater to activate the facade), for acoustical reasons (using the rainwater to dilute exterior noise in certain areas of the building), or for thermal reasons (using the rainwater to mitigate heat gain/loss in the facade). Allowing the collection of rainwater to assist in the design of a mass, rather than leaving it as an afterthought the way traditional rainwater collection systems do allows the WDR to become a more integral part of the architecture. As with other passively sustainable systems these criteria need to be evaluated early on in the design process, too often designers use technology to mediate the effects of poor decisions and this tool hopes to help them visualize the impacts of one aspect of massing design. The visual impact of a facade designed to collect rainwater could also create more awareness and afford the client the ability to equate their brand with a desire to minimize their impact on the environment.

Implementation

The collection and reuse of rainwater could begin to reshape the architectural landscape of our cities. As computational manufacturing and prefabricated building components become more pervasively used, buildings formal typologies are continuing to expand into ever more complex shapes. These complex forms are often driven by arbitrary and stylistic ideas (or lack thereof). We envision a series of tools integrated for evaluation of multivariate performative criteria integrating forms and orientation choices driven by Kaktos with other evaluative tools, based on solar radiation, day lighting and structural performance, amongst others.

The integration of facade based water collection systems with other infrastructure could help create more resilient neighborhoods and cities. Various cities have infrastructure in place which could supplement the performance of such systems, for instance in New York City most buildings already have water towers to create pressure for the water systems in the building, but could also serve as the location for storing greywater. This would require a transfer system for creating pressure in both systems within the same tower, but is a reasonable use for existing technology. New York as well as other cities could respond better to heavy storm water scenarios by collecting rather than shedding the WDR which hits buildings.

Furthermore, these localized systems could provide backup for disaster scenarios such as the Haitian Earthquake, whose impact was exacerbated by the lack of clean drinking water following the earthquake. If the infrastructure were in place to collect water on buildings that remained through the event, there would have been hyper local sources for relatively safe water following the event.

Additionally, envelope designs and their capabilities continue to evolve becoming ever more complex. They are at the same time more water resistant while still affording for the building to exchange air with the outside. These envelopes could supplement their performance by integrating collected rainwater as a cooling system for the envelope as well as to create light shelves or other day lighting controls with or containing the forms of the WDR collection gutter system. Once collected, there are multitudes of ways to reuse the water: irrigation, fire protection, toilet flushing, and heating/cooling systems. Some uses require the filtration and sanitization of the rainwater. The Center for Architecture, Science, and Ecology (CASE) at Rensselaer Polytechnic Institute (RPI) has developed a facade system of modular glass blocks in which greywater flows through channels that are embedded in the surface of the block. By tilting each module for maximum exposure to solar energy, the grey water is purified for reuse via sunlight. Using one natural resource to purify another creates the potential for an unlimited resource, which requires little if any large-scale infrastructure. This hyper local water source could create cities which are much more resilient to the ever changing and intensifying climate in which we live, in the twenty-first Century.

Conclusion

The design for Kaktos is intended to create an integrated method for designers to analyze and test massing studies based on wind driven rainwater collection. Challenging current design standards, we have developed a method and tool to calculate the amount of rainfall that can be collected on all surfaces of a building design. In a typical office building, 90% of the water consumed is for non-potable uses. As the need for water in urban areas predictably increases especially in sensitive areas, over the coming decades, tools like Kaktos will provide the ability for designers to factor rainwater collection in as one of the primary schematic conditions for successful building design.

References

Beorkrem, C., McGregor, M., Polyakov, I., Desimini, N.: Sphere mapping: a method for responsive surface rationalization. Int. J. Archit. Comput. **11**(3), 319–330 (2013)

Blockena, B., Carmelieta, J.: A review of wind-driven rain research in building science. J. Wind Eng. Ind. Aerodyn. 92:1079–1130 (2004)

Hoppestad, S., Slagregn, I.: Norge (in Norwegian), Norwegian Building Research Institute, rapport Nr. 13, Oslo (1955)
Roberts, A.G.: Predicting the future of global water stress. MIT News. MIT Joint Program on the Science and Policy of Global Change. 09 Jan 2014

Redundancy and Resilience in Reciprocal Systems

Joseph Benedetti[1,2(✉)], Pierre André[3], Audrey Aquaronne[3], and Olivier Baverel[4,5]

[1] F-O-R-T Ingénierie, 412 rue Léon Gambetta, 59000 Lille, France
joseph@f-o-r-t.com
[2] Ecole Nationale Supérieure d'Architecture et de Paysage de Lille, 2 rue Verte, 59650 Villeneuve d'Ascq, France
[3] Ecole des Ponts ParisTech, Cité Descartes, 77455 Champs-sur-Marne, France
[4] Laboratoire Navier (UMR CNRS 8205), Ecole des Ponts ParisTech, Cité Descartes, 77455 Champs-sur-Marne, France
[5] Ecole Nationale Supérieure d'Architecture de Grenoble, 60 Avenue de Constantine, 38036 Grenoble, France

Abstract. This work presents some new methods for the comprehension of reciprocal frames and interlocked elements in order to design more resilient systems. First the authors propose a new method and mathematical notation based on graph theory to describe the topology of such interlocking systems. We then give an alternate construction of interlocked element with horizontal contact surface, minimizing thrust, so that any given topology can be applied to a reciprocal system. We choose three different topologies with a level of redundancy, and apply the method in order to design three different interlocked elements and reciprocal systems. At last the elements are 3d-printed, and we analyse the first experimental results and the failure mechanism observed and then build failure graphs that will help analysing the resilience of these interlocking systems.

Keywords: Reciprocal system · Graph theory · 3d printing

Introduction

Reciprocal structural systems—frames or planar vaults—allow us to multiply the span of structures with simply supported small elements. In 1699 Joseph Abeille and Jean Truchet both presented new "planar vault" systems where each voussoir supports 2 adjacent voussoirs and is supported by two other adjacent voussoirs (see Fig. 1).

However, systems like Abeille vault or nexorades as defined in Baverel (2000)—when the voussoirs are replaced by linear elements—cannot allow for the loss of one single element, since two of its neighbours are then only supported by one element and will then be overthrown. Some actual vaults will still stand still due to high friction between voussoirs and prestress, but as a system the Abeille vault lacks resiliency. Such resilient systems could be well used in the construction of rock fall barriers wherever it is not possible to tense nets as in Fig. 2 or put wide embankments as in

K. De Rycke et al., *Humanizing Digital Reality*,
https://doi.org/10.1007/978-981-10-6611-5_51

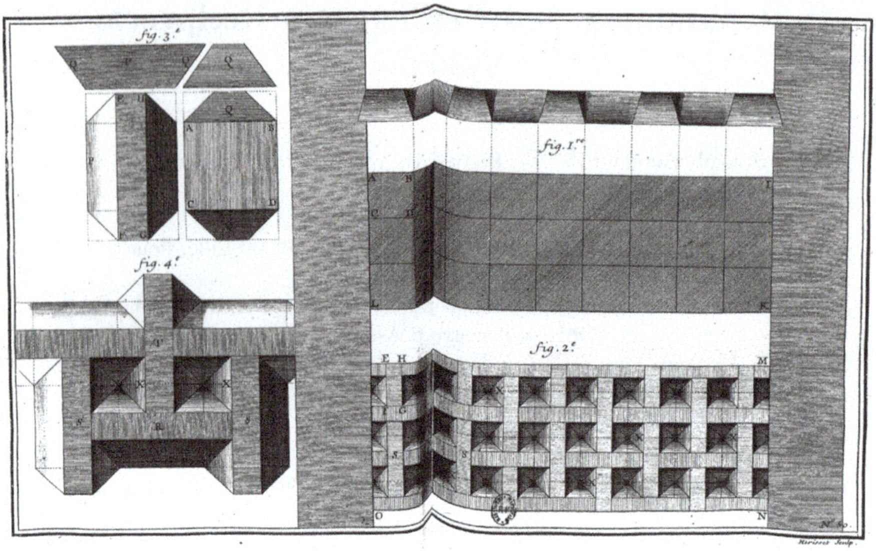

Fig. 1. Machines et inventions approuvées par l'Académie royale des Sciences, tome I, 1735. *Left* is Abeille planar vault and *right* is Truchet planar vault

Fig. 2. Steel nets as rock fall barrier. © Maccaferri

Fig. 3. Embankment used as rock fall energy dissipator. © Maccaferri

Fig. 3. Interlocking elements can be put in place without the need for cement and the energy of the rock fall is dissipated by the multiple cracks in the elements.

Therefore, it is natural to study extensions of these systems with more connections between elements, allowing the system to stand still even after the loss of one, two, or more elements. By doing so, we should expect some kind of pseudo-ductile behavior even with elements made of fragile material, the level of redundancy being connected to the maximum number of allowable cracks in the elements. However, since Abeille vaults or systems as developed by Weizman et al. (1999) or Kanel-Belov et al. (2008) have a hybrid behavior, as the inclined contact surface between elements allows horizontal thrust due to high friction, we need to develop a new strategy for designing the elements geometry. Moreover, the design method based on polyhedra leads to a limited number of systems and could be expanded.

This article follows this idea by first introducing some new mathematical notation based on graph theory in order to generalize the design method, provide tools for classifying the topology of such interlocking systems. It then describes an alternate construction for element assembly that follows a given topology which minimize thrust, by designing only horizontal contact surface between elements. We then choose three different but close topologies and build the three different assemblies, have some first experimental results that show the failure mechanism and then build the failure graphs that can help analyzing the resilience of the interlocking system.

Redundancy: A Topological Approach

Graph Notation

Weizman et al. (1999) and Kanel-Belov et al. (2008) use an arrow notation to help visualizing the orientation in which the vertical planes between elements are tilted and explain the geometric construction of their interlocking systems. However, it is possible to interpret the arrow just as a mathematical symbol representing the order relation between two elements—which one is the support and which one is supported. We develop this notation in this article in which the topology of a reciprocal system is mathematically represented in a unique way by a directed graph that shows how each element is supported by each other.

A directed graph is just a collection of vertices and arrows between two vertices. The directed graph is easily built following these two rules:

- Each interlocked element is a vertex of the graph
- Each support relation between two elements is a directed edge, the arrow going from the supported element to the support element.

We just need to add the last vertices and arrows corresponding to the fixed supports in order to finish the graph. As an illustration, the graph of the historical Abeille Vault is given in Fig. 4.

The black and red dots represent the structural elements while the green dots represent the fixed supports. The central red dot is supported by its upper and lower

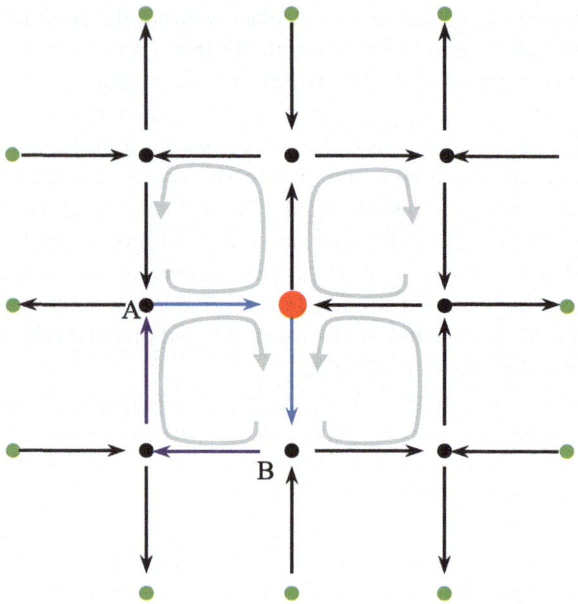

Fig. 4. Graph associated to the Abeille planar vault

neighbors and supports its left and right neighbors. With this representation, the authors propose the following definition:

Definition A *reciprocal system* is a system in which we can flip the direction of each arrow in its graph notation and get another stable structure.

This graph notation allows us to underline some topological properties of a reciprocal system:

Property The directed graph of a reciprocal system is cyclic and more precisely strongly connected.

The cyclic property is highlighted by the grey arrows in Fig. 4: from each (non-terminal) vertex you can loop back to it. The strongly connected property means that for each pair of vertices, there exists at least two paths, one from each vertex to the other: for example, connecting vertices A and B there is the blue path from A to B and the violet path from B to A.

A regular graph is a graph where each vertex is connected to the same number of other vertices: each vertex has the same degree. A regular directed graph must also satisfy the stronger condition that the indegree and outdegree of each vertex are equal to each other.

In Fig. 4 the shortest loop length is 4. If you don't consider the terminal green vertices, it is a regular graph of degree 4 (4 arrows by vertex) and also a regular directed graph of degree 2: 2 arrows coming in and 2 arrows going out each vertex.

Finally, the authors propose a last definition:

Definition The *level of redundancy* of a reciprocal system is the lower bound of the highest number of arrows you can discard for each vertex while retaining a stable structure. Rising the level of redundancy means that you rise the degree of its graph.

This graph theory representation will be further expanded with network theory based on Shai and Preiss work (1999). Represented this way, force equilibrium is essentially the same as Kirchhoff laws in electricity and we will be able to use network theory algorithm which are well documented and efficient to solve the systems.

From Directed Graphs to Interlocking Element Geometry

Redundancy and First Construction

Since we want to build redundant reciprocal systems, we need to rise the degree of its graph. The simplest way to rise the degree is to start with a triangular grid instead of a quadrangular grid: the degree of a vertex in a triangular grid is six instead of four.

The most symmetric way to choose the direction of the arrows is to have alternating outward and inward arrows as in Fig. 5. We call it Type 1.

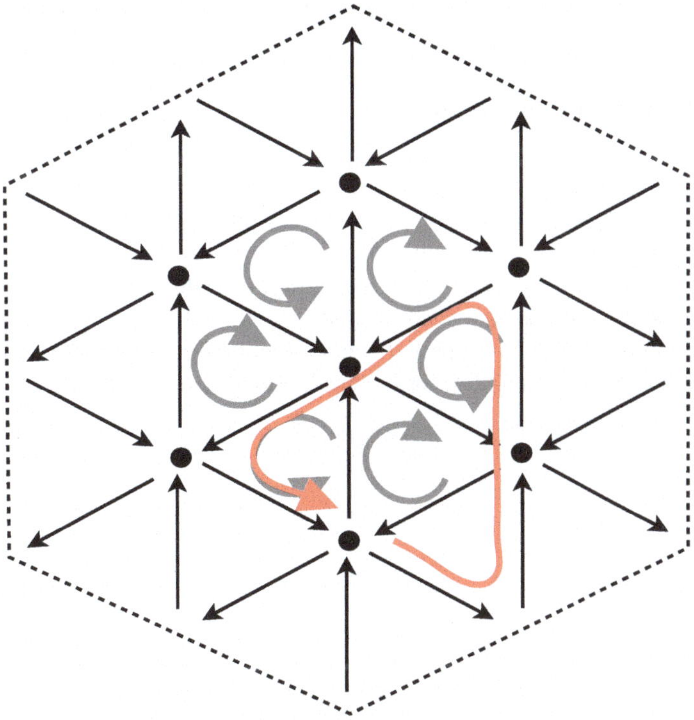

Fig. 5. Type 1: (3,3)-regular directed graph

Each element is then supported by three elements and supports three other elements, forming a (3,3) regular directed graph. The shortest loop length is three, shorter than for the Abeille system, and each vertex is the beginning of six different such loops. Moreover, contrary to the Abeille system, here the arrows that are in the same line all have the same orientation, therefore there are loops of length 6 such as the one highlighted in red, loops of length 9, 12,…: the type 1 graph is strongly connected with a lot of redundancy.

Given this reciprocal graph, we design a compatible element geometry. However, contrary to Weizmann et al. (1999), the elements are built in a way that the least thrust is generated: the supporting faces need to be horizontal instead of tilted so that we can expect a different behavior.

The idea is to start as in Fig. 7 with a hexagonal prism—the hexagon having the same number of edges than the degree of the graph. More generally, for a k-regular graphs, we can start with k-sided prisms. The prism can be thought as having two levels: the lower level which will be the supporting level (intrados) and the upper level which will be the supported level (extrados). We then extend the upper (supported) level following the outward arrows, forming cornices, and extend the lower (supporting) level on the inward arrows, forming corbels. Figure 6 shows the operation for the upper level: a cornice (in yellow) of an element is supported by the corresponding corbel (in grey) of its neighbor.

We can then finish the design of the element by filling the remaining gap as in Fig. 7.

Two Closely Related Constructions

We can follow the same method with two other graphs built on the same triangular grid but with different choices in the direction of arrows.

The first one is given by this other (3,3)-regular directed graph shown in Fig. 8. It can also be thought of an Abeille system (the black arrows) with one more support/supported pair of arrows, which can be left or right-handed: the blue ones are left-handed and the red ones are right-handed.

The last one is a 6-regular graph but is not a regular directed graph: vertices in and out degrees alternate between (2,4) for the blue vertices and (4,2) for the red vertices. It

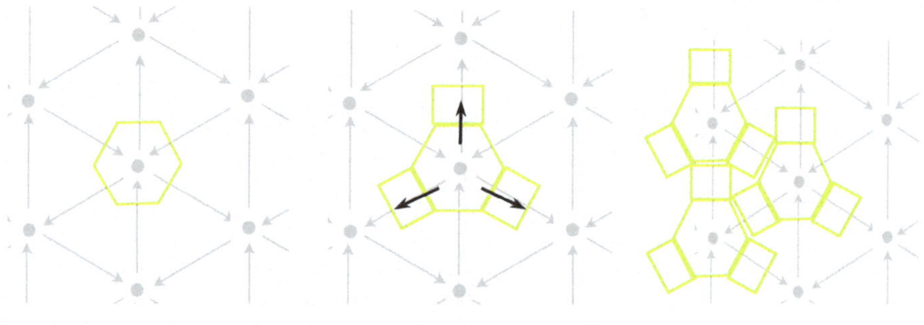

Fig. 6. An element construction

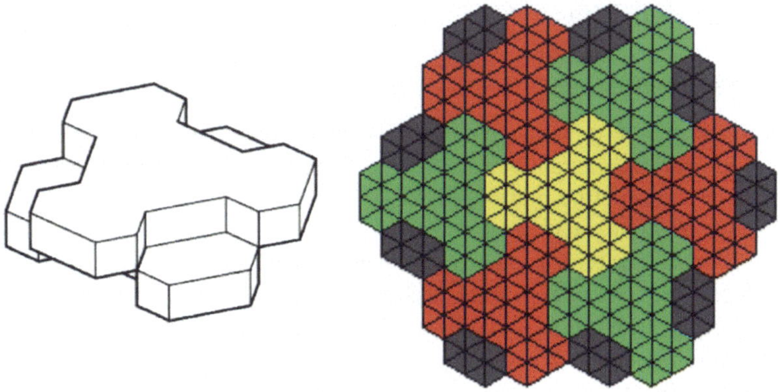

Fig. 7. The finished element and an assembly

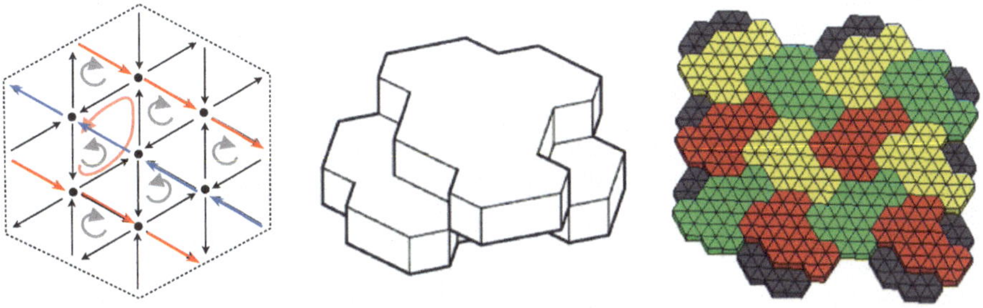

Fig. 8. Type 2: another 3-regular directed graph where the elements are chiral. Each vertex has only 3 different order 3 loops but also loops of order 4

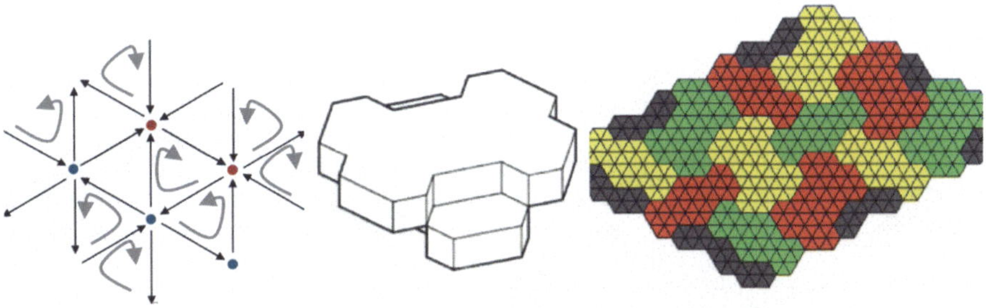

Fig. 9. Type 3: A more complex system. *Blue vertices* are (2,4) and *red vertices* are (4,2). However, there is just one type of elements: it is just flipped upside down

is as Type 2 an expansion of the Abeille type in which the added pair of arrows goes in for the red vertices or out for the blue vertices. See Fig. 9.

These three construction are interesting in that they all have redundant arrows (in and out degrees >2) but in different topological ways. We have thus shown that the notation and method can lead to a wide number of different interlocking systems configurations.

Experimentation Phase

Geometry Definition

After having selected our 3 test candidates we need to perform some experiments. We decided to 3d-print the elements in PLA (PolyLactic Acid), even though PLA is not a fragile material, considering that 3D-printing would help us build the high number of required elements in an easy and automated way.

Figure 10 shows a typical element with its dimensions. The distance between 2 elements is 28 mm. The overall thickness is 12 mm.

It was decided to build a global hexagonal plate with 5 elements on each side so that the overall diameter is 28 cm.

The edge element geometry was adapted in order to fill a hexagonal test frame made by two MDF plates (Medium Density Fiberboard) as can be seen in Fig. 11.

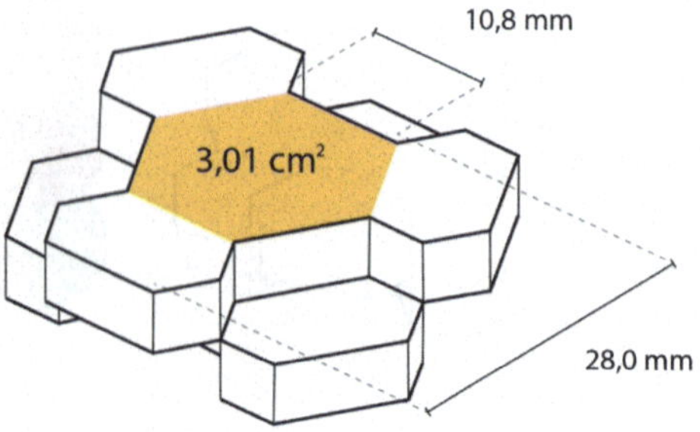

Fig. 10. An element dimensions

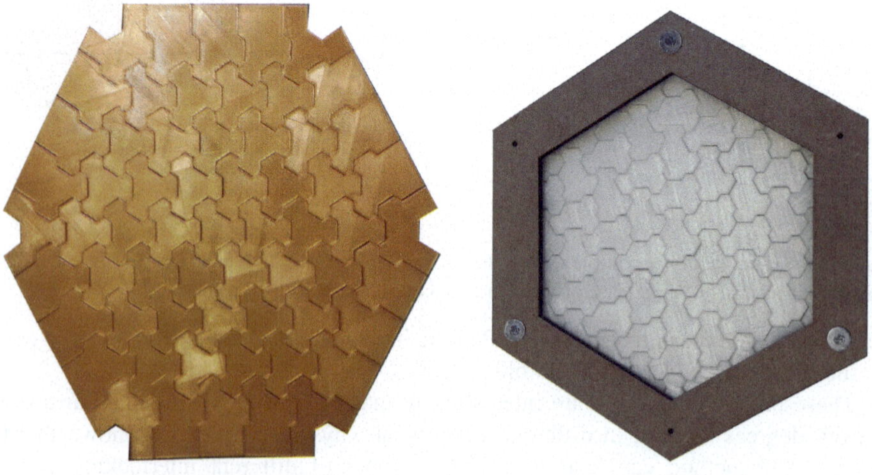

Fig. 11. Type 1 plate (*left*). Type 2 plate in its test frame (*right*)

Tolerance and Play

During the impression phase, we had to manage the relation between the necessary play between elements and the impression tolerance, so that the 3 types were not printed with the same rules, as shown in Fig. 12.

Test Definition

We then conducted a series of displacement-driven tests on a compression machine until overall failure, three for each type.

The load was applied through a hexagonal support to the seven central elements, as can be seen in Fig. 13.

Results

Observations and Qualitative Results

We observed that the failure comes from horizontal cracks, showing shear failure: most of the failing elements are like the one shown in Fig. 14a. This tends to prove that the horizontal support geometry was a good candidate to avoid any thrust. However, since the systems are flexible, and the material is ductile, second order efforts become more and more important. Figure 14b shows also a failed element with a strong plastic deformation.

The systems are not stiff and allow for big deformation. The max. displacement before failure is higher than 20 mm: a deformation of more than 10%! This high level of deformability comes from the fact that the pieces can shift a lot while still staying interlocked: we can see on Fig. 15 that the interval between two pieces increases, but this interval is not a crack and cannot propagate and lead to failure. After the break of a number of pieces—two or three, mostly near the load application—other pieces start to

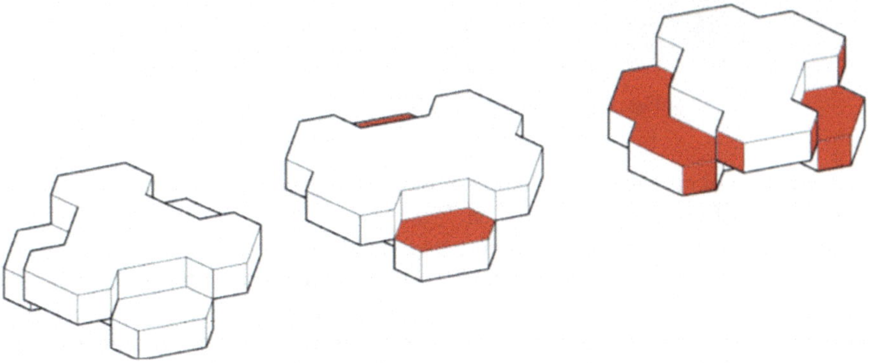

Fig. 12. 0.1 mm planar offset to account for the impression tolerance and allow for more play between the pieces

Fig. 13. Loading system and testing configuration

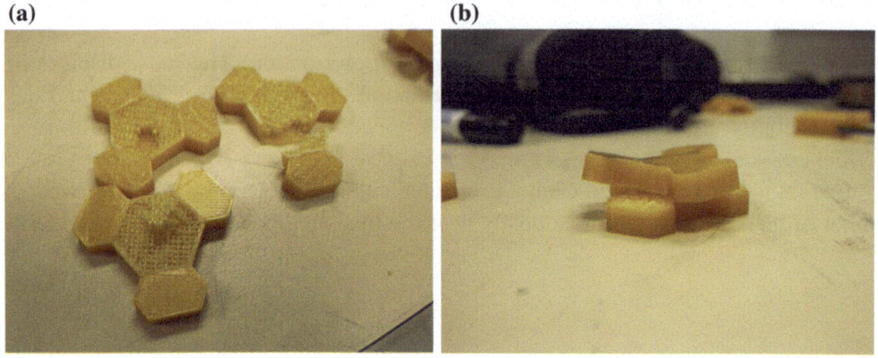

Fig. 14. a Shear crack. b Plastic deformation

slide down, provoking the overall failure of the system (see Fig. 15 which is a selection of video frames from the second test on Type 1).

The pieces broke near the center for type 2 and type 3 systems and the second test of type 1: it shows that the system behavior is close to nexorades (where the maximum shear is near the center) than to continuous plate (where the maximum shear is near the support). For the first test of type 1 experiment, there was not any allowable play between the pieces, the pieces broke near the support (see Fig. 16). Since we used the same pieces for the second test, some plastic deformation made for the allowable play.

In conclusion, as was predicted in the theoretical part of this article, the systems we tested did show some resiliency.

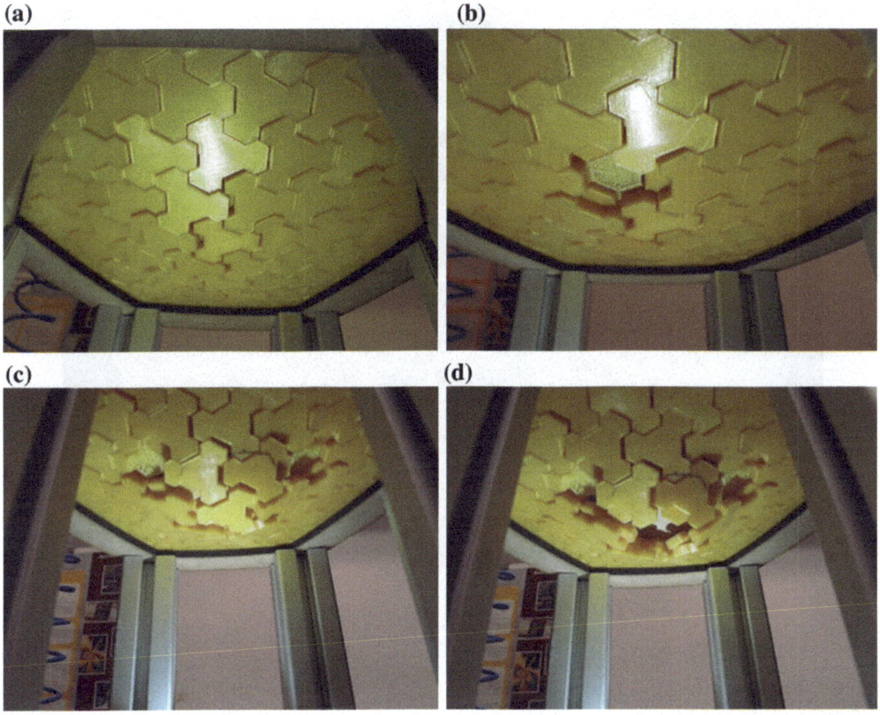

Fig. 15. **a** Before the first break. **b** The first element breaks. **c** A second element breaks. **d** The central element slides down

Quantitative Results

Quantitative results are shown in Fig. 17.

It is remarkable that the interlocked plate can allow for such a huge displacement before failure. Few elements break before overall failure due to too much shift. Since the tests are made consecutively with only the broken pieces changed, the first tests always show a stiffer behavior. It appears that during the first tests some small plastic deformations occur (see Fig. 14), so that—even with type 1 subsequent tests—there are more and more allowable play between the pieces.

Eventually the last test on type 3 shows that the system could withstand even more load even after the first elements broke, proving that our initial expectation that redundancy can lead to resilience.

Exploitation of the Results: Failure Graphs

Given the way an element fails based on the experiment, we can create new directed graphs with one or more failed elements and then analyse the stability of the system. Figure 18 is an example based on Type 1 where the red element breaks as in Fig. 14a. The remaining part is only supported by its neighbours and cannot support any

Fig. 16. Failure during the first test of Type 1

element. The fact that the paths are longer means that the system is more deformable, but for Type 1 the increase is small: the system shows its resiliency.

Conclusion

In this article, we have first shown a new way to represent reciprocal systems using graph and network theory, then build interesting new interlocked geometry of simply supported elements with the use of horizontal supports, conducted some experiments to highlight the failure mechanisms of such systems and represented the failure mechanism also using graph theory so that it will be possible to analyse in further computer simulations and using network theory algorithms.

For planar graphs, Type 1 graph is the most strongly connected and the corresponding reciprocal system has the higher redundancy. It is also the one allowing the highest load before failure, so it seems as if maximising graph connexion and redundancy is a good design goal for reciprocal systems. However further developments including experiments with the use of a more brittle material are still needed in order to experiment failure mechanisms on a larger scale and be able to better distinguish the behaviour of the different types.

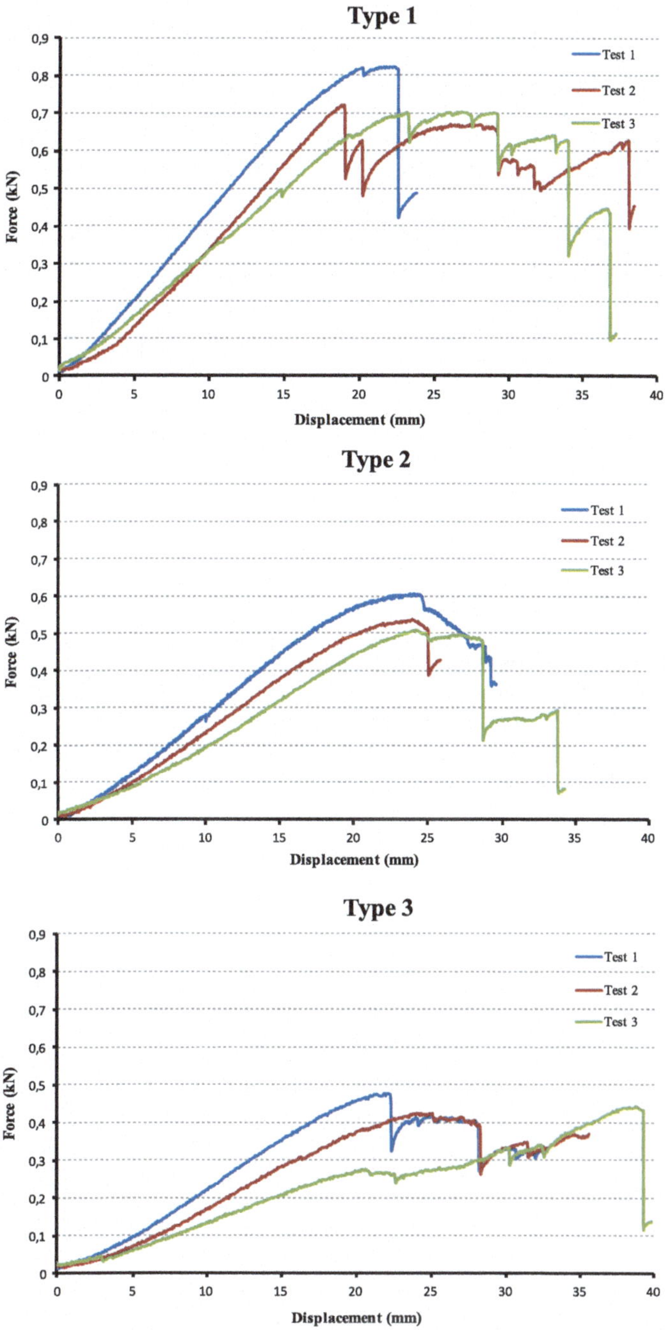

Fig. 17. Force-displacement graphs for each 3 types and tests

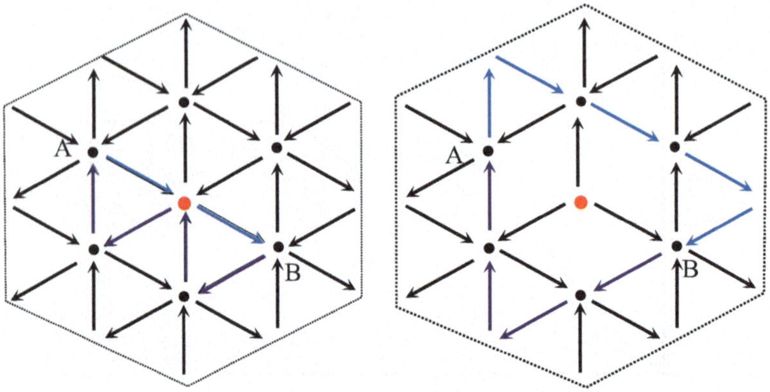

Fig. 18. Failure graph. The *red vertex* is not cyclic anymore, the graph without the *red vertex* is still strongly connected, but the path from A to B is longer

References

Baverel, O., et al.: Nexorades. Int. J. Space Struct. **15**(2), 155–159 (2000)

Kanel-Belov, J., et al.: Interlocking of convex polyhedra: towards a geometric theory of fragmented solids. ArXiv08125089 Math (2008)

Shaia, O., Preiss, K.: Graph theory representations of engineering systems and their embedded knowledge. Artif. Intell. Eng. **13**, 273–285 (1999)

Weizmann, M., et al.: Topological interlocking in buildings: a case for the design and construction of floors. Autom. Constr. **72**(1), 18–25 (1999)

Negotiating Sound Performance and Advanced Manufacturing in Complex Architectural Design Modelling

Pantea Alambeigi[1], Canhui Chen[1(✉)], and Jane Burry[2]

[1] Royal Melbourne Institute of Technology, Melbourne, VIC 3000, Australia
canhui.chen@rmit.edu.au
[2] Swinburne University of Technology, Melbourne, VIC 3122, Australia

Abstract. This paper will explore ways that digital design modelling can support the human facility for reconciling multiple, often conflicting requirements and goals. We detail a project that explores the design application and advanced construction of highly custom architecture for tuning architectural shape and surfaces to manage sound in open work areas. The research focuses on the way in which digital fabrication and advanced manufacturing techniques can support this highly custom approach to the combination of architectural surface, materiality and auditory perception. A set of evaluation criteria is selected by the users and it is organised in a hierarchy to determine the design outcome.

Keywords: Digital modelling · Sound performance · Acoustic simulation · Fabrication and automation

Introduction

Architectural design projects must frequently respond to multiple goals and systems of requirements (Szalapaj 2014). This requires a level of complexity that designers are uniquely equipped to balance.

The research project is the second iteration in response to a brief that calls for a bespoke spatial enclosure in the form of a private meeting space in an open plan office. The first enduring prototype 'room' in this series accomplished excellent precision and finish in the prototype by combining digital fabrication and a high level of craft in the assembly. The aim in this iteration is to (a) increase automation in assembly, migrate some of the craft in construction to a manufacturing paradigm, without losing the extreme variability and organic nature of the design system and (b) introduce new findings regarding room shape and acoustic performance, and an increased number of parameters related to auditory experience. Although a seemingly simple brief, an important provision was for an enclosure open to the main ceiling that nevertheless maintained speech privacy. Two parallel yet distinct digital modelling investigations were developed to maintain material and geometrical complexity and responsiveness to sound performance while fulfilling the functional brief, site conditions and architectural intent: (1) a thorough study of the relationship between architectural geometries and

K. De Rycke et al., *Humanizing Digital Reality*,
https://doi.org/10.1007/978-981-10-6611-5_52

human sound perception through parametric digital modelling and acoustic simulation; and (2) a tectonic system that aims to combine various materials and different types of digital manufacturing technique in order to form the structural logic of the project. The separate but simultaneous establishment of the two systems provides a set of both specific constraints and design possibilities which later informs the development of a holistic parametric modelling system.

General Framework

As illustrated in Fig. 1 the development of the individual systems presents a complex network of interrelated constraints. Through interpretation and evaluation, these parameters are synthesised into a new algorithmic framework that negotiates and transforms these constraints into active design drivers for further exploration.

The parametric model consists of three major sections:

1. Overall geometric exploration: a segmented double skin form that consists of internally and externally concave shells fulfills both the original architectural formal preference, whilst providing significantly improved sound performance. The underlying geometric rules of the form comprise a sphere intersecting method, which not only provides a consistent and gentle curvature for tessellation and planarisation but also creates planar intersections at the segment junctions which

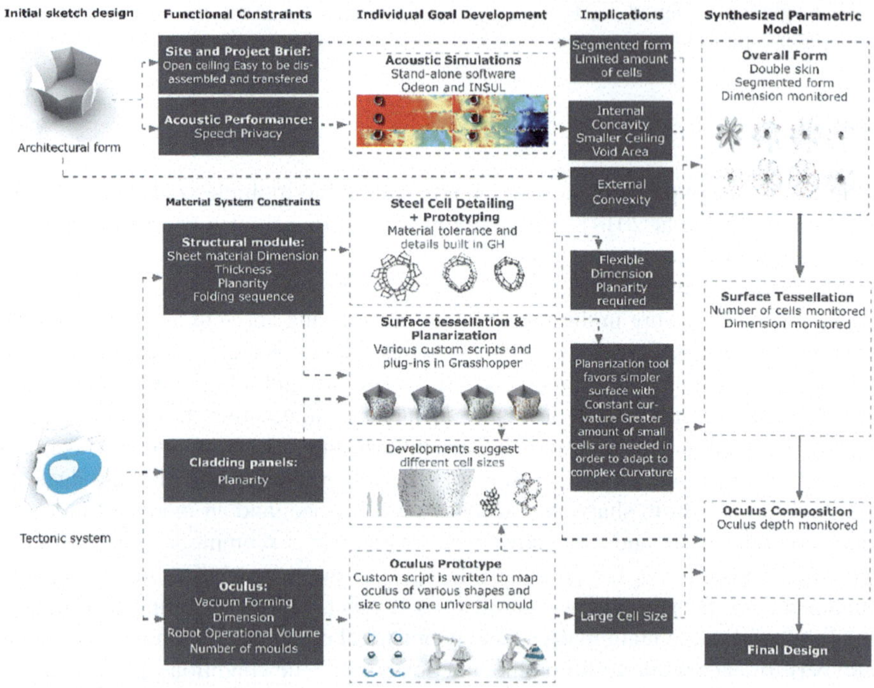

Fig. 1. Orchestrating a synthesised algorithmic framework

reduces the complexity at the seams, allowing for easier assembly and disassembly (see Section "Overall Geometric Exploration").
2. Tectonic system: the materiality, thickness and layer composition were fully studied with acoustic digital modelling to provide an integrated acoustic solution which can immediately respond to human auditory requirements (see Section "Tectonic System").
3. Fabrication system: an interactive mesh wrapping section of the script was developed for the cell pattern tessellation on the overall form. It allows the user to compose the pattern and cell sizes on a flat plane by simply manipulating a few input points. The input changes are instantly reflected on the form. The script is intentionally lightweight to allow designers to have the ability to focus on designing the composition of cells whilst easily maintaining control over the number and size of the cells. The final model combines the composition of the oculi, adds the steel cell module details, door, and other architectural expressions, integrating construction information for various forms of CNC manufacturing (see Section "Fabrication System").

This paper aims to introduce a specific architectural approach to the design development which responds to the design parameters and constraints for controlling human auditory environment through a combination of simulation and parametric digital modelling.

Overall Geometric Exploration

In order to respond optimally to the performance brief, the geometric design is guided by acoustic simulations. A user-experienced centred architectural approach to digital sound modelling was applied. Sound as a design driver has been integrated into the digital modelling and its interactive influence on other design imperatives has been investigated and analysed. A thorough digital modelling study was performed to explore geometry and materiality in two CAD modelling acoustic software, Odeon and Insul, and the data obtained provided reliable ground to control the human auditory environment of the designed meeting space.

Aesthetic and Brief Requirements

The design is an evolution of a previously constructed pod named FabPod at the Design Hub, Royal Melbourne Institute of Technology (RMIT) University in Melbourne, Australia. The brief called for a semi-enclosed meeting space within an open plan office which can accommodate 8 people for a semi-private conversation (Williams et al. 2013). The sound behaviour of the first prototype was measured objectively and subject of subjective experiments to both validate digital simulation and feedback into the new pod design. Although the spiral plan shape of the FabPod was designed to control the sound transmission from the inside to the open plan office, it was found that the entrance was the main source of sound leakage (Alambeigi et al. 2016). Therefore, the new FabPod included a door in the design. The new strategy assisted in providing a

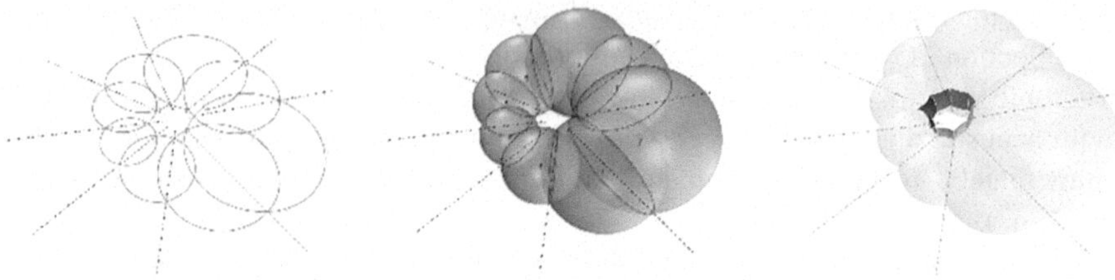

Fig. 2. Intersection based spherical composition

smooth sound propagation with a better auditory environment by preventing the direct sound traveling from the pod into the open plan office. Reflected sound could then be controlled by geometric strategies.

A segmented form composed of intersecting spheres was developed for a number of reasons. The segmentation allows offsite manufacturing and assembly facilitating transportation, onsite installation and post-construction relocation. In addition, spheres can create planar wall intersections to reduce complexity at the seams and allow easier assembly and disassembly.

As illustrated in Fig. 2, a set of lines is set out to determine the location of the wall segments' seams and the ratio of each segment. Based on the intersection information, a set of matching circles and spheres were generated for defining floor layout and wall tilt angles. The radii of the spheres were created to respond to both architectural aesthetic requirements and functional needs such as the minimum curvature, maximum radius and absence of the tilt angle for the door.

Acoustic Requirements

Odeon, the room acoustic software, was adopted to explore the design of the overall geometry and the impact of room's shape on increasing speech privacy of the space.

It was found that the most influential geometry is an internally concave form (Alambeigi et al. 2017a), which produces an external convex shape that is aesthetically undesirable. The parametric digital modelling allowed us to quickly examine various iterations to find a solution that fulfilled both architectural and acoustical requirements (Alambeigi et al. 2017b). A double-skin structure has been developed through parametric modelling to generate a bi-concave overall form which responds to aesthetic and performance needs simultaneously. A 55% increase in speech privacy is suggested by purely adopting a bi-concave overall geometry in the design process (Alambeigi et al. 2017a).

As demonstrated in the Fig. 3 a set of spheres is calculated and generated based on the positions and angles of the intersection planes of the external walls to produce a bi-concave geometry. The diverging interior and exterior surfaces produce an internal gap within the pod's structure. Acoustic simulations suggest that the gap has a positive influence on the acoustic performance of the pod and by trapping the reflected sound from the ceiling can increase the speech privacy in the open plan office (Alambeigi et al. 2017b).

Tectonic System

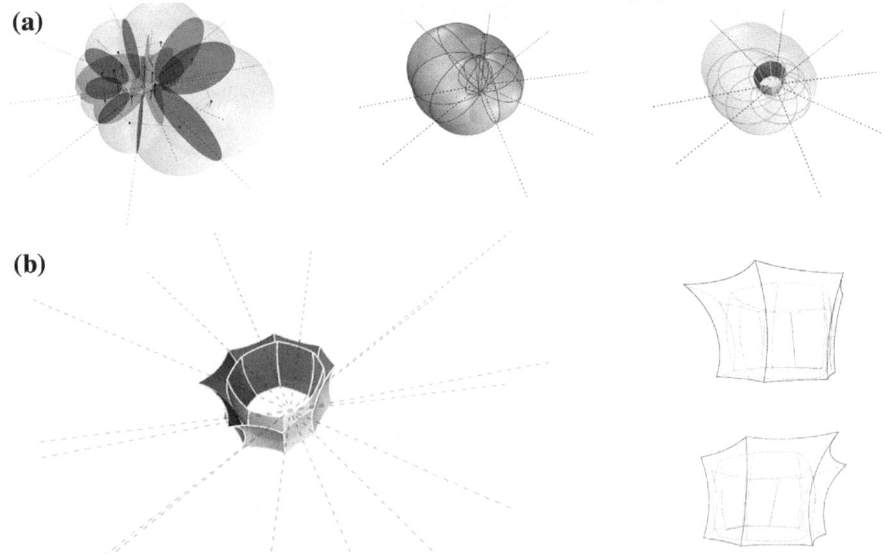

Fig. 3. **a** Intersection planes of the external spheres determine the ratio between the internal spheres. **b** Resulting double skin with coplanar seams

Materiality

For a better understanding and representation of the materiality, thickness and composition, a quick iterative simulation was carried out with INSUL software. The iterations were carefully chosen based on the market and economic constraints. It was found that the most effective material composition comprises two layers of 6 mm Medium Density Fiber (MDF) bolted to the steel structure from each side with the final finishing layers of a 7 mm acoustic felt product, Echopanel, attached to the MDF (Fig. 4).

Figure 5 compares different structures and materials in terms of Weighted Sound Reduction Index (R_W). The R_W is a measure to rate the soundproofing effectiveness of a system or a material. It can be seen from the graphs that Echopanel has the lowest R_W while it has the highest sound absorption coefficient. Therefore, it is suggested to attach the Echopanel to Medium Density Fiber (MDF) which can increase the R_W from 10 to 38. This can enhance the efficiency of the structure considerably. Incorporating an oculus in one module the R_W decreases but only slightly to 34. Given the advantages that the oculus brings to the design (see Section "Oculus"), this marginal reduction in R_W is negligible.

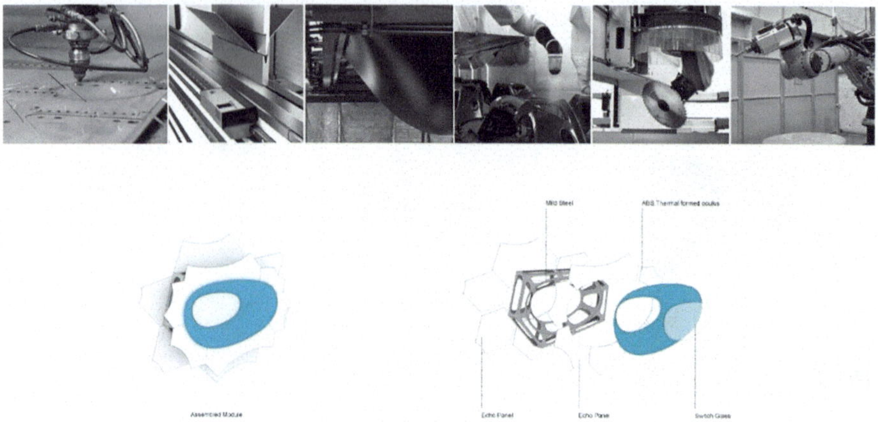

Fig. 4. Proposed materials and tectonic system

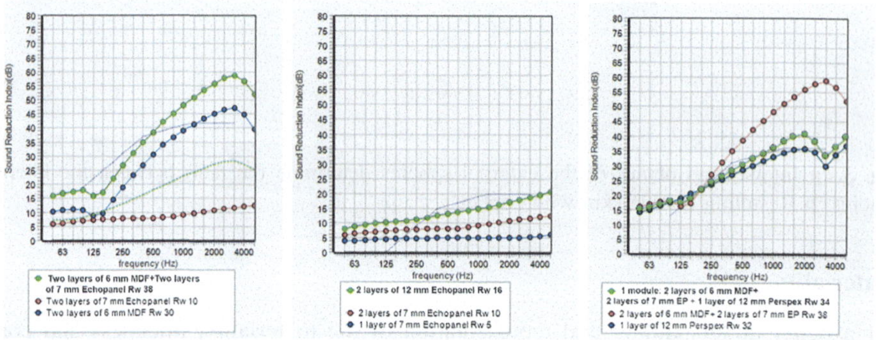

Fig. 5. The INSUL result, comparing sound transmission from different materials and compositions in terms of Rw

Metal Cells

A 3-dimensional form was required for each individual module to ensure a self-supporting stable structure. A foldable steel sheet was selected to provide a geometric volume for each module rather than constructing every single face out of a flat sheet. Moreover, all the construction fixing details can be embedded into the cell structure to increase the construction speed as well as the automation.

The air gap in each and every module acoustically adds an extra layer to increase the R_W and consequently to decrease the sound transmission through the structure (Fig. 6). Furthermore, by cutting out the unnecessary materials and creating hollow metal cells, less sound vibration is transferred from one layer to the other. Similarly, for attaching MDF to Echopanel on each side of the pod a dual lock tape is applied which provides an approximately 6 mm gap between the layers to prevent transfer of the sound vibration from one layer to the other. By integrating acoustic and fabrication

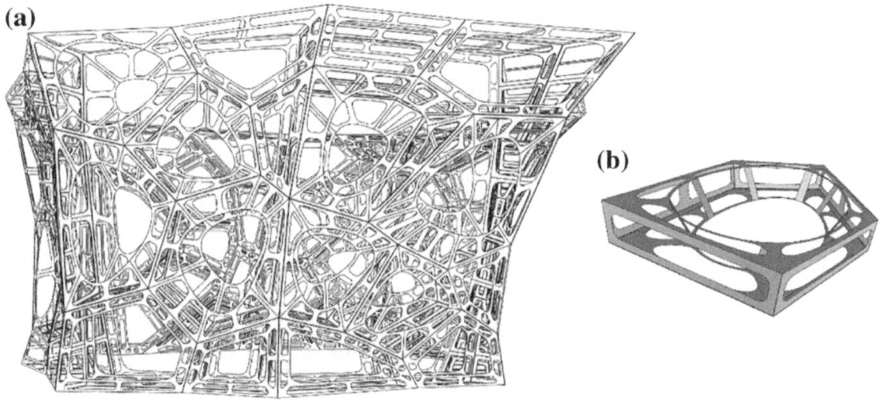

Fig. 6. **a** Steel light structure, **b** hollow single module with at least 100 mm air gap

strategies in designing each module, the structure can respond to multiple goals and requirements at the same time.

Oculus

Architecturally, a space with openings provides a more pleasant environment for occupants. Oculi bring light and visual connection. For a meeting space, oculi may affect occupants' perception of the privacy as well. It is commonly believed in open plan offices that if occupants become aware of their environment, they reduce their vocal effort to secure more speech privacy. Although this might be true for open layouts, the subjective experiments on the first prototype of this semi-enclosed meeting pod suggested a reverse trend. That is to say, with a less visual connection to the outside, occupants reduced their vocal effort. Hypothetically, the reason behind this unconscious phenomenon is the human spatial impression. In a less visually private space, occupants' unconscious perception of sound propagation makes them increase their vocal effort for sound loss compensation. Further details on this subjective experiment are beyond the scope of this paper. The design solution for providing light, visual connection and a comfortable human environment yet inhibiting unconscious vocal effort increase is to apply the switch glass in oculi and to place the oculi above the human sitting height. This allows the users to turn the oculi opaque when they have a more private conversation by switching on the lights and leave it transparent when they need to have a visual connection.

Fabrication System

5-1-Geometric Tessellation

For tessellating the geometry, a Voronoi-based discretisation strategy is applied, similar to the first prototype. The base pattern is generated through a series of points on the XY

Fig. 7. Pattern generation and mapping

plane within a grid which approximates the area of the pod surface. This methodology was adopted to allow the designer to iterate various design options through simple manipulation of points on a flat plane. This variation can be adapted to different contexts in future. More elaborations on the iterations and variations are well beyond the scope and limit of this paper.

The generated pattern is then remapped on internal and external pod surfaces that are linked to define each three-dimensional cell through a particle-spring network and is optimized by using the physics-based simulation plugin, Kangaroo II, developed by Daniel Piker in grasshopper (Fig. 7). The generated pattern is further detailed to design, fabricate and assemble the pod's entrance. The leading edge of the surface adjacent to the door is altered to become perpendicular to the ground for providing a hinging edge for the door swing (Fig. 8).

Construction and Folding Methods

Efficiency and automation in cell construction are one of the primary aims of this project. This is achievable by various means. Each three-dimensional cell is formed from a folded piece of metal, which creates the internal and external faces of the pod. The modularity and foldable cells significantly reduce assembly and construction time. Cells' clustering and assembly were tested and subsequently, the folding configuration was iteratively explored and prototyped. Although folding strategy helps in creating each cell out of one single sheet, the folding sequence needs to be considered carefully.

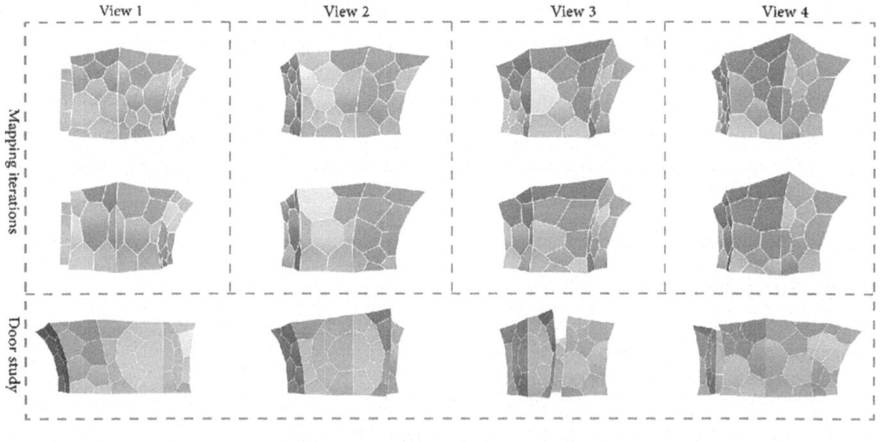

Fig. 8. Example of pattern iterations and door study

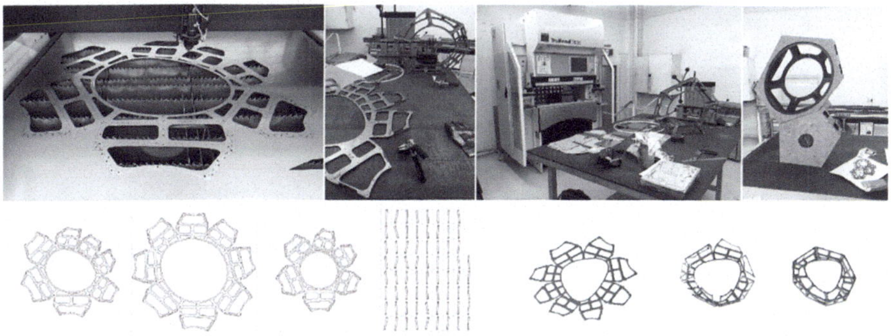

Fig. 9. Steel cell prototype

The angle of each bend should be engraved on each corresponding cell face (Fig. 9), which is then shaped with a CNC folding machine.

Oculi Fabrication

Materiality and fabrication strategy. The fabrication process of the oculi is developed in response to the dynamic and varied module tessellation considering fabrication and cost efficiency. Conical surfaces are selected as the base geometry for the oculi since they satisfy the required variability in spatial depth while remaining developable geometry. A series of developable surfaces are produced with Acrylonitrile Butadiene Styrene (ABS) plastic thermoforming to use as the base for the oculi that can be later trimmed into the desired shape. Since adopting multiple moulds for creating a wide range of oculi, different in size and depth, is economically inefficient, an algorithm is developed to exploit the flexibility of the ABS.

With a given conical surface, the sum of the unrolled ruling lines' angles is constant and predictable (Fig. 10). The data obtained can be applied as a parameter in the custom algorithm to generate various oculi sizes from one basic cone. It also allows the designer to shape the cones' outlines in the late stages of the design since for all cones only one thermoformed mould is adopted. The direction of the cone can also be manipulated and is described by a hypothesized axis in the algorithm.

Fig. 10. Conical surface study

Fig. 11. Oculus form finding through angle matching

One axis is assigned to a cone outline and a set of ruling lines are connected between the curve and a point on the axis. This point is moved iteratively along the axis until the sum of the ruling lines' angles reaches the desired number that is 180° in this project to ensure a 60° cone. As a result, all the oculi generated with this method have the same degree but different in size and depth (Fig. 11).

Prototyping and design possibilities. The base surface of the 60-degree cone is 700 mm high with 1000 mm diameter that is determined from the maximum operational volume of the robotic arms. Fabrication information such as the trimming axis is extracted from the ruling lines of the oculi, mapped back and translated into machining code (Fig. 12a). A prototype fabrication process is illustrated in Fig. 12b and confirms the accuracy of the algorithm.

The advantages of adopting this method are not only the considerable reduction in the fabrication cost by moulding many cones from a single mould, but it can also bring a wide range of design opportunities for designers. As part of the design system, fabrication information and constraints are integrated into the digital modelling that allows the designers to have a real-time feedback and to adjust the design to a holistic solution (Figs. 13 and 14).

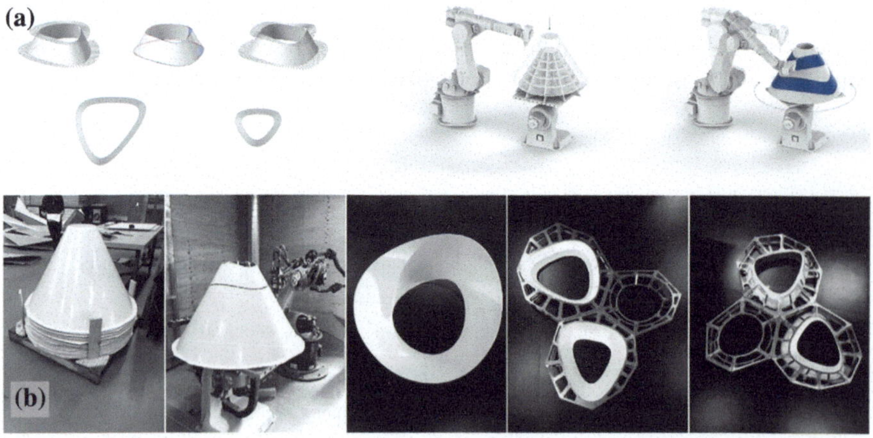

Fig. 12. a Obtain fabrication information from oculus. b Oculi prototypes

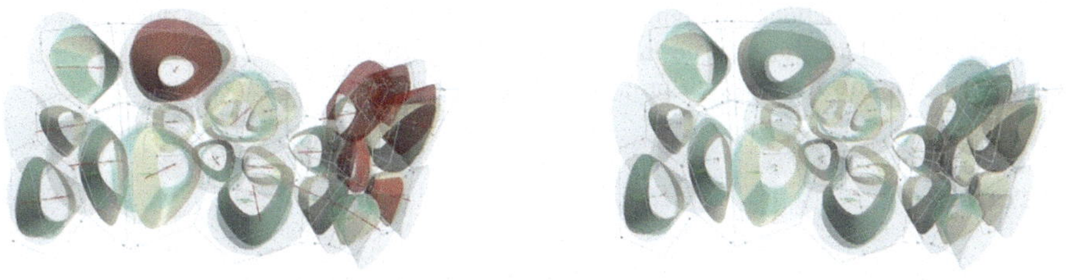

Fig. 13. Fabrication constraints embedded in design model: *Red* oculi are outside of material boundary; this can be rectified by adjusting the axis to ensure all the oculi can be fabricated (*green*)

Fig. 14. Design outcome and fabrication layout is monitored

Conclusion

The aim of the model design is to allow designers to shift their focus from technical performance requirements back to 'design' which is sometimes hindered by the proscriptive aspects of a project.

Functional and Fabrication design constraints can be recognised by digital modelling and simulations and turned into a design opportunity by integrating the solutions at early stages of the design.

Nesting multiple complex systems according to the requirements of a functional brief, as discussed in this paper, through the analysis of an architectural meeting space, highlights that human control of digital design modelling can provide designers with the tools that embrace and amplify their own particular facility to balance complexity and respond to multiple goals and system requirements. By utilising a synthesised parametric system, the architect is able to assert architectural expression in the design with technical brief criteria integrated into the established workflow. The increased human interaction in the modelling system leads to increased automation and predictability in the construction. Ready-made solutions can be avoided by supporting the design process with appropriate digital tools and simulations.

Acknowledgements. The project received funding from RMIT Property Services and is supported by the Australian Research Council (ARC) grant through 'The Sound of Space: Architecture for improved auditory performance in the age of digital manufacturing' project. The authors acknowledge Nicholas Williams' design input to the FabPod II project during his time at RMIT University, Mark Burry as co-investigator in the ARC Linkage Project and Haworth as Partner Organisation in the research project.

References

Alambeigi, P., Burry, J., Cheng, E.: Investigating the effects of the geometry on speech privacy of semi-enclosed meeting spaces. In: Symposium on Simulation for Architecture and Urban Design, Proceedings, pp. 9–16. Toronto (2017a)

Alambeigi, P., Chen, C., Burry, J.: Shape the Design with Sound Performance Prediction; A Case Study for Exploring the Impact of Early Sound Performance Prediction on Architectural Design. In: CAAD Futures, Proceedings, Future Trajectories of Computation in Design, Istanbul (2017b)

Alambeigi, P., Zhao, S., Burry, J., Qui, X.: Complex human auditory perception and simulated sound performance prediction. In: Conference on Computer-Aided Architectural Design Research in Asia (CAADRIA), Proceedings, pp. 631–40. Melbourne (2016)

Szalapaj, P.: Contemporary Architecture and the Digital Design Process. Routledge, New York (2014)

Williams, N., Davis, D., Peters, B., De Leon, A. P., Burry, J., Burry, M.: FABPOD.: an open design-to-fabrication system. In: Open Systems, Proceedings of the 18th International Conference on Computer-Aided Architectural Design Research in Asia, pp. 251–260. Hong Kong (2013)

Pneu and Shell: Constructing Free-Form Concrete Shells with a Pneumatic Formwork

Maxie Schneider[1,2(✉)]

[1] University of Arts Berlin, Berlin, Germany
maxieschneider@udk-berlin.de
[2] Springer Heidelberg, Tiergartenstr. 17, 69121 Heidelberg, Germany

Abstract. This paper presents a computationally driven form-finding methodology for the creation of surface-active structural systems. The case study outlines a feasible construction technique that uses fabric formwork and textile reinforcement as an innovative integrated solution for the fabrication of free-form concrete shells. Extensive developments in mineral and polymer materials such as ultra-high performance concrete (UHPC) and carbon fiber reinforcement may facilitate larger spans of composite shells in future. The research focuses on the aesthetic and structural potentials of the transforming from tension based inflatable structures into compression based thin shells. Pneumatic formwork allows for large span structures by a minimal material use and is less laborious due to prefabrication and on site casting, hereby offering critical advantages to conventional casting procedures. The presented physical prototype demonstrates that the shape of an inflated formwork can be manipulated in order to improve the aesthetical and structural behavior of the casted shell. With the use of a network of restraining ropes a ribbed shell can be built in order to utilize the shell for freeform geometry and prevent buckling. This method of a pneumatic restraint cast (PneuReCa) is effective for fabricating complex geometries at low cost while taking advantage of a digitally-driven, topologically-optimized design process.

Keywords: Pneumatic structures · Textile form-work · Thin shells · Carbon fiber textile reinforcement

Introduction

Free-form concrete shell structures are equilibrium shapes: their shape balances loads through membrane stresses. Graphic statics and physical models such as those introduced by Heinz Isler and Frei Otto since the 1950s were a key to controlling the force flow in order to reach out to a structurally efficient design (Otto 1983). Nowadays computer aided design (CAD) replicates these analogue form-finding methods and is instrumental in optimizing material and economic efficiency (Figs. 1 and 2).

A variety of advanced theories of static optimization for thin shells have been introduced in the past. The ribbed concrete structures of Pier Luigi Nervi, the translation of stress trajectories into the rebar pattern of Eduardo Torroja or the unequotial

K. De Rycke et al., *Humanizing Digital Reality*,
https://doi.org/10.1007/978-981-10-6611-5_53

Fig. 1. Inflatocookbook, Ant Farm, 1971

Fig. 2. Binishell, State Library of NSW, 1974

physics of Heinz Isler's suspended models all show that the structural approach can be of morphological, material or physical nature.

However, as shells are constructed using timber formwork, they are typically considered complex, time-consuming, laborious and therefore financially unattractive. Another method of lightweight construction appeared in the beginning of the 1960s, which has since been enhanced by the development of synthetic fibers and PVC-coated fabric entering the marked. Inflatable structures use an overpressure of air to rigidize membranes, a principle that is suitable for erecting desirable wide span concrete shell structures via a fabric formwork. The potentials of fabric formwork in general can be seen in various technologies that show their advantages in terms of aesthetics and efficiency (West 2016).

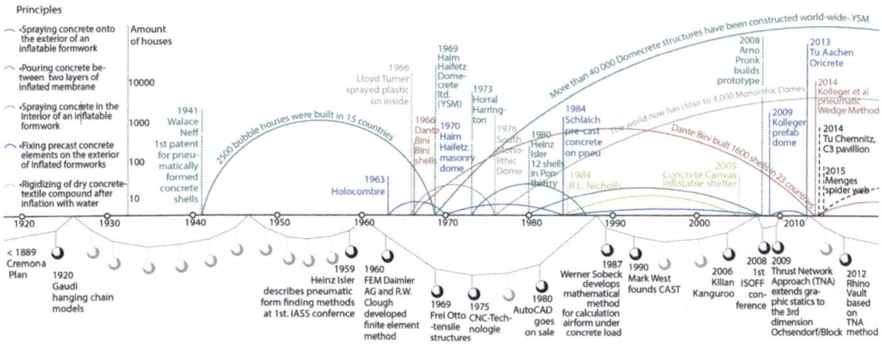

Fig. 3. A century of pneumatic formwork for concrete shells. Chronogram made by the author

Among them, the *pneuform* is considered to be a feasible technique for casting concrete shells. From the early developments of pneumatic formwork-used first as drainage cast, later employed by Wallace Neff for his bubble houses in the 1940s and recently advanced by Benjamin Kollegger et al. at the TU Vienna–pneumatically formed concrete shells have been developed using various methods in order to reduce construction time and costs of concrete shells (Kromoser and Kollegger 2015) (Fig. 3).

With over 70,000 buildings worldwide, pneumatic construction has already proven its economic relevance (Van Hennik and Houtman 2008). Indifferent to their construction method, build examples have shown significant disadvantages, such as critical placement of openings and aesthetic flaws. To date, only regular domes with a synclastic shape have been built with inflatable formwork. It is clear that designers are lacking a framework for embedding pneumatic design methods in their workflow. In fact, it is one of the most promising technique when it comes to fabric formwork, as curvature and a smooth surface seem to come so effortlessly (Sobek (1987).

Research Objective

By studying the history of tectonics of different pneumatic casting methods for concrete shells, it becomes quite clear that the economic advantages are very high. Concrete shells can be erected very quickly and in many cases the formwork can be reused. This research project aims to develop a prefabricated, textile formwork system, which will be easy to transport and handle, include reinforcement, be tensioned on site and can be reused after casting. It is suggested that the usage of pneumatic formwork would create new opportunities to apply ornamentation, geometry and texture onto large surface structures by harnessing the geometric potential of CAD/CAM technologies (Bechthold 2008).

The workflow presented here, which spans between the computational and the material realm, strives for an integrated method that allows for an accurate placement of material, allows design of free form geometry, integrated openings in relation to surface and stress flows, structural optimization and will therefore use minimal resources. Carbon fiber textile reinforcement, in combination with a reusable formwork will reduce waste and increase the economy of concrete structures by creating more efficient and lighter cross sections. Alongside the challenging structural questions that need to be addressed, the objective is to enable a design method in which the architect is in control of authorship. Given the character of the employed composite materials, the design scale extends down to the constructing material.

Methodology

The methodology that is followed here is induced by the detailed inspection of the Bini system, specifically the local surface morphologies of a Binishell shortly after construction (Fig. 2). The cast shell reveals an enormously woven ornament on the inside as a result of the reinforcement wires restraining an inflated formwork (Bini 2014). This observation stimulates the idea that the ribbed surface is both the physical representation of the building process and the shell's structural optimization. Furthermore, it creates a dialogue between the employed construction materials, between formwork and form, and between textile and concrete. Translating this observation into a morphological and topological method means to create an actual structural system that can be described on three scales:

- Global equilibrium geometry (suspended model)
- Adaptive morphological optimization (restrained formwork creates a ribbed shell)
- Material structure (UHPC and carbon fiber reinforcement)

The form-finding process is based on three major steps:

1. modelling of the desired geometry
2. optimization of force distribution
3. transformation from force pattern to structural elements

The suspended target geometry of the shell is structurally analyzed according to its force flow then restraints are placed onto a pneumatic formwork according to the force pattern. As the casted material fills up the constricted area, a rib is created which adds structural fitness to the shell in order to obtain larger resistance against buckling (Fig. 4). These states of modelling compose an interactive workflow in which the designed spatial geometry drives the optimization and fabrication processes (Fig. 5). Testing these hybrid tectonics and aesthetic requires building a physical prototype.

Applied Computational Workflow

A multi-step digital workflow is required to convert tension-based inflatable structures into compression-based thin shells. The computational form-finding switches between different states of modelling that follow different physical principles (Fig. 6).

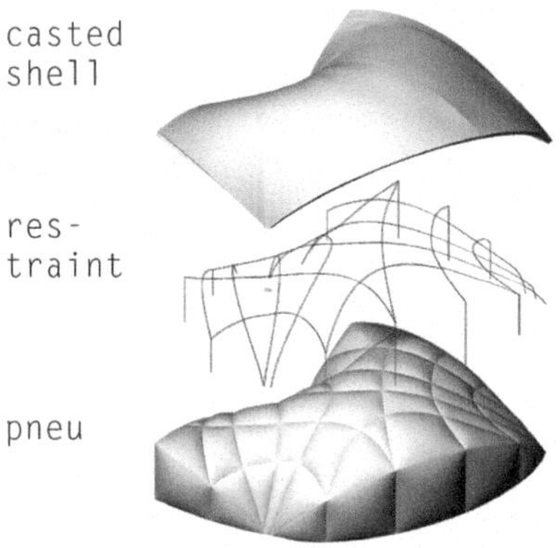

Fig. 4. Modelling principle

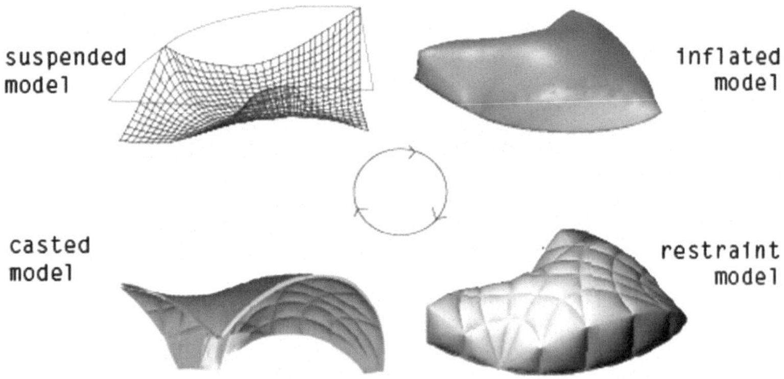

Fig. 5. Feedback loop between the software-workflow

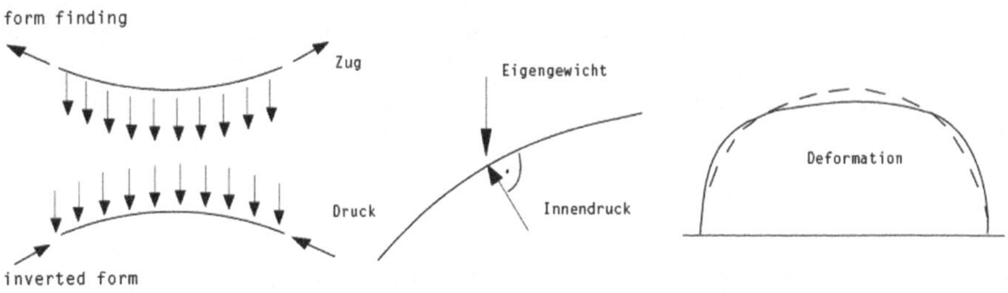

Fig. 6. Suspended principle, inflated principle, loaded principle

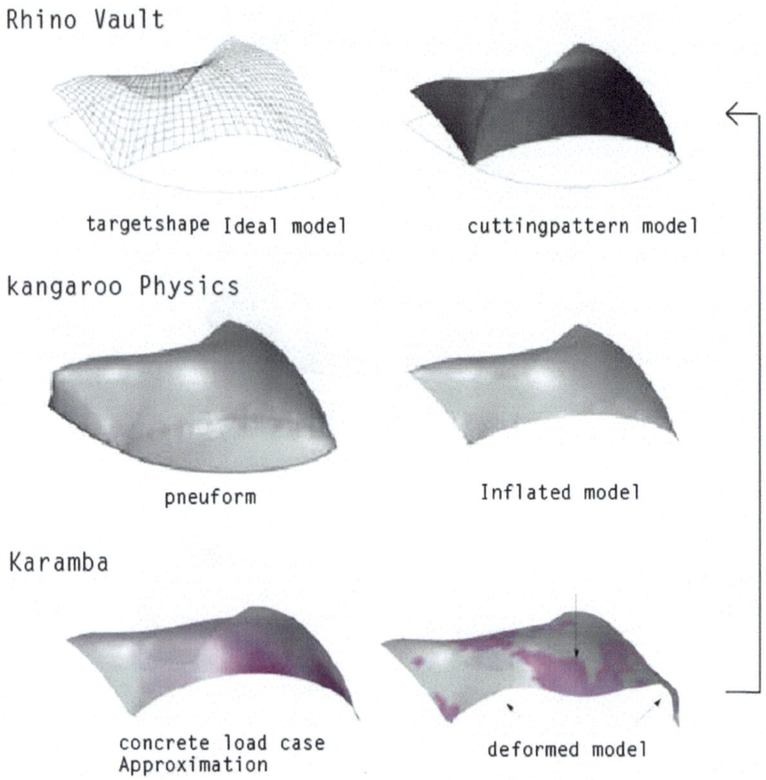

Fig. 7. Digital chain, feedback loops between the models

- The suspended model, representing the ideal design
- The inflated model, representing the casting technique
- The loaded model, representing the shell

Structural optimization, with finite element analyses as a tool to inform the logic of the systems internal organization, enables control over the structural behavior, the material properties and the aesthetic appearance. Computation works as a communicative tool between the physical and the digital model, establishing a dialogue between the material assembly and a digital fabrication (Fig. 9).

Suspended Model

Since suspended surfaces have only compressive strength, it is the easiest way to balance the structure and spread loads evenly to the support points. By using the powerful plug-in Rhino Vault for Rhinoceros, an equilibrium shell was designed that approaches free curvature, large span, integrated openings, different shapes of foot points and even cantilevering parts (Rippmann et al. 2012). This is described as the target geometry. The shell refers to an analytical shape designed to show the rich

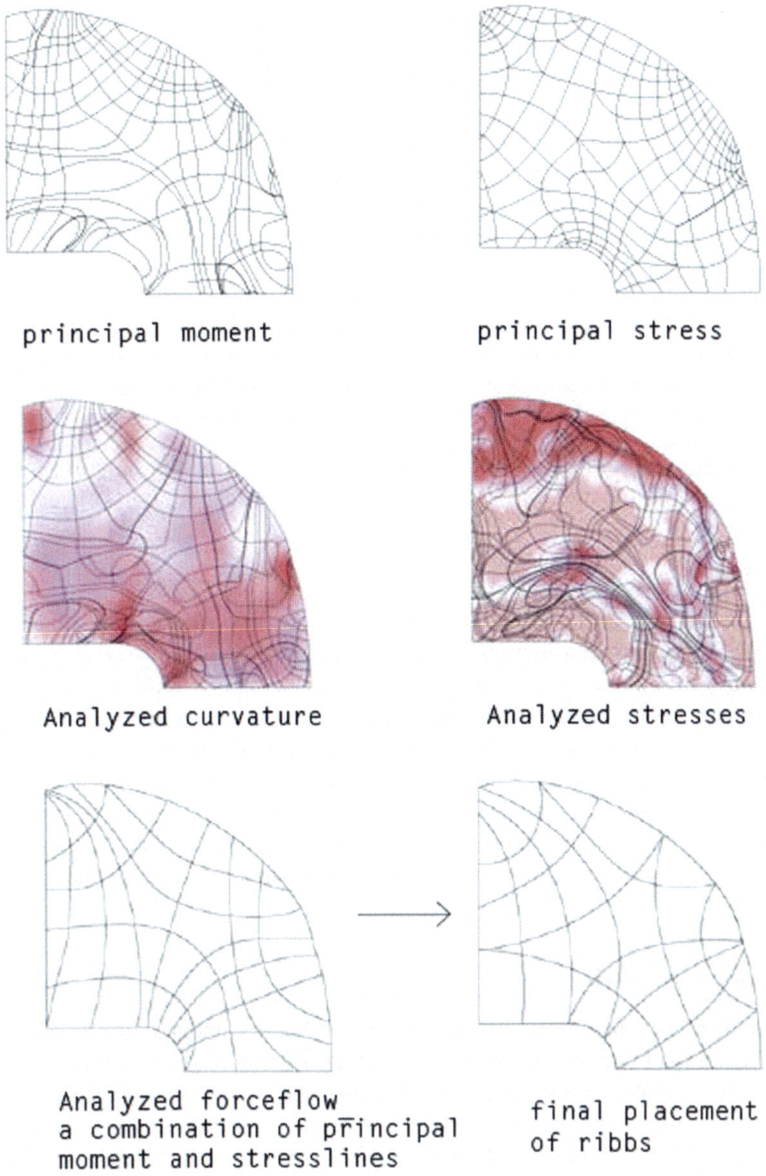

Fig. 8. Force flow integration for fabrication

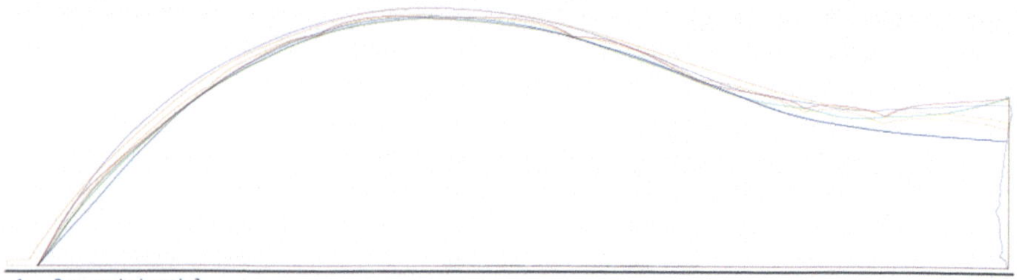

Fig. 9. Section of stares of modelling, deformation diagram

repertoire of possible morphologies. As seen in Isler's shells, the hanging models are particularly suitable for buildings with free edges and a few supports (Isler 1961).

Inflated Model

By using the particle spring simulation software Kangaroo physics—a plug-in for the 3D Software Rhinoceros/Grasshopper—it is possible to simulate physical behavior like deformation from inflation (Piker 2013). Since it is the natural tendency of every pneu to push spherically outwards, it is required to optimize the geometry to the inflated target surface. Simulation must be made with an inextensible material under the required pressure. Wrinkles are an indicator for uneven shape, appearing on parts where it will not be cast is considered to be unproblematic. In the second iteration the constraint model can be achieved by projecting the force pattern back to the surface and adapting the mesh. Placing anchor points along the vertices hinders the membrane from moving outwards.

Finite Element Analysis

If shells fail, they either crack under bending or buckle under compression. In thin shells the dead load is comparatively low, so that the live load (e.g. wind or snow), if not distributed evenly, can produce considerable second stresses (Fig. 7). These circumstances require a stiffening network of ribs that is consequently placed according to the force flow. The FE plug-in Karamba describes principal stress lines and principal moment lines as pairs of orthogonal curves that indicate trajectories of internal forces. Due to the fact that not only the stress lines play a role on structural shell optimization, but also the occurring bending, it was the objective to formulate a strategy to integrate both principal stress and principal moment patterns into one pattern. As a combination of these two loadcases it is optimized to sustain buckling in regions of lower curvature and breaking cases resulting from high stresses in the regions of foot points. Additionally, the resulting pattern was simplified and optimized for fabrication (Fig. 8).

Advantageous of stress line interpretation is the idealized paths of material continuity. The usage of stress trajectories barely found application jet, as a lack of standardization and parameterization of the process for generating and interpreting stress lines can be seen. Here further work is needed to provide a ruleset for implementation in design strategies (Tam and Mueller 2015). Common FE programs normally do not accommodate different thicknesses in one shell. That means that two different systems need to be modelled in order to gain information about the final deflections: one part with a solid minimal thickness and another trustpath-like discrete structure for the ribs. This ribbed shell structurally is a hybrid between an isotropic shell and a discrete structure and can eventually be described as discrete shell (Fig. 9).

Pneumatic Restraint Cast (PneuReCa)

Prototype

The prototype is not only a physical verification of the presented system but also an iterative and physical investigative tool to extract further information on materialization and assembly that cannot be gained out of the digital model. A translation of the idea of the discrete shell into a fabrication process for pneumatic formworks is explained in following (Fig. 10).

The realized structure spans a length of 2.5 m, covers an area of 4.5 m^2 and weighs 300 kg. The Thickness of the shell lies by about 3 cm (in the area of the ribs 6 cm). The demonstrator was manufactured in situ in the following steps:

1. Pieces of TPU-coated nylon were cut according to the pattern of the pneuform.
2. The pneu was sewed with a double sewn folded seam and welded afterwards.
3. The prefabricated formwork was placed and attached to a platform.
4. The ropes were placed and connected with ring nuts to the pneuform and were tied to the platform.
5. Two layers of carbon fiber reinforcement spaced with distance holders got attached to the knots at every crossing.
6. White UHPC was mixed to a clay-like state and laminated on to the exterior of the formwork.
7. After two days of hardening by constant air pressure the formwork could be removed.

Pneuform

The pneuform consist of an inflated membrane and exterior restraining ropes that are directly fixed to it via ring nuts. The inflated formwork is designed on three scales.

- On a fabric (micro) scale, parameters for the textile needs to fit the requirements of the structure, i.e., it is airtight, strong in tension and alkali resistant. A 170 g/m^2 TPU-coated nylon with a breaking strength of 600 N/50 mm has been used.
- The cutting pattern (meso) is unrolled according to the designed force pattern of the resulting shell (Fig. 13).
- In the inflated state the overall geometry (macro) needs to fit with the target shape.

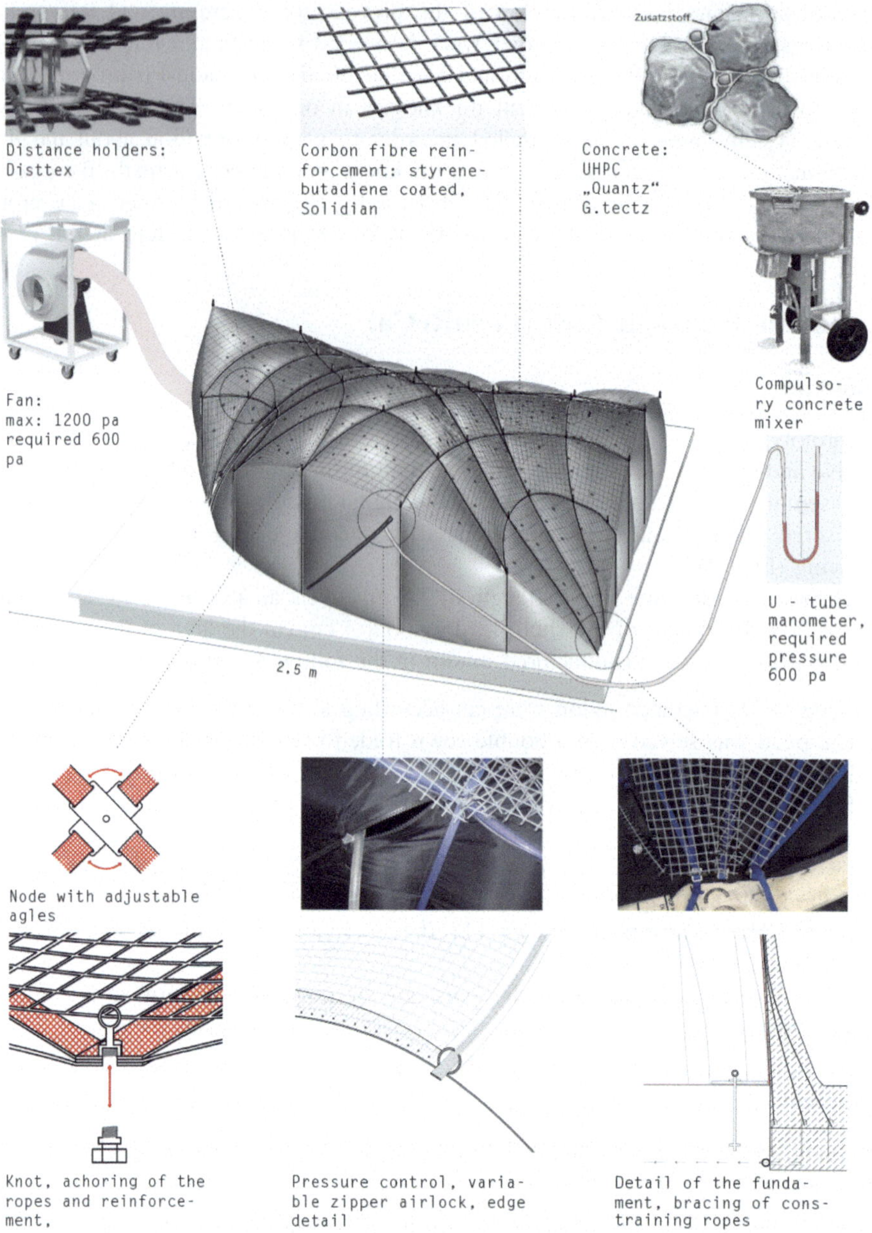

Fig. 10. Physical setup of fabrication

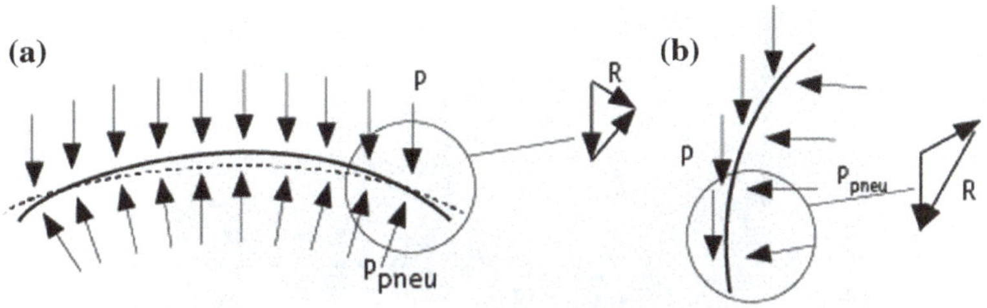

Fig. 11. Resultants from load (**a**) slanting (**b**)

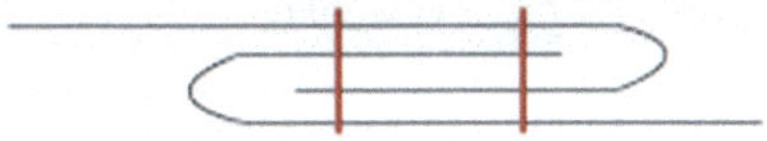

Fig. 12. Double-sewn, folded seam

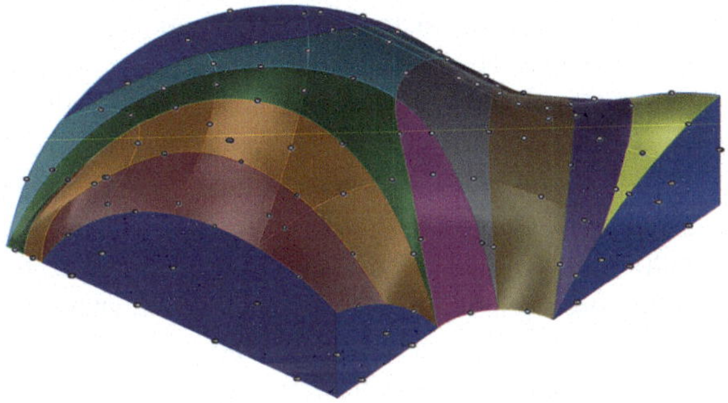

Fig. 13. Cutting pattern, sewing plan

The tensile force in the membrane is directly dependent on the internal pressure of the formwork and the radius of curvature (Fig. 11). The resulting shell will impart a weight of 50 kg/m^2 onto the inflated membrane, which needs to be compensated by an inner overpressure of at least 500 pa. A radial fan with a capacity of 1200 pa has been chosen for safety reasons. The pressure can be regulated with an airlock additionally providing an entrance for fixing and striping. The structure is only as strong as the weakest link, i.e., the yarn. Its strength and type has to be chosen according to the tension in the membrane (Fig. 12). A double-sewn folded seam is a strong connection that becomes airtight by welding it after sewing (Ermolov 1979). A net of ropes is spanned onto the inflated membrane, effecting high strength and can be adjusted. The ropes are placed exactly above the seams, so that the seam is protected from concreting (Fig. 14). Adding local curvature will transmit membrane stresses onto the cables, which can cause local wrinkles. No holes should be located where the concrete is applied since released air can cause damage to the fluid concrete (Fig. 13).

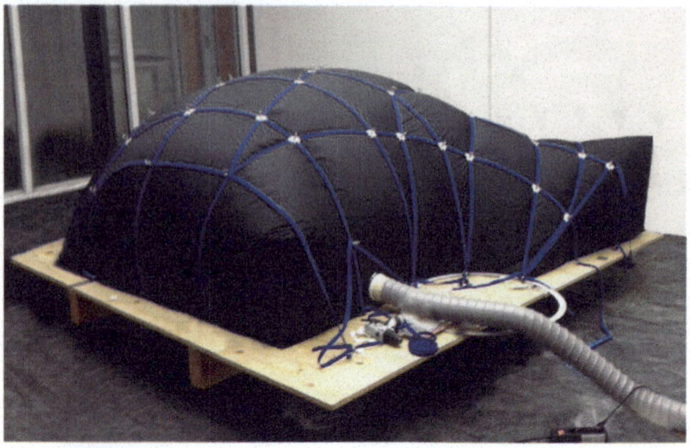

Fig. 14. Inflated, restraint pneuform

Reinforcement

The reinforcement of the structure consists of two elements (a) the woven carbon fiber textile mesh and (b) the distance holders. Carbon fiber reinforcement is used for its non-corrosive properties, which allow for thinner, slender and lighter designs. During the construction process it demonstrates favorable material properties, most notably its lightness and flexibility. For better friction and a stronger compound the carbon rovings are coated with a styrene-butadiene resin, an impregnation that gives bonding at a tensile strength of 2000 mpa (Kulas 2013). The system is attached in two layers on-to the pneu and fixed at the ring nuts (Fig. 15). The minimal thickness of 3 cm and a span of the shell of 2.5 m make a double layered compound of reinforcement possible. The investigations so far show that the position of the textile reinforcement during concreting can be ensured by using special distance holders (Disttex) (Walther et al. 2014). They are placed in relation to the curvature and ensure a distance of 1 cm between each layer and the formwork (Fig. 16). The distance holders are relatively new to the market

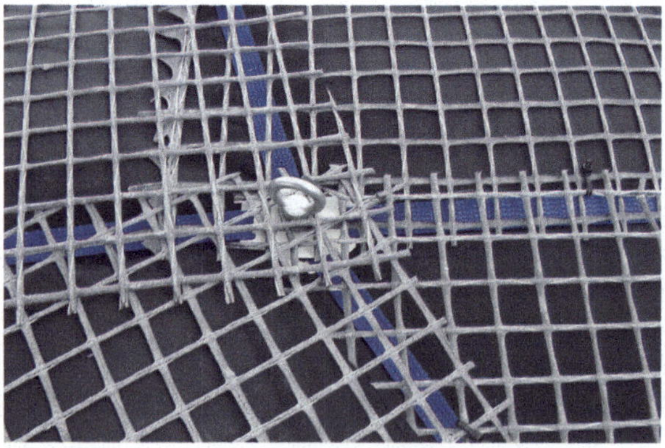

Fig. 15. The knot

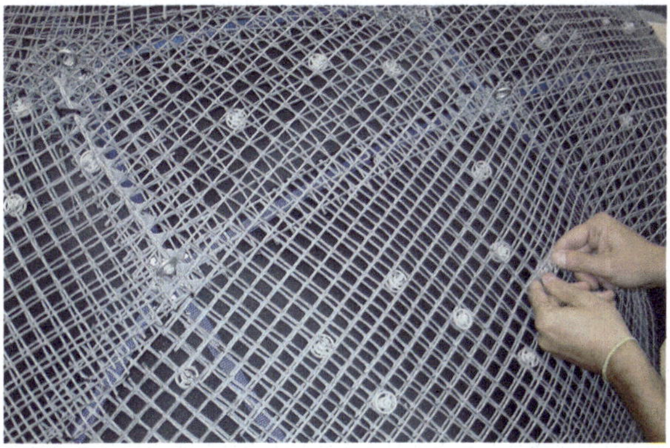

Fig. 16. Fixing the second layer with distance holders

and in this case adapted yet for fabric formwork. That is why they are rather pointy and leave an imprint at the inner surface of the shell.

More design freedom can be achieved by this measure, since the reinforcement transfers tensile stresses that result from the membrane and bending action, as well as the tensile stresses due to transverse and membrane shear. Furthermore, it provides a distribution net for local concentrated loads and the grid reduces shrinkage and temperature effects. Local bending can be reduced, which minimizes the chances of instability due to buckling of the shell.

Concrete

Spraying at the exterior of the shell is a feasible technique for placing the concrete in a minimal time onto large surfaces (Fig. 17). The disadvantage is the higher costs involved in compare to conventional placement, as well as the dust formation when

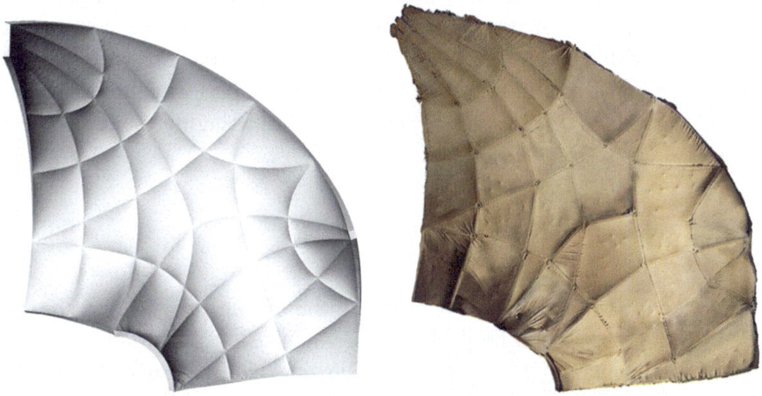

Fig. 17. Pattern of ribs, simulation and photograph

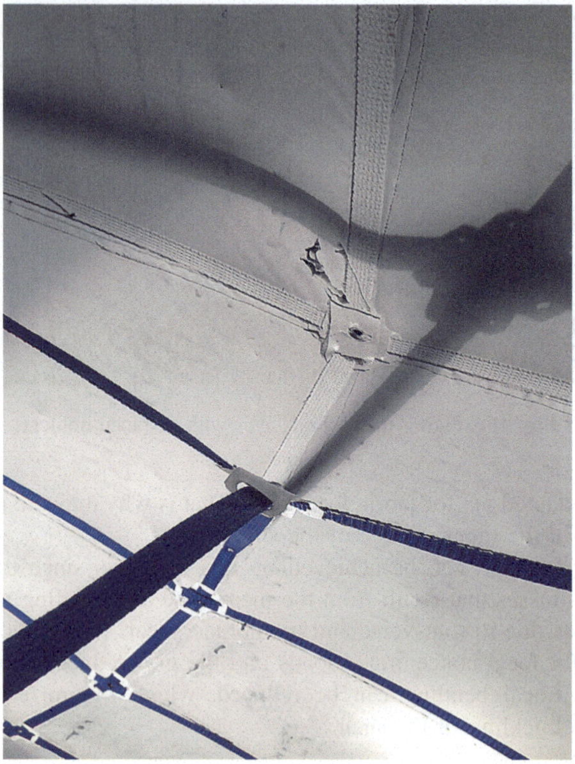

Fig. 18. Stripping the formwork, revealing the cast

using the dry method (Teichmann and Schmidt 2002). By using a tailored concrete mixture of UHPC (Quantz), the curing time can be adapted with plasticizers. Both the high compressive strength and the improved durability of UHPC are based on the principles of a very low water-cement-ratio minimizing the capillary pores and a high packing density especially of the fine grains, which results in a very strong structure. The final mixture has a proper density and bonding character with relatively high compression strength of minimum to 80 N/mm^2. UHPC performs high water tightness and little shrinkage, its good flowability is suitable for passaging the reinforcement and generates precise casts (Fig. 18).

Discussion

In general, deflections of the membrane during the curing process and the associated weakening of the concrete can be problematic when using pneumatic formwork. The lamination of the concrete was rather difficult in steep regions of the formwork. Spraying concrete is recommended instead. Deformations can be seen in the areas of lower curvature of the pneuform. By adapting the pneuform, this deviation can be compensated in the next step of iteration. In order to keep the deviations as small as possible further work should focus on finding the closest possible equilibrium surface to a given target surface by anticipating the later deformations through the loading conditions like inflation and weight of concrete and exterior forces. Also further prototype load-testing is necessary to gain relevant, comparable data about the strength of the compound, its structural safety and scalability of the construction system for wider architectural applications.

Conclusion

A holistic approach was required to iterate the design process between geometry, inner force flow and fabrication method. This offers new design freedom for double-curved shells that is barely feasible to realize with traditional formwork. It could be shown that with the PneuReCa strategy it is possible to create a structure, which enables free curvature, integrated openings and even cantilevering parts (Figs. 19 and 20). In that, both the problems and solutions encountered within the planning and fabrication

Fig. 19. Inside of the casted shell

Fig. 20. Elevations of the prototype

process, such as the attempt to develop a knot connection, may be applicable for further testing. As a result, the prototype presented here is the first pneumatic formwork restrained by a cable-net and casted with a carbon fiber reinforced UHPC. As opposed to their present applications, largely being used under their capacities in facade panels or secondary structures, this showcases potentials for load-bearing, more challenging applications of composites through structural activated shell structures. Apart from this specific contribution to the closer field of pneumatically erected cast structures, the project attempts to showcase potentials of how computationally driven design and manufacturing processes may develop autonomous aesthetics through the incorporation of structural capacities, material behavior and fabrication logics.

Acknowledgements. This Research was developed as part of the authors diploma thesis at the University of Arts, berlin. The Thesis was advised by Prof. C. Gengnagel and Prof. N. Palz. The author thanks the following companies for sponsoring material for the prototype: G.tecz Engeneering, viz. Gregor Zimmermann, Solidian, viz. Mr. Pfaff and Mr. Kulas, Distex, viz. Frank Schladitz. Special thanks go to my colleagues and family for their great support building the Prototype.

References

Bechthold, M.: Innovative Surface Structures: Technologies and Applications. Taylor & Francis, New York (2008)

Bini, D.: Building with Air. Bibliotheque McLean, London (2014)

Ermolov, V.V.: Zur Problematik von luftgestützten Membranschalen großer Spannweite (1979)

Isler, H.: New Shapes for Shells. IASS1961, Madrid (1961)

Kromoser, B., Kollegger, J.: Application areas for pneumatic forming of hardened concrete. J. IASS **56**(3), 187–198 (2015). doi:10.1002/stab.201410211

Kulas, C.H.: Zum Tragverhalten getränkter textiler Bewehrungselemente für Betonbauteile. Dissertation (2013)

Otto, F.O.: Air Hall Handbook. Institut für leichte Flächentragwerke, Universität Stuttgart, Stuttgart (1983)

Piker, D.P.: Kangaroo: Form Finding with Computational Physics. Archit. Design **83**, 136–137 (2013)

Rippmann, M., Lachauer, L., Block, P.: Interactive Vault Design. Int. J. Space Struct. **27**(4), 219–230 (2012)

Sobek, W.: Auf pneumatisch gestützten Schalungen hergestellte Betonschalen. 1. Aufl., Stuttgart (1987)

Tam, K.M.M., Mueller, C.: Stress line generation for structurally performative architectural design. In: Combs, L., Perry, C. (eds.) Computational Ecologies: Proceedings of the 35th Annual Conference of the Association for Computer Aided Design in Architecture, 22–24 October 2015. University of Cincinnati School of Architecture and Interior Design, Cincinnati (2015)

Teichmann, T., Schmidt, M.: Mix Design and Durability of Ultra High Performance Concrete (UHPC); 4th International Ph.D. Symposium in Civil Engeneering, Munich (2002)

Van Hennik, P.C., Houtman, R.: Pneumatic formwork for irregular curved thin shells. Text. Compos. Inflat. Struct. II **8**, 99–116 (2008)

Walther, T., Schladitz, F., Curbach, M.: Textilbetonherstellung im Gießverfahren mit Hilfe von Abstandhaltern. Beton- und Stahlbetonbau **109**, 216–222 (2014). Patent: DE102015104830

West, M.: The Fabric Formwork Book: Methods for Building New Architectural and Structural Forms in Concrete. Routledge, London (2016)

A Computational Approach to Methodologies of Landscape Design

Ignacio López Busón[1], Mary Polites[2], Miguel Vidal Calvet[3,4](✉), and Han Yu[5,6]

[1] AA MA Landscape Urbanism, MAPS, AAVS Shanghai and Turenscape Academy, Shanghai, China
[2] AA MArch Emergent Technologies, MAPS, D&I College, Biomimetic Design Lab (BiDL), Tongji University, Shanghai, China
[3] Foster + Partners, London, England
mvidalc@faculty.ie.edu
[4] Computational Design, Master in Architectural Management and Design (MAMD), IE University, Segovia, Spain
[5] MLA, Landscape Architect, Kuala Lumpur, Malaysia
[6] SOM, Shanghai, China

Abstract. This work provides a conceptual approach for determining eco-regions linked with Landscape Design, computational techniques, and remote sensing technologies. The methods explored develop a computational model for eco-regions in order to assist Architects in the evaluation and design for the environment with a main focus on promoting natural elements. The outcomes bring the natural and the built environment to the same level in order to extrapolate and analyze relationships using a data overlay methodology. This work takes methods used dominantly in large scale territorial design and applies similar techniques to the small scale contexts which are largely unexplored in terms of resolution. The paper is a first step of a longer-term research on computational landscape design. Some topics are outlined and will be developed in future research, such as point clouds and sensors.

Keyword: Computational landscape design

Introduction

This paper proposes a methodology for landscape design in small scale contexts that merges together GIS techniques and environmental analysis by means of computational tools. Our aim is to go beyond the generic large scale approach of landscape design typically prevalent in contemporary cities design and advocate for an environmentally specific and ecologically conscious strategy.

The present text is an extended version of the short paper proceedings published in PROTOCOLS, FLOWS AND GLITCHES, Proceedings of the 22nd International Conference on Computer-Aided Architectural Design Research in Asia. CAADRIA. Suzhou Liton Digital printing Co., LTD 2017.

K. De Rycke et al., *Humanizing Digital Reality*,
https://doi.org/10.1007/978-981-10-6611-5_54

The concept of *eco-region*, a "unit of land or water containing a geographically distinct assemblage of species, natural communities, and environmental conditions", is taken as a starting point for this research. By exploring the emergence of local ecoregions influenced by the built environment we aim to determine the small-scale environmental context that affects plants and other living systems in contemporary cities. This methodology suggests that by understanding these constraints and thereby designing landscapes upon them we will be promoting a new city ecology that will truly integrate nature within the urban context. The research will be applied in the city of Shanghai whose government approved an initiative in 2014 to create up to 2000 landscape bodies before 2040 as a means of greening the city.

Towards an Integrated Urban—Natural Model Through Computation

It is the aim of this paper to introduce a design methodology that blurs the boundaries between the urban and the natural and explores the potential of digital tools in the location, generation, visualization and construction of urban green spaces in contemporary cities. This proposal suggests that computation, with an interdisciplinary integrative approach, can be considered as the main tool to seamlessly articulate the integration of natural systems within cities. However, along with its opportunities, we would also like to highlight some its potential risks.

Although computational technologies have been extensively implemented on territorial analysis through GIS platforms since the 1960s, small-scale landscape design has yet to benefit from the potential of computational tools, especially in terms of intersecting environmental analysis with biological and ecological strategies, data management and data visualization techniques. Ian McHarg's seminal book Design with Nature linked regional planning and landscape architecture, creating a corpus for the integration of urban and natural systems, which were later to become GIS techniques. McHarg's approach (the layer-cake method) was based on an organized way of qualifying the existing conditions on site, translating them into opportunities and constraints, and eventually overlaying them to test alternative options in order to find the best possible solution.

The Local Eco-Region Model

Eco-regions are defined by the WWF as "large areas of relatively uniform climate that harbour a characteristic set of species and ecological communities." However, as the concept of size remains relative, other definitions exist for areas of different extent that link climate to ecology, such as biomes, eco-zones or bio-regions. Therefore, the abstraction of the term eco-region is used as the main driver of this proposal by assuming that such definition would be still ecologically relevant in a small scale context.

Shanghai Case Study: Guannong Park

As of 2016 China is the most populous country in the world whose total population has reached approximately 1.36 billion. As a result of an unprecedented migration phenomenon and rapid urbanization processes, it is estimated that 50–75% of Chinese population will be living in cities by 2050. In order to contextualize this research, China's largest city has been selected as a case study for this paper. Currently in Shanghai, as part of a governmental incentive, a series of community gardens are inserted throughout the city. Just recently, Shanghai′s government has released a renewed vision for greening the city included in the Shanghai Masterplan 2016–2040. For this research, the site selected represents the current policy measures, scale, style, program and typical use of local residents for neighborhood parks in Shanghai. Guannong Park is located at the northeast corner of Putuo District, with Guannong road on the east and Huning Railway on the north. The total area of the park is 12,500 m^2, including 6060 m^2 of green space and 170 m^2 water surface.

Proposed Methodology

The methodological approach introduced in this paper tries to break away from the standard engineering approach in which data collection the real breakthrough to a more design, anthropocentric perspective, where the data is accessible and abundant and we have the technology and ability to truly be critical about it and the design outcome of the built environment is the real focus. A conventional standard GIS workflow could be summarised into four main phases: capture (data is collected from several sources: personal devices, infrastructures, remote sensing technologies…), processing/management (data is filtered, classified and stored), analysis (data is deconstructed, combined and evaluated) and visualization (data is visualized and shared).

However, this paper claims that such approach is shortsighted and lacks ecological and environmental contextualization, which would potentially have an impact in shortening species lifespan and the eventual degradation and even disappearance of the green space. The implementation of the four phases of the proposed workflow, namely **Mapping, Analysis, Overlay and Eco-modelling**, will be described in Table 1.

Stage 1—Mapping: Guannong Park Digital Model

The data collection for the digital modelling was based on the use of an UAV, DJI's Mavic Pro model. The Pix4d Capture app (https://pix4d.com/product/pix4dcapture/) was installed on a Galaxy Note 4 mobile phone to define a flight plan the drone would follow to take aerial photographs. The software chosen for the photogrammetry process was Agisoft Photoscan (http://www.agisoft.com/) (Fig. 1).

Table 1. Summary of eco-model design methodology

Phase	Objective	Toolkit	Data sources
Mapping	Creation of a comprehensive digital model of the site including physical constraints, environmental parameters and existing plant species	UAVs (DJI Mavic) PIX4D Capture Agisoft Photoscan Rhinoceros 3d	Google Earth Openstreetmaps EPWmap
Environmental simulation and analysis	Analysis and simulations of physical and environmental constraints in the digital model: solar radiation, wind and water accumulation. Quantitative data is obtained	Rhinoceros 3d Grasshopper (with Ladybug) Excel	EPWmap
Data overlay	Organization, reclassification and overlay of the different analysis layers in order to define the climate regions on site	Rhinoceros 3d Grasshopper	
Eco-model design	Linking of climate regions with plant species that match each area's attributes	Rhinoceros 3d Grasshopper Excel	

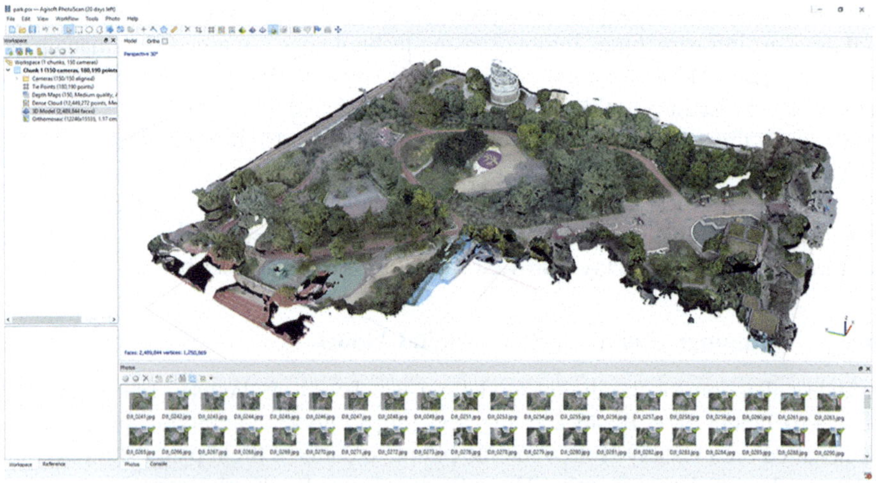

Fig. 1. Screenshot of the Agisoft photoscan interface

Stage 2: Site Analysis

Planting Analysis

A visual mapping study was firstly carried out to catalog the dominant tree species within the park and the documentation was conducted using images. Whereas the site visit allowed for the identification of the main plant species in the park, the orthomosaic generated by Agisoft Photoscan helped map the different species clusters. In Rhinoceros, the plant clusters were traced as closed boundaries and labeled accordingly. The species were also included in a spreadsheet and evaluated according to their attributes. Each plant was qualified in terms of their tolerance/affinity to shade, drought, flooding, wind and high and low temperatures. The measurements were constrained to low, medium and high tolerance. This simplification was decided for the sake of clarity and would not reduce the model's goal. As a result no logic could be observed as far as environmental constraints are concerned. Although all the species identified match the standard Shanghai catalogue plants, it seems that their location on site may only respond to aesthetic reasons and composition rules. A certain degree of freedom could be assumed regarding design at such small scale, it is the aim of this paper to search for a layer of environmental hyper-specificity that could not only raise ecological standards but also add unexpected value to landscape design (Figs. 2, 3 and 4; Tables 2 and 3).

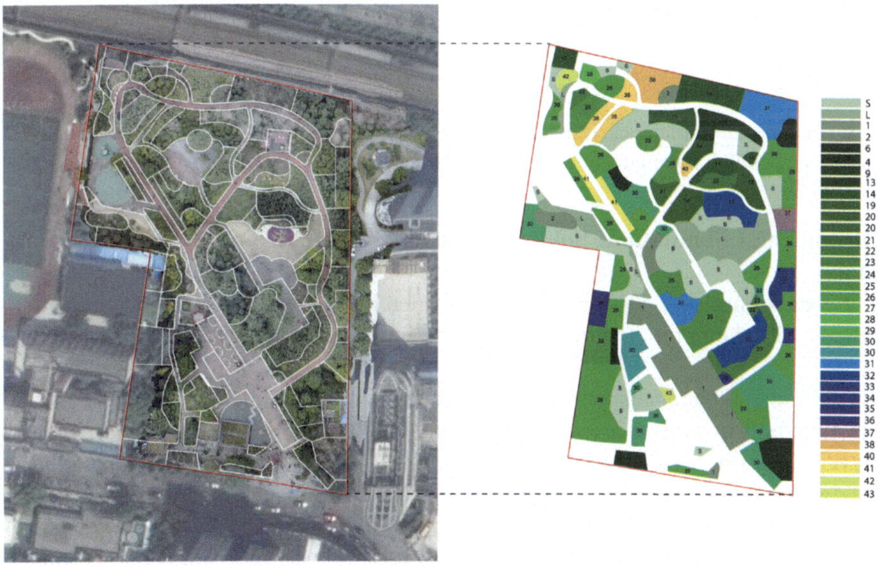

Fig. 2. Plan view from visual planting analysis

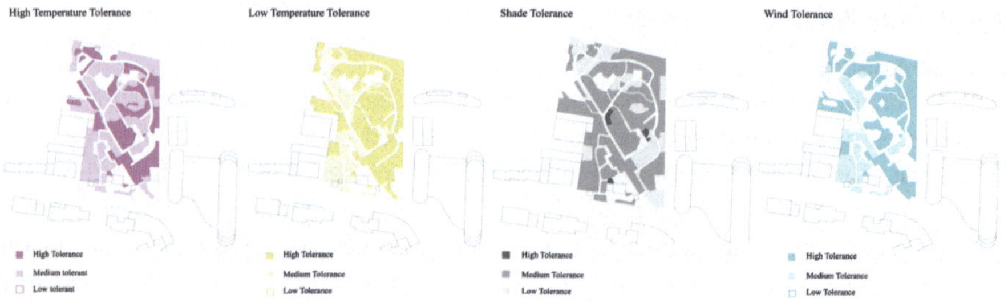

Fig. 3. Species tolerance map for each parameter: high temperature, low temperature, shade and wind

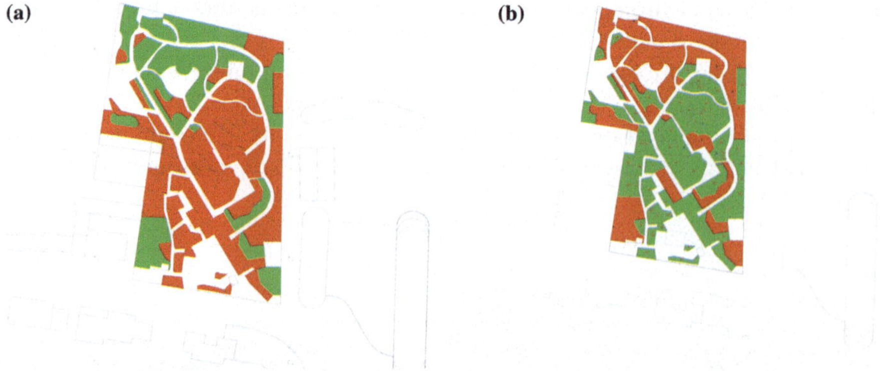

Fig. 4. Plants *shade* tolerance test for winter (*left*) and summer (*right*) conditions. In *green*, plants clusters which have adequate shading tolerances; in *red*, plants clusters which do not

Environmental Simulation: Solar Incidence, Wind Flow and Water Runoffs

For the environmental simulation, a simplified 3D model of the plot surrounding buildings with the point cloud from the drone in Rhinoceros was used as shown below (Fig. 5).

Solar radiation analysis has been carried out both in summer and winter conditions using Ladybug in Grasshopper as explained in previous sections with climatic data retrieved from Energy Plus database (https://energyplus.net/weather). Energy Plus is a building simulation program funded by the U.S. Department of Energy's (DOE). The solar radiation is modelled in these analysis as sunlight hours as shown and the periods used for the simulation is hourly data from 8 h to 18 h as follows: (Figs. 6 and 7).

As for the wind simulation, it was carried out with a simplified 3D model of the plot and surroundings as shown and taking the terrain as a flat surface. The computational fluid dynamics (CFD) software used is Rhino CFD by CHAM in beta (http://www.cham.co.uk/rhinoCFD.php). The orientations and intensities used in the simulations are as follows: Winter period: orientation, North East (NE); intensity, 5.50 (m/s); Summer

Table 2. Guannong Park species and classification from visual analysis

No.	Chinese	Latin name	Shade	Low temp.	High temp.	Drought	Flooding	Wind
1	悬铃木	*Platanus orientalis*	1	2	3	2	1	1
2	合欢	*Albizzia julibrissin*	1	3	3	3	1	3
3	樱花	*Cerasus yedoensis*	1	3	1	1	1	1
4	水杉	*Metasequoia glyptostroboides*	1	3	2	1	3	2
5	无患子	*Sapindus mukorossi*	2	3	2	3	1	3
6	栾树	*Koelreuteria paniculata*	2	3	2	3	2	3
7	喜树	*Camptotheca Acuminata*	2	1	2	1	2	2
8	臭椿	*Ailanthus altissima*	1	3	2	3	1	3
9	女贞	*Ligustrum lucidum*	2	3	2	2	2	1
10	榉树	*Zelkova serrata*	1	3	2	1	2	3
11	苦楝	*Melia azedarach*	1	2	2	3	2	3
12	玉兰	*Magnolia denudata*	1	2	2	3	1	2
13	银杏	*Ginkgo biloba*	1	3	2	1	1	3
14	三角枫	*Acer buergerianum*	2	3	2	3	2	3
15	马褂木	*Liriodendron chinense*	2	3	2	1	1	3
16	国槐	*Sophora japonica*	2	3	2	3	1	3
17	垂柳	*Salix babylonica*	1	2	2	3	3	1
18	落羽杉	*Taxodium distichum*	1	3	2	3	3	2
19	垂丝海棠	*Malus Halliana*	1	2	2	2	1	1
20	鸡爪槭	*Acer palmatum cv. dissectum*	2	3	1	3	1	1
21	紫薇	*Lagerstroemia indica*	2	3	2	3	1	3
22	腊梅	*Chimonanthus praecox*	2	3	2	3	1	1
23	紫叶李	*Prunus Cerasifera*	2	3	2	1	2	1
24	木芙蓉	*Hibiscus mutabilis*	2	1	3	1	2	2
25	雪松	Cedrus deodara	2	3	3	3	1	1
26	香樟	*Cinnamomum camphora*	2	3	3	1	2	3
27	柳杉	*Cryptomeria fortunei*	2	3	2	1	1	1
28	杜英	*Elaeocarpus decipiens*	2	2	2	1	1	3
29	龙柏	*Juniperus chinensis 'Kaizuka'*	2	3	2	3	1	3
30	广玉兰	*Magnolia grandiflora*	2	3	2	1	1	3
31	桂花	Osmanthus fragrans	3	3	3	2	1	3
32	罗汉松	*Podocarpus macrophyllus*	3	2	2	2	1	1

(*continued*)

Table 2. (*continued*)

No.	Chinese	Latin name	Shade	Low temp.	High temp.	Drought	Flooding	Wind
33	棕榈	*Trachycarpus fortunei*	2	3	3	3	2	1
34	熊掌木	*Fatshedera lizei*	3	3	1	1	1	3
35	盘槐	*Sophora japonica (cultivated)*	2	3	2	3	1	3
36	夹竹桃	*Nerium indicum*	3	2	2	3	1	3
37	石楠	*Photinia* spp.	3	3	2	2	1	3
38	孝顺竹	*Bambusa multiplex*	3	2	2	3	1	3
39	火棘	*Pyracantha fortuneana*	1	2	2	3	1	3
40	紫荆	*Cercis chinensis*	2	3	2	2	1	2
41	青枫	*Acer palmatum*	2	3	1	2	1	2
42	梅花	*Armeniaca mume*	1	3	3	2	1	2
43	桃花	*Amygdalus persica*	1	3	3	3	1	2
S	*shrubs*	*shrubs*	2	2	2	2	2	2
L	*lawn*	*lawn*	1	1	1	1	1	3

Table 3. Species tolerance legend

Shade	3	Shade tolerant
	2	Semi shade tolerant
	1	Not tolerant to shady condition
Low temp	3	Low temperature tolerant (−5° and below)
	2	Low temperature Semi tolerant (−5° to 5°)
	1	Not tolerant to Low temperature (above 5°)
High temp	3	Fine when above 35°
	2	Best not to exceed 30°–35°
	1	Best not to exceed 25°–30°
Drought	3	Drought tolerant
	2	Semi drought tolerant
	1	Not tolerant to drought
Flooding	3	Tolerant to short term flooding
	2	Tolerant to damp soil
	1	Not tolerant to constant damp soil
Wind	3	Strong wind tolerant
	2	Semi wind tolerant
	1	Not tolerant to windy condition

period: orientation, South West (SW); intensity, 4.50 (m/s). The bounding box of the analysis and a coloured visualization is shown in Fig. 8.

For the water parameter, a map for the plot's runoffs has been modelled in order to show the potential water paths in the zone which would provide the likelihood of water

Fig. 5. Axonometric view of the simplified 3D model used

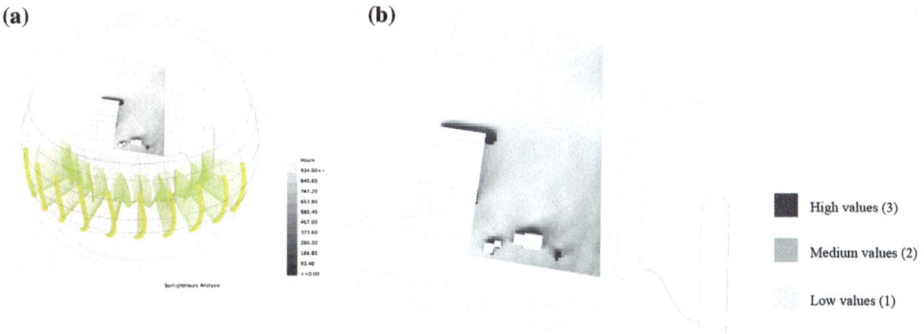

Fig. 6. Winter Sunlight hours simulation with sun vectors projection—Site Plan and Plot. Winter period: from the winter equinox (21 December) to the spring solstice (20 March)

accumulation in the plot. In order to extract these paths, the software MOSQUITO for Grasshopper by Studio Smuts has been used (http://www.food4rhino.com/app/mosquito-media-4-grasshopper). The measuring approach we took was to divide the runoffs curves in 10 equal parts and then measuring amount of points in a 20 m radius from each plot sample point (Fig. 9).

Although our ambition is to be as close as possible to reality and to aim for a contextualized mapping of the site, we are aware of the limitations of such digital explorations. By no means we claim that these results are absolutely accurate but good approximations of what may occur in reality. The goal is also to show the relevance of bringing these technologies in Landscape Design on a regular basis.

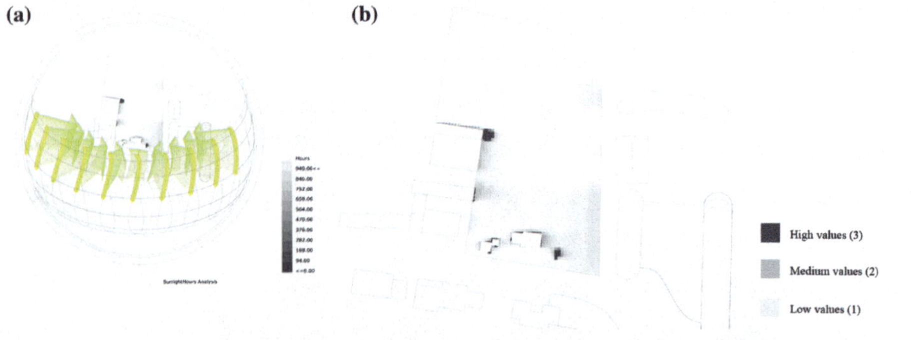

Fig. 7. Summer Sunlight hours simulation with projection of sun vectors—Site Plan and Plot View. Summer period: from the summer equinox (21 June) to the autumn solstice (22 September)

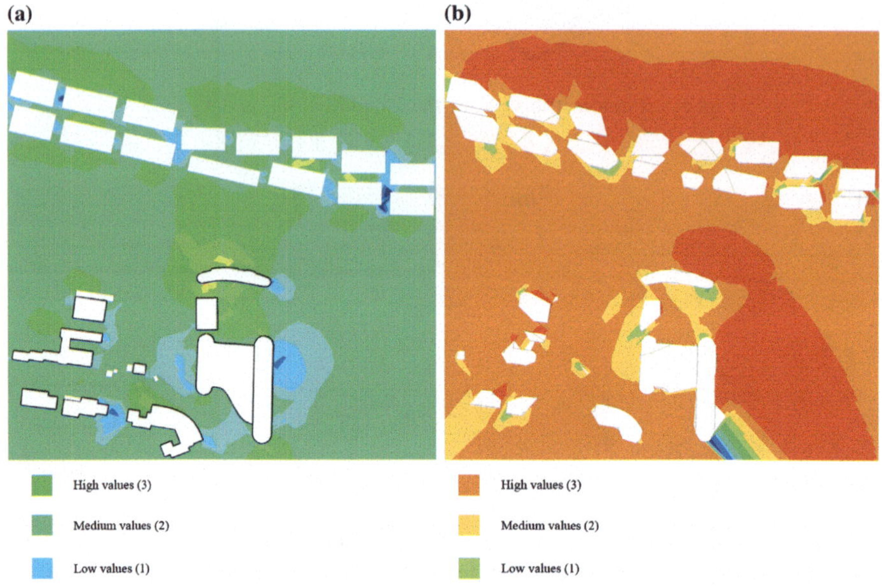

Fig. 8. Wind CFD Simulation for winter (*left*) and summer (*right*) conditions

Stage 3: Data Overlay

The data overlay step was carried out following Ian McHarg's layer-cake methodology as mentioned in previous sections overlapping the planting and thermodynamic site analysis. In order for all these parameters to be intersected two procedures were followed:

- Firstly, the analysis values were projected on the same analysis grid. The plot has been divided in a 2 m dense flat grid, 4558 points and by means of this discretization samples of the three parameters analysed can be stored and compared (Fig. 10).

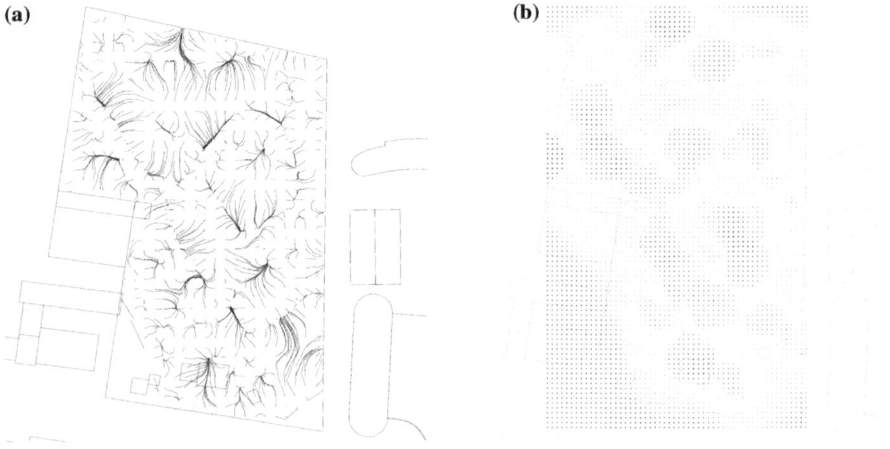

Fig. 9. Water runoffs visualization: paths (*left*) and influence points (*right*)—Plot plan view

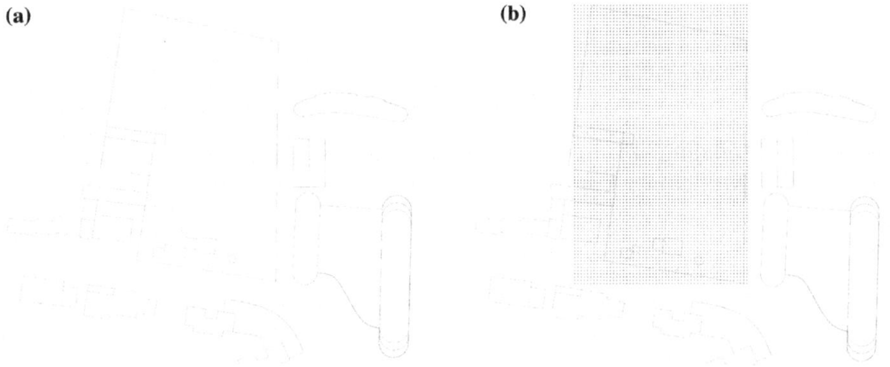

Fig. 10. Plot view (*left*) and Grid plan view (*right*)

– Secondly, the environmental analysis values in each of their units (number of hours for sunlight, m/s for wind and number of points for water runoffs) were remapped to integer numbers onto a (Biomes 2017; Society 2011) domain so they can also be compared with the planting analysis.

In the below visualizations, a data overlay strategy has been taken using the a primary color (red, green and blue) for each environmental parameter (sun, wind, water runoffs) which has been represented as a height field with their values. When overlayed those colours, represent the importance of either sun, wind or water which characterize the eco-regions definition (Fig. 11).

(a) (b)

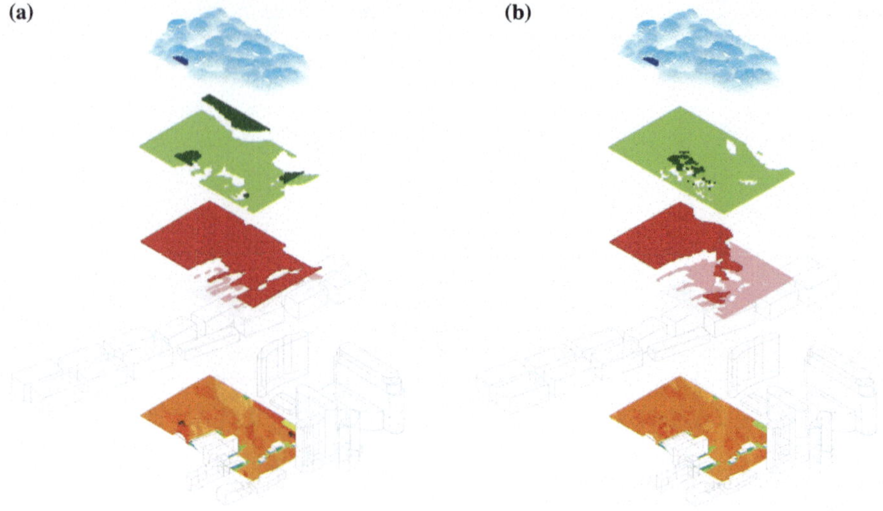

Fig. 11. *Left*: Axonometric Data Overlay visualization of three climatic parameters in Winter conditions: solar incidence (*red*), wind (*green*) and water runoffs (*blue*). *Right*: Axonometric Data Overlay visualization of three climatic parameters in Summer conditions: solar incidence (*red*), wind (*green*) and water runoffs (*blue*)

Stage 4 Eco-Modeling: From Micro-Climates to Eco-Climates. A Parametric Modelling Approach to Landscape Design

Following the environmental simulation remapped as 1, 2 and 3 values (low, medium and high) for the analysed values: sunlight hours, wind incidence and water runoffs, the eco-regions are numbered from 1-less tolerant in each value 1, 1, 1 for all the parameters—to 27—more tolerant 3, 3, 3—s A computational plants tolerance test has been carried out as shown in Fig. 12.

Theoretically there could exist 27 potential eco-region types both for summer and for winter as the permutations with repeated values of 1, 2 and 3 intensity for the each of the environmental analysis. The lower the environmental values are the bigger variety of plants could be suitable to live in that local eco-region and vice versa: the higher those values are the less tolerant plant species were to grow in that space. This criteria would eliminate species that were suitable as we go up in the list.

Conclusions

The aim of this work is to go beyond the large scale approach to landscape design typically prevalent in contemporary cities and advocate for ecological specificity and a environmentally conscious methodology. We would also like to criticise the apparent randomness in the planting strategy in urban parks and we believe that a data-backed approach such as the one that was presented here is a valid proposal for contemporary landscape design. The application of this approach holds the potential not only on the

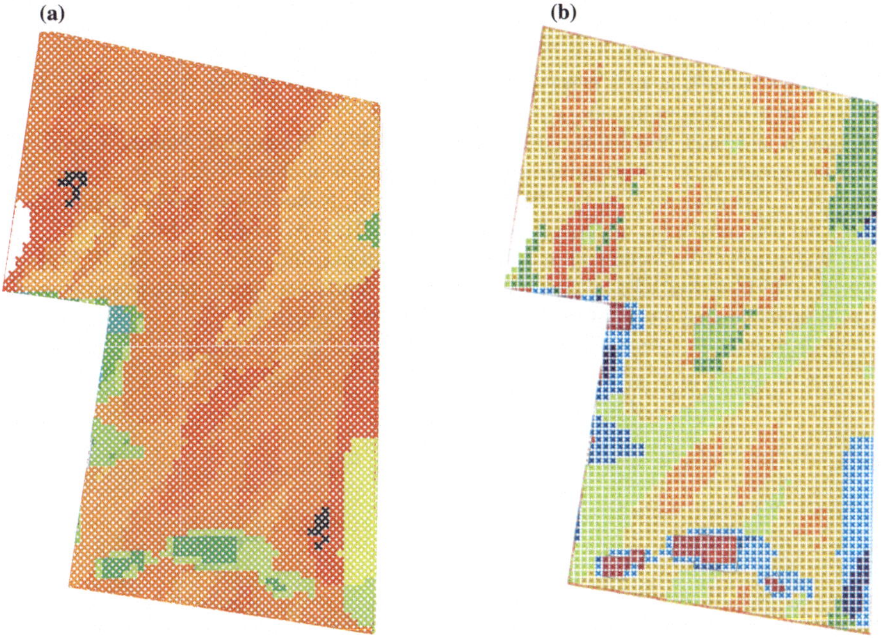

Fig. 12. Final Eco-Regions for Winter (*left*) and summer (*right*) conditions

design profession but has the capability of activating a knowledge data-base that could offer practical outcomes as well as insight for a wide range of contributors. The potential for this work is that it has the possibility of becoming a practice-oriented information base or a smart advisor, to the design process. Such information could offer a high level of practical information necessary to execute contextual work based on simple computational instructions that is typically lacking without the expertise of consultants.

On the other hand, from a research perspective, the possibility to add upon a detailed base of information, through refinements and additional systemic information would be of great use to continuing research in the field of mapping, landscape and urbanism. Each method needs to have the ability to continuously be refined towards more specificity and accuracy. In addition, contemporary design tools and techniques such as parametric and algorithmic design represent an interesting toolkit capable of dealing, embracing uncertainty, while preserving specificity.

References

Biomes | Conserving Biomes | WWF [WWW Document], n.d. . World Wildlife Fund. https://www.worldwildlife.org/biomes. Accessed 19 May 2017

China: urban and rural population 2015 | Statistic [WWW Document]. Statista. https://www.statista.com/statistics/278566/urban-and-rural-population-of-china/. Accessed 19 May 2017

Society, N.G., Society, N.G.: biome [WWW Document]. National Geographic Society. http://www.nationalgeographic.org/encyclopedia/biome/ (2011). Accessed 19 May 2017

上海总规2040 [WWW Document], n.d. http://www.supdri.com/2040/. Accessed 19 May 2017

管弄公园_普陀区管弄公园旅游指南 [WWW Document], n.d. URL http://wap.bytravel.cn/Landscape/43/guannonggongyuan.html. Accessed 19 May 2017

Design Tools and Workflows for Braided Structures

Petras Vestartas[1(✉)], Mary Katherine Heinrich[1], Mateusz Zwierzycki[1], David Andres Leon[1], Ashkan Cheheltan[1,2], Riccardo La Magna[3], and Phil Ayres[1]

[1] Centre for Information Technology and Architecture, School of Architecture, The Royal Danish Academy of Fine Arts, 1435 Copenhagen, Denmark
petrasvestartas@gmail.com
[2] Berlin University of the Arts, 10623 Berlin, Germany
[3] Konstruktives Entwerfen Und Tragwerksplanung, Berlin University of the Arts, 10623 Berlin, Germany

Abstract. This paper presents design workflows for the representation, analysis and fabrication of braided structures. The workflows employ a braid pattern and simulation method which extends the state-of-the-art in the following ways: by supporting the braid design of both pre-determined target shapes and exploratory, generative, or evolved designs; by incorporating material and fabrication constraints generalised for both hand and machine; by providing a greater degree of design agency and supporting real-time modification of braid topologies. The paper first introduces braid as a technique, stating the objectives and motivation for our exploration of braid within an architectural context and highlighting both the relevance of braid and current lack of suitable design modelling tools to support our approach. We briefly introduce the state-of-the-art in braid representation and present the characteristics and merits of our method, demonstrated though four example design and analysis workflows. The workflows frame specific aspects of enquiry for the ongoing research project *flora robotica*. These include modelling target geometries, automatically producing instructions for fabrication, conducting structural analysis, and supporting generative design. We then evaluate the performance and generalisability of the modelling against criteria of geometric similarity and simulation performance. To conclude the paper we discuss future developments of the work.

Keywords: Braided structures · Braid representation · Computational design

Introduction—Context, Motivation and Objectives

Braid technique is based on a principle of oblique interlacing of three or more strands of yarn, filament or strapping (Rana and Fangueiro 2015). The technique is used to produce artifacts that are larger, stronger, more resilient and, often, more aesthetically charged than the original material. Braid offers tremendous versatility in terms of material, scale of artifact, and method of production (Fig. 1). As such, braid is applied across a broad range of industries, including medicine (stents), agriculture (industrial

K. De Rycke et al., *Humanizing Digital Reality*,
https://doi.org/10.1007/978-981-10-6611-5_55

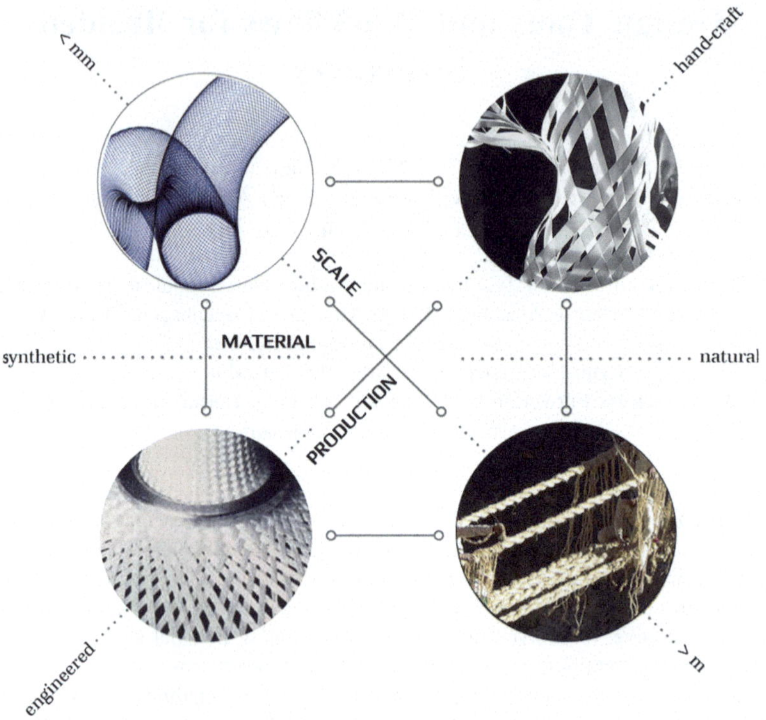

Fig. 1. The versatility of braid technique spans across scales, materials and production processes

hoses) and leisure equipment (ropes). Braid is also used to produce preforms for advanced composite manufacture (Potluri et al. 2003). The resulting composites are amongst the lightest yet strongest components currently produced and are utilised in high-performance arenas such as cycling, racing and aeronautics (Rana and Fangueiro 2015). Despite the versatility, resilience, and scalability demonstrated by braid across many industries and functional uses, it is not a commonly used or discussed technique within contemporary architecture. In parallel to the limited use of braid in architecture, there is also a lack of suitable design modelling tools for freely exploring complex braid topologies against architectural design, analysis, and fabrication considerations. In the work presented here, we concentrate on exploring hollow tubular braids [typically referred to as '2D braid' in industrial contexts (Ayranci and Carey 2008)]. Tubular braids offer a rich space of topological freedom, as well as structural potential when approaching architectural scale.

Our motivation for exploring braid derives from the on-going *flora robotica* project (Hamann et al. 2015), focusing on symbiotic relationships between biological plants and distributed robotics, for the purpose of constructing architectural artifacts and spaces. The project pursues braided structures as adaptive mechanical scaffolds, on which biological (plant) and artificial (robot) materials can grow and act (Fig. 2, left). Spatial and topological organisation of the bio-hybrid system is steered over time by a

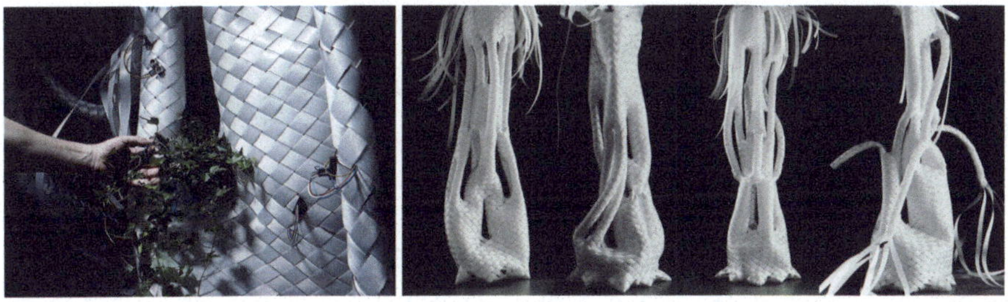

Fig. 2. (*left*) Braided scaffolds supporting plants and robotic elements in the *flora robotica* project. (*right*) Hand braided examples of self-supporting complex braid topologies

combination of high-level design objectives, distributed robot controllers, and user interaction. Achieving this broader goal requires that the braided scaffolds have the ability to reorganise continuously in situ, through a distributed construction process. The scaffolds may be collaboratively fabricated by stationary centrally-controlled braid machines (Richardson 1993), swarms of distributed mobile braiding robots (Heinrich et al. 2016), or manual braiding by users (Fig. 2, right). To achieve these aims, it has been necessary to develop tools and workflows that can support design speculation and specification, and provide an adequate representation of braid in the context of diverse and often independent design and evaluation tasks. Within *flora robotica*, these include the modelling of complex braids produced by hand, the generation of fabrication instruction for both hand and automated methods, the analysis of mechanical performance of braids, and the adaptive generation of braid morphology.

Braid Representation Method

This section briefly reviews our developed braid representation method (Heinrich et al. 2016). In general, the method uses a precomputed set of tiles, which, in their underlying logic, combine different approaches seen in the literature (e.g., Mercat 2001; Kaplan and Cohen 2003; Akleman et al. 2009). The method works directly on polygon meshes that can be modified in real-time.

The first step in the method is tiling. Following the approach defined by Mercat (2001), a predefined tile dictionary provides a complete description of possible braid strip organizations (Fig. 3). In the tile dictionary utilised in this paper, there are three possible relationships between neighboring tiles: no connection (green colour in Fig. 3); connection with two separated strips (yellow); connection with two crossing strips (blue). To be able to simulate the physical characteristics of the braiding pattern, it is necessary to translate the tile notation into geometry. The method uses a point grid, with each strip declared as a series of grid-based coordinates on the mesh.

The next step is relaxation. In evaluating the physical properties of the digital model, the braid cannot be approximated to a grid-shell due to torsion occurring in the flat strips. The mesh topology is therefore constructed from triangle meshes with varying density. The constraint-based geometry solver Kangaroo 2 [update to Piker

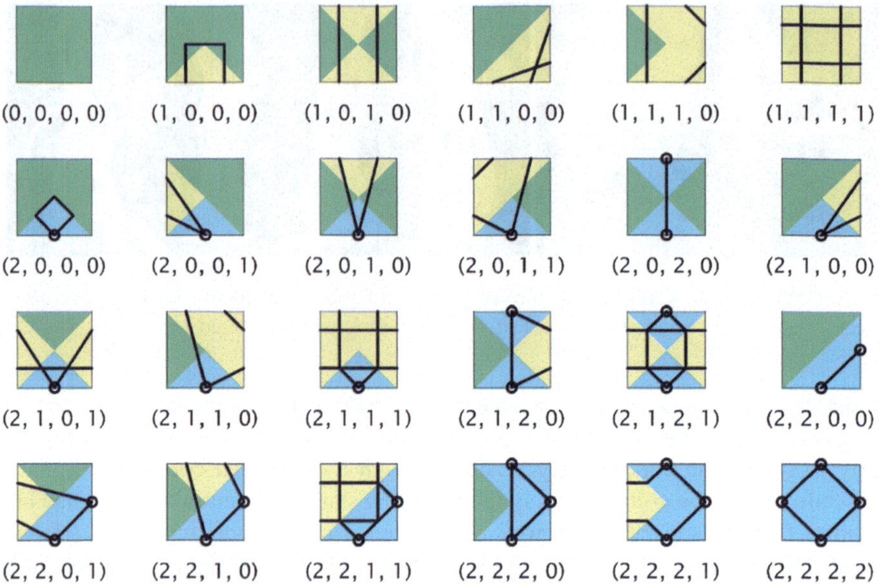

Fig. 3. An example of a complete set of tiles for 3 colours and 4-sided polygons. 24 tiles guarantee all the combinations of colours

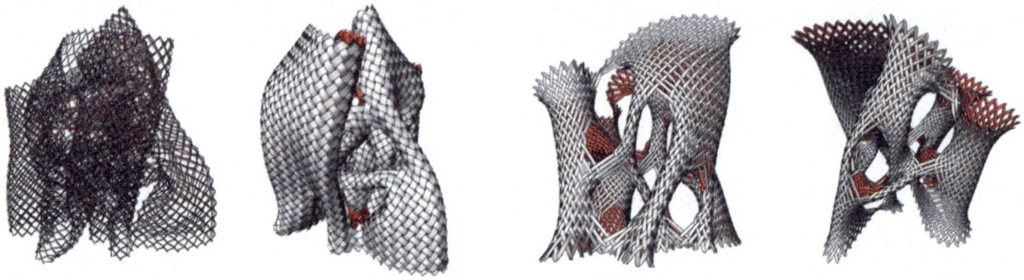

Fig. 4. Initial geometry (*left*) and three example relaxation results when changing target mesh edge length

(2013)] is used to perform relaxation, with objectives to equalise mesh edges, detect collisions for zero length mesh manifolds, and add shell-like behaviour. The problem of 'overshooting' may occur when mesh faces are undesirably stretched or compressed by large percentages, so a dynamic constraint incrementally increases initial edge lengths to reach the target. This results in longer calculation time, but achieves tight braids with strips that, if unrolled, approximate the straightness of physical strips (Fig. 4).

Design Workflows

In this section, we describe four examples that integrate the braid representation method with possible design workflows (Fig. 5). As outlined in the previous section, these workflows support design efforts of the *flora robotica* project.

In the first workflow (modelling target geometries), we compare our method to existing state-of-the-art approaches by tiling the input of a manually modelled mesh. In the second (generating fabrication instructions), we interpret a model to provide the output of textual instructions for hand braiding. Third (analysing mechanical properties), we address calibrated simulations for the output of assessing structural performance. Fourth (generative design), we generate braids from the input of an environmentally responsive robotic controller. In combination, these workflows show that our braid representation is sufficiently generalised that it is able to both receive multiple inputs (workflow 1 and 4) and be interpreted for multiple outputs (workflow 2 and 3).

Workflow 1: Modelling Target Geometries

In this workflow we demonstrate modelling to pre-defined design targets, such as those shown in Fig. 2. In these cases, material geometry is a hard constraint that must be

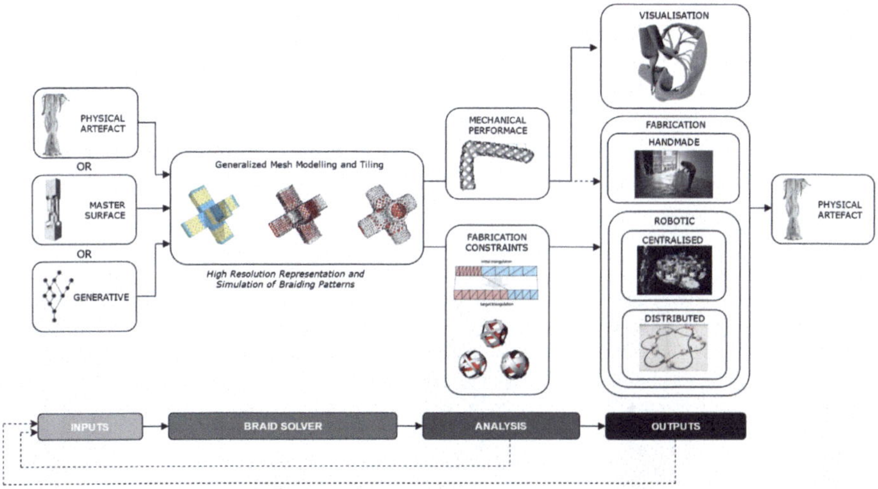

Fig. 5. Overview of the workflows supported by the generalised modelling approach for braided artifacts. The generalised modelling method can receive one of the following inputs: target physical artifact (Fig. 2), target master surface (workflow 1), or generative input (workflow 4). The mesh resulting from the generalised modelling approach can be interpreted in multiple complementary ways, such that it can be analysed (workflow 3), visualized (Figs. 4, 6), and used to generate instructions for hand braiding or braiding robots (workflow 2)

considered to ensure an adequate representational approximation. In addition, conforming to local braid conditions such as asymmetric bifurcations (in terms of strip numbers), inversion and 'cornering' displayed by these physical prototypes present further modelling challenges. This workflow has five stages. First, a low-poly quad mesh model is created of the physical prototype. Next, quad mesh edges are equalised by using a custom Kangaroo 2 solver with spherical collision and equal length line constraints in a parallel thread for faster calculation time. Thirdly, all braid conditions are specified by the application of tile colours to the quad mesh. The braid tiles are then applied from the existing tile dictionary (Fig. 3). Finally, strip-strip simulation is used to achieve a tight and realistic braid model. Within *flora robotica* the making of physical prototypes is an essential mode of exploration, as is the ability to accurately represent these complex morphologies once produced (Fig. 6).

Workflow 2: Generating Fabrication Instructions

In this workflow, a braid result from the solver is used to automatically generate fabrication instructions for distributed robots, centralised robots, or hand fabrication. In this example, we look at hand fabrication of a simple 12-strip bifurcating braid. We developed a method, firstly, to generate a low-level graph representation of the braid's organisation and, secondly, to interpret that graph into instructions using a string replacement dictionary. The low-level graph represents the braid as a series of connections, in which each intersection of braid strips receives a unique ID, and the under or over condition of a strip is indicated by a sign attached to the intersection ID. In order to generate this low-level graph, an agent starting at the beginning of each strip walks it until it finds a new strip intersection point. Once it has found an intersection, it waits there until it sees that its neighbors have also found intersections. At this time, the agents each log their newly found connection, and resume walking along their assigned strips. This process is continued until the agents all find the end of their assigned

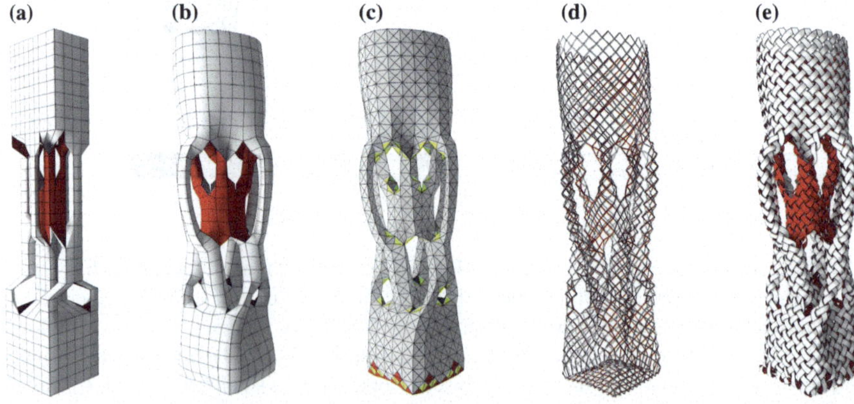

Fig. 6. 3D modelling braid typology informed by the hand-braided structures. Modelling steps from *left* to *right* are: **a** 3D mesh modelling; **b** equalise edge lengths; **c** assign colours to mesh edges; **d** apply tiling; **e** strip–strip relaxation

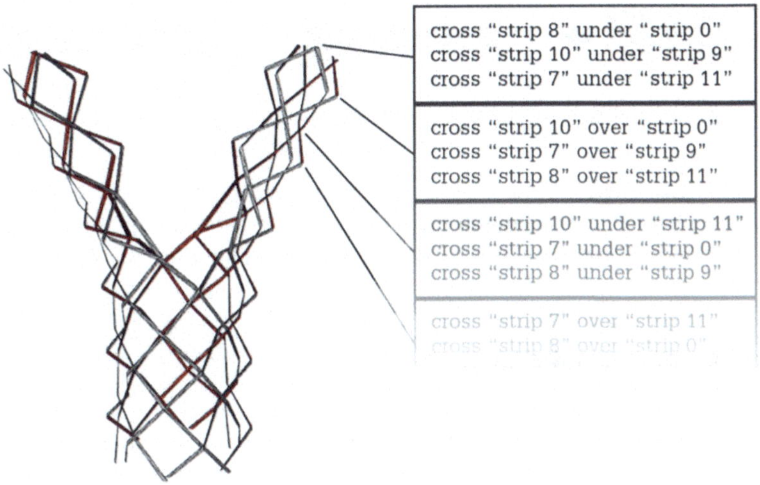

Fig. 7. Example instructions for hand braiding, generated using a low-level graph representation interpreted through an instruction dictionary

strip. Conceptually, this method could be extended to be used in real-time with a continuous fabrication process. After the low-level graph is produced, their connections are interpreted through a dictionary into instructions for hand braiding (Fig. 7). These instructions could potentially be implemented into a user interface and combined with a braid visualisation to interactively show the intended result of each instruction step.

Workflow 3: Analysing Mechanical Properties

Explicit and dynamic formulations of structural elements are particularly attractive for the simulation of large deformations and nonlinear phenomena. Especially in the context of form-finding, new and alternative approaches have recently emerged which are well-suited for these types of problems. The results shown in this paper clearly demonstrate the capacity and versatility of such formulations. Modern nonlinear Finite Element packages, the de facto standard in engineering simulation, are completely equipped to perform simulations of complex mechanical systems and accurately describe their behaviour, but it still requires a certain effort to organise entire simulation routines for large design explorations in Finite Elements environments. For this reason, analysis and evaluation of mechanical performance of braided systems could be divided into the following two steps:

1. Strip-strip interaction (form-finding and system generation, as developed so far)
2. Shell-like behavior (subsequent Finite Element Analysis for evaluation of system stiffness and buckling behavior, through further developments).

As in most cases, pre-stressing effects emerging from the deformation of thin and slender elements can be safely disregarded. The geometry emerging from the form-finding step of this paper could therefore represent the direct input for FEM

Fig. 8. Strip–strip interaction for testing physical properties of braid (tension)

analysis. This two-step workflow aims to explore and analyse the characteristics of mechanical performance, focusing in particular on the assessment of axial, bending and torsional stiffness, along with potential buckling behaviour of the braided systems (Fig. 8).

Workflow 4: Generative Design

In this workflow, a generative input is given to the braid solver. As an example of such an input, we use a low-level robotic controller from the literature—the Vascular Morphogenesis Controller (VMC) (Zahadat et al. 2017)—to supply a macro scale graph that grows over time based on environmental conditions. In a case where a braided artifact is manufactured robotically in situ and sensors are embedded in the physical braid, a controller like the VMC could be used to guide the shape of the braid in a way that is adaptive to dynamics of the environment [and provides behavior diversity (Zahadat and Schmickl 2014)]. Therefore, in this workflow, we use a simulated VMC graph output to generate a mesh topology for the braid solver (Fig. 9).

Unlike the inputs of a physical target or a master surface, which can be manually defined, a generative input necessitates a solver to automate the integration of fabrication constraints. To guarantee that a generated braiding pattern can be fabricated by hand or with a braiding robot, the underlying mesh has to have all the faces marked with a direction pointing in the fabrication direction. A macro scale directed graph approach was developed to generate meshes following this constraint (Fig. 10). The graph serves as a scaffold for the mesh tiling, and has to comply with several constraints to satisfy the fabrication method and mesh generation routine.

Evaluation

We evaluate the performance and generalisability of the method against geometric similarity and simulation performance. To assess simulation performance, we compare approaches to collision detection within the braid relaxation phase of the method to extrapolate the limits on complexity of currently achievable models. To assess geometric similarity we compare geometries of simulated results against physical examples.

Simulation performance is a limiting factor to the complexity that can be represented, with the relaxation phase being the most computationally demanding due to monitoring line-to-line collision detection in the mesh. We tested several methods to evaluate collision detection performance. Using an input mesh with 11,105 edges we obtained the following results:

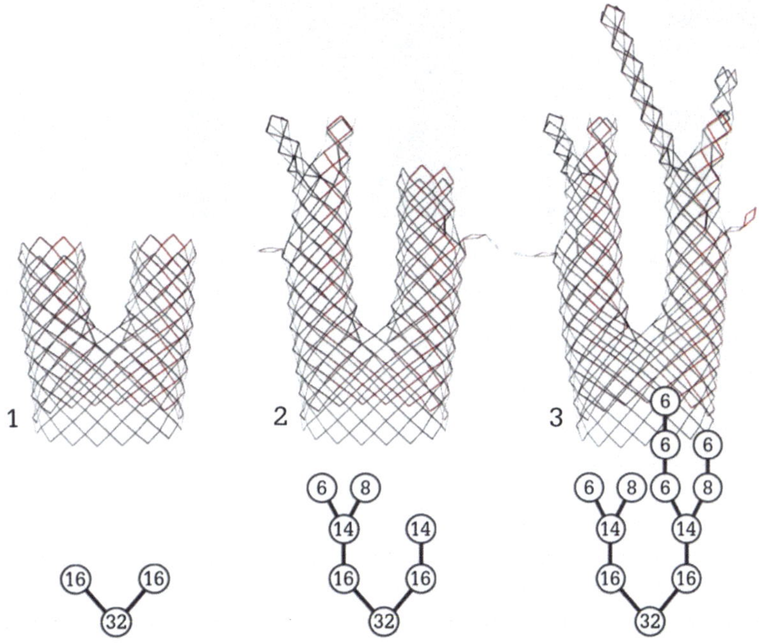

Fig. 9. Three example time-steps from the simulated growth of a VMC graph, and the braided artifacts resulting from those graphs

Fig. 10. Example mesh solution (*center*) from a weighted macro scale directed graph topology (*left*) and the eventual result from the braid solver (*right*)

- using a Spatial-Grid method (Teschner et al. 2003) without multi-threading results in a running speed of 60–100 ms per frame;
- using an R-Tree search method results in 80–200 ms per frame;
- using a conventional line-to-line constraint when calculating collision between all possible pairs runs at 6.1–7.5 s per frame and runs out of memory for larger models.

The geometric similarity of relaxed braid meshes is visually assessed. In Fig. 11, two features are physically prototyped, and then modelled for comparison. In both cases, the macro geometry of the model conforms closely to the physical prototype,

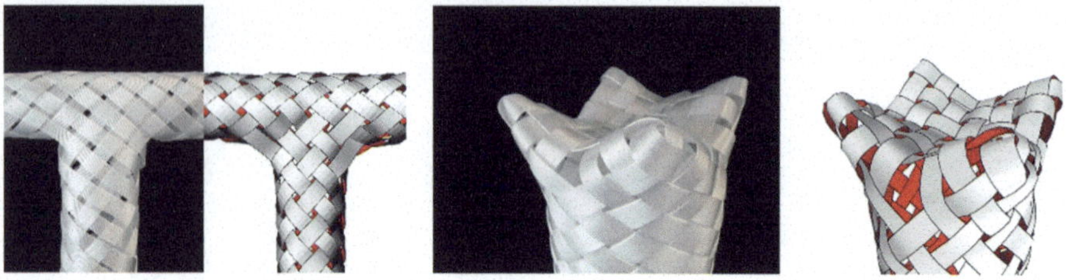

Fig. 11. Evaluation of the method by visual comparison to physical prototypes

whilst yarn-to-yarn relations show some geometric deviation. The modelled braid appears looser around regions of large geometric transition (Fig. 11, right).

Conclusion and Further Work

This paper has reported on a method and its integration within design workflows that address a gap in suitable tools for braid pattern generation and physics-based simulation. We have described and demonstrated workflows that target the exploration of generative or target based geometry, the production of fabrication instructions, and the conducting of structural analysis. These workflows support current research efforts in the *flora robotica* project, but are sufficiently generalised for the exploration of braid patterns beyond this project.

Further work aims to develop calibrated modelling of mechanical performance of modelled braids, and to fully test fabrication workflows using braiding machines and distributed robotic methods currently under development within the *flora robotica* project.

Acknowledgements. We would like to thank Payam Zahadat for feedback on the generative section and for providing us with example VMC graphs, from simulations tailored to braided structures.

Project *flora robotica* has received funding from the European Union's Horizon 2020 research and innovation program under the FET Grant Agreement, No. 640959.

References

Akleman, E., Chen, J., Xing, Q., Gross, J.L.: Cyclic plain-weaving on polygonal mesh surfaces with graph rotation systems. ACM Trans. Graph. **28**(3), 1 (2009)

Ayranci, C., Carey, J.: 2D braided composites: a review for stiffness critical applications. Compos Struct **85**(1), 43–58 (2008)

Hamann, H., Wahby, M., Schmickl, T., Zahadat, P., Hofstadler, D., Stoy, K., Risi, S., Faina, A., Veenstra, F., Kernbach, S., Kuksin, I., Kernback, O., Ayres, P., Wojtaszek, P.: Flora robotica —mixed societies of symbiotic robot-plant bio-hybrids. In: IEEE Symposium Series on Computational Intelligence, Cape Town, pp. 1102–1109 (2015)

Heinrich, M.K., Wahby, M., Soorati, M., Hofstadler, D., Zahadat, P., Ayres, P., Stoy, K., Hamann, H.: Self-organized construction with continuous building material: higher flexibility based on braided structures. In: IEEE 1st International Workshops on Foundations and Applications of Self* Systems (FAS*W), Augsburg, pp. 154–159 (2016)

Heinrich, M.K., Zwierzycki, M., Vestartas, P., Zahadat, P., Leon, D.A., Cheheltan, A., Evers, H. L., Hansson, P.K., Wahby, M., Hofstadler, D., Soorati, M., Hamann, H., Schmickl, T., Ayres, P.: Representations and Design Rules—Deliverable D3.1—EU-H2020 FET project 'flora robotica'. https://zenodo.org/record/816270/export/hx#.WV8-g4SGNJE. Verified 07 July 2017

Kaplan, M., Cohen, E.: Computer generated celtic design. In: Proceedings of the 14th Eurographics Workshop on Rendering, vol. 44, pp. 9–19 (2003)

Mercat, C.: Les entrelacs des enluminure celtes. Dossier Pour La Sci. **15** (2001)

Piker, D.: Kangaroo: form finding with computational physics. Archit. Des. **83**(2), 136–137 (2013). Updated to Kangaroo 2. http://www.food4rhino.com/app/kangaroo-physics

Potluri, P., Rawal, A., Rivaldi, M., Porat, I.: Geometrical modeling and control of a triaxial braiding machine for producing 3D preforms. Compos. A Appl. Sci. Manuf. **34**(6), 481–492 (2003)

Rana, S., Fangueiro, R.: Braided Structures and Composites: Production, Properties, Mechanics, and Technical Applications. CRC Press, Boca Raton (2015)

Richardson, D.: U.S. Patent No. 5,257,571. U.S. Patent and Trademark Office, Washington, DC (1993)

Teschner, M., Heidelberger, B., Müller, M., Pomerantes, D., Gross, M.H.: Optimized spatial hashing for collision detection of deformable objects. Vmv **3**, 47–54 (2003)

Zahadat, P., Hofstadler, D.N., Schmickl, T.: Vascular morphogenesis controller: a generative model for developing morphology of artificial structures. In: Proceedings of Genetic and Evolutionary Computation Conference (GECCO 2017), Berlin (2017)

Zahadat, P., Schmickl, T.: Generation of diversity in a reaction-diffusion based controller. Artif. Life **20**(3), 319–342 (2014)

Erratum to: Humanizing Digital Reality

Klaas De Rycke, Christoph Gengnagel, Olivier Baverel, Jane Burry,
Caitlin Mueller, Minh Man Nguyen, Philippe Rahm,
and Mette Ramsgaard Thomsen

Erratum to:
K. De Rycke et al., *Humanizing Digital Reality*, https://doi.org/10.1007/978-981-10-6611-5

In the original version of the book, the following belated corrections had been incorporated:

The author name "Mariana Popscu" has been changed to "Mariana Popescu" in Table of Contents, Contributors, Chapter 24 and Author Index.

The author name "Tom van Mele" has been changed to "Tom Van Mele" in Chapter 42.

In FM Acknowledgement, author names "Betty Pinks", "Adreas Kofler" and "Kombini" have been changed as "Bettina Pinks", "Andreas Kofler" and "Konbini", respectively.

In Back Cover, the term "Sixth" has been changed as "6th".

The updated original online version of this book can be found at https://doi.org/10.1007/978-981-10-6611-5.

K. De Rycke et al., *Humanizing Digital Reality*,
https://doi.org/10.1007/978-981-10-6611-5_56

Author Index

K. De Rycke et al., *Humanizing Digital Reality*,
https://doi.org/10.1007/978-981-10-6611-5

The manufacturer's authorised representative in the EU is Springer Nature Customer Service Centre GmbH, Europaplatz 3, 69115 Heidelberg, Germany. If you have any concerns regarding our products, please contact ProductSafety@springernature.com

Printed and bound by CPI Group (UK) Ltd, Croydon, CR0 4YY
06/10/2026
02228868-0007